Theory and Methods of Statistics

Theory and Methods of Statistics

P. K. Bhattacharya
Prabir Burman

ELSEVIER

AMSTERDAM • BOSTON • HEIDELBERG • LONDON
NEW YORK • OXFORD • PARIS • SAN DIEGO
SAN FRANCISCO • SINGAPORE • SYDNEY • TOKYO
Academic Press is an imprint of Elsevier

Academic Press is an imprint of Elsevier
32 Jamestown Road, London NW1 7BY, UK
525 B Street, Suite 1800, San Diego, CA 92101-4495, USA
50 Hampshire Street, 5th Floor, Cambridge, MA 02139, USA
The Boulevard, Langford Lane, Kidlington, Oxford OX5 1GB, UK

British Library Cataloguing in Publication Data
A catalogue record for this book is available from the British Library

Library of Congress Cataloging-in-Publication Data
A catalog record for this book is available from the Library of Congress

ISBN: 978-0-12-802440-9

For information on all Academic Press publications
visit our website at http://www.elsevier.com/

Publisher: Nikki Levy
Acquisition Editor: Graham Nisbet
Editorial Project Manager: Susan Ikeda
Production Project Manager: Poulouse Joseph
Designer: Victoria Pearson

Typeset by SPi Global, India

To my wife, Srilekha Bhattacharya.
P. K. Bhattacharya

To my mother, Shanti Lata Barman.
Prabir Burman

Contents

Preface

This book has grown out of lecture notes in courses in Mathematical Statistics, Linear Models, Multivariate Analysis, and Time Series Analysis taught by the authors over many years at UC Davis. Of these, Mathematical Statistics (Chapters 1–10) and Linear Models (Chapter 11) are core courses for PhD students in Statistics and Biostatistics at most institutions. Courses in Multivariate Analysis and Time Series Analysis are also taken by many students in these programs. A good background in Advanced Calculus and Matrix Analysis is a prerequisite for these courses.

Although most students in these courses would have taken an intermediate level course in Probability Theory, we have included such material in Chapters 1 and 2 for a review. Chapter 3 introducers various modes of convergence of an infinite sequence of random variables, which may be deferred until the study of asymptotic theory begins in Chapter 7. In Chapter 4, we outline the main problems of statistical inference and various approaches to optimality in the decision theoretic framework before treating Point Estimation, Hypothesis Testing, and Confidence Sets in Chapters 5 and 6. Methods based on Likelihood, Distribution-free Tests, and Curve Estimation are treated in Chapters 7–9, and Chapter 10 deals with Statistical Functionals. Gauss-Markov Models, topics in Model selection and Linear Mixed Models are the main focus of Chapter 11. Chapter 12 deals with Multivariate Analysis covering many of the standard methods used in practice. An introduction to Time Series Analysis is given in Chapter 13 covering aspects of ARMA modeling and Spectral Analysis. These chapters mostly concentrate on asymptotic properties of these methodologies, using the material in Chapter 3. Some technical results are included in Appendices A and B.

Throughout the book we have restricted to discrete and absolutely continuous random variables. Except for using the concept of a σ-field of subsets and in particular the Borel σ-field in Euclidean spaces, we have avoided all technicalities of measure theory. On the other hand, we have used Stieltjes integrals, providing a quick introduction to these integrals in Appendix A. Basic results from matrix algebra and distribution of quadratic forms are included in Appendix B.

We have not been able to cover everything in Chapters 1 through 10 and Appendix A in a three-quarter course with three lectures and one discussion per week in Mathematical Statistics. The following possibility of slightly trimming this material is suggested so as to fit into a three-quarter framework.

(i) Give a quick review of Chapters 1 and 2, leaving out most of the materials for self-study.
(ii) Introduce as many topics as possible out of locally best tests (Section 6.8), SPRT (Section 6.11), locally most powerful rank tests (Section 8.2), and curve estimation (Chapter 9).
(iii) Leave the details of most of the proofs in Sections 6.7–6.9, 7.2, 7.4, and 10.6 for self-study, going over only the main ideas in class-room presentation.

We also suggest the following outline for a full academic year Master's level course in Probability and Mathematical Statistics.

(i) Chapters 1–3 omitting Theorems 3.2.5 parts IX–XI; Theorems 3.2.3–3.2.5.
(ii) Omit the following: proofs in Sections 4.6.5 and 4.6.6; Ancillarity and Completeness; proof in Section 5.2.3; Equivariance in its generality; Section 6.8 and Section 6.11; proofs in Sections 6.12.2, 7.1.1, and 7.1.2; proofs of Theorems 7.2.1, 7.2.2, and 7.4.1; Section 8.2; proofs in Section 8.3; proofs in Chapter 9 and the entire Chapter 10.

Chapters 5 and 6 are influenced by Ferguson [1] and Lehmann [2, 3], and Chapter 8 relies heavily on Hájek and Šidák [4].

The material in Chapter 11 can be fitted in a two-quarter course in Linear Models with minor modifications; introduce as many topics as possible from Model Selection omitting some of the proofs, using examples to clarify the concepts.

Most of the material in Chapter 12 should be accessible to doctoral students along with the examples. An introduction to Time Series (Chapter 13) can be done by focusing on the main ideas and examples. The material in these two chapters can be fitted in two one-quarter courses for PhD students.

Chapters 11 and 12 use the material provided in Appendix B.

Our work on this book was initiated at the insistence of Jerome Braun, who took the Mathematical Statistics course from one of us. Thank you Jerome! The authors are also grateful to Ms Christine Cai for typing some parts of the book. We are indebted to the National Science Foundation for a grant (DMS 09-07622) for support of research in writing this book.

P.K. Bhattacharya, Prabir Burman
Davis, California

1

Probability Theory

1.1 Random Experiments and Their Outcomes

Probability theory is the systematic study of outcomes of a random experiment such as the roll of a die, or a bridge hand dealt from a thoroughly shuffled deck of cards, or the life of an electric bulb, or the minimum and the maximum temperatures in a city on a certain day, etc. The very first step in such a study is to *visualize* all possible outcomes of the experiment in question, and then to *realize* that the actual outcome of such an experiment is not predictable in advance. However, from the nature of the experiment, or from our experience with the past results of the experiment (if available), we may be able to assign probabilities to the possible outcomes or sets thereof.

For example, in the roll of a balanced, six-faced die, the possible outcomes are $\{1, 2, 3, 4, 5, 6\}$ to each of which we may assign a probability of $\frac{1}{6}$ (ie, in many repetitions of the trial, we expect each of these outcomes to occur $\frac{1}{6}$ of the times). From this, we can also conclude the outcome to be an even number with probability $\frac{3}{6} = \frac{1}{2}$.

Similarly, the possible outcomes of a bridge hand dealt from a standard deck are all the $\binom{52}{13} = \frac{52!}{13!39!}$ combinations of 13 cards from a 52-card deck, each carrying a probability of $1/\binom{52}{13}$.

The possible outcomes of the life of an electric bulb (in hours) is the set $[0, \infty)$ and the possible outcomes of the minimum and maximum temperatures in Philadelphia on a certain day in future (in °C) is $\{(x, y)\colon -273 < x < y < \infty\}$. In these last two examples, the sets of possible outcomes may seem to be unrealistically large, but this is because we cannot put an upper limit to the life of an electric bulb, or lower and upper limits to the temperatures in a city. However, from the past performance of many electric bulbs of a certain make, or from the meteorological records over many years in the past, we may be able to assign probabilities to various outcomes in these two examples. Such probabilities are empirical in nature.

1.2 Set Theory

The collection of all possible outcomes of a random experiment and various sub-collections of these outcomes are the entities to which we want to assign probabilities.

Theory and Methods of Statistics. http://dx.doi.org/10.1016/B978-0-12-802440-9.00001-1

These sub-collections are *sets* in the *space* of all possible outcomes. Our aim is to develop a logical system which would enable us to calculate the probabilities of sets of complicated nature from the probabilities of sets whose probabilities are more clearly understood. For this, we need to have an understanding of some basic facts about the theory of sets.

We denote by S the space of all possible outcomes of a random experiment, consisting of elements (particular outcomes) $s \in S$. In S, we would be interested in various sets $A \subset S$ and their combinations of different types.

Definition 1.2.1.

(i) The entire space is S consisting of elements $s \in S$, and $\emptyset$ is the empty set which contains no element of S.
(ii) $A^c = \{s \in S: s \notin A\}$ is the complement of $A \subset S$.
(iii) $A_1 \cap A_2 = A_1A_2 = \{s \in S: s \in A_1 \text{ and } s \in A_2\}$ is the intersection of A_1 and A_2.
(iv) $A_1 \cup A_2 = \{s \in S: s \in \text{at least one of } A_1, A_2\}$ is the union of A_1 and A_2.
(v) $A_1 \subset A_2$ if $s \in A_1$ implies $s \in A_2$; $A_1 = A_2$ if $A_1 \subset A_2$ and $A_2 \subset A_1$.

The following are immediate from the above definitions:

(i) $\emptyset = S^c$. **(ii)** $A_1 \cup A_2 = (A_1A_2^c) \cup (A_1^cA_2) \cup (A_1A_2)$. **(iii)** $(A^c)^c = A$.
(iv) $(A_1 \cup A_2)^c = A_1^c \cap A_2^c$ and more generally $\left(\bigcup_{i=1}^n A_i\right)^c = \bigcap_{i=1}^n A_i^c$.
(v) $(A_1 \cap A_2)^c = A_1^c \cup A_2^c$ and more generally $\left(\bigcap_{i=1}^n A_i\right)^c = \bigcup_{i=1}^n A_i^c$.
(vi) $A \cap (B \cup C) = (A \cap B) \cup (A \cap C)$ and $A \cup (B \cap C) = (A \cup B) \cap (A \cup C)$.
(vii) If $A_1 \subset A_2$, then $A_1 \cup A_2 = A_2$ and $A_1 \cap A_2 = A_1$.

Note. (iv) and (v) are called DeMorgan's rules.

Definition 1.2.2. Sets $A_1, A_2, \ldots$ are disjoint (or mutually exclusive) if $A_i \cap A_j = \emptyset$ for all $i \neq j$.

Definition 1.2.3.

(i) If the sets $A_1, A_2, \ldots$ are such that $\bigcup_{n=1}^\infty A_n = S$, then the collection $\{A_1, A_2, \ldots\}$ forms a covering of S.
(ii) If the sets $A_1, A_2, \ldots$ are disjoint and $\bigcup_{n=1}^\infty A_n = S$, then the collection $\{A_1, A_2, \ldots\}$ forms a partition of S.

Note. To avoid triviality, the sets in a covering or a partition of S are nonempty.

From any covering $\{A_1, A_2, \ldots\}$ of S we can construct a partition $\{B_1, B_2, \ldots\}$ of S by letting $B_1 = A_1, B_2 = A_1^cA_2, \ldots, B_n = A_1^cA_2^c \cdots A_{n-1}^cA_n, \ldots$.

Clearly, $B_1, B_2, \ldots$ are disjoint and $\cup_{n=1}^\infty B_n = \cup_{n=1}^\infty A_n = S$. In particular, if $A_1 \subset A_2 \subset \cdots \subset A_n \subset \cdots$ and $\cup_{n=1}^\infty A_n = S$, then $\{B_1, B_2, \ldots\}$ forms a partition of S if $B_1 = A_1, B_2 = A_1^cA_2, \ldots, B_n = A_{n-1}^cA_n, \ldots$.

1.3 Axiomatic Definition of Probability

A *probability space* is described by a triple $(S, \mathcal{A}, P)$ where S is an arbitrary space consisting of elements $s \in S$, $\mathcal{A}$ is a collection of sets $A \subset S$, called *events*, with the properties:

(Ei)$S \in \mathcal{A}$. (Eii) $A \in \mathcal{A} \Rightarrow A^c \in \mathcal{A}$. (Eiii) $A_1, A_2, \ldots \in \mathcal{A} \Rightarrow \bigcup_{n=1}^{\infty} A_n \in \mathcal{A}$,

and P: $\mathcal{A} \to [0, 1]$ is a function on $\mathcal{A}$ with the properties:

(Pi)$P[S] = 1$. (Pii) for disjoint sets $A_1, A_2, \ldots \in A$, $P\left[\bigcup_{n=1}^{\infty} A_n\right] = \sum_{n=1}^{\infty} P[A_n]$.

A collection of sets of S with properties E(i, ii, iii) is called a *σ-field of subsets of* S and a function on $\mathcal{A}$ with the property P(ii) is called a *a countably additive set function.* The property P(i) makes such a set function a *Probability.*

Proposition 1.3.1 (Continuity Property of Probability). *From the axioms P(i, ii), it follows that*

(Piii) *if* $A_1 \supset A_2 \supset A_3 \supset \cdots \supset A_n \supset \cdots$ *and* $\lim_{n\to\infty} A_n = \bigcap_{n=1}^{\infty} A_n = \emptyset$, *then* $\lim_{n\to\infty} P[A_n] = 0$, *or equivalently,*

(Piv) *if* $A_1 \subset A_2 \subset A_3 \subset \cdots \subset A_n \subset \cdots$ *and* $\lim_{n\to\infty} A_n = \bigcup_{n=1}^{\infty} A_n = S$, *then* $\lim_{n\to\infty} P[A_n] = 1$.

Conversely, either P(i, iii) or P(i, iv) implies P(i, ii).

Proof of the equivalence of P(i, ii), P(i, iii), and P(i, iv) is left as an exercise.

1.4 Some Simple Propositions

Proposition 1.4.1.

(i) $P[A^c] = 1 - P[A]$, $P[\emptyset] = 0$.

(ii) *If* $A \subset B$, *then* $P[A] \le P[B]$.

(iii) $P[A \cup B] = P[A] + P[B] - P[AB]$.

(iv) $P\left[\bigcup_{i=1}^{n} A_i\right] \le \sum_{i=1}^{n} P[A_i]$.

Proof.

(i) A and A^c are disjoint and $A \cup A^c = S$. Hence
$1 = P[S] = P[A \cup A^c] = P[A] + P[A^c]$, so $P[A^c] = 1 - P[A]$.
In particular, $\emptyset = S^c$ and therefore, $P[\emptyset] = 1 - P[S] = 1 - 1 = 0$.

(ii) If $A \subset B$, then $B = A \cup (A^c B)$, where A and $A^c B$ are disjoint. Hence
$P[B] = P[A] + P[A^c B] \ge P[A]$, because $P[A^c B] \ge 0$.

(iii) $A = (AB) \cup (AB^c)$, $B = (AB) \cup (A^c B)$, $A \cup B = (AB^c) \cup (A^c B) \cup (AB)$, and the sets AB^c, $A^c B$, AB are disjoint. Thus

$$\begin{aligned} P[A \cup B] &= P[AB^c] + P[A^c B] + P[AB] \\ &= \{P[AB] + P[AB^c]\} + \{P[AB] + P[A^c B]\} - P[AB] \\ &= P[A] + P[B] - P[AB]. \end{aligned}$$

(iv) Let $B_1 = A_1, B_2 = A_1^c A_2, \ldots, B_n = A_1^c \cdots A_{n-1}^c A_n$. Then
(a) $\bigcup_{i=1}^n A_i = \bigcup_{i=1}^n B_i$,
(b) $B_1, B_2, \ldots, B_n$ are disjoint, and
(c) $B_i \subset A_i, i = 1, \ldots, n$.
Hence $P[\bigcup_{i=1}^n A_i] = P[\bigcup_{i=1}^n B_i] = \sum_{i=1}^n P[B_i] \leq \sum_{i=1}^n P[A_i]$ by (ii).

□

Proposition 1.4.2.

$$P[A_1 \cup \cdots \cup A_n] = \sum_{i=1}^{n} P[A_i] - \sum_{1 \leq i_1 < i_2 \leq n} P[A_{i_1} A_{i_2}] + \cdots + (-1)^{r+1} \sum_{1 \leq i_1 < i_2 < \cdots < i_r \leq n} P[A_{i_1} A_{i_2} \cdots A_{i_r}] + \cdots + (-1)^{n+1} P[A_1 A_2 \cdots A_n]$$

The proof, which follows by induction, starting from Proposition 1.4.1(iii), is left as an exercise.

Proposition 1.4.3 (Bonferroni's Inequality). $P[A_1 A_2 \cdots A_n] \geq P[A_1] + P[A_2] + \cdots + P[A_n] - 1$.

For $n = 2$, $P[A_1 A_2] \geq P[A_1] + P[A_2] - 1$ is proved as follows: $1 \geq P[A_1 \cup A_2] = P[A_1] + P[A_2] - P[A_1 A_2]$. Hence the result.The proof of the general case follows by induction, and is left as an exercise.

This inequality may be useful in some situations to get some idea about $P[A_1 A_2 \cdots A_n]$ where actual evaluation may be difficult.

1.5 Equally Likely Outcomes in Finite Sample Space

Suppose that S consists of N elements, $S = \{s_1, \ldots, s_N\}$, which are equally likely. Then the total probability of 1 is shared equally by these N elements, so that $P[\{s_i\}] = \frac{1}{N}$ for each i. Consequently, for any event $A = \{s_{i_1}, \ldots, s_{i_K}\}$, $P[A] = \frac{K}{N} = \frac{\#(A)}{\#(S)}$, where $\#(E) \doteq$ number of elements in E.

Evaluation of probabilities of events in such a setting, essentially reduces to counting problems which are often of a combinatorial nature.

Example 1.5.1. Two balanced dice with 1, 2, 3, 4, 5, 6 on the six faces are rolled. Find the probabilities of the events: (a) the numbers on the two dice are equal, (b) the sum of the numbers on the two dice is 10, (c) the number on at least one die is 6, and (d) none of the numbers is a 6.

Solution.

(a) $\#(S) = 6 \cdot 6 = 36$ and if we let A_1 be the event in question, then $A_1 = \{(1,1), (2,2), \ldots, (6,6)\}$ and $\#(A_1) = 6$. Thus $P[A_1] = \frac{\#(A_1)}{\#(S)} = \frac{6}{36} = \frac{1}{6}$.
(b) Here the event $A_2 = \{(4,6), (5,5), (6,4)\}$, so $\#(A_2) = 3$. Thus $P[A_2] = \frac{\#(A_2)}{\#(S)} = \frac{3}{36} = \frac{1}{12}$.

(d) We answer (d) first. None of the numbers is a 6, means no 6 on the first, with probability $\left(1-\frac{1}{6}\right)$ and no 6 on the second, with probability $\left(1-\frac{1}{6}\right)$. Hence the probability of no 6 on any of the two dice is the probability of the event A_4, $P[A_4]=(1-\frac{1}{6})(1-\frac{1}{6})=\frac{5}{6}\cdot\frac{5}{6}=\frac{25}{36}$, because the roll of the first die has nothing to do with the roll of the second die, that is, they are "independent," which is a concept to be introduced in the sequel. In a direct counting approach, $A_4=\{(x,y)\colon x=1,2,3,4,5 \text{ and } y=1,2,3,4,5\}$. Thus $\#(A_4)=5\cdot 5=25$, so $P[A_4]=\frac{\#(A_4)}{\#(S)}=\frac{25}{36}$, which agrees with our calculation based on the concept of independence.

(c) "At least one 6" is the complement of "no 6". Hence $P[A_3]=P[A_4^c]=1-P[A_4]=1-\frac{25}{36}=\frac{11}{36}$.

Example 1.5.2. What is the probability that a bridge hand dealt from a well-shuffled standard deck of cards will contain 2 Aces and 2 Kings?

Solution. Here $\#(S)\doteq$ number of ways to choose 13 cards from a deck of 52 cards = $\binom{52}{13}$, and if A denotes the event in question, then $\#(A)\doteq$ number of ways to choose 2 out of 4 Aces, 2 out of 4 Kings, and 9 out of the remaining 44 cards = $\binom{4}{2}\binom{4}{2}\binom{44}{9}$. Thus

$$P[A]=\frac{\binom{4}{2}\binom{4}{2}\binom{44}{9}}{\binom{52}{13}}\simeq 0.04.$$

1.6 Conditional Probability and Independence

Definition 1.6.1. For events A and B with $P[A]>0$, the conditional probability of B given A is defined to be

$$P[B|A]=\frac{P[AB]}{P[A]}. \tag{1}$$

This is the probability that we assign to the event B if we know that the event A has occurred. Now if the knowledge of occurrence of A does not change the probability of B (ie, if $P[B|A]=P[B]$), then we say that *B is independent of A*. But this holds if $P[AB]=P[A]P[B]$. Now if $P[B]>0$, then $P[AB]=P[A]P[B]$ also implies

$$P[A|B]=\frac{P[AB]}{P[B]}=P[A],$$

so that *A is independent of B*. This leads to the following definition.

Definition 1.6.2. Events A, B are independent if

$$P[AB]=P[A]P[B]. \tag{2a}$$

More generally, events $A_1,\ldots,A_n$ are independent if for every subset $\{i_1,\ldots,i_r\}$ of $\{1,2,\ldots,n\}$, $2\le r\le n$,

$$P[A_{i_1}A_{i_2}\cdots A_{i_r}]=P[A_{i_1}]P[A_{i_2}]\cdots P[A_{i_r}]. \tag{2b}$$

From the definition of $P[B|A]$ in Eq. (1), we have

$$P[AB] = P[A]P[B|A]. \tag{3a}$$

More generally, we have

$$P[A_1A_2\cdots A_n] = P[A_1]P[A_2|A_1]P[A_3|A_1A_2]\cdots P[A_n|A_1\cdots A_{n-1}]. \tag{3b}$$

This is proved by applying the above formula repeatedly to the right side to see that

$$\begin{aligned} &P[A_1]P[A_2|A_1]P[A_3|A_1A_2]\cdots P[A_n|A_1\cdots A_{n-1}] \\ &\quad = P[A_1A_2]P[A_3|A_1A_2]\cdots P[A_n|A_1\cdots A_{n-1}] \\ &\quad = P[A_1A_2A_3]\cdots P[A_n|A_1\cdots A_{n-1}] \\ &\quad = \cdots = P[A_1A_2\cdots A_n]. \end{aligned}$$

Formulas (2a) and (2b) are special cases of Eqs. (3a) and (3b) when the events are independent.

Proposition 1.6.1. *[Bayes Formula] If the collection of events $\{A_1, A_2, \ldots\}$ forms a partition of S, that is, if $A_1, A_2, \ldots \in \mathcal{A}$ are disjoint and $\bigcup_{j=1}^{\infty} A_j = S$, then*

$$P[A_i|B] = \frac{P[A_iB]}{P[B]} = \frac{P[A_i]P[B|A_i]}{\sum_{j=1}^{\infty} P[A_j]P[B|A_j]}.$$

Proof. We only have to observe that by virtue of $A_1, A_2, \ldots$ being a partition of S, $B = \bigcup_{j=1}^{\infty} A_jB$, and the events $A_1B, A_2B, \ldots$ are disjoint. Hence $P[B] = \sum_{j=1}^{\infty} P[A_jB] = \sum_{j=1}^{\infty} P[A_j]P[B|A_j]$ by P(ii) and Eq. (3a). □

For any event B with $P[B] > 0$, $P[A|B]$ as a function of $A \in \mathcal{A}$ is a probability, that is, $0 \leq P[A|B] \leq 1$ for all $A \in \mathcal{A}$ and $P[\cdot|B]$ satisfies P(i, ii). The verification of these facts and the proof of the following proposition are left as exercises.

Proposition 1.6.2. *If A and B are independent, then the pairs A and B^c, A^c and B, A^c and B^c are also independent.*

Example 1.6.1. There are 20 balls in a bag, of which 8 are white and 12 are black. We draw three balls at random from the bag. Find the probability that all three of these balls are white if (a) each ball drawn is replaced before the next draw, (b) the balls are drawn without replacement.

Solution.

(a) Here the outcomes of the three draws are independent events, because the composition of the bag is unchanged after a draw due to replacement. Let A_i denote the event of a white ball in ith draw, $i = 1, 2, 3$. Then $P[A_i] = \frac{8}{20}$ for each i and A_1, A_2, A_3 are independent. Hence P(All three balls are white) $= P[A_1A_2A_3] = P[A_1]$ $P[A_2]P[A_3] = \left(\frac{8}{20}\right)^3 = \left(\frac{2}{5}\right)^3 = \frac{8}{125}$.

(b) When the draws are without replacement, the events A_1, A_2, A_3 are not independent. Here we can approach the problem in two ways:

(i) By direct counting, we have $\#(A_1A_2A_3) = \binom{8}{3} = 56$ and $\#(S) = \binom{20}{3} = 1140$. Hence $P[A_1A_2A_3] = \frac{56}{1140} = \frac{14}{285}$.

(ii) Arguing with conditional probabilities, by looking at the composition of the bag conditional upon the outcomes of previous draws we have $P[A_1] = \frac{8}{20}, P[A_2|A_1] = \frac{7}{19}, P[A_3|A_1A_2] = \frac{6}{18}$.

Hence $P[A_1A_2A_3] = P[A_1]P[A_2|A_1]P[A_3|A_1A_2] = \frac{8}{20} \cdot \frac{7}{19} \cdot \frac{6}{18} = \frac{14}{285}$.

Example 1.6.2. There are three boxes of which Box 1 contains tickets numbered $1, \ldots, 5$; Box 2 contains tickets numbered $1, \ldots, 10$; and Box 3 contains tickets numbered $1, \ldots, 20$. One of the three boxes is chosen at random and then a ticket is drawn at random from the chosen box. Find the probability that the ticket is from Box 2 if (a) the number on the ticket is 4, (b) the number on the ticket is 7.

Solution. Let A_i be the event that Box i is chosen, B the event that the number on the ticket is 4 and C the event that the number on the ticket is 7. Then $P[A_i] = \frac{1}{3}, i = 1, 2, 3$; $P[B|A_1] = \frac{1}{5}, P[B|A_2] = \frac{1}{10}, P[B|A_3] = \frac{1}{20}$; $P[C|A_1] = 0, P[C|A_2] = \frac{1}{10}, P[C|A_3] = \frac{1}{20}$.

(a) Here we need $P[A_2|B]$. By Bayes' formula,

$$P[A_2|B] = \frac{P[A_2]P[B|A_2]}{\sum_{j=1}^{3} P[A_j]P[B|A_j]} = \frac{\frac{1}{3} \cdot \frac{1}{10}}{\frac{1}{3} \cdot \frac{1}{5} + \frac{1}{3} \cdot \frac{1}{10} + \frac{1}{3} \cdot \frac{1}{20}} = \frac{2}{7}.$$

(b) Here we need $P[A_2|C]$. By Bayes' formula,

$$P[A_2|C] = \frac{P[A_2]P[C|A_2]}{\sum_{j=1}^{3} P[A_j]P[C|A_j]} = \frac{\frac{1}{3} \cdot \frac{1}{10}}{\frac{1}{3} \cdot 0 + \frac{1}{3} \cdot \frac{1}{10} + \frac{1}{3} \cdot \frac{1}{20}} = \frac{2}{3}.$$

1.7 Random Variables and Their Distributions

Definition 1.7.1. A real-valued function $X\colon S \to \mathbb{R}$ on a probability space $(S, \mathcal{A}, P)$ is a random variable (*rv*) if for all $a \in \mathbb{R}$,

$$\{s \in S\colon X(s) \le a\} \in \mathcal{A}. \tag{4a}$$

More generally, $\boldsymbol{X} = (X_1, \ldots, X_k)\colon S \to \mathbb{R}^k$ on a probability space $(S, \mathcal{A}, P)$ is a k-dimensional (k-dim) random vector (*rv*) if for all $(a_1, \ldots, a_k) \in \mathbb{R}^k$,

$$\left\{s \in S\colon X_1(s) \le a_1, \ldots, X_k(s) \le a_k\right\} \in \mathcal{A}. \tag{4b}$$

Definition 1.7.2. The cumulative distribution function (cdf) of a 1-dim rv X is defined as

$$F_X(a) = P[X \le a] = P[\{s\colon X(s) \le a\}] \quad \text{for all } a \in \mathbb{R}. \tag{5a}$$

Note. We shall use P as a set function $P\colon \mathcal{A} \to [0, 1]$ as in P(i, ii) as well as in a more informal way as in $P[X \le a]$ to indicate the probability of $X \le a$.

Remark 1.7.1. We are using the abbreviation rv to indicate both a random variable and a random vector. The meaning of one or the other should be clear in the context.

Proposition 1.7.1 (Properties of a cdf $F_X(x)$).

(i) $F_X(a) \leq F_X(b)$ *for all* $a \leq b$.
(ii) $\lim_{h\downarrow 0} F_X(x+h) = F_X(x)$ *for all* x.
(iii) $\lim_{x\to-\infty} F_X(x) = 0$.
(iv) $\lim_{x\to\infty} F_X(x) = 1$.

Property (i) says that F_X is monotone nondecreasing (because $(-\infty, a] \subset (-\infty, b]$ for all $a \leq b$) and Property (ii) says that F_X is right continuous. Properties (ii), (iii), and (iv) follow from the continuity property of probability, the proofs of which are left as exercises.

For any rv X,

$$\begin{aligned} P[X = a] &= P[X \leq a] - P[X < a] = F_X(a) - \lim_{h\uparrow 0} F_X(a+h) \\ &= \text{magnitude of jump of } F_X \text{ at } a \text{ (if any).} \end{aligned}$$

We shall consider only two types of rv's, discrete and continuous. A rv X is *discrete* if its cdf F_X *is a step-function* (ie, increases only by jumps and stays constant elsewhere) and is *continuous* if its cdf F_X *is differentiable everywhere.* Since a cdf F_X can have only a finite or a countably infinite number of jumps, a discrete rv can have only a finite or countably infinite number of possible values $\{x_i\}$.

Definition 1.7.3. The probability mass function (pmf) of a discrete rv X is defined as

$$f_X(x_i) = P[X = x_i], \quad i = 1, 2, \ldots \quad \text{and} \quad f_X(x) = 0 \ \text{ for all } x \notin \{x_i\}. \tag{6a}$$

For any set $B \subset \mathbb{R}$, $P[X \in B] = \sum_{x\in B} f_X(x)$ and $\sum_{i=1}^{\infty} f_X(x_i) = P[X \in \mathbb{R}] = 1$.

Definition 1.7.4. The probability density function (pdf) of a continuous rv X is defined as

$$f_X(x) = F'_X(x) \geq 0 \quad \text{for all } x \in \mathbb{R}. \tag{7a}$$

Clearly, $P[X \leq a] = F_X(a) = \int_{-\infty}^{a} f_X(x)\,dx$, and $\int_{-\infty}^{\infty} f_X(x)\,dx = \lim_{x\to\infty} F_X(x) = 1$.

Note. Since the cdf of a continuous rv X is differentiable everywhere, it has no jumps. Hence $P[X = a] = 0$ for all a.

On the real line $\mathbb{R}$, the smallest σ-field of subsets which includes all intervals, is called the σ*-field of Borel sets and is denoted by* $\mathbb{B}$. For each $B \in \mathbb{B}$,

$$P[X \in B] = \int_B f_X(x)\,dx,$$

when this integral is defined in an appropriate manner.

The cdf of a k-dim rv $\boldsymbol{X} = (X_1, \ldots, X_k)$ is defined as

$$F_{\boldsymbol{X}}(a_1, \ldots, a_k) = P\big[X_1 \leq a_1, \ldots, X_k \leq a_k\big] \quad \text{for all } (a_1, \ldots, a_k) \in \mathbb{R}^k. \tag{5b}$$

The marginal cdf of X_i is obtained as $F_{X_i}(a) = F_{\boldsymbol{X}}(\infty, \ldots, \infty, a, \infty, \ldots, \infty)$, that is, by taking $a_i = a$, and $a_j = \infty$ for all $j \neq i$ in $F_{\boldsymbol{X}}(a_1, \ldots, a_k)$. Again we consider the discrete and continuous cases, and define

$$f_{\boldsymbol{X}}(x_1, \ldots, x_k) \\ = \begin{cases} P[X_1 = x_1, \ldots, X_k = x_k] & \text{if } (x_1, \ldots, x_k) \in \text{ set of possible values of } \boldsymbol{X} \\ 0 & \text{otherwise} \end{cases} \tag{6b}$$

as the pmf of $\boldsymbol{X}$ in the discrete case, and

$$f_{\boldsymbol{X}}(x_1, \ldots, x_k) = \frac{\partial^k}{\partial x_1 \cdots \partial x_k} F_{\boldsymbol{X}}(x_1, \ldots, x_k) \tag{7b}$$

as the pdf of $\boldsymbol{X}$ in the continuous case.

The various marginal pmf's and pdf's are obtained by summing or integrating over the other variables in Eq. (6b) or (7b).

For random variables defined by (4a, b), the cdf defined by (5a, b), the pmf/pdf defined by (6a, b) or (7a, b) contain all the information.

Example 1.7.1. Players 1 and 2 each rolls a balanced six-faced die. Let (s_1, s_2) be the outcome of the rolls, s_i the number on the die rolled by Player i. On the sample space of $\{(s_1, s_2): s_i = 1, 2, 3, 4, 5, 6,\ i = 1, 2\}$, we define $X(s_1, s_2) = s_1 - s_2$ as the amount which Player 1 wins from Player 2, a negative win being interpreted as a loss. Find the pmf of X.

Solution. The possible values of the rv X are 0, ± 1, ± 2, ± 3, ± 4, and ± 5. For $x \geq 0$, $\#\{(s_1, s_2): s_1 - s_2 = x\} = \#\{(1 + x, 1), \ldots, (6, 6 - x)\} = 6 - x$, each carrying a probability of $\frac{1}{36}$. Hence for $x \geq 0$,

$$f_X(x) = P[X = x] = \frac{6 - x}{36}, \quad x = 0, 1, \ldots, 5.$$

Again for $x < 0$, $\#\{(s_1, s_2): s_1 - s_2 = x\} = \#\{(s_1, s_2): s_2 - s_1 = |x|\} = \#\{(1, 1 + |x|), \ldots, (6 - |x|, 6)\} = 6 - |x|$, each carrying a probability of $\frac{1}{36}$. Hence for $x < 0$,

$$f_X(x) = P[X = x] = \frac{6 - |x|}{36}, \quad x = -5, \ldots, -1.$$

Thus, $f_X(x) = \frac{6-|x|}{36}$, $x = 0, \pm 1, \pm 2, \pm 3, \pm 4, \pm 5$.

Example 1.7.2. A point s is selected at random from the unit circle with center at the origin (0, 0) and radius 1 and

$$X(s) = \text{distance of } s \text{ from the center.}$$

Find the cdf and the pdf of X and the probability that $\frac{1}{2} \leq X \leq \frac{3}{4}$.

Solution. The cdf of X is

$$\begin{aligned} F_X(x) &= P[\{s: X(s) \leq x\}] \\ &= P\left[s \in \text{a circle of radius } x \text{ centered at origin}\right] \\ &= \frac{\text{Area of a circle of radius } x}{\text{Area of unit circle}} \\ &= \frac{\pi x^2}{\pi 1^2} = x^2, \quad 0 \leq x \leq 1. \end{aligned}$$

This is because, for a randomly selected point s from a set S in the plane,

$$P[s \in A] = \frac{\text{Area of } A}{\text{Area of } S}, \quad \text{for all } A \subset S$$

assuming that the areas of S and of A are well-defined.

Now the pdf of X is

$$f_X(x) = F_X'(x) = \frac{d}{dx}(x^2) = 2x, \quad 0 \leq x \leq 1.$$

Finally,

$$P\left[\frac{1}{2} \leq X \leq \frac{3}{4}\right] = F_X\left(\frac{3}{4}\right) - F_X\left(\frac{1}{2}\right) = \left(\frac{3}{4}\right)^2 - \left(\frac{1}{2}\right)^2 = \frac{5}{16}.$$

Alternatively,

$$P\left[\frac{1}{2} \leq X \leq \frac{3}{4}\right] = \int_{\frac{1}{2}}^{\frac{3}{4}} f_X(x)\, dx = \int_{\frac{1}{2}}^{\frac{3}{4}} 2x\, dx = x^2\Big|_{\frac{1}{2}}^{\frac{3}{4}} = \frac{5}{16}.$$

1.8 Expected Value, Variance, Covariance, and Correlation Coefficient

Definition 1.8.1. The expected value or the mean of an rv X is defined as

$$\mu_X = \mathrm{E}[X] = \begin{cases} \sum_{i=1}^{\infty} x_i f_X(x_i) & \text{if } X \text{ is discrete} \\ \int_{-\infty}^{\infty} x f_X(x)\, dx & \text{if } X \text{ is continuous,} \end{cases}$$

provided that the sum or the integral exists and is finite. Otherwise, we say that $\mathrm{E}[X]$ does not exist.

More generally, we define

$$\mathrm{E}[g(X)] = \begin{cases} \sum_{i=1}^{\infty} g(x_i) f_X(x_i) & \text{if } X \text{ is discrete} \\ \int_{-\infty}^{\infty} g(x) f_X(x)\, dx & \text{if } X \text{ is continuous.} \end{cases}$$

Note. $\mathrm{E}[X]$ exists if and only if $\mathrm{E}[|X|] < \infty$.

For a k-dim random vector $\boldsymbol{X} = (X_1, \ldots, X_k)$, we define

$$\mathrm{E}[g(X_1, \ldots, X_k)] = \begin{cases} \sum_{\text{all } x_1,\ldots,x_k} g(x_1, \ldots, x_k) f_X(x_1, \ldots, x_k) & \text{if } X \text{ is discrete} \\ \int_{-\infty}^{\infty} \cdots \int_{-\infty}^{\infty} g(x_1, \ldots, x_k) f_X(x_1, \ldots, x_k)\, dx_1 \cdots dx_k & \text{if } X \text{ is continuous.} \end{cases}$$

Note. The term *Expectation* is often used for *Expected Value.*

Proposition 1.8.1 (Properties of Expectation).

(i) *If* $P[X = c] = 1$, *then* $E[X] = c$.
(ii) *If* $P[X \geq 0] = 1$ *with probability 1, then* $E[X] \geq 0$.
(iii) *If* $I_A(x) = 1$ *for* $x \in A$ *and* $I_A(x) = 0$ *for* $x \notin A$, *then* $E[I_A(x)] = P[X \in A]$.
(iv) $E\left[aX + b\right] = aE[X] + b$.
(v) $E\left[\sum_{i=1}^{k} a_i X_i\right] = \sum_{i=1}^{k} a_i E[X_i]$.
(vi) *If* $P[X_1 \geq X_2] = 1$, *then* $E[X_1] \geq E[X_2]$.

The proof is omitted.

Definition 1.8.2. The variance of an rv X is defined as

$$\sigma_X^2 = \mathrm{Var}[X] = \mathrm{E}\left[\{X - \mathrm{E}[X]\}^2\right] = \mathrm{E}\left[X^2\right] - \{\mathrm{E}[X]\}^2.$$

The last equality follows by expanding $\{X - \mathrm{E}[X]\}^2$ and then using the properties of expectation in Proposition 1.8.1, remembering that $\mathrm{E}[X] = \mu_X$ is a constant. The *standard deviation* of X is $\sqrt{\mathrm{Var}[X]} = \sigma_X$.

Definition 1.8.3. The covariance between rv's X and Y is defined as

$$\sigma_{XY} = \mathrm{Cov}[X, Y] = \mathrm{E}[\{X - \mathrm{E}[X]\}\{Y - \mathrm{E}[Y]\}] = \mathrm{E}[XY] - \mathrm{E}[X]\mathrm{E}[Y].$$

Again the last equality follows by expanding $\{X - \mathrm{E}[X]\}\{Y - \mathrm{E}[Y]\}$ and using the properties of expectation. Obviously,

$$\mathrm{Cov}[X, X] = \mathrm{Var}[X].$$

Note. If $\boldsymbol{X}^\top = (X_1, \ldots, X_k)$ is a k-dim rv, then $\boldsymbol{\Sigma} = ((\mathrm{Cov}\left[X_i, X_j\right]_{i,j}))$ is called the *covariance matrix of* $\boldsymbol{X}$.

Proposition 1.8.2 (Properties of Variance and Covariance).

(i) $Var\left[aX + b\right] = a^2\sigma_X^2$.
(ii) $Var\left[\sum_{i=1}^{k} a_i X_i\right] = \sum_{i=1}^{k} a_i^2 \sigma_{X_i}^2 + 2\sum_{1 \leq i < j \leq k} a_i a_j \, \sigma_{X_i X_j}$.
(iii) $Cov\left[aX + b, cY + d\right] = ac\,\sigma_{X,Y}$.
(iv) $Cov\left[\sum_{i=1}^{m} a_i X_i + \sum_{j=1}^{n} b_j Y_j\right] = \sum_{i=1}^{m}\sum_{j=1}^{n} a_i b_j \, \sigma_{X_i Y_j} = \boldsymbol{a}^\top \boldsymbol{\Sigma} \boldsymbol{b}$,
where $\boldsymbol{a}^\top = (a_1, \ldots, a_m)$, $\boldsymbol{b}^\top = (b_1, \ldots, b_n)$, *and* $\boldsymbol{\Sigma} = ((\sigma_{(X_i, Y_j)}))$.
(v) *If* $\boldsymbol{X}$ *is a k-dim rv, then* $Cov[\boldsymbol{X}, \boldsymbol{X}] = E\left[(\boldsymbol{X} - \mu_{\boldsymbol{X}})(\boldsymbol{X} - \mu_{\boldsymbol{X}})^\top\right]$.
(vi) *If* $\boldsymbol{X}$ *is a k-dim rv and* $\boldsymbol{A}$ *is a* $k \times k$ *matrix, then* $Cov[\boldsymbol{AX}, \boldsymbol{AX}] = \boldsymbol{A}Cov[\boldsymbol{X}, \boldsymbol{X}]\boldsymbol{A}^\top$.

The proof is omitted.

Question. Why is the mean μ_X called the "expected value" of X?

The answer is statistical in nature. Suppose we want to *predict* the value of the rv X *before it is actually observed*, by making a guess a and paying a penalty $(X - a)^2$, which is the square of the discrepancy between our guess a and the actual observation X. Then the average penalty is

$$\begin{aligned}
\mathrm{E}\left[(X-a)^2\right] &= \mathrm{E}\left[\{(X-\mu_X)-(a-\mu_X)\}^2\right] \\
&= \mathrm{E}\left[(X-\mu_X)^2\right] + (a-\mu_X)^2 = \sigma_X^2 + (a-\mu_X)^2,
\end{aligned}$$

(using the fact $\mathrm{E}[X-\mu_X]=0$), which is minimized by taking $a=\mu_X$. Thus the mean μ_X is the best guess for an unobserved X under the rule of squared-error penalty and that is why $\mu_X = \mathrm{E}[X]$ is called the expected value of X. In a variation of the above game of prediction, what would be the best guess if the penalty is the absolute error $|X-a|$? Here we want to minimize $\mathrm{E}[|X-a|]$ by our choice of a. For simplicity, suppose F_X is continuous and strictly increasing and let $m = m_X = F_X^{-1}(\frac{1}{2})$ be the solution of the equation $F_X(m) = \frac{1}{2}$. Then under the rule of absolute error penalty, m is the best guess for an unobserved X. The quantity m_X is called the *median of* X, which has some advantages over the mean μ_X, as will be seen later. For one thing, m_X always exists even when μ_X may not.

We now prove an important inequality.

Proposition 1.8.3 (Cauchy-Schwarz Inequality). *Let X, Y be rv's for which $\mathrm{E}[X^2] < \infty$ and $\mathrm{E}[Y^2] < \infty$. Then*

$$\{E|XY|\}^2 \le E\left[X^2\right]E\left[Y^2\right].$$

Proof. The function

$$h(t) = \mathrm{E}\left[(tX-Y)^2\right] = t^2\mathrm{E}\left[X^2\right] - 2t\mathrm{E}[XY] + \mathrm{E}\left[Y^2\right] \ge 0 \quad \text{for all } t,$$

so the quadratic equation $h(t) = 0$ can have *at most one real root*. But the roots of this equation are

$$t = \frac{2\mathrm{E}[XY] \pm \sqrt{4\{\mathrm{E}[XY]\}^2 - 4\mathrm{E}\left[X^2\right]\mathrm{E}\left[Y^2\right]}}{2\mathrm{E}\left[X^2\right]}.$$

Hence the expression under the radical must be ≤ 0. □

Replacing X and Y by $X-\mu_X$ and $Y-\mu_Y$, respectively, in the above inequality, we have

$$\mathrm{Cov}^2[X,Y] \le \mathrm{Var}[X]\,\mathrm{Var}[Y], \quad \text{ie, } \sigma_{XY}^2 \le \sigma_X^2\sigma_Y^2.$$

Definition 1.8.4. The correlation coefficient between X and Y is defined as

$$\rho_{XY} = \frac{\mathrm{Cov}[X,Y]}{\sqrt{\mathrm{Var}[X]\mathrm{Var}[X]}} = \frac{\sigma_{XY}}{\sigma_X\sigma_Y}.$$

From *Cauchy-Schwarz inequality*, it follows that $\rho_{XY}^2 \le 1$, that is, $-1 \le \rho_{XY} \le 1$. Moreover, $\rho_{XY} = +1$ or $\rho_{XY} = -1$, that is, $\rho_{XY}^2 = 1$ *iff* the equation

$$h(t) = \mathrm{E}\left[\{t(X-\mu_X) - (Y-\mu_Y)\}^2\right] = 0$$

holds for exactly one real value of t. But

$$\begin{aligned}
&\mathrm{E}\left[\{t(X-\mu_X) - (Y-\mu_Y)\}^2\right] = 0 \\
&\Leftrightarrow Y-\mu_Y = t(X-\mu_X) \quad \text{w.p. 1 for some } t \ne 0.
\end{aligned}$$

($t = 0 \Rightarrow Y - \mu_Y = 0$ w.p. 1 $\Rightarrow \sigma_Y^2 = 0$, and we exclude this degenerate case from our consideration.) Thus $\rho_{XY} = +1$ or -1 *iff* Y is a linear function of X w.p. 1, the slope having the same sign as ρ_{XY}.

1.9 Moments and the Moment Generating Function

Definition 1.9.1. The rth moment of an rv X is defined as

$$\mu_{r,X} = \mathrm{E}[X^r], \quad r = 1, 2, \ldots$$

assuming existence, and $\mu_{0,X} = 1$. In particular, the mean $\mu_X = \mu_{1,X}$, and $\sigma_X^2 = \mu_{2,X} - \mu_{1,X}^2$.

Definition 1.9.2. The moment generating function (mgf) of X is defined as $M_X(t) = \mathrm{E}[\mathrm{e}^{tX}]$, provided that the expectation exists.

Differentiating under the integral,

$$\mu_X^{(r)}(0) = \frac{d^r}{dt^r} M_X(t)\bigg|_{t=0} = \mathrm{E}\left[X^r \mathrm{e}^{tX}\right]_{t=0} = E[X^r] = \mu_r,$$

$$M_X(t) = \sum_{r=0}^{\infty} \frac{t^r}{r!} \mu_r.$$

Hence the name moment generating function. For a linear function $a + bX$, $M_{a+bX}(t) = \mathrm{E}\left[\mathrm{e}^{t(a+bX)}\right] = \mathrm{e}^{at} M_X(bt)$.

1.10 Independent Random Variables and Conditioning When There Is Dependence

Random variables $X_1, \ldots, X_k$ are said to be mutually independent if

$$F_{X_1,\ldots,X_k}(x_1, \ldots, x_k) = \prod_{i=1}^{k} F_{X_i}(x_i) \quad \text{for all } x_1, \ldots, x_k. \tag{8a}$$

Equivalently, for mutually independent rv's,

$$f_{X_1,\ldots,X_k}(x_1, \ldots, x_k) = \prod_{i=1}^{k} f_{X_i}(x_i) \quad \text{for all } x_1, \ldots, x_k, \tag{8b}$$

holds for their joint pdf or joint pmf. The conditional pmf or pdf of $X_{r+1}, \ldots, X_k$ given $(X_1, \ldots, X_r) = (x_1, \ldots, x_r)$ when $f_{X_1,\ldots,X_r}(x_1, \ldots, x_r) > 0$ is

$$f_{(X_{r+1},\ldots,X_k)|(X_1,\ldots,X_r)}(x_{r+1}, \ldots, x_k | x_1, \ldots, x_r) = \frac{f_{X_1,\ldots,X_k}(x_1, \ldots, x_k)}{f_{X_1,\ldots,X_r}(x_1, \ldots, x_r)}.$$

In particular, for (X, Y) having joint pmf/pdf $f_{XY}(x, y)$,

$$f_{Y|X}(y|x) = \frac{f_{XY}(x,y)}{f_X(x)} = \frac{f_{XY}(x,y)}{\int_{-\infty}^{\infty} f_{XY}(x,y)\, dy}, \quad \text{when } f_X(x) > 0.$$

On the other hand, from the marginal pdf f_X of X and the conditional pdf $f_{Y|X}$ of Y given X, the joint pdf of (X, Y) is obtained as

$$f_{XY}(x,y) = f_X(x) f_{Y|X}(y|x). \tag{9}$$

If $X_1, \ldots, X_k$ are mutually independent rv's, then

$$\mathrm{E}\big[g_1(X_1)\cdots g_k(X_k)\big] = \prod_{i=1}^{k} \mathrm{E}\big[g_i(X_i)\big]$$

provided that $\mathrm{E}\big[g_i(X_i)\big]$, $i = 1, \ldots, k$ exist. This follows immediately from the definition of independence (8a, b). It follows that if X, Y are independent, then $\sigma_{XY} = \mathrm{Cov}[X, Y] = 0$, and $\rho_{XY} = 0$, and therefore,

$$\sigma^2_{aX+bY} = a^2\sigma^2_X + b^2\sigma^2_Y.$$

However, $\rho_{XY} = 0$ does not imply that X, Y are independent.

Suppose that $X_1, \ldots, X_n$ are mutually independent and each X_i is distributed as X. We then say that $X_1, \ldots, X_n$ are independent and identically distributed (iid) as X, and if $\bar{X} = \frac{1}{n}\sum_{i=1}^{n} X_i$, then

$$\mu_{\bar{X}} = \mu_X \text{ and } \sigma^2_{\bar{X}} = \frac{\sigma^2_X}{n}.$$

If $X_1, \ldots, X_k$ are independent rv's and if the mgf $M_{X_i}(t)$ exists for each X_i, then

$$M_{X_1+\cdots+X_k}(t) = \mathrm{E}\left[e^{t\sum_{i=1}^{k} X_i}\right] = \prod_{i=1}^{k} \mathrm{E}\left[e^{tX_i}\right] = \prod_{i=1}^{k} M_{X_i}(t).$$

Going back to the general case of (X, Y) with joint pdf $f_{XY}(x, y)$, let

$$m(x) = \mathrm{E}\big[g(Y)|X = x\big] = \int g(y) f_{Y|X}(y|x)\, dy.$$

We now denote by $m(X) = \mathrm{E}\big[g(Y)|X\big]$ the conditional expectation of $g(Y)$ given X. This conditional expectation is a function of the rv X and therefore, is itself an rv, which takes the value $m(x)$ when $X = x$. Hence

$$\begin{aligned}
\mathrm{E}\big[\mathrm{E}\big[g(Y)|X\big]\big] = E[m(X)] &= \int m(x) f_X(x)\, dx \\
&= \int\left[\int g(y) f_{Y|X}(y|x)\, dy\right] f_X(x)\, dx \\
&= \int g(y) f_Y(y)\, dy = E\big[g(Y)\big],
\end{aligned}$$

because $\int f_{Y|X}(y|x) f_X(x)\, dx = \int f_{XY}(x,y)\, dx = f_Y(y)$ using Eq. (9). Next consider

$$\begin{aligned} \mathrm{E}\big[h(X)g(Y)|X=x\big] &= \int h(x)g(y)f_{Y|X}(y|x)\, dy \\ &= h(x)\int g(y)f_{Y|X}(y|x)\, dy \\ &= h(x)\,\mathrm{E}\left[g(Y)|X=x\right] \quad \text{for each } x. \end{aligned}$$

Hence $\mathrm{E}\big[h(X)g(Y)|X\big] = h(X)\mathrm{E}\big[g(Y)|X\big]$. We thus have the following important results:

$$\mathrm{EE}\big[g(Y)|X\big] = \mathrm{E}\big[g(Y)\big] \text{ and } \mathrm{E}\big[h(X)g(Y)|X\big] = h(X)\mathrm{E}\big[g(Y)|X\big].$$

We next consider the conditional properties of variance.

$$\begin{aligned} \mathrm{Var}[Y] &= \mathrm{E}\Big[\{Y-\mathrm{E}[Y]\}^2\Big] \\ &= \mathrm{EE}\Big[\{(Y-\mathrm{E}[Y|X]) + (\mathrm{E}[Y|X]-\mathrm{E}[Y])\}^2|X\Big] \\ &= \mathrm{EE}\Big[(Y-\mathrm{E}[Y|X])^2|X\Big] + \mathrm{EE}\Big[(\mathrm{E}[Y|X]-\mathrm{E}[Y])^2|X\Big] \\ &\quad + 2\mathrm{EE}[(Y-\mathrm{E}[Y|X])(\mathrm{E}[Y|X]-\mathrm{E}[Y])|X]. \end{aligned}$$

The three terms in the last expression are

$$\mathrm{EE}\Big[(Y-\mathrm{E}[Y|X])^2|X\Big] = \mathrm{E}[\mathrm{Var}[Y|X]],$$

since $\mathrm{E}\big[(Y-\mathrm{E}[Y|X])^2|X\big] = \mathrm{Var}[Y|X]$,

$$\mathrm{EE}\Big[(\mathrm{E}[Y|X]-\mathrm{E}[Y])^2|X\Big] = \mathrm{E}\Big[(\mathrm{E}[Y|X]-\mathrm{E}[Y])^2\Big] = \mathrm{Var}[\mathrm{E}[Y|X]],$$

and the third term is 0, using $\mathrm{E}\big[h(X)g(Y)|X\big] = h(X)\mathrm{E}\big[g(Y)|X\big]$.

Summary. Besides all the properties analogous to the properties of expectation, we have proved the following important properties of conditional expectation.

Proposition 1.10.1.

(i) $E\big[E\big[g(Y)|X\big]\big] = E\big[g(Y)\big]$.
(ii) $E\big[h(X)g(Y)|X\big] = h(X)E\big[g(Y)|X\big]$.
(iii) $Var[Y] = E[Var[Y|X]] + Var[E[Y|X]]$.

Definition 1.10.1. The function $m(x) = \mathrm{E}[Y|X=x]$ is called the regression function of Y on X. In particular, if $m(x)$ is a linear function of x, then we can represent Y as

$$Y = \alpha + \beta X + \varepsilon \text{ with } \mathrm{E}[\varepsilon|X] = 0,$$

and if ε is independent of X, then this is called the *linear regression model* and $\mathrm{Var}[\varepsilon]$, if it exists, is the *residual variance.* More generally, if the dependence of Y on a k-dim rv $\boldsymbol{X} = (X_1, \ldots, X_k)$ is such that $m(x_1, \ldots, x_k) = \mathrm{E}[Y|\boldsymbol{X} = \boldsymbol{x}]$ is a linear function of $(x_1, \ldots, x_k)$, then

$$Y = \alpha + \beta_1 X_1 + \cdots + \beta_k X_k + \varepsilon \text{ with } \mathrm{E}[\varepsilon|\boldsymbol{X}] = 0,$$

and $(\boldsymbol{X}, Y)$ is said to follow a *multiple linear regression model* if ε is independent of $\boldsymbol{X}$.

Example 1.10.1. The joint pdf of (X, Y) is

$$f_{XY}(x,y) = \begin{cases} C(x^2+2y^2) & 0 < x, y < 1 \\ 0 & \text{otherwise.} \end{cases}$$

Find the constant C, the marginal pdf's of X and Y, the conditional pdf of Y given $X = x$ and then find the means μ_X, μ_Y, the variances σ_X^2, σ_Y^2, the correlation coefficient ρ_{XY} and the conditional expectation $\mathrm{E}[Y|X = x]$. Also find $P[X > Y]$.

Solution.

$$1 = C\int_0^1\int_0^1 (x^2+2y^2)\,dx\,dy = C\left[1\int_0^1 x^2\,dx + 2\int_0^1 y^2\,dy\right] = C\left[\frac{1}{3}+\frac{2}{3}\right].$$

Thus $C = 1$ and $f_{XY}(x,y) = x^2+2y^2$, $0 < x < 1$, $0 < y < 1$. Now

$$f_X(x) = \int_0^1 (x^2+2y^2)\,dy = x^2+\frac{2}{3}, \quad 0 < x < 1,$$

$$f_Y(y) = \int_0^1 (x^2+2y^2)\,dx = 2y^2+\frac{1}{3}, \quad 0 < y < 1.$$

The conditional pdf of Y given $X = x$ is

$$f_{Y|X}(y|x) = \frac{f_{XY}(x,y)}{f_X(x)} = \frac{x^2+2y^2}{x^2+\frac{2}{3}}, \quad 0 < y < 1 \text{ for } 0 < x < 1.$$

Next, we evaluate the means, the variances, and the correlation coefficient:

$$\begin{aligned}
\mu_X &= \int_0^1 x f_X(x)\,dx = \int_0^1 x\left(x^2+\frac{2}{3}\right)dx \\
&= \frac{1}{4}+\frac{2}{3}\cdot\frac{1}{2} = \frac{7}{12}, \\
\mu_Y &= \int_0^1 y f_Y(y)\,dy = \int_0^1 y\left(2y^2+\frac{1}{3}\right)dy \\
&= 2\cdot\frac{1}{4}+\frac{1}{3}\cdot\frac{1}{2} = \frac{2}{3}, \\
\sigma_X^2 &= \mathrm{E}\left[X^2\right]-\mu_X^2 = \int_0^1 x^2\left(x^2+\frac{2}{3}\right)dx - \left(\frac{7}{12}\right)^2 \\
&= \left(\frac{1}{5}+\frac{2}{3}\cdot\frac{1}{3}\right)-\left(\frac{7}{12}\right)^2 = \frac{19}{45}-\frac{49}{144} = \frac{59}{720}, \\
\sigma_Y^2 &= \mathrm{E}\left[Y^2\right]-\mu_Y^2 = \int_0^1 y^2\left(2y^2+\frac{1}{3}\right)dy - \left(\frac{2}{3}\right)^2 \\
&= \left(\frac{2}{5}+\frac{1}{3}\cdot\frac{1}{3}\right)-\left(\frac{2}{3}\right)^2 = \frac{23}{45}-\frac{4}{9} = \frac{1}{15}, \\
\sigma_{XY} &= \mathrm{E}[XY]-\mu_X\mu_Y = \int_0^1\int_0^1 xy(x^2+2y^2)\,dx\,dy - \frac{7}{12}\cdot\frac{2}{3}
\end{aligned}$$

$$= \left(\frac{1}{4}\cdot\frac{1}{2}+2\cdot\frac{1}{2}\cdot\frac{1}{4}\right)-\frac{7}{12}\cdot\frac{2}{3}=\frac{3}{8}-\frac{7}{18}=-\frac{1}{72},$$

$$\rho_{XY}=\frac{\sigma_{XY}}{\sigma_X\sigma_Y}=\frac{-\frac{1}{72}}{\sqrt{\frac{59}{720}\cdot\frac{1}{15}}}=-\frac{1}{72}\sqrt{\frac{10{,}800}{59}}=-0.1879.$$

The conditional expectation of Y given $X=x$ is

$$\begin{aligned}
\mathrm{E}[Y|X=x] &= \int_0^1 y f_{Y|X}(y|x)\,dy = \int_0^1 y\,\frac{x^2+2y^2}{x^2+\frac{2}{3}}\,dy \\
&= \frac{1}{x^2+\frac{2}{3}}\int_0^1 (x^2y+2y^3)\,dy = \frac{x^2\frac{1}{2}+2\cdot\frac{1}{4}}{x^2+\frac{2}{3}} \\
&= \frac{x^2+1}{2x^2+\frac{4}{3}}.
\end{aligned}$$

Note that

$$\begin{aligned}
\mathrm{EE}[Y|X] &= \int_0^1 \mathrm{E}[Y|X=x]f_X(x)\,dx = \int_0^1 \frac{x^2+1}{2x^2+\frac{4}{3}}\left(x^2+\frac{2}{3}\right)dx \\
&= \int_0^1 \frac{1}{2}(x^2+1)\,dx = \frac{1}{2}\left(\frac{1}{3}+1\right)=\frac{2}{3}=\mu_Y, \text{ as it should be.}
\end{aligned}$$

Finally,

$$P[X>Y]=\int_0^1\left[\int_0^x (x^2+2y^2)\,dy\right]dx=\int_0^1\left(x^2+2\frac{x^3}{3}\right)dx=\frac{5}{3}\cdot\frac{1}{4}=\frac{5}{12}.$$

1.11 Transforms of Random Variables and Their Distributions

We start with some simple transforms. In our following discussion, we shall often work with continuous rv's and their pdf's, of which (i), (ii), and (iii) will remain valid for discrete rv's and their pmf's by replacing $\int\ dx$ by $\sum_x$.

(i) For $b>0$, $F_{a+bX}(z)=P[a+bX\le z]=P[X\le\frac{z-a}{b}]=F_X(\frac{z-a}{b})$. Hence $f_{a+bX}(z)=b^{-1}f_X(\frac{z-a}{b})$. More generally,

$$\text{for } b\neq 0,\ f_{a+bX}(z)=|b|^{-1}f_X\left(\frac{z-a}{b}\right). \tag{10}$$

In the discrete case, $f_{a+bX}(z)=f_X(\frac{z-a}{b})$.

(ii) $F_{X^2}(z)=P[-\sqrt{z}\le X\le\sqrt{z}]=F_X(\sqrt{z})-F_X(-\sqrt{z})$ for $z>0$. Hence

$$\begin{aligned}
f_{X^2}(z) &= \frac{1}{2\sqrt{z}}f_X(\sqrt{z})-\left(-\frac{1}{2\sqrt{z}}\right)f_X(-\sqrt{z}) \qquad (11a)\\
&= \frac{1}{2\sqrt{z}}[f_X(\sqrt{z})+f_X(-\sqrt{z})] \text{ for } z>0.
\end{aligned}$$

If X is distributed symmetrically about 0, then $f_X(\sqrt{z})=f_X(-\sqrt{z})$, so that

$$f_{X^2}(z) = \frac{1}{\sqrt{z}} f_X(\sqrt{z}), \quad z > 0 \tag{11b}$$

for symmetrically distributed X. In the discrete case,

$$f_{X^2}(z) = f_X(\sqrt{z}) + f_X(-\sqrt{z})$$

and $f_{X^2}(z) = 2f_X(\sqrt{z})$ if X is symmetric.

(iii) If X_1, X_2 are independent, then

$$\begin{aligned} F_{X_1+X_2}(z) &= \int\int_{x_1+x_2 \le z} f_{X_1}(x_1) f_{X_2}(x_2)\, dx_1\, dx_2 \\ &= \int_{x_1=-\infty}^{\infty} \int_{x_2=-\infty}^{z-x_1} f_{X_1}(x_1) f_{X_2}(x_2)\, dx_1\, dx_2. \end{aligned}$$

Differentiating under the integral, we have

$$\begin{aligned} f_{X_1+X_2}(z) &= \int_{-\infty}^{\infty} f_{X_1}(x_1) \left[\frac{d}{dz} \int_{-\infty}^{z-x_1} f_{X_2}(x_2)\, dx_2 \right] dx_1 \\ &= \int_{-\infty}^{\infty} f_{X_1}(x) f_{X_2}(z-x)\, dx. \end{aligned}$$

If X_1, X_2 are nonnegative, then $f_{X_1}(x) = 0$ for $x < 0$ and $f_{X_2}(z-x) = 0$ for $x > z$. In this case the above formula becomes

$$f_{X_1+X_2}(z) = \int_0^z f_{X_1}(x) f_{X_2}(z-x)\, dx. \tag{12a}$$

For independent discrete rv's,

$$f_{X_1+X_2}(z) = \sum_x f_{X_1}(x) f_{X_2}(z-x). \tag{12b}$$

(iv) If X_1, X_2 are independent with $P[X_2 > 0] = 1$, then

$$F_{\frac{X_1}{X_2}}(z) = \int_{x_2=0}^{\infty} \left[\int_{x_1=-\infty}^{x_2 z} f_{X_1}(x_1)\, dx_1 \right] f_{X_2}(x_2)\, dx_2$$

and differentiating under the integral, we have

$$f_{\frac{X_1}{X_2}}(z) = \int_0^{\infty} x f_{X_1}(xz) f_{X_2}(x)\, dx. \tag{13}$$

(v) For mutually independent $X_1, \ldots, X_k$, let $Y = \min(X_1, \ldots, X_k)$ and $Z = \max(X_1, \ldots, X_k)$. Then

$$\begin{aligned} 1 - F_Y(y) &= P\big[\min(X_1, \ldots, X_k) > y\big] \\ &= \prod_{i=1}^{k} P[X_i > y] = \prod_{i=1}^{k} \{1 - F_{X_i}(y)\} \text{ and} \\ F_Z(z) &= P\big[\max(X_1, \ldots, X_k) \le z\big] \\ &= \prod_{i=1}^{k} P[X_i \le z] = \prod_{i=1}^{k} F_{X_i}(z). \end{aligned}$$

Hence

$$f_Y(y) = -\frac{d}{dy}\prod_{i=1}^{k}\{1 - F_{X_i}(y)\} = \sum_{i=1}^{k} f_{X_i}(y)\prod_{j\neq i}\left[1 - F_{X_j}(y)\right]$$

$$f_Z(z) = \frac{d}{dz}\prod_{i=1}^{k} F_{X_i}(z) = \sum_{i=1}^{k} f_{X_i}(z)\prod_{j\neq i=1}^{k} F_{X_j}(z).$$

We now consider a transformation of

$$\boldsymbol{X} = (X_1, \ldots, X_k) \xrightarrow{g} (Y_1, \ldots, Y_k) = \boldsymbol{Y}$$

(ie, $Y_i = g_i(X_1, \ldots, X_k)$), where g is one to one with continuous partial derivatives and a nonvanishing Jacobian of transformation in the continuous case. Let g^{-1} denote the inverse transformation, that is,

$$g^{-1}(\boldsymbol{y}) = g^{-1}(y_1, \ldots, y_k) := (h_1(y_1, \ldots, y_k), \ldots, h_k(y_1, \ldots, y_k)).$$

In the discrete case,

$$f_{\boldsymbol{Y}}(y_1, \ldots, y_k) = f_{\boldsymbol{X}}(h_1(y_1, \ldots, y_k), \ldots, h_k(y_1, \ldots, y_k)) = f_{\boldsymbol{X}}\left(g^{-1}(\boldsymbol{y})\right).$$

In the continuous case, for all events B, $P[\boldsymbol{Y} \in B] = P\left[\boldsymbol{X} \in g^{-1}(B)\right]$, where $g^{-1}(B) = \left\{\boldsymbol{x}\colon g(\boldsymbol{x}) \in B\right\}$. Thus

$$\begin{aligned}
&\int_B f_{\boldsymbol{Y}}(y_1, \ldots, y_k)\, dy_1 \cdots dy_k \\
&= \int_{g^{-1}(B)} f_{\boldsymbol{X}}(x_1, \ldots, x_k)\, dx_1 \cdots dx_k \\
&= \int_{g(g^{-1}(B))} f_{\boldsymbol{X}}\left(g^{-1}(\boldsymbol{y})\right)\left|\det\begin{bmatrix} \frac{\partial x_1}{\partial y_1} & \cdots & \frac{\partial x_k}{\partial y_1} \\ \vdots & \ddots & \vdots \\ \frac{\partial x_1}{\partial y_k} & \cdots & \frac{\partial x_k}{\partial y_k} \end{bmatrix}\right| dy_1 \cdots dy_k \\
&= \int_B f_{\boldsymbol{X}}\left(g^{-1}(\boldsymbol{y})\right)|J_{g^{-1}(\boldsymbol{y})}|\, dy_1 \cdots dy_k,
\end{aligned}$$

where $J_{g^{-1}(\boldsymbol{y})} = \det\begin{bmatrix} \frac{\partial h_1}{\partial y_1} & \cdots & \frac{\partial h_k}{\partial y_1} \\ \vdots & \ddots & \vdots \\ \frac{\partial h_1}{\partial y_k} & \cdots & \frac{\partial h_k}{\partial y_k} \end{bmatrix}$, and note that $J_{g^{-1}(\boldsymbol{y})} = \left[J_g(g^{-1}(\boldsymbol{y}))\right]^{-1}$, so we compute the one which is easier. Thus

$$f_{\boldsymbol{Y}}(\boldsymbol{y}) = f_{\boldsymbol{X}}\left(g^{-1}(\boldsymbol{y})\right)|J_{g^{-1}(\boldsymbol{y})}| = f_{\boldsymbol{X}}\left(g^{-1}(\boldsymbol{y})\right)\frac{1}{|J_g(g^{-1}(\boldsymbol{y}))|}.$$

An Extension

Let $S_1, \ldots, S_l$ be disjoint open sets in $\mathbb{R}^k$ with $\sum_{j=1}^{l} P[\boldsymbol{X} \in S_j] = 1$. Let g: $\bigcup_{j=1}^{l} S_j \to \mathbb{R}^k$, $\boldsymbol{Y} = g(\boldsymbol{X})$, where for each j, the restriction g_j of g on S_j is one to one with continuous partial derivatives and nonvanishing J_{g_j}. Then for all events B,

$$\begin{aligned}
\int_B f_{\boldsymbol{Y}}(\boldsymbol{y})\, d\boldsymbol{y} = P[\boldsymbol{Y} \in B] &= \sum_{j=1}^{l} P\left[\boldsymbol{Y} \in B, \boldsymbol{X} \in S_j\right] \\
&= \sum_{j=1}^{l} P\left[\boldsymbol{X} \in g_j^{-1}(B),\ \boldsymbol{X} \in S_j\right] \\
&= \sum_{j=1}^{l} \int_{g^{-1}B \cap S_j} f_{\boldsymbol{X}}(\boldsymbol{x})\, d\boldsymbol{x} \\
&= \sum_{j=1}^{l} \int_{B \cap g_j(S_j)} f_{\boldsymbol{X}}\left(g^{-1}(\boldsymbol{y})\right) |J_{g^{-1}(\boldsymbol{y})}|\, d\boldsymbol{y} \\
&= \int_B \sum_{j=1}^{l} f_{\boldsymbol{X}}\left(g^{-1}(\boldsymbol{y})\right) |J_{g^{-1}(\boldsymbol{y})}| I_{g(S_j)}(\boldsymbol{y})\, d\boldsymbol{y},
\end{aligned}$$

where $I_A(\boldsymbol{y}) = 1$ if $\boldsymbol{y} \in A$ and $= 0$ if $\boldsymbol{y} \notin A$. Hence

$$f_{\boldsymbol{Y}}(\boldsymbol{y}) = \sum_{j=1}^{l} f_{\boldsymbol{X}}\left(g^{-1}(\boldsymbol{y})\right) |J_{g^{-1}(\boldsymbol{y})}| I_{g(S_j)}(\boldsymbol{y}).$$

Joint Distribution of Order Statistics

Let $X_1, \ldots, X_n$ be iid real-valued rv's with common pdf f. Define

$$Y_i = X_{n:i}, \quad i = 1, \ldots, n,$$

where $X_{n:1} < \cdots < X_{n:n}$ are the order statistics, that is, the ordered values of $X_1, \ldots, X_n$. (The X_i's are all distinct with probability 1.) The joint pdf of $(X_1, \ldots, X_n)$ is $f_{\boldsymbol{X}}(\boldsymbol{x}) = \prod_{i=1}^{n} f(x_i)$. For the $n!$ permutations $(j_1, \ldots, j_n)$ of $(1, \ldots, n)$, let

$$S_{j_1 \cdots j_n} = \left\{(x_1, \ldots, x_n) \colon x_{j_1} < \cdots < x_{j_n}\right\}$$

and define $\boldsymbol{Y} = (Y_1, \ldots, Y_n) = g_{j_1,\ldots,j_n}(X_1, \ldots, X_n) = (X_{j_1}, \ldots, X_{j_n})$ on $S_{j_1,\ldots,j_n}$ for each $(j_1, \ldots, j_n)$. Then $Y_1 < \cdots < Y_n$ with probability 1 and $g_{j_1,\ldots,j_n}$ is one to one with $\left|J_{g^{-1}_{j_1,\ldots,j_n}}(\boldsymbol{y})\right| = 1$ for each permutation $(j_1, \ldots, j_n)$ of $(1, \ldots, n)$. Hence

$$f_{\boldsymbol{Y}}(\boldsymbol{y}) = \sum_{j_1,\ldots,j_n} f_{\boldsymbol{X}}\left(g_{j_1,\ldots,j_n}^{-1}(\boldsymbol{y})\right) = n! \prod_{i=1}^{n} f(y_i) \quad \text{for } y_1 < \cdots < y_n$$

and $f_{\boldsymbol{Y}}(\boldsymbol{y}) = 0$ otherwise.

Linear Transformation

As a special case of g: $\boldsymbol{X} \to \boldsymbol{Y}$ from $\mathbb{R}^k \to \mathbb{R}^k$, consider a linear transformation $\boldsymbol{Y} = \boldsymbol{AX}$ where $\boldsymbol{A}$ is a nonsingular $k \times k$ matrix. Then the inverse transformation is $\boldsymbol{X} = \boldsymbol{A}^{-1}\boldsymbol{Y} = g^{-1}(\boldsymbol{Y})$, so $|J_{g^{-1}}(\boldsymbol{y})| = |\det(\boldsymbol{A}^{-1})| = \frac{1}{|\det(\boldsymbol{A})|}$. Hence

$$f_{\boldsymbol{Y}}(\boldsymbol{y}) = f_{\boldsymbol{X}}(\boldsymbol{A}^{-1}\boldsymbol{y})\frac{1}{|\det(\boldsymbol{A})|}. \tag{14}$$

In particular, if A is orthonomial, that is, $\boldsymbol{AA}^{\mathsf{T}} = \boldsymbol{A}^{\mathsf{T}}\boldsymbol{A} = \boldsymbol{I}$, then $|\det(\boldsymbol{A})| = 1$, so that $f_{\boldsymbol{Y}}(\boldsymbol{y}) = f_{\boldsymbol{X}}(\boldsymbol{A}^{-1}\boldsymbol{y})$.

Probability Integral Transform

Suppose X has cdf F which is continuous and strictly increasing. Then F^{-1} is uniquely defined as

$$F^{-1}(u) = x \text{ iff } F(x) = u \text{ for } 0 < u < 1.$$

Then the cdf of $Y = F(X)$ at $u \in (0, 1)$ is

$$F_Y(u) = P[F(x) \leq u] = P\left[X \leq F^{-1}(u)\right] = F\left(F^{-1}(u)\right) = u.$$

Thus $f_Y(u) = 1$ for $0 < u < 1$ and $f_Y(u) = 0$ for $u \notin (0, 1)$, because $0 < Y = F(X) < 1$ with probability 1. In other words, if X has a continuous and strictly increasing cdf F, then $Y = F(X)$ is distributed with pdf

$$f_Y(u) = \begin{cases} 1 & \text{if } 0 < u < 1, \\ 0 & \text{otherwise.} \end{cases}$$

A rv with this pdf is said to be a Uniform(0, 1) rv. Conversely, if U is Uniform(0, 1), then $X = F^{-1}(U)$ has cdf F. This fact is useful in generating random samples (ie, iid rv's) with cdf F by first generating random samples $U_1, U_2, \ldots$ from Uniform(0, 1), which is easy, and then transforming $U_1, U_2, \ldots$ to $X_1 = F^{-1}(U_1), X_2 = F^{-1}(U_2), \ldots$.

Exercises

1.1. Prove that the axioms P(i, ii), P(i, iii), and P(i, iv) are equivalent.

1.2. Prove Proposition 1.4.2 by induction, starting from Proposition 1.4.1(iii).

1.3. Five cards with numbers $1, \ldots, 5$ are arranged in random order. A person claiming psychic power, declares the arrangement. Assuming that the numbers are declared purely at random, what is the probability that
(a) All are correct? (b) Exactly 4 are correct?

1.4. From a well-shuffled standard deck of cards, eight cards are dealt.
(a) What is the probability that the hand contains two trios (exactly three cards of the same denomination such as three kings)?

(b) What is the probability that the hand contains a run of five cards (exactly five cards of consecutive denominations such as $8, 9, 10, J, Q$ irrespective of suits, $10, J, Q, K, A$ included)?

1.5. A poker hand consists of five cards dealt from a well-shuffled standard deck of cards. A hand is a

(a) *straight* if the cards are of consecutive denominations (including $A2345$ and $AKQJ10$), but not of the same suit,

(b) *flush* if all five cards are of the same suit but not a straight,

(c) *full house* if three cards are of the same denomination and the other two cards are also of the same denomination such as three 10s and two Queens.

Find the probabilities of a hand being a straight, a flush, and a full house.

1.6. (*The birthday problem*) In an assembly of n people, what is the probability that no two have the same birthday? Find the smallest n for which this probability is less than $1/2$. [For simplicity, ignore leap-years and assume that the birthdays are equally likely to be any of the 365 days of a year.]

1.7. Show that for any event B with $P[B] > 0$, $P[A|B]$ as a function of $A \in \mathcal{A}$ is a probability.

1.8. Prove Proposition 1.6.2.

1.9. A couple has two children. If one child is a boy, what is the probability that the other child is a girl?

1.10. Two shooters A and B can hit a target with probabilities 0.8 and 0.7, respectively. They shoot alternatively and the one who first makes three hits wins. Find the probability that A wins if

(a) A shoots first, (b) B shoots first, (c) the one who shoots first is decided by the toss of a fair coin.

1.11. There are two dice in a bag both of which are balanced, but one has $1, 2, 3, 4, 5, 6$ and the other has $1, 1, 2, 3, 4, 5$ on the six faces. One of the dice is selected at random and rolled.

(a) What is the probability that the outcome is an even number?

(b) If the outcome is an even number, what is the probability that the standard die was rolled?

1.12. In a city, in the month of July, the maximum temperature reaches 100°F or above with probability 0.2 on any day and with probability 0.8 if the previous day was 100°F or above. What is the probability that there will be exactly 5 consecutive days of 100°F or above starting on July 15?

1.13. Prove Proposition 1.7.1.

1.14. There are N tickets in a bag, numbered $1, \ldots, N$, from which n tickets are drawn. Let X be the largest number drawn. Find the pmf and the cdf of X if the tickets are drawn at random (a) with replacement, (b) without replacement.

1.15. The lifetime of an equipment (in hours) is a random variable X with pdf

$$f(x) = \begin{cases} c & 0 < x \le 50 \\ c\exp(-(x-50)) & x > 50. \end{cases}$$

Find (a) the constant c, (b) the cdf of X, (c) $P[10 < X \le 100]$.

1.16. Let X be a random variable with pmf

$$f(x) = c/2^x, \quad x = 1, 2, \ldots.$$

Find (a) the constant c, (b) the cdf of X, (c) $P[X \ge 5]$.

1.17. A game consists of tossing a fair coin and then drawing a random number from $0, 1, \ldots, 9$ with equal probabilities if head comes up, or spinning a wheel to choose a random number in the interval $(0, 10]$ if tail comes up. Let X be the outcome of this game, which is either an integer between 0 and 9 or a number in the interval $(0, 10]$.

(a) Find the cdf of X and $P[3 \le X \le 5]$. Note that the rv X is neither discrete nor continuous, but a mixture.

(b) Express X as a mixture, $X = \alpha U + (1-\alpha)V$, where U is discrete and V is continuous, giving the pmf of U and the pdf of V. [Interpretation of mixture is $F_X = \alpha F_U + (1-\alpha)F_V$.]

1.18. (*Censoring: another example of a mixture*) Let T be a nonnegative rv (life of an equipment or survival time of a patient undergoing a certain treatment) and $t_c > 0$ is a time at which observation stops. So we observe T if $0 \le T < t_c$ and t_c if $T \ge t_c$. The resulting observation is $Y = \min(T, t_c)$. Find the cdf of Y which is a mixture.

1.19. Verify (i)–(vi) of Proposition 1.8.1.

1.20. Verify (i)–(vi) of Proposition 1.8.2.

1.21. Let (X, Y) be a two-dimensional rv with joint pdf

$$f(x, y) = c(x + 2y), \quad 0 \le x, \; y \le 1.$$

(a) Find the constant c and $P[X \le Y]$.

(b) Find the marginal pdf's f_X of X and f_Y of Y.

(c) Calculate μ_X, μ_Y, σ_X^2, and σ_Y^2. Also find the medians of X and Y.

(d) Find Cov$[X, Y]$ and ρ_{XY}.

1.22. The joint pdf of (X, Y) is

$$f(x, y) = c(2x^2 + xy), \quad 0 \le x \le 1, \; 0 \le y \le 2.$$

Find

(a) the constant c, (b) the marginal pdf f_X of X, (c) the conditional pdf $f_{Y|X}(y|x)$ of Y given $X = x$, (d) the regression function $m(x) = \mathrm{E}[Y|X = x]$.

1.23. The pdf of X and the conditional pdf of Y given $X = x$ are given by

$$f_X(x) = \exp(-x), \; 0 < x < \infty, \text{ and } f_{Y|X}(y|x) = x\exp(-xy), \; 0 < y < \infty.$$

(a) Find the joint pdf of (X, Y), the marginal pdf of Y, and the conditional pdf of X given $Y = y$.

(b) Find the regression function $m(x) = \mathrm{E}[Y|X = x]$.

1.24. The joint pdf of (X, Y) is

$$f(x, y) = cx\exp(-x), \quad 0 < y < x < \infty.$$

Find

(a) the constant c and the marginal pdf's f_X and f_Y,

(b) $\mathrm{E}[X], \mathrm{Var}[X], \mathrm{E}[Y], \mathrm{Var}[Y], \mathrm{Cov}[X, Y]$, and ρ_{XY},

(c) the conditional pdf of Y given $X = x$, $\mathrm{E}[Y|X = x]$ and $\mathrm{Var}[Y|X = x]$, and

(d) verify: $\mathrm{E}[Y[= \mathrm{E}[\mathrm{E}(Y|X)]$ and $\mathrm{Var}[Y] = \mathrm{E}[\mathrm{Var}(Y|X)] + \mathrm{Var}[\mathrm{E}(Y|X)]$.

2

Some Common Probability Distributions

2.1 Discrete Distributions

We shall start with the simplest nontrivial rv X which can take only one of two values, say 0 and 1, with $P[X = 0] = 1 - p = q$ and $P[X = 1] = p$ where $0 < p < 1$. This arises in the context of a random experiment resulting in success or failure, such as the toss of a coin resulting in head (success) or tail (failure), or a medicine having favorable effect (success) or not (failure) on a patient, etc., with $P[\text{success}] = p$ and $P\left[\text{failure}\right] = q = 1 - p$. The rv X takes the values 1 or 0 for the outcomes success or failure, respectively. Such an rv is called a *Bernoulli*(p) rv and we write the pmf of X as

$$f_X(x) = p^x(1-p)^{1-x}, \quad x = 0, 1. \tag{1}$$

2.1.1 Binomial Distribution *Bin*(n, p)

Suppose $X_1, \ldots, X_n$ are independent *Bernoulli*(p) rv's as described by Eq. (1) and let $X = X_1 + \cdots + X_n$, that is, X = total number of successes in n independent experiments with $P[\text{success}] = p$ in each experiment. The pmf of X is obtained by noting that each sequence of n outcomes with x successes and $n - x$ failures has probability $= p^x(1-p)^{n-x}$, and the number of such sequences is the number of ways in which x trials (resulting in successes) can be chosen from n trials, which is $\binom{n}{x} = \frac{n!}{x!(n-x)!}$. Hence the pmf of X is

$$f_X(x) = \binom{n}{x} p^x(1-p)^{n-x}, \quad x = 0, 1, \ldots, n.$$

Since $\mathrm{E}[X_1] = 1 \cdot p + 0 \cdot (1-p) = p$ and

$$\begin{aligned} \mathrm{Var}[X_1] &= \mathrm{E}\left[X_1^2\right] - (\mathrm{E}[X_1])^2 \\ &= \left[1^2 \cdot p + 0^2 \cdot (1-p)\right] - p^2 \\ &= p - p^2 = p(1-p), \end{aligned}$$

we have

$$\mathrm{E}[X] = n\mathrm{E}[X_1] = np \text{ and } \mathrm{Var}[X] = n\,\mathrm{Var}[X_1] = np(1-p). \tag{2}$$

Theory and Methods of Statistics. http://dx.doi.org/10.1016/B978-0-12-802440-9.00002-3

Example 2.1.1. An equipment with n components needs at least 3 to function properly for its overall effectiveness. If each of these components function with probability 4/5, independently of one another, what should be the minimum n for the equipment to remain effective with probability 0.9 or more?

Solution.

$$\begin{aligned} P[\text{Not Effective}] &= \sum_{x=0}^{2} \binom{n}{x} \left(\frac{4}{5}\right)^{x} \left(\frac{1}{5}\right)^{n-x} \\ &= \frac{1}{5^n}\left[\binom{n}{0}4^0 + \binom{n}{1}4^1 + \binom{n}{2}4^2\right] \\ &= \frac{1}{5^n}[1 + 4n + 8n(n-1)] \\ &= \frac{1}{5^n}\left[1 - 4n + 8n^2\right]. \end{aligned}$$

We want the smallest n so that $\frac{1}{5^n}\left[1 - 4n + 8n^2\right] \leq \frac{1}{10}$.

n	$\left(1 - 4n + 8n^2\right)/5^n$
3	$(1 - 12 + 72)/5^3 = 0.488$
4	$(1 - 16 + 128)/5^4 = 0.1808$
5	$(1 - 20 + 200)/5^5 = 0.05792$

so the minimum n to keep the equipment effective, $n_{\min} = 5$.

2.1.2 Multinomial Distribution $Multi(n; p_1, \ldots, p_k)$

Extending the concept of the *Binomial*(n, p) distribution from the outcomes of n independent trials, each with two possible outcomes, consider n independent trials with k possible outcomes $1, \ldots, k$ with

$$P[i\text{th outcome}] = p_i > 0, \quad \sum_{i=1}^{k} p_i = 1.$$

Let $X = (X_1, \ldots, X_k)$ with X_i = number of trials resulting in the ith outcome. Then any sequence of n outcomes with x_i = number of trials with the ith outcome, $i = 1, \ldots, k$ has probability

$$p_1^{x_1} \cdots p_k^{x_k} \text{ if } x_i \geq 0, \quad \sum_{i=1}^{k} x_i = n \text{ and } 0 \text{ otherwise},$$

and the number of such sequences is the number of ways in which such $(x_1, \ldots, x_k)$ can be chosen from the n trials. This number is $n!/(x_1!, \ldots, x_k!)$, the multinomial coefficient. Thus, the joint pmf of $X = (X_1, \ldots, X_k)$ is

$$f_X(x_1,\ldots,x_k) = \frac{n!}{x_1!\cdots x_k!}p_1^{x_1}\cdots p_k^{x_k}$$

$$\text{if } x_1,\ldots,x_k \geq 0 \text{ and } \sum_{i=1}^{k} x_i = n, \text{ and 0 otherwise.}$$

The multinomial coefficient is obtained by multiplying the number of successive choices of x_1 out of n trials, followed by x_2 out of $n - x_1$ trials and so on, and finally, x_{k-1} out of $n - (x_1 + \cdots + x_{k-2})$. Thus, using $x_k = n - \sum_{i=1}^{k-1} x_i$, we have

$$\begin{aligned}&\binom{n}{x_1}\binom{n-x_1}{x_2}\cdots\binom{n-(x_1+\cdots+x_{k-2})}{x_{k-1}}\\ &\quad= \frac{n!}{x_1!(n-x_1)!}\cdot\frac{(n-x_1)!}{x_2!(n-x_1-x_2)!}\cdots\frac{(n-x_1-\cdots-x_{k-2})!}{x_{k-1}!x_k!}.\end{aligned}$$

By cancellation, this yields the multinomial coefficient. To find $\mathrm{E}[X_i], \mathrm{Var}[X_i]$, and $\mathrm{Cov}[X_i, X_j]$ for any $i \neq j$, note that each $X_i \sim Bin(n, p_i)$ and $X_j|X_i \sim Bin(n - X_i,\ p_j/(1-p_i))$. Hence

$$\mathrm{E}[X_i] = np_i, \quad \mathrm{Var}[X_i] = np_i(1-p_i) \text{ and}$$

as in Eq. (2), and using Proposition 1.10.1,

$$\begin{aligned}\mathrm{Cov}[X_i, X_j] &= \mathrm{E}[X_iX_j] - \mathrm{E}[X_i]\mathrm{E}[X_j]\\ &= \mathrm{E}[X_i\mathrm{E}(X_j|X_i)] - n^2p_ip_j\\ &= \mathrm{E}[X_i(n-X_i)p_j/(1-p_i)] - n^2p_ip_j\\ &= \frac{p_j}{1-p_i}\left[n\cdot np_i - \{np_i(1-p_i) + n^2p_i^2\}\right] - n^2p_ip_j\\ &= -np_ip_j\end{aligned}$$

by algebraic simplification.

2.1.3 Geometric Distribution *Geom* (*p*)

In a sequence of independent trials, each resulting in success with probability p or failure with probability $1 - p = q$, let X be the number of trials needed for the first success to occur. Then the pmf of X is

$$\begin{aligned}f_X(n) &= P[\text{Failures in the first } n-1 \text{ trials and success in the } n\text{th trial}]\\ &= q^{n-1}p, \quad n = 1, 2, \ldots\end{aligned}$$

Then $\sum_{n=1}^{\infty} f_X(n) = \sum_{n=1}^{\infty} q^{n-1}p = p\sum_{s=0}^{\infty} q^s = p(1-q)^{-1} = 1$, and

$$\begin{aligned}
\mathrm{E}[X] &= \sum_{n=1}^{\infty} n f_X(n) = \sum_{n=1}^{\infty} n q^{n-1} p = p \frac{d}{dq} \sum_{s=0}^{\infty} q^s \\
&= p \cdot \frac{d}{dq}(1-q)^{-1} = p(1-q)^{-2} = \frac{1}{p}
\end{aligned}$$

$$\begin{aligned}
\mathrm{E}\left[X^2\right] &= \sum_{n=1}^{\infty} n^2 f_X(n) = \sum_{n=1}^{\infty} n^2 q^{n-1} p = p \sum_{n=1}^{\infty} [n + n(n-1)] q^{n-1} \\
&= p \sum_{n=1}^{\infty} n q^{n-1} + pq \sum_{n=1}^{\infty} n(n-1) q^{n-2} = \mathrm{E}[X] + pq \frac{d^2}{dq^2} \sum_{s=0}^{\infty} q^s \\
&= \frac{1}{p} + pq \frac{d^2}{dq^2}(1-q)^{-1} = \frac{1}{p} + pq \cdot 2(1-q)^{-3} = \frac{2}{p^2} - \frac{1}{p}.
\end{aligned}$$

$$\mathrm{Var}[X] = \mathrm{E}\left[X^2\right] - (\mathrm{E}[X])^2 = \left(\frac{2}{p^2} - \frac{1}{p}\right) - \frac{1}{p^2} = \frac{q}{p^2}.$$

Remark 2.1.1. Memoryless property of the Geometric distribution. If $X \sim Geom(p)$, then $P[X > k] = p \sum_{n=k+1}^{\infty} q^{n-1} = pq^k \sum_{s=0}^{\infty} q^s = pq^k (1-q)^{-1} = q^k$. More directly, $P[X > k] = P[\text{all failures in the first } k \text{ trials}] = q^k$. So

$$P[X = k + n | X > k] = \frac{P[X = k + n]}{P[X > k]} = \frac{q^{k+n-1} p}{q^k} = q^{n-1} p = P[X = n].$$

2.1.4 Negative Binomial Distribution $N\ Bin(r, p)$

In a sequence of independent trials resulting in success or failure with probabilities p and $q = 1 - p$ respectively, let X be the number of trials needed for the rth success to occur. Then

$$\begin{aligned}
f_X(n) &= P[(r-1) \text{ successes in the first } (n-1) \text{ trials and success on the } n\text{th trial}] \\
&= \binom{n-1}{r-1} p^{r-1} q^{n-r}, \quad n = r, r+1, \ldots
\end{aligned}$$

Note that this X can be expressed as $X = Y_1 + \cdots + Y_r$, where $Y_1, \ldots, Y_r$ are independent $Geom(p)$. Hence

$$\mathrm{E}[X] = r\mathrm{E}[Y_1] = r/p \text{ and } \mathrm{Var}[X] = r\ \mathrm{Var}[Y_1] = rq/p^2.$$

Note. A binomial rv is the number of successes in a given number of trials, whereas, a negative binomial rv is the number of trials needed for a given number of successes.

Example 2.1.2. A target-shooter can hit the bull's eye once in three attempts on average, that is, with probability 1/3. In a tournament with an entry fee of \$10, there are prizes of \$100, \$50, and \$20 if one can hit the bull's eye three times in 3, 4, or 5 attempts, respectively. Should our target-shooter enter this tournament paying the fee?

Solution. Let N = number of attempts needed to hit the bull's eyes $r = 3$ times, where the probability of a hit is $p = 1/3$ and the results of the attempts are mutually independent, and for $N = n$, let $X = X(n)$ be the prize won. Then $X(3) = \$100$, $X(4) = \$50$, $X(5) = \$20$, $X(n) = 0$ for $n > 5$, and N is $NB(r = 3,\ p = 1/3)$. Since

$$P[N=n] = \binom{n-1}{r-1} p^r (1-p)^{n-r} = \binom{n-1}{2}\left(\frac{1}{3}\right)^3\left(\frac{2}{3}\right)^{n-3},$$

$$\begin{aligned}
P[N=3] &= \binom{2}{2}\left(\frac{1}{3}\right)^3\left(\frac{2}{3}\right)^0 = \frac{1}{27}\\
P[N=4] &= \binom{3}{2}\left(\frac{1}{3}\right)^3\left(\frac{2}{3}\right)^1 = \frac{2}{27}\\
P[N=5] &= \binom{4}{2}\left(\frac{1}{3}\right)^3\left(\frac{2}{3}\right)^2 = \frac{8}{81}.
\end{aligned}$$

So $\mathrm{E}[X] = \$100 \cdot \frac{1}{27} + \$50 \cdot \frac{2}{27} + \$20 \cdot \frac{8}{81} = \$\frac{760}{81} \approx \$9.38$. Since the entry fee of \$10 is greater than the expected value of prize to be won, which is \$9.38, this game is not in favor of our target-shooter, who (strictly speaking) should not enter unless for fun, because the expected loss is very little.

2.1.5 Hypergeometric Distribution (n, N, m)

Suppose that from a box containing N balls of which m are white and $N - m$ are red, n balls are drawn at random *without replacement*. Then the number of white balls in the sample is an rv X with pmf

$$f_X(x) = \binom{m}{x}\binom{N-m}{n-x}\Big/\binom{N}{n}, \quad x = 0, \ldots, n,$$

with the convention $\binom{r}{k} = 0$ for $k < 0$ or $k > r$. Thus the effective range of X is $\max(0,\ m+n-N) \le x \le \min(m, n)$. To find the mean and variance of X, we calculate $\mathrm{E}\left[X^k\right]$ for $k = 1$ and $k = 2$, using $x\binom{m}{x} = m\binom{m-1}{x-1}$ and $n\binom{N}{n} = N\binom{N-1}{n-1}$ as shown below.

$$\begin{aligned}
\mathrm{E}\left[X^k\right] &= \sum_{x=0}^{n} x^k \cdot \frac{\binom{m}{x}\binom{N-m}{n-x}}{\binom{N}{n}}\\
&= \sum_{x=1}^{n} x^{k-1} \cdot \frac{m\binom{m-1}{x-1}\binom{N-m}{n-x}}{\frac{N}{n}\binom{N-1}{n-1}}\\
&= \frac{mn}{N}\sum_{x-1=0}^{n-1} x^{k-1} \cdot \frac{\binom{m-1}{x-1}\binom{N-m}{n-x}}{\binom{N-1}{n-1}}
\end{aligned}$$

$$= \frac{mn}{N} \sum_{y=0}^{n-1} (y+1)^{k-1} \cdot \frac{\binom{m-1}{y}\binom{N-m}{n-1-y}}{\binom{N-1}{n-1}}$$

$$= \frac{mn}{N} \mathrm{E}\left[(Y+1)^{k-1}\right],$$

where Y is Hypergeometric $(n-1, N-1, m-1)$. Hence

$$\mathrm{E}[X] = \frac{mn}{N} \mathrm{E}\left[(Y+1)^0\right] = \frac{mn}{N} \cdot 1 = \frac{mn}{N} = np,$$

where $p = m/N$ = proportion of white balls in the box, and

$$\begin{aligned}
\mathrm{Var}[X] &= \mathrm{E}\left[X^2\right] - (\mathrm{E}[X])^2 \\
&= np\{\mathrm{E}[Y] + 1\} - (np)^2 \\
&= np\left\{\frac{(m-1)(n-1)}{N-1} + 1\right\} - (np)^2 \\
&= \frac{N-n}{N-1} np(1-p)
\end{aligned}$$

by algebraic simplification.

Remark 2.1.2.

1. If in the above set-up, n balls are drawn at random with replacement , then the number of white balls in the sample is distributed as $Bin(n, p)$. In the without replacement case, the mean of the hypergeometric rv remains the same as in $Bin(n, p)$, but the variance is reduced by a factor of $\frac{N-n}{N-1}$.
2. For any n, the factor $\frac{N-n}{N-1} \to 1$ as $N \to \infty$, which agrees with the common sense that for sampling from a very large collection with $\frac{m}{N} = p$, sampling with or without replacement are practically the same.

2.1.6 Poisson Distribution *Poi*(λ)

The Poisson Distribution is an approximation of the Binomial Distribution $Bin(n, p)$, where n is large, p is small, and np is of moderate magnitude. So let $n \to \infty$, $p \to 0$ in such a manner that $np \to \lambda \in (0, \infty)$, and approximate the pmf of $Bin(n, p)$ for such n, p as follows:

$$\begin{aligned}
f_X(x; n, p) &= \frac{n(n-1)\cdots(n-x+1)}{x!} p^x (1-p)^{n-x} \\
&= \frac{(np)^x}{x!}\left[\left(1-\frac{1}{n}\right)\left(1-\frac{2}{n}\right)\cdots\left(1-\frac{x-1}{n}\right)\right][1-p]^{-x}\left[1-\frac{np}{n}\right]^n \\
&:= A_{1n}A_{2n}A_{3n}A_{4n},
\end{aligned}$$

where $A_{1n} \to \lambda^{\lambda}/x!$, $A_{2n} \to 1$, $A_{3n} \to 1$ and since $(1-\frac{u}{n})^n \to \mathrm{e}^{-a}$, which extends to $(1-\frac{a_n}{n})^n \to \mathrm{e}^{-a}$, if $a_n \to a$, we see that $A_{4n} \to \mathrm{e}^{-\lambda}$. Thus the pmf of $Bin(n, p)$, as $n \to \infty$

and $p \to 0$ in the above manner, tends to the following:

$$f_X(x; n, p) \to f_X(x) = \frac{\lambda^x}{x!} \cdot 1 \cdot 1 \cdot e^{-\lambda} = e^{-\lambda}\frac{\lambda^x}{x!}, \quad x = 0, 1, 2, \ldots$$

This limiting pmf of $Bin(n, p)$ as $n \to \infty, p \to 0$ so that $np \to \lambda \in (0, \infty)$ is the pmf of the Poisson distribution $Poi(\lambda)$:

$$f_X(x) = e^{-\lambda}\frac{\lambda^x}{x!}, \quad x = 0, 1, 2, \ldots$$

and

$$\sum_{x=0}^{\infty} f_X(x) = e^{-\lambda}\sum_{x=0}^{\infty}\frac{\lambda^x}{x!} = e^{-\lambda}e^{\lambda} = 1.$$

The mean and the variance of $Poi(\lambda)$ are

$$\begin{aligned} E[X] &= \sum_{x=0}^{\infty} x e^{-\lambda}\frac{\lambda^x}{x!} = \sum_{x=1}^{\infty} e^{-\lambda}\frac{\lambda^x}{(x-1)!} \\ &= \lambda\sum_{y=0}^{\infty} e^{-\lambda}\frac{\lambda^y}{y!} = \lambda \cdot 1 = \lambda \text{ and,} \\ E[X(X-1)] &= \sum_{x=0}^{\infty} x(x-1)e^{-\lambda}\frac{\lambda^x}{x!} = \sum_{x=2}^{\infty} e^{-\lambda}\frac{\lambda^x}{(x-2)!} \\ &= \lambda^2\sum_{y=0}^{\infty} e^{-\lambda}\frac{\lambda^y}{y!} = \lambda^2 \cdot 1 = \lambda^2, \text{ so} \\ Var[X] &= E\left[X^2\right] - (E[X])^2 = E[X(X-1)] + E[X] - (E[X])^2 \\ &= \lambda^2 + \lambda - \lambda^2 = \lambda. \end{aligned}$$

Some Properties of $Poi(\lambda)$.

1. Let X_1, X_2 be independent, $X_i \sim Poi(\lambda_i)$. Then $X = X_1 + X_2 \sim Poi(\lambda_1 + \lambda_2)$, by Eq. (12b) of Chapter 1,

$$\begin{aligned} f_X(x) &= \sum_{x_1=0}^{x} f_{X_1}(x_1)f_{X_2}(x - x_1) = e^{-(\lambda_1+\lambda_2)}\sum_{x_1=0}^{x}\frac{\lambda_1^{x_1}}{x_1!}\frac{\lambda_2^{x-x_1}}{(x-x_1)!} \\ &= e^{-(\lambda_1+\lambda_2)} \cdot \frac{(\lambda_1+\lambda_2)^x}{x!}\sum_{x_1=0}^{x}\binom{x}{x_1}\left(\frac{\lambda_1}{\lambda_1+\lambda_2}\right)^{x_1}\left(\frac{\lambda_2}{\lambda_1+\lambda_2}\right)^{x-x_1} \\ &= e^{-(\lambda_1+\lambda_2)} \cdot \frac{(\lambda_1+\lambda_2)^x}{x!} \cdot 1, \end{aligned}$$

which is the pmf of $Poi(\lambda_1 + \lambda_2)$. More generally, $X_1, \ldots, X_k$ independent with $X_i \sim Poi(\lambda_i)$ implies $\sum_{i=1}^{k} X_i \sim Poi\left(\sum_{i=1}^{k}\lambda_i\right)$.

2. If X_1, X_2 are independent with $X_i \sim Poi(\lambda_i)$, then conditionally, given $X_1 + X_2 = n$, $X_1 \sim Bin(n, \lambda_1/(\lambda_1 + \lambda_2))$.
Proof.

$$\begin{aligned} P[X_1 = x | X_1 + X_2 = n] &= \frac{P[X_1 = x, X_2 = n - x]}{P[X_1 + X_2 = n]} \\ &= \frac{e^{-(\lambda_1+\lambda_2)} \frac{\lambda_1^x}{x!} \frac{\lambda_2^{n-x}}{(n-x)!}}{e^{-(\lambda_1+\lambda_2)} \frac{(\lambda_1+\lambda_2)^n}{n!}} \\ &= \frac{n!}{x!(n-x)!} \left(\frac{\lambda_1}{\lambda_1 + \lambda_2}\right)^x \left(\frac{\lambda_2}{\lambda_1 + \lambda_2}\right)^{n-x}, \end{aligned}$$

which is the pmf of $Bin(n, \lambda_1/(\lambda_1 + \lambda_2))$. □

More generally, if $X_i \sim Poi(\lambda_i)$ $i = 1, \ldots, k$ are independent, then conditionally, given $X_1 + \cdots + X_k = n$, $(X_1, \ldots, X_k) \sim Mult(n; p_1, \ldots, p_k)$ with $p_i = \lambda_i/(\lambda_1 + \cdots + \lambda_k)$.

3. If $N \sim Poi(\lambda)$ and conditionally, given $N = n$, $X_1, \ldots, X_n$ are independent *Bernoulli*(p), then $X_1 + \cdots + X_N = X \sim Poi(\lambda p)$.
Proof.

$$\begin{aligned} P[X = x] &= \sum_{n=x}^{\infty} P[N = n, X_1 + \cdots + X_n = x] \\ &= \sum_{n=x}^{\infty} \frac{e^{-\lambda}\lambda^n}{n!} \frac{n!}{x!(n-x)!} p^x (1-p)^{n-x} \\ &= e^{-\lambda} \cdot \frac{(\lambda p)^x}{x!} \sum_{n=x}^{\infty} \frac{\{\lambda(1-p)\}^{n-x}}{(n-x)!} \\ &= e^{-\lambda} \frac{(\lambda p)^x}{x!} \cdot e^{\lambda(1-p)} = e^{-\lambda p} \frac{(\lambda p)^x}{x!}. \end{aligned}$$

□

Example 2.1.3. An insurance agent has sold fire insurance policies to 200 homeowners in a town. If the probability of a fire in a house during a year is 1/250 in this town, what is the probability that this agent will have to handle two or more claims in a year?

Solution. The number of policies $n = 200$ (large) and the probability of a fire is $p = 1/250$ (small), so $\lambda = np = 0.8$. Assuming independence, X = number of claims = number of fires, is approximately $Poi(\lambda = 0.8)$. Hence

$$\begin{aligned} P[X \geq 2] &= 1 - (P[X = 0] + P[X = 1]) = 1 - e^{-0.8}[1 + 0.8] \\ &= 1 - 0.4493 \cdot 1.8 = 1 - 0.8088 = 0.1912. \end{aligned}$$

Example 2.1.4. If the number of tropical storms on the Gulf of Mexico during September follows a Poisson distribution with mean $\lambda = 5$ and if each of these storms can become a hurricane with probability $p = 1/4$, what is the probability of at least one hurricane in September?

Solution. N = Number of storms $\sim Poi(\lambda = 5)$, and $X_1, \ldots, X_N$ given N are iid *Bernoulli*$(p = 1/4)$, where $X_i = 1$ or 0 according as the ith storm becomes a hurricane or not. Hence X = number of hurricanes $= X_1 + \cdots + X_N \sim Poi(\lambda p = 1.25)$, and so

$$
\begin{aligned}
P[\text{At least one hurricane}] &= P[X \geq 1] \\
&= 1 - P[X = 0] = 1 - \mathrm{e}^{-1.25} \\
&= 1 - 0.2865 = 0.7135.
\end{aligned}
$$

2.2 Continuous Distributions

2.2.1 The Gamma and Beta Functions

These two functions are essential in connection with many distributions and their properties discussed here.

Definition 2.2.1. The gamma function is defined as

$$\Gamma(\alpha) = \int_0^\infty \mathrm{e}^{-u} u^{\alpha-1}\, du, \quad \alpha > 0$$

and the beta function is defined as

$$Be(\alpha_1, \alpha_2) = \int_0^1 t^{\alpha_1 - 1}(1-t)^{\alpha_2 - 1}\, dt, \quad \alpha_1, \alpha_2 > 0.$$

Properties of the Gamma and Beta Functions.

(i) $\Gamma(\alpha+1) = \alpha\Gamma(\alpha)$ for $\alpha > 0$, so for a positive integer α, $\Gamma(\alpha+1) = \alpha!$.
Proof. Integrating by parts,

$$
\begin{aligned}
\Gamma(\alpha+1) &= \int_0^\infty \mathrm{e}^{-u} u^{\alpha}\, du = -\int_0^\infty u^\alpha d(\mathrm{e}^{-u}) \\
&= \int_0^\infty \mathrm{e}^{-u} d(u^\alpha) = \alpha \int_0^\infty \mathrm{e}^{-u} u^{\alpha-1}\, du \\
&= \alpha\Gamma(\alpha),
\end{aligned}
$$

because $\mathrm{e}^{-u}u^\alpha = 0$ at $u = 0$ and $\to 0$ as $u \to \infty$. □

(ii) $\Gamma\left(\frac{1}{2}\right) = \sqrt{\pi}$.
Proof.

$$
\begin{aligned}
\left\{\Gamma\left(\frac{1}{2}\right)\right\}^2 &= \int_0^\infty \int_0^\infty \mathrm{e}^{-u+v} u^{-1/2} v^{-1/2}\, du\, dv \\
&= \int_0^\infty \int_0^1 \mathrm{e}^{-s}(st)^{-1/2}\{s(1-t)\}^{-1/2} \cdot s\, ds\, dt,
\end{aligned}
$$

with $u + v = s$, $u/(u+v) = t$, so that $du\, dv = s\, ds\, dt$. Next let $\sin^2\theta = t$, so that $dt = 2\sin\theta\cos\theta\, d\theta$ and $\{t(1-t)\}^{1/2} = \sin\theta\cos\theta$. Hence

$$
\begin{aligned}
\left\{\Gamma\left(\frac{1}{2}\right)\right\}^2 &= \int_0^\infty \mathrm{e}^{-s}\, ds \cdot \int_0^1 \{t(1-t)\}^{-1/2}\, dt \\
&= 1 \cdot \int_0^{\pi/2} \frac{2\sin\theta\cos\theta}{\sin\theta\cos\theta}\, d\theta \\
&= 2 \cdot \frac{\pi}{2} = \pi.
\end{aligned}
$$

□

(iii) $Be(\alpha_1, \alpha_2) = \frac{\Gamma(\alpha_1)\Gamma(\alpha_2)}{\Gamma(\alpha_1+\alpha_2)}$. This will follow as part of the proof of Proposition 2.2.1.

We now introduce some important continuous distributions and examine some of their properties.

2.2.2 Uniform Distribution *Unif*(*a*, *b*)

Definition 2.2.2. X is a $Unif(a, b)$ rv if $f_X(x) = I_{[a,b]}(x)/(b-a)$.

Of course, $\int_a^b (b-a)^{-1}\, dx = (b-a)^{-1}(b-a) = 1$.

$$\mathrm{E}[X] = \int_a^b \frac{x}{b-a}\, dx = \frac{1}{b-a} \cdot \frac{b^2-a^2}{2} = \frac{a+b}{2},$$

$$\mathrm{E}\left[X^2\right] = \int_a^b \frac{x^2}{b-a}\, dx = \frac{b^3-a^3}{3(b-a)} = \frac{a^2+ab+b^2}{3}, \text{ and}$$

$$\mathrm{Var}[X] = \frac{a^2+ab+b^2}{3} - \frac{a^2+2ab+b^2}{4} = \frac{(b-a)^2}{12}.$$

In particular, if $X \sim Unif(0,1)$, then $f_X(x) = I_{[0,1]}(x)$, $\mathrm{E}[X] = \frac{1}{2}$, and $\mathrm{Var}[X] = \frac{1}{12}$.

2.2.3 Gamma and Beta distributions

Definition 2.2.3. X is a $Gamma(\alpha, \beta)$ rv if the pdf of X is

$$f_X(x) = \frac{1}{\Gamma(\alpha)\beta^\alpha} e^{-x/\beta} x^{\alpha-1} I_{(0,\infty)}(x), \quad \alpha > 0, \ \beta > 0.$$

The mean and variance of a $Gamma(\alpha, \beta)$ rv are

$$\begin{aligned}
\mathrm{E}[X] &= \frac{1}{\Gamma(\alpha)\beta^\alpha} \int_0^\infty x e^{-x/\beta} x^{\alpha-1}\, dx = \frac{\beta}{\Gamma(\alpha)} \int_0^\infty e^{-u} u^\alpha\, du \\
&= \frac{\beta}{\Gamma(\alpha)} \Gamma(\alpha+1) = \alpha\beta, \\
\mathrm{Var}[X] &= \mathrm{E}\left[X^2\right] - (\alpha\beta)^2 = \frac{\beta^2}{\Gamma(\alpha)} \int_0^\infty e^{-u} u^{\alpha+1}\, du - (\alpha\beta)^2 \\
&= \frac{\beta^2}{\Gamma(\alpha)} \Gamma(\alpha+2) - (\alpha\beta)^2 \\
&= \alpha(\alpha+1)\beta^2 - \alpha^2\beta^2 = \alpha\beta^2.
\end{aligned}$$

Proposition 2.2.1. *If X_i, $i = 1, 2$ are independent $Gamma(\alpha_i, \beta)$ rv's, then $X_1 + X_2$ is a $Gamma(\alpha_1 + \alpha_2, \beta)$ rv.*

Proof. By Eq. (12a) in Chapter 1,

$$
\begin{aligned}
f_{X_1+X_2}(z) &= \frac{1}{\Gamma(\alpha_1)\Gamma(\alpha_2)\beta^{\alpha_1+\alpha_2}} \int_0^z e^{-x/\beta} x^{\alpha_1-1} e^{-(z-x)/\beta}(z-x)^{\alpha_2-1}\,dx \\
&= \frac{e^{-z/\beta} z^{\alpha_1+\alpha_2-1}}{\Gamma(\alpha_1)\Gamma(\alpha_2)\beta^{\alpha_1+\alpha_2}} \int_0^z \left(\frac{x}{z}\right)^{\alpha_1-1}\left(1-\frac{x}{z}\right)^{\alpha_2-1} \frac{dx}{z} \\
&= \frac{e^{-z/\beta} z^{\alpha_1+\alpha_2-1}}{\Gamma(\alpha_1)\Gamma(\alpha_2)\beta^{\alpha_1+\alpha_2}} \int_0^1 t^{\alpha_1-1}(1-t)^{\alpha_2-1}\,dt \\
&= \frac{e^{-z/\beta} z^{\alpha_1+\alpha_2-1}}{\Gamma(\alpha_1)\Gamma(\alpha_2)\beta^{\alpha_1+\alpha_2}} \cdot Be(\alpha_1,\alpha_2), \quad z>0.
\end{aligned}
$$

But we must have

$$
\begin{aligned}
1 &= \int_0^\infty f_{X_1+X_2}(z)\,dz \\
&= \frac{Be(\alpha_1,\alpha_2)}{\Gamma(\alpha_1)\Gamma(\alpha_2)\beta^{\alpha_1+\alpha_2}} \int_0^\infty e^{-z/\beta} z^{\alpha_1+\alpha_2-1}\,dz \\
&= \frac{Be(\alpha_1,\alpha_2)}{\Gamma(\alpha_1)\Gamma(\alpha_2)} \Gamma(\alpha_1+\alpha_2).
\end{aligned}
$$

Hence $Be(\alpha_1,\alpha_2) = \frac{\Gamma(\alpha_1)\Gamma(\alpha_2)}{\Gamma(\alpha_1+\alpha_2)}$ as stated earlier, and therefore,

$$
f_{X_1+X_2}(z) = \frac{1}{\Gamma(\alpha_1+\alpha_2)\beta^{\alpha_1+\alpha_2}} e^{-z/\beta} z^{\alpha_1+\alpha_2-1} I_{0,\infty}(z).
$$

□

By induction, if X_i, $i = 1,\ldots,k$ are independent $Gamma(\alpha_i,\beta)$, then $\sum_{i=1}^k X_i \sim Gamma\left(\sum_{i=1}^k \alpha_i, \beta\right)$.

Definition 2.2.4. T is a $Be(\alpha_1,\alpha_2)$ rv if the pdf of T is

$$
f_T(t) = \frac{\Gamma(\alpha_1+\alpha_2)}{\Gamma(\alpha_1)\Gamma(\alpha_2)} t^{\alpha_1-1}(1-t)^{\alpha_2-1} I_{(0,1)}(t), \quad \alpha_1,\alpha_2>0.
$$

The mean and variance of a $Beta(\alpha_1,\alpha_2)$ rv are

$$
\begin{aligned}
E[T] &= \frac{\Gamma(\alpha_1+\alpha_2)}{\Gamma(\alpha_1)\Gamma(\alpha_2)} \int_0^1 t^{\alpha_1}(1-t)^{\alpha_2-1}\,dt \\
&= \frac{\Gamma(\alpha_1+\alpha_2)}{\Gamma(\alpha_1)\Gamma(\alpha_2)} Be(\alpha_1+1,\alpha_2) \\
&= \frac{\Gamma(\alpha_1+\alpha_2)}{\Gamma(\alpha_1)\Gamma(\alpha_2)} \cdot \frac{\Gamma(\alpha_1+1)\Gamma(\alpha_2)}{\Gamma(\alpha_1+\alpha_2+1)} \\
&= \frac{\alpha_1}{\alpha_1+\alpha_2},
\end{aligned}
$$

$$
\mathrm{Var}[T] = E\left[T^2\right] - \left(\frac{\alpha_1}{\alpha_1+\alpha_2}\right)^2
$$

$$= \frac{\alpha_1(\alpha_1+1)}{(\alpha_1+\alpha_2)(\alpha_1+\alpha_2+1)} - \left(\frac{\alpha_1}{\alpha_1+\alpha_2}\right)^2$$
$$= \frac{\alpha_1\alpha_2}{(\alpha_1+\alpha_2)^2(\alpha_1+\alpha_2+1)}.$$

Proposition 2.2.2. *If $X_i \sim Gamma(\alpha_i, \beta)$, $i = 1, 2$ are independent, then $Z = X_1 + X_2$ and $T = X_1/(X_1 + X_2)$ are independent, $Z \sim Gamma(\alpha_1 + \alpha_2, \beta)$ and $T \sim Be(\alpha_1, \alpha_2)$.*

Proof. Rewrite the transformation as: $X_1 = TZ$, $X_2 = (1-T)Z$. Then the Jacobian of the transformation is $|J| = z$, and

$$f_{Z,T}(z,t) = \frac{1}{\Gamma(\alpha_1)\Gamma(\alpha_2)\beta^{\alpha_1+\alpha_2}} e^{-z/\beta}(tz)^{\alpha_1-1}\{(1-t)z\}^{\alpha_2-1} z I_{(0,\infty)}(z) I_{(0,1)}(t)$$
$$= \frac{e^{-z/\beta} z^{\alpha_1+\alpha_2-1}}{\Gamma(\alpha_1+\alpha_2)\beta^{\alpha_1+\alpha_2}} I_{(0,\infty)}(z) \cdot \frac{\Gamma(\alpha_1+\alpha_2)}{\Gamma(\alpha_1)\Gamma(\alpha_2)} t^{\alpha_1-1}(1-t)^{\alpha_2-1} I_{(0,1)}(t),$$

as was to be shown. □

2.2.4 Exponential Distribution

Definition 2.2.5. A rv X is distributed as Exp(θ) if X has pdf

$$f_X(x) = \frac{1}{\theta} e^{-x/\theta}, \quad \theta > 0,$$

and cdf

$$F_X(t) = \frac{1}{\theta}\int_0^t e^{-x/\theta}\, dx = 1 - e^{-t/\theta}.$$

Clearly, Exp(θ) is a special case of $Gamma(\alpha, \beta)$ with $\alpha = 1$ and $\beta = \theta$. Hence, $E[X] = \theta$ and $Var[X] = \theta^2$.

The exponential distribution shares with geometric distribution, the memoryless property:

$$P[X > t+s | X > t] = \frac{P[X > t+s]}{P[X > t]} = \frac{e^{-(t+s)/\theta}}{e^{-t/\theta}} = e^{-s/\theta} = P[X > s].$$

2.2.5 Normal Distribution $N(\mu, \sigma^2)$

Definition 2.2.6. X is an $N(\mu, \sigma^2)$ rv if X has pdf

$$f_X(x) = \frac{1}{\sqrt{2\pi}\sigma} e^{-\frac{1}{2\sigma^2}(x-\mu)^2}, \quad -\infty < x < \infty.$$

We first verify that $f_X(x)$ is indeed a pdf. Since $f_X(x) \geq 0$, we only need to check that $\int_{-\infty}^{\infty} f_X(x)\, dx = 1$. By symmetry,

$$\int_{-\infty}^{\infty} f_X(x)\, dx = \frac{1}{\sqrt{2\pi}}\int_{-\infty}^{\infty} e^{-z^2/2}\, dz = \frac{2}{\sqrt{2\pi}}\int_0^{\infty} e^{-z^2/2}\, dz$$
$$= \frac{2}{\sqrt{2\pi}}\int_0^{\infty} e^{-u} \cdot \frac{1}{\sqrt{2}} u^{-1/2}\, du \quad \left(\text{with } u = \frac{z^2}{2},\ dz = (2u)^{-1/2}\, du\right)$$

$$= \frac{1}{\sqrt{\pi}} \int_0^\infty e^{-u} u^{-1/2}\, du = \frac{1}{\sqrt{\pi}} \cdot \Gamma\left(\frac{1}{2}\right)$$
$$= 1, \text{ since } \Gamma\left(\frac{1}{2}\right) = \sqrt{\pi}.$$

The mean and variance of $N(\mu, \sigma^2)$ are

$$\begin{aligned}
\mathrm{E}[X] &= \int_{-\infty}^{\infty} x f_X(x)\, dx = \int_{-\infty}^{\infty} (x-\mu) f_X(x)\, dx + \mu \cdot 1 \\
&= \int_{-\infty}^{\infty} \sigma z \cdot \frac{1}{\sqrt{2\pi}} e^{-z^2/2}\, dz + \mu \\
&= \mu, \text{ because } \int_{-\infty}^{\infty} z e^{-z^2/2}\, dz = 0 \text{ by symmetry.} \\
\mathrm{Var}[X] &= \mathrm{E}[(X-\mu)]^2 \\
&= \int_{-\infty}^{\infty} (x-\mu)^2 \frac{1}{\sqrt{2\pi}\sigma} e^{-\frac{(x-\mu)^2}{2\sigma^2}}\, dx = \frac{2\sigma^2}{\sqrt{2\pi}} \int_0^\infty z^2 e^{-z^2/2}\, dz \\
&= \frac{2\sigma^2}{\sqrt{2\pi}} \int_0^\infty 2u e^{-u} \cdot (2u)^{-1/2}\, du \quad \left(\text{with } u = \frac{z^2}{2} \text{ as above}\right) \\
&= \frac{2\sigma^2}{\sqrt{\pi}} \int_0^\infty e^{-u} u^{1/2}\, du = \frac{2\sigma^2}{\sqrt{\pi}} \cdot \Gamma\left(\frac{3}{2}\right) \\
&= \frac{2\sigma^2}{\sqrt{\pi}} \cdot \frac{1}{2} \Gamma\left(\frac{1}{2}\right) = \sigma^2.
\end{aligned}$$

For this reason, $N(\mu, \sigma^2)$ is called a normal distribution with mean μ and variance σ^2. For $\mu = 0$ and $\sigma = 1$, $N(0,1)$ *is called the standard normal distribution.*

Example 2.2.1. The life of a tire on freeway is a Normal rv with mean $\mu = 25{,}000$ miles and sd $\sigma = 1{,}000$ miles. If a car goes on a coast-to-coast round-trip in the United States, which we can assume is 6000 miles, what is the probability that none of the four tires will need replacement during the 6000 miles trip, given that the tires were already used for 17,500 miles?

Solution. Note that

$$\begin{aligned}
P[\min(X_1, \ldots, X_4) > 23{,}500] &= \{P[X_1 > 23{,}500]\}^4 \\
&= \left\{P\left[\frac{X_1 - \mu}{\sigma} = Z > \frac{23{,}500 - 25{,}000}{1000}\right]\right\}^4 \\
&= \{P[Z > -1.5]\}^4 = \{P[Z < 1.5]\}^4 \\
&= \{\Phi(1.5)\}^4 = \{1 - 0.0668\}^4 = 0.7584,
\end{aligned}$$

using the normal table.

Strictly speaking, in this problem, we should have considered

$$P\left[\min(X_1, \ldots, X_4) > 23{,}500 | X_i > 17{,}500,\ i = 1, \ldots, 4\right],$$

but

$$P[X_i > 17{,}500] = P\left[Z > \frac{17{,}500 - 25{,}000}{1000}\right] = P[Z > -7.5] \approx 1,$$

so this conditionality does not matter.

pdf and cdf of $N(0,1)$. Let $Z \sim N(0,1)$. Then the pdf of Z is

$$f_Z(x) = \frac{1}{\sqrt{2\pi}} e^{-x^2/2}$$

which we shall denote by $\varphi(x)$. The cdf of Z is $F_Z(x) = \int_{-\infty}^{x} \varphi(t)\, dt$, which we shall denote by $\Phi(x)$. Since $\varphi(x)$ is symmetric about 0, we have

$$\Phi(-x) = 1 - \Phi(x) \text{ and } P[|Z| > x] = 2[1 - \Phi(x)].$$

The cdf $\Phi(x)$ cannot be calculated analytically. There are tables for $\Phi(x)$. However, the following are useful upper and lower bounds for $1 - \Phi(x)$ for all $x > 0$,

$$x^{-1}\left(1 - x^{-2}\right)\varphi(x) < 1 - \Phi(x) < x^{-1}\varphi(x).$$

By a change of variable $y = x^2/2$,

$$1 - \Phi(x) = \int_x^{\infty} \varphi(t)\, dt = (1/2)\pi^{-1/2} \int_{x^2/2}^{\infty} y^{-1/2} e^{-y}\, dy.$$

For notational convenience use $c = x^2/2$. The upper bound is obtained by noting that

$$(1/2)\pi^{-1/2} \int_c^{\infty} y^{-1/2} e^{-y}\, dy < (1/2)\pi^{-1/2} c^{-1/2} \int_c^{\infty} e^{-y}\, dy = (1/2)\pi^{-1/2} c^{-1/2} e^{-c} = x^{-1}\varphi(x).$$

To get the lower bound, use the integration by parts formula

$$\begin{aligned}
(1/2)\pi^{-1/2} \int_c^{\infty} y^{-1/2} e^{-y}\, dy &= (1/2)\pi^{-1/2}\left[c^{-1/2} e^{-c} - \frac{1}{2}\int_c^{\infty} y^{-3/2} e^{-y}\, dy\right] \\
&> (1/2)\pi^{-1/2}\left[c^{-1/2} e^{-c} - \frac{1}{2} c^{-3/2} \int_c^{\infty} e^{-c}\right] \\
&= (1/2)\pi^{-1/2}\left[c^{-1/2} e^{-c} - \frac{1}{2} c^{-3/2} e^{-c}\right] \\
&= (1/2)\pi^{-1/2} c^{-1/2}\left[1 - \frac{1}{2} c^{-1}\right] e^{-c} \\
&= x^{-1}\left(1 - x^{-2}\right)\varphi(x).
\end{aligned}$$

Normal Approximation to the Binomial Distribution: De Moivre-Laplace Theorem. If $S_n = X_1 + \cdots + X_n$, where $X_1, \ldots, X_n$ are independent *Bernoulli*(p), that is, $S_n \sim Bin(n, p)$, then for large n and for any $a < b$,

$$P\left[a \leq \frac{S_n - np}{\sqrt{np(1-p)}} \leq b\right] \approx \Phi(b) - \Phi(a).$$

This is a special case of the Central Limit Theorem which holds for sums of iid rv's $S_n = X_1 + \cdots + X_n$ with finite variance.

Some Properties of $N(\mu, \sigma^2)$.

1. If $X \sim N(\mu, \sigma^2)$, then $a + bX \sim N(a + b\mu, b^2\sigma^2)$ follows by using Eq. (10) in Chapter 1. In particular, $Z = (X - \mu)/\sigma \sim N(0, 1)$.
2. If $X_i \sim N(\mu_i, \sigma_i^2)$, $i = 1, 2$ are independent, then $X_1 + X_2 \sim N(\mu_1 + \mu_2, \sigma_1^2 + \sigma_2^2)$.
   Proof. $Z_i = (X_i - \mu_i)/\sigma_i$, $i = 1, 2$ are independent $N(0, 1)$, and
   $X_1 + X_2 = (\mu_1 + \mu_2) + (\sigma_1 Z_1 + \sigma_2 Z_2)$. Transform $\begin{pmatrix} Y_1 \\ Y_2 \end{pmatrix} = \begin{pmatrix} \sigma_1 & \sigma_2 \\ \sigma_2 & -\sigma_1 \end{pmatrix} \cdot \begin{pmatrix} Z_1 \\ Z_2 \end{pmatrix}$. Then
   $\left|\det\begin{bmatrix} \sigma_1 & \sigma_2 \\ \sigma_2 & -\sigma_1 \end{bmatrix}\right| = \sigma_1^2 + \sigma_2^2$, and $\begin{bmatrix} Z_1 \\ Z_2 \end{bmatrix} = \frac{1}{\sigma_1^2+\sigma_2^2}\begin{bmatrix} \sigma_1 & \sigma_2 \\ \sigma_2 & -\sigma_1 \end{bmatrix} \cdot \begin{bmatrix} Y_1 \\ Y_2 \end{bmatrix}$, and the joint pdf of $Y_1 = \sigma_1 Z_1 + \sigma_2 Z_2$ and $Y_2 = \sigma_2 Z_1 - \sigma_1 Z_2$ is

$$
\begin{aligned}
&f_{Y_1,Y_2}(y_1, y_2) \\
&= \frac{1}{2\pi\left(\sigma_1^2 + \sigma_2^2\right)} \exp\left[-\frac{1}{2\left(\sigma_1^2 + \sigma_2^2\right)^2}\left\{(\sigma_1 y_1 + \sigma_2 y_2)^2 + (\sigma_2 y_1 - \sigma_1 y_2)^2\right\}\right] \\
&= \frac{1}{2\pi\left(\sigma_1^2 + \sigma_2^2\right)} \exp\left[-\frac{1}{2\left(\sigma_1^2 + \sigma_2^2\right)^2}\left\{\left(\sigma_1^2 + \sigma_2^2\right)y_1^2 + \left(\sigma_1^2 + \sigma_2^2\right)y_2^2\right\}\right] \\
&= \frac{1}{2\pi\left(\sigma_1^2 + \sigma_2^2\right)} \exp\left[-\frac{y_1^2}{2\left(\sigma_1^2 + \sigma_2^2\right)} - \frac{y_2^2}{2\left(\sigma_1^2 + \sigma_2^2\right)}\right]
\end{aligned}
$$

   Thus Y_1, Y_2 are independent $N(0, \sigma_1^2 + \sigma_2^2)$, so

$$
\begin{aligned}
X_1 + X_2 &= (\mu_1 + \mu_2) + (\sigma_1 Z_1 + \sigma_2 Z_2) \\
&= (\mu_1 + \mu_2) + Y_1 \\
&\sim N\left(\mu_1 + \mu_2, \sigma_1^2 + \sigma_2^2\right).
\end{aligned}
$$

□

   By induction, if $X_i \sim N(\mu_i, \sigma_i^2)$, $i = 1, \ldots, k$ are mutually independent, then $\sum_{i=1}^k X_i \sim N\left(\sum_{i=1}^k \mu_i, \sum_{i=1}^k \sigma_i^2\right)$. In particular, if $X_1, \ldots, X_n$ are independent $N(\mu, \sigma^2)$, then $\sum_{i=1}^n X_i \sim N(n\mu, n\sigma^2)$, and $\bar{X}_n = \frac{1}{n}\sum_{i=1}^n X_i \sim N(\mu, \sigma^2/n)$.
3. If $Z \sim N(0, 1)$, then $W = Z^2$ has pdf

$$
f_W(w) = \frac{1}{2^{1/2}\Gamma\left(\frac{1}{2}\right)} e^{-w/2} w^{-1/2} I_{(0,\infty)}(w). \tag{3}
$$

   Proof. $W = Z^2$ has cdf (by Eq. (11b) in Chapter 1)

$$
F_{Z^2}(w) = P\left[-\sqrt{w} \le Z \le \sqrt{w}\right] = 2 \cdot P\left[0 \le Z \le \sqrt{w}\right]
$$

$$= 2\int_0^{\sqrt{w}} \frac{1}{\sqrt{2\pi}} e^{-z^2/2}\, dz,$$

so

$$f_{Z^2}(w) = \sqrt{\frac{2}{\pi}} e^{-w/2} \cdot \frac{1}{2} w^{-1/2} = \frac{1}{2^{1/2}\Gamma\left(\frac{1}{2}\right)} e^{-w/2} w^{-1/2}, \quad w > 0.$$

□

2.2.6 Chi-Square Distribution

If $Z \sim N(0,1)$, then from the pdf of Z^2 derived in Eq. (3) we see that Z^2 is a $Gamma\left(\frac{1}{2}, 2\right)$ rv and from the additive property of independent Gamma rv's with common β in Proposition 2.2.1, the following proposition is an immediate consequence.

Proposition 2.2.3. *If $Z_1, \ldots, Z_k$ are independent $N(0,1)$, then $W = \sum_{i=1}^{k} Z_i^2$ is a $Gamma(k/2, 2)$ rv with pdf:*

$$f_W(w) = \frac{1}{2^{k/2}\Gamma\left(\frac{k}{2}\right)} e^{-w/2} w^{k/2-1} I_{(0,\infty)}(w).$$

This rv is known as χ^2 *with k degrees of freedom (df)*, denoted by χ_k^2. Clearly, the sum of independent χ^2 rv's is a χ^2 rv, the df of which is the sum of the df's of its components.

Proposition 2.2.4. *If $\chi_{k_1}^2, \ldots, \chi_{k_n}^2$ are mutually independent, then by the additive property of independent Gamma rv's derived in Proposition 2.2.1, $\chi_{k_1}^2 + \cdots + \chi_{k_n}^2$ is distributed as $\chi_{k_1+\cdots+k_n}^2$.*

Now suppose that $X \sim N(\mu, 1)$, then by Eq. (11a) of Chapter 1,

$$f_{X^2}(w) = \frac{1}{2\sqrt{w}} \cdot \frac{1}{\sqrt{2\pi}} \left[e^{-\frac{1}{2}(\sqrt{w}-\mu)^2} + e^{-\frac{1}{2}(-\sqrt{w}-\mu)^2} \right]$$
$$= \frac{1}{2\sqrt{w}} \cdot \frac{1}{\sqrt{2\pi}} e^{-\frac{1}{2}\mu^2 - \frac{1}{2}w} \left[e^{\mu\sqrt{w}} + e^{-\mu\sqrt{w}} \right].$$

Using the properties of Gamma function and remembering $\Gamma\left(\frac{1}{2}\right) = \sqrt{\pi}$,

$$e^a + e^{-a} = 2\sum_{r=0}^{\infty} \frac{a^{2r}}{(2r)!} = 2\sum_{r=0}^{\infty} \frac{\left(a^2\right)^r}{2^r r![1\cdot 3\cdots(2r-1)]}$$
$$= \sqrt{\pi}\sum_{r=0}^{\infty} \frac{\left(a^2\right)^r/2^{2r-1}}{r!\left[\frac{1}{2}\cdot\frac{3}{2}\cdots\left(r-\frac{3}{2}\right)\cdot\left(r-\frac{1}{2}\right)\right]\Gamma\left(\frac{1}{2}\right)}$$
$$= \sqrt{\pi}\sum_{r=0}^{\infty} \frac{\left(a^2\right)^r/2^{2r-1}}{r!\,\Gamma\left(r+\frac{1}{2}\right)}.$$

We thus arrive at the *Noncentral* χ^2 *distribution with 1 df* and noncentrality parameter $\frac{1}{2}\mu^2$:

$$f_{X^2}(w) = \frac{1}{2\sqrt{w}} \cdot \frac{1}{\sqrt{2\pi}} e^{-\frac{1}{2}\mu^2 - \frac{1}{2}w} \cdot \sqrt{\pi} \sum_{r=0}^{\infty} \frac{\left(\mu^2\right)^r w^r / 2^{2r-1}}{r!\Gamma\left(r+\frac{1}{2}\right)}$$

$$= \sum_{r=0}^{\infty} \left(\frac{e^{-\frac{1}{2}\mu^2}\left(\frac{1}{2}\mu^2\right)^r}{r!} \right) \cdot \frac{1}{\Gamma\left(\frac{2r+1}{2}\right) 2^{\frac{2r+1}{2}}} e^{-w/2} w^{\frac{2r+1}{2}-1}.$$

Let N be a Poisson rv with mean $\frac{1}{2}\mu^2$. Then the rth term in the above sum is

$$f_N\left(r, \frac{1}{2}\mu^2\right) \cdot f_{\chi^2_{1+2r}}(w).$$

Hence, $f_{X^2}(w) = \mathrm{E}\left[f_{\chi^2_{1+2N}}(w)\right]$, where N is a $Poi\left(\frac{1}{2}\mu^2\right)$ rv. This is the pdf of a noncentral χ^2 rv with 1 df and noncentrality parameter $\frac{1}{2}\mu^2$ denoted by $\chi^2_1\left(\frac{1}{2}\mu^2\right)$.

2.2.7 Sampling From a Normal Distribution, Sample Mean and Sample Variance

Let $X_1, \ldots, X_n$ be n independent rv's, each distributed as $N(\mu, \sigma^2)$. Such a collection of rv's will be called "a random sample of n observations from the normal population $N(\mu, \sigma^2)$." The concepts of population and sample will be discussed in Chapter 4.

We have seen that μ is the mean and σ^2 is the variance of $N(\mu, \sigma^2)$. The analogs of μ and σ^2 in the sample $(X_1, \ldots, X_n)$ are, respectively, the sample mean and sample variance

$$\bar{X} = \frac{1}{n} \sum_{i=1}^{n} X_i \text{ and } s^2 = \frac{1}{n-1} \sum_{i=1}^{n} (X_i - \bar{X})^2.$$

(The reason for the factor $\frac{1}{n-1}$ instead of $\frac{1}{n}$ in s^2 will be made clear later in Chapter 5). In the following proposition we derive the joint distribution of $\bar{X}$ and s^2.

Proposition 2.2.5. *Let $\bar{X}$ and s^2 be the sample mean and sample variance, respectively, in a random sample $(X_1, \ldots, X_n)$ from $N(\mu, \sigma^2)$. Then*

(i) $\bar{X} \sim N(\mu, \sigma^2/n)$,
(ii) $(n-1)s^2/\sigma^2 \sim \chi^2_{n-1}$, *and*
(iii) $\bar{X}$ *and* s^2 *are independent.*

Proof. Let $W_i = X_i - \mu$. Then $W_1, \ldots, W_n$ are iid $N(0, \sigma^2)$, and letting $\boldsymbol{W}^{\mathsf{T}} = (W_1, \ldots, W_n)$, $\boldsymbol{w}^{\mathsf{T}} = (w_1, \ldots, w_n)$, the joint pdf of $(W_1, \ldots, W_n)$ is expressed as

$$f_{\boldsymbol{W}}(\boldsymbol{w}) = \frac{1}{\left(\sqrt{2\pi}\sigma\right)^n} \exp\left[-\frac{1}{2\sigma^2}\boldsymbol{w}^{\mathsf{T}}\boldsymbol{w}\right]$$

Transform $\boldsymbol{Y} = \boldsymbol{A}\boldsymbol{W}$, where $\boldsymbol{A} = ((a_{ij}))$ with $\boldsymbol{a}_i^{\mathsf{T}} = (a_{i1}, \ldots, a_{in}),\ i = 1, \ldots, n$ satisfying

$$\boldsymbol{a}_1^{\mathsf{T}} = n^{-1/2}(1, \ldots, 1),\ \boldsymbol{a}_i^{\mathsf{T}}\boldsymbol{a}_i = 1, \text{and } \boldsymbol{a}_i^{\mathsf{T}}\boldsymbol{a}_j = 0 \text{ for all } i \neq j$$

so that $\boldsymbol{A}$ is orthonormal. (In particular,

$$\boldsymbol{a}_j^{\mathsf{T}} = \{(j-1)j\}^{-1/2}(1, \ldots, 1, j-1, 0, \ldots, 0), \quad j = 2, \ldots, n$$

will do.) Then

$$Y_1 = \boldsymbol{a}_1^{\mathsf{T}}\boldsymbol{W} = n^{-1/2}\sum_{i=1}^{n} W_i = \sqrt{n}\bar{W} = \sqrt{n}(\bar{X} - \mu),$$

where $\bar{W} = n^{-1}\sum_{i=1}^{n} W_i$, and $\boldsymbol{Y}^{\mathsf{T}}\boldsymbol{Y} = \boldsymbol{W}^{\mathsf{T}}\left(\boldsymbol{A}^{\mathsf{T}}\boldsymbol{A}\right)\boldsymbol{W} = \boldsymbol{W}^{\mathsf{T}}\boldsymbol{W}$. Hence

$$\sum_{i=2}^{n} Y_i^2 = \boldsymbol{Y}^{\mathsf{T}}\boldsymbol{Y} - Y_1^2 = \boldsymbol{W}^{\mathsf{T}}\boldsymbol{W} - (\sqrt{n}\bar{W})^2 = \sum_{i=1}^{n}(W_i - \bar{W})^2 = \sum_{i=1}^{n}(X_i - \bar{X})^2.$$

Now by Eq. (14) of Chapter 1, the joint pdf of $(Y_1, \ldots, Y_n)$ is

$$\begin{aligned} f_{\boldsymbol{Y}}(\boldsymbol{y}) = f_{\boldsymbol{W}}\left(\boldsymbol{A}^{-1}\boldsymbol{y}\right) &= \frac{1}{\left(\sqrt{2\pi}\sigma\right)^n}\exp\left[-\frac{1}{2\sigma^2}\left(\boldsymbol{A}^{-1}\boldsymbol{y}\right)^{\mathsf{T}}\left(\boldsymbol{A}^{-1}\boldsymbol{y}\right)\right] \\ &= \frac{1}{\left(\sqrt{2\pi}\sigma\right)^n}\exp\left[-\frac{1}{2\sigma^2}\boldsymbol{y}^{\mathsf{T}}\boldsymbol{y}\right], \end{aligned}$$

that is, $Y_1, \ldots, Y_n$ are iid $N(0, \sigma^2)$. Thus

(i) $\sqrt{n}(\bar{X} - \mu) = \boldsymbol{a}_1^{\mathsf{T}}\boldsymbol{W} = Y_1$ is $N(0, \sigma^2)$, that is, $\bar{X}$ is $N(\mu, \sigma^2/n)$,
(ii) $\sum_{i=1}^{n}(X_i - \bar{X})^2 = \sum_{i=1}^{n}(W_i - \bar{W})^2 = \sum_{i=2}^{n} Y_i^2$ is $\sigma^2\chi_{n-1}^2$, that is, $(n-1)s^2/\sigma^2$ is χ_{n-1}^2,
(iii) $\bar{X} = \mu + n^{-1/2}Y_1$ and $s^2 = (n-1)\sum_{i=2}^{n} Y_i^2$ are independent.

□

2.2.8 *t* and *F* Distributions

Definition 2.2.7. A rv $t_k = Z/\sqrt{W/k}$, where $Z \sim N(0, 1)$ and $W \sim \chi_k^2$ are independent, is called a *t* rv with *k* df. It follows from Proposition 2.2.5 that

$$\frac{\sqrt{n}(\bar{X} - \mu)}{s} = \frac{\sqrt{n}(\bar{X} - \mu)/\sigma}{\sqrt{\{(n-1)s^2/\sigma^2\}/(n-1)}}$$

is a *t* rv with $(n-1)$ df.
This rv is commonly referred to as Student's *t* ("Student," [5]).

Proposition 2.2.6. *If $Z \sim N(0,1)$ and $W \sim \chi_k^2$ are independent, then $T \sim t_k = Z/\sqrt{W/k}$ has pdf*

$$f_T(t) = \frac{\Gamma\big((k+1)/2\big)}{\Gamma(1/2)\Gamma(k/2)}\frac{1}{\sqrt{k}}\left(1+t^2/k\right)^{-(k+1)/2}, \quad -\infty < t < \infty.$$

A distribution with this pdf is called the t distribution with k df.

Proof. The cdf of $T \sim t_k$ is

$$F_T(t) = P\Big[Z/\sqrt{W/k} \le t\Big] = \int_{w=0}^{\infty}\int_{z=-\infty}^{t\sqrt{w/k}} f_Z(z)f_W(w)\,dz\,dw, \text{ and}$$

$$\begin{aligned}
f_T(t) &= \int_0^{\infty} f_W(w)\sqrt{w/k}\,f_Z\big(t\sqrt{w/k}\big)\,dw \text{ (by Eq. (13) in Chapter 1)}\\
&= \int_0^{\infty}\left(\frac{1}{\Gamma(k/2)}2^{-k/2}\mathrm{e}^{-w/2}w^{k/2-1}\right)\sqrt{\frac{w}{k}}\left(\frac{1}{\sqrt{2\pi}}\mathrm{e}^{-(1/2)t^2w/k}\right)dw\\
&= \frac{(1/2)^{(k+1)/2}}{\Gamma(1/2)\Gamma(k/2)}\frac{1}{\sqrt{k}}\int_0^{\infty}\mathrm{e}^{-(1/2)(1+t^2/k)w}w^{(k+1)/2-1}\,dw\\
&= \frac{\Gamma\big((k+1)/2\big)}{\Gamma(1/2)\Gamma(k/2)}\frac{1}{\sqrt{k}}\left(1+t^2/k\right)^{-(k+1)/2}, \quad -\infty < t < \infty,
\end{aligned}$$

recognizing the last integral as $\Gamma\big((k+1)/2\big)$ by appropriate normalization. □

Using Stirling's approximation, $\Gamma(p) \approx \sqrt{2\pi}e^{-(p-1)}(p-1)^{p-1/2}$ for large p, we see that the pdf $f_T(t)$ converges to the pdf of $N(0,1)$ at each t, as $k \to \infty$.

Another important rv is

$$F_{k_1,k_2} = \frac{\chi_{k_1}^2/k_1}{\chi_{k_2}^2/k_2},$$

where $\chi_{k_1}^2$ and $\chi_{k_2}^2$ are independent χ^2 rv's with k_1 df and k_2 df, respectively.

To find the pdf of F_{k_1,k_2}, we first use Eq. (13) and then Eq. (10) in Chapter 1 for the pdf of the ratio of two independent rv's and then for the pdf of the multiple of an rv. In this way, we arrive at

$$f_{F_{k_1,k_2}}(w) = \frac{\Gamma\big((k_1+k_2)/2\big)}{\Gamma(k_1/2)\Gamma(k_2/2)}\left(\frac{k_1}{k_2}\right)^{k_1/2}\frac{w^{k_1/2-1}}{\left[1+(k_1/k_2)w\right]^{(k_1+k_2)/2}}, \quad w > 0,$$

the verification of which is left as an exercise.

2.2.9 Noncentral χ^2 and F Distributions

In Section 2.2.5 we introduced the noncentral χ^2 rv with 1 df and noncentrality parameter $\mu^2/2$ as the rv X^2 where $X \sim N(\mu, 1)$. The pdf of this rv was expressed as $\mathrm{E}\Big[f_{\chi_{1+2N}^2}(w)\Big]$ where N is a $Poi(\mu^2/2)$ rv. We now extend this by considering the distribution of $\sum_{i=1}^k X_i^2$ where $X_1,\ldots,X_k$ are independent $N(\mu_i, 1)$.

As in the proof of the Proposition 2.2.5, transform $\boldsymbol{Y} = \boldsymbol{A}\boldsymbol{X}$ where $\boldsymbol{A} = ((a_{ij}))$ with $\boldsymbol{a}_i^{\mathsf{T}} = (a_{i1}, \ldots, a_{ik})$ satisfying

$$\boldsymbol{a}_1^{\mathsf{T}} = \frac{1}{||\boldsymbol{\mu}||}(\mu_1, \ldots, \mu_k) \text{ and } \boldsymbol{a}_i^{\mathsf{T}}\boldsymbol{a}_j = \begin{cases} 1 & \text{if } i = j \\ 0 & \text{if } i \neq j \end{cases},$$

where $\|\boldsymbol{\mu}\|^2 = \sum_{i=1}^k \mu_i^2$. Then $\boldsymbol{A}$ is orthonormal and $\sum_{i=1}^k X_i^2 = \sum_{i=1}^k Y_i^2$. Moreover, since $Y_1 = \boldsymbol{a}_1^{\mathsf{T}}\boldsymbol{X} = \sum_{i=1}^k \mu_i X_i / \|\boldsymbol{\mu}\|$, we have

$$\begin{aligned}\left(\boldsymbol{A}^{-1}\boldsymbol{Y} - \boldsymbol{\mu}\right)^{\mathsf{T}}\left(\boldsymbol{A}^{-1}\boldsymbol{Y} - \boldsymbol{\mu}\right) &= (\boldsymbol{X} - \boldsymbol{\mu})^{\mathsf{T}}(\boldsymbol{X} - \boldsymbol{\mu}) = \sum_{i=1}^k (X_i - \mu_i)^2 \\ &= \sum_{i=1}^k X_i^2 - 2\sum_{i=1}^k \mu_i X_i + \sum_{i=1}^k \mu_i^2 = \sum_{i=1}^k Y_i^2 - 2\|\boldsymbol{\mu}\| Y_1 + \|\boldsymbol{\mu}\|^2 \\ &= (Y_1 - \|\boldsymbol{\mu}\|)^2 + \sum_{i=2}^k Y_i^2,\end{aligned}$$

and the joint pdf of $(Y_1, \ldots, Y_k)$ is

$$\begin{aligned}f_{\boldsymbol{Y}}(\boldsymbol{y}) = f_{\boldsymbol{X}}\left(\boldsymbol{A}^{-1}\boldsymbol{y}\right) &= \frac{1}{\left(\sqrt{2\pi}\right)^n} \exp\left[-\frac{1}{2}\left(\boldsymbol{A}^{-1}\boldsymbol{Y} - \boldsymbol{\mu}\right)^{\mathsf{T}}\left(\boldsymbol{A}^{-1}\boldsymbol{Y} - \boldsymbol{\mu}\right)\right] \\ &= \frac{1}{\left(\sqrt{2\pi}\right)^n} \exp\left[-\frac{1}{2}(y_1 - \|\boldsymbol{\mu}\|)^2 - \frac{1}{2}\sum_{i=2}^k y_i^2\right].\end{aligned}$$

It follows that $Y_1 \sim N(\|\boldsymbol{\mu}\|, 1)$ and $\sum_{i=2}^k Y_i^2 \sim \chi_{k-1}^2$, which are independent. Now

$$\sum_{i=1}^k X_i^2 = Y_1^2 + \sum_{i=2}^k Y_i^2 \sim \chi_{1+2N}^2 + \chi_{k-1}^2,$$

where the two χ^2 rv's are independent. Using additive property of independent χ^2 rv's, we see that $W = \sum_{i=1}^k X_i^2 = \chi_{k+2N}^2$ has the pdf

$$f_W(w) = \sum_{r=0}^{\infty} f_N\left(r, \|\boldsymbol{\mu}\|^2/2\right) f_{\chi_{k+2r}^2}(w) = \mathrm{E}\left[f_{\chi_{k+2N}^2}(w)\right],$$

$N \sim Poi(\|\mu\|^2/2)$. We denote this by writing $W \sim \chi_k^2(\|\mu\|^2/2)$, which is a χ^2 rv with k df and noncentrality parameter $\|\boldsymbol{\mu}\|^2/2$.

A rv of the form $V = \frac{W_1/m}{W_2/n}$, where $W_1 \sim \chi_m^2(\delta^2)$ and $W_2 \sim \chi_n^2$ are independent, is said to be distributed as F with numerator df m, denominator df n and noncentrality parameter δ^2, denoted by $F_{m,n}(\delta^2)$.

Since $W_1 \sim \chi^2_{m+2N}$ with $N \sim Poi(\delta^2/2)$, we can write $V = F_{m+2N,n}$, where $F_{m+2r,n}$ is F with numerator df $m+2r$ and denominator df n. In particular, if $X_1, \ldots, X_n$ are independent $N(\mu, \sigma^2)$, then

$$T = \frac{\sqrt{n}\bar{X}}{s} = \frac{\sqrt{n}\bar{X}/\sigma}{\sqrt{\{(n-1)s^2/\sigma^2\}/(n-1)}}$$

is distributed as

$$\frac{N(\sqrt{n}\mu/\sigma, 1)}{\sqrt{\chi^2_{n-1}/(n-1)}},$$

the numerator and denominator being independent (by Proposition 2.2.5), so that

$$T^2 \sim \frac{\chi_1^2\left(n\mu^2/\sigma^2\right)}{\chi^2_{n-1}/(n-1)} = F_{1,n-1}\left(\frac{n\mu^2}{\sigma^2}\right).$$

The noncentral χ^2 and F distributions are used in evaluating "power properties" of tests of significance of various hypotheses about means and variances of normal populations and in tests of hypotheses in Linear Models, as will be seen later. The noncentrality parameter is a measure of departure from "null hypothesis," and for fixed c, k, m, n, $P[\chi_k^2(\delta^2) \geq c]$ and $P[F_{m,n}(\delta^2) \geq c]$ are increasing functions of δ^2. Proofs of these monotonicity properties are left as exercises.

2.2.10 Cauchy Distribution

Definition 2.2.8. A rv X is distributed as *Cauchy*(θ) if X has pdf

$$f_X(x) = \frac{1}{\pi}\frac{1}{1+(x-\theta)^2}, \quad -\infty < x < \infty.$$

We first verify that f_X is indeed a pdf. Since $f_X(x) > 0$ for all x, we only need to check

$$\begin{aligned}\int_{-\infty}^{\infty} f_X(x)\,dx &= \frac{1}{\pi}\int_{-\infty}^{\infty}\frac{1}{1+(x-\theta)^2}\,dx = \frac{1}{\pi}\int_{-\infty}^{\infty}\frac{1}{1+y^2}\,dy \\ &= \frac{2}{\pi}\int_0^{\infty}\frac{1}{1+y^2}\,dy = \frac{2}{\pi}\int_0^{\pi/2}\frac{1+\tan^2\varphi}{1+\tan^2\varphi}\,d\varphi \\ &= \frac{2}{\pi}\int_0^{\pi/2} d\varphi = 1,\end{aligned}$$

using the transformation $\varphi = \tan^{-1} y$, that is, $y = \tan\varphi$. To find the mean of X, note that $X = Y + \theta$ where $Y = X - \theta$ has pdf

$$f_Y(y) = \frac{1}{\pi}\frac{1}{1+y^2}, \quad -\infty < y < \infty.$$

Now

$$\mathrm{E}[|Y|] = \frac{1}{\pi}\int_{-\infty}^{\infty} \frac{|y|}{1+y^2}\,dy = \frac{2}{\pi}\int_{0}^{\infty} \frac{y}{1+y^2}\,dy$$
$$= \frac{1}{\pi}\int_{1}^{\infty} \frac{1}{t}\,dt \left(\text{with } t = 1+y^2\right),$$

which diverges. Hence $\mathrm{E}[Y]$ does not exist, and therefore, $\mathrm{E}[X]$ does not exist.

2.2.11 Multivariate Normal Distribution

Definition 2.2.9. If $\boldsymbol{X}^{\mathsf{T}} = (X_1, \ldots, X_p)$ is a p-dim rv which can be written as $\boldsymbol{X} = \boldsymbol{\mu} + \boldsymbol{BZ}$, where $\boldsymbol{\mu}$ is a p-dim vector in $\mathbb{R}^p$, $\boldsymbol{B}$ is a $p \times k$ matrix and $\boldsymbol{Z}$ is a k-dim vector of iid $N(0, 1)$ rv's, then $\boldsymbol{X}$ is said to follow a p-variate normal distribution.

Since $\mathrm{E}[\boldsymbol{X}] = \boldsymbol{\mu}$ and $\mathrm{Cov}[\boldsymbol{X}, \boldsymbol{X}] = \mathrm{E}\left[(\boldsymbol{X}-\boldsymbol{\mu})(\boldsymbol{X}-\boldsymbol{\mu})^{\mathsf{T}}\right] = \boldsymbol{BB}^{\mathsf{T}} := \boldsymbol{\Sigma}$, we call $\boldsymbol{X}$ a p-variate normal rv with mean vector $\boldsymbol{\mu}$ and covariance matrix $\boldsymbol{\Sigma}$, and write $\boldsymbol{X} \sim N_p(\boldsymbol{\mu}, \boldsymbol{\Sigma})$.

We now list the important properties of p-variate normal distribution in the following propositions.

Proposition 2.2.7. *If* $\boldsymbol{X} \sim N_p(\boldsymbol{\mu}, \boldsymbol{\Sigma})$ *and* $\boldsymbol{Y} = \boldsymbol{c} + \boldsymbol{AX}$, *where* $\boldsymbol{c}$ *is in* $\mathbb{R}^r$ *and* $\boldsymbol{A}$ *is a* $r \times p$ *matrix, then* $\boldsymbol{Y} \sim N_r\left(\boldsymbol{c} + \boldsymbol{A\mu}, \boldsymbol{A\Sigma A}^{\mathsf{T}}\right)$.

It follows that the vector formed by any subset of r coordinates, $1 \le r \le p-1$, follows an r-variate normal distribution with appropriate mean vector and covariance matrix. In particular, each $X_i \sim N(\mu_i, \sigma_{ii})$.

Proof. Since $\boldsymbol{X} = \boldsymbol{\mu} + \boldsymbol{BZ}$, the transform

$$\boldsymbol{Y} = \boldsymbol{c} + \boldsymbol{A}(\boldsymbol{\mu} + \boldsymbol{BZ}) = (\boldsymbol{c} + \boldsymbol{A\mu}) + (\boldsymbol{AB})\boldsymbol{Z},$$

and $(\boldsymbol{AB})(\boldsymbol{AB})^{\mathsf{T}} = \boldsymbol{A}\left(\boldsymbol{BB}^{\mathsf{T}}\right)\boldsymbol{A}^{\mathsf{T}} = \boldsymbol{A\Sigma A}^{\mathsf{T}}$. Hence, $\boldsymbol{Y} \sim N_r\left(\boldsymbol{c} + \boldsymbol{A\mu}, \boldsymbol{A\Sigma A}^{\mathsf{T}}\right)$. □

Proposition 2.2.8. *If* $p \le k$ *and rank*$(\boldsymbol{B}) = p$, *then* $\boldsymbol{\Sigma} = \boldsymbol{BB}^{\mathsf{T}}$ *is positive definite, and* $\boldsymbol{X} = \boldsymbol{\mu} + \boldsymbol{BZ}$ *has pdf*

$$f_{\boldsymbol{X}}(\boldsymbol{x}; \boldsymbol{\mu}, \boldsymbol{\Sigma}) = \frac{1}{(2\pi)^{p/2}|\boldsymbol{\Sigma}|^{1/2}} \exp\left[-\frac{1}{2}(\boldsymbol{x}-\boldsymbol{\mu})^{\mathsf{T}}\boldsymbol{\Sigma}^{-1}(\boldsymbol{x}-\boldsymbol{\mu})\right], \quad \boldsymbol{x} \in \mathbb{R}^p.$$

Conversely, if $\boldsymbol{X}$ *has pdf* $f_{\boldsymbol{X}}(\boldsymbol{x}; \boldsymbol{\mu}, \boldsymbol{\Sigma})$ *given above, for some* $\boldsymbol{\mu}$ *and positive definite* $\boldsymbol{\Sigma}$, *then* $\boldsymbol{X}$ *can be expressed as* $\boldsymbol{X} = \boldsymbol{\mu} + \boldsymbol{BZ}$, $\boldsymbol{Z}^{\mathsf{T}} = (Z_1, \ldots, Z_p)$ *with* $Z_1, \ldots, Z_p$ *iid as* $N(0, 1)$.

Proof. Since the rank of $\boldsymbol{B}$ is $p \le k$, $\boldsymbol{B}^{\mathsf{T}}\boldsymbol{a} \ne \boldsymbol{0}$ for any $\boldsymbol{a} \ne \boldsymbol{0}$ in $\mathbb{R}^p$, so that $\boldsymbol{\Sigma} = \boldsymbol{BB}^{\mathsf{T}}$ is positive definite. Next augment $\boldsymbol{B}$ and $\boldsymbol{\mu}$ with a $(k-p) \times k$ matrix $\boldsymbol{C}$ and a $k-p$ vector of zeros, respectively, so that

$$\boldsymbol{B}^* = \begin{bmatrix} \boldsymbol{B} \\ \boldsymbol{C} \end{bmatrix} \text{and } \boldsymbol{\mu}^* = \begin{bmatrix} \boldsymbol{\mu} \\ \boldsymbol{0} \end{bmatrix},$$

where the rows of $\boldsymbol{C}$ are of unit length, mutually orthogonal, and orthogonal to the rows of $\boldsymbol{B}$. Note that $\boldsymbol{B}^*$ is a square matrix of order k and $\boldsymbol{\mu}^*$ is a k-dim vector. Then letting $\boldsymbol{Y} = \boldsymbol{CZ}$, we have

$$\boldsymbol{X}^* = \begin{bmatrix} \boldsymbol{X} \\ \boldsymbol{Y} \end{bmatrix} = \begin{bmatrix} \boldsymbol{\mu} + \boldsymbol{BZ} \\ \boldsymbol{CZ} \end{bmatrix}$$

as a one-to-one transformation from $\boldsymbol{Z}$. We now find the pdf of $\boldsymbol{X}^*$ from the pdf of $\boldsymbol{Z}$ and integrate out $\boldsymbol{Y}$. The details are left as an exercise.

To prove the converse, find $\boldsymbol{B}$ so that $\boldsymbol{\Sigma} = \boldsymbol{BB}^{\mathsf{T}}$ ($\boldsymbol{B}$ is not unique) and transform $\boldsymbol{Z} = \boldsymbol{B}^{-1}(\boldsymbol{X} - \boldsymbol{\mu})$. Then $\boldsymbol{Z}$ has the desired property. □

Remark 2.2.1. If $\boldsymbol{X}_1$ and $\boldsymbol{X}_2$ are p- and q-dimensional rv's, then independence of $\boldsymbol{X}_1$ and $\boldsymbol{X}_2$ implies $\boldsymbol{\Sigma}_{12} = \mathrm{Cov}[\boldsymbol{X}_1, \boldsymbol{X}_2] = 0$. However, the converse is not true in general. In the case of normal distributions the converse is also true. More precisely, if $\boldsymbol{X} = \begin{bmatrix} \boldsymbol{X}_1 \\ \boldsymbol{X}_2 \end{bmatrix} \sim N_{p+q}(\boldsymbol{\mu}, \boldsymbol{\Sigma})$, then $\boldsymbol{X}_1$ and $\boldsymbol{X}_2$ are independent iff $\boldsymbol{\Sigma}_{12} = \mathrm{Cov}[\boldsymbol{X}_1, \boldsymbol{X}_2] = 0$. The proof can be easily seen when $\boldsymbol{\Sigma}$ is positive definite. If $\boldsymbol{\Sigma}_{12} = 0$, then $\boldsymbol{\Sigma}$ is a block diagonal matrix of the form $\begin{bmatrix} \boldsymbol{\Sigma}_{11} & \boldsymbol{0} \\ \boldsymbol{0}^{\mathsf{T}} & \boldsymbol{\Sigma}_{22} \end{bmatrix}$ and its inverse is also block diagonal. This results in the joint pdf of $\boldsymbol{X}_1$ and $\boldsymbol{X}_2$ being the product of the pdf's of $\boldsymbol{X}_1$ and $\boldsymbol{X}_2$.

Proposition 2.2.9. *Suppose* $\boldsymbol{X} \sim N_{p+q}(\boldsymbol{\mu}, \boldsymbol{\Sigma})$ *where* $\boldsymbol{\Sigma}$ *is positive definite. Let*

$$\boldsymbol{X} = \begin{bmatrix} \boldsymbol{X}_1 \\ \boldsymbol{X}_2 \end{bmatrix}, \ \boldsymbol{\mu} = \begin{bmatrix} \boldsymbol{\mu}_1 \\ \boldsymbol{\mu}_2 \end{bmatrix}, \ \boldsymbol{\Sigma} = \begin{bmatrix} \boldsymbol{\Sigma}_{11} & \boldsymbol{\Sigma}_{12} \\ \boldsymbol{\Sigma}_{21} & \boldsymbol{\Sigma}_{22} \end{bmatrix},$$

where $\boldsymbol{X}_1$, $\boldsymbol{\mu}_1$ *and* $\boldsymbol{X}_2$, $\boldsymbol{\mu}_2$ *denote the first p and last q coordinates of* $\boldsymbol{X}$ *and* $\boldsymbol{\mu}$, *respectively, and* $\boldsymbol{\Sigma}_{11} = \mathit{Cov}[\boldsymbol{X}_1, \boldsymbol{X}_1]$, $\boldsymbol{\Sigma}_{22} = \mathit{Cov}[\boldsymbol{X}_2, \boldsymbol{X}_2]$, $\boldsymbol{\Sigma}_{21}^{\mathsf{T}} = \boldsymbol{\Sigma}_{12} = \mathit{Cov}[\boldsymbol{X}_1, \boldsymbol{X}_2]$, *where* $\mathit{Cov}\left[\boldsymbol{X}_i, \boldsymbol{X}_j\right] = E\left[(\boldsymbol{X}_i - \boldsymbol{\mu}_i)(\boldsymbol{X}_j - \boldsymbol{\mu}_j)^{\mathsf{T}}\right]$. *Then the conditional distribution of* $\boldsymbol{X}_2$ *given* $\boldsymbol{X}_1 = \boldsymbol{x}_1$ *is* $N_q\left(\boldsymbol{\mu}_2 + \boldsymbol{\Sigma}_{21}\boldsymbol{\Sigma}_{11}^{-1}(\boldsymbol{x}_1 - \boldsymbol{\mu}_1), \ \boldsymbol{\Sigma}_{22} - \boldsymbol{\Sigma}_{21}\boldsymbol{\Sigma}_{11}^{-1}\boldsymbol{\Sigma}_{12}\right)$.

Proof. Write $f_{\boldsymbol{X}_2|\boldsymbol{X}_1}(\boldsymbol{x}_2|\boldsymbol{x}_1) = f_{\boldsymbol{X}_1,\boldsymbol{X}_2}(\boldsymbol{x}_1, \boldsymbol{x}_2)/f_{\boldsymbol{X}_1}(\boldsymbol{x}_1)$, where

$$f_{\boldsymbol{X}_1,\boldsymbol{X}_2}(\boldsymbol{x}_1, \boldsymbol{x}_2) = \frac{1}{(2\pi)^{(p+q)/2}\begin{vmatrix} \boldsymbol{\Sigma}_{11} & \boldsymbol{\Sigma}_{12} \\ \boldsymbol{\Sigma}_{21} & \boldsymbol{\Sigma}_{22} \end{vmatrix}^{1/2}} \times \exp\left\{-\frac{1}{2}\left[(\boldsymbol{x}_1 - \boldsymbol{\mu}_1)^{\mathsf{T}}, (\boldsymbol{x}_2 - \boldsymbol{\mu}_2)^{\mathsf{T}}\right]\begin{bmatrix} \boldsymbol{\Sigma}_{11} & \boldsymbol{\Sigma}_{12} \\ \boldsymbol{\Sigma}_{21} & \boldsymbol{\Sigma}_{22} \end{bmatrix}^{-1}\begin{bmatrix} \boldsymbol{x}_1 - \boldsymbol{\mu}_1 \\ \boldsymbol{x}_2 - \boldsymbol{\mu}_2 \end{bmatrix}\right\}$$

and

$$f_{\boldsymbol{X}_1}(\boldsymbol{x}_1) = \frac{1}{(2\pi)^{p/2}|\boldsymbol{\Sigma}_{11}|^{1/2}} \exp\left[-\frac{1}{2}(\boldsymbol{x}_1 - \boldsymbol{\mu}_1)^{\mathsf{T}}\boldsymbol{\Sigma}_{11}^{-1}(\boldsymbol{x}_1 - \boldsymbol{\mu}_1)\right].$$

To calculate the ratio of the last two terms, the main task is to find the inverse and the determinant of the partitioned matrix involved in $f_{\boldsymbol{X}_1,\boldsymbol{X}_2}(\boldsymbol{x}_1, \boldsymbol{x}_2)$. For this we need the following argument. Let

$$\boldsymbol{A} = \begin{bmatrix} \boldsymbol{A}_{11} & \boldsymbol{A}_{12} \\ \boldsymbol{A}_{21} & \boldsymbol{A}_{22} \end{bmatrix} \text{ with } \boldsymbol{A}_{21} = \boldsymbol{A}_{12}^{\mathsf{T}} \text{ and } \boldsymbol{B} = \begin{bmatrix} \boldsymbol{I} & \boldsymbol{0} \\ -\boldsymbol{A}_{21}^{\mathsf{T}}\boldsymbol{A}_{11}^{-1} & \boldsymbol{I} \end{bmatrix}.$$

Then

$$D = BAB^{\mathsf{T}} = \begin{bmatrix} A_{11} & \mathbf{0} \\ \mathbf{0} & A_{22} - A_{12}^{\mathsf{T}} A_{11}^{-1} A_{12} \end{bmatrix},$$

so $D^{-1} = \begin{bmatrix} A_{11}^{-1} & \mathbf{0} \\ \mathbf{0} & \left(A_{22} - A_{12}^{\mathsf{T}} A_{11}^{-1} A_{12}\right)^{-1} \end{bmatrix}$. This implies $A = B^{-1} D \left(B^{\mathsf{T}}\right)^{-1}$ and therefore, denoting $\left(A_{22} - A_{12}^{\mathsf{T}} A_{11}^{-1} A_{12}\right)^{-1}$ by H,

$$A^{-1} = B^{\mathsf{T}} D^{-1} B = \begin{bmatrix} A_{11}^{-1} + A_{11}^{-1} A_{12} H A_{12}^{\mathsf{T}} A_{11}^{-1} & -A_{11}^{-1} A_{12} H \\ -H A_{12}^{\mathsf{T}} A_{11}^{-1} & H \end{bmatrix}.$$

Finally,

$$|A| = |B|^{-1} |D| |B^{\mathsf{T}}|^{-1} = |D| = |A_{11}| |A_{22} - A_{12}^{\mathsf{T}} A_{11}^{-1} A_{12}|,$$

because $|B| = |B^{\mathsf{T}}| = 1$. Using these results on $\begin{bmatrix} \Sigma_{11} & \Sigma_{12} \\ \Sigma_{21} & \Sigma_{22} \end{bmatrix}^{-1}$ and $\begin{vmatrix} \Sigma_{11} & \Sigma_{12} \\ \Sigma_{21} & \Sigma_{22} \end{vmatrix}$, we obtain after some algebraic simplification,

$$\begin{aligned} f_{X_2|X_1}(x_2|x_1) &= f_{X_1,X_2}(x_1, x_2)/f_{X_1}(x_1) \\ &= \frac{1}{(2\pi)^{q/2} |\Sigma_{22.1}|^{1/2}} \exp\left[-\frac{1}{2}(x_2 - \mu_{2.1})^{\mathsf{T}} \Sigma_{22.1}^{-1} (x_2 - \mu_{2.1})\right], \end{aligned}$$

where $\mu_{2.1} = \mu_2 + \Sigma_{21}\Sigma_{11}^{-1}(x_1 - \mu_1)$ and $\Sigma_{22.1} = \Sigma_{22} - \Sigma_{21}\Sigma_{11}^{-1}\Sigma_{12}$. The expression displayed above is the pdf of $N_q(\mu_{2.1}, \Sigma_{22.1})$. □

Remark 2.2.2. Special Cases.

(i) For $q = 1$, write $\sigma_{2(1)}^{\mathsf{T}} = (\sigma_{1,p+1}, \ldots, \sigma_{p,p+1})$ where $\sigma_{ij} = \text{Cov}[X_i, X_j]$. Then

$$\Sigma_{22} - \Sigma_{21}\Sigma_{11}^{-1}\Sigma_{12} = \sigma_{p+1}^2 - \sigma_{2(1)}^{\mathsf{T}} \Sigma_{11}^{-1} \sigma_{2(1)} := \sigma_{(p+1)\cdot(12\cdots p)}^2, \text{ and}$$

$$\begin{aligned} \mu_2 + \Sigma_{21}\Sigma_{11}^{-1}(x_1 - \mu_1) &= \left(\mu_{p+1} - \sigma_{2(1)}^{\mathsf{T}} \Sigma_{11}^{-1} \mu_1\right) + \sigma_{2(1)}^{\mathsf{T}} \Sigma_{11}^{-1} x_1 \\ &:= \alpha + \beta_1 x_1 + \cdots + \beta_p x_p, \end{aligned}$$

where $\beta^{\mathsf{T}} = (\beta_1, \ldots, \beta_p) = \sigma_{2(1)}^{\mathsf{T}} \Sigma_{11}^{-1}$ and $\alpha = \mu_{p+1} - (\beta_1\mu_1 + \cdots + \beta_p\mu_p)$. (Remember: In the above, $x_1^{\mathsf{T}} = (x_1, \ldots, x_p)$ and $\mu_1^{\mathsf{T}} = (\mu_1, \ldots, \mu_p)$.) Thus the conditional distribution of X_{p+1} given $(X_1, \ldots, X_p) = (x_1, \ldots, x_p)$ is $N\left(\alpha + \beta_1 x_1 + \cdots + \beta_p x_p,\ \sigma_{(p+1)\cdot(12\cdots p)}^2\right)$ where the conditional mean $\alpha + \beta_1 x_1 + \cdots + \beta_p x_p$ and the conditional variance $\sigma_{(p+1)\cdot(12\cdots p)}^2$ are given by the above formulas.

(ii) For $p = q = 1$, $\Sigma = \begin{bmatrix} \sigma_1^2 & \rho\sigma_1\sigma_2 \\ \rho\sigma_1\sigma_2 & \sigma_2^2 \end{bmatrix}$ and therefore,

$$|\Sigma|^{1/2} = \sigma_1\sigma_2\sqrt{1 - \rho^2} \text{ and } \Sigma^{-1} = \left(1 - \rho^2\right)^{-1} \begin{bmatrix} 1/\sigma_1^2 & -\rho/(\sigma_1\sigma_2) \\ -\rho/(\sigma_1\sigma_2) & 1/\sigma_2^2 \end{bmatrix}.$$

Thus for $-1 < \rho < 1$,

$$f_{X_1,X_2}(x_1, x_2) = \frac{1}{2\pi\sigma_1\sigma_2\sqrt{1-\rho^2}} \exp\left[-\frac{1}{2(1-\rho^2)}\left\{\left(\frac{x_1-\mu_1}{\sigma_1}\right)^2 \right.\right.$$
$$\left.\left. -2\rho\left(\frac{x_1-\mu_1}{\sigma_1}\right)\left(\frac{x_2-\mu_2}{\sigma_2}\right) + \left(\frac{x_2-\mu_2}{\sigma_2}\right)^2\right\}\right],$$

which is the pdf of bivariate normal distribution $N_2(\mu_1, \mu_2, \sigma_1^2, \sigma_2^2, \rho)$ with mean vector $(\mu_1, \mu_2)^\top$ and covariance matrix $\begin{bmatrix} \sigma_1^2 & \rho\sigma_1\sigma_2 \\ \rho\sigma_1\sigma_2 & \sigma_2^2 \end{bmatrix}$. The marginal distribution of X_i is $N(\mu_i, \sigma_i^2)$ and the conditional distribution of X_2 given $X_1 = x_1$ is $N(\alpha + \beta x_1, \sigma_{2.1}^2)$, where $\beta = \rho\sigma_2/\sigma_1$, $\alpha = \mu_2 - \beta\mu_1$, and $\sigma_{2.1}^2 = \sigma_2^2(1-\rho^2)$. Conversely, if $X_1 \sim N(\mu_1, \sigma_1^2)$ and the conditional distribution of X_2 given X_1 is $N(\alpha + \beta X_1, \tau^2)$, then $(X_1, X_2) \sim N_2(\mu_1, \mu_2, \sigma_1^2, \sigma_2^2, \rho)$, where $\mu_2 = \alpha + \beta\mu_1$, $\sigma_2^2 = \tau^2 + \beta^2\sigma_1^2$, and $\rho = \beta\sigma_1/\sqrt{\tau^2 + \beta^2\sigma_1^2}$.

Moreover, if $(X_1, X_2) \sim N_2(\mu_1, \mu_2, \sigma_1^2, \sigma_2^2, \rho)$ and $\boldsymbol{A} = \begin{bmatrix} a_{11} & a_{12} \\ a_{21} & a_{22} \end{bmatrix}$ is nonsingular, then $Y_1 = a_{11}X_1 + a_{12}X_2$ and $Y_2 = a_{21}X_1 + a_{22}X_2$ are jointly bivariate normal with appropriate parameters. In particular $X_1 + X_2$ and $X_1 - X_2$ are independent *iff* $\sigma_1 = \sigma_2$.

Proposition 2.2.10. *Suppose* $\boldsymbol{X} \sim N_p(\boldsymbol{\mu}, \boldsymbol{\Sigma})$ *with positive definite* $\boldsymbol{\Sigma}$. *Then* $(\boldsymbol{X} - \boldsymbol{\mu})^\top \boldsymbol{\Sigma}^{-1} (\boldsymbol{X} - \boldsymbol{\mu}) \sim \chi_p^2$.

Proof. Find $\boldsymbol{B}$ such that $\boldsymbol{B}\boldsymbol{B}^\top = \boldsymbol{\Sigma}$ and write $\boldsymbol{X} = \boldsymbol{\mu} + \boldsymbol{B}\boldsymbol{Z}$ with $\boldsymbol{Z} \sim N_p(\boldsymbol{0}, \boldsymbol{I})$, as in the proof of the converse part of Proposition 2.2.8. Then

$$(\boldsymbol{X} - \boldsymbol{\mu})^\top \boldsymbol{\Sigma}^{-1} (\boldsymbol{X} - \boldsymbol{\mu}) = \boldsymbol{Z}^\top \boldsymbol{B}^\top \left(\boldsymbol{B}\boldsymbol{B}^\top\right)^{-1} \boldsymbol{B}\boldsymbol{Z} = \boldsymbol{Z}^\top \boldsymbol{Z} = \sum_{i=1}^{p} Z_i^2 \sim \chi_p^2.$$

□

2.2.12 Exponential Family of Distributions

A family of pdf's or pmf's is said to be an exponential family if it is of the form

$$f(x, \boldsymbol{\theta}) = c(\boldsymbol{\theta}) \exp\left[\sum_{j=1}^{k} Q_j(\boldsymbol{\theta}) T_j(x)\right] r(x), \quad \boldsymbol{\theta} \in \Theta,$$

where $r(x) > 0$, $c(\boldsymbol{\theta}) > 0$, and each $T_j(x)$ is a real-valued function of x, while each $Q_j(\boldsymbol{\theta})$ is a real-valued function of $\boldsymbol{\theta}$.

Note. The set $\{x: f(x, \boldsymbol{\theta}) > 0\} = \{x: r(x) > 0\}$ does not depend on $\boldsymbol{\theta}$.

Many distributions including most of those discussed above belong to this family.

Example 2.2.2. The pmf of the Binomial distribution $Bin(n, p)$ is

$$f(x, p) = \binom{n}{x} \exp[x\log(p) + (n - x)\log(1 - p)]$$
$$= \binom{n}{x}(1 - p)^n \exp\left[x \log\left(\frac{p}{1 - p}\right)\right],$$

$x = 0, 1, \ldots, n$. Thus $f(x, p)$ is of the desired form with $\theta = p$, $c(\theta) = (1 - \theta)^n$, $r(x) = \binom{n}{x}$, $Q(\theta) = \log(\theta/(1 - \theta))$, and $T(x) = x$.

Example 2.2.3. The pmf of $Poi(\lambda)$ is

$$f(x, \lambda) = e^{-\lambda}\frac{\lambda^x}{x!} = e^{-\lambda}\frac{1}{x!}\exp(x \log \lambda), \quad x = 0, 1, \ldots$$

This pmf is of the desired form with $\theta = \lambda$, $r(x) = 1/x!$, $c(\theta) = e^{-\theta}$, $Q(\theta) = \log\theta$, and $T(x) = x$.

Example 2.2.4. The pdf of $N(\mu, \sigma^2)$ is

$$f\left(x; \mu, \sigma^2\right) = \frac{1}{\sqrt{2\pi}\sigma}\exp\left[-\frac{1}{2\sigma^2}(x - \mu)^2\right]$$
$$= \left(\frac{1}{\sqrt{2\pi}\sigma}e^{-\mu^2/(2\sigma^2)}\right)\exp\left[\frac{\mu}{\sigma^2}x + \left(-\frac{x^2}{2\sigma^2}\right)\right].$$

This pdf is of the desired form with $\boldsymbol{\theta} = \left(\mu, \sigma^2\right)$, $r(x) = 1$, $c(\boldsymbol{\theta}) = \frac{1}{\sqrt{2\pi}\sigma}e^{-\mu^2/(2\sigma^2)}$, $Q_1(\boldsymbol{\theta}) = \mu/\sigma^2$, $T_1(x) = x$, $Q_2(\boldsymbol{\theta}) = -1/(2\sigma^2)$, and $T_2(x) = x^2$.

The pdf of $Unif(0, \theta)$ is $f(x, \theta) = I_{(0,\theta)}(x)$. Here, the set $\{x: f(x, \theta) > 0\} = (0, \theta)$ depends on θ. Therefore, this family of pdf's $\{f(x, \theta),\ \theta > 0\}$ is not an exponential family.

Natural Parameters. In the form of $f(x; \boldsymbol{\theta})$ given in the definition of an exponential family, if we reparameterize by letting

$$\theta_j = Q_j(\boldsymbol{\theta}),\ j = 1, \ldots, k, \text{ and let } c(\boldsymbol{\theta}) = c(\theta_1, \ldots, \theta_k),$$

then the k-parameter exponential family of pdf's or pmf's is represented as

$$f(x; \boldsymbol{\theta}) = c(\boldsymbol{\theta})\exp\left[\sum_{j=1}^{k}\theta_j T_j(x)\right]r(x) = c(\boldsymbol{\theta})\exp[\langle\boldsymbol{\theta}, \boldsymbol{T}(x)\rangle]r(x),$$

where the parameters $\theta_1, \ldots, \theta_k$ are called *natural parameters* and Θ^* is called the natural parameter space defined as

$$\Theta^* = \left\{\boldsymbol{\theta} = (\theta_1, \ldots, \theta_k): \int_{-\infty}^{\infty}\exp\left[\sum_{j=1}^{k}\theta_j T_j(x)\right]r(x)\,dx < \infty\right\},$$

where $\mathcal{X} = \mathbb{R}$ (and analogously for more general $\mathcal{X}$). Also, the constant $c(\boldsymbol{\theta})$ is given by

$$c(\boldsymbol{\theta}) = \left\{\int_{-\infty}^{\infty}\exp\left[\sum_{j=1}^{k}\theta_j T_j(x)\right]r(x)\,dx\right\}^{-1}.$$

With the reparameterization $\theta_1 = \mu/\sigma^2$, $\theta_2 = -1/(2\sigma^2)$, the pdf's of the family of normal distributions can be written as

$$f(x;\theta_1,\theta_2) = c(\boldsymbol{\theta})\exp\left[\theta_1 x + \theta_2 x^2\right].$$

Proposition 2.2.11. *The natural parameter space* Θ^* *of a k-parameter exponential family of distributions is a convex set in* $\mathbb{R}^k$*; that is, if* $\boldsymbol{\theta}$ *and* $\boldsymbol{\theta}'$ *are in* Θ^* *then* $\alpha\boldsymbol{\theta} + (1-\alpha)\boldsymbol{\theta}'^*$ *for all* $0 \le \alpha \le 1$.

Proof. We shall use *Hölder's Inequality*: if f and g are real-valued functions on $\mathbb{R}^n$ for which $\int |f|^p$ and $\int |g|^q$ are finite for $p > 1$, $q > 1$ with $1/p + 1/q = 1$, then $\int |fg| \le \left(\int |f|^p\right)^{1/p}\left(\int |g|^q\right)^{1/q}$ (see Section A.2).

To prove the proposition, let $p = 1/\alpha$, $q = 1/(1-\alpha)$ for $0 < \alpha < 1$, so that p and q are larger than 1 with $1/p + 1/q = 1$. Then by Hölder's inequality,

$$\int \exp[\langle\boldsymbol{\theta}, \boldsymbol{T}(x)\rangle]r(x)\,dx < \infty \text{ and } \int \exp[\langle\boldsymbol{\theta}', \boldsymbol{T}(x)\rangle]r(x)\,dx < \infty$$

and hence

$$\begin{aligned}
&\int \exp[\langle\alpha\boldsymbol{\theta} + (1-\alpha)\boldsymbol{\theta}', \boldsymbol{T}(x)\rangle]r(x)\,dx \\
&\quad = \int \left\{\exp[\alpha\langle\boldsymbol{\theta}, \boldsymbol{T}(x)\rangle]r^{\alpha}(x)\right\}\left\{\exp[(1-\alpha)\langle\boldsymbol{\theta}', \boldsymbol{T}(x)\rangle]r^{1-\alpha}(x)\right\} dx \\
&\quad \le \left[\int \left\{\exp[\alpha\langle\boldsymbol{\theta}, \boldsymbol{T}(x)\rangle]r^{\alpha}(x)\right\}^{1/\alpha} dx\right]^{\alpha} \\
&\qquad \times \left[\int \left\{\exp[(1-\alpha)\langle\boldsymbol{\theta}', \boldsymbol{T}(x)\rangle]r^{1-\alpha}(x)\right\}^{1/(1-\alpha)} dx\right]^{1-\alpha} \\
&\quad = \left[\int \exp[\langle\boldsymbol{\theta}, \boldsymbol{T}(x)\rangle]r(x)\,dx\right]^{\alpha}\left[\int \exp[\langle\boldsymbol{\theta}', \boldsymbol{T}(x)\rangle]r(x)\,dx\right]^{1-\alpha} \\
&\quad < \infty.
\end{aligned}$$

□

Exercises

2.1. A multiple choice exam consists of 20 questions, each with four possible answers and carries four points. If a question is attempted, then the score is 4 for a correct answer and 0 otherwise, while a score of 1 is awarded for a question which is not attempted. A student knows the answers to eight questions, but has no idea about six questions and is 50% sure about the answers of the other six questions. It takes 50 points to pass the exam. Find the probability of passing for each of the following strategies and choose the best one.

(a) Answer the 8 sure ones and guess at random the answers of the other 12.

(b) Answer the 8 sure ones, choose the answers of the 6 that you are 50 sure about and guess at random the other 6.

(c) Answer the 8 sure ones and do not answer the other 12.

2.2. Samples from a large lot of manufactured items are being inspected to determine whether the proportion of defectives $p = 0.1$ or less as claimed by the supplier. The following sampling plan is under consideration:

Take a sample of 15 items and accept the lot if all are good, reject the lot if 2 or more are defectives and take another sample of 15 items if there is 1 defective in the sample. Accept the lot if all 15 in the second sample are good and reject the lot otherwise.

Find the probabilities that this sampling plan will result in

(a) rejection of a lot with $p = 0.1$,

(b) acceptance of a lot with $p = 0.15$.

2.3. An insurance company writes automobile insurance policies in an area where 1 out of 200 drivers causes accidents in a year according to past record. If the company writes 500 policies, find

(a) the probability that there will be no more than two claims in a year,

(b) the expected amount of claims to be settled if the average claim is \$1500.

2.4. On a certain segment of a freeway, accidents happen during the rush hour at the rate of two per hour, following the Poisson distribution. Find the probabilities of

(a) at least one accident in an hour,

(b) if there was one accident in an hour, the accident occurred during the first half hour.

2.5. A radioactive source emits particles at a rate of 0.4/s. Suppose that the number of particles emitted per second is a Poisson rv.

(a) Find the probability that two or more particles will be emitted in 3 s.

(b) A counter registers an emitted particle with probability 0.75. What is the probability that two or more particles will be registered on the counter in 3 s?

2.6. Three players A, B, and C will play a table tennis match. In each game, A defeats B with probability 0.6, C defeats B with probability 0.6, and A and C are evenly matched. First, A plays against B and the one who wins three games, plays against C. Then the one who wins three games wins the match.

(a) Find the probability that A wins the first round and the probability that the first round is settled in four games.

(b) Find the probabilities of A, B, and C winning the match and the probability that the entire match is settled in nine games.

(c) Find the expected number of games to settle the match.

2.7. Find the moment generating function (mgf) of each of the following rv's:

(a) $X \sim Poi(\lambda)$, (b) $Y \sim Geom(p)$, (c) $T \sim \text{Exp}(\theta)$, (d) $W \sim Gamma(\alpha, \beta)$.

Also find the means and variances of these rv's from their mgf's.

2.8. Show that the pdf of $F_{k_1,k_2} = \left(\chi^2_{k_1}/k_1\right)/\left(\chi^2_{k_2}/k_2\right)$ where $\chi^2_{k_1}$ and $\chi^2_{k_2}$ are independent chi-square rv's with df's k_1 and k_2, respectively, is as given in the text.

2.9. As in Section 2.2.8, let $\chi_k^2(\delta^2)$ and $F_{m,n}(\delta^2)$ denote the noncentral χ_k^2 and $F_{m,n}$ rv's with noncentrality parameter δ^2. Show that for fixed c, k, m, n, $P[\chi_k^2(\delta^2) \geq c]$ and $P[F_{m,n}(\delta^2) \geq c]$ are increasing functions of δ^2.

2.10. Verify the details of the proof of Proposition 2.2.8.

2.11. Let $X_1, \ldots, X_n$ be independent $N(0, \sigma^2)$ rv's. Then $Y_n = (X_1^2 + \cdots + X_n^2)^{1/2}$ is distributed as $\sigma\sqrt{\chi_n^2}$. Find the pdf of Y_n. [This distribution is called Raleigh distribution for $n = 2$ and Maxwell distribution for $n = 3$.]

2.12. Let X_1, X_2 be independent $N(0, \sigma^2)$ rv's. Find the pdf's of $Y_1 = X_1/X_2$, $Y_2 = X_1/|X_2|$, $Y_3 = |X_1|/|X_2|$.

2.13. Let $V_1, \ldots, V_{n+1}$ be independent Exp(1) rv's and let $S_k = V_1 + \cdots + V_k$, $k = 1, \ldots, n+1$. Let $U_{(1)} < \cdots < U_{(n)}$ be the order statistics in a random sample $(U_1, \ldots, U_n)$ from *Unif* (0, 1).

(a) Show that the joint distribution of $(U_{(1)}, \ldots, U_{(n)})$ is the same as that of $(S_1/S_{n+1}, \ldots, S_n/S_{n+1})$.

(b) Use this result and Proposition 2.2.2 to find the pdf's of (i) $U_{(k)}$ and (ii) $U_{(l)} - U_{(k)}$ for $1 \leq k < l \leq n$.

2.14. Let $X_{(1)} < \cdots < X_{(n)}$ be the order statistics in a random sample $(X_1, \ldots, X_n)$ from Exp(1). Show that $nX_{(1)}, (n-1)(X_{(2)} - X_{(1)}), (n-2)(X_{(3)} - X_{(2)}), \ldots, (X_{(n)} - X_{(n-1)})$ are iid Exp(1).

2.15. **(a)** Show that if X has a continuous strictly increasing cdf F, then $F(X)$ is *Unif* (0, 1).

(b) If $T_1, \ldots, T_n$ are iid Exp(mean θ), then $2(T_1 + \cdots + T_n)/\theta$ is χ_{2n}^2.

(c) Let $(Z_1, Z_2) = (R\cos\theta, R\sin\theta)$ define a one-to-one map between (Z_1, Z_2) and (R, θ). What is the joint distribution of (R, θ) when Z_1, Z_2 are independent $N(0, 1)$?

(d) Let Z_1, Z_2 be independent $N(0, 1)$ and let

$$X_1 = a_1 + b_{11}Z_1 + b_{12}Z_2, \quad X_2 = a_2 + b_{21}Z_1 + b_{22}Z_2$$

Find the constants $a_1, a_2, b_{11}, b_{12}, b_{21}, b_{22}$ so that (X_1, X_2) follows a bivariate normal distribution with mean vector (μ_1, μ_2) and covariance matrix $\begin{bmatrix} \sigma_1^2 & \rho\sigma_1\sigma_2 \\ \rho\sigma_1\sigma_2 & \sigma_2^2 \end{bmatrix}$.

2.16. Suppose that you have a computer program to generate *Unif* (0, 1) rv's $U_1, U_2, \ldots$. How would you use these rv's to generate independent rv's distributed as

(a) $T \sim$ Exp(1), (b) $V \sim \chi_{10}^2$, (c) $W \sim$ *Cauchy*(0, 1), (d) $Z \sim N(0, 1)$,

(e) $(X_1, X_2) \sim N_2(\boldsymbol{\mu}, \boldsymbol{\Sigma})$ where $\boldsymbol{\mu} = (\mu_1, \mu_2)$ and $\boldsymbol{\Sigma} = \begin{bmatrix} \sigma_1^2 & \rho\sigma_1\sigma_2 \\ \rho\sigma_1\sigma_2 & \sigma_2^2 \end{bmatrix}$

Use the results in Exercise 2.15.

2.17. Suppose that (X_1, X_2, X_3) follows a 3-dim normal distribution with the mean vector (μ_1, μ_2, μ_3) and a positive definite covariance matrix $\boldsymbol{\Sigma} = ((\sigma_{ij}))$. Let $X_{2\cdot 1} = X_2 - \mathrm{E}[X_2|X_1]$, $X_{3\cdot 1} = X_3 - \mathrm{E}[X_3|X_1]$, and $X_{3\cdot 12} = X_3 - \mathrm{E}[X_3|X_1, X_2]$.

(a) Show that $X_1, X_{2\cdot 1}$ and $X_{3\cdot 12}$ are mutually independent

(b) The partial correlation coefficient between X_2, X_3 given X_1 is

$$\rho_{23\cdot 1} = \mathrm{Corr}\left[X_{2\cdot 1}, X_{3\cdot 1}\right] = \frac{\mathrm{Cov}\left[X_{2\cdot 1}, X_{3\cdot 1}\right]}{\sqrt{\mathrm{Var}\left[X_{2\cdot 1}\right]\mathrm{Var}\left[X_{3\cdot 1}\right]}}$$

Express $\rho_{23\cdot 1}$ in terms of ρ_{12}, ρ_{13} and ρ_{23} where $\rho_{ij} = \mathrm{Corr}\left[X_i, X_j\right]$.

(c) The multiple correlation of X_3 on (X_1, X_2) is $\rho_{3\cdot 12} = \mathrm{Corr}[X_3, \mathrm{E}(X_3|X_1, X_2)]$. Show that $\rho_{3\cdot 12} \geq 0$ and $\mathrm{Var}[X_{3\cdot 12}] = \sigma_{33}\left(1 - \rho_{3\cdot 12}^2\right)$.

(d) Show that $1 - \rho_{3\cdot 12}^2 = \left(1 - \rho_{13}^2\right)\left(1 - \rho_{23\cdot 1}^2\right)$.

3

Infinite Sequences of Random Variables and Their Convergence Properties

3.1 Introduction

Let T be a statistic based on the data consisting of rv's $X_1, \ldots, X_n$. Most statistical methods (such as estimation and hypothesis testing discussed in subsequent chapters) use a statistic appropriate for the problem at hand. It is of interest to know how T behaves as we have more and more data in order to understand the behavior of the procedures based on T with large data sets. For a rigorous examination of this question, we need proper notation for T based on $(X_1, \ldots, X_n)$ as $n \to \infty$ and the concept of how T behaves as $n \to \infty$ should be made more precise.

3.2 Modes of Convergence

Let T be a statistic based on the data consisting of rv's $X_1, \ldots, X_n$. Most statistical methods (such as estimation and hypothesis testing discussed in subsequent chapters) use a statistic appropriate for the problem at hand. It is of interest to know how T behaves as we have more and more data in order to understand the behavior of the procedures based on T with large data sets. For a rigorous examination of this question, we need proper notation for T based on $(X_1, \ldots, X_n)$ as $n \to \infty$ and the concept of how T behaves as $n \to \infty$ should be made more precise.

Definition 3.2.1. Let $X_1, \ldots, X_n$ be iid rv's with common cdf $F \in \mathcal{F}$, where $\mathcal{F}$ is the class of all cdf's on $\mathbb{R}$. Then the function

$$F_n(x) = n^{-1} \sum_{i=1}^{n} I_{(-\infty,x]}(X_i)$$

is called the empirical distribution function (edf).

Definition 3.2.2. Let $\boldsymbol{T}: \mathcal{F} \to \mathbb{R}^k$. Then $\boldsymbol{T}_n = \left(T_{n1}, \ldots, T_{nk}\right) = \boldsymbol{T}(F_n)$ is the k-dim statistic $\boldsymbol{T}$ based on $(X_1, \ldots, X_n)$, that is, $\boldsymbol{T}_n = \boldsymbol{T}(X_1, \ldots, X_n)$.

We now define various modes of convergence.

Theory and Methods of Statistics. http://dx.doi.org/10.1016/B978-0-12-802440-9.00003-5

Definition 3.2.3. The various modes of convergence are defined as:

1. Convergence in Probability: $X_n \xrightarrow{P} a$ if $\lim_{n\to\infty} P[|X_n - a| > \varepsilon] = 0$ for all $\varepsilon > 0$ and $X_n \xrightarrow{P} X$ if $|X_n - X| \xrightarrow{P} 0$, denoted by $X_n = a + o_P(1)$ and $X_n = X + o_P(1)$, respectively.
2. Convergence in Law: $X_n \xrightarrow{\mathcal{L}} X$ if $P[X_n \in A] \to P[X \in A]$ for all A for which $P[X \in \partial A] = 0$ where ∂A is the boundary of A. For 1-dim rv's, $F_{X_n}(x) \to F_X(x)$ at all continuity points of F_X is enough.
3. Convergence in Quadratic Mean: $X_n \xrightarrow{q.m.} a$ if $\mathrm{E}\big[(X_n - a)^2\big] \to 0$ and $X_n \xrightarrow{q.m.} X$ if $\mathrm{E}\big[(X_n - X)^2\big] \to 0$. More generally, $X_n \xrightarrow{r} X$ if $\mathrm{E}\big[|X_n - X|^r\big] \to 0$ for X_n, X such that $\mathrm{E}\big[|X_n|^r\big] < \infty$ and $\mathrm{E}\big[|X|^r\big] < \infty$.
4. Bounded in Probability: $\{X_n\}$ is bounded in probability, denoted by $X_n = O_P(1)$, if for any $\varepsilon > 0$ there exists a positive constant M_ε such that $P[|X_n| \le M_\varepsilon] > 1 - \varepsilon$ for all n.
5. Almost Sure Convergence: $X_n \xrightarrow{a.s.} a$ if $P\big[\lim_{n\to\infty} X_n = a\big] = 1$.
6. Uniform Integrability: $\{X_n\}$ is uniformly integrable if

$$\lim_{k\to\infty} \overline{\lim_{n\to\infty}} \mathrm{E}\big[|X_n| I_{[k,\infty)}(|X_n|)\big] = 0.$$

Compare the stochastic order relations o_P and O_P with the usual order relations o and O for sequence of real numbers:

1. If $\lim_{n\to\infty} x_n = 0$, then $x_n = o(1)$.
2. If $\{x_n\}$ is bounded, then $x_n = O(1)$.

Extensions of the above definitions. Let $\{r_n\}$ be a sequence of real numbers. Then $X_n = o_P(r_n)$ *iff* $X_n/r_n = o_P(1)$, $X_n = O_P(r_n)$ *iff* $X_n/r_n = O_P(1)$, $x_n = o(r_n)$ *iff* $x_n/r_n = o(1)$, $x_n = O(r_n)$ *iff* $X_n/r_n = O(1)$. For example, $X_n = \theta + O_P(1/\sqrt{n})$ *iff* $\sqrt{n}(X_n - \theta) = O_P(1)$.

Convergence Properties of Sample Means of iid Random Variables and Random Vectors. Suppose $\{X_n\}$ is a sequence of iid rv's and $\bar{X}_n = n^{-1}\sum_{i=1}^n X_i$.

Theorem 3.2.1 (Weak Law of Large Numbers (WLLN, Khinchine)). *If* $E[X_1] = \mu$ *exists, then* $\bar{X}_n \xrightarrow{P} \mu$.

Theorem 3.2.2 (Central Limit Theorem (CLT, Lindeberg-Lévy)). *If* $E[X_1] = \mu$ *and* $Var[X_1] = \sigma^2 > 0$ *exist, then*

$$\sqrt{n}\frac{\bar{X}_n - \mu}{\sigma} \xrightarrow{\mathcal{L}} Z \sim N(0,1).$$

Theorem 3.2.3 (Multivariate CLT). *Let* $\{X_i\}$ *be k-dim iid random vectors with mean* μ *and covariance matrix* Σ, *then*

$$\sqrt{n}(\bar{X}_n - \mu) \xrightarrow{\mathcal{L}} Z \sim N_k(0, \Sigma).$$

Theorem 3.2.4 (Strong Law of Large Numbers (SLLN, [6])). *If* $E[X_1] = \mu$ *exists, then* $\bar{X}_n \xrightarrow{a.s.} \mu$.

The proofs of Theorems 3.2.1–3.2.3 using characteristic functions, are given in Section A.4.

Two proofs of Theorem 3.2.4 will be given in Section 3.3, one under a stronger condition that X_1 has a finite fourth moment and the other assuming finite variance.

We conclude this section with a number of basic facts which follow from the definitions of various modes of convergence. We shall sketch the proofs of some of these, while the proofs of others are left as exercises. These facts will be used in the sequel without further explanation.

Theorem 3.2.5. *The following are true.*

I. *Convergence in quadratic mean implies convergence in probability.*

II. *If* $X_n \xrightarrow{P} X$ *and* g *is continuous, then* $g(X_n) \xrightarrow{P} g(X)$.

III. *If* $X_n \xrightarrow{\mathcal{L}} X$ *and* g *is continuous (more generally, if* $P[X \in D_g] = 0$*, where* D_g *is the set of discontinuity points of* g*), then* $g(X_n) \xrightarrow{\mathcal{L}} g(X)$.

IV. (a) $X_n \xrightarrow{\mathcal{L}} X$ *implies* $X_n = O_P(1)$.

(b) $X_n \xrightarrow{\mathcal{L}} 0$ *implies* $X_n = o_P(1)$.

V. $X_n \xrightarrow{\mathcal{L}} X$ *and* $Y_n \xrightarrow{P} c$ *implies* $X_n + Y_n \xrightarrow{\mathcal{L}} X + c$, $X_nY_n \xrightarrow{\mathcal{L}} cX$, *and if* $c \neq 0$, $X_n/Y_n \xrightarrow{\mathcal{L}} X/c$.

VI. (a) *If* $X_n = O_P(1)$ *and* $Y_n = o_P(1)$, *then* $X_nY_n = o_P(1)$ *and* $X_n + Y_n = O_P(1)$.

(b) *If* $X_n = o_P(1)$ *and* $Y_n = o_P(1)$, *then* $X_nY_n = o_P(1)$ *and* $X_n + Y_n = o_P(1)$.

VII. *Slutsky's Theorem:* $X_n \xrightarrow{\mathcal{L}} X$ *and* $Y_n = o_P(1)$ *implies* $X_n + Y_n \xrightarrow{\mathcal{L}} X$.

VIII. *Polya's Theorem: If* F_n *and* F *are cdf's,* F *is continuous, and* $F_n(x) \to F(x)$ *for all* x, *then the convergence is uniform.*

IX. *Borel-Cantelli Lemma: If* $\sum_{n=1}^{\infty} P[|X_n - a| > \varepsilon] < \infty$ *for all* $\varepsilon > 0$, *then* $X_n \xrightarrow{a.s.} a$. *(The converse also holds under further conditions.)*

X. (a) *If* $X_n \xrightarrow{a.s.} a$, *then* $X_n \xrightarrow{P} a$.

(b) *If* $X_n \xrightarrow{P} a$, *then there exists a sequence* $\{n_j\}$ *such that the subsequence* $X_{n_j} \xrightarrow{a.s.} a$.

Outlines of some proofs.

I. If $E[(X_n - a)^2] \to 0$ as $n \to \infty$, then by Tchebyshev's inequality (to be proved in the next section), for any $\varepsilon > 0$,

$$P[|X_n - a| > \varepsilon] \leq \frac{E[|X_n - a|^2]}{\varepsilon^2} \to 0, \text{ as } n \to \infty.$$

II. (i) For $\delta > 0$, there exists $M > 0$ such that $P[|X| > M] \leq \delta$.

(ii) For $\delta > 0$, there exists a positive integer $n(\delta)$ such that $P[|X_n - X| > \varepsilon] \leq \delta$ for all $n \geq n(\delta)$.

(iii) By (i) and (ii), for $\delta > 0$, there exist $M > 0$ and $n(\delta)$, so that for all $n \geq n(\delta)$,

$$P[|X| \leq M,\ |X_n - X| \leq \varepsilon,\ |X_n| \leq M + \varepsilon] \geq 1 - 2\delta,$$

(iv) g is continuous on $\mathbb{R}$, so g is uniformly continuous on $[-M - \varepsilon, M + \varepsilon]$. Therefore, for $\varepsilon_1 > 0$, there exists $\varepsilon_2 > 0$, so that for $x, x' \in [M - \varepsilon, M + \varepsilon]$, $|x - x'| < \varepsilon_2$ implies $|g(x) - g(x')| < \varepsilon_1$.

(v) In (iii), choose $\varepsilon = \varepsilon_2$ and choose δ accordingly. Then

$$P\big[|g(X_n) - g(X)| < \varepsilon_1\big] \geq 1 - 2\delta \text{ for } n \geq n(\delta).$$

III. The proof is in Section A.4.

IV. The proof of part (a) is in (iii) of II above. To prove part (b), note that $X_n \xrightarrow{\mathcal{L}} 0$ means $X_n \xrightarrow{\mathcal{L}} X$ where $P[X = 0] = 1$. Now taking $A = (-\varepsilon, \varepsilon)$ in the definition of $X_n \xrightarrow{\mathcal{L}} X$, we have

$$P[|X_n| < \varepsilon] = P[X_n \in A] \to P[X \in A] = P[|X| < \varepsilon] = 1.$$

V. For any constant a

$$P[X_n + Y_n \leq a] = P[X_n + Y_n \leq a, |Y_n - c| \leq \varepsilon] + P[X_n + Y_n \leq a, |Y_n - c| > \varepsilon].$$

Hence as $n \to \infty$,

$$P[X_n + Y_n \leq a] \leq P[X_n \leq a - c + \varepsilon] + P[|Y_n - c| > \varepsilon] \to P[X \leq a - c + \varepsilon].$$

On the other hand

$$\begin{aligned} P[X_n + Y_n \leq a] &\geq P[X_n \leq a - c - \varepsilon, |Y_n - c| \leq \varepsilon] + 0 \\ &\geq P[X_n \leq a - c - \varepsilon] - P[|Y_n - c| > \varepsilon] \\ &\to P[X \leq a - c + \varepsilon]. \end{aligned}$$

Thus for any $\varepsilon > 0$,

$$P[X \leq a - c - \varepsilon] \leq \lim_{n\to\infty} P[X_n + Y_n \leq a] \leq P[X \leq a - c + \varepsilon].$$

Hence

$$\lim_{n\to\infty} P[X_n + Y_n \leq a] = P[X \leq a - c] = P[X + c \leq a]$$

at all continuity points a of $X + c$. Hence $X_n + Y_n \xrightarrow{\mathcal{L}} X + c$.

VII. Slutsky's Theorem is a special case of (V) for $c = 0$.

VIII. To prove Polya's Theorem, choose $x_1 < \cdots < x_k$ such that $F(x_1) < \varepsilon/2$, $1 - F(x_k) < \varepsilon/2$, and $F(x_{i+1}) - F(x_i) < \varepsilon/2$, $i = 1, \ldots, k - 1$. Now find N such that $|F_n(x_i) - F(x_i)| < \varepsilon/2$, $i = 1, \ldots, k$, for $n \geq N$. The proof follows from these.

IX. Almost Sure Convergence $\left(\xrightarrow{a.s.}\right)$ takes place in the context of infinite sequences $\{X_n,\ n = 1, 2, \ldots\}$. For this, the appropriate probability space is constructed by extending the probability spaces of $(X_1, \ldots, X_n)$, $n = 1, 2, \ldots$ in a suitable manner.

Now $\lim_{n\to\infty} X_n$ is to be thought of in the space $(\Omega, \mathcal{F}, P)$, where Ω is the space of all infinite sequences $\omega = (x_1, x_2, \ldots)$, $\mathcal{F}$ is a suitable σ -field consisting of all sets $C_n(B_n) = \{\omega: (x_1, x_2, \ldots): (x_1, \ldots, x_n) \in B_n\}$, $n = 1, 2, \ldots$ and all Borel sets B_n in $\mathbb{R}^n$ with $P[C_n(B_n)] = P[(X_1, \ldots, X_n) \in B_n]$.

Almost sure convergence $X_n \stackrel{a.s.}{\to} a$ has the following meaning in this space: for every $\varepsilon > 0$ and for $\omega \notin N$ where $P[N] = 0$, there exists $n(\omega, \varepsilon)$ such that $|X_n(\omega) - a| < \varepsilon$ for all $n \geq n(\omega, \varepsilon)$. If there is no such $n(\omega, \varepsilon)$, then the event $A_n(\varepsilon) = \{\omega: |X_n(\omega) - a| \geq \varepsilon\}$ must occur infinity often (i.o.). Thus $X_n \stackrel{a.s.}{\to} a$, that is, $P[\lim_{n\to\infty} X_n = a] = 1$ if $P[A_n(\varepsilon) \text{ i.o.}] = 0$ for all $\varepsilon > 0$.

Now $\{A_n(\varepsilon) \text{ i.o.}\}^c$ means that $A_n(\varepsilon)^c$ must occur for all $n \geq k$, for some positive integer k, that is,

$$\{A_n(\varepsilon) \text{ i.o.}\}^c = \bigcup_{k=1}^{\infty} \bigcap_{n=k}^{\infty} A_n^c(\varepsilon), \text{ so}$$

$$\{A_n(\varepsilon) \text{ i.o.}\} = \left[\bigcup_{k=1}^{\infty} \bigcap_{n=k}^{\infty} A_n^c(\varepsilon)\right]^c = \bigcap_{k=1}^{\infty} \bigcup_{n=k}^{\infty} A_n(\varepsilon)$$

by De Morgan's rule (Chapter 1). Since $\left\{\bigcup_{n=k}^{\infty} A_n(\varepsilon)\right\}$ is a decreasing sequence of sets,

$$P[|X_n - a| \geq \varepsilon \text{ i.o.}] = P[A_n(\varepsilon) \text{ i.o.}] = \lim_{k\to\infty} P\left[\bigcup_{n=k}^{\infty} A_n(\varepsilon)\right] \leq \lim_{k\to\infty} \sum_{n=k}^{\infty} P[A_n(\varepsilon)].$$

Hence $\sum_{n=1}^{\infty} P[|X_n - a| \geq \varepsilon] = \sum_{n=1}^{\infty} P[A_n(\varepsilon)] < \infty$ implies $\lim_{k\to\infty} \sum_{n=k}^{\infty}$ $P[A_n(\varepsilon)]$, so

$$P[|X_n - a| \geq \varepsilon \text{ i.o.}] = 0 \text{ for all } \varepsilon > 0,$$

that is, $X_n \stackrel{a.s.}{\to} a$. This proves the direct part of Borel-Cantelli Lemma.

X(a). Note that $X_n \stackrel{a.s.}{\to} a$ *iff* $\lim_{k\to\infty} P\left[\bigcup_{n=k}^{\infty} A_n(\varepsilon)\right] = 0$ for all $\varepsilon > 0$, which implies $\lim_{k\to\infty} P[A_k(\varepsilon)] = \lim_{k\to\infty} P[|X_k - a| \geq \varepsilon] = 0$ for all $\varepsilon > 0$, that is, $X_k \stackrel{P}{\to} a$.

Algebra of o_P, O_P, o, *and* O. The properties (VIa, b) given above can be stated as:

$$O_P(1) + o_P(1) = O_P(1),\ O_P(1)o_P(1) = o_P(1),$$
$$o_P(1) + o_P(1) = o_P(1), \text{ and } o_P(1)o_P(1) = o_P(1).$$

The following also hold

$$O_P(1) + o(1) = O_P(1),\ O(1) + o_P(1) = O_P(1),\ O_P(1)o(1) = o_P(1),$$
$$O(1)o_P(1) = o_P(1),\ o_P(1) + o(1) = o_P(1), \text{ and } o_P(1)o(1) = o_P(1).$$

Theorem 3.2.6 (Delta Method). *Suppose that in* $\mathbb{R}^k$, $\sqrt{n}(\mathbf{X}_n - \boldsymbol{\mu}) \xrightarrow{\mathcal{L}} \mathbf{Z} \sim N_k(\mathbf{0}, \boldsymbol{\Sigma})$ *and* $g\colon \mathbb{R}^k \to \mathbb{R}$ *has continuous first partial derivatives. Then*

$$\sqrt{n}[g(\mathbf{X}_n) - g(\boldsymbol{\mu})] \xrightarrow{\mathcal{L}} W \sim N\left(0, [\nabla g(\boldsymbol{\mu})]^{\mathsf{T}} \boldsymbol{\Sigma} [\nabla g(\boldsymbol{\mu})]\right).$$

The proof of this theorem is left as an exercise.

In many instances of asymptotic normality, the variance of the asymptotic normal distribution is a function of its mean. This poses a problem in the construction of a large sample confidence interval for the mean, which is the parameter of interest. In such cases, it is convenient to make a transformation so that in the asymptotic distribution of the transform, the variance is a constant. These are called *variance-stabilizing transformations* of which some well-known examples are given below.

Example 3.2.1. If X is $Poi(\mu)$, then for large μ, $(X - \mu)$ is asymptotically $N(0, \mu)$. Here $\left(\sqrt{X} - \sqrt{\mu}\right)$ is asymptotically $N(0, 1/4)$.

Example 3.2.2. If $X_1, \ldots, X_n$ are iid $Bernoulli(p)$ and $\hat{p}_n = n^{-1}\sum_{i=1}^{\infty} X_i$, then $\sqrt{n}(\hat{p}_n - p) \xrightarrow{\mathcal{L}} N(0, p(1-p))$. Here $\sqrt{n}\left(\arcsin\sqrt{\hat{p}_n} - \arcsin\sqrt{p}\right) \xrightarrow{\mathcal{L}} N(0, 1/4)$.

Example 3.2.3. Let $S_n^2 = (n-1)^{-1}\sum_{i=1}^{\infty}(X_i - \bar{X}_n)^2$, where $X_1, \ldots, X_n$ are iid $N(\mu, \sigma^2)$. Then $\sqrt{n}(S_n^2 - \sigma^2) \xrightarrow{\mathcal{L}} N(0, 2\sigma^4)$. Here $\sqrt{n}(\log S_n^2 - \log\sigma^2) \xrightarrow{\mathcal{L}} N(0, 2)$.

Example 3.2.4. If $(X_1, Y_1), \ldots, (X_n, Y_n)$ are iid bivariate normal with $\mathrm{Corr}[X_i, Y_i] = \rho \in (-1, 1)$ and if r_n is the sample correlation coefficient, then $\sqrt{n}(r_n - \rho) \xrightarrow{\mathcal{L}} N\left(0, (1-\rho^2)^2\right)$. Here $\sqrt{n}\left(\tanh^{-1} r_n - \tanh^{-1}\rho\right) \xrightarrow{\mathcal{L}} N(0, 1)$ $\left(\text{recall that } \tanh^{-1} x = (1/2)\log[(1+x)/(1-x)]\right)$. Actually, using $\sqrt{n-3}\left(\tanh^{-1} r_n - \tanh^{-1}\rho\right)$ as an asymptotically $N(0, 1)$ rv for the purpose of constructing confidence intervals for ρ results in a better approximation.

To prove these results, find a transformation g in each case, such that $\sqrt{n}(X_n - \mu) \xrightarrow{\mathcal{L}} N(0, \sigma^2(\mu))$ leads to $\sqrt{n}[g(X_n) - g(\mu)] \xrightarrow{\mathcal{L}} N\left(0, \{g'(\mu)\}^2\sigma^2(\mu)\right)$, where $\{g'(\mu)\}^2\sigma^2(\mu)$ is a constant.

3.3 Probability Inequalities

In the previous section we have seen that the proofs of many "in probability" or "almost sure" convergence results require good upper bounds for tail probabilities of deviations of rv's from their means, such $P[|\bar{X}_n - \mu| \geq a]$ where $\bar{X}_n$ is the mean of iid $X_1, \ldots, X_n$ with mean μ. The following probability inequalities are useful for this purpose.

Theorem 3.3.1 (Markov Inequality). *If* X *is an rv with* $P[X \geq 0] = 1$, *then*

$$P[X \geq a] \leq \frac{E[X]}{a}, \quad \text{for any } a > 0.$$

Proof. Note that for any $a > 0$, $I_{[a,\infty)}(x) \le x/a$ and hence

$$P[X \ge a] = \mathrm{E}\big[I_{[a,\infty)}(X)\big] \le \frac{\mathrm{E}[X]}{a}.$$

(See Proposition 1.8.1(iii and vi).) □

Tchebyshev's Inequality. *If X is an rv with mean μ and variance σ^2, then*

$$P[|X-\mu| \ge t] \le \frac{\sigma^2}{t^2}, \quad \textit{for any } t > 0.$$

Proof. Using Markov's inequality

$$P[|X-\mu| \ge t] = P\Big[|X-\mu|^2 \ge t^2\Big] \le \frac{\mathrm{E}\big[|X-\mu|^2\big]}{t^2} = \frac{\sigma^2}{t^2}.$$

□

Generalization. It follows in the same way, that if X is an rv with mean μ and $(2r)$th central moment $\mu_{2r} = \mathrm{E}\big[|X-\mu|^{2r}\big]$, then

$$P[|X-\mu| \ge t] \le \frac{\mu_{2r}}{t^{2r}}, \quad \text{for any } t > 0.$$

Applications.

1. A simple proof of the WLLN assuming finite variance:
   WLLN. If $\bar{X}_n$ is the sample mean of iid rv's $X_1, \ldots, X_n$ with mean μ and finite variance σ^2, then by Tchebyshev's inequality

$$\lim_{n\to\infty} P\big[|\bar{X}_n - \mu| \ge \varepsilon\big] \le \lim_{n\to\infty} \frac{\mathrm{Var}\big[\bar{X}_n\big]}{\varepsilon^2} = \lim_{n\to\infty} \frac{\sigma^2}{n\varepsilon^2} = 0.$$

2. A simple proof of the SLLN assuming finite fourth moment:
   SLLN. If $\bar{X}_n$ is the sample mean of iid rv's $X_1, \ldots, X_n$ with mean μ and finite fourth central moment τ^4, then by generalized Tchebyshev's inequality;

$$P\big[|\bar{X}_n - \mu| \ge \varepsilon\big] = P\big[|S_n| \ge n\varepsilon\big] \le \frac{\mathrm{E}\big[S_n^4\big]}{(n\varepsilon)^4},$$

where $S_n = \sum_{i=1}^{\infty}(X_i - \mu) = \sum_{i=1}^{\infty} Y_i$, where $Y_1, \ldots, Y_n$ are iid with mean 0 and finite fourth moment $\mathrm{E}\big[Y_i^4\big] = \tau^4$. Therefore

$$\begin{aligned}\mathrm{E}\Big[S_n^4\Big] &= \mathrm{E}\Big[(S_{n-1}+Y_n)^4\Big] = \mathrm{E}\Big[S_{n-1}^4\Big] + 6(n-1)\sigma^4 + \tau^4 = \cdots \\ &= \mathrm{E}\Big[S_1^4\Big] + 6\{1+2+\cdots+(n-1)\}\sigma^4 + (n-1)\tau^4 \\ &= 3n(n-1)\sigma^4 + n\tau^4.\end{aligned}$$

Thus

$$P\left[|\bar{X}_n - \mu| \geq \varepsilon\right] \leq \frac{3n(n-1)\sigma^4 + n\tau^4}{(n\varepsilon)^4} < \frac{3\sigma^4 + \tau^4}{n^2\varepsilon^4},$$

so that

$$\sum_{n=1}^{\infty} P\left[|\bar{X}_n - \mu| \geq \varepsilon\right] \leq \frac{3\sigma^4 + \tau^4}{\varepsilon^4} \sum_{n=1}^{\infty} \frac{1}{n^2} < \infty,$$

and now $\bar{X}_n \overset{a.s.}{\rightarrow} \mu$ by the Borel-Cantelli Lemma.

This covers the case of many standard distributions where the SLLN holds. Another proof of the SLLN, assuming only finite variance, will be given later in this section, using a much stronger probability inequality.

We now obtain two more powerful inequalities which provide *exponential bounds* for tail probabilities. Of these, the first holds for bounded rv's and the second requires some moment conditions, providing a sharper bound. However, the first inequality due to Hoeffding has the advantage of simplicity and is very useful in many situations.

Theorem 3.3.2 (Hoeffding's Inequality [7]). *If* $X_1, \ldots, X_n$ *are independent with* $P[0 \leq X_i \leq 1] = 1$ *for all* i, $S_n = X_1 + \cdots + X_n$, $\bar{X}_n = S_n/n$, *and* $\mu = E[\bar{X}_n] = E[S_n]/n$, *then for* $0 \leq t \leq 1 - \mu$, $P[\bar{X}_n - \mu \geq t] \leq e^{-2nt^2}$.

Proof. Note that

$$\begin{aligned}
P[\bar{X}_n - \mu \geq t] &= P[S_n - n\mu - nt \geq 0] = \mathrm{E}[I_{[0,\infty)}(S_n - n\mu - nt)] \\
&\leq \mathrm{E}\left[\mathrm{e}^{h(S_n - n\mu - nt)}\right], \quad \text{for } h > 0 \\
&= \mathrm{e}^{-nh(\mu+t)} \prod_{i=1}^{n} \mathrm{E}\left[\mathrm{e}^{hX_i}\right] \leq \mathrm{e}^{-nh(\mu+t)} \prod_{i=1}^{n} \mathrm{E}\left[(1 - X_i)\mathrm{e}^0 + X_i \mathrm{e}^h\right] \\
&= \mathrm{e}^{-nh(\mu+t)} \prod_{i=1}^{n} \left[1 - \mu_i + \mu_i \mathrm{e}^h\right] \leq \mathrm{e}^{-nh(\mu+t)} \left[n^{-1} \sum_{i=1}^{n} \left(1 - \mu_i + \mu_i \mathrm{e}^h\right)\right]^n \\
&= \mathrm{e}^{-nh(\mu+t)} \left(1 - \mu + \mu \mathrm{e}^h\right)^n \\
&= \mathrm{e}^{nL(h)}, \text{ with } L(h) = \log\left(1 - \mu + \mu \mathrm{e}^h\right) - h(\mu + t),
\end{aligned}$$

using the convexity of e^{hx} and because geometric mean is no larger than the arithmetic mean. Next, we want to choose $h > 0$ so that $L(h) < 0$. This is possible because $L'(0) = -t$ and $L''(h) \leq 1/4$ for all h, as can be easily verified, so that

$$\begin{aligned}
L(h) &= L(0) + hL'(0) + \frac{1}{2}h^2 L''(h^*) = 0 - ht + \frac{1}{2}h^2 L''(h^*) \\
&\leq -ht + \frac{h^2}{8} = -2t^2 \text{ for } h = 4t.
\end{aligned}$$

Hence we get the inequality by taking $h = 4t$. □

Theorem 3.3.3 (Bernstein's Inequality). *If* $X_1, \ldots, X_n$ *are independent with* $E[X_i] = 0$, $E[X_i^2] = b_i$, $B_n = b_1 + \cdots + b_n$, *and for* $r > 2$,

$$E[|X_i|^r] \leq \frac{1}{2} r! b_i c^{r-2}, \quad i = 1, \ldots, n,$$

where c is a constant, then

$$P[|X_1 + \cdots + X_n| > t] \leq 2 \exp\left[-\frac{t^2}{2B_n + 2ct}\right].$$

Proof. For an outline of the proof, see Uspensky [8, p. 204–5]. □

Corollary. *If* $P[|X_i| \leq M] = 1$, *then for* $r > 2$,

$$\begin{aligned} E[|X_i|^r] &= E\left[X_i^2 |X_i|^{r-2}\right] \leq b_i M^{r-2} \\ &< \frac{b_i M^{r-2} r!}{2 \cdot 3^{r-2}} = \frac{1}{2} r! b_i \left(\frac{M}{3}\right)^{r-2}, \end{aligned}$$

because $r!/(2 \cdot 3^{r-2}) > 1$. *In Bernstein's inequality we can now take* $c = M/3$, *resulting in*

$$P[|X_1 + \cdots + X_n| > t] \leq 2 \exp\left[-\frac{t^2}{2B_n + 2Mt/3}\right].$$

In particular, if $X_1, \ldots, X_n$ *are iid Bernoulli*(p), *then* $M = \max(p, 1-p)$, *and we have*

$$P\left[|\bar{X}_n - p| > t\right] \leq 2 \exp\left[-\frac{1}{2} \cdot \frac{nt^2}{p(1-p) + \frac{t}{3}\max(p, 1-p)}\right].$$

Finally we obtain two bounds for tail probabilities of the maximum of cumulative sums of independent rv's. For this reason, they are called maximal inequalities.

In what follows, $X_1, \ldots, X_n$, is a sequence of independent rv's with $\mathrm{E}[X_i] = 0$, $\mathrm{E}[X_i^2] = \sigma_i^2$, $S_k = X_1 + \cdots + X_k$, and $\varepsilon > 0$.

Theorem 3.3.4 (Kolmogorov's Inequality).

$$P\left[\max_{1 \leq k \leq n} |S_k| \geq \varepsilon\right] \leq \frac{1}{\varepsilon^2} \sum_{1 \leq k \leq n} \sigma_k^2.$$

Proof. In the space of infinite sequences $\omega = (x_1, x_2, \ldots)$, let $A_1 = \{\omega\colon |X_1| \geq \varepsilon\}$ and $A_r\{\omega\colon |S_k| < \varepsilon,\ k = 1, \ldots, r-1,\ |S_r| \geq \varepsilon\}$, for $r \geq 2$, that is, A_r is the set of all ω for which the cumulative sums $S_1, S_2, \ldots$ go beyond $\pm\varepsilon$ for the first time. Then $A_1, A_2, \ldots$ are mutually exclusive and

$$\left\{\omega\colon \max_{1 \leq k \leq n} |S_k| \geq \varepsilon\right\} = \bigcup_{k=1}^{n} A_k \quad \text{so } P\left[\max_{1 \leq k \leq n} |S_k| \geq \varepsilon\right] = \sum_{k=1}^{n} P[A_k].$$

Now

$$\begin{aligned}
\sum_{k=1}^{n}\sigma_k^2 &= \mathrm{E}\left[(X_1+\cdots+X_n)^2\right] \geq \sum_{k=1}^{n} P[A_k]\mathrm{E}\left[(X_1+\cdots+X_n)^2|A_k\right] \\
&= \sum_{k=1}^{n} P[A_k]\Bigg\{\mathrm{E}\left[(X_1+\cdots+X_k)^2|A_k\right]+\mathrm{E}\left[(X_{k+1}+\cdots+X_n)^2|A_k\right] \\
&\quad +2\sum_{i=1}^{k}\sum_{j=k+1}^{n}\mathrm{E}\left[X_iX_j|A_k\right]\Bigg\} \\
&\geq \sum_{k=1}^{n} P[A_k]\left\{\varepsilon^2+0+0\right\} = \varepsilon^2\sum_{k=1}^{n}P[A_k] = \varepsilon^2 P\left[\max_{1\leq k\leq n}|S_k|\geq\varepsilon\right].
\end{aligned}$$

Hence the inequality follows. In the above argument,

(i) $\mathrm{E}\left[(X_1+\cdots+X_k)^2|A_k\right]\geq\varepsilon^2$ because A_k is defined that way, and
(ii) for $1\leq i\leq k$, $k+1\leq j\leq n$,

$$\begin{aligned}
\mathrm{E}\left[X_iX_j\middle|A_k\right] &= \mathrm{E}\left[\mathrm{E}\left(X_iX_j\middle|X_1,\ldots,X_k\right)\middle|A_k\right] \\
&= \mathrm{E}\left[X_i\mathrm{E}\left(X_j\middle|X_1,\ldots,X_k\right)\middle|A_k\right] \\
&= \mathrm{E}\left[X_i\mathrm{E}\left(X_j\right)\middle|A_k\right] = 0,
\end{aligned}$$

since the X_i's are independent and $\mathrm{E}(X_j)=0$. □

Remark 3.3.1. In this proof, independence is used only in (ii) which holds more generally, so long as $\mathrm{E}[X_j|X_1,\ldots,X_k]=0$ for $j\geq k+1$. Let $\mathcal{F}_k$ be the σ-field generated by $X_1,\ldots,X_k$. Then $\mathcal{F}_1\subset\mathcal{F}_2\subset\cdots$ and $\mathrm{E}[S_j|\mathcal{F}_k]=S_k$ for $j\geq k+1$. This property of $(S_k,\mathcal{F}_k)_{1\leq k<\infty}$ is called the *martingale property*, which is enough for this theorem (see [9, p. 105]).

Theorem 3.3.5 (Hájek-Rényi Inequality). *Let $c_1,c_2,\ldots$ be a nonincreasing sequence of positive numbers. Then for any two positive integers $m<n$,*

$$P\left[\max_{m\leq k\leq n}c_k|S_k|\geq\varepsilon\right]\leq\frac{1}{\varepsilon^2}\left[c_m^2\sum_{k=1}^{m}\sigma_k^2+\sum_{k=m+1}^{n}c_k^2\sigma_k^2\right].$$

(For $m=1$ and $c_1=\cdots=c_n$, this reduces to the Kolmogorov inequality.)

Proof. Generalizing the definition of $\{A_k\}$ in the proof of the Kolmogorov inequality, let

$$A_m=\{\omega\colon c_m|S_m|\geq\varepsilon\}\text{ and }A_r\{\omega\colon c_k|S_k|<\varepsilon,\ m\leq k\leq r-1,\ c_r|S_r|\geq\varepsilon\}$$

for $r\geq m+1$, and as a generalization of $(X_1+\cdots+X_n)^2$, let

$$Z=\sum_{k=m}^{n-1}\left(c_k^2-c_{k+1}^2\right)S_k^2+c_n^2.$$

Then

$$\begin{aligned}
\mathrm{E}[Z] &= \sum_{k=m}^{n-1}\left(c_k^2 - c_{k+1}^2\right)\left(\sigma_1^2 + \cdots + \sigma_k^2\right) + c_n^2\left(\sigma_1^2 + \cdots + \sigma_n^2\right) \\
&= \sum_{k=m}^{n-1} c_k^2 \sum_{j=1}^{k} \sigma_j^2 - \sum_{k=m+1}^{n} c_k^2 \sum_{j=1}^{k-1} \sigma_j^2 + c_n^2 \sum_{j=1}^{n} \sigma_j^2 \\
&= \left\{ c_m^2 \sum_{j=1}^{m} \sigma_j^2 + \sum_{k=m+1}^{n-1} c_k^2 \sum_{j=1}^{k} \sigma_j^2 \right\} - \left\{ \sum_{k=m+1}^{n-1} c_k^2 \sum_{j=1}^{k-1} \sigma_j^2 + c_n^2 \sum_{j=1}^{n-1} \sigma_j^2 \right\} \\
&\quad + c_n^2 \sum_{j=1}^{n} \sigma_j^2 \\
&= c_m^2 \sum_{k=1}^{m} \sigma_k^2 + \sum_{k=m+1}^{n} c_k^2 \sigma_k^2, \text{ and}
\end{aligned}$$

$$P\left[\max_{m \le k \le n} c_k |S_k| \ge \varepsilon\right] = \sum_{k=m}^{n} P[A_k].$$

We now proceed as in the previous proof:

$$\begin{aligned}
c_m^2 \sum_{k=1}^{m} \sigma_k^2 + \sum_{k=m+1}^{n} c_k^2 \sigma_k^2 = \mathrm{E}[Z] &\ge \sum_{k=m}^{n} P[A_k]\mathrm{E}[Z|A_k] \\
&= \sum_{k=m}^{n} P[A_k]\mathrm{E}\left[\sum_{j=m}^{n-1}\left(c_j^2 - c_{j+1}^2\right)S_j^2 + c_n^2 S_n^2 | A_k\right] \\
&\ge \sum_{k=m}^{n} P[A_k]\mathrm{E}\left[\sum_{j=k}^{n-1}\left(c_j^2 - c_{j+1}^2\right)\left\{S_k^2 + \left(S_j - S_k\right)^2 + 2\sum_{r=1}^{k}\sum_{s=k+1}^{j} X_r X_s\right\}\right. \\
&\quad \left. + c_n^2\left\{S_k^2 + (S_n - S_k)^2 + 2\sum_{r=1}^{k}\sum_{s=k+1}^{n} X_r X_s\right\} \Bigg| A_k\right] \\
&\ge \sum_{k=m}^{n} P[A_k]\left[\sum_{j=k}^{n-1}\left(c_j^2 - c_{j+1}^2\right)\frac{\varepsilon^2}{c_k^2} + \frac{c_n^2 \varepsilon^2}{c_k^2}\right] \\
&= \varepsilon^2 \sum_{k=m}^{n} P[A_k] = \varepsilon^2 P\left[\max_{m \le k \le n} c_k |S_k| \ge \varepsilon\right].
\end{aligned}$$

□

Application of Hájek-Rényi Inequality. A simple proof of the SLLN assuming finite variance is given here. Using the notations used in the above proof we see that for $c_k = 1/k$, $c_k|S_k| = |\bar{X}_k|$. Hence by the H-R inequality,

$$P\left[\max_{k\ge m}|\bar{X}_k| = \max_{k\ge m} c_k|S_k| \ge \varepsilon\right] = \lim_{n\to\infty} P\left[\max_{m\le k\le n} c_k|S_k| \ge \varepsilon\right]$$
$$\le \lim_{n\to\infty}\frac{1}{\varepsilon^2}\left[c_m^2\sum_{k=1}^{m}\sigma_k^2 + \sum_{k=m+1}^{n} c_k^2\sigma_k^2\right]$$
$$= \frac{1}{\varepsilon^2}\left[c_m^2\sum_{k=1}^{m}\sigma_k^2 + \sum_{k=m+1}^{\infty} c_k^2\sigma_k^2\right]$$
$$= \frac{1}{\varepsilon^2}\left[m^{-2}\sum_{k=1}^{m}\sigma_k^2 + \sum_{k=m+1}^{\infty}\frac{\sigma_k^2}{k^2}\right].$$

Now suppose that $\mathrm{E}\left[X_k^2\right] = \sigma^2$ for all k, that is, $X_1, X_2, \ldots$ have finite variance σ^2. Then

$$\lim_{m\to\infty}\left[m^{-2}\sum_{k=1}^{m}\sigma_k^2 + \sum_{k=m+1}^{\infty}\frac{\sigma_k^2}{k^2}\right] = \lim_{m\to\infty}\sigma^2\left[\frac{m}{m^2} + \sum_{k=m+1}^{\infty}\frac{1}{k^2}\right] = 0.$$

Hence $\lim_{m\to\infty} P\left[\max_{k\ge m}|\bar{X}_k| \ge \varepsilon\right] = 0$.

But the events

$$\left\{\omega\colon \max_{k\ge m}|\bar{X}_k| \ge \varepsilon\right\} = \bigcup_{k=m}^{\infty}\{\omega\colon |\bar{X}_k| \ge \varepsilon\} = \bigcup_{k=m}^{\infty} B_k(\varepsilon)$$

are nonincreasing as m increases. Therefore

$$0 = \lim_{m\to\infty} P\left[\max_{k\ge m}|\bar{X}_k| \ge \varepsilon\right] = P\left[\bigcap_{m=1}^{\infty}\bigcup_{k=m}^{\infty} B_k(\varepsilon)\right]$$
$$= P\left[B_k(\varepsilon)\ \text{i.o.}\right] = P\left[|\bar{X}_k| \ge \varepsilon\ \text{i.o.}\right] = P\left[\bar{X}_k \nrightarrow 0\right].$$

Thus $\bar{X}_k \overset{a.s.}{\to} 0$ if $\{X_k\}$ is a sequence of independent rv's with mean 0 and finite variance σ^2.

3.4 Asymptotic Normality: The Central Limit Theorem and Its Generalizations

The basic Lindeberg-Lévy CLT has been stated in Theorem 3.2.2 and its proof, using characteristic functions is given in Section A.4. The Lindeberg-Lévy CLT is restricted to sample means of iid rv's with finite variance. However, in statistical inference based on large samples, we often have to deal with sample means of *independent but nonidentically distributed* rv's and in some situations, even the distribution of the n independent rv's changes from one n to another, as well as the sample size n. In the following, we state two generalizations of Theorem 3.2.2 without proof which address theses generalities.

Theorem 3.4.1 (Lindeberg-Feller Theorem). *Let $X_1, X_2, \ldots$ be independent rv's with $E[X_i] = \mu_i$, $Var[X_i] = \sigma_i^2 < \infty$, and suppose $B_n^2 = \sigma_1^2 + \cdots + \sigma_n^2$ satisfies $\lim_{n\to\infty} B_n = \infty$ and $\lim_{n\to\infty}(\sigma_n^2/B_n^2) = 0$. Then*

$$B_n^{-1} \sum_{i=1}^{n} (X_i - \mu_i) \xrightarrow{\mathcal{L}} Z \sim N(0,1) \textit{ iff}$$

$$\lim_{n\to\infty} B_n^{-2} \sum_{i=1}^{n} E\Big[I_{(\varepsilon B_n,\infty)}(|X_i - \mu_i|)(X_i - \mu_i)^2\Big] = 0 \quad \textit{for all } \varepsilon > 0.$$

Triangular Arrays. Here we think of a double array of rv's in which each row consists of independent rv's

$$\{(X_{11}, \ldots, X_{1k_1}), (X_{21}, \ldots, X_{2k_2}), \ldots, (X_{n1}, \ldots, X_{nk_n}), \ldots\}$$

with $E[X_{ni}] = \mu_{ni}$, $\mathrm{Var}(X_{ni}) = \sigma_{ni}^2 < \infty$, and $B_n^2 = \sigma_{n1}^2 + \cdots + \sigma_{nk_n}^2$.

Theorem 3.4.2 (Lindeberg-Liapounov Theorem). *The Lindeberg condition*

$$\lim_{n\to\infty} B_n^{-2} \sum_{i=1}^{k_n} E\Big[I_{(\varepsilon B_n,\infty)}(|X_{ni} - \mu_{ni}|)(X_{ni} - \mu_{ni})^2\Big] = 0 \quad \textit{for all } \varepsilon > 0$$

implies $B_n^{-1}\sum_{1\le i\le k_n}(X_{ni} - \mu_{ni}) \xrightarrow{\mathcal{L}} Z \sim N(0,1)$. The above convergence in law is also implied by the Liapounov condition

$$\lim_{n\to\infty} \frac{\rho_n^3}{B_n^3} = 0, \textit{ where } \rho_n^3 = \sum_{i=1}^{k_n} E\Big[|X_{ni} - \mu_{ni}|^3\Big] < \infty \textit{ and } B_n^3 = \left(\sum_{i=1}^{k_n} \sigma_{ni}^2\right)^{3/2}.$$

Finally, we state another generalization of Theorem 3.2.2 for sequences of rv's with limited dependence.

Definition 3.4.1. A sequence of rv's $\{X_1, X_2, \ldots\}$ is said to be m-dependent if $(X_1, \ldots, X_r)$ is independent of $(X_s, X_{s+1}, \ldots)$ whenever $s - r > m$.

Let $A_i = 2\sum_{j=0}^{m-1} \mathrm{Cov}\big[X_{i+j}, X_{i+m}\big] + \mathrm{Var}\big[X_{i+m}\big]$.

Theorem 3.4.3 (Hoeffding and Robbins [10]). *If for an m dependent sequence $\{X_i\}$ with $E[X_i] = 0$, $Var\big[|X_i|^3\big] \le K < \infty$, and $\lim_{r\to\infty} r^{-1}\sum_{j=1}^{r} A_{i+j} = A > 0$ exists uniformly for all i, then $n^{-1/2}\sum_{i=1}^{n} X_i \xrightarrow{\mathcal{L}} N(0,A)$.*

In particular, if $\{X_i\}$ is stationary m-dependent sequence with $E[X_i] = \mu$, $E\big[|X_i|^3\big] < \infty$ and $A = Var[X_1] + 2\sum_{i=2}^{m+1} Cov[X_1, X_i] > 0$, then $\sqrt{n}(\bar{X}_n - \mu) \xrightarrow{\mathcal{L}} N(0,A)$.

Exercises

For all the problems below, when there are observations $X_1, \ldots, X_n$, it is understood that

$$\bar{X}_n = n^{-1}\sum_{i=1}^{n} X_i,\ s_n^2 = (n-1)^{-1}\sum_{i=1}^{n}(X_i - \bar{X}_n)^2,$$
$$X_{n:1} = \min(X_1, \ldots, X_n),\ X_{n:n} = \max(X_1, \ldots, X_n).$$

3.1. Prove VI(a,b) of Theorem 3.2.5.

3.2. Prove X(b) of Theorem 3.2.5.

3.3. Prove Theorem 3.2.6, the Delta Method.

3.4. Prove the results stated in the examples on applications of the Delta Method.

3.5. Let $X_1, \ldots, X_n$ be iid with $\mathrm{E}[X_i] = \mu \neq 0$, $\mathrm{Var}[X_i] = \sigma^2$ and $\mathrm{E}[X_i^4] < \infty$. Find the asymptotic distribution of $\sqrt{n}(s_n/\bar{X}_n - \sigma/\mu)$ as $n \to \infty$.

3.6. Let $B_{m,n}$ denote a *Beta(m, n)* rv. Show that if $m, n \to \infty$ in such a way that $m/(m+n) \to \alpha \in (0,1)$, then $\sqrt{m+n}(B_{m,n} - m/(m+n))/\sqrt{\alpha(1-\alpha)}$ converges in distribution to $Z \sim N(0,1)$. [First show that if $X_1, \ldots, X_m, Y_1, \ldots, Y_n$ are iid Exp(1), then $R_{m,n}(1+R_{m,n})^{-1} \stackrel{\mathcal{D}}{=} B_{m,n}$, where $R_{m,n} = (m\bar{X}_m)/(n\bar{Y}_n)$.]

3.7. Let $\{X_i\}$ be a sequence of rv's with $\mathrm{E}[X_i] = 0$, $\mathrm{Var}1X_i] = 1$ and $\mathrm{Cov}[X_i, X_j] = 0$ for $|i - j| \geq k$, where k is a fixed positive integer. Show that $\bar{X}_n \xrightarrow{P} 0$.

3.8. Let $\{X_n\}$ be a sequence of rv's' with $\mathrm{E}[X_n] = \mu_n$ and $\mathrm{Var}[X_n] = \sigma_n^2$. Show that if $\mu_n \to 0$ and $\sigma_n^2 \to 0$ as $n \to \infty$, then $X_n \xrightarrow{P} 0$.

3.9. Let $\{X_n\}$ be a sequence of rv's such that X_n is distributed as Z with probability p_n and as $\sigma_n Z$ with probability $1 - p_n$ where $Z \sim N(0,1)$, $p_n \to p \in [0,1]$ and $\sigma_n \to \infty$.

(a) Show that $X_n \xrightarrow{\mathcal{L}} X$ if and only if $p = 1$. What is the distribution of X?

(b) $\lim_{n\to\infty} \mathrm{Var}[X_n]$ is not necessarily the same as $\mathrm{Var}[X]$. Find the possible values of $\lim_{n\to\infty} \mathrm{Var}[X_n]$.

3.10. Let $\{(x_{ni}, Y_{ni}), i = 1, \ldots, n\}_{n=1}^{\infty}$ be a triangular array following a simple linear regression model $Y_{ni} = \alpha + \beta x_{ni} + Z_{ni}$, where $Z_{n1}, \ldots, Z_{nn}$ are iid with mean 0 and variance σ^2. Assume that the common distribution of Z_{ni}'s is the same for all n. Let

$$\hat{\beta}_n = \sum_{i=1}^{n}(x_{ni} - \bar{x}_n)Y_{ni} \Big/ \sum_{i=1}^{n}(x_{ni} - \bar{x}_n)^2, \quad \bar{x}_n = n^{-1}\sum_{i=1}^{n} x_{ni}$$

denote the least squares estimator of β. Use Lindeberg's condition to establish the asymptotic normality of $\sqrt{n}\left(\hat{\beta}_n - \beta\right)$, making appropriate assumptions on $\{x_{ni}\}$.

4

Basic Concepts of Statistical Inference

4.1 Population and Random Samples

The term "Statistics" is commonly used as a synonym for data, but "Statistical Inference" is the science of analyzing data to probe into where the data came from. The data is a "sample" and where the data came from is the "population" having some unknown characteristics in which we are interested. The sample has to be a "random sample" for the sake of objectivity and thus randomness brings probability into the picture.

A population can be thought of as a concrete, finite collection, such as individuals in a city, from which a random sample is drawn and some characteristic of each of these individuals, such as opinion on a certain issue, or income, etc., is recorded. This is the data, from which a summary measure of the characteristic in the entire population must be inferred. This is the framework of Survey Sampling.

Here, on the other hand, we think of a population consisting of observations on the outcomes of infinite repetitions of an experiment, such as survival times of cancer patients on a certain drug, or the number of defective items in lots of N items coming out of a production line, or the number of accidents during certain hours on a particular stretch of a freeway. Clearly, these observations vary from one experiment to another and they vary in a random manner. The collection of possible outcomes of such an experiment is a probability space $(S, \mathcal{A}, P)$ and the observation on an outcome, $X = X(s)$ is a random variable or more generally, a random vector. From n independent repetitions of such an experiment, we observe independent random variables $X_1, \ldots, X_n$. This is the data.

The probability distribution of X should be denoted by P^X in strict notation (to distinguish it from the set function P: $\mathcal{A} \to [0, 1]$ in the basic probability space), defined as $P^X[(-\infty, a]] = P[\{s: X(s) \leq a\}]$ and more generally by $P^X[B] = P[\{s: X(s) \in B\}]$ for all Borel sets in $\mathbb{R}$. However, to keep the notation simple, we shall use P instead of P^X to denote the probability distribution of X when there is no chance of confusion. These notations will extend to the case of multidimensional rv's in an obvious manner. We can now say that our observations $X_1, X_2, \ldots$ are independent rv's with common probability distribution P which is an unknown element of a family of distributions $\mathbb{P}$.

Theory and Methods of Statistics. http://dx.doi.org/10.1016/B978-0-12-802440-9.00004-7

4.2 Parametric and Nonparametric Models

Instead of leaving the family $\mathbb{P}$ wide open, we shall assume that $\mathbb{P}$ is either a *parametric family*

$$\mathbb{P} = \{P_{\boldsymbol{\theta}},\ \boldsymbol{\theta} \in \Theta\},$$

where $\boldsymbol{\theta} = (\theta_1, \ldots, \theta_k)$ is a k-dim parameter vector belonging to the parameter space Θ and $P_{\boldsymbol{\theta}}$ is known for any given $\boldsymbol{\theta} \in \Theta$, or a *nonparametric family* $\mathbb{P}$ whose members cannot be identified by a finite number of parameters.

Examples of parametric family are

P(i) *Bernoulli* (θ), $0 < \theta < 1$,
P(ii) *Poisson* (λ), $\lambda > 0$,
P(iii) *Normal* (μ, σ^2), $-\infty < \mu < \infty$ and $\sigma^2 > 0$,
P(iv) *Gamma* (α, β), $\alpha > 0$ and $\beta > 0$,
P(v) Linear Regression Model $(\alpha, \beta, \sigma^2)$, etc.

The families P(i)–P(iv) have been discussed in Chapter 2 and P(v) has been introduced in Chapter 1, Section 1.10. In P(v), the distributions of X and ε are not fully specified, but we still can do a lot within the model as it is and much more conditionally given $X = x$ if $\varepsilon \sim N(0, \sigma^2)$.

Examples of nonparametric family are

NP(i) All probability distributions on the real line.
NP(ii) All probability distributions on the real line with pdf's satisfying some smoothness conditions.
NP(iii) The family described in NP(ii) with the further restriction that the pdf's are symmetric about some $\theta \in \mathbb{R}$.
NP(iv) All probability distribution of X_1 and X_2 on the real line which are independent and

$$f_{X_2}(x) = f_{X_1}(x - \theta) \quad \text{for all } x \text{ and for some } \theta \in \mathbb{R}.$$

NP(v) All probability distributions of $(X_1, \ldots, X_k, Y)$ on $\mathbb{R}^{k+1}$ with $f_{X_1,\ldots,X_k}$ and $m(x_1, \ldots, x_k) = \mathrm{E}[Y|\boldsymbol{X} = \boldsymbol{x}]$, satisfying some smoothness conditions.

4.3 Problems of Statistical Inference

Based on observed data consisting of random samples $(X_1, \ldots, X_n)$ or $(X_{11}, \ldots, X_{1n_1}; X_{21}, \ldots, X_{2n_2})$ or $((X_1, Y_1), \ldots, (X_n, Y_n))$, etc., from an unknown probability distribution $P \in \mathbb{P}$ in a parametric or nonparametric model, we have to make inference about some unknown features of P. The three main types of inference that the statistical science has been concerned with from its very inception are *Point Estimation*, *Hypothesis Testing*, and

Confidence Sets. We briefly describe these three types here by means of some examples before taking them up in more details in subsequent chapters.

Point Estimation

Here we want to make a guess about a function $g(\theta)$ of the unknown $\theta \in \Theta$ in a parametric model, or construct a pdf or a regression function with prescribed properties as our guess for an unknown pdf f or an unknown regression function m in a nonparametric model. In a parametric model, we may want to estimate the mean λ of a Poisson distribution or the slope β of a linear regression function or $P[X \leq a]$ of X distributed as $N(\mu, \sigma^2)$.

Hypothesis Testing

In a parametric model, this involves deciding whether $\theta \in \Theta_0$ or $\theta \in \Theta_1$ where Θ_0 and Θ_1 are disjoint sets in the parameter space Θ, such as deciding in the context of $N(\mu, \sigma^2)$ with σ^2 known or unknown, whether $\mu \leq 0$ or $\mu > 0$. We call H_0: $\theta \in \Theta_0$ the *null hypothesis* which we are inclined to believe unless the data provides *significant* evidence in favor of H_1: $\theta \in \Theta_1$ which we call the *alternative hypothesis*. So there is an asymmetry in this problem due to our attitude toward H_0 and H_1. For this reason, this is a problem of testing H_0 against H_1. An example in a nonparametric model is to decide whether or not $\theta = 0$ (ie, testing H_0: $\theta = 0$ against H_1: $\theta \neq 0$) in the nonparametric family NP(iv).

Confidence Sets

Unlike point estimation, here we want to *construct a set* in the parameter space to which we guess the unknown θ to belong. In a parametric model, this may be an interval which includes the mean μ of a normal distribution $N(\mu, \sigma^2)$ with σ^2 known or unknown, while in a nonparametric model this may be a band in which we guess that an unknown cdf F belongs.

These three problems will be treated in Chapters 5 and 6 within parametric families and in Chapter 8 within nonparametric families.

4.4 Statistical Decision Functions

Problems of statistical inference in a parametric model such as Point Estimation, Hypothesis Testing, and many others can be fitted in a general framework of decision making based on a random sample X from $(\mathfrak{X}, \mathbb{B}, P_\theta)$ with unknown $\theta \in \Theta$. The decision consists of taking an action $a \in A$, where A is the set of available actions which we call the *action space*. This action has to be a function of the observed value x of X. We call this function a *Decision Function* or *Decision Rule*:

$$d: \mathfrak{X} \to A \text{ with } d(x) = a \in A \text{ when } X = x.$$

Now there are good decisions and bad decisions for any given $\theta \in \Theta$, and for making a bad decision one has to pay a penalty, or take a loss, which is the consequence of taking the action $a = d(x)$ when the observed value x of X was generated from P_θ. This leads to the definition of a *Loss Function*: L: $\Theta \times A \to \mathbb{R}$ with $L(\theta, a) :=$ loss due to action a when θ is the true value of the parameter. Then for $X = x$, $L(\theta, d(x)) =$ loss due to action $d(x)$ when $X = x$ is generated by $P(\theta)$.

Since the data is the rv X, we should be thinking of the loss as an rv $L(\theta, d(X))$ where $X \sim P_\theta$. Note that the θ in P_θ and the θ in $L(\theta, d(X))$ is the same unknown element of Θ (ie, the true value of the parameter).

A statistical decision problem is thus described by the triple $(\{P_\theta: \theta \in \Theta\}, A, L)$, and the overall performance of a decision function d is measured by its risk

$$R(\theta, d) = \mathrm{E}_\theta\left[L(\theta, d(X))\right] = \begin{cases} \int_{\mathfrak{X}} L(\theta, d(x)) f(x, \theta)\, dx & \text{in the continuous case,} \\ \sum_{x \in \mathfrak{X}} L(\theta, d(x)) f(x, \theta) & \text{in the discrete case,} \end{cases}$$

assuming that the $\int$ or the $\sum$ defining $R(\theta, d)$ exists. This holds if $\mathfrak{X}$ is finite and more generally, if $L(\cdot, \cdot)$ is bounded below, although the $\int$ or the $\sum$ may be $+\infty$.

Definition 4.4.1. The function $R(\cdot, d)$ is called the risk function of the decision rule d.

Ideally, we should be using a decision rule d^* for which $R(\theta, d^*) \leq R(\theta, d)$, for all $\theta \in \Theta$ and for all decision rules d.

Unfortunately, such a decision rule does not exist (except in some trivial cases). This is illustrated by the following examples. Before introducing these examples, we enlarge the class of decision rules by allowing randomization over $a \in A$ for each x.

Definition 4.4.2. A *behavioral decision rule* δ consists of a probability distribution $\delta(\cdot|x)$: $\mathcal{C} \to [0, 1]$ for each $x \in \mathfrak{X}$, where $\mathcal{C}$ is a σ-field of subsets of the action space A.

If A is countable, then $\mathcal{C}$ could be the class of all subsets of A. For $A = \mathbb{R}^k$ or some subsets of $\mathbb{R}^k$, $\mathcal{C}$ could be the Borel sets and $\delta(\cdot|x)$ would have a density. The risk of a behavioral decision rule $\delta(\cdot|x)$ is given by:

$$R(\theta, \delta) = \int_{\mathfrak{X}} \int_A L(\theta, a)\, d\delta(a|x) f(x, \theta)\, dx,$$

where $\int_A L(\theta, a)\, d\delta(a|x) = \sum_{a \in A} L(\theta, a)\delta(a|x)$ if A is countable and

$$\int_A L(\theta, a)\, d\delta(a|x) = \int_A L(\theta, a) p(a|x)\, da$$

when $A = \mathbb{R}^k$ is some subset of $\mathbb{R}^k$, $p(\cdot|x)$ being the pdf of $\delta(\cdot|x)$.

Example 4.4.1. Estimating a Bernoulli parameter.

Let $X_1, \ldots, X_n$ be iid with

$$P_\theta(X_i = x) = \theta^x (1 - \theta)^{1-x}, \quad x = 0, 1$$
$$A = [0, 1] = \Theta, \quad L(\theta, a) = (\theta - a)^2.$$

This defines a statistical decision problem, which is the problem of estimating θ under squared-error loss. Consider a class of estimators (ie, functions of the data taking values in Θ) denoted by $d_\gamma(\boldsymbol{x}) = \gamma\bar{x}$, $0 \le \gamma \le 1$, where $\bar{x} = n^{-1}\sum_{i=1}^n x_i$. The usual estimator $d_1(\boldsymbol{x}) = \bar{x}$ belongs to this class. Is there an estimator $d_{\gamma^*}(\boldsymbol{x})$ in this class which is better than all others?

Solution. Note that

$$\begin{aligned} R(\theta, d_\gamma) &= \mathrm{E}_\theta\left[(\gamma\bar{X} - \theta)^2\right] = \mathrm{Var}_\theta[\gamma\bar{X}] + \{\mathrm{E}_\theta[\gamma\bar{X}] - \theta\}^2 \\ &= \frac{\gamma^2\theta(1-\theta)}{n} + (\gamma-1)^2\theta^2. \end{aligned}$$

So $R(\theta, d_\gamma)/R(\theta, d_1) = \gamma^2 + n\theta(1-\gamma)^2/(1-\theta)$, which is $>, =,$ or < 1 according as $\theta >, =,$ or $< (1+\gamma)/[n(1-\gamma) + (1+\gamma)]$. For example, taking $\gamma = 3/4$, d_γ is better or worse than d_1 according as $\theta <$ or $> 7/(n+7)$. This shows that there is no clear winner in the class of $d_\gamma, 0 \le \gamma \le 1$.

Example 4.4.2. Choosing between two values of θ.

Let $X \sim P_\theta$, $\Theta = \{\theta_0, \theta_1\}$, $A = \{a_0, a_1\}$ where a_i is to choose θ_i as the true value of θ, and let $L(\theta_0, a_0) = L(\theta_1, a_1) = 0$ and $L(\theta_0, a_1) = L(\theta_1, a_0) = 1$. A typical behavioral decision rule is described by a function $\varphi\colon \mathfrak{X} \to [0, 1]$, so that

$$\delta_\varphi(a_1|x) = \varphi(x) \text{ and } \delta_\varphi(a_0|x) = 1 - \varphi(x),$$

that is, δ_φ chooses actions a_1, a_0 with probabilities $\varphi(x)$ and $1-\varphi(x)$, respectively. Is there a best decision rule in this class?

Solution. The risk function of δ_φ is given by

$$\begin{aligned} R(\theta_0, \delta_\varphi) &= \int_{\mathfrak{X}} [L(\theta_0, a_0)\{1-\varphi(x)\} + L(\theta_0, a_1)\varphi(x)]f(x, \theta_0)\, dx \\ &= \mathrm{E}_{\theta_0}[\varphi(X)], \\ R(\theta_1, \delta_\varphi) &= \int_{\mathfrak{X}} [L(\theta_1, a_0)\{1-\varphi(x)\} + L(\theta_1, a_1)\varphi(x)]f(x, \theta_1)\, dx \\ &= 1 - \mathrm{E}_{\theta_1}[\varphi(X)]. \end{aligned}$$

The *risk set* $S = \left\{\left(R(\theta_0, \delta_\varphi), R(\theta_1, \delta_\varphi)\right),\ \varphi\colon \mathfrak{X} \to [0, 1]\right\}$ has the following features:

- **(i)** $(0, 1) \in S$ corresponding to $\varphi(x) \equiv 0$.
- **(ii)** $(1, 0) \in S$ corresponding to $\varphi(x) \equiv 1$.
- **(iii)** S is convex, because for any φ_1, φ_2, $\lambda\varphi_1 + (1-\lambda)\varphi_2$ for $0 \le \lambda \le 1$, has the property:

$$\begin{aligned} &\left(R(\theta_0, \delta_{\lambda\varphi_1 + (1-\lambda)\varphi_2}), R(\theta_1, \delta_{\lambda\varphi_1+(1-\lambda)\varphi_2})\right) \\ &\quad = \lambda\left(R(\theta_0, \delta_{\varphi_1}), R(\theta_1, \delta_{\varphi_1})\right) + (1-\lambda)\left(R(\theta_0, \delta_{\varphi_2}), R(\theta_1, \delta_{\varphi_2})\right). \end{aligned}$$

- **(iv)** $(\alpha, \beta) \in S$ implies $(1-\alpha, 1-\beta) \in S$ (ie, S is symmetric about $(1/2, 1/2)$). (To see this, consider $\psi = 1 - \varphi$ for any φ.)
- **(v)** S can be shown to be closed. We omit the proof.

Consider any two points on the lower boundary of S. Among the φ's corresponding to two such points, there is no clear winner. However, for any $s \in S$, the point s^* on the lower boundary which is on the vertical strip through s is decidedly better than s. For this reason, it is enough to restrict our attention to the φ's corresponding to the points on the lower boundary of S.

4.5 Sufficient Statistics

Consider a problem of statistical inference in the framework of a parametric family with pdf or pmf $\{f(\boldsymbol{x}, \theta),\ \theta \in \Theta\}$ based on a random sample $(X_1, \ldots, X_n)$. The unknown $\theta \in \Theta$, which we want to estimate or test a hypothesis about, is of a finite dimension, often of only one or two dimensions, while the number of observations in the sample, n, may be quite large. Do we really need to carry all the n observations in the data in our search for a good estimate or test or whatever else to decide about?

Definition 4.5.1. A function $T\colon \mathfrak{X}^n \to \mathbb{R}$ of the sample observations is said to be a statistic if

$$\{(x_1, \ldots, x_n)\colon T(x_1, \ldots, x_n) \le a\} \in \mathbb{B}^n \quad \text{for all } a \in \mathbb{R},$$

where $\mathbb{B}^n$ is the class of all Borel sets in $\mathfrak{X}^n$. A vector $\boldsymbol{T} = (T_1, \ldots, T_k)$ is said to be a k-dim statistic.

If we could extract all the relevant information about θ in a *fixed-dimensional* statistic $T = T(X_1, \ldots, X_n)$, then we can concentrate our efforts in search of a good procedure based on T without carrying the burden of the entire sample $(X_1, \ldots, X_n)$. If we have such a statistic, then it can rightly be called a *Sufficient Statistic.*

Definition 4.5.2. A statistic $T = T(X_1, \ldots, X_n)$ is said to be sufficient for $\theta \in \Theta$ in $\boldsymbol{X}$ if the conditional distribution of $\boldsymbol{X}$ given T is independent of θ.

This means the following:

Discrete Case. Here the pmf of T is $g(t, \theta) = \sum_{\{\boldsymbol{x}:\ T(\boldsymbol{x})=t\}} f(\boldsymbol{x}, \theta)$, and the conditional distribution of $\boldsymbol{X}$ given $T = t$ is given by

$$h_\theta(\boldsymbol{x}|t) = \frac{P_\theta[\boldsymbol{X} = \boldsymbol{x}, T = t]}{P_\theta[T = t]} = \begin{cases} f(\boldsymbol{x}, \theta)/g(t, \theta) & \text{if } T(x_1, \ldots, x_n) = t, \\ 0 & \text{otherwise} \end{cases}$$

for all t with $g(t, \theta) > 0$. Sufficiency requires $h_\theta(\boldsymbol{x}|t) = h(\boldsymbol{x}|t)$, independent of θ.

Continuous Case. For simplicity of notations, suppose $\theta \in \mathbb{R}$ and let $T = T(X_1, \ldots, X_n)$ be a 1-dim statistic. (Generalization to higher dimension is routine.) Suppose that $U_i = U_i(X_1, \ldots, X_n)$, $1 \le i \le n-1$ is any collection of $n-1$ other statistics such that $\varphi\colon (X_1, \ldots, X_n) \leftrightarrow (T, U_1, \ldots, U_{n-1})$ is one-to-one with continuous first partials. Then the joint pdf of $(T, U_1, \ldots, U_{n-1})$ is

$$f(\varphi^{-1}(t, u_1, \ldots, u_{n-1}), \theta)\left|J_\varphi(\varphi^{-1}(t, u_1, \ldots, u_{n-1}))\right|^{-1}.$$

From this, the pdf of $g(t, \theta)$ of T is obtained by integrating out $u_1, \ldots, u_{n-1}$ and the conditional pdf of $(U_1, \ldots, U_{n-1})$ given $T = t$ is

$$h_\theta(u_1, \ldots, u_{n-1}|t) = \frac{f(\varphi^{-1}(t, u_1, \ldots, u_{n-1}), \theta)\left|J_\varphi(\varphi^{-1}(t, u_1, \ldots, u_{n-1}))\right|^{-1}}{g(t, \theta)}.$$

Sufficiency of T requires $h_\theta(u_1, \ldots, u_{n-1}|t)$ to be independent of θ. Since $J_\varphi(\varphi^{-1}(t, u_1, \ldots, u_{n-1}))$ does not involve θ anyway, and since

$$f\left(\varphi^{-1}(T(\boldsymbol{x}), u_1(\boldsymbol{x}), \ldots, u_{n-1}(\boldsymbol{x})), \theta\right) = f(\boldsymbol{x}, \theta),$$

$h_\theta(u_1, \ldots, u_{n-1}|t)$ being independent of θ is equivalent to $f(\boldsymbol{x}, \theta)/g(T(\boldsymbol{x}), \theta)$ being independent of θ. Thus in both the discrete and continuous case, T is sufficient for θ in $\boldsymbol{X}$ if $f(\boldsymbol{x}, \theta)/g(T(\boldsymbol{x}), \theta)$ is independent of θ.

If $\boldsymbol{T}(\boldsymbol{x}) = (T_1(\boldsymbol{x}), \ldots, T_k(\boldsymbol{x}))$ has the above property (using $U_1, \ldots, U_{n-k}$ to construct a one-to-one map), then $(T_1, \ldots, T_k)$ are jointly sufficient for θ, usually when $\boldsymbol{\theta} = (\theta_1, \ldots, \theta_k)$.

Suppose T is sufficient for θ in $\boldsymbol{X}$ and let ψ be a real-valued function of $\boldsymbol{X}$ with $\mathrm{E}_\theta[\psi(\boldsymbol{X})]$ finite. Let $U_1, \ldots, U_{n-1}$ be as described above, and $\psi^*(T, U_1, \ldots, U_{n-1}) = \psi \circ \varphi^{-1}(T, U_1, \ldots, U_{n-1}.)$ Then

$$\begin{aligned}\mathrm{E}_\theta[\psi(\boldsymbol{X})|T = t] &= \mathrm{E}_\theta[\psi^*(T, U_1, \ldots, U_{n-1})|T = t] \\ &= \mathrm{E}_\theta[\psi^*(t, U_1, \ldots, U_{n-1})|T = t]\end{aligned}$$

is independent of θ.

Now consider an arbitrary decision problem described by (Θ, A, L) in the context of $(\mathfrak{X}, \mathbb{B}, \{P_\theta, \theta \in \Theta\})$. For simplicity of discussion, let $\mathfrak{X}$ and A be countable and let $\delta(\cdot|\boldsymbol{x})$ be an arbitrary behavioral rule. Suppose T is sufficient for θ in $\boldsymbol{X}$ and let $\mathcal{T} = T(\mathfrak{X})$ and $\mathfrak{X}_t = \{\boldsymbol{x} \in \mathfrak{X}\colon T(\boldsymbol{x}) = t\}$. Then the risk of δ is

$$\begin{aligned}R(\theta, \delta) &= \sum_{\boldsymbol{x} \in \mathfrak{X}}\left(\sum_{a \in A} L(\theta, a)\delta(a|\boldsymbol{x})\right)f(\boldsymbol{x}, \theta) \\ &= \sum_{t \in \mathcal{T}}\sum_{\boldsymbol{x} \in \mathfrak{X}_t}\left(\sum_{a \in A} L(\theta, a)\delta(a|\boldsymbol{x})\right)g(t, \theta)h(\boldsymbol{x}|t) \\ &= \sum_{t \in \mathcal{T}}\left[\sum_{a \in A} L(\theta, a)\left(\sum_{\boldsymbol{x} \in \mathfrak{X}_t} \delta(a|\boldsymbol{x})h(\boldsymbol{x}, t)\right)\right]g(t, \theta) \\ &= \sum_{t \in \mathcal{T}}\left(\sum_{a \in A} L(\theta, a)\delta^*(a|t)\right)g(t, \theta) \\ &= R(\theta, \delta^*),\end{aligned}$$

where δ^* is defined by

$$\delta^*(a|\boldsymbol{x}) = \delta^*(a|t) = \sum_{\boldsymbol{x}\in\mathfrak{X}_t} \delta(a|\boldsymbol{x})h(\boldsymbol{x}, t) \quad \text{for all } \boldsymbol{x} \in \mathfrak{X}_t,$$

which depends on $\boldsymbol{x}$ only through $T(\boldsymbol{x})$ and so we write $\delta^*(a|\boldsymbol{x}) = \delta^*(a|t)$. In other words, $\delta^*(a|t) = \mathrm{E}[\delta(a|\boldsymbol{X})|T = t]$. We thus have for arbitrary δ, an equivalent decision rule δ^* which uses the data $\boldsymbol{x}$ only through the summary provided by the statistic T. This justifies the term sufficient statistic. In any decision problem, a sufficient statistic is used for data reduction without losing anything that could be achieved by the full data. We now state and prove the key theorem about sufficient statistic.

Theorem 4.5.1 (Factorization Theorem). *A statistic $T = T(\boldsymbol{X})$ is sufficient for θ in $\boldsymbol{X}$ iff there exist functions $g(t, \theta)$ and $h(\boldsymbol{x})$ so that*

$$f(\boldsymbol{x}, \theta) = g(T(\boldsymbol{x}), \theta)h(\boldsymbol{x}) \quad \textit{for all } \boldsymbol{x}, \theta,$$

where for every fixed $t = T(\boldsymbol{x})$, the function $h(\boldsymbol{x})$ is independent of θ.

The proof follows from our discussion above.

Example 4.5.1. In Section 2.2.11 we defined an exponential family of distributions of $\boldsymbol{X}$ without specifying $\mathfrak{X}$ in which $\boldsymbol{X}$ takes its values. We now let $\boldsymbol{X} = (X_1, \ldots, X_n)$ be a random sample from a k-parameter regular exponential family. Then

$$f_{\boldsymbol{X}}(\boldsymbol{x}, \theta) = \{c(\theta)\}^n \exp\left[\sum_{j=1}^{k} Q_j(\theta) \sum_{i=1}^{n} T_j(x_i)\right] \prod_{i=1}^{n} r(x_i).$$

Thus $(T_1, \ldots, T_k) = \sum_{i=1}^{n}(T_1(x_i), \ldots, T_k(x_i))$ are jointly sufficient for θ in $\boldsymbol{X}$.

4.6 Optimal Decision Rules

Definition 4.6.1. A decision rule δ_1 is said to be (i) as good as δ_2, (ii) better than δ_2, or (iii) equivalent to δ_2, according as

(i) $R(\theta, \delta_1) \le R(\theta, \delta_2)$ for all $\theta \in \Theta$,
(ii) $R(\theta, \delta_1) \le R(\theta, \delta_2)$ for all $\theta \in \Theta$, with strict inequality for some θ,
(iii) $R(\theta, \delta_1) = R(\theta, \delta_2)$ for all $\theta \in \Theta$.

If there exists a decision rule δ_0 which is as good as any other δ, then of course δ_0 is optimal, but typically, such a δ_0 does not exist as seen in Examples 4.4.1 and 4.4.2. The concept of optimality therefore needs adjustment. Two general approaches are taken for this purpose:

(a) restricting the class of decision rule to choose from, or
(b) ordering the decision rules in a less stringent manner.

4.6.1 Restrictions Used in the Estimation Problem

Unbiasedness

Restrict attention to only those estimators $\hat{\theta} = d(X)$ of θ, which satisfies

$$\mathrm{E}_\theta\left[\hat{\theta}\right] = \theta \quad \text{for all } \theta \in \Theta.$$

In the example of estimating a Bernoulli parameter, all d_γ with $\gamma \neq 1$ are now ruled out. In a large class of estimation problems with squared-error loss, there exists a best estimator in the class of unbiased estimators. These are the uniformly minimum variance unbiased estimators (UMVUE).

Equivariance

Suppose that $f(x, \theta) = g(x - \theta)$ where g is a known pdf and we want to estimate θ, which is called a location parameter, subject to $L(\theta, a) = (a - \theta)^2$. If $X_1, \ldots, X_n$ are iid with pdf $f(\cdot, \theta)$, then $X_1 + c, \ldots, X_n + c$ are iid with pdf $f(\cdot, \theta + c)$. Moreover, $L(\theta + c, a + c) = L(\theta, a) = (a - \theta)^2$. Thus the problem of estimating $\theta + c$ from $X_1 + c, \ldots, X_n + c$ is the same as that of estimating θ from $X_1, \ldots, X_n$; that is, the problem of estimating a location parameter under squared-error loss is *invariant under location*, that is, under the transformations

$$g_c\colon \mathfrak{X} \to \mathfrak{X} \text{ and its corresponding } \bar{g}_c\colon \Theta \to \Theta, \quad c \in \mathbb{R},$$

defined by $g_c(x) = x + c$ and $\bar{g}_c(\theta) = \theta + c$. It is therefore reasonable to restrict our estimator by the requirement: $d(x_1 + c, \ldots, x_n + c) = d(x_1, \ldots, x_n) + c$ for all x, c. Such estimators are called *equivariant under location*. Among equivariant estimators in this, and many more general invariant problems, there exists a best estimator, known as minimum risk equivariant estimator (MRE).

Optimal estimators under these restrictions will be discussed in Chapter 5.

4.6.2 Restriction Used in the Two-Decision Problem

As mentioned earlier, the two-decision problem is treated as a problem of testing a hypothesis $H_0\colon \theta \in \Theta_0$ against $H_1\colon \theta \in \Theta_1$ by introducing an asymmetry between the two hypotheses. We are *inclined to accept* H_0 unless there is *significant evidence provided by the data to reject* H_0 *in favor of* H_1. For this reason, we call H_0 the *null hypothesis* and the error in rejecting H_0 when it is true is called a *Type I error*, while H_1 is called the *alternative hypothesis* and the error in accepting H_0 when H_0 is not true (ie, H_1 is true) is called a *Type II error*.

We illustrate the asymmetric nature of H_0 and H_1 by the following example. Suppose for a certain disease there is a drug on the market with an effectiveness of $p_0 = 0.40$ established over a long period of time. Now a new drug which claims to be better has been shown to be effective in $\hat{p}_1 = 0.45$ in a clinical study based on $n = 100$ cases. To test the validity of this claim, we have to choose between the actions:

$$a_0 = \{p_1 \leq 0.40\} \text{ and } a_1 = \{p_1 > 0.40\},$$

where p_1 = the unknown effectiveness of the new drug. Here H_0: $p_0 = 0.40$ is based on a long record, while H_1: $p_1 > p_0 = 0.40$ needs to be substantiated by significant (ie, overwhelming) evidence provided by the data in favor of H_1. This is why H_0 *is called the null hypothesis*, giving it a special role, while H_1 *is called an alternative hypothesis* which is to be accepted only if the evidence provided by the data in its favor is significant enough. Therefore, we want to control the probability, $P_{H_0}[\text{Reject } H_0] \le \alpha$ (preassigned) and with that restriction, want to maximize $P_{H_1}[\text{Reject } H_0]$. We now introduce the restrictions in the two-decision problem.

Prescribed Type I Error Probability

Recall the discussion about the lower boundary of the risk set in Example 4.4.2. Here if we require

$$R(\theta_0, \delta_\varphi) = \mathrm{E}_{\theta_0}[\varphi(X)] = P_{\theta_0}[\text{Decide } \theta = \theta_1] = \alpha,$$

where $0 < \alpha < 1$ is given, then we are restricted to a vertical strip in the risk set, and in this restricted set, the φ^* which corresponds to the point s^* on the lower boundary in that vertical strip is the best (ie, this φ^* minimizes the Type II Error Probability $R(\theta_1, \delta_\varphi) = 1 - \mathrm{E}_{\theta_1}[\varphi(X)] = P_{\theta_1}[\text{Decide } \theta = \theta_0]$ subject to the requirement that the Type I Error Probability $R(\theta_0, \delta_\varphi) = P_{\theta_0}[\text{Decide } \theta = \theta_1] = \alpha$).

Unbiased Tests

The problem of testing H_0: $\theta \in \Theta_0$ against H_1: $\theta \in \Theta_1$ where Θ_0 and Θ_1 are disjoint sets in Θ is more complicated than testing H_0: $\theta = \theta_0$ against H_1: $\theta = \theta_1$. Here we restrict to those δ_φ for which

(i) $\sup_{\theta \in \Theta_0} \mathrm{E}_\theta[\varphi(X)] = \alpha$ (these are tests of level α),
(ii) $\mathrm{E}_\theta[\varphi(X)] \ge \alpha$ for all $\theta \in \Theta_1$ (these are unbiased tests of level α),

and then among tests satisfying (i) and (ii) search for φ^* in this class for which

(iii) $\mathrm{E}_\theta[\varphi^*(X)] \ge \mathrm{E}_\theta[\varphi(X)]$, for all φ, and for all $\theta \in \Theta_1$.

In many situations such a φ^* exists and is called the *Uniformly Most Powerful (UMP) Unbiased* level α test.

Note. If a test φ does not satisfy condition (ii) for unbiasedness, then $\mathrm{E}_{\theta_1}[\varphi(X)] = \alpha - \varepsilon$ for some $\theta_1 \in \Theta_1$ and some $\varepsilon > 0$, while $\mathrm{E}_{\theta_0}[\varphi(X)] > \alpha - \varepsilon$ for some $\theta_0 \in \Theta_0$, so that φ rejects H_0 when $\theta = \theta_0$ and H_0 is true with a larger probability than when $\theta = \theta_1$ and H_0 is not true. The condition of unbiasedness does not allow such undesirable decisions.

Optimum tests under these restrictions will be discussed in Chapter 6.

4.6.3 Suitable Ordering of Decision Rules

The stringent ordering by the entire risk function can be replaced by invoking other principles of ordering. The following are two such important principles.

The Bayes Principle

Let τ denote a probability distribution on the parameter space Θ. Actually, we should introduce a suitable class of events $\mathcal{C}$ in Θ at this point, on which τ is defined, but when $\Theta = \mathbb{R}^k$ or a subset thereof (as in many situations), we shall simply use the events $\mathbb{B}$ (introduced earlier) for this purpose. Moreover, with a slight abuse of notation, we shall let τ denote the pdf of this distribution and define the Bayes risk of a behavioral decision rule δ with respect to τ as:

$$r(\tau, \delta) = \int_{\Theta} R(\theta, \delta)\tau(\theta)\, d\theta$$

with $R(\theta, \delta)$ as defined in Section 4.4.

Definition 4.6.2. The probability distribution τ is called the *prior distribution of* θ. A decision rule δ_o is said to be *Bayes* with respect to the prior distribution τ if

$$r(\tau, \delta_o) = \inf_{\delta} r(\tau, \delta).$$

There may be many Bayes rules with respect to a τ.

Sometimes it is useful to work with a function τ on Θ which is *not a pdf* and define $r(\tau, \delta)$ for such a τ in a formal way, and then minimize it with respect to δ.

Definition 4.6.3. A decision rule δ_0 is said to be

(a) generalized Bayes if there exists $\tau(\theta) \geq 0$ but $\int_{\Theta} \tau(\theta)\, d\theta = \infty$ (ie, τ is not a pdf), and

$$\int_{\Theta} R(\theta, \delta_o)\tau(\theta)\, d\theta \leq \int_{\Theta} R(\theta, \delta)\tau(\theta)\, d\theta \quad \text{for all } \delta,$$

(b) ε-Bayes with respect to τ if $r(\tau, \delta_0) \leq \inf_{\delta} r(\tau, \delta) + \varepsilon$, $\varepsilon > 0$, and

(c) extended Bayes if δ_0 is ε-Bayes for every $\varepsilon > 0$ (ie, for every $\varepsilon > 0$, there exists a prior distribution τ_ε such that δ_0 is ε-Bayes with respect to τ_ε).

The Minimax Principle

Definition 4.6.4. A decision rule δ_0 is said to be *minimax* if

$$\sup_{\theta} R(\theta, \delta_0) = \inf_{\delta} \sup_{\theta} R(\theta, \delta).$$

There may be many minimax rules.

The minimax principle summarizes the performance of each δ by $\sup_\theta R(\theta, \delta)$ (ie, judges a rule δ by its worst performance). A minimax rule may, therefore, have quite weak overall performance.

Admissibility

Finally, we introduce another property of a decision rule which requires that it cannot be improved upon.

Definition 4.6.5. A decision rule δ_0 is said to be *inadmissible* if there is a rule δ_1 which is better than δ_0 (ie, $R(\theta, \delta_1) \leq R(\theta, \delta_0)$ for all $\theta \in \Theta$ with strict inequality for some $\theta \in \Theta$).

A decision rule δ_0 is said to be *admissible* if there is no decision rule that is better than δ_0, that is, for every rule δ, $R(\theta_1, \delta) < R(\theta_1, \delta_0)$ for some $\theta_1 \in \Theta$ implies that there exists $\theta_2 \in \Theta$ such that $R(\theta_2, \delta) > R(\theta_2, \delta_0)$.

Admissibility is in no sense an indicator of high performance of a decision rule. It merely says that there is none that is uniformly better. On the other hand, inadmissibility indicates that there is a uniformly better rule that we may want to search for. However, there are rules with many nice properties which are inadmissible. A prime example is the sample mean vector $\bar{\boldsymbol{X}}_n$ as an estimator of the mean vector of $N_p(\boldsymbol{\mu}, I)$ with the loss function

$$L(\boldsymbol{\mu}, \boldsymbol{a}) = \|\boldsymbol{a} - \boldsymbol{\mu}\|^2 = (\boldsymbol{a} - \boldsymbol{\mu})^T(\boldsymbol{a} - \boldsymbol{\mu})$$

for $p \geq 3$, which is strictly improved upon by the shrinkage estimator

$$\delta_{\boldsymbol{c}}(\boldsymbol{X}) = \bar{\boldsymbol{X}}_n - \frac{p-2}{\|\bar{\boldsymbol{X}}_n - \boldsymbol{c}\|^2}\left(\bar{\boldsymbol{X}}_n - \boldsymbol{c}\right), \quad \boldsymbol{c} \in R^p$$

due to Stein [11]. Unfortunately, $\boldsymbol{\delta}_c$ is also inadmissible, because it can be improved upon by (see [3, p. 302–3])

$$\delta_c^+(\boldsymbol{X}) = \bar{\boldsymbol{X}}_n - \min\left\{1, \frac{p-2}{\|\bar{\boldsymbol{X}}_n - \boldsymbol{c}\|^2}\right\}\left(\bar{\boldsymbol{X}}_n - \boldsymbol{c}\right).$$

But $\boldsymbol{\delta}_c^+$ is also inadmissible.

4.6.4 Finding Bayes Rules: Prior to Posterior

The idea in the Bayes Principle is to judge a δ by its average performance with weights assigned by τ, which reflects the likelihood of various values of θ in the statistician's assessment based on prior experience (before obtaining the data). The key step in the calculation of the Bayes rules is to incorporate the data $X = x$ in the prior distribution τ to obtain the *posterior distribution of* θ, which is the continuous analog of the Bayes Formula in Chapter 1, Proposition 1.6.1. Write $f(x, \theta) = f(x|\theta)$, treating $f(x, \theta)$ as the conditional pdf of X given θ. Then proceeding as in Section 1.10, we have

$$\begin{aligned}
r(\tau, d) &= \int_\Theta \left[\int_{\mathfrak{X}} L(\theta, d(x)) f(x|\theta)\, dx\right]\tau(\theta)\, d\theta \\
&= \int_{\mathfrak{X}}\left[\int_\Theta L(\theta, d(x))\frac{f(x|\theta)\tau(\theta)}{\int_\Theta f(x|\theta)\tau(\theta)\, d\theta}\, d\theta\right]\left(\int_\Theta f(x|\theta)\tau(\theta)\, d\theta\right) dx \\
&= \int_{\mathfrak{X}}\left[\int_\Theta L(\theta, d(x)) g(\theta|x)\, d\theta\right] f(x)\, dx \\
&= \int_{\mathfrak{X}} \mathrm{E}[L(\theta, d(x))|X = x] f(x)\, dx,
\end{aligned}$$

where $f(x) = \int_\Theta f(x|\theta)\tau(\theta)\, d\theta$ is the *marginal* pdf of X, and $g(\theta|x) = [f(x|\theta)\tau(\theta)]/\int_\Theta f(x|\theta)\tau(\theta)\, d\theta$ is the *posterior* pdf of θ given $X = x$. The minimization of $r(\tau, d)$ with respect to d is achieved by choosing $d(x) = d^*(x) \in A$ for each x so that

$$E\left[L(\theta, d^*(X))|X = x\right] \le E[L(\theta, a)|X = x] \quad \text{for all } a \in A.$$

Interchanging the order of integration over Θ and $\mathfrak{X}$ is justified if $L(\theta, a)$ is bounded below. The case when X is discrete is treated in exactly the same way, replacing $\int_{\mathfrak{X}} dx$ by $\sum_{\mathfrak{X}}$.

Example 4.6.1. Estimating the mean of $N(\theta, \sigma^2)$ under squared-error loss.

Our data consists of $X_1, \ldots, X_n$ iid as $N(\theta, \sigma^2)$ with σ^2 known, and $\Theta = A = \mathbb{R}$, $L(\theta, a) = (a - \theta)^2$. Since $\bar{X}_n$ is sufficient for θ in $\boldsymbol{X}$, it is enough to restrict attention to decision rules based on $\bar{X}_n$ having pdf

$$f(\bar{x}_n, \theta) = \frac{\sqrt{n}}{\sqrt{2\pi}\sigma} \exp\left[-\frac{n}{2\sigma^2}(\bar{x}_n - \theta)^2\right], \quad \text{ie, } \bar{X}_n \sim N\left(\theta, \frac{\sigma^2}{n}\right).$$

First consider the prior τ:

$$\tau(\theta) = \frac{1}{\sqrt{2\pi}\gamma} \exp\left[-\frac{1}{2\gamma^2}(\theta - \mu)^2\right],$$

that is, $\theta \sim N(\mu, \gamma^2)$, and $\bar{X}_n|\theta \sim N\left(\theta, \frac{\sigma^2}{n}\right)$. Using the properties of bivariate normal distribution (see Section 2.2.10), we have

$$\theta|\bar{X}_n \sim N\left(\frac{w_1\mu + w_2\bar{X}_n}{w_1 + w_2}, \frac{1}{w_1 + w_2}\right)$$

as the posterior distribution, where $w_1 = 1/\gamma^2$ and $w_2 = n/\sigma^2$, that is, $\mathrm{E}\left[\theta|\bar{X}_n\right]$ is the weighted average of the mean μ of the prior and the sample mean $\bar{X}_n$ with weights inversely proportional to the variance of the prior distribution for θ and the conditional variance of $\bar{X}_n$ given θ for $\bar{X}_n$.

Remark 4.6.1.

1. This effect of the prior distribution on $\bar{X}_n$ is called: "a shrinkage of the sample mean toward the prior mean."
2. On the other hand, we can also interpret this as "overcoming a prior belief by observed data."
3. Note that the weight $1/\gamma^2$ attached to the prior remains fixed, but the weight n/σ^2 attached to $\bar{X}_n$ keep increasing with n and $1/\sigma^2$ determines the rate of increase. Thus a bad prior cannot hurt much once the sample size gets large.

Finally, the Bayes estimator of θ under squared-error loss, using this prior is:

$$\mathrm{E}\left[\theta|\bar{X}_n\right] = \frac{n\gamma^2\bar{X}_n + \sigma^2\mu}{n\gamma^2 + \sigma^2}.$$

For the special case of $\sigma^2 = 1$, $n = 1$, and $\mu = 0$, $\mathrm{E}[\theta|X] = \frac{\gamma^2}{1+\gamma^2}X$.

4.6.5 Solving for Minimax Rules

In this section we discuss two methods of finding minimax rules.

Definition 4.6.6. A prior distribution τ_0 is said to be *least favorable* if

$$\inf_{\delta} r(\tau_0, \delta) = \sup_{\tau} \inf_{\delta} r(\tau, \delta).$$

Next note that

$$\underline{V} = \sup_{\tau} \inf_{\delta} r(\tau, \delta) \le \inf_{\delta} \sup_{\tau} r(\tau, \delta) = \overline{V},$$

where $\underline{V}$ and $\overline{V}$ are, respectively, the lower and upper value of the statistical decision problem which is viewed as a game between nature who chooses θ randomly according to the distribution τ and the statistician who chooses a decision rule δ, resulting in a pay-off by the statistician of the quantity $r(\tau, \delta)$. Also note that

$$\sup_{\theta} R(\theta, \delta) = \sup_{\tau} r(\tau, \delta) \quad \text{for all } \delta.$$

We say that the game of a statistical decision problem has a value if $\underline{V} = \overline{V}$.

We now describe the *first method* for finding a minimax rule which involves guessing a τ_0 as least favorable, finding a rule δ_0 which is Bayes with respect to τ_0 and checking that δ_0 is indeed minimax. The actual procedure is described in Theorem 4.6.1.

In many situations, our guess of the least favorable τ_0 is not a pdf (ie, $\int \tau(\theta)\, d\theta = \infty$), in which case δ_0 , which is formally Bayes with respect to τ_0, is not really a Bayes rule. This needs a modification of the above method, which is described in Theorem 4.6.2.

Theorem 4.6.1. *If δ_0 is Bayes with respect to τ_0 and*

$$R(\theta, \delta_0) \le r(\tau_0, \delta_0) \quad \textit{for all } \theta,$$

then δ_0 is minimax and τ_0 is least favorable.

Proof. Note that

$$\begin{aligned}\overline{V} &= \inf_{\delta} \sup_{\tau} r(\tau, \delta) = \inf_{\delta} \sup_{\theta} R(\theta, \delta) \le \sup_{\theta} R(\theta, \delta_0) \le r(\tau_0, \delta_0) \\ &= \inf_{\delta} r(\tau_0, \delta) \le \sup_{\tau} \inf_{\delta} r(\tau, \delta) = \underline{V}.\end{aligned}$$

Therefore, all inequalities are equalities and therefore,

$$\inf_{\delta} r(\tau_0, \delta) = \sup_{\tau} \inf_{\delta} r(\tau, \delta),$$

that is, τ_0 is least favorable, and

$$\sup_{\theta} R(\theta, \delta_0) = \inf_{\delta} \sup_{\theta} R(\theta, \delta),$$

which proves that δ_0 is minimax. □

Theorem 4.6.2. *Let $\{\tau_n\}$ be a sequence of distributions of θ and let δ_n be Bayes with respect to τ_n. If*

$$R(\theta, \delta_0) \le \overline{\lim}_n r(\tau_n, \delta_n) \quad \textit{for all } \theta,$$

then δ_0 is minimax.

Proof. For arbitrary δ and for all n,

$$\int R(\theta,\delta)\tau_n(\theta)\,d\theta = r(\tau_n,\delta) \geq r(\tau_n,\delta_n),$$

implying $\sup_\theta R(\theta,\delta) \geq r(\tau_n,\delta_n)$. Therefore, for all δ,

$$\sup_\theta R(\theta,\delta) \geq \overline{\lim}_n r(\tau_n,\delta_n) \geq \sup_\theta R(\theta,\delta_0),$$

proving that δ_0 is minimax. □

Remark 4.6.2. Often $c = \lim_n r(\tau_n,\delta_n)$ exists, in which case we simply have to check that $R(\theta,\delta_0) \leq c$ for all θ.

Definition 4.6.7. A decision rule is an equalizer rule if $R(\theta,\delta_0)$ is constant for all θ.

The second method of finding a minimax rule is to look for an equalizer rule that is Bayes or extended Bayes.

Theorem 4.6.3. *If an equalizer rule is extended Bayes, then it is minimax.*

Proof. Since $R(\theta,\delta_0) = c$ (constant) for all θ, $r(\tau,\delta_0) = c$ for all τ, and since δ_0 is extended Bayes, there exists a sequence $\{\tau_n\}$ with Bayes rules $\{\delta_n\}$ such that

$$c = r(\tau_n,\delta_0) \leq r(\tau_n,\delta_n) + 1/n \quad \text{for all } n.$$

Hence for all θ,

$$R(\theta,\delta_0) = c \leq \overline{\lim}_n r(\tau_n,\delta_n).$$

It now follows from Theorem 4.6.2 that δ_0 is minimax. □

4.6.6 Conditions for Admissibility

The simplest (and obvious) condition for admissibility is the following.

Theorem 4.6.4. *If δ_0 is unique Bayes with respect to some τ, then δ_0 is admissible.*

Proof. Since δ_0 is unique Bayes with respect to τ, $r(\tau,\delta_0) < r(\tau,\delta_1)$ for all $\delta_1 \neq \delta_0$. Hence there exists θ for any $\delta_1 \neq \delta_0$ such that $R(\theta,\delta_0) < R(\theta,\delta_1)$. □

The requirement of uniqueness in the above theorem can be dispensed with if the risk function $R(\theta,\delta)$ for every δ is continuous in θ and if the τ with respect to which δ_0 is Bayes has plenty of support.

Theorem 4.6.5. *Suppose Θ is an interval in $\mathbb{R}$, the risk function $R(\theta,\delta)$ for every decision rule δ is continuous in θ and τ is a probability distribution on Θ such that $\tau(I) > 0$ for all nondegenerate intervals $I \subset \Theta$. Then a Bayes rule δ_0 with respect to τ is admissible.*

Proof. If δ_0 is inadmissible, then there is a rule δ_1 such that $R(\theta,\delta_1) \leq R(\theta,\delta_0)$ for all $\theta \in \Theta$, and $R(\theta_0,\delta_1) < R(\theta_0,\delta_0)$ for some $\theta_0 \in \Theta$. Since $R(\theta,\delta_0)$ and $R(\theta,\delta_1)$ are both continuous, there exist $\varepsilon > 0$ and a neighborhood I of θ_0 such that $R(\theta,\delta_0) - R(\theta,\delta_1) > \varepsilon$ for all $\theta \in I$. Hence

$$r(\tau,\delta_0) - r(\tau,\delta_1) \geq \int_I [R(\theta,\delta_0) - R(\theta,\delta_1)]\tau(\theta)\,d\theta > \varepsilon\tau(I) > 0,$$

contradicting the hypothesis of δ_0 being Bayes with respect to τ. □

This theorem is now extended to cover some familiar situations where δ_0 is extended Bayes and has constant risk.

Theorem 4.6.6. *Suppose that a decision rule δ_0 satisfies the following conditions in relation to a sequence of probability distributions $\{\tau_n\}$ on $\Theta \subset \mathbb{R}$ with respect to which $\{\delta_n\}$ are Bayes rules:*

(i) $R(\theta, \delta_0) = c$ *(constant) for all θ,*
(ii) $\lim_n r(\tau_n, \delta_n) = c$,
(iii) $R(\theta, \delta)$ *is continuous in θ for all δ,*
(iv) $\tau_n(I) > 0$ *for all nondegenerate intervals $I \subset \Theta$, and*
(v) $\lim_n \tau_n(I)/[c - r(\tau_n, \delta_n)] = \infty$.

Then δ_0 is admissible.

Proof. Note that

$$c - r(\tau_n, \delta_n) = r(\tau_n, \delta_0) - r(\tau_n, \delta_n) \geq 0 \quad \text{for all } n.$$

If equality holds for some n, then δ_0 is Bayes with respect to that τ_n which makes it admissible by Theorem 4.6.5. Therefore, assume that $c - r(\tau_n, \delta_n) > 0$ for all n and suppose that δ_0 is inadmissible. Then there is a rule δ' such that $c = R(\theta, \delta_0) \geq R(\theta, \delta')$ for all $\theta \in \Theta$, and $R(\theta_0, \delta_0) > R(\theta_0, \delta')$ for some $\theta_0 \in \Theta$. We can now find $\varepsilon > 0$ and a neighborhood I of θ_0 so that $c - r(\tau_n, \delta') > \varepsilon\tau_n(I)$ as in the proof of Theorem 4.6.5. Thus

$$\frac{c - r(\tau_n, \delta')}{c - r(\tau_n, \delta_n)} > \frac{\varepsilon\tau_n(I)}{c - r(\tau_n, \delta_n)} > 1 \quad \text{for large } n,$$

since by Condition (v), the middle quantity in the last display tends to ∞ as $n \to \infty$. This contradicts the hypothesis that δ_n is Bayes with respect to τ_n. □

Theorems 4.6.5 and 4.6.6 depend on the key condition that the risk $R(\theta, \delta)$ is continuous in θ for all decision rules δ, so the applicability of these theorems depends on the verification of this condition. The following theorem from Ferguson [1] serves that purpose for one-parameter exponential families.

Theorem 4.6.7. *Let Θ be the real line. Suppose that*

(a) *there exist nonnegative functions $B_1(\theta_1, \theta_2)$ and $B_2(\theta_1, \theta_2)$ bounded on compact sets of $\Theta \times \Theta$ such that*

$$|L(\theta_2, a)| \leq B_1(\theta_1, \theta_2)|L(\theta_1, a)| + B_2(\theta_1, \theta_2), \quad \textit{for all } a \in A;$$

(b) $L(\theta, a)$ *is continuous in θ for each $a \in A$; and*
(c) $f(x, \theta) = c(\theta)h(x)\exp\left[Q(\theta)T(x)\right]$ *where $Q(\theta)$ is a continuous increasing function.*

Then for all nonrandomized decision rules d, the risk function $R(\theta, d)$ is continuous in θ.

Proof. See Ferguson [1, p. 139–40]. □

Other methods of proving admissibility have been developed by Hodges and Lehmann [12], Karlin [13], and Stein [14].

Example 4.6.2. An admissible minimax estimator of a Bernoulli proportion under squared-error loss.

Let X be the number of successes in n independent Bernoulli trials with probability of success θ. We want to estimate θ under squared-error loss from the data X. Here $\Theta = [0,1] = A$ and $L(\theta, a) = (a - \theta)^2$.

Solution. We shall apply Theorem 4.6.3, looking for an equalizer rule which is Bayes. This will give us a minimax rule. We look among Bayes rules with respect to $\tau_{a,b} = Beta(a,b)$, that is, Beta distributions with parameters a, b with pdf

$$\tau_{a,b}(\theta) = \frac{\Gamma(a+b)}{\Gamma(a)\Gamma(b)}\theta^{a-1}(1-\theta)^{b-1}I_{(0,1)}(\theta).$$

The Bayes rule with respect to $\tau_{a,b}$ is $d_{a,b}(x) = (x+a)/(n+a+b)$ (the verification of which is left as an exercise), with risk function

$$\begin{aligned} R(\theta, d_{a,b}) &= E_\theta\left[d_{a,b}(X) - \theta\right]^2 = \mathrm{Var}_\theta\left[d_{a,b}(X)\right] + \left\{\mathrm{E}_\theta\left[d_{a,b}(X)\right] - \theta\right\}^2 \\ &= \frac{n\theta(1-\theta)}{(n+a+b)^2} + \left[\frac{n\theta + a}{n+a+b} - \theta\right]^2 \\ &= \frac{\left[(a+b)^2 - n\right]\theta^2 - \left[2a(a+b) - n\right]\theta + a^2}{(n+a+b)^2}. \end{aligned}$$

To make this a constant in θ, take $(a+b)^2 = 2a(a+b) = n$; leading to $a = b = \sqrt{n}/2$. The resulting Bayes rule $d_{\sqrt{n}/2,\sqrt{n}/2}(X) = (X + \sqrt{n}/2)/(n+\sqrt{n})$ is minimax.

Since this rule is unique Bayes with respect to $Beta(\sqrt{n}/2, \sqrt{n}/2)$, it is admissible by Theorem 4.6.4. Thus $\hat{\theta} = (X + \sqrt{n}/2)/(n + \sqrt{n})$ is an admissible minimax estimator of θ.

Example 4.6.3. Admissibility of $\bar{X}_n$ as an estimator of mean θ of $N(\theta, \sigma^2)$ under squared-error loss.

Under squared-error loss, the Bayes estimator of θ based on a random sample $\boldsymbol{X} = (X_1, \ldots, X_n)$ from $N(\theta, \sigma^2)$, with respect to $\tau_k = N(0, k)$ is

$$d_k(\boldsymbol{X}) = \frac{nk}{nk + \sigma^2}\bar{X}_n, \quad \text{where } \bar{X}_n = n^{-1}\sum_{i=1}^{n} X_i.$$

This result is derived in Section 4.6.4. The Bayes risk of d_k is $r(\tau_k, d_k) = (k\sigma^2)/(nk + \sigma^2)$, the proof of which is left an exercise. We shall use Theorems 4.6.6 and 4.6.7 to show that the decision rule $d_0(\boldsymbol{X}) = \bar{X}_n$ is admissible.

Solution. Since

$$R(\theta, d_0) = \sigma^2/n = \lim_{k\to\infty} r(\tau_k, d_k)$$

is constant, and $\tau_k(I) > 0$ for all nondegenerate intervals, Conditions (i), (ii), and (iv) of Theorem 4.6.6 are satisfied, while Theorem 4.6.7 provides justification for Condition (iii) that $R(\theta, d)$ is continuous for all d. So we only need to check Condition (v). Now

$$\frac{\tau_k([a,b])}{c - r(\tau_k, d_k)} = \frac{\Phi\left(b/\sqrt{k}\right) - \Phi\left(a/\sqrt{k}\right)}{\sigma^2/n - (k\sigma^2)/(nk+\sigma^2)} = \frac{n\left(nk+\sigma^2\right)}{\sigma^4}\left[\Phi\left(b/\sqrt{k}\right) - \Phi\left(a/\sqrt{k}\right)\right].$$

Since $\Phi\left(b/\sqrt{k}\right) - \Phi\left(a/\sqrt{k}\right) = \frac{b-a}{\sqrt{2\pi k}}[1+o(1)]$ as $k \to \infty$ (see Section 2.2.4), the above tends to ∞ as $k \to \infty$, justifying Condition (v). This proves the admissibility of $\bar{X}_n$.

Exercises

4.1. Suppose that X is distributed with a pdf $f(x,\theta)$ where θ is an unknown real parameter. Consider the two-decision problem of choosing between the hypotheses H_0: $\theta \le \theta_0$ and H_1: $\theta > \theta_0$ (for a given θ_0) with the loss function

$$L(\theta, a_0) = (\theta - \theta_0)_+, \quad L(\theta, a_1) = (\theta_0 - \theta)_+,$$

where for any real number x, x_+ denotes $\max(x, 0)$, and a_i is the action to accept H_i, $i = 0, 1$. Show that the Bayes rule with respect to a prior cdf G of θ rejects H_0 if and only if $E_G[\theta|X=x] > \theta_0$.

4.2. Under $0-1$ loss function in the problem of testing H_0: $\theta = \theta_0$ vs H_1: $\theta = \theta_1$, the risk set is

$$\begin{aligned} S &= \left\{(R(\theta_0, \phi), R(\theta_1, \phi)) \colon \phi \text{ maps } \mathcal{X} \to [0,1]\right\} \\ &= \left\{(E_{\theta_0}[\phi(X)], 1 - E_{\theta_1}[\phi(X)]) \colon \phi \text{ maps } \mathcal{X} \to [0,1]\right\}, \end{aligned}$$

where X denotes the data. Let $f(x,\theta) = \binom{2}{x}\theta^x(1-\theta)^{2-x}$, $x = 0, 1, 2$, and consider H_0: $\theta = 1/2$ vs H_1: $\theta = 2/3$.

(a) Plot the risk set S.

(b) Find a minimax test.

4.3. Let X follow a binomial distribution $Bin(n,\theta)$, $0 < \theta < 1$.

(a) Show that $d_0(x) = x/n$ is a minimax estimator of θ with constant risk $1/n$ under the loss function $L(\theta, a) = (\theta - a)^2/\{\theta(1-\theta)\}$.

(b) Show that $d_0(x) = x/n$ is not minimax under the loss function $L(\theta, a) = (\theta - a)^2$. [Consider δ_ε^* defined as: $\delta_\varepsilon^*(x) = d_0(x) = x/n$ with probability $1-\varepsilon$ and $\delta_\varepsilon^*(x) = d_1(x) \equiv 1/2$ with probability ε. Examine the risk function of δ_ε^* with $\varepsilon = (n+1)^{-1}$.]

(c) Show that $d_0(x)$ is not a Bayes rule, but it is an extended Bayes rule (ie, ε-Bayes for every $\varepsilon > 0$). [Try Beta priors.]

(d) Show that d_0 is Bayes with respect to *Unif* $(0,1)$ prior under $L(\theta, a)$ given in (a).

4.4. Let X follow the binomial $Bin(n,\theta)$ distribution with pdf

$$f(x,\theta) = \binom{n}{x}\theta^x(1-\theta)^{n-x}, \quad x = 0, 1, \ldots, n,$$

where $\theta \in \Theta = (0,1)$ is to be estimated in the action space $A = [0,1]$ under the loss function $L(\theta, a) = (a - \theta)^2$.

(a) Show that the estimator $d_0(x) = x/n$ is admissible.
(b) If the parameter space is $\Theta = [1/3, 2/3]$ and everything else is the same as above, then show that $d_0(x) = x/n$ is inadmissible. [*Hint*: Think of a sensible estimator.]

4.5. Let X be an rv with pdf

$$f(x, \theta) = \theta^{-1} \exp(-x/\theta), \quad x > 0,\ \theta = 1 \text{ or } 2.$$

Consider the two-decision problem with action space $A = \{a_1, a_2\}$ with $a_i = i$ and the loss function $L(\theta, a) = I(\theta \neq a)$, based on a single observation X.
(a) Let S denote the risk set consisting of points $(R(1, \delta), R(2, \delta))$ corresponding to all behavioral rules δ. Determine the function $\beta(\alpha)$ describing the points $(\alpha, \beta(\alpha))$ on the lower boundary of S.
(b) Find a minimax rule.
(c) Consider a prior g_τ with probabilities τ for $\theta = 1$ and $1 - \tau$ for $\theta = 2$. Show that the Bayes rule with respect to the prior g_τ will always take action a_2 (irrespective of the observed values of X) if $\tau < 1/3$.

4.6. Show that the sample mean $\bar{X}$ is an admissible estimator of the mean θ of a normal distribution $N(\theta, \sigma^2)$ under the absolute error loss.

4.7. Suppose we have one observation X from $N(\theta, 1)$ on the basis of which we have to take action a_0 to decide $\theta = 0$ or a_1 to decide $\theta \neq 0$ subject the loss function

$$L(0, a_0) = 0, \quad L(\theta, a_1) = 0 \text{ for } \theta \neq 0,$$
$$L(0, a_1) = 1, \quad L(\theta, a_0) = 1 \text{ for } \theta \neq 0.$$

Consider a prior distribution τ which assigns probability p to $\{\theta = 0\}$ and distributes the remaining probability $1 - p$ on $(-\infty, \infty)$ according to $N(\mu, \sigma^2)$, that is

$$\tau(\theta = 0) = p \text{ and } \tau[\theta \in B] = (1 - p) \int_B \sigma^{-1} \phi((\theta - \mu)/\sigma)\, d\theta \quad \text{if } 0 \notin B,$$

where ϕ is the pdf of the standard normal distribution. Find the Bayes rule with respect to τ based on a sample X of size 1. Give a common sense interpretation of the Bayes rule.

4.8. Let $X_1, \ldots, X_n$ be a random sample from *Unif* $(0, \theta)$, $\theta > 0$ and the prior of θ is $\tau(\theta) = \theta \exp(-\theta)$, $\theta > 0$.
(a) Find the posterior distribution of θ given $X_1, \ldots, X_n$.
(b) Find the Bayes estimators of θ under the loss functions $L_1(\theta, a) = |\theta - a|$ and $L_2(\theta, a) = (\theta - a)^2$.

4.9. We want to estimate the mean θ of a Poisson distribution on the basis of a random sample $X_1, \ldots, X_n$ subject to the loss function $L(\theta, a) = (\theta - a)^2$, $\theta \in \Theta = (0, \infty)$ and $a \in A = [0, \infty)$. Assume that θ has a Gamma(α, β) prior (ie, it has the pdf)

$$\tau_{\alpha,\beta}(\theta) = \{\Gamma(\alpha)\beta^\alpha\}^{-1} \theta^{\alpha-1} \exp(-\theta/\beta), \theta > 0, \text{ with}$$
$$\mathrm{E}_{\alpha,\beta}(\theta) = \alpha\beta,\ \mathrm{E}_{\alpha,\beta}\left(\theta^2\right) = \alpha(\alpha + 1)\beta^2.$$

(a) Find the posterior distribution of θ given $X_1, \ldots, X_n$ and the Bayes rule with respect to $\tau_{\alpha,\beta}$.

(b) Show that $d_0(x) = \bar{X}_n$ is not a Bayes rule with respect to any prior on $\Theta = (0, \infty)$. However, d_0 is a Bayes rule if $\Theta = [0, \infty)$.

(c) Show that d_0 is

(i) a limit of Bayes rules, (ii) a generalized Bayes rule with respect to $\tau(\theta) = \theta^{-1}, \theta > 0$, which is not a pdf, and (iii) an extended Bayes rule.

4.10. Let X and Y be independent binomial $Bin(n, \theta_1)$ and $Bin(n, \theta_2)$ rv's. We want to estimate $\theta_1 - \theta_2$ subject to the squared error loss $L((\theta_1, \theta_2), a) = ((\theta_1 - \theta_2) - a)^2$, $|a| \leq 1$. Find the Bayes rule with respect to the prior with pdf $\tau(\theta_1, \theta_2) = I_{(0,1)}(\theta_1)I_{(0,1)}(\theta_2)$ (ie, θ_1 and θ_2 have independent uniform distributions on $(0, 1)$).

4.11. Suppose that X has a pdf $f(x, \theta)$, T is a sufficient statistics for θ, and that the factorization theorem holds. Show that all Bayes rules are functions of T.

4.12. Let $X_1, \ldots, X_n$ be a random sample from a Weibull distribution with pdf

$$f(x, \theta) = \theta a x^{a-1} \exp(-\theta x^a), \quad x > 0, \ \theta > 0,$$

where $a > 0$ is known. Find a sufficient statistic for θ.

4.13. Let $X_1, \ldots, X_n$ be a random sample from a distribution with pdf

$$f(x, \boldsymbol{\theta}) = a(\boldsymbol{\theta})h(x)I(\theta_1 \leq x \leq \theta_2), \quad \boldsymbol{\theta} = (\theta_1, \theta_2), \ -\infty < \theta_1 < \theta_2 < \infty,$$

where $h(x) > 0$ is a known function with $\int_{-\infty}^{\infty} h(x)\, dx = 1$ and $a(\boldsymbol{\theta}) = 1 \big/ \int_{\theta_1}^{\theta_2} h(x)\, dx$ for $\boldsymbol{\theta} = (\theta_1, \theta_2)$. Find a two-dimensional sufficient statistics for $\boldsymbol{\theta}$ and apply your result to the special case of uniform distribution on $[\theta_1, \theta_2]$.

5

Point Estimation in Parametric Models

5.1 Optimality Under Unbiasedness, Squared-Error Loss, UMVUE

In the general framework of statistical decision theory (see Section 4.4), the estimation problem is described by the triple $(\{P_\theta, \theta \in \Theta\}, \mathcal{A}, L)$ where $\mathcal{A} = \{g(\theta) : \theta \in \Theta\}$ and $L(\theta, a)$ = loss due to estimating $g(\theta)$ by a. For a decision rule d, which is an estimator $T = T(X_1, \ldots, X_n)$ based on the data $\boldsymbol{X} = (X_1, \ldots, X_n)$, the risk is

$$R(\theta, T) = \mathrm{E}_\theta[L(\theta, T(\boldsymbol{X}))] = \int_{\mathcal{X}} L(\theta, T(\boldsymbol{x})) f(\boldsymbol{x}; \theta)\, d\boldsymbol{x}$$

in the continuous case and analogously in the discrete case.

The concept of unbiasedness in estimation has been introduced in Section 4.6.1. In a parametric family $\{(\mathcal{X}, \mathcal{A}, P_\theta), \theta \in \Theta\}$, a statistic T based on a random sample $\boldsymbol{X}$ is an unbiased estimator of $g(\theta)$ if $\mathrm{E}_\theta[T(\boldsymbol{X})] = g(\theta)$ for all $\theta \in \Theta$.

Example 5.1.1. Let $(X_1, \ldots, X_n)$ be a random sample from *Bernoulli*(θ) and let $g(\theta) = \theta(1-\theta)$. Let $T = X_1(1 - X_2)$. Then

$$\mathrm{E}_\theta[T] = \mathrm{E}_\theta\left[X_1(1 - X_2)\right] = \theta(1-\theta) = g(\theta).$$

Example 5.1.2. Let $(X_1, \ldots, X_n)$ be a random sample from *Unif*(θ) and let $g(\theta) = \theta$. Let $T = 2X_1$. Then

$$\mathrm{E}_\theta[T] = \mathrm{E}_\theta\left[2X_1\right] = 2\mathrm{E}_\theta\left[X_1\right] = 2 \cdot \frac{\theta}{2} = \theta = g(\theta).$$

Example 5.1.3. Let $(X_1, \ldots, X_n)$ be a random sample from $N(\mu, \sigma^2)$, and let $g_1(\mu, \sigma) = \mu$, and $g_2(\mu, \sigma) = \sigma^2$. Let $T_1 = \bar{X} = n^{-1}\sum_{i=1}^n X_i$ and $T_2 = (n-1)^{-1}\sum_{i=1}^n (X_i - \bar{X})^2$. Then

$$\mathrm{E}_{\mu,\sigma}\left[T_1\right] = n^{-1}\mathrm{E}_{\mu,\sigma}\left[\sum_{i=1}^n X_i\right] = n^{-1} n\mu = \mu, \text{ and}$$

$$\mathrm{E}_{\mu,\sigma}\left[T_2\right] = (n-1)^{-1}\mathrm{E}_{\mu,\sigma}\left[\sum_{i=1}^n (X_i - \bar{X})^2\right]$$

Theory and Methods of Statistics. http://dx.doi.org/10.1016/B978-0-12-802440-9.00005-9

$$= (n-1)^{-1}\mathrm{E}_{\mu,\sigma}\left[\sum_{i=1}^{n}(X_i-\mu)^2 - n(\bar{X}-\mu)^2\right]$$
$$= (n-1)^{-1}\left[n\sigma^2 - n\frac{\sigma^2}{n}\right] = \sigma^2.$$

Thus in these examples, $X_1(1-X_2)$ is an unbiased estimator of $\theta(1-\theta)$, $2X_1$ is an unbiased estimator of θ, and $\bar{X}$ and $(n-1)^{-1}\sum_{i=1}^{n}(X_i-\bar{X})^2$ are unbiased estimators of μ and σ^2, respectively.

Theorem 5.1.1 (Rao-Blackwell Theorem). *Let A be a convex set in $\mathbb{R}^k$, $L(\theta, a)$ convex in $a \in A$ for each $\theta \in \Theta$ and T sufficient for θ in $\mathbf{X}$. Then for any nonrandomized decision rule d, the nonrandomized decision rule $d^\star$ based on T, defined as*

$$d^\star(t) = E\big[d(\mathbf{X})|T = t\big], \tag{1}$$

assuming that the expectation exists, is at least as good as d (ie, $R(\theta, d^\star)$ is either $< R(\theta, d)$ or $= R(\theta, d)$ for all θ).

Proof. Sufficiency of T makes $d^\star(T) = \mathrm{E}\big[d(\mathbf{X})|T\big]$ free of θ so that $d^\star(T)$ is a bona fide estimator. By Jensen's inequality (see Section A.2), for fixed t,

$$\mathrm{E}\big[L(\theta, d(\mathbf{X}))|T = t\big] \ge L(\theta, E\big[d(\mathbf{X})|T = t\big]) = L(\theta, d^\star(t)) \quad \text{for all } \theta.$$

Hence, for each θ,

$$\begin{aligned} R(\theta, d) &= \mathrm{E}_\theta\big[L(\theta, d(\mathbf{X}))\big] = \mathrm{E}_\theta \mathrm{E}\big[L(\theta, d(\mathbf{X}))|T\big] \\ &\ge E_\theta\big[L(\theta, d^\star(T))\big] = R(\theta, d^\star) \end{aligned}$$

showing that $d^\star$ is "at least as good" as d. □

Remark 5.1.1. If d is an unbiased estimator of $g(\theta)$ (ie, $\mathrm{E}_\theta\big[d(\mathbf{X})\big] = g(\theta)$ for all $\theta \in \Theta$), then

$$\mathrm{E}_\theta\big[d^\star(T)\big] = \mathrm{E}_\theta\big[\mathrm{E}(d(\mathbf{X})|T)\big] = \mathrm{E}_\theta\big[d(\mathbf{X})\big] = g(\theta) \quad \text{for all } \theta.$$

Thus, $d^\star$ is also an unbiased estimator of $g(\theta)$.

Remark 5.1.2. If d is an unbiased estimator of $g(\theta)$ then, for $L(\theta, a) = \{a - g(\theta)\}^2$,

$$R(\theta, d) = \mathrm{E}_\theta\Big[\{d(\mathbf{X}) - g(\theta)\}^2\Big] = \mathrm{Var}_\theta\big[d(\mathbf{X})\big],$$

and $R(\theta, d^\star)$ is also $\mathrm{Var}_\theta\big[d^\star(T)\big]$. Hence, if d is an unbiased estimator of $g(\theta)$, then $d^\star$ is also an unbiased estimator of $g(\theta)$, the variance of which is uniformly bounded above by the variance of d.

The Rao-Blackwell Theorem is often stated in this form.

Theorem 5.1.2 (Rao-Blackwell Theorem for Squared-Error Loss). *If d is an unbiased estimator of $g(\theta)$ based on a sample $\mathbf{X}$ from P_θ, $\theta \in \Theta$, and if T is sufficient for θ in $\mathbf{X}$, then $d^\star$, defined by*

$$d^\star(t) = E\big[d(\mathbf{X})|T = t\big],$$

is also an unbiased estimator of $g(\theta)$*, and*

$$Var_\theta\left[d^\star(T)\right] \le Var_\theta\left[d(\boldsymbol{X})\right] \quad \textit{for all } \theta \in \Theta.$$

(See [15, 16].)

Under squared-error loss, the Rao-Blackwell Theorem provides us with a method for "improving" upon an unbiased estimator, using a sufficient statistic, by constructing an unbiased estimator with uniformly smaller (or equal) variance. However, we do not know whether there is an unbiased estimator with even smaller variance than this "improved" estimator. More to the point, among many sufficient statistics, which one should we use in the Rao-Blackwell formula to find the best unbiased estimator? Such an estimator will be called the Uniformly Minimum Variance Unbiased Estimator (UMVUE).

Before searching for the UMVUE, we shall first show that it is unique and find a characterization for the UMVUE.

Theorem 5.1.3. *The UMVUE is unique.*

Proof. Suppose T_1 and T_2 are two distinct UMVUEs of $g(\theta)$, that is

$$\begin{aligned} \mathrm{E}_\theta\left[T_1\right] &= \mathrm{E}_\theta\left[T_2\right] = g(\theta), \text{ and} \\ \mathrm{Var}_\theta\left[T_1\right] &= \mathrm{Var}_\theta\left[T_2\right] := \sigma^2(\theta) \quad \text{for all } \theta. \end{aligned}$$

Then, by Cauchy-Schwartz,

$$\mathrm{Cov}_\theta\left[T_1, T_2\right] \le \left\{\mathrm{Var}_\theta\left[T_1\right]\mathrm{Var}_\theta\left[T_2\right]\right\}^{1/2} = \sigma^2(\theta).$$

Now let $\bar{T} = (T_1 + T_2)/2$. Then, $\bar{T}$ is also an unbiased estimator of $g(\theta)$ and

$$\mathrm{Var}_\theta\left[\bar{T}\right] = \frac{1}{4}\left\{\mathrm{Var}_\theta\left[T_1\right] + \mathrm{Var}_\theta\left[T_2\right] + 2\mathrm{Cov}_\theta\left[T_1, T_2\right]\right\} \le \sigma^2(\theta) = \mathrm{Var}_\theta\left[T_1\right].$$

If this inequality is strict, then the UMVUE property of T_1 is violated. On the other hand, equality holds only if we have equality in the Cauchy-Schwartz, for which we need $T_2 = a(\theta) + b(\theta)T_1$, but then,

$$\sigma^2(\theta) = \mathrm{Cov}_\theta\left[T_1, T_2\right] = \mathrm{Cov}_\theta\left[T_1, a(\theta) + b(\theta)T_1\right] = b(\theta)\sigma^2(\theta),$$

so $b(\theta) = 1$ and $a(\theta) = 0$ since $\mathrm{E}_\theta[T_1] = \mathrm{E}_\theta[T_2] = g(\theta)$. Thus, $T_1 = T_2$ showing that T_1 is unique. □

Definition 5.1.1. Let $\mathcal{U}$ denote the class of estimators U with $\mathrm{E}_\theta[U] = 0$ and $\mathrm{E}_\theta\left[U^2\right] < \infty$ for all θ.

Theorem 5.1.4. *If* T *is an unbiased estimator of* $g(\theta)$*, then a necessary and sufficient condition for* U *to be the UMVUE of* $g(\theta)$ *is* $Cov_\theta[T, U] = E_\theta[TU] = 0$ *for all* $U \in \mathcal{U}$.

Proof (Necessity). Let T be UMVUE of $g(\theta)$. Then, for arbitrary $U \in \mathcal{U}$ and $\lambda \in \mathbb{R}$, $T + \lambda U$ is also an unbiased estimator of $g(\theta)$. Hence,

$$\mathrm{Var}_\theta[T] \le \mathrm{Var}_\theta[T + \lambda U] = \mathrm{Var}_\theta[T] + \lambda^2\mathrm{Var}_\theta[U] + 2\lambda\mathrm{Cov}_\theta[T, U] \quad \text{for all } \lambda,$$

that is, $f(\lambda) = \lambda^2 \mathrm{Var}_\theta[U] + 2\lambda \mathrm{Cov}_\theta[T, U] \geq 0$ for all λ. But if $\mathrm{Cov}_\theta[T, U] \neq 0$, then

$$\min_\lambda f(\lambda) = -\frac{\{\mathrm{Cov}_\theta[T, U]\}^2}{\mathrm{Var}_\theta[U]} < 0.$$

To avoid contradiction, we therefore need $\mathrm{Cov}_\theta[T, U] = 0$.

(Sufficiency). Suppose $\mathrm{Cov}_\theta[T, U] = \mathrm{E}_\theta[TU] = 0$ for all $U \in \mathcal{U}$. We shall show that T is UMVUE of $g(\theta)$. Let T' be an arbitrary unbiased estimator of $g(\theta)$ with $\mathrm{Var}_\theta[T'] < \infty$. Then, $T - T' \in \mathcal{U}$, so that $\mathrm{E}_\theta[T(T - T')] = 0$ (ie, $\mathrm{E}_\theta[T^2] = \mathrm{E}_\theta[TT']$). Since $\mathrm{E}_\theta[T] = g(\theta)$, it follows that

$$\begin{aligned}\mathrm{Var}_\theta[T] &= \mathrm{E}_\theta\left[T^2\right] - \{g(\theta)\}^2 = \mathrm{E}_\theta[TT'] - \{g(\theta)\}^2 \\ &= \mathrm{Cov}_\theta[T, T'] \leq \{\mathrm{Var}_\theta[T]\mathrm{Var}_\theta[T']\}^{1/2},\end{aligned}$$

which implies $\mathrm{Var}_\theta[T] \leq \mathrm{Var}_\theta[T']$.

The main steps in finding the UMVUE of $g(\theta)$ are

(i) finding an unbiased estimator of $g(\theta)$, and

(ii) using an *appropriate* sufficient statistic T to use in Eq. (1), the Rao-Blackwell formula $d^\star(T) = \mathrm{E}[d(\mathbf{X})|T]$.

The following example will illustrate these issues.

Example 5.1.4. Let $\mathbf{X} = (X_1, \ldots, X_n)$ be a random sample from *Bernoulli*(θ), $0 < \theta < 1$. We want to estimate $g(\theta) = \theta^2$.

It is easy to see that $d(\mathbf{X}) = X_1 X_2$ is an unbiased estimator of θ^2:

$$\mathrm{E}_\theta[X_1 X_2] = \mathrm{E}_\theta[X_1]\mathrm{E}_\theta[X_2] = \theta^2 \quad \text{for all } \theta.$$

In many situations, as in this example, finding an unbiased estimator of $g(\theta)$ is easy. Getting into the question of the choice of a sufficient statistic (Step (ii)), first note that there are many sufficient statistics of which we consider the following three:

$$T_1 = (X_1 + X_2, X_3 + \cdots + X_n),\ T_2 = (X_1 + X_n, X_2 + \cdots + X_{n-1}),\ \text{and}$$
$$T_3 = X_1 + \ldots + X_n.$$

It is easy to see that they are all sufficient statistics, which we leave as exercises. We now use the Rao-Blackwell formula, conditioning $d(\mathbf{X}) = X_1 X_2$ by T_1, T_2, T_3 given above:

(a) We first note that

$$\begin{aligned}\mathrm{E}[d(\mathbf{X})|T_1] &= \mathrm{E}[X_1 X_2 | X_1 + X_2, X_3, \ldots, X_n] \\ &= \mathrm{E}[X_1 X_2 | X_1 + X_2] = X_1 X_2,\end{aligned}$$

because

$$\begin{aligned}&\mathrm{E}[X_1 X_2 | X_1 + X_2 = 0] = 0 = X_1 X_2, \\ &\mathrm{E}[X_1 X_2 | X_1 + X_2 = 1] = 0 = X_1 X_2,\ \text{and} \\ &\mathrm{E}[X_1 X_2 | X_1 + X_2 = 2] = 1 = X_1 X_2.\end{aligned}$$

(b) To find $\mathrm{E}\big[d(\boldsymbol{X})|T_2\big]$, we first note that

$$\begin{aligned}
&\mathrm{E}\big[X_1X_2|X_1+X_n=r, X_2+\cdots+X_{n-1}=s\big]\\
&\qquad = P\big[X_1=X_2=1|X_1+X_n=r, X_2+\cdots+X_{n-1}=s\big]\\
&\qquad := P_{rs}.
\end{aligned}$$

Obviously, $P_{rs}=0$ if $r=0$ and/or $s=0$. Next, for $r=1,2$ and $s=1,\ldots,n-2$,

$$\begin{aligned}
P_{rs} &= \frac{P\big[X_1=X_2=1, X_n=r-1, X_3+\cdots+X_{n-1}=s-1\big]}{P\big[X_1+X_n=r, X_2+\cdots+X_{n-1}=s\big]}\\
&= \frac{\theta^2\theta^{r-1}(1-\theta)^{2-r}\binom{n-3}{s-1}\theta^{s-1}(1-\theta)^{n-s-2}}{\binom{2}{r}\theta^r(1-\theta)^{2-r}\binom{n-2}{s}\theta^s(1-\theta)^{n-s-2}}\\
&= \frac{rs}{2(n-2)},
\end{aligned}$$

after simplification, since $\binom{2}{r}=2/r$ for $r=1$ or $r=2$. Thus,

$$\mathrm{E}\big[d(\boldsymbol{X})|T_2\big] = \frac{(X_1+X_n)(X_2+\cdots+X_{n-1})}{2(n-2)}.$$

(c) Note that

$$\begin{aligned}
\mathrm{E}\big[d(\boldsymbol{X})|T_3=t\big] &= P\big[X_1=X_2=1|X_1+\cdots+X_n=t\big]\\
&= \frac{P\big[X_1=X_2=1, X_3+\cdots+X_n=t-2\big]}{P\big[X_1+\cdots+X_n=t\big]}\\
&= \frac{\theta^2\binom{n-2}{t-2}\theta^{t-2}(1-\theta)^{n-t}}{\binom{n}{t}\theta^t(1-\theta)^{n-t}}\\
&= \frac{\binom{n-2}{t-2}}{\binom{n}{t}} = \frac{t(t-1)}{n(n-1)}.
\end{aligned}$$

Thus,

$$\mathrm{E}\big[d(\boldsymbol{X})|T_3\big] = \frac{(\sum_{i=1}^n X_i)(\sum_{i=1}^n X_i - 1)}{n(n-1)}.$$

In applying the Rao-Blackwell method on $d(\boldsymbol{X}) = X_1X_2$, using the sufficient statistics T_1, T_2, and T_3, we have seen that $\mathrm{E}\big[d(\boldsymbol{X})|T_1\big] = d(\boldsymbol{X})$, so the method provides an "improvement" in a trivial sense. On the other hand, $\mathrm{E}\big[d(\boldsymbol{X})|T_2\big]$ provides a real improvement, because $d(\boldsymbol{X})$ is not a function of T_2, so the conditional distribution of $d(\boldsymbol{X})$ given T_2 is nondegenerate. Also, by direct calculation, we can verify that

$$\mathrm{Var}_\theta \mathrm{E}\big[d(\boldsymbol{X})|T_2\big] = \frac{\theta^2(1-\theta)}{2(n-2)}\{1+(n-1)\theta\} < \theta^2(1-\theta)^2 = \mathrm{Var}_\theta\big[d(\boldsymbol{X})\big]$$

for $n \geq 3$. Since $d(\boldsymbol{X})$ is not a function of T_3, the estimator $\mathrm{E}\big[d(\boldsymbol{X})|T_3\big]$ also provides a real improvement over $d(\boldsymbol{X})$, although a direct calculation of the variance of $\mathrm{E}\big[d(\boldsymbol{X})|T_3\big]$ is somewhat messy.

Finally, we use the Rao-Blackwell method on the estimator $\mathrm{E}\big[d(\mathbf{X})|T_2\big]$, using the sufficient statistic T_3. This leads to

$$\begin{aligned}
&\mathrm{E}\left[\frac{(X_1+X_n)(X_2+\cdots+X_{n-1})}{2(n-2)}\middle|X_1+\cdots+X_n=t\right]\\
&=\sum_{r=1}^{2}\frac{r(t-r)P\big[X_1+X_n=r,X_2+\cdots+X_{n-1}=t-r\big]}{2(n-2)\binom{n}{t}\theta^t(1-\theta)^{n-t}}\\
&=\sum_{r=1}^{2}\frac{r(t-r)\binom{2}{r}\theta^r(1-\theta)^{2-r}\binom{n-2}{t-r}\theta^{t-r}(1-\theta)^{n-t+r-2}}{2(n-2)\binom{n}{t}\theta^t(1-\theta)^{n-t}}\\
&=\sum_{r=1}^{2}\frac{r(t-r)\binom{2}{r}\binom{n-2}{t-r}}{2(n-2)\binom{n}{t}}=\frac{t(t-1)}{n(n-1)}
\end{aligned}$$

since the numerator simplifies to $2(n-2)\binom{n-2}{t-2}$.

In summary, we have seen that starting with an unbiased estimator $d(\mathbf{X}) = X_1X_2$ of $g(\theta)=\theta^2$ and using the Rao-Blackwell method with sufficient statistics $T_1=(X_1+X_2,X_3+\cdots+X_n)$, $T_2=(X_1+X_n,X_2+\cdots+X_{n-1})$, and $T_3=X_1+\cdots+X_n$, we obtained:

(a) $\mathrm{E}\big[d(\mathbf{X})|T_1\big]=d(\mathbf{X})$, which is an "improvement" in a trivial sense,
(b) $\mathrm{E}\big[d(\mathbf{X})|T_2\big]=\frac{(X_1+X_n)(X_2+\cdots+X_{n-1})}{2(n-2)}$, which is a real improvement,
(c) $\mathrm{E}\big[d(\mathbf{X})|T_3\big]=\frac{(\sum_{i=1}^n X_i)(\sum_{i=1}^n X_i-1)}{n(n-1)}$, which is also an improvement, and moreover,
(d) $\mathrm{E}\big[\mathrm{E}[d(\mathbf{X})|T_2]|T_3\big]=\mathrm{E}\big[d(\mathbf{X})|T_3\big]$.

Thus, $\mathrm{E}\big[d(\mathbf{X})|T_2\big]$ and $\mathrm{E}\big[d(\mathbf{X})|T_3\big]$ are successive improvements on $d(\mathbf{X})$.

The question still remains: can $\mathrm{E}\big[d(\mathbf{X})|T_3\big]$ be further improved? The key property of a sufficient statistic addressing this question is the property of completeness, as defined below.

Definition 5.1.2. A sufficient statistic T for a parameter $\theta\in\Theta$ is said to be complete (or boundedly complete) if for every real-valued (bounded real-valued) function φ, $\mathrm{E}_\theta[\varphi(T)]=0$ for all θ implies $\varphi(t)=0$ for all $t\notin N$ where $P_\theta[T\in N]=0$.

Completeness requires the family of distributions $\{P_\theta^T\}$ of T to be sufficiently rich, so that the condition $\mathrm{E}_\theta[\varphi(T)]=0$ for all θ forces $\varphi(t)$ to be identically zero for all practical purposes. One can think of it as a condition on the pdf's $\{f_T(\cdot,\theta),\theta\in\Theta\}$ of T having full rank, so that

$$\mathrm{E}_\theta[\varphi(T)]=\int\varphi(t)f_T(t,\theta)\,dt=0\quad\text{for all }\theta,$$

that is, $\varphi(\cdot)$ is "orthogonal" to $f_T(\cdot,\theta)$ for all $\theta\in\Theta\Rightarrow\varphi(\cdot)$ must be a zero function with probability 1.

How to check that a sufficient statistic is complete?

1. If a convergent power series $\sum_n a_nz^n=0$ for all z in some open interval, then each coefficient a_n must be zero. This fact can be used to prove completeness of some important families of discrete distributions such as $\{Bin(n, p),\ n \text{ fixed},\ 0 < p < 1\}$, $\{Poi(\theta),\ \theta > 0\}$, etc.

2. If $X_1, \ldots, X_n$ are iid as $Unif(0, \theta)$, then $T = \max\{X_1, \ldots, X_n\}$ is sufficient for θ in $\boldsymbol{X} = (X_1, \ldots, X_n)$. The pdf of T corresponding to θ is

$$f_T(t, \theta) = n\theta^{-n} t^{n-1} I_{(0,\theta)}(t).$$

   Hence, $\mathrm{E}_\theta[\varphi(T)] = 0$ for all θ implies $\int_0^\theta \varphi(t) t^{n-1}\, dt = 0$ for all θ. If φ is continuous, then $\varphi(t) = 0$ for all t by the Fundamental Theorem of Calculus, but even without continuity, the result holds with probability 1, of which the proof needs more advanced analysis. This shows that $T = \max\{X_1, \ldots, X_n\}$ is a complete sufficient statistic for θ in $\boldsymbol{X}$.

3. If $X_1, \ldots, X_n$ is a random sample from a regular k -parameter exponential family:

$$f(x, \theta) = \exp\left[\sum_{j=1}^{k} \theta_j T_j(\boldsymbol{x}) + S(\boldsymbol{x}) + d(\theta)\right] I_A(\boldsymbol{x}),$$

   where A is the support of the distribution, then

$$T_j = \sum_{i=1}^{n} T_j(x_i), \quad j = 1, \ldots, k$$

   are jointly sufficient for θ in $\boldsymbol{X} = (X_1, \ldots, X_n)$ and the joint distribution of $\boldsymbol{T} = (T_1, \ldots, T_k)$ also belongs to a regular k-parameter exponential family. If $\Theta = \left\{\boldsymbol{\theta} \in \mathbb{R}^k\colon \int_{\mathcal{X}} f(\boldsymbol{x}, \boldsymbol{\theta})\, d\boldsymbol{x} < \infty\right\}$ contains a k -dimensional open rectangle, then $\boldsymbol{T}$ is a complete sufficient statistic for $\boldsymbol{\theta}$ in $\boldsymbol{X}$. This takes care of many important situations.

With the concept of completeness, we now know how to choose the *appropriate sufficient statistic* in the Rao-Blackwell formula (1) that will lead to the UMVUE.

Theorem 5.1.5 (Lehmann-Scheffé). *If $d(\boldsymbol{X})$ is an unbiased estimator of $g(\theta)$ and T is a complete sufficient statistic for θ in $\boldsymbol{X}$, then*

$$d^\star(T) = E\big[d(\boldsymbol{X})|T\big]$$

is the (essentially unique) UMVUE of $g(\theta)$.

Proof. For any other unbiased estimator $d_1(\boldsymbol{X})$ of $g(\theta)$, consider the unbiased estimator $d_1^\star(T) = \mathrm{E}\big[d_1(\boldsymbol{X})|T\big]$ which must have variance $\leq$ that of d_1 by the Rao-Blackwell Theorem, but by completeness, $d_1^\star(T) = d^\star(T)$ with probability 1. Therefore,

$$\mathrm{Var}_\theta\big[d_1(\boldsymbol{X})\big] \geq \mathrm{Var}_\theta\big[d_1^\star(T)\big] = \mathrm{Var}_\theta\big[d^\star(T)\big]. \qquad \square$$

In the above example, $T_3 = \sum_{i=1}^n X_i$ is a complete sufficient statistic. Hence,

$$\mathrm{E}\big[d(\boldsymbol{X})|T_3\big] = \frac{(\sum_{i=1}^n X_i)(\sum_{i=1}^n X_i - 1)}{n(n-1)}$$

is the UMVUE of $g(\theta) = \theta^2$. Also, note that the sufficient statistic $T_2 = (X_1 + X_n,\ X_2 + \cdots + X_{n-1})$ is *not complete*, because the expectation of

$$\varphi(T_2) = \frac{1}{2}(X_1 + X_n) - \frac{1}{n-2}(X_2 + \cdots + X_{n-1})$$

is 0 for all $\theta \in (0, 1)$. We end this section with a very interesting property of completeness.

Ancillarity and Completeness

Definition 5.1.3. A statistic whose distribution does not depend on θ is called an ancillary statistic.

Theorem 5.1.6 (Basu's Theorem [17]). *Suppose that the distribution of X belongs to the family $\{P_\theta, \theta \in \Theta\}$. If $T = T(X)$ is a complete sufficient statistic for θ and if $V = V(X)$ is an ancillary statistic, then V is independent of T.*

Proof. The probability $p_A = P_\theta[V \in A]$ is independent of θ for all A by ancillarity of V. Let $\eta_A(t) = P[V \in A|T = t]$, which is also independent of θ since T is sufficient, and $\mathrm{E}_\theta\left[\eta_A(T)\right] = \mathrm{E}_\theta P[V \in A|T] = P_\theta[V \in A] = p_A$. Thus

$$\mathrm{E}_\theta\left[\eta_A(T) - p_A\right] = 0 \quad \text{for all } \theta.$$

Since T is complete, this implies that for all A, $\eta_A(T) = p_A$, that is

$$P[V \in A|T] = P[V \in A] \text{ w.p. } 1.$$

Hence, V is independent of T. □

5.2 Lower Bound for the Variance of an Unbiased Estimator

5.2.1 The Information Inequality: Cramèr-Rao Lower Bound

For brevity of notation, we shall write

$$\frac{\partial f(x,\theta)}{\partial \theta} = \dot{f}(x,\theta), \quad \frac{\partial^2 f(x,\theta)}{\partial \theta^2} = \ddot{f}(x,\theta),$$

$$\log f(x,\theta) = \ell(x,\theta), \quad \frac{\partial \ell(x,\theta)}{\partial \theta} = \dot{\ell}(x,\theta), \quad \frac{\partial^2 \ell(x,\theta)}{\partial \theta^2} = \ddot{\ell}(x,\theta).$$

These notations will also be used in all subsequent discussions.

Regularity conditions on $\{f(x,\theta),\ \theta \in \Theta \subset \mathbb{R}\}$:

1. The parameter space Θ is an open interval and the set $S = \{x\colon f(x,\theta) > 0\}$ does not depend upon θ.
2. For all $x \in S$ and $\theta \in \Theta$, $\dot{\ell}(x,\theta)$ exists and is finite.
3. For any statistic T such that $\mathrm{E}_\theta[|T|] < \infty$ for all $\theta \in \Theta$,

$$\begin{aligned}\frac{d}{d\theta} \int T(x) f(x,\theta)\, dx \bigg|_{\theta = \theta_0} &= \int T(x) \dot{f}(x,\theta_0)\, dx \\ &= \int T(x) \dot{\ell}(x,\theta_0) f(x,\theta_0)\, dx\end{aligned}$$

whenever the right-hand side is finite. In other words,

$$\frac{d}{d\theta}\mathrm{E}_\theta[T(X)]\bigg|_{\theta=\theta_0} = \mathrm{E}_{\theta_0}\big[T(X)\dot{\ell}(X,\theta_0)\big].$$

These conditions are satisfied in a regular exponential family.

Remark 5.2.1. Write $\mathrm{E}_\theta[T(X)] = g(\theta) + b(\theta)$, where $b(\theta)$ is the bias of T at θ as an estimator of $g(\theta)$. Then, Condition 3 becomes

$$g'(\theta) + b'(\theta) = \mathrm{E}_\theta\big[T(X)\dot{\ell}(X,\theta)\big].$$

Remark 5.2.2.

(a) Taking $T(x) \equiv 1$ in Condition 3, we have $\mathrm{E}_\theta\big[\dot{\ell}(X,\theta)\big] = 0$.

(b) Sometimes (as in regular exponential families), a stronger version of Condition 3,

$$\frac{d^2}{d\theta^2}\int T(x)f(x,\theta)\,dx\bigg|_{\theta=\theta_0} = \int T(x)\ddot{f}(x,\theta_0)\,dx$$

holds for $T(x) \equiv 1$, so that $\int \ddot{f}(x,\theta)\,dx = 0$. Since

$$\ddot{f}(x,\theta) = \Big[\ddot{\ell}(x,\theta) + \big\{\dot{\ell}(x,\theta)\big\}^2\Big]f(x,\theta),$$

this implies that

$$\mathrm{Var}_\theta\big[\dot{\ell}(X,\theta)\big] = \mathrm{E}_\theta\Big[\big\{\dot{\ell}(X,\theta)\big\}^2\Big] = -\,\mathrm{E}_\theta\big[\ddot{\ell}(X,\theta)\big].$$

Definition 5.2.1 (Fisher-Information). Condition 2 allows us to define

$$I(\theta_0) = \mathrm{E}_{\theta_0}\Big[\big\{\dot{\ell}(X,\theta_0)\big\}^2\Big], \tag{2}$$

which is called the Fisher-Information in X for the family $\big\{f(x,\theta),\ \theta \in \Theta\big\}$ at θ_0.

Remark 5.2.3. Note that $0 \le I(\theta) \le \infty$. By Condition 3,

$$I(\theta) = \mathrm{Var}_\theta\big[\dot{\ell}(X,\theta)\big],$$

and the stronger version of Condition 3 implies

$$I(\theta) = \mathrm{E}_\theta\Big[\big\{\dot{\ell}(X,\theta)\big\}^2\Big] = \mathrm{Var}_\theta\big[\dot{\ell}(X,\theta)\big] = \mathrm{E}_\theta\big[-\ddot{\ell}(X,\theta)\big].$$

Theorem 5.2.1 (Cramér-Rao Inequality [15, 18]). *Suppose $T = T(X)$ has $\mathrm{E}_\theta[T] = g(\theta) + b(\theta)$ and $Var_\theta[T] < \infty$. Then, under the regularity Conditions 1, 2, and 3, we have*

$$Var_\theta[T] \ge \frac{\{g'(\theta) + b'(\theta)\}^2}{I(\theta)}.$$

The right-hand side of this inequality is known as the *Cramér-Rao lower bound.*

Proof. Fix $\theta \in \Theta$ and let $S = S(X,\theta) = \dot{\ell}(X,\theta)$. By Remark 5.2.2, $\mathrm{E}_\theta[S] = 0$ and $\mathrm{Var}_\theta[S] = I(\theta)$. Hence,

$$\mathrm{Cov}_\theta\big[T,S\big] = \mathrm{E}_\theta\big[TS\big] = \mathrm{E}_\theta\big[T(X)\dot{\ell}(X,\theta)\big] = g'(\theta) + b'(\theta),$$

as shown earlier. But $\text{Cov}_\theta^2[T, S] \le \text{Var}_\theta[T]\text{Var}_\theta[S]$. Thus,

$$\text{Var}_\theta[T] \ge \frac{\text{Cov}_\theta^2[T, S]}{\text{Var}_\theta[S]} = \frac{\{g'(\theta) + b'(\theta)\}^2}{I(\theta)}. \qquad \square$$

Remark 5.2.4.

(a) If $\boldsymbol{X} = (X_1, \ldots, X_n)$ is a random sample of size n from a population with pdf (or pmf) $f(x, \theta)$, then $\dot{\ell}(\boldsymbol{X}, \theta) = \sum_{i=1}^n \dot{\ell}(X_i, \theta)$. Hence, the Fisher-information in $\boldsymbol{X} = (X_1, \ldots, X_n)$ for the family $\{f(\boldsymbol{x}, \theta)\colon \theta \in \Theta\}$ is

$$\text{Var}_\theta\left[\sum_{i=1}^n \dot{\ell}(X_i, \theta)\right] = n\text{Var}_\theta\left[\dot{\ell}(X_1, \theta)\right] = nI(\theta).$$

Thus the information inequality for $\text{Var}_\theta[T]$ of an estimator T based on a random sample of size n from $f(x, \theta)$ is

$$\text{Var}_\theta[T] \ge \frac{\{g'(\theta) + b'(\theta)\}^2}{nI(\theta)}.$$

(b) If T is an unbiased estimator of θ based on a random sample of size n, then

$$\text{Var}_\theta[T] \ge \frac{1}{nI(\theta)}.$$

5.2.2 Effect of Reparametrization on Fisher-Information in Exponential Families

Consider the exponential family of pdf's (pmf's):

$$f(x, \theta) = \exp\left[c(\theta)T(x) + d(\theta) + h(x)\right]I_A(x),$$

where c is one-to-one and twice differentiable. If $f(x, \theta)$ is reparametrized by $\eta - c(\theta)$ and rewritten as

$$g(x, \eta) = \exp\left[\eta T(x) + d_0(\eta) + h(x)\right]I_A(x)$$

with $d_0(\eta) = d\left[c^{-1}(\eta)\right]$, then the Fisher-Information for the family $\{g(x, \eta)\colon \eta \in c(\Theta)\}$ is $I_g(\eta) = -d_0''(\eta)$. How does this I_g relate to the Fisher-Information $I_f(\theta) = \text{E}_\theta\left[-\ddot{\ell}(X, \theta)\right]$ for the original family? The answer is:

$$I_f(\theta) = \left(d_0''(\eta)\big|_{\eta=c(\theta)}\right)\{c'(\theta)\}^2,$$

the proof of which is left as an exercise.

5.2.3 Information Inequality in Multiparameter Families

We now consider the multiparameter case in which we observe a random variable X following a distribution with pdf (or pmf) belonging to a family $\{f(x,\boldsymbol{\theta})\colon \boldsymbol{\theta} \in \Theta\}$, where $\boldsymbol{\theta} = (\theta_1,\ldots,\theta_k)$ and $\Theta \subset \mathbb{R}^k$. Our aim is to estimate a function $g(\boldsymbol{\theta})$ of the parameter vector $\boldsymbol{\theta}$ such as $g(\boldsymbol{\theta}) = \theta_r$, or, as in the pdf of $N(\theta_1,\theta_2)$ with $\theta_1 \neq 0$, $g(\boldsymbol{\theta}) = \sqrt{\theta_2}/\theta_1$, the coefficient of variation. For a statistic $T = T(X)$, estimating $g(\boldsymbol{\theta})$ with a bias $b(\boldsymbol{\theta})$ (ie, $\mathrm{E}_{\boldsymbol{\theta}}[T(X)] = g(\boldsymbol{\theta})+b(\boldsymbol{\theta})$), we now look for a lower bound of $\mathrm{Var}_{\boldsymbol{\theta}}[T]$. The notations introduced earlier are now extended as:

$$\frac{\partial f(x,\boldsymbol{\theta})}{\partial\theta_r} = \dot{f}_r(x,\boldsymbol{\theta}), \quad \frac{\partial^2 f(x,\boldsymbol{\theta})}{\partial\theta_r\partial\theta_s} = \ddot{f}_{rs}(x,\boldsymbol{\theta}),$$

$$\log f(x,\boldsymbol{\theta}) = \ell(x,\boldsymbol{\theta}) \text{ as before}, \quad \frac{\partial \ell(x,\boldsymbol{\theta})}{\partial\theta_r} = \dot{\ell}_r(x,\boldsymbol{\theta}), \quad \frac{\partial^2 \ell(x,\boldsymbol{\theta})}{\partial\theta_r\partial\theta_s} = \ddot{\ell}_{rs}(x,\boldsymbol{\theta})$$

for $r, s = 1,\ldots,k$.

The Cramèr-Rao Information Inequality is simply a restatement of the inequality $\rho^2(U,V) \leq 1$, where $\rho(U,V)$ is the correlation between U and V with $U = T(X)$ and $V = \dot{\ell}(X,\boldsymbol{\theta})$.

In the multiparameter case, we extend the concept of correlation between U and V to the

$$\text{“correlation” between } U = T(X) \text{ and } \boldsymbol{V}^{\mathsf{T}} = (\dot{\ell}_1(X,\boldsymbol{\theta}),\ldots,\dot{\ell}_k(X,\boldsymbol{\theta}))$$

which we shall call the multiple correlation of U on $\boldsymbol{V}$.

Let $\boldsymbol{V}^{\mathsf{T}} = (V_1,\ldots,V_k)$ be a k-dim random vector with covariance matrix

$$\boldsymbol{\Sigma} = \mathrm{Cov}[\boldsymbol{V},\boldsymbol{V}] = \mathrm{E}\Big[\{\boldsymbol{V} - \mathrm{E}(\boldsymbol{V})\}\{\boldsymbol{V} - \mathrm{E}(\boldsymbol{V})\}^{\mathsf{T}}\Big],$$

and let U be a random variable with finite second moment,

$$\gamma_i = \mathrm{Cov}\big[U,V_i\big],\ i = 1,\ldots,k, \text{ and } \boldsymbol{\gamma}^{\mathsf{T}} = (\gamma_1,\ldots,\gamma_k).$$

Definition 5.2.2. If $\boldsymbol{\Sigma} = \mathrm{Cov}[\boldsymbol{V},\boldsymbol{V}]$ is positive definite, then the multiple correlation of U on $\boldsymbol{V}$ is defined as:

$$\rho^\star = \rho^\star_{U\cdot\boldsymbol{V}} = \max_{a_1,\ldots,a_k} \mathrm{Corr}\left[U, \sum_{i=1}^k a_i V_i\right] = \max_{a_1,\ldots,a_k} \frac{\boldsymbol{a}^{\mathsf{T}}\boldsymbol{\gamma}}{\{\mathrm{Var}[U]\cdot \boldsymbol{a}^{\mathsf{T}}\boldsymbol{\Sigma}\boldsymbol{a}\}^{1/2}}.$$

Note that

(i) when $\boldsymbol{\Sigma}$ is singular and is of rank r, we can always find a r-dim random vector $\boldsymbol{V}^*$, which is a linear function of $\boldsymbol{V}$, whose covariance matrix is positive definite, and

(ii) we can assume without loss of generality that $\boldsymbol{a}^{\mathsf{T}}\boldsymbol{\Sigma}\boldsymbol{a} = 1$.

Justification of (i). By spectral decomposition, $\boldsymbol{\Sigma} = \boldsymbol{Q}\boldsymbol{\Lambda}\boldsymbol{Q}^{\mathsf{T}}$ where $\boldsymbol{Q}$ is an orthogonal matrix with columns $\boldsymbol{q}_1,\ldots,\boldsymbol{q}_k$ (the eigenvectors of $\boldsymbol{\Sigma}$) and $\boldsymbol{\Lambda}$ is a diagonal matrix whose diagonal elements $\lambda_1 \geq \cdots \geq \lambda_k$ are the eigenvalues of $\boldsymbol{\Sigma}$. If $\boldsymbol{\Sigma}$ is of rank $r < k$, then only the first

r diagonal elements of $\boldsymbol{\Lambda}$ are nonzero and the rest are zeros. Let $\boldsymbol{V}^* = \left[\boldsymbol{q}_1^\top \boldsymbol{V}, \ldots, \boldsymbol{q}_r^\top \boldsymbol{V}\right]^\top = \boldsymbol{Q}_0^\top \boldsymbol{V}$, where $\boldsymbol{Q}_0 = \left[\boldsymbol{q}_1, \ldots, \boldsymbol{q}_r\right]$ is a $k \times r$ matrix. Then,

$$\text{Cov}\left[V_i^*, V_j^*\right] = \text{Cov}\left[\boldsymbol{q}_i^\top \boldsymbol{V}, \boldsymbol{q}_j^\top \boldsymbol{V}\right] = \boldsymbol{q}_i^\top \boldsymbol{\Sigma} \boldsymbol{q}_j = \begin{cases} \lambda_i & i = j \\ 0 & i \neq j \end{cases}$$

Since $\boldsymbol{\Sigma}$ is of rank r, $\lambda_1 \geq \cdots \geq \lambda_r > 0$ and $\lambda_{r+1} = \cdots = \lambda_k = 0$. Consequently, the covariance matrix $\boldsymbol{\Sigma}^*$ of $\boldsymbol{V}^* = \boldsymbol{Q}_0^\top \boldsymbol{V}$ is a diagonal matrix with diagonal elements $\lambda_1, \ldots, \lambda_k$ (ie, $\boldsymbol{\Sigma}^*$ is positive definite).

Justification of (ii). Correlation coefficient is invariant under scale change. Hence, for any vector $\boldsymbol{a} \in \mathbb{R}^k$, if we take $\boldsymbol{a}_0 = \boldsymbol{a}/\sqrt{\boldsymbol{a}^\top \boldsymbol{\Sigma} \boldsymbol{a}}$, then

$$\text{Var}\left[\boldsymbol{a}_0^\top \boldsymbol{V}\right] = 1 \text{ and } \text{Corr}\left[U, \boldsymbol{a}^\top \boldsymbol{V}\right] = \text{Corr}\left[U, \boldsymbol{a}_0^\top \boldsymbol{V}\right].$$

Proposition 5.2.1. *Let U and $(V_1, \ldots, V_k)$ have finite second moments and let $\gamma_i = Cov[U, V_i]$, $i = 1, \ldots, k$, and $\boldsymbol{\Sigma} = Cov[\boldsymbol{V}, \boldsymbol{V}]$. Then,*

$$\rho^{\star 2} = \frac{\boldsymbol{\gamma}^\top \Sigma^{-1} \boldsymbol{\gamma}}{Var[U]}.$$

Proof. Since $\rho^\star = \max_{a_1, \ldots, a_k} \frac{\boldsymbol{a}^\top \boldsymbol{\gamma}}{\text{Var}[U]}$ and $\boldsymbol{a}^\top \boldsymbol{\Sigma} \boldsymbol{a} = 1$, we need to maximize $\boldsymbol{a}^\top \boldsymbol{\gamma}$ subject to the condition $\boldsymbol{a}^\top \boldsymbol{\Sigma} \boldsymbol{a} = 1$. Using Lagrange's undetermined multiplier, we maximize

$$\boldsymbol{a}^\top \boldsymbol{\gamma} - \frac{1}{2} \lambda \boldsymbol{a}^\top \boldsymbol{\Sigma} \boldsymbol{a}$$

and then find λ using $\boldsymbol{a}^\top \boldsymbol{\Sigma} \boldsymbol{a} = 1$. To this end,

$$\begin{aligned} 0 &= \frac{\partial}{\partial a_r} \left[\sum_{r=1}^{k} a_r \gamma_r - \frac{1}{2} \lambda \sum_{r=1}^{k} \sum_{s=1}^{k} a_r a_s \sigma_{rs} \right] \\ &= \gamma_r - \frac{1}{2} \lambda \left[2 a_r \sigma_{rr} + 2 \sum_{r \neq s = 1}^{k} a_s \sigma_{rs} \right] \\ &= \gamma_r - \lambda \sum_{s=1}^{k} a_s \sigma_{rs}, \quad r = 1, \ldots, k \end{aligned}$$

implies $\boldsymbol{\Sigma} \boldsymbol{a} = \lambda^{-1} \boldsymbol{\gamma}$ (ie, $\boldsymbol{a} = \lambda^{-1} \boldsymbol{\Sigma}^{-1} \boldsymbol{\gamma}$).

To find λ, we now have the equation

$$1 = \boldsymbol{a}^\top \boldsymbol{\Sigma} \boldsymbol{a} = \lambda^{-2} (\boldsymbol{\gamma}^\top \boldsymbol{\Sigma}^{-1}) \boldsymbol{\Sigma} (\boldsymbol{\Sigma}^{-1} \boldsymbol{\gamma}) = \lambda^{-2} \boldsymbol{\gamma}^\top \boldsymbol{\Sigma}^{-1} \boldsymbol{\gamma},$$

with solutions: $\lambda = \pm (\boldsymbol{\gamma}^\top \boldsymbol{\Sigma}^{-1} \boldsymbol{\gamma})^{1/2}$. Hence,

$$\boldsymbol{a}^\star = \lambda^{-1} \boldsymbol{\Sigma}^{-1} \boldsymbol{\gamma} = \frac{\pm \boldsymbol{\Sigma}^{-1} \boldsymbol{\gamma}}{\sqrt{\boldsymbol{\gamma}^\top \boldsymbol{\Sigma}^{-1} \boldsymbol{\gamma}}} \text{ and}$$

$$\boldsymbol{a}^{\star\mathsf{T}}\boldsymbol{\gamma} = \pm\frac{\boldsymbol{\gamma}^{\mathsf{T}}\boldsymbol{\Sigma}^{-1}\boldsymbol{\gamma}}{\sqrt{\boldsymbol{\gamma}^{\mathsf{T}}\boldsymbol{\Sigma}^{-1}\boldsymbol{\gamma}}} = \pm\sqrt{\boldsymbol{\gamma}^{\mathsf{T}}\boldsymbol{\Sigma}^{-1}\boldsymbol{\gamma}}$$

are the maximizers of $\boldsymbol{a}^{\mathsf{T}}\boldsymbol{\gamma}$ subject to $\boldsymbol{a}^{\mathsf{T}}\boldsymbol{\Sigma}\boldsymbol{a} = 1$. Hence,

$$\begin{aligned}\rho^{\star 2} &= \max_{\{a_1,\ldots,a_k\},\boldsymbol{a}^{\mathsf{T}}\boldsymbol{\Sigma}\boldsymbol{a}=1} \operatorname{Corr}^2\left[U, \sum_{i=1}^{k} a_i V_i\right] = \max_{a_1,\ldots,a_k}\left(\frac{\boldsymbol{a}^{\mathsf{T}}\boldsymbol{\gamma}}{\sqrt{\operatorname{Var}[U]}}\right)^2 \\ &= \frac{(\boldsymbol{a}^{\star\mathsf{T}}\boldsymbol{\gamma})^2}{\operatorname{Var}[U]} = \frac{\boldsymbol{\gamma}^{\mathsf{T}}\boldsymbol{\Sigma}^{-1}\boldsymbol{\gamma}}{\operatorname{Var}[U]}.\end{aligned}$$

□

Getting back to $U = T(X)$ and $\boldsymbol{V}^{\mathsf{T}} = \left(\dot{\ell}_1(X,\boldsymbol{\theta}), \ldots, \dot{\ell}_k(X,\boldsymbol{\theta})\right)$, and recognizing that the multiple correlation $\rho^{\star}_{U\cdot\boldsymbol{V}}$, like a correlation lies between -1 and $+1$, so that

$$\rho^{\star 2} = \frac{\boldsymbol{\gamma}^{\mathsf{T}}\boldsymbol{\Sigma}^{-1}\boldsymbol{\gamma}}{\operatorname{Var}[U]} \le 1, \quad \text{ie, } \operatorname{Var}[U] \ge \boldsymbol{\gamma}^{\mathsf{T}}\boldsymbol{\Sigma}^{-1}\boldsymbol{\gamma},$$

we arrive at a lower bound for $\operatorname{Var}[U] = \operatorname{Var}[T(X)]$. To make this lower bound explicit, we need to find $\boldsymbol{\gamma}^{\mathsf{T}} = (\gamma_1, \ldots, \gamma_k)$, where $\gamma_j = \operatorname{Cov}_\theta\left[T(X), \dot{l}_j(X,\boldsymbol{\theta})\right]$, and $\boldsymbol{\Sigma} = \left(\left(\operatorname{Cov}_{\boldsymbol{\theta}}\left[\dot{l}_r(X,\boldsymbol{\theta}), \dot{l}_s(X,\boldsymbol{\theta})\right]\right)\right)$.

As in the single-parameter case, here also we assume that the following regularity conditions hold for the family $\left\{f(x,\boldsymbol{\theta}), \boldsymbol{\theta} \in \Theta \subset \mathbb{R}^k\right\}$:

1. The parameter space Θ is an open interval and the set $S = \left\{x: f(x,\boldsymbol{\theta}) > 0\right\}$ does not depend upon $\boldsymbol{\theta}$.
2. For all $x \in S$ and $\boldsymbol{\theta} \in \Theta$, $\dot{\ell}_r(x,\boldsymbol{\theta})$ exists and is finite for $r = 1, \ldots, k$.
3. For any statistic T such that $\mathrm{E}_{\boldsymbol{\theta}}[|T|] < \infty$ for all $\boldsymbol{\theta} \in \Theta$,

$$\begin{aligned}\frac{\partial}{\partial\theta_r}\int T(x) f(x,\boldsymbol{\theta})\, dx\bigg|_{\boldsymbol{\theta}=\boldsymbol{\theta}_0} &= \int T(x)\dot{f}_r(x,\boldsymbol{\theta}_0)\, dx \\ &= \int T(x)\dot{\ell}_r(x,\boldsymbol{\theta}_0) f(x,\boldsymbol{\theta}_0)\, dx\end{aligned}$$

for $r = 1, \ldots, k$ whenever the right-hand side is finite. In other words,

$$\frac{\partial}{\partial\theta_r}\mathrm{E}_{\boldsymbol{\theta}}[T(X)]\bigg|_{\boldsymbol{\theta}=\boldsymbol{\theta}_0} = \mathrm{E}_{\boldsymbol{\theta}_0}\left[T(X)\dot{\ell}_r(X,\boldsymbol{\theta}_0)\right], \quad r = 1, \ldots, k.$$

As in the single-parameter case, these conditions are also satisfied in a regular k-parameter exponential family.

Condition 3 becomes

$$\mathrm{E}_{\boldsymbol{\theta}}\left[T(X)\dot{\ell}_r(X,\boldsymbol{\theta})\right] = \frac{\partial}{\partial\theta_r}\left[g(\boldsymbol{\theta}) + b(\boldsymbol{\theta})\right],$$

when $T(X)$ is an estimator of $g(\boldsymbol{\theta})$ with bias $b(\boldsymbol{\theta})$. Taking $T(x) \equiv 1$ in Condition 3, here we have $\mathrm{E}_{\boldsymbol{\theta}}\left[\dot{\ell}_r(X,\boldsymbol{\theta})\right] = 0$, $r = 1, \ldots, k$, and a stronger version of Condition 3 (which holds in exponential families), namely,

$$\frac{\partial^2}{\partial\theta_r\partial\theta_s}\int T(x)f(x,\boldsymbol{\theta})\,dx\bigg|_{\boldsymbol{\theta}=\boldsymbol{\theta}_0} = \int T(x)\ddot{f}_{rs}(x,\boldsymbol{\theta})\,dx$$

holds for $T(x) \equiv 1$, so that $\int \ddot{f}_{rs}(x,\boldsymbol{\theta})\,dx = 0$. Since

$$\ddot{f}_{rs}(x,\boldsymbol{\theta}) = \left[\ddot{\ell}_{rs}(x,\boldsymbol{\theta}) + \dot{\ell}_r(x,\boldsymbol{\theta})\dot{\ell}_s(x,\boldsymbol{\theta})\right]f(x,\boldsymbol{\theta}),$$

this implies

$$\mathrm{Cov}_{\boldsymbol{\theta}}\left[\dot{\ell}_r(X,\boldsymbol{\theta}), \dot{\ell}_s(X,\boldsymbol{\theta})\right] = \mathrm{E}_{\boldsymbol{\theta}}\left[\dot{\ell}_r(X,\boldsymbol{\theta})\dot{\ell}_s(X,\boldsymbol{\theta})\right] = -\mathrm{E}_{\boldsymbol{\theta}}\left[\ddot{\ell}_{rs}(X,\boldsymbol{\theta})\right].$$

Definition 5.2.3 (Information Matrix). By Condition 2, $\dot{\ell}_r(x,\boldsymbol{\theta})$ exist and is finite for $r = 1, \ldots, k$, so we can define the $k \times k$ matrix $I(\boldsymbol{\theta}_0) = ((I_{rs}(\boldsymbol{\theta}_0)))$ where $I_{rs}(\boldsymbol{\theta}_0) = \mathrm{E}_{\boldsymbol{\theta}_0}\left[\dot{\ell}_r(X,\boldsymbol{\theta}_0)\dot{\ell}_s(X,\boldsymbol{\theta}_0)\right]$, which is called the Information Matrix in X for the family $\left\{f(x,\boldsymbol{\theta}),\ \boldsymbol{\theta} \in \Theta \subset \mathbb{R}^k\right\}$ at $\boldsymbol{\theta}_0$.

By Condition 3, $I_{rs}(\boldsymbol{\theta}) = \mathrm{Cov}_{\boldsymbol{\theta}}\left[\dot{\ell}_r(X,\boldsymbol{\theta}), \dot{\ell}_s(X,\boldsymbol{\theta})\right]$ and the stronger version of Condition 3 implies $I_{rs}(\boldsymbol{\theta}) = \mathrm{E}_{\boldsymbol{\theta}}\left[-\ddot{\ell}_{rs}(X,\boldsymbol{\theta})\right]$ as shown above.

We now go back to the inequality $\mathrm{Var}[T(X)] \geq \boldsymbol{\gamma}^{\mathsf{T}}\boldsymbol{\Sigma}^{-1}\boldsymbol{\gamma}$, where

$$\begin{aligned}\gamma^{\mathsf{T}} &= \left(\mathrm{Cov}_{\boldsymbol{\theta}}\left[T(X), \dot{\ell}_1(X,\boldsymbol{\theta})\right], \ldots,\ \mathrm{Cov}_{\boldsymbol{\theta}}\left[T(X), \dot{\ell}_k(X,\boldsymbol{\theta})\right]\right)\\ &= \left(\frac{\partial}{\partial\theta_1}\mathrm{E}_{\boldsymbol{\theta}}[T(X)], \ldots, \frac{\partial}{\partial\theta_r}\mathrm{E}_{\boldsymbol{\theta}}[T(X)]\right)\\ &= \nabla^{\mathsf{T}}\left[g(\boldsymbol{\theta}) + b(\boldsymbol{\theta})\right].\end{aligned}$$

by Condition 3. We thus arrive at the following theorem.

Theorem 5.2.2 (Information Lower Bound in Multiparameter Case). *Suppose $T = T(X)$ is an estimator of $g(\boldsymbol{\theta})$ with bias $b(\boldsymbol{\theta})$ and finite variance, that is, $E_{\boldsymbol{\theta}}[T] = g(\boldsymbol{\theta}) + b(\boldsymbol{\theta})$ and $Var_{\boldsymbol{\theta}}[T] < \infty$ where the family $\left\{f(x,\boldsymbol{\theta}),\ \boldsymbol{\theta} \in \Theta \subset \mathbb{R}^k\right\}$ satisfies regularity Conditions 1, 2, and 3. Then,*

$$Var_{\boldsymbol{\theta}}[T] \geq \left\{\nabla^{\mathsf{T}}\left[g(\boldsymbol{\theta}) + b(\boldsymbol{\theta})\right]\right\}I(\boldsymbol{\theta})^{-1}\left\{\nabla\left[g(\boldsymbol{\theta}) + b(\boldsymbol{\theta})\right]\right\}.$$

Example 5.2.1. In a random sample $(X_1, \ldots, X_n)$ from $N(\mu, \sigma^2)$, let

$$\bar{X} = n^{-1}\sum_{i=1}^{n} X_i, \quad W = n^{-1}\sum_{i=1}^{n}(X_i - \mu)^2, \text{ and}$$

$$S^2 = (n-1)^{-1}\sum_{i=1}^{n}(X_i - \bar{X})^2.$$

We want UMVUEs of σ^2 and σ when μ is known and when μ is unknown.

Solution. When μ is known, W is a complete sufficient statistic for σ^2, and when μ is unknown, $(\bar{X}, S^2)$ is a complete sufficient statistic for (μ, σ^2) in $(X_1, \ldots, X_n)$. Therefore, it is enough to find unbiased estimators of σ^2 and σ, which are functions of W when μ is known and functions of $(\bar{X}, S^2)$ when μ is unknown, because these estimators would then be UMVUEs by the Lehmann-Scheffé Theorem. We now obtain such unbiased estimators.

When μ is known,

(a) W is an unbiased estimator of σ^2, since $\mathrm{E}_\sigma\left[(X_i - \mu)^2\right] = \sigma^2$, and
(b) starting with $\sqrt{W}$ as a natural estimator of σ, we find

$$\mathrm{E}_\sigma\left[\sqrt{W}\right] = \sigma n^{-1/2}\mathrm{E}\left[(\chi_n^2)^{1/2}\right] = \frac{\sigma}{\sqrt{n}} \frac{\sqrt{2}\Gamma\left(\frac{n+1}{2}\right)}{\Gamma\left(\frac{n}{2}\right)},$$

so that

$$T_0 = c_n\sqrt{W}, \text{ with } c_n = \sqrt{\frac{n}{2}}\frac{\Gamma\left(\frac{n}{2}\right)}{\Gamma\left(\frac{n+1}{2}\right)},$$

is the UMVUE of σ.

When μ is unknown,

(a) recall from Proposition 2.2.5 that

$$\frac{(n-1)S^2}{\sigma^2} = \frac{1}{\sigma^2}\sum_{i=1}^{n}(X_i - \bar{X})^2 \sim \chi_{n-1}^2,$$

so that $\mathrm{E}_\sigma\left[S^2\right] = \frac{\sigma^2}{n-1}\mathrm{E}\left[\chi_{n-1}^2\right] = \sigma^2$, making S^2 the UMVUE of σ^2, and
(b) starting with S as an estimator of σ, we see that

$$\mathrm{E}_\sigma\left[S\right] = \frac{\sigma}{\sqrt{n-1}}\mathrm{E}\left[(\chi_{n-1}^2)^{1/2}\right] = \frac{\sigma}{\sqrt{n-1}}\frac{\sqrt{2}\Gamma\left(\frac{n}{2}\right)}{\Gamma\left(\frac{n-1}{2}\right)},$$

so that

$$T_1 = c_{n-1}S, \text{ with } c_{n-1} = \sqrt{\frac{n-1}{2}}\frac{\Gamma\left(\frac{n-1}{2}\right)}{\Gamma\left(\frac{n}{2}\right)},$$

is the UMVUE of σ.

In summary, we have

1. W and $T_0 = c_n\sqrt{W}$ are the UMVUEs of σ^2 and σ, respectively, when μ is known, and
2. S^2 and $T_1 = c_{n-1}S$ are the UMVUEs of σ^2 and σ, respectively, when μ is unknown, where

$$c_n = \sqrt{\frac{n}{2}}\frac{\Gamma\left(\frac{n}{2}\right)}{\Gamma\left(\frac{n+1}{2}\right)}.$$

Example 5.2.1 (Continuation). Find the Information Lower Bounds for the variances of unbiased estimators of σ^2 and σ when μ is known and when μ is unknown. Do the variances of the UMVUEs obtained above attain these lower bounds?

Solution. To find these lower bounds when μ is known, we write $\log f(X;\sigma^2)$ as

$$\ell(X;\theta) = \begin{cases} -\frac{1}{2}\log(2\pi) - \frac{1}{2}\log\theta - \frac{1}{2\theta}(X-\mu)^2 & \text{with } \theta = \sigma^2 \\ -\frac{1}{2}\log(2\pi) - \log\theta - \frac{1}{2\theta^2}(X-\mu)^2 & \text{with } \theta = \sigma \end{cases},$$

leading to

$$\dot{\ell}(X;\theta) = \begin{cases} \frac{1}{2\theta^2}\left[(X-\mu)^2 - \theta\right] & \text{for } \theta = \sigma^2 \\ \frac{1}{\theta^3}\left[(X-\mu)^2 - \theta^2\right] & \text{for } \theta = \sigma \end{cases}.$$

Hence,

$$I(\theta) = \mathrm{E}_\theta\left[\dot{\ell}^2(X;\theta)\right] = \begin{cases} \frac{1}{2\theta^2} = \frac{1}{2\sigma^4} & \text{for } \theta = \sigma^2 \\ \frac{2}{\theta^2} = \frac{2}{\sigma^2} & \text{for } \theta = \sigma \end{cases},$$

so the information lower bounds for variances of the unbiased estimators of σ^2 and σ are, respectively,

$$\frac{1}{nI(\sigma^2)} = \frac{2\sigma^4}{n} \text{ and } \frac{1}{nI(\sigma)} = \frac{\sigma^2}{2n}.$$

When μ is unknown, we write $\log f(X;\mu,\sigma^2)$ as

$$\ell(X;\theta_1,\theta_2) = \begin{cases} -\frac{1}{2}\log(2\pi) - \frac{1}{2}\log\theta_2 - \frac{1}{2\theta_2}(X-\theta_1)^2 & \text{with } (\theta_1,\theta_2) = (\mu,\sigma^2) \\ -\frac{1}{2}\log(2\pi) - \log\theta_2 - \frac{1}{2\theta_2^2}(X-\theta_1)^2 & \text{with } (\theta_1,\theta_2) = (\mu,\sigma) \end{cases},$$

leading to

$$\dot{\ell}^{\mathsf{T}}(X;\theta_1,\theta_2) = \begin{cases} \left(\frac{1}{\theta_2}[X-\theta_1],\ -\frac{1}{2\theta_2} + \frac{1}{2\theta_2^2}[X-\theta_1]^2\right) & \text{for } (\theta_1,\theta_2) = (\mu,\sigma^2) \\ \left(\frac{1}{\theta_2^2}[X-\theta_1],\ -\frac{1}{\theta_2} + \frac{1}{\theta_2^3}[X-\theta_1]^2\right) & \text{for } (\theta_1,\theta_2) = (\mu,\sigma) \end{cases}.$$

Hence, the information matrix

$$\boldsymbol{I}(\theta) = \mathrm{E}_\theta\left[\dot{\ell}(X;\theta_1,\theta_2)\dot{\ell}^{\mathsf{T}}(X;\theta_1,\theta_2)\right]$$

is obtained as

$$\boldsymbol{I}(\theta) = \begin{bmatrix} \frac{1}{\theta_2} & 0 \\ 0 & \frac{1}{2\theta_2^2} \end{bmatrix} = \begin{bmatrix} \frac{1}{\sigma^2} & 0 \\ 0 & \frac{1}{2\sigma^4} \end{bmatrix} \text{ for } (\theta_1,\theta_2) = (\mu,\sigma^2), \text{ and}$$

$$\boldsymbol{I}(\theta) = \begin{bmatrix} \frac{1}{\theta_2^2} & 0 \\ 0 & \frac{2}{\theta_2^2} \end{bmatrix} = \begin{bmatrix} \frac{1}{\sigma^2} & 0 \\ 0 & \frac{2}{\sigma^2} \end{bmatrix} \text{ for } (\theta_1,\theta_2) = (\mu,\sigma),$$

with inverses

$$\boldsymbol{I}^{-1}(\theta) := \begin{bmatrix} I^{11} & I^{12} \\ I^{21} & I^{22} \end{bmatrix},$$

where

$$I^{22} = \begin{cases} 2\sigma^4 & \text{for } (\theta_1, \theta_2) = (\mu, \sigma^2) \\ \sigma^2/2 & \text{for } (\theta_1, \theta_2) = (\mu, \sigma) \end{cases}.$$

Thus the information lower bounds $n^{-1}I^{22}$ for variances of unbiased estimators of σ^2 and σ are $2\sigma^4/n$ and $\sigma^2/(2n)$, respectively.

We now compare these lower bounds with the variances of the UMVUEs obtained above:

$$\begin{aligned} \mathrm{Var}_\sigma[W] &= n^{-1}\mathrm{Var}_\sigma\left[(X_i - \mu)^2\right] \\ &= n^{-1}\left\{\mathrm{E}_\sigma\left[(X_i - \mu)^4\right] - \mathrm{E}_\sigma^2\left[(X_i - \mu)^2\right]\right\} \\ &= n^{-1}\left[3\sigma^4 - \sigma^4\right] = \frac{2\sigma^4}{n}, \end{aligned}$$

which equals the information lower bound for this case.

$$\begin{aligned} \mathrm{Var}_\sigma[T_0] &= c_n^2\mathrm{Var}_\sigma\left[\sqrt{W}\right] = c_n^2\left\{\mathrm{E}_\sigma[W] - \mathrm{E}_\sigma^2\left[\sqrt{W}\right]\right\} \\ &= c_n^2\left[\sigma^2 - \left(\frac{\sigma}{c_n}\right)^2\right] = \sigma^2(c_n^2 - 1), \end{aligned}$$

where $c_n = \sqrt{\frac{n}{2}}\frac{\Gamma(\frac{n}{2})}{\Gamma(\frac{n+1}{2})}$, while the corresponding lower bound is $\sigma^2/(2n)$.

Next, we have

$$\mathrm{Var}_\sigma\left[S^2\right] = \mathrm{Var}\left[(n-1)^{-1}\sigma^2\chi_{n-1}^2\right] = \frac{2\sigma^4}{n-1},$$

which is greater than the corresponding lower bound $\frac{2\sigma^4}{n}$. Finally,

$$\begin{aligned} \mathrm{Var}_\sigma[T_1] &= c_{n-1}^2\mathrm{Var}_\sigma[S] \\ &= c_{n-1}^2\left\{\mathrm{E}_\sigma\left[S^2\right] - \mathrm{E}_\sigma^2[S]\right\} \\ &= c_{n-1}^2\left\{\sigma^2 - \left(\frac{\sigma}{c_{n-1}}\right)^2\right\} = \sigma^2(c_{n-1}^2 - 1), \end{aligned}$$

while the corresponding lower bound is $\sigma^2/(2n)$.

Thus the variance of the UMVUE of σ^2 when μ is known (ie, $\mathrm{Var}_{\sigma^2}[W]$ attains the information lower bound), whereas the other three UMVUEs do not attain their corresponding lower bounds. In particular,

$$\frac{\mathrm{Var}_\sigma\left[S^2\right]}{2\sigma^4/n} = \frac{n}{n-1}, \quad \frac{\mathrm{Var}_\sigma\left[T_0\right]}{\sigma^2/(2n)} = 2n(c_n^2 - 1), \text{ and}$$
$$\frac{\mathrm{Var}_\sigma\left[T_1\right]}{\sigma^2/(2n)} = 2n(c_{n-1}^2 - 1).$$

However, all three of these ratios are nearly equal to 1 for moderately large n. For example, for $n = 20$,

$$\frac{n}{n-1} = 1.0526, \; 2n(c_n^2 - 1) = 1.0144, \text{ and } 2n(c_{n-1}^2 - 1) = 1.0630.$$

5.3 Equivariance

Like unbiasedness, equivariance is another property which narrows down the class of estimators to only those with constant risk functions, allowing us to look for the best among them.

We first briefly discuss equivariance in the general context of a decision problem specified by

$$(\mathcal{X}, \mathcal{B}, \{P_\theta, \theta \in \Theta\}), A, L$$

in the light of a group $\mathcal{G}$ of one-to-one transformations of $\mathcal{X} \overset{\text{onto}}{\to} \mathcal{X}$, the group operation being composition, that is,

$$g_2 \cdot g_1(x) = g_2\left[g_1(x)\right].$$

The identity transformation is $e(x) = x$ for all x, the inverse of g is the usual inverse transformation g^{-1} defined by $g^{-1}\left[g(x)\right] = x$ for all x. We assume that the family $\mathcal{P} = \{P_\theta, \theta \in \Theta\}$ is identifiable (ie, if $\theta \neq \theta'$, then $P_\theta \neq P_{\theta'}$).

The family of probabilities $\mathcal{P} = \{P_\theta, \theta \in \Theta\}$ is said to be *invariant under* $\mathcal{G}$ if for every $g \in \mathcal{G}$ and for every $\theta \in \Theta$, there exists $\theta' \in \Theta$ such that if $X \sim P_\theta$, then $g(X) \sim P_{\theta'}$. Since the family $\{P_\theta, \theta \in \Theta\}$ is identifiable, such θ' is unique and we denote it by $\bar{g}(\theta)$. Thus for any B,

$$P_\theta\left[g^{-1}(B)\right] = P_\theta\left[X \in g^{-1}(B)\right] = P_\theta\left[g(X) \in B\right] = P_{\bar{g}(\theta)}[B].$$

Let $\bar{\mathcal{G}} = \left\{\bar{g} \colon g \in \mathcal{G}\right\}$. Then $\bar{\mathcal{G}}$ is a group of transformations from $\Theta \to \Theta$, all $\bar{g} \in \bar{\mathcal{G}}$ being one-to-one and onto. [To check that $\bar{\mathcal{G}}$ is a group, verify (i) $\bar{g}_2 \cdot \bar{g}_1 = \overline{g_2 \cdot g_1}$, (ii) $\bar{e}$ is the identity element of $\bar{\mathcal{G}}$, and (iii) $\bar{g}^{-1} = \overline{g^{-1}}$.]

Definition 5.3.1. A statistical decision problem specified by

$$(\mathcal{X}, \mathcal{B}, \{P_\theta, \theta \in \Theta\}), \; A, \; L$$

is said to be invariant under a group $\mathcal{G}$ of transformations from $\mathcal{X} \to \mathcal{X}$ if

(i) $\mathcal{P} = \{P_\theta, \theta \in \Theta\}$ is invariant under $\mathcal{G}$ with the associated group of transformations $\bar{\mathcal{G}}$ from $\Theta \to \Theta$, and

(ii) for every $\bar{g} \in \bar{\mathcal{G}}$ and for every $a \in A$, there is a unique $a' \in A$ denoted by $a' = \tilde{g}(a)$ such that for all $\theta \in \Theta$

$$L(\theta, a) = L(\bar{g}(\theta), a') = L(\bar{g}(\theta), \tilde{g}(a)).$$

Remark 5.3.1. If for a given $\bar{g} \in \bar{\mathcal{G}}$ and $a \in A$, (ii) is satisfied by more than one a', then we can remove all such candidates except one and call this $a' = \tilde{g}(a)$, thereby achieving uniqueness of a' as specified in (ii).

Let $\tilde{\mathcal{G}} = \{\tilde{g}: g \in \mathcal{G}\}$. Then $\tilde{\mathcal{G}}$ is a group of transformations from $A \to A$. All $\tilde{g}$ are one-to-one and onto. [Verify that (i) $\tilde{g}_2 \cdot \tilde{g}_1 = \widetilde{g_2 \cdot g_1}$, (ii) $\tilde{e}$ is the identity transformation of $\tilde{\mathcal{G}}$, and (iii) $\tilde{g}^{-1} = \widetilde{g^{-1}}$.]

Definition 5.3.2. If a decision problem is invariant under $\mathcal{G}$, then

(i) a nonrandomized decision rule d is equivariant under $\mathcal{G}$ if for all $x \in \mathcal{X}$ and $g \in \mathcal{G}$, $d(g(x)) = \tilde{g}(d(x))$, and
(ii) a behavioral decision rule δ is equivariant under $\mathcal{G}$ if $\delta(S|g(x)) = \delta(\tilde{g}^{-1}(S)|x)$ for all "events" $S \subset A$, for all $x \in \mathcal{X}$ and $g \in \mathcal{G}$.

Definition 5.3.3.

(a) For $\theta_1, \theta_2 \in \Theta$, we say $\theta_1 \equiv \theta_2$ if there exists $\bar{g} \in \bar{\mathcal{G}}$ such that $\theta_2 = \bar{g}(\theta_1)$, $\equiv$ being an equivalence relation;
(b) let Θ^* denote the collection of equivalence classes of Θ; and
(c) the equivalence classes are called orbits, Θ^* being the set of all orbits.

For the rest of this section, we consider only nonrandomized decision rules.

Theorem 5.3.1. *For any (nonrandomized) equivariant decision rule d, the risk function is constant on each orbit.*

Proof. Note that

$$\begin{aligned} R(\theta, d) &= \mathrm{E}_\theta\left[L(\theta, d(X))\right] = \mathrm{E}_\theta\left[L(\bar{g}(\theta), \tilde{g}(d(X)))\right] \\ &= \mathrm{E}_\theta\left[L(\bar{g}(\theta), d(g(X)))\right] = \mathrm{E}_{\bar{g}(\theta)}\left[L(\bar{g}(\theta), d(Y))\right] = R(\bar{g}(\theta), d), \end{aligned}$$

using (i) $L(\theta, a) = L(\bar{g}(\theta), \tilde{g}(a))$, (ii) $\tilde{g}(d(x)) = d(g(x))$ since d is equivariant, and (iii) $X \sim P_\theta$ implies $Y = g(X) \sim P_{\bar{g}(\theta)}$. □

This theorem shows how equivariance simplifies the search for an optimal rule. If there is a single orbit, then there is an optimal equivariant rule, which is called the Minimum Risk Equivariant (MRE) estimator.

5.3.1 Location Equivariance in a Location Family

Suppose $\boldsymbol{X} = (X_1, \ldots, X_n)$ has pdf $f(\boldsymbol{x}; \theta) = f(x_1 - \theta, \ldots, x_n - \theta)$ where θ is the unknown location parameter to be estimated.

Let $\boldsymbol{Y} = g_c(\boldsymbol{X}) = (X_1 + c, \ldots, X_n + c) := (\boldsymbol{X} + c\mathbf{1})$ where $\mathbf{1}^\mathsf{T} = (1, \ldots, 1)$, and let $\bar{g}_c(\theta) = \theta + c$. Then $\boldsymbol{Y}$ has pdf

$$f([y_1 - c] - \theta, \ldots, [y_n - c] - \theta) = f(\boldsymbol{y} - [\theta + c]\mathbf{1}) = f(\boldsymbol{y}; \bar{g}_c(\theta)).$$

For $a \in \mathbb{R}$, taking $\tilde{g}_c(a) = a + c$, suppose

$$L(\tilde{g}_c(\theta), \tilde{g}_c(a)) = L(\theta + c, a + c) = L(\theta, a) \quad \text{for all } \theta,\ a,\ c.$$

Then with $c = -\theta$,

$$L(\theta, a) = L(0, a - \theta) := \rho(a - \theta).$$

Thus the problem is invariant under $\mathcal{G} = \{g_c, c \in \mathbb{R}\}$ and a location equivariant rule d must satisfy: $d(\boldsymbol{x} + c\mathbf{1}) = d(\boldsymbol{x}) + c$. All the usual estimators of a location parameter such as mean, median, etc., are location equivariant.

In this set-up, there is only one orbit in $\Theta = \mathbb{R}$ and therefore, the risk function is constant in θ for equivariant rules d satisfying $d(\boldsymbol{x} + c\mathbf{1}) = d(\boldsymbol{x}) + c$. Indeed, for such a rule,

$$\begin{aligned} R(\theta, d) &= \int L(\theta, d(\boldsymbol{x}))f(\boldsymbol{x} - \theta\mathbf{1})\, d\boldsymbol{x} \\ &= \int L(\theta, d(\boldsymbol{y} + \theta\mathbf{1}))f(\boldsymbol{y})\, d\boldsymbol{y} \\ &= \int L(\theta, d(\boldsymbol{y}) + \theta)f(\boldsymbol{y})\, d\boldsymbol{y} \\ &= \int L(0, d(\boldsymbol{y}))f(\boldsymbol{y})\, d\boldsymbol{y} = R(0, d). \end{aligned}$$

We now find the MRE estimator of θ. We first characterize a location equivariant estimator.

Lemma 5.3.1. *d is location equivariant iff $d(\boldsymbol{x}) = d_0(\boldsymbol{x}) + u(\boldsymbol{x})$ where d_0 is an arbitrary location equivariant estimator and $u(\boldsymbol{x} + c\mathbf{1}) = u(\boldsymbol{x})$ for all $\boldsymbol{x}$ and $c \in \mathbb{R}$.*

Proof. If d is of the above form, then

$$\begin{aligned} d(\boldsymbol{x} + c\mathbf{1}) &= d_0(\boldsymbol{x} + c\mathbf{1}) + u(\boldsymbol{x} + c\mathbf{1}) \\ &= \{d_0(\boldsymbol{x}) + c\} + u(\boldsymbol{x}) \\ &= \{d_0(\boldsymbol{x}) + u(\boldsymbol{x})\} + c = d(\boldsymbol{x}) + c \end{aligned}$$

showing that d is equivariant. Conversely, if d is equivariant, then taking $u(\boldsymbol{x}) = d(\boldsymbol{x}) - d_0(\boldsymbol{x})$, we have $d(\boldsymbol{x}) = d_0(\boldsymbol{x}) + u(\boldsymbol{x})$ and

$$\begin{aligned} u(\boldsymbol{x} + c\mathbf{1}) &= d(\boldsymbol{x} + c\mathbf{1}) - d_0(\boldsymbol{x} + c\mathbf{1}) \\ &= \{d(\boldsymbol{x}) + c\} - \{d_0(\boldsymbol{x}) + c\} \\ &= d(\boldsymbol{x}) - d_0(\boldsymbol{x}) = u(\boldsymbol{x}). \end{aligned}$$ □

To find the MRE estimator,

1. For $n = 1$, take $d_0(x) = x$ which is location equivariant, and

$$u(x + c) = u(x) \quad \text{for all } x \text{ and } c \in \mathbb{R}$$

implies $u(x) = \text{constant} := -b$. Hence, all location equivariant estimators are of the form: $d(x) = x - b := d_b(x)$. Now the MRE estimator is obtained by minimizing

$$R(0, d_b) = \mathrm{E}_0\left[\rho(X - b)\right]$$

with respect to b.

2. For $n \geq 2$, let $\boldsymbol{y} = (y_1, \ldots, y_{n-1}) = (x_1 - x_n, \ldots, x_{n-1} - x_n)$. Then $u(\boldsymbol{x} + c\mathbf{1}) = u(\boldsymbol{x})$ for all $\boldsymbol{x}$ and $c \in \mathbb{R}$ iff $u(\boldsymbol{x}) = v(\boldsymbol{y})$ for some function v.

Proof. $u(\boldsymbol{x}) = v(\boldsymbol{y})$ means $u(x_1, \ldots, x_n) = v(x_1 - x_n, \ldots, x_{n-1} - x_n)$. Therefore, $u(\boldsymbol{x} + c\mathbf{1}) = v(x_1 - x_n, \ldots, x_{n-1} - x_n) = u(\boldsymbol{x})$. Conversely, if $u(\boldsymbol{x} + c\mathbf{1}) = u(\boldsymbol{x})$, then with $c = -x_n$,

$$u(\boldsymbol{x}) = u(x_1 - x_n, \ldots, x_{n-1} - x_n, 0) = u(y_1, \ldots, y_{n-1}, 0) := v(\boldsymbol{y}). \qquad \square$$

Thus location equivariant estimators are of the form $d(\boldsymbol{x}) = d_0(\boldsymbol{x}) - v(\boldsymbol{y})$, so choose v so that

$$R(0, d) = \mathrm{E}_0\left[\rho(d_0(\boldsymbol{X}) - v(\boldsymbol{Y}))\right] = \mathrm{E}_0\mathrm{E}_0\left[\rho(d_0(\boldsymbol{X}) - v(\boldsymbol{Y}))|\boldsymbol{Y}\right]$$

is minimum, which is achieved by minimizing $\mathrm{E}_0\left[\rho(d_0(\boldsymbol{X}) - v(\boldsymbol{y}))|\boldsymbol{Y} = \boldsymbol{y}\right]$ for each $\boldsymbol{y}$. Therefore, to find the MRE estimator, choose $v(\boldsymbol{y}) = v^*(\boldsymbol{y})$ for each $\boldsymbol{y}$, so that

$$\mathrm{E}_0\left[\rho(d_0(\boldsymbol{X}) - v^*(\boldsymbol{y}))|\boldsymbol{Y} = \boldsymbol{y}\right] \leq \mathrm{E}_0\left[\rho(d_0(\boldsymbol{X}) - b)|\boldsymbol{Y} = \boldsymbol{y}\right] \quad \text{for all } b.$$

The above development is summarized in the following theorem.

Theorem 5.3.2. *For $n = 1$, the MRE estimator of a location parameter is given by $X - b^*$ where*

$$\mathrm{E}_0\left[\rho(X - b^*)\right] \leq \mathrm{E}_0\left[\rho(X - b)\right] \quad \textit{for all } b.$$

For $n \geq 2$, let $\boldsymbol{Y} = (Y_1, \ldots, Y_{n-1}) = (X_1 - X_n, \ldots, X_{n-1} - X_n)$ and let $d_0(\boldsymbol{X})$ be an arbitrary location equivariant estimator with finite risk. Then the MRE estimator is given by $d_0(\boldsymbol{X}) - v^*(\boldsymbol{Y})$ where for each $\boldsymbol{y}$,

$$\mathrm{E}_0\left[\rho(d_0(\boldsymbol{X}) - v^*(\boldsymbol{y}))|\boldsymbol{Y} = \boldsymbol{y}\right] \leq \mathrm{E}_0\left[\rho(d_0(\boldsymbol{X}) - b)|\boldsymbol{Y} = \boldsymbol{y}\right] \quad \textit{for all } b.$$

In particular, for squared-error loss $L(\theta, a) = (a - \theta)^2$, the MRE estimator is:

$$d^*(x) = x - \mathrm{E}_0[X] \text{ for } n = 1 \text{ and}$$
$$d^*(\boldsymbol{x}) = d_0(\boldsymbol{x}) - \mathrm{E}_0\left[d_0(\boldsymbol{X})|\boldsymbol{Y} = \boldsymbol{y}\right] \text{ for } n \geq 2,$$

because $\mathrm{E}_0\left[(X - b)^2\right]$ and $\mathrm{E}_0\left[\left\{d_0(\boldsymbol{X}) - b\right\}^2|\boldsymbol{Y} = \boldsymbol{y}\right]$ are minimized at $b = \mathrm{E}_0[X]$ and $b = \mathrm{E}_0\left[d_0(\boldsymbol{X})|\boldsymbol{Y} = \boldsymbol{y}\right]$, respectively.

Likewise, for absolute error loss $L(\theta, a) = |a - \theta|$, the MRE estimator is:

$$d^*(x) = x - \mathrm{median}_0[X] \text{ for } n = 1 \text{ and}$$
$$d^*(\boldsymbol{x}) = d_0(\boldsymbol{x}) - \mathrm{median}_0\left[d_0(\boldsymbol{X})|\boldsymbol{Y} = \boldsymbol{y}\right] \text{ for } n \geq 2.$$

Suppose $L(\theta, a) = \rho(a - \theta)$ where ρ is convex and even. Suppose further that f is symmetric about 0. Then for $n = 1$, the MRE estimator is $d^*(x) = x$. This is because

$$\begin{aligned} \mathrm{E}_0\big[\rho(X-b)\big] &= \int_{-\infty}^{0} \rho(x-b)f(x)\,dx + \int_0^\infty \rho(x-b)f(x)\,dx \\ &= \int_0^\infty \rho(-x-b)f(-x)\,dx + \int_0^\infty \rho(x-b)f(x)\,dx \\ &= \int_0^\infty \big[\rho(x+b) + \rho(x-b)\big]f(x)\,dx, \end{aligned}$$

so that

$$\mathrm{E}_0\big[\rho(X-b)\big] - \mathrm{E}_0[\rho(X)] = 2\int_0^\infty \left[\frac{1}{2}\{\rho(x+b)+\rho(x-b)\} - \rho(x)\right] f(x)\,dx \geq 0$$

for all b.

Example 5.3.1. Let $(X_1, \dots, X_n)$ be a random sample from $\mathcal{N}(\theta, \sigma^2)$ where σ^2 is known. Take $d_0(\boldsymbol{X}) = \bar{X}$ which is complete sufficient, while

$$\boldsymbol{Y} = (X_1 - X_n, \dots, X_{n-1} - X_n)$$

is an *ancillary statistic* (ie, its distribution does not depend on θ). By Basu's Theorem 5.1.6, it follows that $d_0(\boldsymbol{X}) = \bar{X}$ is independent of $\boldsymbol{Y}$. Hence, for each $\boldsymbol{y}$,

$$\mathrm{E}_0\big[\rho(d_0(\boldsymbol{X}) - b)|\boldsymbol{Y} = \boldsymbol{y}\big] = \mathrm{E}_0\big[\rho(\bar{X} - b)|\boldsymbol{Y} = \boldsymbol{y}\big] = \mathrm{E}_0\big[\rho(\bar{X} - b)\big].$$

For $\theta = 0$, the pdf of $\bar{X}$ is symmetric about 0, so that for all convex and even ρ, $\mathrm{E}_0\big[\rho(\bar{X} - b)\big]$ is minimized for $b = 0$. Thus the MRE estimator is $d^*(\boldsymbol{X}) = d_0(\boldsymbol{X}) - 0 = \bar{X}$.

Theorem 5.3.3. *Under squared-error loss, the MRE estimator*

$$d^*(\boldsymbol{x}) = d_0(\boldsymbol{x}) - \mathrm{E}_0\big[d_0(\boldsymbol{X})|\boldsymbol{Y} = \boldsymbol{y}\big]$$

can be expressed as

$$d^*(\boldsymbol{x}) = \frac{\int_{-\infty}^{\infty} u f(x_1 - u, \dots, x_n - u)\,du}{\int_{-\infty}^{\infty} f(x_1 - u, \dots, x_n - u)\,du}.$$

In this form it is known as the Pitman estimator of θ [19], which is also the generalized Bayes estimator with respect to the "improper prior having uniform distribution over the entire real line."

Proof. Take $d_0(\boldsymbol{X}) = X_n$ and compute $\mathrm{E}_0\big[d_0(\boldsymbol{X})|\boldsymbol{Y}\big] = \mathrm{E}_0[X_n|\boldsymbol{Y}]$ by augmenting the transformation $\boldsymbol{x} \to \boldsymbol{y}$ in the following manner:

$$y_1 = x_1 - x_n, \dots, y_{n-1} = x_{n-1} - x_n, \quad y_n = x_n.$$

The Jacobian of this transformation is 1 and the joint pdf of

$$\boldsymbol{Y}^* = (Y_1, \dots, Y_{n-1}, Y_n)$$

for $\theta = 0$ is

$$\begin{aligned} f_{\boldsymbol{Y}^*}(y_1, \dots, y_{n-1}, y_n) &= f_{\boldsymbol{X}}(y_1 + y_n, \dots, y_{n-1} + y_n, y_n) \\ &= f(y_1 + y_n, \dots, y_{n-1} + y_n, y_n). \end{aligned}$$

The conditional pdf of $Y_n = X_n$ given $\boldsymbol{Y} = (Y_1, \ldots, Y_{n-1})$ is

$$\frac{f_{\boldsymbol{Y}^*}(y_1, \ldots, y_n)}{f_{\boldsymbol{Y}}(y_1, \ldots, y_{n-1})} = \frac{f(y_1 + y_n, \ldots, y_{n-1} + y_n, y_n)}{\int_{-\infty}^{\infty} f(y_1 + t, \ldots, y_{n-1} + t, t)\, dt}.$$

Hence, for $\boldsymbol{y} = (x_1 - x_n, \ldots, x_{n-1} - x_n)$,

$$\begin{aligned}
\mathrm{E}_0\left[X_n | \boldsymbol{Y} = \boldsymbol{y}\right] &= \mathrm{E}_0\left[Y_n | \boldsymbol{Y} = \boldsymbol{y}\right] \\
&= \frac{\int_{-\infty}^{\infty} t f(y_1 + t, \ldots, y_{n-1} + t, t)\, dt}{\int_{-\infty}^{\infty} f(y_1 + t, \ldots, y_{n-1} + t, t)\, dt} \\
&= \frac{\int_{-\infty}^{\infty} t f(x_1 - x_n + t, \ldots, x_{n-1} - x_n + t, t)\, dt}{\int_{-\infty}^{\infty} f(x_1 - x_n + t, \ldots, x_{n-1} - x_n + t, t)\, dt}, \text{ with } x_n - t = u \\
&= \frac{\int_{-\infty}^{\infty} (x_n - u) f(x_1 - u, \ldots, x_{n-1} - u, x_n - u)\, du}{\int_{-\infty}^{\infty} f(x_1 - u, \ldots, x_{n-1} - u, x_n - u)\, du} \\
&= x_n - \frac{\int_{-\infty}^{\infty} u f(x_1 - u, \ldots, x_n - u)\, du}{\int_{-\infty}^{\infty} f(x_1 - u, \ldots, x_n - u)\, du}
\end{aligned}$$

showing that $d^*(\boldsymbol{x}) = x_n - \mathrm{E}_0\left[X_n | \boldsymbol{Y} = \boldsymbol{y}\right]$ has the desired form. □

5.3.2 Scale Equivariance in a Scale Family

Suppose $\boldsymbol{X} = (X_1, \ldots, X_n)$ has pdf $f(\boldsymbol{x}; \sigma) = \frac{1}{\sigma^n} f\left(\frac{x_1}{\sigma}, \ldots, \frac{x_n}{\sigma}\right)$ where $\sigma \in \mathbb{R}^+$ is the unknown scale parameter to be estimated.

Let $\boldsymbol{Y} = g_c(\boldsymbol{X}) = c\boldsymbol{X}$ and $\bar{g}_c(\sigma) = c\sigma$. Then $\boldsymbol{Y} = g_c(\boldsymbol{X})$ has pdf

$$\frac{1}{(c\sigma)^n} f\left(\frac{\boldsymbol{y}}{c\sigma}\right) = f(\boldsymbol{y}; \bar{g}_c(\sigma)),$$

that is, $\boldsymbol{Y} = g_c(\boldsymbol{X})$ has pdf $f(\boldsymbol{y}; \bar{g}_c(\sigma))$.

For $a \in A = \mathbb{R}^+$, taking $\tilde{g}_c(a) = ca$, suppose

$$L(\bar{g}_c(\sigma), \tilde{g}_c(a)) = L(c\sigma, ca) = L(\sigma, a) \quad \text{for all } \sigma,\ a,\ c.$$

Then with $c = 1/\sigma$, $L(\sigma, a) = L(1, a/\sigma) := \rho(a/\sigma)$.

Thus the problem is invariant under $\mathcal{G} = \left\{g_c, c \in \mathbb{R}^+\right\}$ and an equivariant rule d must satisfy: $d(c\boldsymbol{x}) = cd(\boldsymbol{x})$. As in the case of location family, here also there is only one orbit in $\Theta = \mathbb{R}^+$ and therefore, the risk function is constant for equivariant rules.

A rule d is scale equivariant iff $d(\boldsymbol{x}) = u(\boldsymbol{x}) d_0(\boldsymbol{x})$, where d_0 is an arbitrary scale equivariant estimator and $u(c\boldsymbol{x}) = u(\boldsymbol{x})$ for all $\boldsymbol{x}$ and $c \in \mathbb{R}^+$. The proof is similar to the location case.

We now find the MRE estimator of σ, proceeding as in the location case:

(i) For $n = 1$, take $d_0(x) = x$ which is scale equivariant. So any scale equivariant d must be of the form: $d(x) = u(x) d_0(x) = u(x)x$, where $u(cx) = u(x)$ for all x and $c \in \mathbb{R}^+$. Hence, $u(x) = b^{-1}$(constant) for all x. Thus all scale equivariant estimators are of the form $d_b(x) = x/b$. To obtain the MRE estimator, we therefore have to minimize the constant risk of d_b, namely,

$$R(1, d_b) = R(1, x/b) = \mathrm{E}_1\left[\rho(X/b)\right]$$

with respect to b.

(ii) For $n \geq 2$, let $\boldsymbol{y} = (y_1, \ldots, y_{n-1}) = (x_1/x_n, \ldots, x_{n-1}/x_n)$. Then

$$u(c\boldsymbol{x}) = u(\boldsymbol{x}) \quad \text{for all } \boldsymbol{x} \text{ and } c \in \mathbb{R}^+ \Leftrightarrow u(\boldsymbol{x}) = v(\boldsymbol{y}) \text{ for some } v,$$

the proof of which is analogous to the corresponding result in the location case. Thus the scale equivariant estimators for $n \geq 2$ are of the form: $d(\boldsymbol{x}) = d_0(\boldsymbol{x})/v(\boldsymbol{y})$. So choose v so that

$$R(1, d) = \mathrm{E}_1\left[\rho(d_0(\boldsymbol{X})/v(\boldsymbol{Y}))\right] = \mathrm{E}_1\mathrm{E}_1\left[\rho(d_0(\boldsymbol{X})/v(\boldsymbol{Y}))|\boldsymbol{Y}\right]$$

is minimized. To find the MRE estimator, choose $v^*(\boldsymbol{y})$ for each $\boldsymbol{y}$ so that

$$\mathrm{E}_1\left[\rho(d_0(\boldsymbol{X})/v^*(\boldsymbol{y}))|\boldsymbol{Y} = \boldsymbol{y}\right] \leq \mathrm{E}_1\left[\rho(d_0(\boldsymbol{X})/b)|\boldsymbol{Y} = \boldsymbol{y}\right] \quad \text{for all } b.$$

The above development is summarized in the following theorem.

Theorem 5.3.4. *For $n = 1$, the MRE estimator of a scale parameter is given by X/b^* where*

$$E_1\left[\rho(X/b^*)\right] \leq E_1\left[\rho(X/b)\right] \quad \textit{for all } b.$$

For $n \geq 2$, let $\boldsymbol{Y} = (Y_1, \ldots, Y_{n-1}) = (X_1/X_n, \ldots, X_{n-1}/X_n)$ and let $d_0(\boldsymbol{X})$ be an arbitrary scale equivariant estimator with finite risk. Then the MRE estimator is given by $d_0(\boldsymbol{X})/v^*(\boldsymbol{Y})$ where for each $\boldsymbol{y}$,

$$E_1\left[\rho(d_0(\boldsymbol{X})/v^*(\boldsymbol{y}))|\boldsymbol{Y} = \boldsymbol{y}\right] \leq E_1\left[\rho(d_0(\boldsymbol{X})/b)|\boldsymbol{Y} = \boldsymbol{y}\right] \quad \textit{for all } b.$$

5.4 Bayesian Estimation Using Conjugate Priors

(I) In Section 4.6.3 we have introduced the Bayes Principle in the general context of a statistical decision problem described by the triple $(\{P_\theta, \theta \in \Theta\}, A, L)$. According to this principle, we choose a decision rule d for which the risk $R(\theta, d)$ averaged with respect to a prior distribution τ over Θ is minimum.

(II) In Section 4.6.4 the general scheme of finding Bayes rules was outlined, which consists of choosing $d(x) = d_\tau(x)$ *for each observed data* x for which the posterior risk is minimized, that is

$$\int_\Theta L(\theta, d_\tau(x))g(\theta|x)\, d\theta \leq \int_\Theta L(\theta, a)g(\theta|x)\, d\theta$$

for all $a \subset A$, where

$$g(\theta|x) = f(x, \theta)\tau(\theta) \Big/ \int_\Theta f(x, u)\tau(u)\, du$$

is the posterior pdf of θ given x.

(III) The method described above is quite straight forward. However, the key step in the method is calculating the posterior distribution $g(\theta|x)$ given the data, from the prior $\tau(\theta)$ over Θ in a family of pdf's $\{f(x,\theta), \theta \in \Theta\}$, which may become very messy. This task becomes very simple in some cases by selecting $\tau(\theta)$ from a family $\{\tau(\theta,\alpha), \alpha \in \Omega\}$ which matches with the family $\{f(x,\theta), \theta \in \Theta\}$ in a manner described below.

Let $\mathcal{F} = \{f(x,\theta), \theta \in \Theta\}$ be a family of pdf's on $\mathfrak{X}$ and let $\boldsymbol{\tau} = \{\tau(\theta,\alpha), \alpha \in \Omega\}$ be a family of priors on Θ. Let

$$g_\alpha(\theta|x) = \frac{f(x,\theta)\tau(\theta,\alpha)}{\int_\Theta f(x,\theta')\tau(\theta',\alpha)\,d\theta'}$$

be the posterior pdf of θ given $X = x$ corresponding to the prior $\tau(\theta,\alpha)$. If there exists ϕ: $\Omega \times \mathfrak{X} \to \Omega$ such that $g_\alpha(\theta|x) = \tau(\theta,\phi(\alpha,x)) = \tau(\theta,\phi_x(\alpha))$, then $\boldsymbol{\tau}$ is said to be a family of conjugate priors for θ. The calculation of posterior from prior in such a situation can be described by the scheme:

$$\boxed{\text{Prior } \tau(\theta,\alpha)} + \boxed{\text{Data } x} = \boxed{\text{Posterior } \tau(\theta,\phi_x(\alpha))}.$$

Consider the special case of a k-parameter regular exponential family:

$$\mathcal{F} = \left\{ f(x,\theta) = \exp\left[\sum_{j=1}^{k} c_j(\theta)\sum_{i=1}^{n} T_j(x_i) + \sum_{i=1}^{n} S(x_i) + nd(\theta)\right] I_A(x), \theta \in \Theta \right\},$$

where $\Theta = \left\{\theta \in \mathbb{R}^k\text{: } \int_{\mathbb{R}^k} f(x,\theta)\,dx = 1\right\}$. Define

$$\omega(\alpha_1,\ldots,\alpha_k,\alpha_{k+1}) = \int_{\theta\in\Theta} \exp\left[\sum_{j=1}^{k}\alpha_j c_j(\theta) + \alpha_{k+1}d(\theta)\right] d\theta,$$

and $\Omega = \left\{(\alpha_1,\ldots,\alpha_k,\alpha_{k+1}) : \omega(\alpha_1,\ldots,\alpha_k,\alpha_{k+1}) < \infty\right\}$. Let

$$\begin{aligned}\tau(\theta;\alpha_1,\ldots,\alpha_{k+1}) &= \exp\left[\sum_{j=1}^{k}\alpha_j c_j(\theta) + \alpha_{k+1}d(\theta) - \log\omega(\alpha_1,\ldots,\alpha_k,\alpha_{k+1})\right] \\ &= \frac{\exp\left[\sum_{j=1}^{k}\alpha_j c_j(\theta) + \alpha_{k+1}d(\theta)\right]}{\omega(\alpha_1,\ldots,\alpha_k,\alpha_{k+1})},\end{aligned} \tag{3}$$

so that $\tau(\theta;\alpha_1,\ldots,\alpha_k,\alpha_{k+1})$ is a pdf for every $(\alpha_1,\ldots,\alpha_{k+1}) \in \Omega$. Then

$$\boldsymbol{\tau} = \left\{\tau(\theta;\alpha_1,\ldots,\alpha_{k+1}),\ (\alpha_1,\ldots,\alpha_k,\alpha_{k+1}) \in \Omega\right\}$$

is a family of conjugate priors for the exponential family $\mathcal{F}$. This is because

$$g_{\alpha_1,\ldots,\alpha_{k+1}}(\theta|x) = \tau(\theta,\phi_x(\alpha_1,\ldots,\alpha_{k+1})), \tag{3a}$$

where

$$\phi_x(\alpha_1,\ldots,\alpha_{k+1}) = \left(\alpha_1 + \sum_{i=1}^{n} T_1(x_i),\ldots,\alpha_k + \sum_{i=1}^{n} T_k(x_i), \alpha_{k+1} + n\right), \tag{3b}$$

as can be easily verified.

Example 5.4.1. Let $(X_1,\ldots,X_n)$ be a random sample from *Bernoulli*(θ). Then

$$\begin{aligned} f(\boldsymbol{x},\theta) &= \theta^{\sum_1^n x_i}(1-\theta)^{n-\sum_1^n x_i} I_A(\boldsymbol{x}) \\ &= \exp\left[\log\left(\frac{\theta}{1-\theta}\right)\sum_{i=1}^{n} x_i + n\log(1-\theta)\right] I_A(\boldsymbol{x}), \end{aligned}$$

where $A = \{0,1\}^n$ and $0 < \theta < 1$. From this we obtain

$$\begin{aligned} \omega(\alpha_1,\alpha_2) &= \int_0^1 \exp\left[\alpha_1\log\left(\frac{\theta}{1-\theta}\right) + \alpha_2\log(1-\theta)\right] d\theta \\ &= \int_0^1 \theta^{\alpha_1}(1-\theta)^{\alpha_2-\alpha_1}\, d\theta \\ &= \Gamma(\alpha_1+1)\Gamma(\alpha_2-\alpha_1+1)/\Gamma(\alpha_2+2), \end{aligned}$$

where $\alpha_1 + 1 > 0$ and $\alpha_2 - \alpha_1 + 1 > 0$. Thus we take

$$\begin{aligned} \tau(\theta;\alpha_1,\alpha_2) &= \frac{\Gamma(\alpha_2+2)}{\Gamma(\alpha_1+1)\Gamma(\alpha_2-\alpha_1+1)} \exp\left[\alpha_1\log\left(\frac{\theta}{1-\theta}\right) + \alpha_2\log(1-\theta)\right] \\ &= \frac{\Gamma(\alpha_2+2)}{\Gamma(\alpha_1+1)\Gamma(\alpha_2-\alpha_1+1)}\theta^{\alpha_1}(1-\theta)^{\alpha_2-1} \end{aligned}$$

on $0 < \theta < 1$, which is *Beta*$(\alpha_1+1, \alpha_2-\alpha_1+1) :=$ *Beta*(a,b) distribution, where $a = \alpha_1 + 1$ and $b = \alpha_2 - \alpha_1 + 1$. Then

$$\begin{aligned} g_{\alpha_1,\alpha_2}(\theta|\boldsymbol{x}) &= \tau\left(\theta;\ \alpha_1 + \sum_{i=1}^{n} x_i,\ \alpha_2 + n\right) \\ &= Beta\left(\alpha_1 + \sum_{i=1}^{n} x_i + 1,\ \alpha_2 + n - \alpha_1 - \sum_{i=1}^{n} x_i + 1\right) \\ &= Beta\left(a + \sum_{i=1}^{n} x_i,\ b + n - x_i\right). \end{aligned}$$

Hence, for the squared-error loss, the Bayes estimator with respect to the prior $\tau(\theta;\alpha_1,\alpha_2)$ is the mean of the distribution *Beta*$\left(a + \sum_{i=1}^n X_i,\ b + n - \sum_{i=1}^n X_i\right)$ which is

$$T = T(\boldsymbol{X}) = \frac{\sum_{i=1}^n X_i + a}{n + a + b}.$$

5.5 Methods of Estimation

We have so far discussed the Rao-Blackwell method of finding UMVUEs and two methods of estimation in the general decision theoretic framework following the Bayesian principle and the principle of equivariance. Three other methods, such as,

(i) the method of maximum likelihood;
(ii) the method of moments; and
(iii) the method of minimum χ^2,

will now be discussed, which are applicable in the context of parametric families.

The method of maximum likelihood was introduced by Fisher [20] who also laid the foundation of the theory of estimation and demonstrated the superiority of this method over the method of moments which was widely used for a long time until then. The method of minimum χ^2 has limited use only in the context of multinomial data to estimate the parameters on which the cell probabilities depend.

5.5.1 The Method of Maximum Likelihood

For a random sample $\boldsymbol{X} = (X_1, \ldots, X_n)$ from a population with pdf/pmf $f(x; \theta_1, \ldots, \theta_k)$, the joint pdf/pmf of the data is:

$$f(\boldsymbol{x}; \theta_1, \ldots, \theta_k) = \prod_{i=1}^{n} f(x_i; \theta_1, \ldots, \theta_k),$$

treating $\boldsymbol{x}$ as variable and $\boldsymbol{\theta} = (\theta_1, \ldots, \theta_k)$ as fixed. We now treat $\boldsymbol{\theta}$ as the variable, since we are searching among all possible $\boldsymbol{\theta} \in \Theta$, and the data $\boldsymbol{x}$ as fixed, because that is what we have in our search for $\boldsymbol{\theta}$. This defines the *likelihood function* of $\boldsymbol{\theta}$ for the data $\boldsymbol{x}$ as:

$$L(\boldsymbol{\theta}|\boldsymbol{x}) = L(\theta_1, \ldots, \theta_k | x_1, \ldots, x_n) = \prod_{i=1}^{n} f(x_i; \theta_1, \ldots, \theta_k).$$

Definition 5.5.1. The maximum likelihood estimate $\hat{\boldsymbol{\theta}}(\boldsymbol{x})$ for a given data $\boldsymbol{x}$ is defined as the value of $\boldsymbol{\theta} \in \Theta$ at which $L(\boldsymbol{\theta}|\boldsymbol{x})$ attains its maximum, that is

$$\hat{\boldsymbol{\theta}}(\boldsymbol{x}) = \arg\max_{\boldsymbol{\theta} \in \Theta} L(\boldsymbol{\theta}|\boldsymbol{x}) \quad \text{for each } \boldsymbol{x}.$$

The maximum likelihood estimator (MLE) of $\boldsymbol{\theta}$ for a random sample $\boldsymbol{X}$ is $\hat{\boldsymbol{\theta}}(\boldsymbol{X})$.

The maximization of the likelihood function $L(\boldsymbol{\theta}|\boldsymbol{x})$ with respect to $\boldsymbol{\theta} \in \Theta$ for a given $\boldsymbol{x}$ is quite straightforward in many situations. Most often, this maximization is achieved by solving the equation:

$$\frac{\partial}{\partial \theta_i} L(\boldsymbol{\theta}|\boldsymbol{x}) = 0, \quad i = 1, \ldots, k$$

and making sure that the extremum is indeed the maximum. In many situations arising in the context of exponential families, it is easier to work with $\log L(\boldsymbol{\theta}|\boldsymbol{x})$, called the log likelihood.

However, various complications may be encountered in the process of maximization of $L(\boldsymbol{\theta}|\boldsymbol{x})$ or $\log L(\boldsymbol{\theta}|\boldsymbol{x})$ such as:

(i) $\frac{\partial}{\partial\theta_i}\log L(\boldsymbol{\theta}|\boldsymbol{x}) = 0$, $i = 1, \ldots, k$ may have several solutions, in which case we have to search for the global maximum among these solutions,

(ii) the parameter space Θ may be restricted and the solution of the likelihood equations may fall outside the restricted part of Θ , in which case we have to modify such a solution appropriately, and

(iii) the equations $\frac{\partial}{\partial\theta_i}\log L(\boldsymbol{\theta}|\boldsymbol{x}) = 0$, $i = 1, \ldots, k$ may not have a closed form solution, so the equation may have to be solved iteratively.

Complications like (i) and (ii) will be dealt with in some examples to be discussed in the sequel and (iii) will be discussed in Chapter 7.

The MLEs have an important property known as *invariance*. This means that if $\eta = g(\theta)$, then the MLE of η based on x is $\hat{\eta}(x) = g(\hat{\theta}(x))$ where $\hat{\theta}(x)$ is the MLE of θ based on x. The proof of this fact is straightforward if $g\colon \Theta \to \mathcal{H}$ is one-to-one, but it needs a little more care if g is not one-to-one.

For given data x, the likelihood function of θ is $L(\theta|x) = f(x, \theta)$. When g is one-to-one and onto, then $\theta = g^{-1}(\eta)$ is well defined for all $\eta \in \mathcal{H}$ and we define $L^*(\eta|x) = L(g^{-1}(\eta)|x)$. In general, when g is onto but not one-to-one, we define $L^*(\eta|x) = \sup_{\{\theta\colon g(\theta)=\eta\}} L(\theta|x)$.

The proof of $\hat{\eta}(x) = g(\hat{\theta}(x))$ is now given for the two cases: g is one-to-one or not.

(i) *g is one-to-one and onto.* If $\hat{\eta}(x)$ is the MLE of η based on x, then

$$L^*(\hat{\eta}(x)|x) = \sup_{\eta\in\mathcal{H}} L^*(\eta|x) \iff L(g^{-1}(\hat{\eta}(x))|x) = \sup_{\eta\in\mathcal{H}} L(g^{-1}(\eta)|x),$$

$$\text{ie, } L(g^{-1}(\hat{\eta}(x))|x) = \sup_{\theta\in\Theta} L(\theta|x).$$

But then $g^{-1}(\hat{\eta}(x)) = \hat{\theta}(x)$, and so $\hat{\eta}(x) = g(\hat{\theta}(x))$.

(ii) *g is onto but not one-to-one.* In this case, if $\hat{\eta}(x)$ is the MLE of η based on x, then

$$\sup_{\{\theta\colon g(\theta)=\hat{\eta}(x)\}} L(\theta|x) = \sup_{\eta\in\mathcal{H}} \sup_{\{\theta\colon g(\theta)=\eta\}} L(\theta|x) = \sup_{\theta\in\Theta} L(\theta|x),$$

since $\bigcup_{\eta\in\mathcal{H}}\{\theta\colon g(\theta) = \eta\} = \Theta$. Hence $\hat{\theta}(x) \in \{\theta\colon g(\theta) = \hat{\eta}(x)\}$, and so $g(\hat{\theta}(x)) = \hat{\eta}(x)$.

The asymptotic properties of the MLEs will be discussed in Chapter 7 along with those of the Likelihood Ratio Tests motivated by the likelihood principle in the context of Hypothesis Testing.

The MLEs are "asymptotically efficient" on one hand, but in finite samples, they may be excessively influenced by a few "outliers." Methods which modify the MLEs against such weakness will be discussed in Chapter 10.

Here we shall only present a heuristic argument to justify the method of maximum likelihood and in the single-parameter context point out a key feature in the model $\{f(\boldsymbol{x}, \theta),\ \theta \in \mathbb{R}\}$ which determines how well the MLE performs.

To emphasize what happens when the sample size $n \to \infty$, we let the log likelihood $\log L$ be subscripted by n and averaged over n to define

$$L_n^*(\theta) = n^{-1} \log L_n(\theta|X_1, \dots, X_n) = n^{-1} \sum_{i=1}^{n} \log f(X_i, \theta),$$

which is maximized with respect to θ.

Let θ_0 denote the true value of θ and suppose that $\mathrm{E}_{\theta_0}[\log f(X, \theta)]$ is finite for all θ in a neighborhood of θ_0. Then as $n \to \infty$,

$$L_n^*(\theta) = n^{-1} \sum_{i=1}^{n} \log f(X_i, \theta) \xrightarrow{p} \mathrm{E}_{\theta_0}[\log f(X, \theta)] := L^*(\theta)$$

at each θ. The MLE $\hat{\theta}_n$ at which $L_n^*(\theta)$ attains its maximum should, therefore, converge in probability to the value of θ at which $L^*(\theta)$ attains its maximum.

Assume that the family $\{f(x, \theta), \theta \in \Theta\}$ satisfies the *identifiability condition* which requires that for $\theta' \neq \theta$, $f(\cdot, \theta')$ is essentially different from $f(\cdot, \theta)$. Then for $\theta \neq \theta_0$, due to strict concavity of the function $\log(\cdot)$, it follows from Jensen's inequality that

$$\begin{aligned}
&\mathrm{E}_{\theta_0}[\log f(X, \theta) - \log f(X, \theta_0)] \\
&= \mathrm{E}_{\theta_0}\left[\log \frac{f(X, \theta)}{f(X, \theta_0)}\right] < \log \mathrm{E}_{\theta_0}\left[\frac{f(X, \theta)}{f(X, \theta_0)}\right] \\
&= \log \int_{\{x:\, f(x, \theta_0) > 0\}} [f(x, \theta)/f(x, \theta_0)] f(x, \theta_0)\, dx \\
&= \log \int_{\{x:\, f(x, \theta_0) > 0\}} f(x, \theta)\, dx \leq \log 1 = 0.
\end{aligned}$$

Thus for $\theta \neq \theta_0$, $L^*(\theta) = \mathrm{E}_{\theta_0}[\log f(X, \theta)] < \mathrm{E}_{\theta_0}[\log f(X, \theta_0)] = L^*(\theta_0)$ (ie, the function $L^*(\theta)$ *has a unique maximum at* θ_0).

Now by the WLLN, the graph of $L_n^*(\theta)$ converges to the graph of $L^*(\theta)$ at each θ with probability tending to 1 as $n \to \infty$. So it seems reasonable that the peaks of the two graphs should approach one another as $n \to \infty$ (ie, $\hat{\theta}_n \to \theta_0$ in probability).

Now for $\Theta = \mathbb{R}$, suppose that the graph of $L^*(\theta)$ has a large curvature at $\theta = \theta_0$. Then it falls off sharply as θ moves away from its peak at θ_0. This increases the tendency of the peak of $L_n^*(\theta)$ to stay close to the peak of $L^*(\theta)$. The geometry of the graph of $L^*(\theta)$ and the convergence property of $L_n^*(\theta)$ together suggest that

1. $\hat{\theta}_n \xrightarrow{P} \theta_0$ as $n \to \infty$, and
2. the precision of $\hat{\theta}_n$ in estimating θ_0 increases with the curvature of $L^*(\theta)$ at $\theta = \theta_0$.

The curvature of $L^*(\theta)$ at $\theta = \theta_0$ is the Fisher-information $I(\theta_0)$ of the family $\{f(x,\theta), \theta \in \mathbb{R}\}$ at θ_0 (defined in Eq. (2)):

$$\begin{aligned}
-\frac{d^2}{d\theta^2} L^*(\theta)\bigg|_{\theta=\theta_0} &= -\frac{d^2}{d\theta^2}\int \{\log f(x,\theta)\} f(x,\theta_0)\,dx\bigg|_{\theta=\theta_0} \\
&= -\int \left(\frac{\partial^2 \log f(x,\theta_0)}{\partial\theta^2}\right) f(x,\theta_0)\,dx \\
&= -\mathrm{E}_{\theta_0}\left[\frac{\partial^2 \log f(X,\theta_0)}{\partial\theta^2}\right] \\
&= \mathrm{E}_{\theta_0}\left[\left(\frac{\partial \log f(X,\theta_0)}{\partial\theta}\right)^2\right] = I(\theta_0)
\end{aligned}$$

under regularity conditions. Note that the Cramér-Rao lower bound is also determined by this curvature, which is an intrinsic measure of how well θ_0 can be estimated.

We now pursue this heuristic argument to examine the asymptotic distribution of $\sqrt{n}(\hat{\theta}_n - \theta_0)$. Assume that the MLE $\hat{\theta}_n$ satisfies

$$n^{-1}\sum_{i=1}^{n} \frac{\partial \log f(X_i, \hat{\theta}_n)}{\partial\theta} = 0, \tag{4}$$

and expand the left-hand side of this equation around θ_0 to see that

$$0 = n^{-1}\sum_{i=1}^{n} \frac{\partial \log f(X_i,\theta_0)}{\partial\theta} + (\hat{\theta}_n - \theta_0)n^{-1}\sum_{i=1}^{n} \frac{\partial^2 \log f\left(X_i, \hat{\hat{\theta}}_n\right)}{\partial\theta^2}$$

with $\hat{\hat{\theta}}_n$ lying between $\hat{\theta}_n$ and θ_0. Thus

$$\sqrt{n}(\hat{\theta}_n - \theta_0) = \frac{n^{-1/2}\sum_{i=1}^{n} \frac{\partial \log f(X_i,\theta_0)}{\partial\theta}}{n^{-1}\sum_{i=1}^{n} -\frac{\partial^2 \log f\left(X_i,\hat{\hat{\theta}}_n\right)}{\partial\theta^2}},$$

of which the numerator $\xrightarrow{\mathcal{L}} \mathcal{N}(0, I(\theta_0))$ by the CLT and the denominator $\xrightarrow{P} I(\theta_0)$ by the WLLN. Hence, $\sqrt{n}(\hat{\theta}_n - \theta_0) \xrightarrow{\mathcal{L}} N(0, 1/I(\theta_0))$ by Slutsky's Theorem.

Remark 5.5.1. The condition that the MLE $\hat{\theta}_n$ satisfies Eq. (4) is essential for the asymptotic normality of $\sqrt{n}(\hat{\theta}_n - \theta_0)$. To see what happens otherwise, let $\boldsymbol{X} = (X_1, \ldots, X_n)$ be a random sample from *Uniform*$(0,\theta)$. Then the MLE $\hat{\theta}_n$ of θ is $X_{n:n} = \max(X_1, \ldots, X_n)$ and $P_{\theta_0}\left[\hat{\theta}_n \leq t\right] = (t/\theta_0)^n$ for $0 < t < \theta_0$ and $= 1$ for $t \geq \theta_0$.

In Chapter 7, the results indicated above by heuristic arguments will be proved more systematically.

5.5.2 The Method of Moments

In a random sample $\boldsymbol{X} = (X_1, \ldots, X_n)$ from a population with pdf/pmf $f(x; \theta_1, \ldots, \theta_k)$, let

$$\mu_j(\theta_1, \ldots, \theta_k) = \mathrm{E}_{\theta_1,\ldots,\theta_k}\left[X_i^j\right] \text{ and } m_{nj} = n^{-1}\sum_{i=1}^{n} X_i^j, \quad j = 1, \ldots, k$$

denote, respectively, the jth population and sample moments. The method of moments estimators (MOME) of $\theta_1, \ldots, \theta_k$ are obtained by equating each of the first k population moments to the corresponding sample moments and solving the system of equations:

$$\mu_j(\theta_1, \ldots, \theta_k) = m_{nj}, \quad j = 1, \ldots, k$$

for $\theta_1, \ldots, \theta_k$. Suppose these equations have unique solutions and let

$$\hat{\theta}_{nj} = g_j(m_{n1}, \ldots, m_{nk}), \quad j = 1, \ldots, k$$

denote the solutions of these equations. If the population distribution has finite $2k$ moments, then $\sqrt{n}(m_{nj} - \mu_j)$, $j = 1, \ldots, k$ will be asymptotically jointly normal and consequently, $\sqrt{n}(\hat{\theta}_{nj} - \theta_j)$, $j = 1, \ldots, k$ will also be asymptotically jointly normal, provided that the functions $g_1, \ldots, g_k$ are sufficiently smooth.

Example 5.5.1. $X_1, \ldots, X_n$ are iid *Bernoulli*(θ).
Here $k = 1$, $\mu(\theta) = \theta$ and $m_n = n^{-1}\sum_1^n X_i = \bar{X}_n$. So we equate $\mu(\theta) = m_n$ to obtain $\hat{\theta}_n = \bar{X}_n$ which is the same as the MLE.

Example 5.5.2. $X_1, \ldots, X_n$ are iid *Uniform*(θ).
Here also $k = 1$, $\mu(\theta) = \theta/2$ and $m_n = \bar{X}_n$; so we equate $\mu(\theta) = m_n$ to obtain $\hat{\theta}_n = 2\bar{X}_n$. Since $\mathrm{E}_\theta[X_i] = \theta/2$ and $\mathrm{Var}_\theta[X_i] = \theta^2/12$, we have $\mathrm{E}_\theta\left[2\bar{X}_n\right] = \theta$ and $\mathrm{Var}_\theta\left[2\bar{X}_n\right] = \theta^2/3$. Hence, $\sqrt{n}(\hat{\theta}_n - \theta) \xrightarrow{\mathcal{L}} \mathcal{N}(0, \theta^2/3)$ which is quite different from the behavior of the MLE. It is left as an exercise to compare the rates at which $|\hat{\theta}_n - \theta| < \epsilon$ converge to 1 for the MOME and the MLE and to note that the convergence is much faster for the MLE.

Remark 5.5.2. Two major drawbacks of the method of moments are:

1. The MOMEs are less efficient than the MLEs in many situations as illustrated in Example 5.5.2.
2. The method is inapplicable if the required number of moments do not exist, as in the case of estimating the median of *Cauchy*(θ).

5.5.3 The Method of Minimum χ^2

Consider a multinomial distribution in m classes with probability $\pi_j(\boldsymbol{\theta})$, $1 \le j \le m$ for the jth class where $\pi_1(\cdot), \ldots, \pi_m(\cdot)$ are known functions of an unknown k-dimensional parameter vector $\boldsymbol{\theta}$, satisfying $\pi_j(\boldsymbol{\theta}) > 0$ for all j and $\sum_{j=1}^m \pi_j(\boldsymbol{\theta}) = 1$. Let $n_1, \ldots, n_m$ denote the observed frequencies in the m classes in a random sample of size n.

In order to put this in a framework suitable for pursuing maximum likelihood estimation of $\boldsymbol{\theta}$, let e_j denote the m-dimensional vector with 1 for the jth coordinate and 0 for all other coordinates. Then $X_i = e_j$ means that in the multinomial sampling, the ith observation is in the jth class. The random vectors $X_1, \ldots, X_n$ are iid with probabilities $P_{\boldsymbol\theta}[X_i = e_j] = f(e_j, \boldsymbol\theta) = \pi_j(\boldsymbol\theta)$, that is, the pmf of X_i is

$$f(x, \boldsymbol\theta) = \begin{cases} \pi_j(\boldsymbol\theta) & \text{if } x = e_j, \quad j = 1, \ldots, m \\ 0 & \text{otherwise} \end{cases}$$

We now have for $x = e_j$,

$$\frac{\partial \log f(x, \boldsymbol\theta)}{\partial \theta_r} = \frac{\partial \log \pi_j(\boldsymbol\theta)}{\partial \theta_r} = \frac{1}{\pi_j(\boldsymbol\theta)} \frac{\partial \pi_j(\boldsymbol\theta)}{\partial \theta_r},$$

and

$$\begin{aligned} S_{nr}(\mathbf{t}) &= \sum_{i=1}^{n} \frac{\partial \log f(X_i, \mathbf{t})}{\partial \theta_r} = \sum_{i=1}^{n} \sum_{j=1}^{m} I\left[X_i = e_j\right] \frac{\partial \log \pi_j(\mathbf{t})}{\partial \theta_r} \\ &= \sum_{j=1}^{m} \left\{ \sum_{i=1}^{n} I\left[X_i = e_j\right] \right\} \frac{1}{\pi_j(\mathbf{t})} \frac{\partial \pi_j(\mathbf{t})}{\partial \theta_r} = \sum_{j=1}^{m} \frac{n_j}{\pi_j(\mathbf{t})} \frac{\partial \pi_j(\mathbf{t})}{\partial \theta_r}, \end{aligned}$$

since $\sum_{i=1}^{n} I[X_i = e_j] = \{\text{number of observations } X_i = e_j\} = n_j$.

The MLEs of $\theta_1, \ldots, \theta_k$ are obtained by solving the likelihood equations: $S_{nr}(\hat{\boldsymbol\theta}_n) = 0$, $1 \le r \le k$, that is

$$\sum_{j=1}^{m} \frac{n_j}{\pi_j(\hat\theta_n)} \frac{\partial \pi_j(\hat\theta_n)}{\partial \theta_r} = 0, \quad 1 \le r \le k.$$

These equations do not have a closed form solution in a typical problem, so they have to be solved iteratively. This issue will be discussed in Chapter 7.

For frequency data described above, one can try to estimate $\boldsymbol\theta$ by minimizing a measure of discrepancy between $(n\pi_1(\theta_n^*), \ldots, n\pi_m(\theta_n^*))$ which are the *expected frequencies* for $(\pi_1(\theta_n^*), \ldots, \pi_m(\theta_n^*))$ and the *observed frequencies* $(n_1, \ldots, n_m)$ given by

$$\chi^2(\theta_n^*) = \sum_{j=1}^{m} \frac{(n_j - n\pi_j(\theta_n^*))^2}{n\pi_j(\theta_n^*)} = \sum_{j=1}^{m} \frac{n_j^2}{n\pi_j(\theta_n^*)} - n$$

with respect to θ_n^*. Such a procedure is known as the *method of minimum* χ^2. The minimizing equations for this method are

$$\sum_{j=1}^{m} \frac{n_j^2}{\pi_j^2(\theta_n^*)} \frac{\partial \pi_j(\boldsymbol\theta_n^*)}{\partial \theta_r} = 0, \quad 1 \le r \le k,$$

whereas the likelihood equations are

$$\sum_{j=1}^{m} \frac{n_j}{\pi_j(\hat\theta_n)} \frac{\partial \pi_j(\dot{\boldsymbol\theta}_n)}{\partial \theta_r} = 0, \quad 1 \le r \le k.$$

The two sets of equations can be shown to be asymptotically equivalent so θ_n^* has the same asymptotic distribution as that of $\hat{\boldsymbol{\theta}}_n$. Thus the estimator θ_n^* obtained by the method of minimum χ^2 is asymptotically efficient.

However, for moderate sample sizes, the MLE $\hat{\boldsymbol{\theta}}_n$ and the minimum χ^2 estimator θ_n^* have been known to perform quite differently. The asymptotics being discussed here do not take into account what happens in respect of terms which are of order of magnitude less than $1/\sqrt{n}$ in probability. What happens beyond this order of magnitude can cause differences in the precision of two estimates which are both "asymptotically efficient" as measured by "first-order efficiency." These concerns have led to several approaches to develop a measure of "second-order efficiency" [21].

Exercises

For all the problems below, when there are observations $X_1, \ldots, X_n$, it is understood that

$$\bar{X}_n = n^{-1}\sum_{i=1}^{n} X_i,\ s_n^2 = (n-1)^{-1}\sum_{i=1}^{n}(X_i - \bar{X}_n)^2,$$
$$X_{n:1} = \min(X_1, \ldots, X_n),\ X_{n:n} = \max(X_1, \ldots, X_n).$$

5.1. Let $X_1, \ldots, X_n$ be a random sample from a population with pdf

$$f(x, \theta) = \theta \exp(-\theta x), \quad x > 0, \ \ \theta > 0.$$

Find the UMVUE of $\gamma(\theta) = \exp(-\theta) = P_\theta[X_i > 1]$.

5.2. Let $X_1, \ldots, X_n$ be a random sample from

$$f(x, \theta) = \exp(-(x - \theta)), \quad x > \theta.$$

(a) Show that $T = \min(X_1, \ldots, X_n)$ is a complete sufficient statistic for θ. [For completeness, you may restrict attention to continuous functions ϕ with $\mathrm{E}_\theta[\phi(T)] = 0$ for all θ.]

(b) Calculate $\mathrm{E}_\theta[T]$ and find a UMVUE of θ.

5.3. Let $X_1, \ldots, X_n$ be a random sample from $N(\mu, \sigma^2)$. Find the UMVUE of σ.

5.4. Let $X_1, \ldots, X_n$ be a random sample from $\mathit{Unif}(\theta_1, \theta_2)$. We want to estimate the mean $\gamma = (\theta_1 + \theta_2)/2$.

(a) Show that $(X_{n:1}, X_{n:n})$ is complete and sufficient for (θ_1, θ_2), and $\bar{X}_n$ is an unbiased estimate of γ.

(b) Find the UMVUE of γ.

5.5. Let $X_1, \ldots, X_n$ be a random sample from $N(\theta, 1)$. Find the UMVUE of $P_\theta[X \geq 0] = \Phi(\theta)$, where Φ is the cdf of the standard normal distribution. [*Hint*: $I_{[0,\infty)}(X_1)$ is unbiased for $\Phi(\theta)$ and $(X_1, \bar{X}_n)$ is bivariate normal.]

5.6. Let $X \sim \mathit{Bin}(n, \theta)$. Show that $X(n - X)/\{n(n-1)\}$ is UMVUE of $\gamma = \theta(1 - \theta)$.

5.7. Suppose that T_1 and T_2 are two UMVUEs of $g(\theta)$ with finite variance. Show that $T_1 = T_2$. [*Hint*: $(T_1 + T_2)/2$ is also unbiased; use the correlation inequality.]

5.8. Show that for an exponential family (in natural form)

$$g(x,\eta) = \exp(\eta T(x) + d_0(\eta) + S(x))I_A(x),$$

the Fisher-information is $I_g(\eta) = -d_0''(\eta)$.

5.9. Consider the exponential family

$$f(x,\theta) = \exp(C(\theta)T(x) + d(\theta) + S(x))I_A(x),$$

where C is one-to-one and twice differentiable. Then we can reparametrize $f(x,\theta)$ by letting $\eta = C(\theta)$ and rewrite $f(x,\theta)$ in the form $g(x,\eta)$ given in Exercise 5.8 above. Show that the Fisher-information in the family $\{f(x,\theta)\}$ is

$$I_f(\theta) = -d_0''(C(\theta)) = -d_0''(\eta)|_{\eta=C(\theta)}\{C'(\theta)\}^2 = I_g(C(\theta))\{C'(\theta)\}^2.$$

5.10. Suppose that $X_1,\ldots,X_n$ is a random sample from the pdf

$$f(x,\theta) = cx^{c-1}\theta\exp(-\theta x^c), \quad x > 0, \quad c > 0 \text{ (known)}, \quad \theta > 0 \text{ (unknown)}.$$

Show that $T = n^{-1}\sum_{i=1}^n X_i^c$ is the UMVUE of $1/\theta$.

5.11. Let X be a random sample from the Cauchy distribution with pdf

$$f(x,\theta) = \frac{1}{\pi\{1+(x-\theta)^2\}}, \quad -\infty < x < \infty.$$

We want to estimate the median of θ under the loss function $L(\theta,a) = \rho(a-\theta)$, where $\rho(t) = I_{(c,\infty)}(|t|)$. Find the MRE estimator of θ based on X.

5.12. Let $X_1,\ldots,X_n$ be a random sample from $Unif(\theta-1/2,\theta+1/2)$. Find the MRE estimator of θ under the loss function $L(\theta,a) = |a-\theta|$.

5.13. Let $X_1,\ldots,X_n$ be a random sample from the half-normal distribution with pdf

$$f(x,\theta) = \sqrt{2/\pi}\exp\left[-(x-\theta)^2/2\right]I_{(0,\infty)}(x).$$

Show that the Pitman estimator of θ is

$$d(\mathbf{x}) = \bar{X}_n - \frac{\exp\left[-n(X_{n:1}-\bar{X}_n)^2/2\right]}{\sqrt{2n\pi}\,\Phi(\sqrt{n}(X_{n:1}-\bar{X}_n)^2)},$$

where Φ is the cdf of $N(0,1)$.

5.14. Let $(X_1,\ldots,X_n)$ have a joint distribution with pdf

$$f(x_1,\ldots,x_n;\theta) = \theta^{-n}g(x_1/\theta,\ldots,x_n/\theta)$$

for some function g which vanishes unless all coordinates are positive, and $\theta > 0$ is an unknown scale parameter. Suppose that the loss function is $L(\theta,a) = \rho(a/\theta)$.

(a) Find the MRE estimator of θ analogous to the MRE estimator in the location problem. [Take $\boldsymbol{Y} = (X_1/X_n,\ldots,X_{n-1}/X_n)$.]

(b) For the loss function $L(\theta, a) = (a/\theta - 1)^2$, show that the MRE estimator is the following analog of the Pitman estimator:

$$d(\boldsymbol{x}) = \frac{\int_0^\infty \theta^{-n-2} f(x_1/\theta, \ldots, x_n/\theta)\, d\theta}{\int_0^\infty \theta^{-n-3} f(x_1/\theta, \ldots, x_n/\theta)\, d\theta}.$$

5.15. **(a)** Express the family of beta distributions $Be(\theta_1, \theta_2)$ as a two-parameter exponential family and find a family of conjugate priors for this family.

(b) Find the prior-to-posterior formula for these conjugate priors. [Use Eqs. (3), (3a), and (3b)]

5.16. Consider the joint pdf $f(\boldsymbol{x}, \boldsymbol{\theta})$, of a random sample $X_1, \ldots, X_n$ from $N(\theta_1, 1/\theta_2)$, where $\boldsymbol{\theta} = (\theta_1, \theta_2)$.

(a) Express $\mathcal{F}_1 = \{f(\boldsymbol{x}, \boldsymbol{\theta})\colon -\infty < \theta_1 < \infty, \theta_2 = 1\}$ as a one-parameter exponential family and find a family of conjugate priors for $\mathcal{F}_1$.

(b) Express $\mathcal{F}_0 = \{f(\boldsymbol{x}, \boldsymbol{\theta})\colon \theta_1 = 0, 0 < \theta_2 < \infty\}$ as a one-parameter exponential family and find a family of conjugate priors for $\mathcal{F}_0$.

(c) Express $\mathcal{F}_2 = \{f(\boldsymbol{x}, \boldsymbol{\theta})\colon -\infty < \theta_1 < \infty, 0 < \theta_2 < \infty\}$ as a two-parameter exponential family and find a family of conjugate priors for $\mathcal{F}_2$.

(d) In each of the cases above, find the prior-to-posterior formula. [Use Eqs. (3), (3a), and (3b).]

5.17. Let $X_1, \ldots, X_n$ be a random sample from a distribution with pdf

$$f(x, \theta) = \theta x^{-\theta-1} I_{(1,\infty)}(x), \quad \theta > 2.$$

(a) Find the method of moment estimator $\tilde{\theta}_n$ of θ.

(b) Determine which of the following is true

i) $\tilde{\theta}_n$ is an unbiased estimator θ,

ii) $\tilde{\theta}_n$ has an upward bias (ie, $E_\theta(\tilde{\theta}_n) > \theta$),

iii) $\tilde{\theta}_n$ has a downward bias. [Write $\tilde{\theta}_n$ in terms of $\bar{X}_n - 1$, find $\mathrm{E}_\theta[\bar{X}_n - 1]$ and use Jensen's inequality.]

(c) Find the asymptotic distribution of $\tilde{\theta}_n$ (suitably normalized) as $n \to \infty$.

5.18. Let $X_1, \ldots, X_n$ be a random sample from a log normal distribution with pdf

$$f(x, \theta) = \frac{1}{\sqrt{2\pi}x} \exp\left[-(\log x - \theta)^2/2\right], \quad x > 0.$$

This means $Y_i = \log X_i$, $i = 1, \ldots, n$, are iid $N(\theta, 1)$, so that $X_i \stackrel{\mathcal{D}}{=} \exp(Z_i + \theta)$ where $Z_1, \ldots, Z_n$ are iid $N(0, 1)$. We want to estimate θ based on $(X_1, \ldots, X_n)$.

(a) Find the MOME $\tilde{\theta}_n$ and its asymptotic distribution.

(b) Find the MLE $\hat{\theta}_n$ and its asymptotic distribution.

(c) Let σ_1^2 and σ_2^2 be the asymptotic variances of $\tilde{\theta}_n$ and $\hat{\theta}_n$, respectively. Find the asymptotic relative efficiency σ_2^2/σ_1^2 of $\tilde{\theta}_n$ with respect to $\hat{\theta}_n$ and comment.

6 Hypothesis Testing

6.1 Early History

Hypothesis testing at its early stage, from the 19th to the early 20th century, was concerned with hypotheses suggested by scientific theories in anthropometry, genetics, etc. The scientists in these disciplines wanted to evaluate the evidence provided by the data in support or against such hypotheses, called the *null hypotheses*.

The extent of departure from such a null hypothesis evidenced by the data was measured by a *test statistic* chosen on an ad hoc basis, which was considered *significant at level* α if the tail probability "beyond" its observed value, under the null hypothesis, called the p-value, fell below α. Whether to use the right tail or the left tail, or both tails for this purpose would be determined by the nature of the problem in a "supposedly obvious" manner. There was no formal basis for the choice of the test statistic or the direction of its tail probability, and the question of optimality of the commonly used test procedures was not addressed until [22] came up with the concept of an *alternative hypothesis* H_1 against which the null hypothesis H_0 was to be tested. This naturally led to the definitions of two types of error which are

(i) deciding in favor of H_1 when H_0 is true (Type I error), and
(ii) deciding in favor of H_0 when H_1 is true (Type II error).

Thus the problem of maximizing the

$$\begin{aligned}\text{Power} &= \text{The probability of } \textit{correctly rejecting } H_0\\ &= 1 - \text{the probability of Type II error}\end{aligned}$$

subject to the condition of the probability of Type I error not exceeding α was defined and the solution to this problem was the *Neyman-Pearson Lemma*.

6.2 Basic Concepts

Let $\{P_\theta,\ \theta \in \Theta\}$ be a family of probabilities on $(\mathfrak{X}, \mathcal{A})$, of which an unknown element P_θ generates a random sample X. We consider two hypotheses, H_0: $\theta \in \Theta_0$ and H_1: $\theta \in \Theta_1$, where Θ_0 and Θ_1 are disjoint subsets of Θ. Based on the observation X, we want to take one

Theory and Methods of Statistics. http://dx.doi.org/10.1016/B978-0-12-802440-9.00006-0

of two actions, a_0: Accept H_0 (ie, decide $\theta \in \Theta_0$) or a_1: Reject H_0 and accept H_1 (ie, decide $\theta \in \Theta_1$). The problem of hypothesis testing was introduced in Chapter 4. In this chapter we shall develop the methods of constructing optimal tests, mostly within the framework of exponential families, subject to the restrictions mentioned in Section 4.6.2.

A nonrandomized decision rule is described by a function $d\colon \mathfrak{X} \to \{a_0, a_1\}$, or equivalently by $C = \{x\colon d(x) = a_1\}$ which is called the *critical region* or *rejection region for* H_0. The complement of C is $A = \{x\colon d(x) = a_0\}$ which is called the *acceptance region for* H_0.

A behavioral decision rule is described by a function $\varphi\colon \mathfrak{X} \to [0, 1]$, where for an x, $\varphi(x)$ is the probability of taking action a_1 (ie, rejecting H_0 when the observed value of X is x). Such a function φ is called a *critical function*. A nonrandomized decision rule d with critical region C is equivalently described by a behavioral decision rule $\varphi = I_C$.

The critical region C of a nonrandomized decision rule and the sets $\{x\colon \varphi(x) \le c\}$ for $c \in [0, 1]$ of the critical function φ of a behavioral decision rule must belong to $\mathcal{A}$.

We consider the $0 - 1$ loss function, that is

$$L(\theta, a_0) = \begin{cases} 0 & \text{if } \theta \in \Theta_0 \\ 1 & \text{if } \theta \in \Theta_1 \end{cases}, \quad \text{and} \quad L(\theta, a_1) = \begin{cases} 1 & \text{if } \theta \in \Theta_0 \\ 0 & \text{if } \theta \in \Theta_1 \end{cases}.$$

Then the risk of a behavioral decision rule φ is

$$R(\theta, \varphi) = \begin{cases} \mathrm{E}_\theta[\varphi(X)] := \text{Probability of Type I Error} & \text{if } \theta \in \Theta_0 \\ 1 - \mathrm{E}_\theta[\varphi(X)] := \text{Probability of Type II Error} & \text{if } \theta \in \Theta_1 \end{cases}.$$

Thus the risk function is described in terms of the function

$$\beta_\varphi(\theta) = \mathrm{E}_\theta[\varphi(X)].$$

On Θ_0, $\beta_\varphi(\theta)$ is a measure of weakness of the test φ (probability of Type I error) and on Θ_1, $\beta_\varphi(\theta)$ is a measure of strength of the test φ (1 – probability of Type II error) which is called the *power*.

In the theory of hypothesis testing, the hypotheses $H_0\colon \theta \in \Theta_0$ and $H_1\colon \theta \in \Theta_1$ are given asymmetric roles, because the Type I error receives more serious consideration than the Type II error. Optimization of φ is formulated as a problem of choosing φ so as to maximize $\beta_\varphi(\theta)$ for $\theta \in \Theta_1$ subject to the condition

$$\sup_{\theta \in \Theta_0} \beta_\varphi(\theta) \le \alpha,$$

where $\alpha \in [0, 1]$ is given. This α is called the *level of significance* and $\sup_{\theta \in \Theta_0} \beta_\varphi(\theta)$ is the *size* of the test φ.

We call "$H_0\colon \theta \in \Theta_0$" the *null hypothesis* and "$H_1\colon \theta \in \Theta_1$" the *alternative hypothesis*. If Θ_0 (and/or Θ_1) is a singleton set, then we call the null hypothesis H_0 (and/or the alternative hypothesis H_1) a *simple hypothesis*, otherwise H_0 (and/or H_1) is composite.

6.3 Simple Null Hypothesis vs Simple Alternative: Neyman-Pearson Lemma

Let P_0 and P_1 be two distinct probability distributions on $(\mathfrak{X}, \mathcal{A})$ with pdf's/pmf's p_0 and p_1, respectively. Let E_i denote E_{P_i}, $i = 0, 1$.

Definition 6.3.1. A test φ is said to be *most powerful (MP)* at level α for testing $H_0: P = P_0$ against $H_1: P = P_1$ if

(a) $\mathrm{E}_0[\varphi(X)] \leq \alpha$; and
(b) if any test Ψ satisfies (a), then $\mathrm{E}_1[\varphi(X)] \geq \mathrm{E}_1[\Psi(X)]$.

Theorem 6.3.1 (Neyman-Pearson Lemma). *Suppose we are testing H_0: $P = P_0$ against H_1: $P = P_1$ and let $0 \leq \alpha \leq 1$.*

(i) *Existence. There exists a test φ and a constant k such that $\mathrm{E}_0[\varphi(X)] = \alpha$ and*

$$\varphi(x) = \begin{cases} 1 & \text{if } p_1(x) > kp_0(x) \\ 0 & \text{if } p_1(x) < kp_0(x) \end{cases}.$$

(ii) *Sufficiency. If a test φ satisfies the above conditions for some k, then it is MP at level α for testing H_0: $P = P_0$ against H_1: $P = P_1$.*
(iii) *Necessity. If φ is MP at level α for testing H_0: $P = P_0$ against H_1: $P = P_1$, then $\varphi(x)$ is of the form given in (i) for some k.*

Proof.

(i) Take $0 < \alpha < 1$ since the proofs for $\alpha = 0$ and 1 are straightforward. Let

$$F(c) = P_0[p_1(X) \leq cp_0(X)] = P_0[p_1(X)/p_0(X) \leq c]$$

be the cdf of the likelihood ratio $T(X) = p_1(X)/p_0(X)$, which is a bona fide rv under P_0 because $P_0[p_0(X) > 0] = 1$. Since F is a cdf, there exists c_0 such that

$$F(c_0 - 0) \leq 1 - \alpha \leq F(c_0).$$

Now define

$$\varphi(x) = \begin{cases} 1 & \text{if } p_1(x) > c_0 p_0(x) \\ \gamma & \text{if } p_1(x) = c_0 p_0(x) \\ 0 & \text{if } p_1(x) < c_0 p_0(x), \end{cases}$$

where γ can be assigned any value in $[0, 1]$, say $\gamma = 1$, if F is continuous at c_0, or else we take

$$\gamma = \frac{F(c_0) - (1 - \alpha)}{F(c_0) - F(c_0 - 0)}.$$

Thus part (i) of the N-P Lemma holds with $k = c_0$ and $0 \leq \gamma \leq 1$ as defined above, because in case of a jump in $F(\cdot)$ at c_0,

$$\begin{aligned} \mathrm{E}_0[\varphi(X)] &= P_0[p_1(X) > c_0 p_0(X)] + \gamma P_0[p_1(X) = c_0 p_0(X)] \\ &= \{1 - F(c_0)\} + \gamma\{F(c_0) - F(c_0 - 0)\} = \alpha, \end{aligned}$$

and the same equality is obvious when α is continuous at c_0.

(ii) Let φ be a test satisfying the conditions in (i) and let Ψ be another level α test for H_0 (ie, $E_0[\Psi(X)] \le \alpha$). Let

$$S^+ = \{x\colon \varphi(x) - \Psi(x) > 0\} \text{ and } S^- = \{x\colon \varphi(x) - \Psi(x) < 0\}.$$

Then for $x \in S^+, \varphi(x) > 0$, so $p_1(x) \ge kp_0(x)$ and for $x \in S^-, \varphi(x) < 1$, so $p_1(x) \le kp_0(x)$. Thus

$$\{\varphi(x) - \Psi(x)\}\{p_1(x) - kp_0(x)\} \begin{cases} \ge 0 \text{ for all } x \in S^+ \cup S^- \text{ and} \\ = 0 \text{ for all other } x. \end{cases}$$

Hence

$$\begin{aligned} E_1[\varphi(X)] - E_1[\Psi(X)] &= \int \{\varphi(x) - \Psi(x)\}p_1(x)\,dx \\ &= \int \{\varphi(x) - \Psi(x)\}\{p_1(x) - kp_0(x)\}\,dx \\ &\quad + k\int \{\varphi(x) - \Psi(x)\}p_0(x)\,dx \\ &\ge 0 + k\{E_0[\varphi(X)] - E_0[\Psi(X)]\} \ge 0. \end{aligned}$$

(iii) Suppose that Ψ is MP at level α for testing $H_0\colon P = P_0$ vs $H_1\colon P = P_1$, and let φ be a test satisfying the conditions in (i) with $k = k^*$. Define S^+ and S^- as in the proof of part (ii). Then $\Psi(x) \ne \varphi(x)$ for $x \in S^+ \cup S^-$. Let

$$\begin{aligned} S &= (S^+ \cup S^-) \cap \{x\colon p_1(x) \ne k^* p_0(x)\} \\ &= \{x\colon \Psi(x) \ne \varphi(x) \text{ and } p_1(x) \ne k^* p_0(x)\}. \end{aligned}$$

Then $P_0(S) = P_1(S) = 0$ implies $\Psi(x) = \varphi(x)$ w.p. 1 under P_0 and P_1 whenever $p_1(x) \ne k^* p_0(x)$, showing that Ψ satisfies condition (i) for $k = k^*$ w.p. 1 under P_0 and P_1, proving part (iii) of the lemma. To prove $P_0(S) = P_1(S) = 0$, it is enough to show that $\int_S dx = 0$. Since $(\varphi - \Psi)(p_1 - k^* p_0) > 0$ on S and $= 0$ on S^c,

$$\int (\varphi - \Psi)(p_1 - k^* p_0) = \int_S (\varphi - \Psi)(p_1 - k^* p_0) > 0, \text{ ie,}$$

$$\int (\varphi - \Psi)p_1 > k^* \int (\varphi - \Psi)p_0,$$

implying $E_1[\varphi(X)] > E_1[\Psi(X)]$, which is a contradiction.

□

Corollary to the N-P Lemma

Let $\beta = E_1[\varphi(X)]$ where φ is the MP test for $H_0\colon P = P_0$ vs $H_1\colon P = P_1$ at level $\alpha < 1$. Then $\beta > \alpha$.

Proof. Consider the test $\varphi_0(x) = \alpha$ for all x. Then $E_0[\varphi_0(X)] = E_1[\varphi_0(X)] = \alpha$. Since φ_0 is a level α test which does not satisfy the necessary condition for an MP test (given by part

(iii) of the N-P Lemma) it is *not* an MP level α test, whereas φ is. Hence

$$\beta = \mathrm{E}_1[\varphi(X)] > \mathrm{E}_1[\varphi_0(X)] = \alpha. \qquad \square$$

Note. $\varphi_0(x) = \alpha$ for all x may satisfy the necessary condition if $p_1(x) \equiv kp_0(x)$ for some k; but then $k = 1$, contradicting $P_0 \neq P_1$.

Remark 6.3.1. For two probabilities P_0, P_1 on $(\mathfrak{X}, \mathcal{A})$, we call $p_1(x)/p_0(x)$ the *likelihood ratio* of P_1 to P_0 at x. The N-P Lemma expresses the MP level α test for H_0: $P = P_0$ vs H_1: $P = P_1$ in terms of the likelihood ratio, or equivalently, in terms of the log likelihood ratio $L(x) = \log(p_1(x)/p_0(x))$

$$\phi(x) = \begin{cases} 0 & \text{if } L(x) < k \\ \gamma & \text{if } L(x) = k \\ 1 & \text{if } L(x) > k, \end{cases}$$

where k and $0 \leq \gamma \leq 1$ are determined by the condition $\mathrm{E}_0[\phi(X)] = \alpha$.

Although the N-P Lemma can be used to construct MP level α tests for arbitrary P_0 vs P_1 such as testing whether the data came from $N(0, 1)$ or from $Cauchy(0, 1)$, our interest mostly lies in testing for one value of the parameter against another *within a parametric family*. We illustrate the use of the N-P Lemma with two examples, namely, testing for the mean θ of $Poi(\theta)$ and the mean θ of $N(\theta, 1)$.

Example 6.3.1. Let $\boldsymbol{X} = (X_1, \ldots, X_n)$ be a random sample from $Poisson(\theta)$, where $\theta > 0$ is unknown. We want the MP level α test for H_0: $\theta = \theta_0$ vs H_1: $\theta = \theta_1$.

Solution. Here the log likelihood ratio is

$$\log \frac{p_{\theta_1}(\boldsymbol{x})}{p_{\theta_0}(\boldsymbol{x})} = -n(\theta_1 - \theta_0) + (\log\theta_1 - \log\theta_0)t_n, \text{ where } t_n = \sum_{i=1}^{n} x_i.$$

(i) Hence the MP level α test for H_0 vs H_1 is

$$\varphi_1(\boldsymbol{x}) = \begin{cases} 0 & \text{if } t_n < c_1 \\ \gamma_1 & \text{if } t_n = c_1 \\ 1 & \text{if } t_n > c_1, \end{cases}$$

where c_1 and $0 \leq \gamma_1 \leq 1$ are determined by $P_{\theta_0}[T_n > c_1] + \gamma_1 P_{\theta_0}[T_n = c_1] = \alpha$ if $\theta_1 > \theta_0$ and

$$\varphi_2(\boldsymbol{x}) = \begin{cases} 0 & \text{if } t_n > c_2 \\ \gamma_2 & \text{if } t_n = c_2 \\ 1 & \text{if } t_n < c_2, \end{cases}$$

where c_2 and $0 \leq \gamma_2 \leq 1$ are determined by $P_{\theta_0}[T_n < c_2] + \gamma_2 P_{\theta_0}[T_n = c_2] = \alpha$, if $\theta_1 < \theta_0$.

(ii) We illustrate the determination of c_1 and γ_1 in the first case by considering H_0: $\theta = \theta_0 = 0.5$ and H_1: $\theta = \theta_1 = 1.0$ for $n = 5$ and $\alpha = 0.5$. Under P_{θ_0}, $T = \sum_{i=1}^{5} X_i \sim Poisson(2.5)$ (ie, $p_{\theta_0}(t) = \mathrm{e}^{-2.5}(2.5)^t/t!$, $t = 0, 1, 2, \ldots$). Calculating these probabilities, we have $P_{\theta_0}[T > 5] = 0.0420$ and $P_{\theta_0}[T = 5] = 0.0668$ and then solve the equation:

$$P_{\theta_0}[T > 5] + \gamma P_{\theta_0}[T = 5] = 0.05.$$

Thus $c_1 = 5$ and $\gamma_1 = 0.12$, so the MP test for H_0: $\theta = 0.5$ vs H_1: $\theta = 1.0$ at level $\alpha = 0.05$ is

$$\varphi_1(\boldsymbol{x}) = \begin{cases} 0 & \text{if } \sum_1^5 x_i < 5 \\ 0.12 & \text{if } \sum_1^5 x_i = 5. \\ 1 & \text{if } \sum_1^5 x_i > 5 \end{cases}$$

(iii) The power function of φ_1 is

$$\beta_1(\theta) = P_\theta[T > 5] + 0.12P_\theta[T = 5] = \sum_{t=6}^{\infty} e^{-\lambda}\lambda^t/t! + 0.12e^{-\lambda}\lambda^5/5!, \text{ where}$$

$$\lambda = 5\theta \text{ and } \beta_1'(\theta) = 5e^{-\lambda}\left[0.12\frac{\lambda^4}{4!} + 0.88\frac{\lambda^5}{5!}\right] > 0.$$

(iv) The MP level α test φ_1 for H_0: $\theta = \theta_0$ vs H_1: $\theta = \theta_1 (> \theta_0)$ does not depend on θ_1, so long as $\theta_1 > \theta_0$. Therefore, the MP level α test for H_0: $\theta = \theta_0$ vs H_1: $\theta = \theta_1 (> \theta_0)$ is the *Uniformly Most powerful (UMP)* level α test for H_0: $\theta = \theta_0$ vs H_1^*: $\theta > \theta_0$. Again, since $\beta_1'(\theta) > 0$, $\beta_1(\theta_0) = \alpha$ implies $\beta_1(\theta) < \alpha$ for all $\theta < \theta_0$. Thus the UMP level α test φ_1 for H_0: $\theta = \theta_0$ vs H_1: $\theta > \theta_0$ is the UMP level α test for the composite null hypothesis H_0^*: $\theta \le \theta_0$ vs the composite alternative H_1^*: $\theta > \theta_0$.

In the same way φ_2 can be shown to be a UMP level α test for H_0^*: $\theta \ge \theta_0$ vs H_1^*: $\theta < \theta_0$.

Example 6.3.2. Let $\boldsymbol{X} = (X_1, \ldots, X_n)$ be a random sample from $N(\theta, \sigma^2)$ where θ is unknown but $\sigma^2 > 0$ is known. We want the MP level α test for H_0: $\theta = \theta_0$ vs H_1: $\theta = \theta_1$.

Solution. Here the log likelihood ratio is

$$\log(p_{\theta_1,\sigma}(\boldsymbol{x})/p_{\theta_0,\sigma}(\boldsymbol{x})) = \frac{n(\theta_1 - \theta_0)}{\sigma^2}\bar{x}_n - \frac{n(\theta_1^2 - \theta_0^2)}{\sigma^2}, \quad \text{where } \bar{x}_n = n^{-1}\sum_{i=1}^{n} x_i.$$

Hence the MP level α test for H_0 vs H_1 is

$$\varphi_1(\boldsymbol{x}) = \begin{cases} 0 & \bar{x}_n < c_1 \\ 1 & \bar{x}_n \ge c_1 \end{cases}, \text{ if } \theta_1 > \theta_0, \text{ and}$$

$$\varphi_2(\boldsymbol{x}) = \begin{cases} 0 & \bar{x}_n > c_2 \\ 1 & \bar{x}_n \le c_2 \end{cases}, \text{ if } \theta_1 < \theta_0,$$

where c_1 is determined by $P_{\theta_0}[\bar{X}_n \ge c_1] = \alpha$ and c_2 is determined by $P_{\theta_0}[\bar{X}_n \le c_2] = \alpha$. The constants c_1 and c_2 are

$$c_1 = \theta_0 + (\sigma/\sqrt{n})\Phi^{-1}(1 - \alpha) \text{ and } c_2 = \theta_0 - (\sigma/\sqrt{n})\Phi^{-1}(1 - \alpha),$$

where Φ is the cdf of $N(0, 1)$.

The power function of φ_1 is

$$\beta_1(\theta) = P_\theta[\bar{X}_n \geq \theta_0 + (\sigma/\sqrt{n})\Phi^{-1}(1-\alpha)] = \Phi(\Phi^{-1}(\alpha) + \sqrt{n}(\theta - \theta_0)/\sigma), \text{ and}$$
$$\beta_1'(\theta) = (\sqrt{n}/\sigma)\Phi'(\Phi^{-1}(\alpha) + \sqrt{n}(\theta - \theta_0)/\sigma) > 0.$$

Using the same argument as in Example 6.3.1, we see that the MP level α test φ_1 for H_0: $\theta = \theta_0$ vs H_1: $\theta = \theta_1 (> \theta_0)$ is also the UMP level α test for the composite null hypothesis H_0^*: $\theta \leq \theta_0$ vs the composite alternative H_1^*: $\theta > \theta_0$.

The MP level α test φ_2 for H_0: $\theta = \theta_0$ vs H_1: $\theta = \theta_1$ if $\theta_1 < \theta_0$, in the same way.

6.4 UMP Tests for One-Sided Hypotheses Against One-Sided Alternatives in Monotone Likelihood Ratio Families

We begin this section with the definition of UMP tests.

Definition 6.4.1. A test φ is a UMP test at level α for H_0: $\theta \in \Theta_0$ vs H_1: $\theta \in \Theta_1$ if

(i) $\sup_{\theta \in \Theta_0} \mathrm{E}_\theta[\varphi(X)] \leq \alpha$, and
(ii) $\mathrm{E}_\theta[\varphi(X)] \geq \mathrm{E}_\theta[\psi(X)]$ for all $\theta \in \Theta_1$, whenever ψ satisfies (i).

Although requirement (ii) is very stringent, UMP tests do exist in certain types of situations. In the previous section we have seen that for *Poisson*(θ), the MP test at level α for H_0: $\theta = \theta_0$ vs H_1: $\theta = \theta_1 (> \theta_0)$ has the stronger property of being the UMP level α test for the composite hypothesis H_0: $\theta \leq \theta_0$ vs the composite alternative H_1: $\theta > \theta_0$. This actually holds in a much wider context.

Definition 6.4.2. Let $\{P_\theta, \theta \in \Theta\}$ be a family of probabilities on $(\mathfrak{X}, \mathcal{A})$ and let p_θ denote the pdf or pmf corresponding to P_θ where $\Theta = \mathbb{R}$ or Θ is an interval in $\mathbb{R}$. Such a family $\{p_\theta\}$ is said to be a monotone likelihood ratio (MLR) family if there exists a real-valued statistic $T(x)$ such that for any $\theta_1 < \theta_2$ in Θ, $p_{\theta_2}(x)/p_{\theta_1}(x)$ is a nondecreasing function of $T(x)$. [If $p_{\theta_1}(x) = 0 < p_{\theta_2}(x)$, define $p_{\theta_2}(x)/p_{\theta_1}(x) = +\infty$.]

Example 6.4.1. Let $p_\theta(x) = c(\theta)\exp[Q(\theta)T(x)]h(x)$ where $Q(\theta)$ is a nondecreasing function. Then $\{p_\theta\}$ is an MLR family. This includes

(a) $p_\theta(\boldsymbol{x}) = (2\pi\sigma^2)^{-n/2}\exp\left[-\frac{1}{2\sigma^2}\sum_1^n (x_i - \theta)^2\right]$, $x_i \in \mathbb{R}, \theta \in \mathbb{R}, \sigma^2$ fixed.

(b) $p_\theta(\boldsymbol{x}) = (2\pi\theta)^{-n/2}\exp\left[-\frac{1}{2\theta}\sum_1^n (x_i - \mu)^2\right]$, $x_i \in \mathbb{R}, \theta \in \mathbb{R}^+, \mu$ fixed.

(c) $p_\theta(\boldsymbol{x}) = \mathrm{e}^{-n\theta}\theta^{\sum_1^n x_i}/\prod_1^n x_i!$, $x_i \in \{0, 1, 2, \ldots\}, \theta \in \mathbb{R}^+$.

(d) $p_\theta(\boldsymbol{x}) = \theta^{\sum_1^n x_i}(1-\theta)^{n - \sum_1^n x_i}$, $x_i \in \{0, 1\}, \theta \in (0, 1)$.

Example 6.4.2. The hypergeometric distribution $\mathcal{H}(n, N, \theta)$ with $p_\theta(\boldsymbol{x}) = \binom{\theta}{x}\binom{N-\theta}{n-x}/\binom{N}{n}$, $x = \max(0, \theta + n - N), \ldots, \min(n, \theta)$.

Example 6.4.3. The family of Cauchy distributions $\mathcal{C}(\theta, 1)$ with $p_\theta(x) = \frac{1}{\pi[1+(x-\theta)^2]}$, $x \in \mathbb{R}, \theta \in \mathbb{R}$ is *not* an MLR family.

Theorem 6.4.1. *Suppose that the family of pdf's or pmf's $\{p_\theta, \theta \in \mathbb{R}\}$ has MLR property in $T(x)$. Then*

(i) *There exists a UMP level α test for $H_0\colon \theta \le \theta_0$ vs $H_1\colon \theta > \theta_0$ given by*

$$\varphi(x) = \begin{cases} 0 & \text{if } T(x) < c \\ \gamma & \text{if } T(x) = c, \text{ where } c \text{ and } 0 \le \gamma \le 1 \text{ are determined by } E_{\theta_0}[\varphi(X)] = \alpha. \\ 1 & \text{if } T(x) > c \end{cases}$$

(ii) *The power function $\beta(\theta) = E_\theta[\varphi(X)]$ is strictly increasing at all θ for which $\beta(\theta) < 1$.*

(iii) *For all θ', this φ is UMP at level $\alpha' = \beta(\theta')$ for $H_0'\colon \theta \le \theta'$ vs $H_1'\colon \theta > \theta'$.*

(iv) *This test minimizes $E_\theta[\psi(X)]$ at all $\theta < \theta_0$ among all tests ψ for which $E_{\theta_0}[\psi(X)] = \alpha$.*

Proof. Parts (i) and (ii). First consider the simple vs simple case with $H_0^*\colon \theta = \theta_0$ vs $H_1^*\colon \theta = \theta_1$ where $\theta_1 > \theta_0$ is fixed. Then by the N-P Lemma, the MP level α test for H_0^* vs H_1^* rejects H_0 for large values of $p_{\theta_1}(x)/p_{\theta_0}(x)$ (ie, for large values of $T(x)$) by the MLR property. Moreover, by the existence part of the N-P Lemma, there exist c and $0 \le \gamma \le 1$ such that the test

$$\varphi(x) = I_{(c,\infty)}[T(X)] + \gamma I_{\{c\}}[T(X)] \text{ satisfies } \mathrm{E}_{\theta_0}[\varphi(X)] = \alpha.$$

Since the forms:

$$\varphi(x) = \begin{cases} 0, & T(x) < c \\ 1, & T(x) > c \end{cases} \text{ and } \varphi(x) = \begin{cases} 0, & p_{\theta''}(x) < kp_{\theta'}(x) \\ 1, & p_{\theta''}(x) > kp_{\theta'}(x) \end{cases}$$

are equivalent for any $\theta' < \theta''$, this test is UMP at level $\alpha' = \beta(\theta')$ for testing $H_0^{**}\colon \theta = \theta'$ vs $H_1^{**}\colon \theta = \theta''$ whenever $\theta' < \theta''$ (by the sufficiency part of the N-P Lemma). Next, note that by the corollary to the N-P Lemma, $\beta(\theta'') > \alpha' = \beta(\theta')$ if $\alpha' < 1$, which proves that $\beta(\theta)$ is strictly increasing, so long as it is < 1. This proves part (ii).

Now note that for this test, $\beta(\theta) = \mathrm{E}_\theta[\varphi(X)] \le \alpha$ for all $\theta \le \theta_0$, which makes φ a level α test for $H_0\colon \theta \le \theta_0$. Let

$$\Psi_\alpha = \left\{\text{all tests } \psi \text{ such that } \sup_{\theta \le \theta_0} \mathrm{E}_\theta[\psi(X)] \le \alpha\right\} \text{ and}$$
$$\Psi_\alpha^* = \{\text{all tests } \psi \text{ such that } \mathrm{E}_{\theta_0}[\psi(X)] \le \alpha\}.$$

Then $\Psi_\alpha \subset \Psi_\alpha^*$. We have shown that

$$\varphi \in \Psi_\alpha \text{ and } \mathrm{E}_{\theta_1}[\varphi(X)] \ge \mathrm{E}_{\theta_1}[\psi(X)] \text{ for all } \psi \in \Psi_\alpha^*.$$

Hence $\mathrm{E}_{\theta_1}[\varphi(X)] \ge \mathrm{E}_{\theta_1}[\psi(X)]$ for all $\psi \in \Psi_\alpha$.

This makes φ the MP test at level α for $H_0\colon \theta \le \theta_0$ vs $H_1\colon \theta > \theta_0$. Finally, since φ does not depend on $\theta_1 > \theta_0$, it is UMP at level α for $H_0\colon \theta \le \theta_0$ vs $H_1\colon \theta > \theta_0$.

The proof of part (iii) is analogous, and part (iv) is proved by observing that power is minimized if all inequalities are reversed in the N-P Lemma. □

6.5 Unbiased Tests

A behavioral test with critical function φ is said to be an unbiased test at level α for H_0: $\theta \in \Theta_0$ vs H_1: $\theta \in \Theta_1$ if (i) $\mathrm{E}_\theta[\varphi(X)] \leq \alpha$ for all $\theta \in \Theta_0$ and (ii) $\mathrm{E}_\theta[\varphi(X)] \geq \alpha$ for all $\theta \in \Theta_1$. An unbiased test rejects the null hypothesis with at least as much probability when it is false as when it is true.

Definition 6.5.1. A test φ is said to be UMP unbiased test at level α for H_0: $\theta \in \Theta_0$ vs H_1: $\theta \in \Theta_1$ if φ is an unbiased test at level α for H_0 vs H_1, and if $\mathrm{E}_\theta[\varphi(X)] \geq \mathrm{E}_\theta[\psi(X)]$ for all $\theta \in \Theta_1$ whenever ψ is also an unbiased level α test for H_0 vs H_1.

For a large class of problems, a UMP test does not exist, but a UMP unbiased test does exist.

Example 6.5.1. Let $X \sim \mathcal{N}(\theta, 1)$, H_0: $\theta = \theta_0$ and H_1: $\theta \neq \theta_0$. Here the tests

$$\varphi_1(x) = I_{[\theta_0+\Phi^{-1}(1-\alpha),\infty)}(x) \text{ and } \varphi_2(x) = I_{(-\infty,\theta_0-\Phi^{-1}(1-\alpha)]}(x),$$

where Φ is the cdf of $\mathcal{N}(0, 1)$, are, respectively, the UMP level α tests for H_0: $\theta = \theta_0$ vs H_1^+: $\theta > \theta_0$ and H_1^-: $\theta < \theta_0$. Moreover, it follows from Theorem 6.4.1, part (ii) that $\mathrm{E}_\theta[\varphi_1(x)] < \alpha$ for all $\theta < \theta_0$ and $\mathrm{E}_\theta[\varphi_2(x)] < \alpha$ for all $\theta > \theta_0$ (because the family $\mathcal{N}(\theta, 1)$ has the MLR property), which shows that a UMP level α test for H_0: $\theta = \theta_0$ vs H_1: $\theta \neq \theta_0$ does not exist.

For testing H_0: $\theta = \theta_0$ vs H_1: $\theta \neq \theta_0$ at level α, an unbiased test φ must satisfy: $\mathrm{E}_{\theta_0}[\varphi(X)] \leq \alpha$ and $\mathrm{E}_\theta[\varphi(X)] \geq \alpha$ for all $\theta \neq \theta_0$, so neither φ_1 nor φ_2 in the above example is an unbiased test. If we restrict to the class of unbiased level α tests then φ_1, φ_2 would not qualify, but in the restricted class a UMP test does exist.

Suppose that the power functions $\beta_\varphi(\theta) = E_\theta[\varphi(X)]$ of all tests φ are differentiable at θ_0 (as in the case of exponential families). Then an unbiased level α test for H_0: $\theta = \theta_0$ vs H_1: $\theta \neq \theta_0$ must satisfy

$$\mathrm{E}_{\theta_0}[\varphi(X)] = \alpha \quad \text{and} \quad \beta'_\varphi(\theta_0) = 0. \tag{1}$$

Let $\mathcal{C}_0$ denote the class of unbiased level α tests for H_0: $\theta = \theta_0$ vs H_1: $\theta \neq \theta_0$ and let $\mathcal{C}_1$ denote the class of all tests satisfying Eq. (1). Then $\mathcal{C}_0 \subset \mathcal{C}_1$ and therefore, if a test $\varphi_0 \in \mathcal{C}_0$ is UMP among all tests in $\mathcal{C}_1$, then it is a UMP unbiased level α test for H_0 vs H_1.

We now examine the case of a single-parameter exponential family $\{P_\theta, \theta \in \Theta \subset \mathbb{R}\}$ with pdf/pmf given by

$$p_X(x; \theta) = C_0(\theta) \exp[\theta T(x)] h_0(x), \quad x \in \mathfrak{X}, \tag{2}$$

and $\boldsymbol{X} = (X_1, \ldots, X_n)$ being a random sample from P_θ. Then $T = \sum_{i=1}^n T(X_i)$ is sufficient for θ in $(X_1, \ldots, X_n)$ and is distributed with pdf/pmf

$$p_T(t; \theta) = C(\theta) \exp[\theta t] h(t), \quad t \in \mathcal{T}. \tag{3}$$

We know that for all tests φ, the power functions

$$\beta_\varphi(\theta) = C(\theta) \int \varphi(t) \exp[\theta t] h(t)\, dt \tag{4}$$

are differentiable at all interior points of the natural parameter space, which is an interval, and if θ_0 is such a point, then the differentiation can be carried out under the integral, that is

$$\begin{aligned}\beta'_\varphi(\theta_0) &= C'(\theta_0)\int \varphi(t)\exp[\theta_0 t]h(t)\,dt + C(\theta_0)\int t\varphi(t)\exp[\theta_0 t]h(t)\,dt\\ &= \frac{C'(\theta_0)}{C(\theta_0)}\mathrm{E}_{\theta_0}[\varphi(T)] + \mathrm{E}_{\theta_0}[T\varphi(T)].\end{aligned}$$

Moreover, for $\varphi(t) \equiv \alpha$, we have $\beta_\varphi(\theta) = \alpha$ for all θ, and therefore, $\beta'_\varphi(\theta_0) = 0$. Thus the above expression becomes: $0 = \frac{C'(\theta_0)}{C(\theta_0)}\alpha + \alpha\mathrm{E}_{\theta_0}[T]$. Hence $C'(\theta_0)/C(\theta_0) = -\mathrm{E}_{\theta_0}[T]$, and the formula for $\beta'_\varphi(\theta_0)$ for an arbitrary level α test becomes:

$$\beta'_\varphi(\theta_0) = -\alpha\mathrm{E}_{\theta_0}[T] + \mathrm{E}_{\theta_0}[T\varphi(T)]. \tag{5}$$

The problem of finding the UMP unbiased level α test for H_0: $\theta = \theta_0$ vs H_1: $\theta \neq \theta_0$ in a single-parameter exponential family, where θ_0 is an interior point of the natural parameter space, now leads to the following restatement of Eq. (1), using the formula (5) for $\beta'_\varphi(\theta_0)$.

Among all tests $\varphi \in \mathcal{C}_1$ satisfying

$$\mathrm{E}_{\theta_0}[\varphi(T)] = \int \varphi(t)f_1(t)\,dt = \alpha, \text{ and} \tag{6}$$

$$\mathrm{E}_{\theta_0}[T\varphi(T)] = \int \varphi(t)f_2(t)\,dt = \alpha\mathrm{E}_{\theta_0}[T] \tag{7}$$

with

$$f_1(t) = C(\theta_0)\exp[\theta_0 t]h(t), \tag{8}$$

$$f_2(t) = C(\theta_0)t\exp[\theta_0 t]h(t) \tag{9}$$

find φ_0 which maximizes $\mathrm{E}_{\theta_1}[\varphi(T)] = \int \varphi(t)f_3(t)\,dt$ with

$$f_3(t) = C(\theta_1)\exp[\theta_1 t]h(t) \tag{10}$$

for a fixed $\theta_1 \neq \theta_0$.

If this φ_0 is in the smaller class $\mathcal{C}_0$ of unbiased level α tests for H_0: $\theta = \theta_0$ vs H_1: $\theta \neq \theta_0$ and if φ_0 does not depend on the specific $\theta_1 \neq \theta_0$ used in the above optimization problem, then φ_0 is a UMP unbiased level α test for H_0 vs H_1.

We next look at the problem of testing H_0: $\theta_1 \leq \theta \leq \theta_2$ vs H_1: $\theta \notin [\theta_1, \theta_2]$ in an exponential family, where θ_1, θ_2 are interior points of the natural parameter space. More generally, first consider the problem of testing H_0: $\theta \in \Theta_0$ vs H_1: $\theta \in \Theta_1$, where Θ_0 and Θ_1 have a common boundary ω. Since the power functions $\beta_\varphi(\theta) = \mathrm{E}_\theta[\varphi(X)]$ of all tests are continuous in θ, $\mathrm{E}_\theta[\varphi(X)] = \alpha$ for all $\theta \in \omega$ must hold for all unbiased level α tests for H_0 vs H_1.

Definition 6.5.2. A test φ satisfying $E_\theta[\varphi(X)] = \alpha$ for all $\theta \in \omega$ is said to be *similar* of size α on ω.

Since the class of similar tests of size α on ω includes the class of unbiased level α tests for H_0: $\theta \in \Theta_0$ vs H_1: $\theta \in \Theta_1$, the following holds.

Lemma 6.5.1. *If the power functions of all tests are continuous in θ and if a test φ_0 is UMP among all similar tests of size α on the common boundary ω of Θ_0 and Θ_1, then φ_0 is UMP unbiased level α test for $H_0\colon \theta \in \Theta_0$ vs $H_1\colon \theta \in \Theta_1$ provided that φ_0 is a level α test.*

Proof. Clearly, φ_0 is UMP on Θ_1 among all unbiased level α tests. Therefore, we only need to verify that φ_0 is an unbiased level α test. For this, we compare φ_0 with $\varphi^*(x) \equiv \alpha$ (which is a similar test of size α), to see that $\mathrm{E}_\theta[\varphi_0(X)] \geq \mathrm{E}_\theta[\varphi^*(X)] = \alpha$ for all $\theta \in \Theta_1$. □

We now consider the problem of testing $H_0\colon\ \theta_1 \leq \theta \leq \theta_2$ vs $H_1\colon\ \theta \notin [\theta_1, \theta_2]$ in an exponential family $\{P_\theta, \theta \in \Theta \subset \mathbb{R}\}$ with pdf/pmf $p_X(x;\theta)$ given by Eq. (2), based on a random sample $\boldsymbol{X} = (X_1, \ldots, X_n)$ in which $T = \sum_{i=1}^n T(X_i)$ is sufficient for θ with pdf/pmf given $f_T(t;\theta)$ given by Eq. (3). We also know that for all tests φ, the power function $\beta_\varphi(\theta)$ given by Eq. (4) is continuous in θ.

To find a UMP unbiased level α test for $H_0\colon \theta \in \Theta_0 = [\theta_1, \theta_2]$ vs $H_1\colon \theta \in \Theta_1 = [\theta_1, \theta_2]^c$, we note that Θ_0 and Θ_1 have a common boundary $\omega = \{\theta_1, \theta_2\}$, and all tests have continuous power functions. So the above Lemma is applicable and we only need to look among tests which are similar of size α on $\omega = \{\theta_1, \theta_2\}$ (ie, among tests satisfying $\mathrm{E}_{\theta_i}[\varphi(T)] = \alpha$, $i = 1, 2$). This leads to the problem of maximizing $\int \varphi(t) f_3(t)\ dt$ subject to the conditions $\int \varphi(t) f_i(t)\ dt = \alpha$, $i = 1, 2$, where

$$f_i(t) = C(\theta_i)\exp\bigl[\theta_i t\bigr]h(t), \quad i = 1, 2, \text{ and} \tag{11}$$

$$f_3(t) = C(\theta)\exp[\theta t]h(t) \quad \text{for } \theta \notin \bigl[\theta_1, \theta_2\bigr]. \tag{12}$$

Thus, our search for UMP unbiased level α tests for $H_0\colon \theta = \theta_0$ vs $H_1\colon \theta \neq \theta_0$ and $H_0\colon \theta_1 \leq \theta \leq \theta_2$ vs $H_1\colon \theta \notin [\theta_1, \theta_2]$ in a single-parameter exponential family, both lead to the problem of maximizing $\int \varphi(t) f_3(t)\ dt$ with respect to φ subject to the conditions $\int \varphi(t) f_i(t)\ dt = c_i$, $i = 1, 2$, where the constants are $c_1 = \alpha$, $c_2 = \alpha \mathrm{E}_{\theta_0}[T]$ and f_1, f_2, f_3 given by Eqs. (8), (9), and (10) in the first problem and $c_1 = c_2 = \alpha$ and f_1, f_2, f_3 given by Eqs. (11) and (12) in the second problem. This calls for a generalization of the Neyman-Pearson Lemma [23].

6.6 Generalized Neyman-Pearson Lemma

Theorem 6.6.1. *Let $f_1, \ldots, f_m, f_{m+1}$ be real-valued functions defined on an Euclidean space $(\mathfrak{X} = \mathbb{R}^n, \mathcal{A})$ for which $\int_{\mathfrak{X}} f_i(x)\ dx$ (or $\sum_{x\in\mathfrak{X}} f_i(x)$ in the discrete case) exist and are finite for $i = 1, \ldots, m+1$. Let $\mathcal{F}$ denote the class of all critical functions (ie $\varphi\colon \mathfrak{X} \to [0, 1]$). Suppose that for given constants $c_1, \ldots, c_m$, there exists $\varphi \in \mathcal{F}$ such that $\int \varphi f_i\ dx = c_i$, $i = 1, \ldots, m$. Let*

$$C = \left\{\varphi \in \mathcal{F}\colon \int \varphi f_i\ dx = c_i, \quad i = 1, \ldots, m\right\}$$

which is nonempty.

(i) *If there exists $\varphi \in C$ such that for some constants $k_1, \ldots k_m$,*

$$\varphi(x) = \begin{cases} 0 & \text{if } f_{m+1}(x) < \sum_{i=1}^m k_i f_i(x) \\ 1 & \text{if } f_{m+1}(x) > \sum_{i=1}^m k_i f_i(x), \end{cases}$$

then $\int \varphi f_{m+1}\,dx \geq \psi f_{m+1}\,dx$ *for all* $\psi \in \mathcal{C}$. *This provides a sufficient condition for maximization of* $\int \varphi f_{m+1}\,dx$ *in* $\mathcal{C}$.

(ii) *If there exists* $\varphi \in \mathcal{C}$ *such that for constants* $k_1, \ldots, k_m \geq 0$, $\varphi(x)$ *is of the form given in (i), then* $\int \varphi f_{m+1}\,dx \geq \int \psi f_{m+1}\,dx$ *for all* $\psi \in \mathcal{C}' = \{\varphi \in \mathcal{F}\colon \int \varphi f_i\,dx \leq c_i,\ i = 1, \ldots, m\}$. *This is an extension of (i).*

Proof. Let $\varphi \in \mathcal{C}$ be of the form given in (i) and $\psi \in \mathcal{F}$. Let

$$S^+ = \left\{x\colon f_{m+1}(x) > \sum_{i=1}^{m} k_i f_i(x)\right\} \text{ and } S^- = \left\{x\colon f_{m+1}(x) < \sum_{i=1}^{m} k_i f_i(x)\right\}.$$

For $x \in S^+$, $\varphi(x) - \psi(x) = 1 - \psi(x) \geq 0$ and $f_{m+1} - \sum_{i=1}^{m} k_i f_i(x) > 0$ and for $x \in S^-$, both of these inequalities are reversed. Hence

$$\{\varphi(x) - \psi(x)\}\left\{f_{m+1}(x) - \sum_{i=1}^{m} k_i f_i(x)\right\} \geq 0 \quad \text{for all } x \in S^+ \cup S^-.$$

Thus,

$$\int_{S^+ \cup S^-} (\varphi - \psi)\left(f_{m+1} - \sum_{i=1}^{m} k_i f_i\right) dx \geq 0, \text{ ie,}$$

$$\int (\varphi - \psi) f_{m+1}\,dx \geq \sum_{i=1}^{m} k_i \int (\varphi - \psi) f_i\,dx.$$

If $\psi \in \mathcal{C}$, then $\sum_{i=1}^{m} k_i \int (\varphi - \psi) f_i\,dx = 0$; if $\psi \in \mathcal{C}'$ and $k_1, \ldots, k_m \geq 0$, then $\sum_{i=1}^{m} k_i \int (\varphi - \psi) f_i\,dx \geq 0$. Hence

$$\int \varphi f_{m+1}\,dx \geq \int \psi f_{m+1}\,dx \text{ if } \psi \in \mathcal{C}, \text{ or if } \psi \in \mathcal{C}' \text{ and } k_1, \ldots, k_m \geq 0. \qquad \square$$

6.7 UMP Unbiased Tests for Two-Sided Problems

We now apply the Generalized N-P Lemma to find unbiased level α tests for the two problems under discussion.

6.7.1 UMP Unbiased Test for $H_0\colon \theta = \theta_0$ vs $H_1\colon \theta \neq \theta_0$ in a Single-Parameter Exponential Family

By the Generalized N-P Lemma, the UMP unbiased level α test for this problem is given by:

$$\varphi_0(t) = \begin{cases} 0 & \text{if } f_3(t) < k_1 f_1(t) + k_2 f_2(t) \\ \gamma(t) & \text{if } f_3(t) = k_1 f_1(t) + k_2 f_2(t) \\ 1 & \text{if } f_3(t) > k_1 f_1(t) + k_2 f_2(t), \end{cases}$$

where f_1, f_2, and f_3 are given by Eqs. (8), (9), and (10) and k_1, k_2, and $0 \leq \gamma(t) \leq 1$ are such that

$$\mathrm{E}_{\theta_0}[\varphi_0(T)] = \alpha \quad \text{and} \quad \mathrm{E}_{\theta_0}[T\varphi_0(T)] = \alpha \mathrm{E}_{\theta_0}[T],$$

as in Eqs. (6) and (7).

The inequality $f_3(t) < k_1 f_1(t) + k_2 f_2(t)$ is equivalent to $\mathrm{e}^{bt} - a_1 - a_2 t < 0$, where $a_i = k_i c(\theta_0)/c(\theta_1)$, $i = 1, 2$, and $b = \theta_1 - \theta_0$ with a fixed $\theta_1 \neq \theta_0$. If the resulting $\varphi_0(t)$ does not depend on the specific $\theta_1 \neq \theta_0$, then it is the desired solution to the problem.

The function $\mathrm{e}^{bt} - a_1 - a_2 t$ has positive second derivative and is therefore convex. Hence $\{t\colon \mathrm{e}^{bt} - a_1 - a_2 t < 0\}$ is either (c, ∞) , or $(-\infty, c')$, or (c_1, c_2). In the first two cases, φ_0 is a one-sided test having monotone power, which contradicts the property of $\beta_{\varphi_0}(\theta) = \mathrm{E}_\theta[\varphi_0(T)]$ being minimized at θ_0. Hence

$$\varphi_0(t) = \begin{cases} 0 & \text{if } c_1 < t < c_2 \\ \gamma_i & \text{if } t = c_i, \quad i = 1, 2 \\ 1 & \text{if } t < c_1 \text{ or } t > c_2, \end{cases}$$

where $c_1 < c_2$ and $0 \leq \gamma_i \leq 1$, $i = 1, 2$ are such that $\mathrm{E}_{\theta_0}[\varphi_0(T)] = \alpha$ and $\mathrm{E}_{\theta_0}[T\varphi_0(T)] = \alpha \mathrm{E}_{\theta_0}[T]$.

To see that φ_0 is an unbiased level α test of $H_0\colon \theta = \theta_0$ vs $H_1\colon \theta \neq \theta_0$, note that $\mathrm{E}_{\theta_0}[\varphi_0(T)] = \alpha$ by construction, and since $\varphi^*(t) \equiv \alpha$ is in the class in which φ_0 maximizes $\beta(\theta)$ for $\theta \neq \theta_0$, we must have $\mathrm{E}_\theta[\varphi_0(T)] \geq \mathrm{E}_\theta[\varphi^*(T)] = \alpha$ for all $\theta \neq \theta_0$, showing that φ_0 is UMP unbiased at level α for H_0 vs H_1.

Remark 6.7.1. If T is symmetrically distributed about c under P_{θ_0}, then choosing a test φ so that $E_{\theta_0}[\varphi(T)] = \alpha$ and φ is symmetric about c, the condition $E_{\theta_0}[T\varphi(T)] = \alpha E_{\theta_0}[T]$ is automatically satisfied.

Proof. Since T is symmetrically distributed about c under P_{θ_0},

$$\begin{aligned} &P_{\theta_0}[T - c < -u] = P_{\theta_0}[T - c > u] \text{ for all } u \in \mathbb{R}, \text{ so that} \\ &\qquad E_{\theta_0}[T - c] = 0, \text{ ie, } E_{\theta_0}[T] = c. \end{aligned}$$

Since φ is symmetric about c, that is, $\varphi(c - u) = \varphi(c + u)$ for all $u \in \mathbb{R}$, letting $T^* = T - c$, we have

$$\begin{aligned} E_{\theta_0}[T\varphi(T)] &= E_{\theta_0}[(T - c)\varphi(T)] + c[\varphi(T)] = E_{\theta_0}[T^*\varphi(T^* + c)] + c\alpha \\ &= \int u\varphi(u + c) p_{T^*}(u; \theta_0)\, du + \alpha E_{\theta_0}[T] = \alpha E_{\theta_0}[T] \end{aligned}$$

due to symmetry of φ and $p_{T^*}(\cdot; \theta_0)$. □

Thus φ_0 is obtained by taking $c_1 = c - k$, $c_2 = c + k$, $\gamma_1 = \gamma_2 = \gamma$, and choosing k and $0 \leq \gamma \leq 1$ so that $\mathrm{E}_{\theta_0}[\varphi_0(T)] = \alpha$.

6.7.2 UMP Unbiased Test for $H_0\colon \theta_1 \leq \theta \leq \theta_2$ vs $H_1\colon \theta \notin [\theta_1, \theta_2]$ in a Single-Parameter Exponential Family

Since all tests have continuous power functions in the context of exponential families, we can use Lemma 6.5.1 to narrow down our search among tests which are similar of size α

on the common boundary of $[\theta_1, \theta_2]$ and $[\theta_1, \theta_2]^c$ which is $\omega = \{\theta_1, \theta_2\}$ (ie, among all tests satisfying $\mathrm{E}_{\theta_i}[\varphi(T)] = \alpha$, $i = 1, 2$).

This leads to the problem of maximizing $\int \varphi(t) f_3(t)\, dt$ with respect to $\varphi \in \mathcal{F}$ subject to the conditions $\int \varphi(t) f_i(t)\, dt = \alpha$, $i = 1, 2$, where f_1, f_2, and f_3 are given in Eqs. (11) and (12).

Again by the Generalized N-P Lemma, the UMP unbiased level α test for H_0 vs H_1 in this problem is found to be *of the same form* as the UMP unbiased level α test for H_0: $\theta = \theta_0$ vs H_1: $\theta \neq \theta_0$, but here c_1, c_2 and $0 \leq \gamma_i \leq 1$ are determined by $\mathrm{E}_{\theta_i}[\varphi_0(T)] = \alpha$, $i = 1, 2$.

6.7.3 Examples

Example 6.7.1. Let $\boldsymbol{X} = (X_1, \ldots, X_n)$ be a random sample from $N(\mu, \sigma^2)$ with σ known. The goal is to find UMP unbiased level α tests for (a) H_0: $\mu = \mu_0$ vs H_1: $\mu \neq \mu_0$, and (b) H_0: $\mu_1 \leq \mu \leq \mu_2$ vs H_1: $\mu \notin [\mu_1, \mu_2]$.

Solution.

(a) Denote $T = \sqrt{n}(\bar{X} - \mu_0)/\sigma$. Then the UMP unbiased level α test is

$$\varphi(\mathbf{x}) = \begin{cases} 0 & \text{if } |T| \leq \Phi^{-1}(1 - \alpha/2) \\ 1 & \text{if } |T| > \Phi^{-1}(1 - \alpha/2), \end{cases}$$

because $\bar{X}$ is sufficient, $\mathrm{E}_{\mu_0}[\varphi(\mathbf{X})] = \alpha$ and $\mathrm{E}_{\mu_0}[(\bar{X} - \mu_0)\varphi(\mathbf{X})] = 0$ holds since $\bar{X} - \mu_0$ is symmetrically distributed about 0.

(b) Transform (μ_1, μ_2) to $(\bar{\mu}, \Delta) = \big((\mu_1 + \mu_2)/2, (\mu_2 - \mu_1)/2\big)$ and let $\theta = \mu - \bar{\mu}$. Also let $Y_i = X_i - \bar{\mu}$, $i = 1, \ldots, n$. Then the problem can be equivalently expressed as that of testing H_0: $-\Delta \leq \theta \leq \Delta$ vs H_1: $|\theta| > \Delta$ based on $Y_1, \ldots, Y_n$ which are iid $N(\theta, \sigma^2)$, and $T = \bar{Y} = n^{-1} \sum_{i=1}^n Y_i$ is sufficient for θ in $(Y_1, \ldots, Y_n)$.
Then the UMP unbiased level α test for H_0 vs H_1 is

$$\varphi_0(t) = \begin{cases} 0 & \text{if } c_1 < t < c_2 \\ 1 & \text{if } t \leq c_1 \text{ or } t \geq c_2, \end{cases}$$

where c_1, c_2 are such that $\mathrm{E}_{\theta=-\Delta}[\varphi_0(T)] = \mathrm{E}_{\theta=\Delta}[\varphi_0(T)] = \alpha$.
Since $T \sim N(\theta, \sigma^2/n)$, choosing $c_1 = -c$ and $c_2 = c$, we have

$$\begin{aligned} \mathrm{E}_{\theta=-\Delta}[\varphi_0(T)] &= P_{-\Delta}[T \leq -c] + P_{-\Delta}[T \geq c] \\ &= \Phi(\sqrt{n}(-c + \Delta)/\sigma) + \{1 - \Phi(\sqrt{n}(c + \Delta)/\sigma)\} \\ &= \{1 - \Phi(\sqrt{n}(c - \Delta)/\sigma)\} + \Phi(\sqrt{n}(-c - \Delta)/\sigma) \\ &= P_{\Delta}[T \geq c] + P_{\Delta}[T \leq -c] = \mathrm{E}_{\theta=\Delta}[\varphi_0(T)]. \end{aligned}$$

Thus, if we choose $c_1 = -c$ and $c_2 = c$ in $\varphi_0(t)$ and let c be such that $\mathrm{E}_{\theta=-\Delta}[\varphi_0(T)] = \alpha$, then $\mathrm{E}_{\theta=\Delta}[\varphi_0(T)] = \alpha$ is automatic.

Example 6.7.2. Let $\boldsymbol{X} = (X_1, \ldots, X_n)$ be a random sample from $\mathrm{Exp}(\theta)$. We will find the UMP unbiased level α test for H_0: $\theta = \theta_0$ vs H_1: $\theta \neq \theta_0$.

Solution. Here $T = \sum_{i=1}^{n} X_i$ is sufficient for θ in $\mathbf{X}$ with $T \mid \theta_0 \sim Gamma(n, 1)$ under H_0. The UMP unbiased level α test is

$$\varphi(t) = \begin{cases} 0 & \text{if } c_1 < t/\theta_0 < c_2 \\ 1 & \text{if } t/\theta_0 \leq c_1 \text{ or } t/\theta_0 \geq c_2, \end{cases}$$

where c_1 and c_2 are determined by

(i) $\int_{c_1}^{c_2} f_n(y)\, dy = 1 - \alpha$, and

(ii) $\int_{c_1}^{c_2} y f_n(y)\, dy = n(1-\alpha)$, where $f_n(y) = y^{n-1} e^{-y}/\Gamma(n)$, $y > 0$, is the pdf of $Gamma(n, 1)$ with mean n. Condition (ii) can be equivalently expressed in any of the following two ways, using integration by part and (i):

(ii)′ $\int_{c_1}^{c_2} f_{n+1}(y)\, dy = 1 - \alpha$ since $y f_n(y) = n f_{n+1}(y)$, or

(ii)″ $e^{-c_1} c_1^n = e^{-c_2} c_2^n$ since

$$\int_{c_1}^{c_2} y f_n(y)\, dy = (e^{-c_1} c_1^n - e^{-c_2} c_2^n)/\Gamma(n) + n(1-\alpha).$$

For moderately large n and for θ_0 neither too large nor too small, CLT provides a reasonable approximation for c_1, c_2 determined by

$$\int_0^{c_1} f_n(y)\, dy = \int_{c_2}^{\infty} f_n(y)\, dy = \alpha/2.$$

Example 6.7.3. Let X be a random sample of size 1 from *Geom*(p) with pmf

$$f(x, p) = pq^{x-1}, \quad x = 1, 2, \ldots, \text{ where } q = 1 - p.$$

We wish to find UMP unbiased test at level α for H_0: $p = p_0$ vs H_1: $p \neq p_0$.

Solution. The UMP unbiased level α test for H_0 vs H_1 is

$$\varphi_0(x) = \begin{cases} 0 & \text{if } c_1 < x < c_2 \\ \gamma_i & \text{if } x = c_i, \quad i = 1, 2 \\ 1 & \text{if } x < c_1 \text{ or } x > c_2, \end{cases}$$

where c_1, c_2, γ_1, and γ_2 are determined by

$$\begin{aligned} E_{p_o}[1 - \varphi_0(X)] &= p_0 q_0^{-1}\left[S_{c_2-1} - S_{c_1} + (1-\gamma_1)q_0^{c_1} + (1-\gamma_2)q_0^{c_2}\right] \\ &= 1 - \alpha, \text{ and} \\ E_{p_o}[X\{1 - \varphi_0(X)\}] &= p_0 q_0^{-1}\left[S^*_{c_2-1} - S^*_{c_1} + (1-\gamma_1)c_1 q_0^{c_1} + (1-\gamma_2)c_2 q_0^{c_2}\right] \\ &= (1-\alpha)p_0^{-1}, \end{aligned}$$

with

$$S_r = \sum_{i=1}^{r} q_0^i = q_0 p_0^{-1}(1 - q_0^r) \text{ and } S_r^* = \sum_{i=1}^{r} i q_0^i = q_0 p_0^{-2}[1 - (1 + rp_0)q_0^r].$$

6.8 Locally Best Tests

We have seen so far that UMP tests exist for one-sided problems of testing H_0: $\theta \le \theta_0$ vs H_1: $\theta > \theta_0$ (or vice versa) in MLR families, and UMP unbiased tests exist for two-sided problems of testing H_0: $\theta = \theta_0$ vs H_1: $\theta \neq \theta_0$ or for testing H_0: $\theta_1 \le \theta \le \theta_2$ vs H_1: $\theta \notin [\theta_1, \theta_2]$ in single-parameter exponential families. In subsequent developments, we are also going to construct UMP unbiased tests for one-sided and two-sided problems concerning any one parameter in multiparameter exponential families when other parameters (called nuisance parameters) are unknown but are of no concern. However, UMP or UMP unbiased tests are not available outside the MLR or the exponential families, so in more general situations, we would have to lower our expectation and settle for some more modest criterion of optimality.

In the one-sided problem, suppose that for every test φ, the power function $\beta_\varphi(\theta)$ has a continuous derivative which can be obtained by differentiating under the integral. In the context of a random sample $\boldsymbol{X} = (X_1, \ldots, X_n)$ from $f(x;\theta)$, this means

$$\begin{aligned}
\beta'_\varphi(\theta) &= \frac{d}{d\theta}\mathrm{E}_\theta[\varphi(\boldsymbol{X})] = \frac{d}{d\theta}\int \varphi(\boldsymbol{x})f(\boldsymbol{x};\theta)d\boldsymbol{x} \\
&= \int \varphi(\boldsymbol{x})\frac{\partial f(\boldsymbol{x};\theta)}{\partial\theta}d\boldsymbol{x} \\
&= \int \varphi(\boldsymbol{x})\left(\sum_{i=1}^{n}\frac{\partial \log f(\boldsymbol{x};\theta)}{\partial\theta}\right)f(\boldsymbol{x};\theta)d\boldsymbol{x} \\
&= \mathrm{E}_\theta\left[\varphi(\boldsymbol{X})\sum_{i=1}^{n}\dot{l}(X_i;\theta)\right],
\end{aligned}$$

where $\dot{l}(x;\theta) = \frac{\partial \log f(x;\theta)}{\partial\theta}$ is the partial derivative of the log likelihood $l(x;\theta) = \log f(x;\theta)$ with respect to θ as in Section 5.2.1.

We shall use the other notations introduced in Section 5.2.1, assuming that the regularity conditions introduced there holds here and use the results obtained under those conditions.

We now formulate our criterion for local optimality in the one-sided problem in a single-parameter family.

Definition 6.8.1. A test φ_0 is said to be a *locally most powerful* (LMP) test at level α for H_0: $\theta \le \theta_0$ vs H_1: $\theta > \theta_0$ if

(i) $\beta_{\varphi_0}(\theta_0) = \mathrm{E}_{\theta_0}[\varphi(\boldsymbol{X})] = \alpha$, and
(ii) $\beta'_{\varphi_0}(\theta_0) \ge \beta'_\varphi(\theta_0)$, that is,

$$\mathrm{E}_{\theta_0}\left[\varphi_0(\boldsymbol{X})\sum_{i=1}^{n}\dot{l}(X_i;\theta_0)\right] > \mathrm{E}_{\theta_0}\left[\varphi(\boldsymbol{X})\sum_{i=1}^{n}\dot{l}(X_i;\theta_0)\right]$$

for all tests φ satisfying (i).

Applying the Generalized N-P Lemma with $m = 1$, $f_1(\boldsymbol{x}) = f(\boldsymbol{x};\theta_0)$ and $f_2(\boldsymbol{x}) = \left\{\sum_{i=1}^n \dot{l}(x_i;\theta_0)\right\} f(\boldsymbol{x};\theta_0)$, we see that any test of the form:

$$\varphi_0(\boldsymbol{x}) = \begin{cases} 0 & \text{if } f_2(\boldsymbol{x}) < kf_1(\boldsymbol{x}) \text{ ie, } \sum_{i=1}^n \dot{l}(x_i;\theta_0) < k \\ \gamma & \text{if } f_2(\boldsymbol{x}) = kf_1(\boldsymbol{x}) \\ 1 & \text{if } f_2(\boldsymbol{x}) > kf_1(\boldsymbol{x}) \end{cases}$$

is an LMP test at level α for H_0 vs H_1 provided that

$$\mathrm{E}_{\theta_0}[\varphi_0(\boldsymbol{X})] = P_{\theta_0}\left[\sum_{i=1}^n \dot{l}(X_i;\theta_0) > k\right] + \gamma P_{\theta_0}\left[\sum_{i=1}^n \dot{l}(X_i;\theta_0) = k\right] = \alpha.$$

For large n, we can find the approximate value of k for a given α by the CLT. Recall that under the regularity conditions in Section 5.2.1,

$$\mathrm{E}_{\theta_0}\left[\dot{l}(X_i;\theta_0)\right] = 0 \quad \text{and} \quad \mathrm{Var}_{\theta_0}\left[\dot{l}(X_i;\theta_0)\right] = I(\theta_0),$$

where $I(\theta_0)$ is the Fisher-information. Thus

$$n^{-1/2}\sum_{i=1}^n \dot{l}(X_i;\theta_0) \xrightarrow{\mathcal{L}} N(0, I(\theta_0)) \text{ under } P_{\theta_0}.$$

Therefore, for large n, the critical value k for a given α can be approximated by

$$k = k_{n,\alpha} \simeq \sqrt{nI(\theta_0)}\Phi^{-1}(1-\alpha).$$

For testing H_0: $\theta = \theta_0$ vs H_1: $\theta \neq \theta_0$, we assume that the power function $\beta_\varphi(\theta)$ of every test φ has two continuous derivatives which can be obtained by differentiating under the integral. Thus for a test φ based on a random sample $\boldsymbol{X} = (X_1, \ldots, X_n)$ from $f(x;\theta)$, we have $\beta'_\varphi(\theta) = \mathrm{E}_\theta[\varphi(\boldsymbol{X})\sum_{i=1}^n \dot{l}(X_i;\theta)]$ as before, and

$$\begin{aligned}\beta''_\varphi(\theta) &= \int \varphi(\boldsymbol{x})\frac{\partial^2 f(\boldsymbol{x};\theta)}{\partial\theta^2}d\boldsymbol{x} \\ &= \mathrm{E}_\theta\left[\varphi(\boldsymbol{X})\left\{\sum_{i=1}^n \ddot{l}(X_i;\theta) + \left(\sum_{i=1}^n \dot{l}(X_i;\theta)\right)^2\right\}\right],\end{aligned}$$

using the identity:

$$\begin{aligned}\frac{\partial^2 f(\boldsymbol{x};\theta)}{\partial\theta^2} &= \left[\frac{\partial^2 \log f(\boldsymbol{x};\theta)}{\partial\theta^2} + \left(\frac{\partial \log f(\boldsymbol{x};\theta)}{\partial\theta}\right)^2\right] \\ &= \left[\sum_{i=1}^n \ddot{l}(x_i;\theta) + \left(\sum_{i=1}^n \dot{l}(x_i;\theta)\right)^2\right] f(\boldsymbol{x};\theta),\end{aligned}$$

where $\dot{l}(x;\theta)$ and $\ddot{l}(x;\theta)$ are the first two partial derivatives of $l(x;\theta) = \log f(x;\theta)$ with respect to θ.

Definition 6.8.2. A test φ_0 is said to be an LMP (locally) unbiased test at level α for $H_0: \theta = \theta_0$ vs $H_1: \theta \neq \theta_0$ if

(i) $\beta_{\varphi_0}(\theta_0) = \mathrm{E}_{\theta_0}[\varphi_0(\boldsymbol{X})] = \alpha$,
(ii) $\beta'_{\varphi_0}(\theta_0) = \mathrm{E}_{\theta_0}\left[\varphi_0(\boldsymbol{X}) \sum_{i=1}^n \dot{l}(X_i;\theta_0)\right] = 0$, and
(iii) $\beta''_{\varphi_0}(\theta_0) \geq \beta''_{\varphi}(\theta_0)$, that is,

$$\mathrm{E}_{\theta_0}\left[\varphi_0(\boldsymbol{X})\left\{\sum_{i=1}^n \ddot{l}(X_i;\theta_0) + \left(\sum_{i=1}^n \dot{l}(X_i;\theta_0)\right)^2\right\}\right]$$
$$\geq \mathrm{E}_{\theta_0}\left[\varphi(\boldsymbol{X})\left\{\sum_{i=1}^n \ddot{l}(X_i;\theta_0) + \left(\sum_{i=1}^n \dot{l}(X_i;\theta_0)\right)^2\right\}\right]$$

for all tests φ satisfying (i) and (ii).

We can now apply the Generalized N-P Lemma with $m = 2$, $f_1(\boldsymbol{x}) = f(\boldsymbol{x};\theta_0)$, $f_2(\boldsymbol{x}) = Uf(\boldsymbol{x};\theta_0)$, $f_3(\boldsymbol{x}) = Vf(\boldsymbol{x};\theta_0)$, where

$$U = \sum_{i=1}^n \dot{l}(x_i;\theta_0) \quad \text{and} \quad V = \sum_{i=1}^n \ddot{l}(x_i;\theta_0) + \left(\sum_{i=1}^n \dot{l}(x_i;\theta_0)\right)^2,$$

to see that any test of the form:

$$\varphi_{k_1,k_2}(\boldsymbol{x}) = \begin{cases} 0 & \text{if } V < k_1 + k_2 U \\ \gamma & \text{if } V = k_1 + k_2 U \\ 1 & \text{if } V > k_1 + k_2 U \end{cases}$$

is an LMP unbiased test for H_0 vs H_1 at level α, provided that k_1 and k_2 are chosen so as to satisfy (i) and (ii).

Finding k_1, k_2 for a given α is difficult in general, but in the special case when $(U, V) \stackrel{\mathcal{D}}{=} (-U, V)$ under P_{θ_0}, the problem is simplified because we can take $k_2 = 0$.

Proposition 6.8.1. *If $(U, V) \stackrel{\mathcal{D}}{=} (-U, V)$ under P_{θ_0}, then for any k_1, condition (ii) holds iff $k_2 = 0$.*

Proof. If $(U, V) \stackrel{\mathcal{D}}{=} (-U, V)$ under P_{θ_0}, then for $k_2 = 0$,

$$\mathrm{E}_{\theta_0}\left[U\varphi_{k_1,0}(\boldsymbol{X})\right] = \mathrm{E}_{\theta_0}\left[UI_{(k_1,\infty)}(V)\right] = \mathrm{E}_{\theta_0}\left[-UI_{(k_1,\infty)}(V)\right]$$
$$= -\mathrm{E}_{\theta_0}\left[U\varphi_{k_1,0}(\boldsymbol{X})\right].$$

Hence $\mathrm{E}_{\theta_0}\left[U\varphi_{k_1,0}(\boldsymbol{X})\right] = 0$. Conversely, if $k_2 > 0$, then

$$\mathrm{E}_{\theta_0}[U\varphi_{k_1,k_2}(\boldsymbol{X})] = \int_{u=-\infty}^{\infty}\int_{v=k_1+k_2 u}^{\infty} uf(u,v;\theta_0)\,du\,dv$$
$$= \sum_{j=1}^{3}\iint_{S_j} uf(u,v;\theta_0)\,du\,dv$$
$$= I_1 + I_2 + I_3, \text{ say}$$

where

$$S_1 = \{(u, v)\colon u \geq 0, k_1 + k_2 u \leq v < \infty\},$$
$$S_2 = \{(u, v)\colon u < 0, k_1 - k_2 u \leq v < \infty\}, \text{ and}$$
$$S_3 = \{(u, v)\colon u < 0, k_1 + k_2 u \leq v < k_1 - k_2 u\}.$$

Since $(u, v) \in S_1 \iff (-u, v) \in S_2$ and $(U, V) \overset{\mathcal{D}}{=} (-U, V)$ under θ_0, it follows that

$$I_2 = \iint_{S_2} uf(u, v; \theta_0)\, du\, dv = -\iint_{S_2} (-u) f(-u, v; \theta_0)\, du\, dv$$
$$= -\iint_{S_1} uf(u, v; \theta_0)\, du\, dv = -I_1.$$

Hence

$$\mathrm{E}_{\theta_0}\left[U\varphi_{k_1,k_2}(\boldsymbol{X})\right] = I_3 = \iint_{S_3} uf(u, v; \theta_0)\, du\, dv < 0$$

because $u < 0$ on S_3. Hence in order to satisfy $\mathrm{E}[U\varphi_{k_1,k_2}(\boldsymbol{X})] = 0$, we must have $P_{\theta_0}[S_3] = 0$. However, if $P_{\theta_0}[S_3] = 0$, then $\mathrm{E}_{\theta_0}[\varphi_{k_1,k_2}(\boldsymbol{X})] = \mathrm{E}_{\theta_0}[\varphi_{k_1,0}(\boldsymbol{X})]$. Thus both (i) and (ii) are satisfied with $k_2 = 0$. The case of $k_2 < 0$ is treated in the same way. □

Example 6.8.1. Let $X = (X_1, \ldots, X_n)$ be a random sample from a logistic distribution with location parameter θ, that is, from

$$f(x, \theta) = \exp[-(x - \theta)]/\{1 + \exp[-(x - \theta)]\}^2, \quad -\infty < x < \infty.$$

We wish to find the locally best

(a) test for $H_0\colon \theta = 0$ vs $H_1\colon \theta > 0$ at level α , and
(b) unbiased test for $H_0\colon \theta = 0$ vs $H_1\colon \theta \neq 0$ at level α.

Solution. Note that

$$l(x, \theta) = \log f(x, \theta) = -(x - \theta) - 2\log\left[1 + \exp\{-(x - \theta)\}\right],$$

$$\dot{l}(x, \theta) = \frac{\partial}{\partial\theta} l(x, \theta) = [\exp(x - \theta) - 1]/[\exp(x - \theta) + 1],$$

$$\ddot{l}(x, \theta) = \frac{\partial^2}{\partial\theta^2} l(x, \theta) = -2\exp(x - \theta)/(\exp(x - \theta) + 1)^2.$$

Thus we have

$$U = \sum_{i=1}^{n} \dot{l}(X_i, 0) = \sum_{i=1}^{n} (\mathrm{e}^{X_i} - 1)/(\mathrm{e}^{X_i} + 1),$$

$$V = \sum_{i=1}^{n} \ddot{l}(X_i, 0) + \left\{\sum_{i=1}^{n} \dot{l}(X_i, 0)\right\}^2 = -2\sum_{i=1}^{n} \mathrm{e}^{X_i}/(1 + \mathrm{e}^{X_i})^2 + U^2.$$

(a) The LMP test at level α for $H_0\colon \theta = 0$ vs $H_1\colon \theta > 0$ is

$$\varphi_1(\boldsymbol{x}) = \begin{cases} 0 & \text{if } U < k \\ \gamma & \text{if } U = k \\ 1 & \text{if } U > k, \end{cases}$$

where $P_0[U > k] + \gamma P_0[U = k] = \alpha$.

(b) The LMP (locally) unbiased test at level α for $H_0\colon \theta = 0$ vs $H_1\colon \theta \neq 0$ is

$$\varphi_0(\boldsymbol{x}) = \begin{cases} 0 & \text{if } V < k_1 + k_2 U \\ \gamma & \text{if } V = k_1 + k_2 U \\ 1 & \text{if } V > k_1 + k_2 U, \end{cases}$$

where $0 \leq \gamma \leq 1$ and k_1, k_2 are determined by (i) $E_0[\varphi_0(\boldsymbol{X})] = \alpha$ and (ii) $E_0[\varphi_0(\boldsymbol{X})U] = 0$.

However, by virtue of symmetry of $f(x, 0)$, $(U, V) \stackrel{\mathcal{D}}{=} (-U, V)$. Hence k_2 can be taken to be equal to 0 and it is enough to choose $0 \leq \gamma \leq 1$ and k_1 to satisfy (i), since (ii) automatically holds.

The condition $(U, V) \stackrel{\mathcal{D}}{=} (-U, V)$ under P_{θ_0} is satisfied in a number of situations such as

(i) $X \sim \mathit{Cauchy}(\theta, 1)$ with $f(x;\theta) = \frac{1}{\pi}\frac{1}{1+(x-\theta)^2}$, $x \in \mathbb{R}$,

(ii) $X \sim$ Double Exponential(θ) with $f(x;\theta) = \frac{1}{2}\exp[-|x-\theta|]$, $x \in \mathbb{R}$, etc.

6.9 UMP Unbiased Tests in the Presence of Nuisance Parameters: Similarity and Completeness

In many situations, the distribution of the observed X depends on several parameters and we want to test a null hypothesis H_0 against an alternative H_1 concerning only one of these parameters. The other parameters are called *nuisance parameters*.

Example 6.9.1. Let $\boldsymbol{X} = (X_1, \ldots, X_n)$ be a random sample from $N(\mu, \sigma^2)$, where both μ, σ^2 are unknown. We want to test $H_1\colon \mu = 0$ vs $H_1\colon \mu \neq 0$. Here σ^2 is a nuisance parameter.

Example 6.9.2. Let $(X_1, \ldots, X_m)$ and $(Y_1, \ldots, Y_n)$ be independent random samples from $\mathit{Poisson}(\lambda)$ and $\mathit{Poisson}(\mu)$, respectively. We want to test $H_0\colon \mu < a\lambda$ vs $H_1\colon \mu > a\lambda$ for a given a . To see how this problem involves testing for one parameter in the presence of a nuisance parameter, first consider the joint distribution of the sufficient statistic $(X, Y) = (\sum_{i=1}^m X_i, \sum_{i=1}^n Y_i)$ given by:

$$f_{X,Y}(x, y; \lambda, \mu) = \mathrm{e}^{-(m\lambda+n\mu)}\frac{(m\lambda)^x(n\mu)^y}{x!y!}, \quad x = 0, 1, 2, \ldots, \; y = 0, 1, 2, \ldots.$$

Now reparametrize: $\pi = n\mu/(m\lambda + n\mu)$, $\xi = m\lambda + n\mu$ and transform the data: $U = Y$, $T = X + Y$. Then the joint distribution of (U, T) is

$$f_{U,T}(u, t; \pi, \xi) = \left\{\binom{t}{u}\pi^u(1-\pi)^{t-u}\right\}\left\{\mathrm{e}^{-\xi}\xi^t/t!\right\},$$

$t = 0, 1, 2, \ldots$ and $u = 0, 1, \ldots, t$. Now H_0: $\mu \le a\lambda \iff \pi \le na/(m + na)$ and H_1: $\mu > a\lambda \iff \pi > na/(m + na)$ while ξ is a nuisance parameter.

Let $\{P^X_{\boldsymbol{\theta}}, \boldsymbol{\theta} \in \Theta\}$ be a family of probabilities on $(\mathfrak{X}, \mathcal{A})$, where $\Theta \subset \mathbb{R}^{k+1}$ and $\boldsymbol{\theta}$ is a $(k + 1)$-dim vector denoted by

$$\boldsymbol{\theta} = (\theta_1, \theta_2, \ldots, \theta_{k+1}) := (\theta, \tau_1, \ldots, \tau_k) := (\theta, \boldsymbol{\tau}) \text{ with}$$
$$\theta = \theta_1 \text{ and } \boldsymbol{\tau} = (\theta_2, \ldots, \theta_{k+1}).$$

Based on a random element X generated by an unknown element $P^X_{\boldsymbol{\theta}}$ of this family, we want to find unbiased level α tests in three situations where the null and the alternative hypotheses are given as:

Problem 1. H_{10}: $\theta \le \theta_0$ vs H_{11}: $\theta > \theta_0$ (testing $\theta \ge \theta_0$ vs $\theta < \theta_0$ is analogous),
Problem 2. H_{20}: $\theta = \theta_0$ vs H_{21}: $\theta \ne \theta_0$,
Problem 3. H_{30}: $\theta_1 \le \theta \le \theta_2$ vs H_{31}: $\theta \notin [\theta_1, \theta_2]$,

in each case, $\boldsymbol{\tau}$ being the nuisance parameter.

A more careful description of the null and the alternative hypotheses in *these three problems* are

$$H_{i0} : (\theta, \boldsymbol{\tau}) \in \Theta_{i0} \text{ vs } H_{i1}: (\theta, \boldsymbol{\tau}) \in \Theta_{i1}, \text{ where}$$
$$\Theta_{10} = \{(\theta, \boldsymbol{\tau}): \theta \le \theta_0,\ \boldsymbol{\tau} \in \Omega\},\ \Theta_{11} = \{(\theta, \boldsymbol{\tau}): \theta > \theta_0,\ \boldsymbol{\tau} \in \Omega\},$$
$$\Theta_{20} = \{(\theta, \boldsymbol{\tau}): \theta = \theta_0,\ \tau \in \Omega\},\ \Theta_{21} = \{(\theta, \boldsymbol{\tau}): \theta \ne \theta_0,\ \boldsymbol{\tau} \in \Omega\},$$
$$\Theta_{30} = \{(\theta, \boldsymbol{\tau}): \theta_1 \le \theta \le \theta_2,\ \boldsymbol{\tau} \in \Omega\},$$
$$\Theta_{31} = \{(\theta, \boldsymbol{\tau}): \theta \notin [\theta_1, \theta_2],\ \boldsymbol{\tau} \in \Omega\}, \text{ where } \Omega \subset \mathbb{R}^k.$$

Now in each of these cases, the null hypothesis and the alternative hypothesis sets Θ_{i0} and Θ_{i1} have a common boundary ω_i, where

$$\omega_1 = \omega_2 = \{\theta_0\} \times \Omega \quad \text{and} \quad \omega_3 = \{\theta_1, \theta_2\} \times \Omega.$$

In the sequel, we assume that the power functions $\beta_\varphi(\theta, \boldsymbol{\tau})$ of all tests φ are continuous in $(\theta, \boldsymbol{\tau})$, so that all unbiased level α tests are similar tests of size α on the boundary ω_i of Θ_{i0} and Θ_{i1}, $i = 1, 2, 3$ in each of the three problems under consideration.

We now omit the subscript i from the boundary ω_i of Θ_{i0} and Θ_{i1}, and continue the development in some generality. Restrict attention to the family of distributions $\mathcal{P}^X_\omega = \{P^X_{\theta,\boldsymbol{\tau}}, (\theta, \boldsymbol{\tau}) \in \omega\}$ and suppose that $\boldsymbol{T}$ is a sufficient statistic for the family $\mathcal{P}^X_\omega$ (ie, the conditional distribution of X given $\boldsymbol{T}$ is the same for all $(\theta, \boldsymbol{\tau}) \in \omega$). Indeed, we should be looking for a sufficient statistic $\boldsymbol{T}$ for $\boldsymbol{\tau} = (\tau_1, \ldots, \tau_k)$ since θ_0 in Problems 1, 2 and θ_1, θ_2 in Problem 3 are fixed and known.

Definition 6.9.1 (Tests of Neyman Structure). Tests satisfying $\mathrm{E}[\varphi(X)|T = t] = \alpha$ for almost all t under $\mathcal{P}^T_\omega$ (ie, for all $t \notin N$, where $P_\theta[T \in N] = 0$ for all $\theta \in \omega$) are called tests of Neyman-structure with respect to the sufficient statistic T.

If a test is of Neyman-structure, then

$$\mathrm{E}_\theta[\varphi(X)] = \mathrm{E}_\theta \mathrm{E}[\varphi(X)|T] = \mathrm{E}_\theta[\alpha] = \alpha \quad \text{for all } \theta \in \omega,$$

that is, φ is similar of size α on ω. Thus all tests of Neyman-structure are similar on ω.

A sufficient statistic T for $\mathcal{P}_\omega^X$ is *complete* if $\int g(t) f_T(t;\theta)\, dt = 0$ for all $\theta \in \omega \Rightarrow g(t) = 0$ for almost all t under $\mathcal{P}_\omega^T$ (ie, $g(t) = 0$ for all $t \notin N$ where $P_\theta[T \in N] = 0$ for all $\theta \in \omega$). The property of completeness of a sufficient statistic has already been used in connection with UMVUE in Section 5.1.

A sufficient statistic T for $\mathcal{P}_\omega^X$ is *boundedly complete* if $\int g(t) f_T(t;\theta)\, dt = 0$ for all $\theta \in \omega$, g bounded $\Rightarrow g(t) = 0$ for almost all t under $\mathcal{P}_\omega^T$. [If T is complete, then it is boundedly complete, but the converse is not true.]

We now have the following theorem.

Theorem 6.9.1. *Suppose that there exists a sufficient statistic T for $\mathcal{P}_\omega^X$. If T is boundedly complete, then every test which is similar on ω, is of Neyman-structure with respect to T.*

Proof. Let φ be a similar test of size α on ω. Then

$$\mathrm{E}_\theta[\varphi(X)] = \alpha \quad \text{for all } \theta \in \omega \Leftrightarrow$$
$$\mathrm{E}_\theta[\varphi(X) - \alpha] = \mathrm{E}_\theta \mathrm{E}[\varphi(X) - \alpha|T] = 0 \quad \text{for all } \theta \in \omega.$$

Since $\psi(t) = \mathrm{E}[\varphi(X) - \alpha|T = t]$ is a bounded function of t,

$$\mathrm{E}_\theta[\psi(t) = 0] \quad \text{for all } \theta \in \omega$$

$\Rightarrow \psi(t) = 0$ for almost all t under $\mathcal{P}_\omega^T \Leftrightarrow \mathrm{E}[\varphi(X)|T = t] = \alpha$ for almost all t under $\mathcal{P}_\omega^T$, that is, φ is a test of Neyman-structure with respect to T. □

The converse (ie, if T is not boundedly complete, then there exists a test which is similar on α but is not of Neyman-structure with respect to T) is also true (see [3, p. 134]), but we do not need it for our purpose.

Let $\boldsymbol{X} = (X_1, \ldots, X_n)$ be a random sample from a $(k+1)$-parameter exponential family with pdf/pmf:

$$f_X(x;\theta,\boldsymbol{\tau}) = C_0(\theta,\boldsymbol{\tau}) \exp\left[\theta U(x) + \sum_{j=1}^{k} \tau_j T_j(x)\right] h_0(x), \quad (\theta,\boldsymbol{\tau}) \in \Theta,$$

where Θ is a convex set in $\mathbb{R}^{k+1}$.

We shall now use the above theorem to find UMP unbiased level α tests in the three problems listed above.

Restrict attention to tests based on

$$U = \sum_{i=1}^{n} U(X_i) \quad \text{and} \quad \boldsymbol{T} = (T_1, \ldots, T_k) = \left(\sum_{i=1}^{n} T_1(X_i), \ldots, \sum_{i=1}^{n} T_k(X_i)\right),$$

since $(U, \boldsymbol{T})$ is sufficient for $(\theta, \boldsymbol{\tau})$ in $\boldsymbol{X}$. The joint pdf/pmf of $(U, \boldsymbol{T})$ is of the form:

$$f_{U,\boldsymbol{T}}(u, \boldsymbol{t}; \theta, \boldsymbol{\tau}) = C(\theta, \boldsymbol{\tau}) \exp\left[\theta u + \sum_{j=1}^{k} \tau_j t_j\right] h(u, \boldsymbol{t}), \quad (\theta, \boldsymbol{\tau}) \in \Theta.$$

We assume that:

(a) $\Theta = I \times \Omega$ where I is an interval in $\mathbb{R}$ and Ω is a convex set in $\mathbb{R}^k$, and recall that the boundaries of the null and the alternative hypotheses in the three problems are

$$\omega_1 = \omega_2 = \{(\theta_0, \boldsymbol{\tau}) \colon \boldsymbol{\tau} \in \Omega\} \text{ in Problems 1 and 2, and}$$
$$\omega_3 = \omega_{31} \cup \omega_{32} = \{(\theta_1, \boldsymbol{\tau}) \colon \boldsymbol{\tau} \in \Omega\} \cup \{(\theta_2, \boldsymbol{\tau}) \colon \boldsymbol{\tau} \in \Omega\} \text{ in Problem 3.}$$

(b) In Problems 1 and 2, θ_0 is an interior point of I and in Problem 3, θ_1 and θ_2 are interior points of I.

(c) Ω contains a nondegenerate k-dim rectangle in $\mathbb{R}^k$.

In each of the three problems under consideration, the statistic $\boldsymbol{T}$ is sufficient for $\mathcal{P}_\omega^{\boldsymbol{X}}$ on the common boundary ω of the null and the alternative hypotheses (ie, for $\mathcal{P}_{\omega_i}^{\boldsymbol{X}}$ in Problems 1, 2, and 3). Also, the marginal distribution of $\boldsymbol{T}$ and the conditional distribution of U given $\boldsymbol{T} = \boldsymbol{t}$ are described by their pdf's/pmf's which are of the form:

$$f_{\boldsymbol{T}}(\boldsymbol{t}; \theta, \boldsymbol{\tau}) = C(\theta, \boldsymbol{\tau}) \exp\left[\sum_{j=1}^{k} \tau_j t_j\right] h_\theta(\boldsymbol{t}), \text{ and}$$
$$f_{U|\boldsymbol{t}}(u; \theta) = C_{\boldsymbol{t}}(\theta) \exp[\theta u] h_{\boldsymbol{t}}(u).$$

[In the discrete case, $h_\theta(\boldsymbol{t}) = \sum_u \exp[\theta u] h(u, \boldsymbol{t})$, $C_{\boldsymbol{t}}(\theta) = 1/h_\theta(\boldsymbol{t})$, $h_{\boldsymbol{t}}(u) = h(u, \boldsymbol{t})$.]

Note that in all three cases, the sufficient statistic $\boldsymbol{T}$ for $\mathcal{P}_\omega^{\boldsymbol{X}}$ is complete by virtue of Assumption (c) and is, therefore, boundedly complete. Hence all tests which are similar on ω_i, $i = 1, 2, 3$ are of Neyman-structure with respect to $\boldsymbol{T}$, which suggests the following strategy:

(i) For each $\boldsymbol{t}$, restrict attention to the family $\{f_{U|\boldsymbol{t}}(u, \theta),\ \theta \in I\}$ and find the UMP level α test for $H_{10}\colon \theta \le \theta_0$ vs $H_1\colon \theta > \theta_0$ in Problem 1, and the UMP unbiased level α tests for $H_{20}\colon \theta = \theta_0$ vs $H_{21}\colon \theta \ne \theta_0$ and for $H_{30}\colon \theta_1 \le \theta \le \theta_2$ vs $H_{31}\colon \theta \notin [\theta_1, \theta_2]$, respectively, in Problems 2 and 3. [We have already obtained the solutions to these problems in Sections 6.4, 6.7.1, and 6.7.2.]

(ii) Put these piecewise optimal tests together and show that the resulting test is an unbiased level α test in each problem.

We now go through the details of each problem.

Problem 1 (UMP Unbiased Level α Test for $H_{10}\colon \theta \le \theta_0$ vs $H_{11}\colon \theta > \theta_0$). We want a test φ such that

(i) $\mathrm{E}_{\theta,\boldsymbol{\tau}}[\varphi(U, \boldsymbol{T})|\boldsymbol{T} = \boldsymbol{t}] = \alpha$ for all $(\theta, \boldsymbol{\tau}) \in \omega_1$, ie, $\mathrm{E}_{\theta_0}[\varphi(U, \boldsymbol{t})|\boldsymbol{T} = \boldsymbol{t}] = \alpha$ for almost all $\boldsymbol{t}$ under $\mathcal{P}_{\omega_1}^{\boldsymbol{T}}$ (written as: a.e. $\mathcal{P}_{\omega_1}^{\boldsymbol{T}}$), and

(ii) $\mathrm{E}_{\theta,\boldsymbol{\tau}}[\varphi(U,\boldsymbol{T})] \geq \mathrm{E}_{\theta,\boldsymbol{\tau}}[\psi(U,\boldsymbol{T})]$ for all $\theta > \theta_0$ and $\boldsymbol{\tau} \in \Omega$ and for every test ψ satisfying (i).

[(i) is the condition of Neyman-structure with respect to $\boldsymbol{T}$ and (ii) ensures the UMP property in this class.]

For every $\boldsymbol{t}$, $\{f_{U|\boldsymbol{t}}(u;\theta) = C_{\boldsymbol{t}}(\theta)\exp[\theta u]h_{\boldsymbol{t}}(u),\ \theta \in I\}$ is an MLR family, so invoking the result in Section 6.4, the piecewise solution to the problem is seen to be

$$\varphi_1(u,\boldsymbol{t}) = \begin{cases} 0 & \text{if } u < c_1(\boldsymbol{t}) \\ \gamma_1(\boldsymbol{t}) & \text{if } u = c_1(\boldsymbol{t}) \\ 1 & \text{if } u > c_1(\boldsymbol{t}), \end{cases}$$

where $c_1(\boldsymbol{t})$ and $0 \leq \gamma_1(\boldsymbol{t}) \leq 1$ are determined by $\mathrm{E}_{\theta_0}[\varphi_1(U,\boldsymbol{t})|\boldsymbol{T}=\boldsymbol{t}] = \alpha$.

To show that φ_1 is a UMP level α test for $H_{10}\colon \theta \leq \theta_0$ vs $H_{11}\colon \theta > \theta_0$, it is enough to check that φ_1 is a level α test for H_{10} vs H_{11}, using Lemma 6.5.1. By the monotone power property of φ_1 in the conditional problem (using part (ii) of Theorem 6.4.1) and the MLR property of $\{f_{U|\boldsymbol{t}}(u;\theta),\ \theta \in I\}$,

$$\mathrm{E}_{\theta_1}\left[\varphi_1(U,\boldsymbol{t})|\boldsymbol{T}=\boldsymbol{t}\right] \leq \mathrm{E}_{\theta_0}\left[\varphi_1(U,\boldsymbol{t})|\boldsymbol{T}=\boldsymbol{t}\right] = \alpha, \text{ a.e. } \mathcal{P}^{\boldsymbol{T}}_{\omega_1} \text{ for all } \theta_1 \leq \theta_0.$$

Hence $\mathrm{E}_{\theta_1,\boldsymbol{\tau}}[\varphi_1(U,\boldsymbol{T})] \leq \alpha$ for all $\theta_1 \leq \theta_0$, showing that φ_1 is a level α test for H_{10} vs H_{11}.

Problem 2 (UMP Unbiased Level α Test for $H_{20}\colon \theta = \theta_0$ vs $H_{21}\colon \theta \neq \theta_0$). We want a test φ such that

(i) $\mathrm{E}_{\theta,\boldsymbol{\tau}}[\varphi(U,\boldsymbol{T})|\boldsymbol{T}=\boldsymbol{t}] = \alpha$ for all $(\theta,\boldsymbol{\tau}) \in \omega_2$, ie, $\mathrm{E}_{\theta_0}[\varphi(U,\boldsymbol{t})|\boldsymbol{T}=\boldsymbol{t}] = \alpha$, a.e. $\mathcal{P}^{\boldsymbol{T}}_{\omega_2}$ (as in Problem 1),

(ii) $\frac{\partial}{\partial\theta}\mathrm{E}_{\theta,\boldsymbol{\tau}}[\varphi(U,\boldsymbol{T})] = 0$ on ω_2, ie, $\frac{\partial}{\partial\theta}\mathrm{E}_{\theta,\boldsymbol{\tau}}[\varphi(U,\boldsymbol{T})]\big|_{\theta=\theta_0,\boldsymbol{\tau}\in\Omega} = 0$, and

(iii) $\mathrm{E}_{\theta,\boldsymbol{\tau}}[\varphi(U,\boldsymbol{T})] \geq \mathrm{E}_{\theta,\boldsymbol{\tau}}[\psi(U,\boldsymbol{T})]$ for all $\theta \neq \theta_0$ and $\boldsymbol{\tau} \in \Omega$, whenever ψ is a test satisfying (i) and (ii).

[(i) is the condition of Neyman-structure with respect to $\boldsymbol{T}$, (ii) is implied by unbiasedness, and (iii) ensures the UMP property in this class.]

Since θ_0 is an interior point of I, $\frac{\partial}{\partial\theta}\mathrm{E}_{\theta,\boldsymbol{\tau}}[\varphi(U,\boldsymbol{T})]$ exists on $\{\theta_0\} \times \Omega$ and can be calculated by differentiating under the integral. Thus

$$\begin{aligned}
&\frac{\partial}{\partial\theta}\mathrm{E}_{\theta,\boldsymbol{\tau}}[\varphi(U,\boldsymbol{T})] \\
&= \frac{\partial}{\partial\theta}\int \varphi(u,\boldsymbol{t})C(\theta,\boldsymbol{\tau})\mathrm{e}^{\theta u+\sum_{j=1}^k \tau_j t_j}h(u,\boldsymbol{t})\,du\,d\boldsymbol{t} \\
&= \int \varphi(u,\boldsymbol{t})\left[\frac{\partial C(\theta,\boldsymbol{\tau})}{\partial\theta}\mathrm{e}^{\theta u+\sum_{j=1}^k \tau_j t_j} + C(\theta,\boldsymbol{\tau})u\mathrm{e}^{\theta u+\sum_{j=1}^k \tau_j t_j}\right]h(u,\boldsymbol{t})\,du\,d\boldsymbol{t} \\
&= \frac{\frac{\partial C(\theta,\boldsymbol{\tau})}{\partial\theta}}{C(\theta,\boldsymbol{\tau})}\mathrm{E}_{\theta,\boldsymbol{\tau}}[\varphi(U,\boldsymbol{T})] + \mathrm{E}_{\theta,\boldsymbol{\tau}}[U\varphi(U,\boldsymbol{T})],
\end{aligned}$$

so condition (ii) amounts to the last expression evaluated at $\theta = \theta_0$ being 0 for all $\boldsymbol{\tau} \in \Omega$.

Since $\varphi^*(u,\boldsymbol{t}) \equiv \alpha$ is an unbiased level α test for H_{20} vs H_{21}, it follows that

$$\frac{\frac{\partial C(\theta_0,\boldsymbol{\tau})}{\partial\theta}}{C(\theta_0,\boldsymbol{\tau})}\alpha + \alpha\mathrm{E}_{\theta_0,\boldsymbol{\tau}}[U] = 0, \text{ ie, } \frac{\frac{\partial C(\theta_0,\boldsymbol{\tau})}{\partial\theta}}{C(\theta_0,\boldsymbol{\tau})} = -\mathrm{E}_{\theta_0,\boldsymbol{\tau}}[U].$$

Condition (ii) thus becomes $E_{\theta_0,\tau}[U\varphi(U,T) - \alpha U] = 0$ for all $\tau \in \Omega$, that is, $\int E_{\theta_0}[U\varphi(U,t) - \alpha U|T = t] f_T(t;\theta_0,\tau)dt = 0$, which implies:

$$E_{\theta_0}[U\varphi(U,t)|T = t] = \alpha E_{\theta_0}[U|T = t], \text{ a.e. } \mathcal{P}_{\omega_2}^{T},$$

since T is complete.

Conditions (i) and (ii) thus become

$$E_{\theta_0}[\varphi(U,t)|T = t] = \alpha \text{ and } E_{\theta_0}[U\varphi(U,t)] = \alpha E_{\theta_0}[U|T = t], \text{ a.e. } \mathcal{P}_{\omega_2}^{T}$$

and subject to these conditions, we now maximize $E_\theta[\varphi(U,t)|T = t]$ for each t, where $\theta \neq \theta_0$.

This conditional problem has already been solved in Section 6.7.1, since for each t, $\{f_{U|t}(u;\theta) = C_t(\theta)\exp[\theta u]h_t(u),\ \theta \in I\}$ is an exponential family. The maximizing test is given as

$$\varphi_2(u,t) = \begin{cases} 0 & \text{if } c_{21}(t) < u < c_{22}(t) \\ \gamma_{2i}(t) & \text{if } u = c_{2i}(t), \quad i = 1,2 \\ 1 & \text{if } u < c_{21}(t) \text{ or } u > c_{22}(t), \end{cases}$$

where $c_{2i}(t)$ and $0 \leq \gamma_{2i}(t) \leq 1$, $i = 1, 2$, for each t are determined by

$$E_{\theta_0}[\varphi_2(U,t)|T = t] = \alpha \text{ and } E_{\theta_0}[U\varphi_2(U,t)|T = t] = \alpha E_{\theta_0}[U|T = t].$$

To show that φ_2 is a UMP level α test for H_{20} vs H_{21}, we only need to verify that φ_2 is a level α test, as shown below:

$$E_{\theta_0}[\varphi_2(U,t)|T = t] = \alpha, \text{ a.e., } \mathcal{P}_{\omega_2}^{T}$$

$$\Rightarrow E_{\theta_0,\tau}[\varphi_2(U,T)] = E_{\theta_0,\tau}E_{\theta_0}[\varphi_2(U,T)|T] = E_{\theta_0,\tau}[\alpha] = \alpha \quad \text{for all } \tau \in \omega_2.$$

Problem 3. UMP Unbiased Level α Test for H_{30}: $\theta_1 \leq \theta \leq \theta_2$ vs H_{31}: $\theta \notin [\theta_1, \theta_2]$.

Here we maximize $E_\theta[\varphi(U,t)|T = t]$, $\theta \notin [\theta_1,\theta_2]$, subject to $E_{\theta_i}[\varphi(U,t)|T = t] = \alpha$, $i = 1, 2$ for each t. As seen in Section 6.7.2, this results in the test:

$$\varphi_3(u,t) = \begin{cases} 0 & \text{if } c_{31}(t) < u < c_{32}(t) \\ \gamma_{3i}(t) & \text{if } u = c_{3i}(t), \ i = 1,2 \\ 1 & \text{if } u < c_{31}(t) \text{ or } u > c_{32}(t), \end{cases}$$

where $c_{3i}(t)$ and $0 \leq \gamma_{3i}(t) \leq 1$, $i = 1, 2$ for each t is determined by

$$E_{\theta_i}[\varphi_3(U,t)|T = t] = \alpha, \quad i = 1,2.$$

Examples of Conditional Tests

Example 6.9.3. Let $X = (X_1, \ldots, X_m)$ and $Y = (Y_1, \ldots, Y_n)$ be independent random samples from *Poisson*(λ) and *Poisson*(μ), respectively. We want to test H_0: $\mu = a\lambda$ vs H_1: $\mu \neq a\lambda$ where $a > 0$ is given. Here $(X, Y) = \left(\sum_{i=1}^m X_i, \sum_{i=1}^n Y_i\right)$ is sufficient for (λ, μ) in (X, Y). As in

Example 6.9.2, reparametrize and transform the data as:

$$\pi = n\mu/(m\lambda + n\mu),\ \xi = m\lambda + n\mu, \text{ and } U = Y, \quad T = X + Y,$$

to arrive at

$$f_{U,T}(u, t; \pi, \xi) = \binom{t}{u}\pi^u(1-\pi)^{t-u}e^{-\xi}\xi^t/t!,$$

$t = 0, 1, \ldots$ and $u = 0, 1, \ldots, t$.

To put the problem in the framework of a two-parameter exponential family, we further reparametrize:

$$\theta = \log\left(\frac{\pi}{1-\pi}\right) = \log\left(\frac{n\mu}{m\lambda}\right) \text{ and } \tau = \log(\xi(1-\pi)) = \log(m\lambda)$$

to write the joint pmf of (U, T) as:

$$f_{U,T}(u, t; \theta, \tau) = \exp\left[-(e^{\tau} + e^{\theta+\tau})\right]e^{\theta u + \tau t}\frac{1}{u!(t-u)!}.$$

The hypotheses H_0 and H_1 now become H_0: $\theta = \theta_0$ and H_1: $\theta \neq \theta_0$, where $\theta_0 = \log(na/m)$.

The UMP unbiased level α test for H_0: $\theta = \theta_0$ vs H_1: $\theta \neq \theta_0$ (τ being a nuisance parameter) is

$$\varphi(u, t) = \begin{cases} 0 & \text{if } c_1(t) < u < c_2(t) \\ \gamma_i(t) & \text{if } u = c_i(t), \quad i = 1, 2 \\ 1 & \text{if } u < c_1(t) \text{ or } u > c_2(t), \end{cases}$$

where $c_i(t)$ and $0 \leq \gamma_i(t) \leq 1$, $i = 1, 2$, are determined by:

$$E_{\theta_0}[\varphi(U, t)|T = t] = \alpha \text{ and } E_{\theta_0}[U\varphi(U, t)|T = t] = \alpha E_{\theta_0}[U|T = t].$$

These equations can be written explicitly, using

$$f_{U|t}(u; \theta_0) = \binom{t}{u}\left(\frac{na}{m+na}\right)^u\left(\frac{m}{m+na}\right)^{t-u}, \quad u = 0, 1, \ldots, t, \text{ and}$$
$$E_{\theta_0}[U|T = t] = t\left(\frac{na}{m+na}\right).$$

The problems of finding unbiased level α tests for testing H_0: $\mu \leq a\lambda$ vs H_1: $\mu > a\lambda$ and for testing H_0: $a_1\lambda \leq \mu \leq a_2\lambda$ vs H_1: $\mu \notin [a_1\lambda, a_2\lambda r]$ for given a or given $a_1 < a_2$ are treated analogously using the methods developed for Problems 1 and 3 of Section 6.9.

Example 6.9.4. Let $X \sim$ *Binomial*(m, π_1) and $Y \sim$ *Binomial*(n, π_2) be independent. We want UMP unbiased level α test for H_0: $\pi_1 = \pi_2$ vs H_1: $\pi_1 \neq \pi_2$. Here,

$$f_{X,Y}(x, y; \pi_1, \pi_2) = \binom{m}{x}\binom{n}{y}(1-\pi_1)^m(1-\pi_2)^n$$
$$\times \exp\left[\left(\log\frac{\pi_2(1-\pi_1)}{\pi_1(1-\pi_2)}\right)y + \left(\log\frac{\pi_1}{1-\pi_1}\right)(x+y)\right].$$

Reparametrizing

$$\theta = \log\frac{\pi_2(1-\pi_1)}{\pi_1(1-\pi_2)}, \quad \tau = \log\frac{\pi_1}{1-\pi_1}$$

and transforming the data to $U = Y$, $T = X + Y$, the model is equivalently expressed as:

$$f_{U,T}(u, t; \theta, \tau) = C(\theta, \tau)\exp[\theta u + \tau t] \quad \text{and} \quad \pi_1 = \pi_2 \Leftrightarrow \theta = 0.$$

The UMP unbiased level α test for H_0: $\theta = 0$ vs H_1: $\theta \neq 0$ is of the form as in Example 6.9.3 with

$$f_{U|t}(u; \theta = 0) = P_{\pi_1=\pi_2}[Y = u|X + Y = t] = \frac{\binom{m}{t-u}\binom{n}{u}}{\binom{m+n}{t}}$$

which is hypergeometric, and $\mathrm{E}_{\theta=0}[U|T = t] = \binom{m}{m+n}t$.

Example 6.9.5. A and B are events in a probability space. The frequencies of AB, A^cB, AB^c, and A^cB^c in n trials are given in Table 6.1, known as a 2×2 contingency table:

Table 6.1 Frequency Distribution in n Trials

Events	A	A^c	Subtotal
B	X	Z	T
B^c	Y	W	T'
Subtotal	S	S'	n

Based on this data, we want to test: H_0: A and B are *independent* or *negatively dependent* (ie, $p_{AB} \leq p_Ap_B$) vs H_1: A and B are *positively dependent* (ie, $p_{AB} > p_Ap_B$). Let $\Delta = (p_{AB^c}p_{A^cB})/(p_{AB}p_{A^cB^c})$. Then the problem can be equivalently described as that of testing H_0: $\Delta \geq 1$ vs H_1: $\Delta < 1$. Reparametrize by transforming

$$(p_{AB}, p_{AB^c}, p_{A^cB}) \rightarrow (\tau_0, \tau_1, \tau_2) \rightarrow (\theta, \tau_1, \tau_2),$$

where

$$\tau_0 = \log(p_{AB}/p_{A^cB^c}), \quad \tau_1 = \log(p_{AB^c}/p_{A^cB^c}),$$
$$\tau_2 = \log(p_{A^cB}/p_{A^cB^c}), \text{ and } \theta = \tau_0 - \tau_1 - \tau_2 = -\log\Delta.$$

Also transform the data from $(X, Y, Z) \rightarrow (X, X+Y, X+Z) = (X, S, T)$. In terms of (θ, τ_1, τ_2), we can write

$$p_{A^cB^c} = (1 + e^{\theta+\tau_1+\tau_2} + e^{\tau_1} + e^{\tau_2})^{-1}, \quad p_{AB} = e^{\theta+\tau_1+\tau_2}p_{A^cB^c},$$
$$p_{AB^c} = e^{\tau_1}p_{A^cB^c}, \quad \text{and} \quad p_{AB^c} = e^{\tau_2}p_{A^cB^c}.$$

The multinomial pmf of (X, Y, Z) can now be rewritten, using $Y = S - X, Z = T - X$ and the above formulas for p_{AB}, p_{AB^c}, p_{A^cB}, and $p_{A^cB^c}$ to express the pmf of (X, S, T) as:

$$f_{X,S,T}(x, s, t; \theta, \tau_1, \tau_2) = C(\theta, \tau_1, \tau_2) \exp[\theta x + \tau_1 s + \tau_2 t] h(x, s, t),$$

where $C(\theta, \tau_1, \tau_2) = [1 + e^{\theta+\tau_1+\tau_2} + e^{\tau_1} + e^{\tau_2}]$ and $h(x, s, t)$ is the multinomial coefficient

$$n!/\{x!(s-x)!(t-x)!(n-s-t+x)!\}.$$

Now the UMP unbiased level α test for H_0: $\Delta \geq 1$ vs H_1: $\Delta < 1$ (ie, for H_0: $\theta \leq 0$ vs H_1: $\theta > 0$) is given by

$$\varphi(x, s, t) = \begin{cases} 0 & \text{if } x < c(s, t) \\ \gamma(s, t) & \text{if } x = c(s, t) \\ 1 & \text{if } x > c(s, t), \end{cases}$$

where $c(s, t)$ and $0 \leq \gamma(s, t) \leq 1$ are determined by

$$\gamma(s, t) P_{\theta=0}[X = c(s, t) | S = s, T = t] + P_{\theta=0}[X > c(s, t) | S = s, T = t] = \alpha.$$

Note that $P_{\theta=0}[X = x | S = s, T = t]$

$$= \begin{cases} \binom{t}{x}\binom{n-t}{s-x}/\binom{n}{s}, & \max(0, s+t-n) \leq x \leq \min(n, s) \\ 0 & \text{otherwise} \end{cases}$$

is the hypergeometric distribution.

The UMP unbiased level α test for H_0: $p_{AB} = p_A p_B$ (independence) vs H_1: $p_{AB} \neq p_A p_B$ (ie, H_0: $\theta = 0$ vs H_1: $\theta \neq 0$) is obtained by the same approach. This is known as the Fisher-Irwin test (also called the "Fisher exact test"), which is formally the same as the test obtained in Example 6.9.2.

Simplified Versions of Conditional Tests

The conditional tests $\varphi_1(u, t)$ for H_{10}: $\theta \leq \theta_0$ vs H_{11}: $\theta > \theta_0$, $\varphi_2(u, t)$ for H_{20}: $\theta = \theta_0$ vs H_{21}: $\theta \neq \theta_0$ and $\varphi_3(u, t)$ for H_{30}: $\theta_1 \leq \theta \leq \theta_2$ vs H_{31}: $\theta \notin [\theta_1, \theta_2]$ are inconvenient in many applications. However, in some situations which include testing problems in the family of normal distributions, these tests can be equivalently expressed in terms of a single statistic $V = g(U, T)$, as shown in the following theorem.

Theorem 6.9.2.

A. *Suppose $V = g(U, T)$ is such that*

(a) *V is independent of T when $\theta = \theta_0$, and*

(b) *$g(u, t)$ is increasing in u for each t.*

Then the UMP unbiased level α test φ_1 for Problem 1 becomes

$$\varphi_1^*(v) = \begin{cases} 0 & \text{if } v < c \\ \gamma & \text{if } v = c \\ 1 & \text{if } v > c, \end{cases}$$

where c and $0 \leq \gamma \leq 1$ are determined by $E_{\theta_0}[\varphi_1^*(V)] = \alpha$.

B. *Suppose $V = g(U, T)$ is such that*
  (a) *V is independent of T when $\theta = \theta_0$, and*
  (b) *$g(u, t) = a(t)u + b(t)$ with $a(t) > 0$.*
  Then the UMP unbiased level α test φ_2 for Problem 2 becomes

$$\varphi_2^*(v) = \begin{cases} 0 & \text{if } c_1 < v < c_2 \\ \gamma_i & \text{if } v = c_i, \quad i = 1, 2 \\ 1 & \text{if } v < c_1 \text{ or } v > c_2, \end{cases}$$

  where c_i and $0 \le \gamma_i \le 1$, $i = 1, 2$ are determined by $E_{\theta_0}[\varphi_2^*(V)] = \alpha$ and $E_{\theta_0}[V\varphi_2^*(V)] = \alpha E_{\theta_0}[V]$.

C. *Suppose $V = g(U, T)$ is such that*
  (a) *V is independent of T when $\theta = \theta_1$ and when $\theta = \theta_2$, and*
  (b) *$g(u, t)$ is increasing in u for each t.*
  Then the UMP unbiased level α test φ_3 in Problem 3 becomes φ_3^* having the same form as φ_2^* (in part B of the theorem), but c_i and $0 \le \gamma_i \le 1$, $i = 1, 2$ are determined by $E_{\theta_i}[\varphi_3^*(V)] = \alpha$, $i = 1, 2$.

Proof of Part A. The UMP unbiased level α test φ_1 in Problem 1 can be equivalently expressed in the form of φ_1^* by condition (b), but the quantities c and $0 \le \gamma \le 1$ may depend on t and are determined by

$$\gamma(t) P_{\theta_0}[V = c(t) | T = t] + P_{\theta_0}[V > c(t) | T = t] = \alpha \quad \text{for all } t.$$

By condition (a), V is independent of T when $\theta = \theta_0$, so $c(t)$ and $0 \le \gamma(t) \le 1$ do not depend on t and are determined by $\mathrm{E}_{\theta_0}[\varphi_1^*(V)] = \alpha$. □

The proof of part C is analogous.

Proof of Part B. Here also, by condition (b), the UMP unbiased level α test φ_2 in Problem 2 can be equivalently expressed in the form φ_2^*, but $c_i(t)$ and $0 \le \gamma_i \le 1$, $i = 1, 2$ may depend on t and are determined by

(i) $\mathrm{E}_{\theta_0}[\varphi_2^*(V, t) | T = t] = \alpha$ and
(ii) $\mathrm{E}_{\theta_0}[U\varphi_2^*(V, t) | T = t] = \alpha \mathrm{E}_{\theta_0}[U | T = t]$ for all t.

Substituting $V = a(t)U + b(t)$ (ie, $U = (V - b(t))/a(t)$ in (ii)), we have

$$a(t)^{-1}\mathrm{E}_{\theta_0}[\{V - b(t)\}\varphi_2^*(V, t) | T = t] = \alpha a(t)^{-1}\mathrm{E}_{\theta_0}[V - b(t) | T = t] \quad \text{for all } t$$

$$\Leftrightarrow \mathrm{E}_{\theta_0}[V\varphi_2^*(V, t) | T = t] - b(t)\mathrm{E}_{\theta_0}[\varphi_2^*(V, t) | T = t] = \alpha \mathrm{E}_{\theta_0}[V | T = t] - \alpha b(t) \quad \text{for all } t$$

$$\Leftrightarrow \mathrm{E}_{\theta_0}[V\varphi_2^*(V, t) | T = t] = \alpha \mathrm{E}_{\theta_0}[V | T = t] \quad \text{for all } t,$$

since $\mathrm{E}_{\theta_0}[\varphi_2^*(V, t) | T = t] = \alpha$ for all t by (i). Now using condition (a), we see that $c_i(t)$ and $0 \le \gamma_i(t) \le 1$, $i = 1, 2$ do not depend on t and are determined by: $\mathrm{E}_{\theta_0}[\varphi_2^*(V)] = \alpha$ and $\mathrm{E}_{\theta_0}[V\varphi_2^*(V)] = \alpha \mathrm{E}_{\theta_0}[V]$. □

In actual applications, condition (b) of this theorem is verified in an obvious manner. The verification of condition (a) often follows from V being an ancillary statistic (ie, one whose distribution does not depend on the nuisance parameter τ) and T being a complete sufficient statistic for τ, using Basu's Theorem 5.1.6.

Examples: Hypothesis Testing With Nuisance Parameters in the Context of Normal Distribution

In each of the following examples, we go through the following steps:

I. The data which is a random sample from the population(s) under consideration are: $(X_1,\ldots,X_n)$, or $(X_1,\ldots,X_m)$ and $(Y_1,\ldots,Y_n)$ or $((X_1,Y_1),\ldots,(X_n,Y_n))$. Also state H_0 and H_1 in terms of the population parameters.

II. The sufficient statistics are (U,T). The conditional distribution of U given $T=t$ is complicated, so the conditional test described earlier in Section 6.9 are impractical.

III. Reparametrize from (μ,σ^2) or (μ_1,μ_2,σ^2), etc., to (θ,τ) and restate H_0 and H_1 in terms of θ, treating τ as the nuisance parameter. In each example, the pdf/pmf of (U,T) belongs to the exponential family with natural parameters (θ,τ).

IV. Transform (U,T) to $V=g(U,T)$ or $(V=g_1(U,T),\ W=g_2(U,T)\,)$ as needed to use Theorem 6.9.2, verifying the conditions.

V. Express the UMP unbiased level α test in terms of V or (V,W) using the distributions of V and W under H_0.

Example 6.9.6.
Testing for μ of $\mathrm{N}(\mu,\sigma^2)$ with σ^2 unknown.

I. On the basis of the data $(X_1,\ldots,X_n)$ test (i) $H_0\colon \mu\le\mu_0$ vs $H_1\colon \mu>\mu_0$ and (ii) $H_0\colon \mu=\mu_0$ vs $H_1\colon \mu\ne\mu_0$ where $\mu_0=0$ (otherwise, replace μ by $\mu-\mu_0$).

II. The sufficient statistics are $(U,T)=\left(\sum_{i=1}^n X_i,\sum_{i=1}^n X_i^2\right)$.

III. Reparametrize $\theta=\mu/\sigma^2$, $\tau=-1/(2\sigma^2)$. In (i), test $H_0\colon \theta\le\theta_0$ vs $H_1\colon \theta>\theta_0$ where $\theta_0=\mu_0/\sigma^2$ and in (ii), test $H_0\colon \theta=\theta_0$ vs $H_1\colon \theta\ne\theta_0$, τ being the nuisance parameter.

IV. Express the UMP unbiased level α tests $\varphi_1(U,T)$ for (i) as $\varphi_1^*(V)$ and $\varphi_2(U,T)$ for (ii) as $\varphi_2^*(W)$, where

$$V=g_1(U,T)=\sqrt{\frac{n-1}{n}}\frac{U}{\sqrt{T-U^2/n}}=\frac{\sqrt{n}\bar{X}}{s}\ \text{and}\ W=g_2(U,T)=U/\sqrt{T}.$$

Verify conditions (a) and (b) of Theorem 6.9.2A for V and W.

V. The test for (i) is $\varphi_1^*(v)=I_{[c,\infty)}(v)$, where c satisfies $P_{\mu=0}[\sqrt{n}\bar{X}/s\ge c]=\alpha$. Under $\mu=0$, $\sqrt{n}\bar{X}/s\sim t_{n-1}$, so $c=t_{n-1}(\alpha)$ where $P[t_{n-1}\ge t_{n-1}(\alpha)]=\alpha$.
The test for (ii) is $\varphi_2^*(w)=I_{[c,\infty)}(|w|)$, using the fact that under $\mu=0$, W is symmetrically distributed about 0, and c satisfies $P_{\mu=0}[|W|\ge c]=\alpha$. Finally, note that

$$V=\sqrt{\frac{n-1}{n}}W\Big/\sqrt{1-W^2/n}=\sqrt{n}\bar{X}/s\sim t_{n-1}\ \text{when}\ \mu=0$$

and $|V|$ is an increasing function of $|W|$. Thus φ_2^* becomes $\varphi_2^*(v)=I_{[t_{n-1}(\alpha/2),\infty)}(|v|)$.

Example 6.9.7. Testing for σ^2 in $\mathrm{N}(\mu,\sigma^2)$ with μ unknown.

I. On the basis of the data $(X_1,\ldots,X_n)$ test (i) $H_0\colon \sigma^2\le\sigma_0^2$ vs $H_1\colon \sigma^2>\sigma_0^2$ and (ii) $H_0\colon \sigma^2=\sigma_0^2$ vs $H_1\colon \sigma^2\ne\sigma_0^2$.

II. Here $(U,T)=\left(\sum_1^n X_i^2,\sum_1^n X_i\right)$ are the sufficient statistics.

III. Reparametrize $\theta = -1/(2\sigma^2)$ and $\tau = \mu/\sigma^2$. In (i), test $H_0\colon \theta \le \theta_0$ vs $H_1\colon \theta > \theta_0$ where $\theta_0 = -1/(2\sigma_0^2)$ and in (ii), test $H_0\colon \theta = \theta_0$ vs $H_1\colon \theta \ne \theta_0$.

IV. Express the UMP unbiased level α tests $\varphi_1(U,T)$ for (i) as $\varphi_1^*(V)$ and $\varphi_2(U,T)$ for (ii) as $\varphi_2^*(V)$, where

$$V = g(U,T) = (1/\sigma_0^2)(U - T^2/n) = \sum_{i=1}^{n}\left(\frac{X_i - \bar{X}}{\sigma_0}\right)^2.$$

Verify conditions (a) and (b) of Theorem 6.9.2A for V.

V. Since $V \sim \chi^2_{n-1}$ under $\sigma^2 = \sigma_0^2$, the UMP level α test for (i) is $\varphi_1^*(v) = I_{[c,\infty)}(v)$ where $c = \chi^2_{n-1}(\alpha)$, $P[\chi^2_{n-1} \ge \chi^2_{n-1}(\alpha)] = \alpha$, and for (ii), $\varphi_2^*(v) = I_{(0,c_1]}(v) + I_{[c_2,\infty)}(v)$ where $c_1 < c_2$ satisfy

$$\int_{c_1}^{c_2} f_{n-1}(y)\,dy = 1 - \alpha \text{ and } \int_{c_1}^{c_2} y f_{n-1}(y)\,dy = (n-1)(1-\alpha),$$

f_{n-1} being the pdf of χ^2_{n-1}. The second condition in the display above can be expressed in any of the following two ways (as in Example 6.7.2):

$$\int_{c_1}^{c_2} f_{n+1}(y)\,dy = 1 - \alpha \text{ since } y f_{n-1}(y) = (n-1) f_{n+1}(y), \text{ or}$$

$$e^{-c_1/2} c_1^{(n-1)/2} = e^{-c_2/2} c_2^{(n-1)/2} \text{ since}$$

$$\int_{c_1}^{c_2} y f_{n-1}(y)\,dy = \frac{1}{2^{(n-3)/2}\Gamma((n-1)/2)}\left(e^{-c_1/2} c_1^{(n-1)/2} - e^{-c_2/2} c_2^{(n-1)/2}\right) + n(n-1)(1-\alpha),$$

using integration by parts and condition (i). For moderately large n and θ_0 neither too small nor too large, CLT provides a reasonably good approximation for c_1 and c_2 determined by

$$\int_0^{c_1} f_{n-1}(y)\,dy = \int_{c_2}^{\infty} f_{n-1}(y)\,dy = \alpha/2.$$

Example 6.9.8. Testing for $\mu_1 = \mu_2$ of $N(\mu_1, \sigma^2)$ and $N(\mu_2, \sigma^2)$ with μ_1, μ_2, σ^2 unknown.

I. On the basis of the data $(X_1, \ldots, X_m)$ and $(Y_1, \ldots, Y_n)$ from $N(\mu_1, \sigma^2)$ and $N(\mu_2, \sigma^2)$, respectively, test
  (i) $H_0\colon \mu_2 - \mu_1 \le 0$ vs $H_1\colon \mu_2 - \mu_1 > 0$ and
  (ii) $H_0\colon \mu_2 - \mu_1 = 0$ vs $H_1\colon \mu_2 - \mu_1 \ne 0$.

II. The sufficient statistics and their one-to-one transforms are

$$\left(\sum_1^n Y_i, \sum_1^m X_i, \sum_1^m X_i^2 + \sum_1^n Y_i^2\right) \text{ and}$$

$$(U, T_1, T_2) = \left(\bar{Y} - \bar{X}, m\bar{X} + n\bar{Y}, \sum_1^m X_i^2 + \sum_1^n Y_i^2\right)$$

III. Reparametrize

$$(\theta, \tau_1, \tau_2) = \left(\frac{\mu_2 - \mu_1}{(m^{-1} + n^{-1})\sigma^2}, \frac{m\mu_1 + n\mu_2}{(m+n)\sigma^2}, -\frac{1}{2\sigma^2}\right).$$

In (i), test $H_0\colon \theta \le \theta_0$ vs $H_1\colon \theta > \theta_0$ where $\theta_0 = 0$ and in (ii), test $H_0\colon \theta = \theta_0$ vs $H_1\colon \theta \ne \theta_0$, τ_1, τ_2 being the nuisance parameters.

IV, V. For testing $H_0\colon \theta \le 0$ vs $H_1\colon \theta > 0$, that is, $H_0\colon \mu_2 \le \mu_1$ vs $H_1\colon \mu_2 > \mu_1$, take

$$V = \sqrt{\frac{mn(m+n-2)}{m+n}} \frac{U}{\sqrt{T_2 - (T_1^2 + mnU^2)/(m+n)}}$$

$$= \frac{\bar{Y} - \bar{X}}{\sqrt{\frac{1}{m} + \frac{1}{n}} \sqrt{\frac{\sum_{i=1}^m (X_i - \bar{X})^2 + \sum_{i=1}^n (Y_i - \bar{Y})^2}{m+n-2}}} := \frac{\bar{Y} - \bar{X}}{\sqrt{\frac{1}{m} + \frac{1}{n}} \cdots}.$$

Then V satisfies both conditions of Theorem 6.9.2A, and the UMP unbiased level α test for $H_0\colon \theta \le 0$ vs $H_1\colon \theta > 0$ is $\varphi_1^*(v) = I_{[c,\infty)}(v)$, where $c = t_{m+n-2}(\alpha)$.

For testing $H_0\colon \theta = 0$ vs $H_1\colon \theta \ne 0$, take

$$W = \frac{U}{\sqrt{T_2 - T_1^2/(m+n)}}$$

$$= \frac{\bar{Y} - \bar{X}}{\sqrt{\sum_1^m X_i^2 + \sum_1^n Y_i^2 - \{\sum_1^m X_i + \sum_1^n Y_i\}^2/(m+n)}}.$$

Then W satisfies both conditions of Theorem 6.9.2B, and the UMP unbiased level α test for $H_0\colon \theta = 0$ vs $H_1\colon \theta \ne 0$ is $\varphi_2^*(w) = I_{[c,\infty)}(|w|)$, where c satisfies $P_{\mu_1=\mu_2}[|W| \ge c] = \alpha$. Now let $V = W/\sqrt{1 - mnW^2/(m+n)}$, which is the same as V defined for problem (i) and $|V|$ is an increasing function of $|W|$. Thus the UMP unbiased level α test for $H_0\colon \mu_1 = \mu_2$ vs $H_1\colon \mu_1 \ne \mu_2$ becomes $\varphi_2^*(v) = I_{[c,\infty)}(|v|)$, where $c = t_{m+n-2}(\alpha/2)$.

Example 6.9.9. Testing for $\sigma_1^2 = \sigma_2^2$ of $N(\mu_1, \sigma_1^2)$ and $N(\mu_2, \sigma_2^2)$, all parameters are unknown.

I. On the basis of the data $(X_1, \ldots, X_m)$ and $(Y_1, \ldots, Y_n)$ from $N(\mu_1, \sigma_1^2)$ and $N(\mu_2, \sigma_2^2)$, respectively, test (for a given $k > 0$):
(i) $H_0\colon \sigma_2^2 \le k\sigma_1^2$ vs $H_1\colon \sigma_2^2 > k\sigma_1^2$ and
(ii) $H_0\colon \sigma_2^2 = k\sigma_1^2$ vs $H_1\colon \sigma_2^2 \ne k\sigma_1^2$.

II. The sufficient statistics and their one-to-one transforms are

$$\left(\sum_1^n Y_i^2, \sum_1^m X_i^2, \sum_1^n Y_i, \sum_1^m X_i\right) \text{ and}$$

$$(U, T_1, T_2, T_3) = \left(\sum_1^n Y_i^2, \sum_1^m X_i^2 + k^{-1}\sum_1^n Y_i^2, \sum_1^n Y_i, \sum_1^m X_i\right).$$

III. Reparametrize

$$(\theta, \tau_1, \tau_2, \tau_3) = \left(-\frac{1}{2\sigma_2^2} + \frac{1}{2k\sigma_1^2}, -\frac{1}{2\sigma_1^2}, \frac{\mu_2}{\sigma_2^2}, \frac{\mu_1}{\sigma_1^2}\right),$$

τ_1, τ_2, τ_3 being the nuisance parameters. In (i), test $H_0\colon \theta \le 0$ vs $H_1\colon \theta > 0$ and in (ii), test $H_0\colon \theta = 0$ vs $H_1\colon \theta \neq 0$.

IV, V. Take

$$V = \frac{\{(n-1)k\}^{-1}\sum_{i=1}^{n}(Y_i - \bar{Y})^2}{(m-1)^{-1}\sum_{i=1}^{m}(X_i - \bar{X})^2}$$

in problem (i). Then both conditions of Theorem 6.9.2A are satisfied and the UMP unbiased level α test in problem (i) is $\varphi_1^*(v) = I_{[c,\infty)}(v)$, where $P[F_{n-1,m-1} \ge c] = \alpha$. In problem (ii), for testing $H_0\colon \sigma_2^2 = k\sigma_1^2$ vs $H_1\colon \sigma_2^2 \neq k\sigma_1^2$, proceed with

$$W = k^{-1}\sum_{i=1}^{n}(Y_i - \bar{Y})^2 \Bigg/ \left[\sum_{i=1}^{m}(X_i - \bar{X})^2 + k^{-1}\sum_{i=1}^{n}(Y_i - \bar{Y})^2\right]$$

to set up the test and then transform to V which increases with W.

Example 6.9.10. Testing for the slope parameter in Simple Linear Regression Model.

I. On the basis of the data $((x_1, Y_1), \ldots, (x_n, Y_n))$ where $x_1, \ldots, x_n$ are given constants and $Y_i = \alpha + \beta x_i + \epsilon_i$, $\epsilon_1, \ldots, \epsilon_n$ being iid $N(0, \sigma^2)$, with α, β, σ^2 all unknown, we want to test

(i) $H_0\colon \beta \le \beta_0$ vs $H_1\colon \beta > \beta_0$ and

(ii) $H_0\colon \beta = \beta_0$ vs $H_1\colon \beta \neq \beta_0$.

Without loss of generality, let $\beta_0 = 0$ (otherwise, replace Y_i by $Y_i - \beta_0 x_i$).

1. Transform the data to reduce the model to its canonical form: Let $Y_i^* = Y_i - \alpha - \beta x_i$, $i = 1, \ldots, n$ and let $\boldsymbol{a}_i^\top = (a_{i1}, \ldots, a_{in})$, $i = 1, \ldots, n$ be n-dim vectors which form an orthonormal basis for $\mathbb{R}^n$ (ie, $\boldsymbol{a}_i^\top \boldsymbol{a}_i = 1$ for all i and $\boldsymbol{a}_i^\top \boldsymbol{a}_j = 0$ for all $i \neq j$). In particular, choose the first two vectors as

$$\boldsymbol{a}_1^\top = n^{-1/2}(1, \ldots, 1) \text{ and } \boldsymbol{a}_2^\top = S_{xx}^{-1/2}(x_1 - \bar{x}, \ldots, x_n - \bar{x}),$$

where $\bar{x} = n^{-1}\sum_1^n x_i$ and $S_{xx} = \sum_{i=1}^n (x_i - \bar{x})^2$. Let $\boldsymbol{Y}^{*^\top} = (Y_1^*, \ldots, Y_n^*)$, and define

$$W_1 = \boldsymbol{a}_1^\top \boldsymbol{Y}^*, W_2 = \boldsymbol{a}_2^\top \boldsymbol{Y}^* \text{and } W_3 = \sum_{i=1}^n a_{3i}Y_i^*, \ldots, W_n^* = \sum_{i=1}^n a_{ni}Y_i^*.$$

Then $W_1, \ldots, W_n$ are iid $N(0, \sigma^2)$ and $\sum_1^n W_i^2 = \sum_1^n Y_i^{*^2}$. Define $S_{xY} = \sum_{i=1}^n (x_i - \bar{x})Y_i$ and $S_{YY} = \sum_{i=1}^n (Y_i - \bar{Y})^2$. It follows that

(a) $\bar{Y} = n^{-1}\sum_1^n Y_i \sim N(\alpha + \beta\bar{x}, \sigma^2/n)$ because $W_1 \sim N(0, \sigma^2)$,

(b) $W_2 \sim N(0, \sigma^2)$ (ie, $B = S_{xY}/S_{xx} \sim N(\beta, \sigma^2/S_{xx})$),

(c) $R = \sum_{i=3}^{n} W_i^2 = \sum_{i=1}^{n} Y_i^{*2} - W_1^2 - W_2^2 = S_{YY} - S_{xY}^2/S_{xx} \sim \sigma^2\chi_{n-2}^2$,

(d) $\bar{Y}$, B, and R are mutually independent.

II. The sufficient statistics and their one-to-one transforms are

$$(B, \bar{Y}, R) \text{ and } (U, T_1, T_2) = (B, \bar{Y}, n\bar{Y}^2 + S_{xx}B^2 + R).$$

III. Reparametrize

$$\theta = S_{xx}\beta/\sigma^2,\ \tau_1 = n(\alpha + \beta\bar{x})/\sigma^2,\ \tau_2 = -1/(2\sigma^2).$$

Starting from the joint distribution of $(\bar{Y}, B, R)$, using (a)-(d) in Step I, the joint pdf of (U, T_1, T_2) can be seen to be of the form

$$f_{U,T_1,T_2}(u, t_1, t_2; \theta, \tau_1, \tau_2) = C(\theta, \tau_1, \tau_2)\exp\left[\theta u + \tau_1 t_1 + \tau_2 t_2\right]h(u, t_1, t_2),$$

and the null and the alternative hypotheses can be restated as $H_0\colon \theta \le 0$ vs $H_1\colon \theta > 0$ in problem (i) and $H_0\colon \theta = 0$ vs $H_1\colon \theta \ne 0$ in problem (ii).

IV, V. In problem (i), take

$$V = g(U, T_1, T_2) = \frac{\sqrt{n-2}U}{\sqrt{(T_2 - nT_1^2)/S_{xx} - U^2}} = \frac{\sqrt{n-2}B}{\sqrt{R/S_{xx}}}.$$

Then V satisfies the conditions of Theorem 6.9.2A and it is distributed as t_{n-2} under H_0. Hence the UMP unbiased level α test for $H_0\colon \theta \le 0$ vs $H_1\colon \theta > 0$ is $\varphi_1^*(v) = I_{[t_{n-2}(\alpha),\infty)}(v)$.

In problem (ii), first take

$$V_1 = g(U, T_1, T_2) = U\Big/\sqrt{(T_2 - nT_1^2)/S_{xx}} = B\Big/\sqrt{B^2 + R/S_{xx}}.$$

Then V_1 satisfies the conditions of Theorem 6.9.2B. Hence the UMP unbiased level α test for $H_0\colon \theta = 0$ vs $H_1\colon \theta \ne 0$ is $\varphi(v_1) = I_{[c,\infty)}(|v_1|)$, where c satisfies $P_{\theta=0}[|V_1| \ge c] = \alpha$. Finally, let

$$V = \frac{\sqrt{n-2}V_1}{\sqrt{1 - V_1^2}} = \frac{\sqrt{n-2}B}{\sqrt{R/S_{xx}}}$$

as in problem (i). Then $V \sim t_{n-2}$ under $\theta = 0$ and $|V|$ is an increasing function of $|V_1|$, so $\varphi(v_1)$ is the same as $\varphi^*(v) = I_{[t_{n-2}(\alpha/2),\infty)}(|v|)$.

Remark 6.9.1. If $\beta_0 \ne 0$, then S_{xY} and S_{YY} are calculated from $\left(x_i, Y_i - \beta_0 x_i\right)$, $i = 1, \ldots, n$ and become

$$S_{xY}^* = S_{xY} - \beta_0 S_{xx} \text{ and } S_{YY}^* = S_{YY} + \beta_0^2 S_{xx} - 2\beta_0 S_{xY},$$

respectively, and B and R are replaced by

$$B^* = B - \beta_0 \text{ and } R^* = S_{YY}^* - (S_{xY}^*)^2/S_{xx} = S_{YY} - S_{xY}^*/S_{xx} = R.$$

As a result, the statistic V is replaced by $V^* = \sqrt{n-2}(B - \beta_0)/\sqrt{R/S_{xx}}$.

6.10 The *p*-Value: Another Way to Report the Result of a Test

In Example 6.7.1(a) of Section 6.7.2, let H_0: $\mu = \mu_0 = 10$ vs H_1: $\mu \neq 10$ with known $\sigma^2 = 4$, and suppose that in a random sample of size $n = 25$, the observed sample mean is $\bar{x} = 10.8$ Then the test statistic $T = \sqrt{n}(\bar{X} - \mu_0)$ has observed value

$$T(\boldsymbol{x}) = \sqrt{n}(\bar{x} - \mu_0)/\sigma = 5(10.8 - 10)/2 = 2.0.$$

Whether to accept or reject H_0, now depends on the choice of α. For $\alpha = 0.05$, $\Phi(1-\alpha/2) = 1.96$ and for $\alpha = 0.01$, $\Phi(1 - \alpha/2) = 2.575$; so for the same data we should reject H_0 at level $\alpha = 0.05$ but accept H_0 at level $\alpha = 0.01$. This raises two concerns:

(i) The choice of α is subjective, contrary to the aim at objectivity in the theory of hypothesis testing.
(ii) Even for a choice of chosen α, the test procedure only tells us whether to "accept H_0" or "reject H_0" without any indication of how strongly the data favors the decision.

To address these concerns, it would be desirable to report not only whether H_0 is rejected or accepted at a preassigned level α by the observed value of the test statistic $T(x)$ but also the smallest level of significance at which $T(x)$ would reject H_0, which is called the *p*-value of $T(x)$.

Definition 6.10.1. For a problem of testing H_0 vs H_1, if $T(x)$ is the appropriate test statistic based on the observed data x, then *p*-value of T = minimum α for which H_0 would be rejected by $T(x)$.

For a one-sided level α test $\phi(x) = I_{[c_\alpha,\infty)}(T(x))$,

$$p\text{-value of } T = \min_{\alpha}\{T(x) \geq c_\alpha\} = P_{H_0}[T \geq T(x)],$$

and for a symmetric two-sided test level α test $\phi(x) = 1 - I_{(-c_\alpha,c_\alpha)}(T(x))$,

$$p\text{-value of } T = \min_{\alpha}\{|T(x)| \geq c_\alpha\} = P_{H_0}[T \geq |T(x)|] + P_{H_0}[T \leq -|T(x)|].$$

In Example 6.7.1(a) discussed above,

$$T = \sqrt{n}(\bar{X} - \mu_0)/\sigma \quad \text{and} \quad T(\boldsymbol{x}) = 2.0.$$

Since $T \sim N(0, 1)$ under H_0,

$$p\text{-value of } T = P_{H_0}[|T| \geq 2.0] = 2\{1 - \Phi(2)\} = (2)(0.0228) = 0.0456.$$

This statement is more informative than the one saying that H_0 is accepted at level $\alpha = 0.01$ but rejected at $\alpha = 0.05$.

6.10.1 Pearson's P_λ Statistic

The *p*-value of a statistic T has so far been discussed in the context of $T(x)$ for an observed sample x. We now look at the *p*-value as a random variable:

$$p\text{-value of } T = P_{H_0}[T \geq T(X)]$$

considering a one-sided test for convenience of discussion.

Let F_T denote the cdf of T under H_0. Then

$$P_{H_0}[T \geq T(X)] = 1 - F_T(T(X)) \sim 1 - U \stackrel{\mathcal{D}}{=} U,$$

where $U \sim \mathit{Uniform}(0, 1)$ as seen in Section 1.11.

In problem 16(a) of Chapter 2, we have seen that

$$W = -2\log U \sim \text{Exp}(1/2) \stackrel{\mathcal{D}}{=} \chi^2_2.$$

Thus the p-value of T for a random sample X being $P = P_{H_0}[T \geq T(X)]$, $W = -2\log P$ is distributed as χ^2_2.

Now suppose that a certain study, such as a trial on the effectiveness of a drug, is carried out independently by several investigators. These studies, being designed differently and based on different sample sizes, may have been analyzed by different test statistics aimed at testing the same null hypothesis H_0, the rejection of which would indicate effectiveness of the drug. Suppose that the results of these studies are summarized by test statistics $T_1, \ldots, T_k$ with observed values $T_1(x_1), \ldots, T_k(x_k)$, and let the observed p-values of these tests (using right-tail tests for simplicity) be denoted by

$$P(x_j) = P_{H_0}[T_j \geq T_j(x_j)], \quad j = 1, \ldots, k.$$

Then these p-values are sample realizations of

$$P_j = P_{H_0}[T_j \geq T_j(X_j)]$$

and therefore, $w_j = -2\log P(x_j)$ are sample realizations of $W_j = -2\log P_j$, which are iid as χ^2_2 under H_0.

Now define $P_\lambda = \sum_{j=1}^{k}[-2\log P_j]$ as a test statistic combining the results of all the k tests, which is distributed as χ^2_{2k} under H_0, and can be used as a test statistic based on the combined evidence provided by the results of all the k investigations.

6.11 Sequential Probability Ratio Test

Traditional statistical inference deals with analysis of a set of data $(X_1, \ldots, X_n)$ to draw some conclusion about a parameter θ involved in the joint pdf (or pmf) $f_n(x_1, \ldots, x_n; \theta)$. The observations $X_1, \ldots, X_n$ are often iid with individual pdf $f(x; \theta)$ (although they need not be so), in which case, $f_n(x_1, \ldots, x_n; \theta) = \prod_{i=1}^{n} f(x_i; \theta)$. But the main thing is that n is fixed and the functional form f_n is fixed (ie, the sample size and the sampling design are predetermined). However, some samples provide more conclusive evidence than others, so it makes sense to take samples gradually (ie, sequentially), and examine the evidence gathered at each stage of sampling to determine whether any more samples are needed, and if not, to draw a final conclusion. This is known as *Sequential Analysis*.

Here we shall only discuss the problem of testing a simple null hypothesis H_0: $\theta = \theta_0$ against a simple alternative H_1: $\theta = \theta_1$ based on iid samples $X_1, X_2, \ldots$ observed sequentially. For this, we need a procedure of the following form:

Step 1. Start with observing X_1. Let S_{10}, S_{11}, and C_1 be a partition of the sample space $\mathcal{X}_1 = \mathcal{X}$ of X_1 such that $S_{10} \cup S_{11} \cup C_1 = \mathcal{X}_1$. If the observed value $x_1 \in S_{10}$, stop sampling and accept H_0; if $x_1 \in S_{11}$, stop sampling and accept H_1; if $x_1 \in C_1$ (continuation region), take another observation X_2.
Step 2. Now the observed data are $(x_1, x_2) \in C_1 \times \mathcal{X} = \mathcal{X}_2$. Let S_{20}, S_{21}, and C_2 be a partition of $\mathcal{X}_2$. If $(x_1, x_2) \in S_{20}$, stop sampling and accept H_0; if $(x_1, x_2) \in S_{21}$, stop sampling and accept H_1; if $(x_1, x_2) \in C_2$ (continuation region), take another observation X_3.
Step n. At the nth stage of sampling, the observed data are $(x_1, \ldots, x_{n-1}, x_n) \in C_{n-1} \times \mathcal{X} = \mathcal{X}_n$. Let S_{n0}, S_{n1}, and C_n be a partition of $\mathcal{X}_n$. If $(x_1, \ldots, x_n) \in S_{n0}$, stop sampling and accept H_0; if $(x_1, \ldots, x_n) \in S_{n1}$, stop sampling and accept H_1; if $(x_1, \ldots, x_n) \in C_n$, observe X_{n+1}.

Sequential Probability Ratio Tests, which we shall refer to as SPRT, are due to Wald [24]. The SPRT is based on the following idea. For a fixed sample size n, the MP test for H_0 vs H_1 at a given level α, accepts H_0 or H_1 according as

$$\lambda_n = \frac{f_n(x_1, \ldots, x_n; \theta_1)}{f_n(x_1, \ldots, x_n; \theta_0)} \text{ is } < k \text{ or } > k,$$

where $k = k_n(\alpha)$ depends on the sample size and the level of significance. [For given α,β, we can also determine the smallest sample size $N(\alpha, \beta)$ such that for $n \geq N(\alpha, \beta)$, the MP level α test based on a sample size n will have the probability of Type II error $\leq \beta$]. In the sequential setting, at the nth stage of sampling, we modify the MP test described above by accepting H_0 if the likelihood ratio $\lambda_n \leq B$, accepting H_1 if $\lambda_n \geq A$, and observing X_{n+1} if $B < \lambda_n < A$.

Definition of *SPRT*(*A*, *B*), *B* < 1 < *A*, Based on iid Observations

Let $Z_i = \log\left[f(X_i, \theta_1)/f(X_i; \theta_0)\right]$, so that the log likelihood ratio at the nth stage of sampling is $\sum_{i=1}^n Z_i$. The procedure at the nth stage of sampling is: accept H_0 if $\sum_{i=1}^n Z_i \leq \log B$, accept H_1 if $\sum_{i=1}^n Z_i \geq \log A$, continue sampling if $\log B < \sum_{i=1}^n Z_i < \log A$, where $0 < B < 1 < A < \infty$ are such that

$$P_{\theta_0}[\text{Accept } H_1] = \alpha \quad \text{and} \quad P_{\theta_1}[\text{ Accept } H_0] = \beta.$$

Note that the actual sample size needed for the SPRT to stop and reach a terminal decision to accept H_0 or H_1 is a random variable. We have to think of this random variable in the context of the sample space of infinite sequences

$$\omega = (z_1, z_2, \ldots), \quad z_i = \log\left[f(x_i, \theta_1)/f(x_i; \theta_0)\right].$$

For each ω the stopping sample size of SPRT is: $N(\omega) = n$ if $\log B < \sum_{i=1}^{j} z_i < \log A$ for $j \le n-1$ and $\sum_{i=1}^{n} z_i$ is either $\le \log B$ or $\ge \log A$.

To facilitate discussion, let us also define the following events in the sample space of ω:

$$S_{n0}^* = \left\{(z_1, z_2, \ldots): \sum_{i=1}^{j} z_i \in (\log B, \log A),\ j \le n-1 \text{ and } \sum_{i=1}^{n} z_i \le \log B\right\},$$

and

$$S_{n1}^* = \left\{(z_1, z_2, \ldots): \sum_{i=1}^{j} z_i \in (\log B, \log A),\ j \le n-1 \text{ and } \sum_{i=1}^{n} z_i \ge \log A\right\}$$

for $n = 1, 2, \ldots$. These events have the following properties:

(i) $S_{n0}^*, S_{n1}^*, n = 1, 2, \ldots$ are disjoint.
(ii) For each n, the events S_{n0}^* and S_{n1}^* are determined by $(Z_1, \ldots, Z_n)$ only.
(iii) *SPRT(A, B)* stops with exactly n observations with acceptance of H_0 if $\omega \in S_{n0}^*$ or with acceptance of H_1 if $\omega \in S_{n1}^*$.

We now ask the following questions:

1. Does the SPRT stop with probability 1, ie, is $P_\theta[N < \infty] = 1$?
2. How are the error probabilities α, β related to the boundaries A, B?
3. The function $L(\theta) = P_\theta[\text{Accept } H_0] = \sum_{1 \le n < \infty} P_\theta(S_{n0}^*)$, called the *operating characteristic* (OC) function, which is simply 1− power function. Obviously, $L(\theta_0) = 1 - \alpha$ and $L(\theta_1) = \beta$. How to calculate $L(\theta)$ for other values of θ?
4. The stopping sample size N is a random variable. It is important to know its average value $E_\theta(N)$ for a given θ. This average value, as a function of θ, is called the *average sampling number* (ASN) function. How to evaluate the ASN function for a given θ?
5. Is the SPRT optimal in any sense?

We now deal with above issues.

6.11.1 SPRT Stops With Probability 1

Theorem 6.11.1. *If $P_\theta[Z_i = 0] < 1$, then there exist $c > 0$ and $0 < r < 1$ such that $P_\theta[N \ge n] \le cr^n$. In such a case, $P_\theta[N < \infty] = 1$.*

Remark 6.11.1. Note that

$$P_\theta[Z_i = 0] = 1 \iff P_\theta[f(X, \theta_0) = f(X, \theta_1)] = 1.$$

If θ_0 and θ_1 are distinguishable, then $P_{\theta_0}[f(X, \theta_0) = f(X, \theta_1)] < 1$ and $P_{\theta_1}[f(X, \theta_0) = f(X, \theta_1)] < 1$.

Proof. Since $P_\theta[Z_i = 0] < 1$, there exist $\varepsilon > 0$ and $\delta > 0$ such that either $P_\theta[Z_i > \varepsilon] = \delta$ or $P_\theta[Z_i < -\varepsilon] = \delta$. We consider the former case (the other case is treated similarly).

Let k be an integer such that $k\varepsilon > \log A - \log B$. We now show that the desired inequality holds with $c = (1-\delta^k)^{-1}$ and $r = (1-\delta^k)^{1/k}$. The idea of the proof is given below.

$N \geq n \iff \sum_{i=1}^{l} Z_i \in (\log B, \log A)$ for all $l \leq n-1 \Longrightarrow \left|\sum_{i=1}^{k} Z_i\right|, \ldots, \left|\sum_{i=(j-1)k+1}^{jk}\right|$ are all $\leq \log A - \log B < k\varepsilon$, where jk is the largest multiple of k which is $\leq n-1$. Thus for $jk+1 \leq n \leq (j+1)k$, using the iid property of the Z_i's, we have

$$\begin{aligned} P_\theta[N \geq n] &\leq \left\{ P_\theta\left[\left|\sum_{i=1}^{k} Z_i\right| < k\varepsilon\right]\right\}^j \leq \left\{ P_\theta\left[\sum_{i=1}^{k} Z_i < k\varepsilon\right]\right\}^j \\ &\leq (1-\delta^k)^j = (1-\delta^k)^{-1}\{(1-\delta^k)^{1/k}\}^{(j+1)k} \leq cr^n, \end{aligned}$$

because $\sum_{i=1}^{k} Z_i < k\varepsilon \Longrightarrow \{Z_1 > \varepsilon, \ldots, Z_k > \varepsilon\}^c$, so that

$$P_\theta\left[\sum_{i=1}^{k} Z_i < k\varepsilon\right] \leq 1 - P_\theta[Z_1 > \varepsilon, \ldots, Z_k > \varepsilon] = 1 - \delta^k.$$

Finally,

$$\begin{aligned} P_\theta[N < \infty] &= \sum_{k=1}^{\infty} P_\theta[N = k] = \lim_{n\to\infty} \sum_{k=1}^{n-1} P_\theta[N = k] \\ &= \lim_{n\to\infty} [1 - P_\theta(N \geq n)] = \lim_{n\to\infty} (1 - cr^n) = 1 \end{aligned}$$

since $0 < r < 1$. □

6.11.2 Error Probabilities of *SPRT*(A, B): Relation Between (A, B) and (α, β)

We will use Theorem 6.11.1 to get

$$\begin{aligned} \alpha &= P_{\theta_0}[\text{Accept } H_1] \\ &= \sum_{n=1}^{\infty} P_{\theta_0}[\text{ Stop with } n \text{ observations and accept } H_1] \\ &= \sum_{n=1}^{\infty} P_{\theta_0}[S_{n1}^*] = \sum_{n=1}^{\infty} \int_{S_{n1}^*} \prod_{i=1}^{n} f(x_i, \theta_0)\, dx_1 \cdots dx_n \\ &= \sum_{n=1}^{\infty} \int_{S_{n1}^*} \left\{\prod_{i=1}^{n} f(x_i, \theta_1)/f(x_i, \theta_0)\right\}^{-1} \prod_{i=1}^{n} f(x_i, \theta_1)\, dx_1 \cdots dx_n \\ &\leq (1/A) \sum_{n=1}^{\infty} \int_{S_{n1}^*} \prod_{i=1}^{n} f(x_i, \theta_1)\, dx_1 \cdots dx_n \\ &= (1/A) P_{\theta_1}[\text{Accept } H_1] = (1/A)\{1 - P_{\theta_1}[\text{ Accept } H_0]\} = \frac{1-\beta}{A}. \end{aligned}$$

Similarly, using the fact that $\lambda_n = \prod_{i=1}^n f(x_i, \theta_1/f(x_i, \theta_0) \leq B$ on S_{n0}^* for all n, we have

$$\begin{aligned} 1-\alpha = P_{\theta_0}[\text{Accept } H_0] &= \sum_{n=1}^{\infty} P_{\theta_0}[S_{n0}^*] \\ &= \sum_{n=1}^{\infty} \int_{S_{n0}^*} \lambda_n^{-1} \prod_{i=1}^n f(x_i, \theta_1)\, dx_1 \cdots dx_n \\ &\geq (1/B) P_{\theta_1}[\text{Accept } H_0] = \beta/B. \end{aligned}$$

Thus the Type I error probability $\alpha(A, B)$ and Type II error probability $\beta(A, B)$ satisfy:

$$\begin{aligned} &\alpha(A,B) \leq [1-\beta(A,b)]/A \text{ and } 1-\alpha(A,B) \geq \beta(A,B)/B, \text{ ie,} \\ &A\alpha(A,B) + \beta(A,B) \leq 1 \text{ and } \alpha(A,B) + \beta(A,B)/B \leq 1. \end{aligned}$$

Suppose for given α^*, β^*, we choose $A' = (1-\beta^*)/\alpha^*$ and $B' = \beta^*/(1-\alpha^*)$.

If the above inequalities were equalities, then we would have $\alpha(A', B') = \alpha^*$ and $\beta(A', B') = \beta^*$ for the resulting $SPRT(A', B')$. This would be the case if $SPRT(A', B')$ always terminated with $\lambda_N = A$ or $\lambda_N = B$ at the stopping time N. But in reality, λ_N will be either $>A$ or $<B$ when the likelihood ratio sequence $\{\lambda_N\}$ first goes out of the continuation region (B, A). Taking these inequalities as equalities amounts to an approximation in which *excess over the boundaries is neglected.* Comparing $(\alpha, \beta) = (\alpha(A', B'), \beta(A', B'))$ with (α^*, β^*), we see

$$\left.\begin{aligned} &(i)\ A'\alpha(A', B') + \beta(A', B') \leq 1 \\ &(ii)\ \alpha(A', B') + \beta(A', B')/B' \leq 1 \end{aligned}\right\} \iff \left.\begin{aligned} (1-\beta^*)\alpha + \alpha^*\beta \leq \alpha^* \\ \beta^*\alpha + (1-\alpha^*)\beta \leq \beta^* \end{aligned}\right\}$$

$$\implies \alpha + \beta \leq \alpha^* + \beta^*.$$

Hence *at most one* of the error probabilities $\alpha(A', B')$ or $\beta(A', B')$ may exceed its intended value α^* or β^* with the above choice of boundaries. Moreover, by (i), $\alpha \leq 1/A' = \alpha^*/(1-\beta^*)$, so $\alpha - \alpha^* \leq \alpha^*/(1-\beta^*) - \alpha^* = \alpha^*\beta^*/(1-\beta^*)$, and by (ii), $\beta \leq B' \leq \beta^*/(1-\alpha^*)$, so $\beta - \beta^* \leq \beta^*/(1-\alpha^*) - \beta^* = \alpha^*\beta^*/(1-\alpha^*)$.

Summary. If for given α^*, β^*, we take $A' = (1-\beta^*)/\alpha^*$ and $B' = \beta^*/(1-\alpha^*)$, then at most one of $\alpha(A', B')$ or $\beta(A', B')$ may exceed α^* or β^*. If $\alpha(A', B') > \alpha^*$, then $\alpha(A', B') - \alpha^* \leq \alpha^*\beta^*/(1-\beta^*)$ and if $\beta(A', B') > \beta^*$, then $\beta(A', B') - \beta^* \leq \alpha^*\beta^*/(1-\alpha^*)$. For example, if $\alpha^* = \beta^* = 0.05$, then $\alpha - \alpha^*$ and $\beta - \beta^*$ are both ≤ 0.002632 and at most one of them is positive.

In what follows, we shall use the approximate formulas: $A \approx (1-\beta)/\alpha$, $B \approx \beta/(1-\alpha)$, $\alpha \approx (1-B)/(A-B)$, and $\beta \approx (A-1)B/(A-B)$.

6.11.3 OC Function

We first prove two lemmas.

Lemma 6.11.1. *Let $M(t)$ be the mgf of an rv Z and let $T = \{t: M(t) < \infty\}$. Then T is an interval containing* 0 *and $M^{(k)}(t_0) = E[Z^k e^{t_0 Z}]$ for all t_0 lying in the interior of T.*

Proof. Since $M(0) = 1$, $0 \in T$ and by Hölder's inequality, t_1 and t_2 in T implies $\lambda t_1 + (1-\lambda)t_2 \in T$ for all $0 < \lambda < 1$. Next note that by Lebesgue Dominated Convergence,

$$\frac{d}{dt}\mathrm{E}[e^{tZ}]|_{t=t_0} = \lim_{\delta \to =0}(1/\delta)\mathrm{E}\left[\int e^{(t_0+\delta)z} f_Z(z)\,dz - \int e^{t_0 z} f_Z(z)\,dz\right]$$
$$= \int \lim_{\delta \to 0} \frac{e^{\delta z} - 1}{\delta} e^{t_0 z} f_Z(z)\,dz = \int z e^{t_0 z} f_Z(z)\,dz = \mathrm{E}[Z e^{t_0 Z}],$$

the dominating function being g such that $|(e^{\delta z} - 1)/\delta| \leq g(z)$ for all z and $\delta \neq 0$ and $\int g(z) e^{t_0 z} f_Z(z)\,dz < \infty$. Such a function exists, because for $|\delta| < \varepsilon$,

$$|(e^{\delta z} - 1)/\delta| \leq \sum_{j=1}^{\infty} |\delta|^{j-1} |z|^j / j! \leq \sum_{j=1}^{\infty} \varepsilon^{j-1} |z|^j / j! < e^{\varepsilon |z|}/\varepsilon = g(z), \text{ say}$$

and

$$E\left[\varepsilon^{-1} e^{\varepsilon |Z|} e^{t_0 Z}\right] \leq \varepsilon^{-1}\left\{\mathrm{E}\left[e^{(t_0+\varepsilon)Z}\right] + \mathrm{E}\left[e^{(t_0-\varepsilon)Z}\right]\right\} < \infty$$

if ε is such that $t_0 + \varepsilon$ and $t_0 - \varepsilon$ are both in T. The proof for the higher derivatives is by induction. □

Lemma 6.11.2. *Suppose that Z satisfies the following conditions:*

(a) $E[Z] \neq 0$,
(b) $M(t) = E[e^{tZ}\} < \infty$ *for all t,*
(c) $P[e^Z > 1 + \delta] > 0$ *and* $P[e^Z < 1 - \delta] > 0$ *for some* $\delta > 0$.

Then there exists a unique nonzero solution of the equation $M(t) = 1$ *(ie, there is a unique* $h \neq 0$ *such that* $M(h) = 1$*).*

Proof. By condition (b) and Lemma 6.11.1, $M''(t) = \mathrm{E}[Z^2 e^{tZ}] > 0$, by condition (a). Hence M is a strictly convex function. Now use condition (c) to see that: since $M(t) \geq (1+\delta)^t P[e^Z > 1+\delta]$ for $t > 0$, $\lim_{t\to\infty} M(t) = \infty$, and since $M(t) \geq (1-\delta)^t P[e^Z < 1-\delta]$ for $t < 0$, $\lim_{t\to-\infty} M(t) = \infty$. Hence $M(t)$ has a unique minimum at some t_0. If $t_0 = 0$, then $0 = M'(0) = E[Z]$ would contradict condition (a). Thus $M(t)$ has the following properties:

(i) $M(0) = 1$,
(ii) $M(t)$ is strictly convex,
(iii) $M(t)$ has a unique minimum at $t_0 \neq 0$.

From these properties it follows that there exists a unique minimum $h \neq 0$ such that $M(h) = 1$. □

Now consider the OC function $L(\theta) = P_\theta[\text{Accept } H_0]$ of $SPRT(A, B)$ for H_0: $\theta = \theta_0$ vs H_1: $\theta = \theta_1$, assuming that the distribution of $Z = \log[f(X, \theta_1)/f(X, \theta_0)]$ satisfies the conditions of Lemma 6.11.2. Then we have the unique $h(\theta) \neq 0$ corresponding to θ, such that

$$1 = \mathrm{E}_\theta[\mathrm{e}^{h(\theta)Z}] = \mathrm{E}_\theta[\{f(X,\theta_1)/f(X;\theta_0)\}^{h(\theta)}]$$
$$= \int \{f(X,\theta_1)/f(X;\theta_0)\}^{h(\theta)} f(x,\theta)\, dx$$

which makes

$$f_1^*(x) = f^*(x,\theta) = \{f(X,\theta_1)/f(X;\theta_0)\}^{h(\theta)} f(x,\theta)$$

a pdf. Now let $f_0^*(x) = f(x,\theta)$ and consider the $SPRT\left(A^{h(\theta)}, B^{h(\theta)}\right)$ for testing $H_0^*\colon X \overset{pdf}{\sim} f_0^*$ vs $H_1\colon X \overset{pdf}{\sim} f_1^*$. If $h(\theta) > 0$, then

$$\lambda_n^* = \prod_{i=1}^n \{f_1^*(x_i)/f_0^*(x_i)\} = \left[\prod_{i=1}^n \{f(x_i,\theta_1)/f(x_i,\theta_0)\}\right]^{h(\theta)}$$

is $\le B^{h(\theta)}$ or $\ge A^{h(\theta)}$ or inside $\left(B^{h(\theta)}, A^{h(\theta)}\right)$ according as $\lambda_n = \prod_{i=1}^n \{f(x_i,\theta_1)/f(x_i,\theta_0)$ is $\le B$ or $\ge A$ or inside (B, A). Hence

$$P_{H_0^*}[SPRT(A^{h(\theta)}, B^{h(\theta)}) \text{ accepts } H_0^*] = P_\theta[SPRT(A, B) \text{ accepts } H_0] = L(\theta).$$

Thus for $h(\theta) > 0$, neglecting excess over boundaries,

$$L(\theta) = 1 - \alpha(A^{h(\theta)}, B^{h(\theta)}) = 1 - (1 - B^{h(\theta)})/(A^{h(\theta)} - B^{h(\theta)})$$
$$= (A^{h(\theta)} - 1)/(A^{h(\theta)} - B^{h(\theta)}).$$

The same formula also holds for $h(\theta) < 0$. To show this, consider the correspondence between $SPRT\left(B^{h(\theta)}, A^{h(\theta)}\right)$ for testing H_0^* vs H_1^* and $SPRT(A, B)$ for testing H_0 vs H_1.

6.11.4 ASN Function

Theorem 6.11.2 (Fundamental Identity of Sequential Analysis). *Let* $S_n = \sum_{i=1}^n Z_i$ *and suppose that* $P_\theta[Z_i = 0] < 1$ *and* $P_\theta[|Z_i| < \infty] = 1$. *Then* $\mathrm{E}_\theta[\mathrm{e}^{tS_N} M(t)^{-N}] = 1$ *for all* t *at which* $M(t) < \infty$, *where* N *is the stopping sample size of* $SPRT(A, B)$.

Remark 6.11.2. For each n,

$$\mathrm{E}_\theta[\mathrm{e}^{tS_n}] = \{\mathrm{E}[\mathrm{e}^{tZ_1}]\}^n = M(t)^n, \quad \text{so } \mathrm{E}_\theta[\mathrm{e}^{tS_n} M(t)^{-n}] = 1.$$

The fundamental identity asserts this fact also for the random stopping time N. We shall see that the proof depends on two properties of N, namely, $P_\theta[N < \infty] = 1$ (Theorem 6.11.1) and the fact that the event $\{N = n\}$ depends only on $Z_1, \ldots, Z_n$.

Remark 6.11.3. Note that $Z_i = \infty$ if $f(X_i,\theta_1) > 0$ and $f(X_i,\theta_0) = 0$, and $Z_i = -\infty$ if $f(X_i,\theta_1) = 0$ and $f(X_i,\theta_0) > 0$. If such Z_i are replaced by $\log A - \log B$ and $-(\log A - \log B)$, respectively, then the crossing behavior of each sample sequence remains unaltered. We can, therefore, assume $P_\theta[|Z_i| < \infty] = 1$, without loss of generality.

Proof. (This proof is due to Bahadur [25].)

Let $T = \{t: M(t) < \infty\}$ and for each $t \in T$, define $p(z|t) = e^{tz}M(t)^{-1}f_Z(z|\theta)$, where $f_Z(z|\theta)$ is the pdf of $Z = \log[f(X, \theta_1)/f(X; \theta_0)]$ induced by the distribution corresponding to $f(x, \theta)$. Then

$$\int p(z|t)\,dz = M(t)^{-1}\int e^{tz} f_Z(z|\theta)\,dz = M(t)^{-1}\mathrm{E}_\theta[e^{tZ}] = 1,$$

showing that $p(z|t)$ is a pdf for each t and $p(z|0) = f_Z(z|\theta)$. Since Z is nondegenerate under $f_Z(z|\theta)$ by virtue of the condition $P_\theta[Z = 0] < 1$, it now follows that Z is nondegenerate under $p(z|t)$ for all $t \in T$. Hence by Theorem 6.11.1, $P_t[N < \infty] = 1$ for all $t \in T$.

We now have, using the fact that the event $\{N = n\}$ depends only on $(Z_1, \ldots, Z_n)$,

$$\begin{aligned}
\mathrm{E}_\theta[e^{tS_N}M(t)^{-N}] &= \sum_{n=1}^{\infty} \mathrm{E}\left[e^{tS_N}M(t)^{-N}I(N = n)\right] \\
&= \sum_{n=1}^{\infty} \int_{N=n} \prod_{i=1}^{n} [e^{tz_i}M(t)^{-1}f_Z(z_i|\theta)]\,dz_1 \cdots dz_n \\
&= \sum_{n=1}^{\infty} \int_{N=n} \prod_{i=1}^{n} p(z_i|t)\,dz_1 \cdots dz_n \\
&= \sum_{n=1}^{\infty} P_t[N = n] = P_t[N < \infty] = 1.
\end{aligned}$$

□

Remark 6.11.4. It can be shown that for t lying in the interior of T, differentiation can be carried under the expectation any number of times in the fundamental identity.

Theorem 6.11.3. *Assume* $P_\theta[Z_i = 0] < 1$, $P_\theta[|Z_i| < \infty] = 1$ *and that* $M(t) = E_\theta[e^{tZ}] < \infty$ *in a neighborhood of zero. Then*

(a) $E_\theta[S_N] = \mu(\theta)E_\theta(N)$,
(b) $E_\theta[\{S_N - N\mu(\theta)\}^2] = \sigma^2(\theta)E_\theta(N)$,

where $\mu(\theta) = E_\theta[Z]$ *and* $\sigma^2(\theta) = Var_\theta(Z)$.

Proof. Differentiating the fundamental identity two times under the expectation sign, we have

$$\begin{aligned}
0 &= \frac{d}{dt}E_\theta\left[e^{tS_N}M(t)^{-N}\right]_{t=0} = E_\theta\left[\left\{S_N - N\frac{M'(t)}{M(t)}\right\}e^{tS_N - N\log M(t)}\right]_{t=0} \\
&= E_\theta[S_N - NE_\theta(Z)] = E_\theta[S_N] - \mu(\theta)E_\theta[N], \text{ and} \\
0 &= \frac{d^2}{dt^2}E_\theta\left[e^{tS_N}M(t)^{-N}\right]_{t=0} \\
&= E_\theta\left[\left\{\left(S_N - N\frac{M'(t)}{M(t)}\right)^2 - N\frac{M(t)M''(t) - M'(t)^2}{M(t)^2}\right\}e^{tS_N - N\log M(t)}\right]_{t=0} \\
&= E_\theta\left[\left[\{S_N - NE_\theta(Z)\}^2\right] - N\mathrm{Var}_\theta(Z)\right] = E_\theta[\{S_N - N\mu(\theta)\}^2] - \sigma^2(\theta)E_\theta[N].
\end{aligned}$$

□

The ASN function is now seen to be

$$\mathrm{E}_\theta[N] = \frac{\mathrm{E}_\theta[S_N]}{\mathrm{E}_\theta[Z]} \approx \frac{L(\theta)\log B + \{1 - L(\theta)\}\log A}{\mathrm{E}_\theta[Z]}, \quad \text{if } \mathrm{E}_\theta[Z] \neq 0,$$

where the approximation is obtained by neglecting excess over boundaries at the stopping time. Finally we state the *Optimally Property of SPRT* without proof.

Among all sequential tests with Type I and Type II error probabilities not exceeding the corresponding error probabilities of *SPRT*(*A*, *B*), the ASN function of *SPRT*(*A*, *B*) has the smallest values at θ_0 and at θ_1.

Example 6.11.1. Suppose $X_1, X_2, \ldots$ are iid $N(\theta, \sigma^2)$; H_0: $\theta = \theta_0$ and H_1: $\theta = \theta_1 > \theta_0$. Then

$$Z_i = \log[f(X_i, \theta_1)/f(X_i; \theta_0)] = \frac{\theta_1 - \theta_0}{\sigma^2} X_i - \frac{\theta_1^2 - \theta_0^2}{2\sigma^2}.$$

At the *n*th stage,

$$\text{accept } H_0 \text{ if } \sum_{i=1}^n X_i \leq \frac{\sigma^2}{\theta_1 - \theta_0}\log B + \frac{\theta_0 + \theta_1}{2} n,$$
$$\text{accept } H_1 \text{ if } \sum_{i=1}^n X_i \geq \frac{\sigma^2}{\theta_1 - \theta_0}\log A + \frac{\theta_0 + \theta_1}{2} n,$$

and continue sampling if $\sum_{i=1}^n X_i$ lies within these boundaries. Here, for $\theta \neq (\theta_0 + \theta_1)/2$,

$$\mathrm{E}_\theta[Z_i] \neq 0, \; h(\theta) = (\theta_1 + \theta_0 - 2\theta)/(\theta_1 - \theta_0) \neq 0,$$

$$L(\theta) = \frac{A^{(\theta_1+\theta_0-2\theta)/(\theta_1-\theta_0)} - 1}{A^{(\theta_1+\theta_0-2\theta)/(\theta_1-\theta_0)} - B^{(\theta_1+\theta_0-2\theta)/(\theta_1-\theta_0)}},$$

whereas, by L'Hôspital's rule, $L\big((\theta_0 + \theta_1)/2\big) = \log A/(\log A - \log B)$.

Example 6.11.2. Suppose $X_1, X_2, \ldots$ are iid taking values 0 and 1 with probabilities θ and $1 - \theta$, respectively; H_0: $\theta = \theta_0$ and H_1: $\theta = \theta_1 > \theta_0$. Using the notations $r_1 = \theta_1/\theta_0$ and $r_2 = (1 - \theta_1)/(1 - \theta_0)$, we have

$$Z_i = X_i \log[r_1/r_2] + \log r_2.$$

At the *n*th stage of sampling, accept H_0 if $\sum_{i=1}^n X_i \leq c_0 \log B + c_1 n$, accept H_1 if $\sum_{i=1}^n X_i \geq c_0 \log A + c_1 n$ and continue sampling if $\sum_{i=1}^n X_i$ lies within these boundaries, where c_0 and c_1 are easily determined. Here

$$\mathrm{E}_\theta[Z_i] \neq 0 \text{ for } \theta \neq -\log r_2/\log[r_1/r_2];$$
$$\mathrm{E}_\theta[e^{hZ_i}] = \theta r_1^h + (1 - \theta) r_2^h = 1 \text{ for } \theta = \frac{1 - r_2^h}{r_1^h - r_2^h}.$$

Plot $(\theta, L(\theta)) = \left(\frac{1-r_2^h}{r_1^h-r_2^h}, \frac{A^h-1}{A^h-B^h}\right)$, using h as a parameter to obtain the OC-curve.

6.12 Confidence Sets

Let $\{P_\theta: \theta \in \Omega\}$ be a family of probability distributions on $(\mathcal{X}, \mathcal{A})$ and let X be an observable rv whose distribution belongs to this family. Let C be a mapping from $\mathcal{X}$ into the class of all subsets of Ω, that is, for each $x \in \mathcal{X}$, $C(x) \subset \Omega$. We call C measurable if for each $\theta \in \Omega$, $\{x: \theta \in C(x)\} \in \mathcal{A}$. [If C is measurable, then "$C(X)$ covers θ" is an event for each $\theta \in \Omega$, so $P_\theta[C(X) \text{ covers } \theta] = P_\theta[\theta \in C(X)]$ is defined.]

A measurable C is said to be a *confidence set with confidence coefficient* $1-\alpha$ if

$$P_\theta[\theta \in C(X)] = P_\theta[\{x: \theta \in C(x)\}] \geq 1-\alpha \quad \text{for all } \theta \in \Omega,$$

that is, $C(X)$ covers the true value of θ with a probability of at least $1-\alpha$.

In particular, if θ is real and if C is such that $C(x) = [\underline{\theta}(x), \infty)$ or $C(x) = (-\infty, \overline{\theta}(x)]$ for each x, then $\underline{\theta}(x)$ and $\overline{\theta}(x)$ are called lower and upper confidence bounds, respectively, for θ with confidence coefficient $1-\alpha$ and we write: $P_\theta[\theta \in C(X)] = P_\theta[\theta \geq \underline{\theta}(X)] \geq 1-\alpha$ for all $\theta \in \Omega$, or $P_\theta[\theta \in C(X)] = P_\theta[\theta \leq \overline{\theta}(X)] \geq 1-\alpha$ for all $\theta \in \Omega$.

A lower confidence bound $\underline{\theta}(X)$ is said to be uniformly most accurate (UMA) with confidence coefficient $1-\alpha$ if

(i) $P_\theta[\theta \geq \underline{\theta}(X)] \geq 1-\alpha$ for all $\theta \in \Omega$, and
(ii) $P_\theta[\theta \geq T(X)] \geq 1-\alpha$ for all $\theta \in \Omega$ implies $P_\theta[\theta' \geq \underline{\theta}(X)] \leq P_\theta[\theta' \geq T(X)]$ for all $\theta' < \theta$,

that is, among all lower confidence bounds with confidence coefficient $1-\alpha$, $[\underline{\theta}(X), \infty)$ includes $\theta' < \theta$ with smallest probability.

UMA upper confidence bounds analogously includes $\theta' > \theta$ with smallest probability among all upper confidence bounds with the same confidence coefficient.

A systematic theory of confidence sets was introduced by Neyman [26]. We now discuss two methods for constructing confidence sets. Of these, the first is based on the concept of pivotal functions and the second uses a duality between acceptance regions of level α tests and confidence sets with confidence coefficient $1-\alpha$.

6.12.1 Methods Based on Pivotal Functions

Definition 6.12.1. A known function $T: \mathcal{X} \times \Omega \to R$ is a pivot if the distribution of $T(X,\theta)$ does not depend on θ (ie, for every $a \in R$, $P_\theta[T(X,\theta) \leq a]$ is the same for all $\theta \in \Omega$).

If $T(X,\theta)$ is a pivot, then its distribution is known. This allows us to find $c_1 < c_2$ for a given $\alpha \in (0,1)$ so that $P_\theta[c_1 \leq T(X,\theta) \leq c_2] \geq 1-\alpha$ for all θ.

Now define $C(x) = \{\theta \in \Omega: c_1 \leq T(x,\theta) \leq c_2\}$. Then C is a confidence set for θ with confidence coefficient $1-\alpha$, because $\theta \in C(x) \iff c_1 \leq T(x,\theta) \leq c_2$, and therefore,

$$P_\theta[\theta \in C(X)] = P_\theta[c_1 \leq T(X,\theta) \leq c_2] \geq 1-\alpha \quad \text{for all } \theta.$$

For real-valued θ, if $T(x,\theta)$ is monotone in θ for each fixed x, then $C(x) = \{\theta: c_1 \leq T(x,\theta) \leq c_2\}$ is an interval of the form $[\underline{\theta}(X), \overline{\theta}(X)]$.

Remark 6.12.1. So far we have considered single-parameter families of probabilities. If the probabilities depend on other nuisance parameters τ in addition to θ, then C is a confidence set for θ with confidence coefficient $1-\alpha$ if $P_{\theta,\tau}[\theta \in C(X)] \geq 1-\alpha$ for all $(\theta,\tau) \in \Omega$.

In such cases, a pivot $T(x,\theta)$ should depend on θ, but not on τ and its distribution must be the same for all $(\theta,\tau) \in \Omega$. Then we can find $c_1 < c_2$ so that $P_{\theta,\tau}[c_1 \leq T(X,\theta) \leq c_2] \geq 1-\alpha$ for all $(\theta,\tau) \in \Omega$, from which $C(x)$ constructed as above is a confidence set with confidence coefficient $1-\alpha$.

Remark 6.12.2. If $T_1,\ldots,T_k$ are pivots which are independent under each P_θ in the family $\{P_\theta : \theta \in \Omega\}$, then any function $g(T_1,\ldots,T_k)$ is also a pivot.

Example 6.12.1 (Location-Scale Family). Let $\boldsymbol{X} = (X_1,\ldots,X_n)$ be a random sample from $\sigma^{-1}f(\sigma^{-1}(x-\mu))$ where $\mu \in R$, $\sigma > 0$ and f is a known pdf.

(a) If σ is known and $\theta = \mu$ is the parameter of interest, then $X_i - \mu$, $i = 1,\ldots,n$ are pivots. So $\bar{X} - \mu$ is also a pivot, from which a confidence interval for μ is obtained as:

$$\begin{aligned} C(\boldsymbol{X}) &= \{\mu : c_1 \leq \bar{X} - \mu \leq c_2\} = \{\mu : \bar{X} - c_2 \leq \mu \leq \bar{X} - c_1\} \\ &= [\bar{X} - c_2, \bar{X} - c_1]. \end{aligned}$$

(b) If μ is known and $\theta = \sigma$ is the parameter of interest, then $(X_i - \mu)/\sigma$, $i = 1,\ldots,n$ are pivots and many other pivots can be constructed as functions of these. In particular, S/σ is a pivot where $S^2 = (n-1)^{-1}\sum_{i=1}^n (X_i - \bar{X})^2$ is the sample variance. Now a confidence interval for σ is obtained as:

$$C(\boldsymbol{X}) = \{\sigma : c_1 \leq S/\sigma \leq c_2\} = \{\sigma : S/c_2 \leq \sigma \leq S/c_1\} = [S/c_2, S/c_1].$$

Here S/σ is still a pivot if μ is unknown and the above confidence interval for σ is still valid.

(c) If $\theta = \mu$ is the parameter of interest and σ is also unknown, then $t(\boldsymbol{X}) = \sqrt{n}(\bar{X} - \mu)/S$, which is called the Studentized Version of $\sqrt{n}(\bar{X} - \mu)/\sigma$, is a pivot, from which a confidence interval for μ is obtained as

$$\begin{aligned} C(\boldsymbol{X}) &= \{\mu : c_1 \leq \sqrt{n}(\bar{X} - \mu)/S \leq c_2\} \\ &= \{\mu : \bar{X} - c_2 S/\sqrt{n} \leq \mu \leq \bar{X} - c_1 S/\sqrt{n}\} \\ &= [\bar{X} - c_2 S/\sqrt{n}, \bar{X} - c_1 S/\sqrt{n}]. \end{aligned}$$

When f is symmetric about 0, one may take $c_1 = -c_2 = -c$ and $C(X)$ is then of the form $[\bar{X} - cS/\sqrt{n}, \bar{X} + cS/\sqrt{n}]$.

Example 6.12.2 (Ratio of Means in Bivariate Normal (Feller's Theorem)). Let $\boldsymbol{X} = \{(X_{11},X_{12}),\ldots,(X_{n1},X_{n2})\}$ be a random sample $N_2(\boldsymbol{\mu},\boldsymbol{\Sigma})$, where $\boldsymbol{\mu}^\mathsf{T} = (\mu_1,\mu_2)$ and $\boldsymbol{\Sigma} = \begin{bmatrix} \sigma_1^2 & \sigma_{12} \\ \sigma_{12} & \sigma_2^2 \end{bmatrix}$. The parameter of interest is $\theta = \mu_2/\mu_1$, $\mu_1 \neq 0$. Let $Y_i(\theta) = X_{i2} - \theta X_{i1}$. Then $Y_i(\theta)$, $i = 1,\ldots,n$, are iid $N(0, \sigma_2^2 - 2\theta\sigma_{12} + \theta^2\sigma_1^2)$, The "sample mean" and "sample variance" of $Y_1(\theta),\ldots,Y_n(\theta)$ are $\bar{Y}(\theta) = \bar{X}_2 - \theta\bar{X}_1$ and

$$S^2(\theta) = (n-1)^{-1}\sum_{i=1}^n (Y_i(\theta) - \bar{Y}(\theta))^2 = S_2^2 - 2\theta S_{12} + \theta^2 S_1^2,$$

where S_1^2, S_2^2, and S_{12} are the sample estimates of σ_1^2, σ_2^2, and σ_{12}, respectively. Hence $\sqrt{n}\bar{Y}(\theta)/S(\theta) \sim t_{n-1}$ and is therefore a pivot and a confidence set for θ is $C(\boldsymbol{X}) = \{\theta\colon n\bar{Y}(\theta)^2/S^2(\theta) \leq t^2_{n-1,\alpha/2}\}$. The actual determination of $C(\boldsymbol{X})$ depends on a quadratic equation in θ.

6.12.2 Inverting Acceptance Regions of Tests

This method is based on a duality between hypothesis testing and confidence sets.

Theorem 6.12.1. *For each θ', let $H_0(\theta')$: $\theta = \theta' \in \Omega$ and $H_1(\theta')$: $\theta \in \Omega_1(\theta')$ where $\Omega_1(\theta') \subset \Omega - \{\theta'\}$. Then the following hold:*

(a) *If for each $\theta' \in \Omega$, $A(\theta')$ is the acceptance region of a test for $H_0(\theta')$ at level α, then $C(x) = \{\theta\colon x \in A(\theta)\}$ is a confidence set for θ with confidence coefficient $1 - \alpha$.*

(b) *If for each $\theta' \in \Omega$, $A(\theta')$ is the acceptance region of a UMP test for $H_0(\theta')$ vs $H_1(\theta')$ at level α, then among all confidence sets for θ with confidence coefficient $1 - \alpha$, $C(x)$ minimizes $P_\theta[\theta' \in C(X)]$ for all $\theta \in \Omega_1(\theta')$.*

Proof. Part (a) follows easily since

$$P_\theta[\theta \in C(X)] = P_\theta[X \in A(\theta)] \geq 1 - \alpha \quad \text{for all } \theta \in \Omega.$$

Now suppose that $C_1(\cdot)$ is also confidence set for θ with confidence coefficient $1 - \alpha$, and let $A_1(\theta) = \{x\colon \theta \in C_1(x)\}$. Then for each $\theta \in \Omega$, $A_1(\theta)$ is the acceptance region of a level α test for $H_0(\theta)$, because

$$P_\theta[X \in A_1(\theta)] = P_\theta[\theta \in C_1(X)] \geq 1 - \alpha \quad \text{for all } \theta.$$

Now,

$$P_\theta[\theta' \in C(X)] = P_\theta[X \in A(\theta')] = \text{Type II error of } A(\theta') \text{ at } \theta,$$
$$P_\theta[\theta' \in C_1(X)] = P_\theta[X \in A_1(\theta')] = \text{Type II error of } A_1(\theta') \text{ at } \theta.$$

Since $A(\theta')$ is the acceptance region of a UMP test for $H_0(\theta')$: $\theta = \theta' \in \Omega$ vs $H_1(\theta')$: $\theta \in \Omega_1(\theta')$, it follows that for all $\theta \in \Omega_1(\theta')$,

$$P_\theta\left[\theta' \in C(X)\right] \leq P_\theta\left[\theta' \in C_1(X)\right],$$

and this proves part (b). □

Corollary. *If $\{P_\theta\}$ is an MLR family in $T(x)$ and if F_θ is the continuous cdf of T under P_θ, then the UMA lower confidence bound for θ with confidence coefficient $1 - \alpha$ is given by $\underline{\theta}(x) = \theta^*(x)$ where $\theta^*(x)$ is the unique solution (in θ) of the equation $F_\theta(T(x)) = 1 - \alpha$.*

Proof. For each θ', the UMP level α test for $H_0(\theta')$: $\theta = \theta' \in \Omega$ vs $H_1(\theta')$: $\theta > \theta'$ has acceptance region $A(\theta') = \{x\colon T(x) \leq k(\theta')\}$ by the MLR property, where

$$P_{\theta'}[T(X) \leq k(\theta')] = F_{\theta'}(k(\theta')) = 1 - \alpha.$$

This defines a function $k(\cdot)$ by the equation $F_\theta(k(\theta)) = 1 - \alpha$. The MLR property also implies that for each k, $F_\theta(k)$ is a strictly decreasing function of θ, because $F_\theta(k)$ is the Type II error of an acceptance region $\{T \le k\}$ for a left-sided null hypothesis against a right-sided alternative. Hence for each k, the set $\{\theta\colon F_\theta(k) \le 1-\alpha\}$ is an interval $[\theta^*(k), \infty)$, where $\theta^*(k)$ is the unique solution of the equation $F_\theta(k) = 1-\alpha$. Now let

$$\begin{aligned} C(x) &= \{\theta\colon x \in A(\theta)\} = \{\theta\colon T(x) \le k(\theta)\} \\ &= \{\theta\colon F_\theta(T(x)) \le F_\theta(k(\theta))\} = \{\theta\colon F_\theta(T(x)) \le 1-\alpha\}. \end{aligned}$$

Then $C(x) = [\underline{\theta}(x), \infty)$, where $\underline{\theta}(x)$ is the unique solution of the equation $F_\theta(T(x)) = 1-\alpha$. By part (b) of the above theorem, it now follows that $\underline{\theta}(x)$ is UMA lower confidence bound for θ with confidence coefficient $1-\alpha$. □

Example 6.12.3. Let $\boldsymbol{X} = (X_1, \ldots, X_n)$ be random sample from $Unif(0, \theta)$. We want to construct a UMA confidence set for θ with confidence coefficient $1-\alpha$.

Solution. We know that the acceptance region of the UMP level α test for $H_0\colon \theta = \theta_0$ vs $H_1(\theta)\colon \theta \ne \theta_0$ is

$$A(\theta_0) = \{x\colon \theta_0 \alpha^{1/n} \le x_{n:n} \le \theta_0\},$$

where $x_{n:n} = \max(x_1, \ldots, x_n)$.

Now

$$\boldsymbol{x} \in A(\theta) \iff \theta\alpha^{1/n} \le x_{n:n} \le \theta \iff x_{n:n} \le \theta \le \alpha^{-1/n} x_{n:n}.$$

It therefore follows from the above theorem that $C(\boldsymbol{x}) = [x_{n:n}, \alpha^{-1/n} x_{n:n}]$ is the UMA confidence interval for θ with confidence coefficient $1-\alpha$.

We now consider the construction of confidence sets for one parameter when there are other nuisance parameters.

Suppose that the distribution of X belongs to the family $\{P_{\theta,\tau}\}$ where θ is real. A confidence set for θ with confidence coefficient $1-\alpha$ must satisfy: $P_{\theta,\tau}[\theta \in C(X)] \ge 1-\alpha$ for all θ and τ.

Definition 6.12.2.

(i) An unbiased lower (or upper) confidence bound $\underline{\theta}(x)$ (or $\overline{\theta}(x)$) for θ with confidence coefficient $1-\alpha$ must satisfy:
$P_{\theta,\iota}[\theta \ge \underline{\theta}(X)] \ge 1-\alpha$ for all θ, τ and
$P_{\theta,\tau}[\theta' \ge \underline{\theta}(X)] \le 1-\alpha$ for all $\theta' < \theta$ and τ,
$P_{\theta,\tau}[\theta \le \overline{\theta}(X)] \ge 1-\alpha$ for all θ, τ and
$P_{\theta,\tau}[\theta' \le \overline{\theta}(X)] \le 1-\alpha$ for all $\theta' > \theta$ and τ.

(ii) An unbiased confidence interval $[\underline{\theta}(X), \overline{\theta}(X)]$ for θ with confidence coefficient $1-\alpha$ must satisfy:
$P_{\theta,\tau}[\underline{\theta}(X) \le \theta \le \overline{\theta}(X)] \ge 1-\alpha$ for all θ, τ and
$P_{\theta,\tau}[\underline{\theta}(X) \le \theta' \le \overline{\theta}(X)] \le 1-\alpha$ for all $\theta' \ne \theta$ and τ.
Subject to these conditions, we minimize
$P_{\theta,\tau}[\theta' \ge \underline{\theta}(X)]$ for all $\theta' < \theta$ and τ, or
$P_{\theta,\tau}[\theta' \le \overline{\theta}(X)]$ for all $\theta' > \theta$ and τ, or
$P_{\theta,\tau}[\underline{\theta}(X) \le \theta' \le \overline{\theta}(X)]$ for all $\theta' \ne \theta$ and τ,

for a UMA unbiased lower confidence bound, or a UMA unbiased upper confidence bound, or a UMA unbiased confidence interval, respectively, for θ with confidence coefficient $1-\alpha$.

Theorem 6.12.2. *If for each θ', $A(\theta')$ is the acceptance region of a UMP unbiased level α test for $H_0\colon \theta = \theta'$ vs $H_1\colon \theta \neq \theta'$, then $C(x) = \{\theta\colon x \in A(\theta)\}$ is a UMA unbiased confidence interval for θ with confidence coefficient $1-\alpha$.*
[Analogous correspondence holds between acceptance regions of UMP unbiased level α tests for $H_0\colon \theta \leq \theta'$ vs $H_1\colon \theta > \theta'$ and UMA unbiased lower confidence bounds with confidence coefficient $1-\alpha$, and similarly for UMA unbiased upper confidence bounds.]

Example 6.12.4. Let $\boldsymbol{X} = (X_1, \ldots, X_m)$ and $\boldsymbol{Y} = (Y_1, \ldots, Y_n)$ be independent random samples from $N(\mu_1, \sigma^2)$ and $N(\mu_1, \sigma^2)$, respectively, where μ_1, μ_2, σ^2 are all unknown. We want to construct a UMA unbiased confidence interval for $\theta = \mu_1 - \mu_2$ with confidence coefficient $1-\alpha$.

Solution. Using the notations of Example 6.9.8 in Section 6.9, note that the acceptance region of the UMP unbiased level α test for $H_0\colon \theta = \theta_0$ vs $H_1\colon \theta \neq \theta_0$ is

$$A(\theta_0) = \left\{(\boldsymbol{x}, \boldsymbol{y})\colon \left|\frac{(\bar{x}-\bar{y}) - \theta_0}{SE}\right| \leq t_{m+n-2,\alpha/2}\right\},$$

where $SE = SE(\bar{X} - \bar{Y}) = s\sqrt{1/m + 1/n}$. Since $(\boldsymbol{x}, \boldsymbol{y}) \in A(\theta_0)$ is equivalent to

$$(\bar{x}-\bar{y}) - t_{m+n-2,\alpha/2}SE \leq \theta_0 \leq (\bar{x}-\bar{y}) + t_{m+n-2,\alpha/2}SE,$$

the UMA unbiased confidence interval for θ with confidence coefficient $1-\alpha$ is

$$C(\boldsymbol{x}, \boldsymbol{y}) = \left[(\bar{x}-\bar{y}) - t_{m+n-2,\alpha/2}SE, (\bar{x}-\bar{y}) + t_{m+n-2,\alpha/2}SE\right].$$

Exercises

6.1. Let $f(\boldsymbol{x}, \theta) = \prod_{i=1}^{n}[c(\theta)h(x_i)I_{(-\infty,\theta)}(x_i)]$ and define $x_{n:n} = \max(x_1, \ldots, x_n)$.

(a) Show that $\{f(\boldsymbol{x}, \theta)\colon \theta \in \mathbb{R}\}$ is an MLR family.

(b) Express the joint pdf of a random sample $\boldsymbol{X} = (X_1, \ldots, X_n)$ from $Unif(0, \theta)$ as a special case of this family.

(c) Show that the test

$$\phi(\boldsymbol{x}) = \begin{cases} 1 & \text{if } x_{n:n} > \theta_0 \\ 0 & \text{if } x_{n:n} \leq \theta_0 \end{cases}$$

is a UMP level α test for $H_0\colon \theta \leq \theta_0$ vs $H_1\colon \theta > \theta_0$, but that ϕ is inadmissible under the $0-1$ loss function (ie, there exists a test ψ whose risk function under the $0-1$ loss function satisfies $R(\theta, \psi) \leq R(\theta, \phi)$ for all θ), with strict inequality holding for some θ.

(d) Show that

$$\phi(\boldsymbol{x}) = \begin{cases} 1 & \text{if } x_{n:n} > \theta_0 \text{ or } x_{n:n} \leq b \\ 0 & \text{if } b < x_{n:n} \leq \theta_0, \end{cases}$$

where $b = \theta_0\alpha^{1/n}$ is a UMP level α test for $H_0\colon \theta = \theta_0$ vs $H_1\colon \theta \neq \theta_0$.

6.2. Let X be the number of successes in n independent trials with $P[\text{success}] = \theta$. Let $\phi(x)$ be the UMP level α test for $H_0\colon \theta \le \theta_0$ vs $H_1\colon \theta > \theta_0$.

(a) For $n = 6$, $\theta_0 = 0.25$ and $\alpha = 0.05, 0.10, 0.20$, find c and γ for ϕ and find the powers of these tests at $\theta = 0.40$.

(b) Use normal approximation in order to find the smallest n required for the UMP test at level $\alpha = 0.05$ for $H_0\colon \theta \le 0.25$ vs $H_1\colon \theta > 0.25$ to attain power $\beta(0.4) = 0.90$.

6.3. Let $X_1, \ldots, X_n$ be a random sample from an exponential distribution with mean θ.

(a) For $n = 5$, find the UMP test for $H_0\colon \theta \le 1$ vs $H_1\colon \theta > 1$ at level $\alpha = 0.1$. [The exponential rv with mean 2 is the same as a χ^2 with 2 df.]

(b) For the test obtained in part (a), calculate the power $\beta(2)$ at $\theta = 2$.

(c) Find the smallest n so that the UMP test at level $\alpha = 0.1$ has $\beta(2) = 0.8$.

[For parts (b) and (c), use integration by parts to show that $\int_c^\infty x^n \exp(-x)\, dx/n! = P[W \le n]$, where $W \sim Poi(c)$.]

6.4. A box contains N manufactured items of which an unknown number θ are defective and the other $N - \theta$ are good. Let X denote the number of defective items in a random sample of n items drawn without replacement from the box. Then X has pmf

$$f(x,\theta) = P_\theta[X = x] = \binom{\theta}{x}\binom{N-\theta}{n-x}\Big/\binom{N}{n}, \quad x = \max(0, n+\theta-N), \ldots, \min(n,\theta).$$

(a) Show that $\{f(x,\theta)\colon \theta = 0, 1, \ldots, N\}$ is an MLR family. [For $\theta_1 < \theta_2$, write $f(x,\theta_2)/f(x,\theta_1) = \prod_{j=\theta_1}^{\theta_2-1}\{f(x,j+1)/f(x,j)\}$ and examine how $f(x,j+1)/f(x,j)$ changes with x.]

(b) For a specified integer θ_0, write down the UMP test at a given level α for $H_0\colon \theta \le \theta_0$ vs $H_1\colon \theta > \theta_0$. Explain how the constants involved in the UMP test are determined.

6.5. Let $X_1, \ldots, X_n$ be a random sample from $Unif(\theta, \theta+1)$.

(a) Show that $(T_1, T_2) = (X_{n:1}, X_{n:n})$ are jointly sufficient for θ and find the joint distribution of (T_1, T_2).

(b) Show that the UMP test at level α for $H_0\colon \theta \le 0$ vs $H_1\colon \theta > 0$ is of the form

$$\phi(t_1,t_2) = \begin{cases} 0 & \text{if } t_1 < k \text{ and } t_2 < 1 \\ 1 & \text{otherwise} \end{cases}, \quad \text{where } k = 1 - \alpha^{1/n}.$$

6.6. Let P_0 and P_1 be two probability distributions with pdf's p_0 and p_1, respectively. Suppose that under P_0, the likelihood ratio $T = p_1(X)/p_0(X)$ has a pdf which is everywhere positive. For $0 < \alpha < 1$, let $\{x\colon T(x) \ge k(\alpha)\}$ denote the critical region of an N-P test of size α for $H_0\colon P = P_0$ vs $H_1\colon P = P_1$ and let $\beta(\alpha) = P_1[T \ge k(\alpha)]$. Show that $\beta'(\alpha) = k(\alpha)$.

6.7. Suppose $P_0 \ne P_1$ are probabilities on $(\mathcal{X}, \mathcal{A})$ and $X_1, \ldots, X_n$ are independent samples from $(\mathcal{X}, \mathcal{A}, P)$ where P is either P_0 or P_1. We want to test $H_0\colon P = P_0$ vs $H_1\colon P = P_1$. Show that there exists a sequence of tests $\{\phi_n\}$, each based on $(X_1, \ldots, X_n)$ such that

$$\lim_{n\to\infty} \mathrm{E}_{P_0}[\phi_n(X_1,\ldots,X_n)] = 0 \text{ and } \lim_{n\to\infty} \mathrm{E}_{P_1}[\phi_n(X_1,\ldots,X_n)) = 1,$$

that is, the Type I error probability converges to zero and the power converges to 1. [*Hint*: Let p_i denote the pdf/pmf corresponding to P_i, $i = 0, 1$. Then for any α, the MP test at level α for H_0 vs H_1 based on $(X_1,\ldots,X_n)$ is roughly (except for discreteness) of the form:

$$\phi_n(X_1,\ldots,X_n) = I_{[k_n,\infty)}\left[n^{-1}\sum_{i=1}^{n}\log(p_1(X_i)/p_0(X_i))\right],$$

where k_n is chosen according to α. Let

$$m_i = \mathrm{E}_{P_i}[\log(p_1(X_i)/p_0(X_i))], \quad i = 0, 1.$$

Use Jensen's inequality to show that $m_0 < 0 < m_1$, choose $k_n = (m_0 + m_1)/2$ in the definition of ϕ_n. Now use the SLLN.]

6.8. Let $X \sim Bin(n, p)$ with $n = 10$ and p unknown. Find the UMP unbiased tests at level $\alpha = 0.10$ for
(a) H_0: $p = 0.2$ vs H_1: $p \neq 0.2$, (b) H_0: $p = 0.5$ vs H_1: $p \neq 0.5$, (c) H_0: $0.4 \le p \le 0.6$ vs H_1: $p \notin [0.4, 0.6]$.

6.9. Let X and Y be independent Poisson rv's with means λ and μ, respectively. Construct UMP unbiased tests at level $\alpha = 0.1$ over the set $\{(x, y)\colon x + y = 8\}$ for
(a) H_0: $\lambda \le \mu$ vs H_1: $\lambda > \mu$, (b) H_0: $\lambda = \mu$ vs H_1: $\lambda \neq \mu$.

6.10. Let $X_1,\ldots,X_n$ be a random sample from an exponential distribution with mean θ. Find the UMP unbiased level α test for H_0: $\theta = 2$ vs H_1: $\theta \neq 2$.

6.11. Let X be a random sample of size 1 from the beta distribution $Be(\theta, 1)$ with the pdf $f(x, \theta) = \theta x^{\theta-1}$, $0 < x < 1$. Find the UMP unbiased test at level $\alpha = 0.1$ for H_0: $\theta = 1$ vs H_1: $\theta \neq 1$ and determine the critical value.

6.12. Use Basu's Theorem to show the following:
(a) If $X_1,\ldots,X_n$ is a random sample from $N(0, \sigma^2)$, then $\sum X_i/\sqrt{\sum X_i^2}$ and $\sum X_i^2$ are independent.
(b) If $(X_1, Y_1),\ldots,(X_n, Y_n)$ are iid $N_2(\boldsymbol{\mu}, \boldsymbol{\Sigma})$, where $\boldsymbol{\mu} = (\mu_1, \mu_2)$ and $\boldsymbol{\Sigma}$ is a diagonal matrix with diagonal entries σ_1^2 and σ_2^2, then $\sum(X_i - \bar{X}_n)^2$, $\sum(Y_i - \bar{Y}_n)^2$ and the sample correlation r are mutually independent.

6.13. Let $\boldsymbol{X} = (X_1,\ldots,X_n)$ be a random sample from $Unif(\theta_1, \theta_2)$. Sufficient statistic for (θ_1, θ_2) in $\boldsymbol{X}$ is $(T_1, T_2) = (X_{n:1}, X_{n:n})$.
(a) Show that T_1 given $T_2 = t_2$ is distributed as $U_{n-1:1}$ where $(U_1,\ldots,U_{n-1})$ is a random sample from $Unif(\theta_1, t_2)$.
(b) Construct a UMP level α test for H_0: $\theta_1 \le 0$ vs H_1: $\theta_1 > 0$ conditionally, given $T_2 = t_2$. Call this test $\phi_{t_2}(t_1)$.
(c) Show that $\phi(t_1, t_2) = \phi_{t_2}(t_1)$ is unconditionally a UMP level α test for H_0 vs H_1.

6.14. Let X and Y be two independent exponential random variables with means $1/\lambda$ and $1/\mu$, respectively. Find UMP unbiased tests at level $\alpha = 0.2$ for

(a) $H_0\colon \lambda \le \mu + 1$ vs $H_1\colon \lambda > \mu + 1$, (b) $H_0\colon \lambda = \mu$ vs $H_1\colon \lambda \ne \mu$, (c) $H_0\colon \lambda \ge 2\mu$ vs $H_1\colon \lambda < 2\mu$.

6.15. Let X_1, X_2 be independent rv's with pmf's

$$f_{X_i}(x, \theta_i) = \theta_i(1-\theta_i)^{x-1}, \quad x = 1, 2, \ldots, \text{ where } 0 < \theta_i < 1.$$

Find UMP unbiased tests at level $\alpha = 0.2$ for

(a) $H_0\colon \theta_1 \le \theta_2$ vs $H_1\colon \theta_1 > \theta_2$, (b) $H_0\colon \theta_1 = \theta_2$ vs $H_1\colon \theta_1 \ne \theta_2$.

6.16. Let $X = (X_1, \ldots, X_m)$ and $Y = (Y_1, \ldots, Y_n)$ be independent samples from $N(\mu_1, \sigma_1^2 = 1)$ and $N(\mu_2, \sigma_2^2 = 2)$, respectively. We want to test $H_0\colon \mu_1 \le \mu_2$ vs $H_1\colon \mu_1 > \mu_2$. Derive the UMP unbiased level α test for H_0 vs H_1 by first expressing it in a conditional form (in terms of appropriate sufficient statistics), and then unconditionally in terms of $\bar{X}_m - \bar{Y}_n$.

6.17. Let $(X_1, Y_1), \ldots, (X_n, Y_n)$ be a random sample from a bivariate normal distribution with $E[X_1] = E[Y_1] = \mu$, $Var[X_1] = Var[Y_1] = \sigma^2$ and $Cov[X_1, Y_1] = \sigma^2/2$, with μ and σ^2 unknown. Find the UMP unbiased level α test for $H_0\colon \mu = 0$ vs $H_1\colon \mu \ne 0$. [Transform: $U_i = (X_i + Y_i)/\sqrt{3}$ and $V_i = X_i - Y_i$.]

6.18. Let $X_1, X_2, \ldots$ be sequentially observed independent samples from a normal distribution with unknown mean θ and known SD $\sigma = 5$. We want to test $H_0\colon \theta = 0$ vs $H_1\colon \theta = 2$ at level $\alpha = 0.01$, holding the probability of Type II error probability at $\beta = 0.05$.

(a) Find the approximate values of A, B in $SPRT(A, B)$ needed for this purpose and describe the procedure in terms of cumulative sums $\sum_{i=1}^{n} X_i$, $n = 1, 2, \ldots$.

(b) Evaluate the OC function $L(\theta) = P_\theta[\text{accept } H_0]$ for $\theta = 3$.

(c) Evaluate the ASN function $E_\theta(N)$ for $\theta = 0$ and $\theta = 2$.

(d) Find the smallest sample size $n(\alpha, \beta)$ needed for a fixed sample size test with $\alpha = 0.01$ and $\beta \le 0.05$ in this problem, and compare $n(\alpha, \beta)$ with the numbers obtained in (c).

6.19. Do problem 18 when the X_i's are sequentially observed independent samples from

(i) Exponential distribution with mean θ,

(ii) Poisson distribution with mean θ, and we want to test $H_0\colon \theta = 1$ vs $H_1\colon \theta = 2$ at level $\alpha = 0.05$, holding $\beta = 0.10$.

6.20. Let $X_1, \ldots, X_{10}$ be a random sample from an exponential distribution with mean θ. Find the UMA lower confidence bound for θ with confidence coefficient $1 - \alpha = 0.95$ based on $(X_1, \ldots, X_{10})$, using the table for the χ^2-distribution.

6.21. Let $X_1, \ldots, X_n$ be iid following the Weibull distribution with the pdf

$$f(x, \lambda) = \lambda c x^{c-1} \exp(-\lambda x^c), \quad x > 0,$$

where c is known but λ is unknown. Show that the UMA upper confidence bound for $\theta = 1/\lambda$ with confidence coefficient $1 - \alpha$ is given by $\bar{\theta} = 2\sum X_i^c/\chi^2_{2n}(\alpha)$ where $P[\chi^2_{2n} \le \chi^2_{2n}(\alpha)] = \alpha$. [Hint: Find the distribution of X_i^c.]

6.22. Let X and Y be independent exponential random variables with means $1/\lambda$ and $1/\mu$, respectively. Construct a UMA unbiased confidence interval with confidence coefficient $1-\alpha$ for $\theta = \lambda/\mu$ by inverting the acceptance regions of UMP unbiased level α tests for $H_0: \theta = \theta_0$ vs $H_1: \theta \neq \theta_0, \theta_0 > 0$.

6.23. Four independent experiments were carried out to test whether a new method of irrigation would provide better yield of a certain crop. The data from all four experiments were analyzed to test the same hypothesis H_0 that the new method is no better than the old vs the alternative H_1 that the new method is better. The results of these experiments are summarized by four t-statistics given below with their respective df's

$$t_1 = 1.50, df = 10;\ t_2 = 2.15, df = 16;\ t_3 = 1.80, df = 19;\ t_4 = 1.30, df = 25.$$

(a) Calculate the p-values of the four statistics

(b) Calculate the P_λ-statistic combining the results of all four experiments.

(c) Comment.

7 Methods Based on Likelihood and Their Asymptotic properties

7.1 Asymptotic Properties of the MLEs: Consistency and Asymptotic Normality

On a historical note, the maximum likelihood estimators (MLEs) were introduced by R.A. Fisher in early 1920s, which he claimed to be "better" than the method of moments estimators used widely, especially by K. Pearson. To justify the superiority of the MLEs, Fisher used the concepts of consistency (converging to the true parameter in probability), sufficiency (capturing everything relevant in the sample), and efficiency (attaining the smallest possible variance among all unbiased estimators, which led to the definition of Fisher-information). This led to the foundation of the theory of statistical inference in the area of estimation. Asymptotic properties of the MLEs will be discussed in this section.

Let $X_1, \ldots, X_n, \ldots$ be iid with pdf/pmf $f(x;\theta_0)$ in the family $\{f(x;\theta)\colon \theta \in \Theta\}$, where θ_0, the unknown value of the parameter is an interior point of Θ. All probability statements (including $\stackrel{P}{\to}$, $\stackrel{\mathcal{L}}{\to}$, o_P, O_P) and all expectations, variances, and covariances are with respect to $f(,\theta_0)$, unless stated otherwise. We assume throughout that the family $\{f(x,\theta)\colon \theta \in \Theta\}$ satisfies the identifiably condition introduced in Section 5.5.1.

The MLE of θ_0 based on $(X_1, \ldots, X_n)$ is denoted by

$$\hat{\theta}_n = \arg\max_{t\in\Theta} \prod_{i=1}^{n} f(X_i, t) = \arg\max_{t\in\Theta} \sum_{i=1}^{n} l(X_i, t), \tag{1}$$

where $l(x,t) = \log f(x,t)$.

Using the notations of Section 5.2.1, we write $\log f(x,t) = l(x,t)$ as in the above paragraph and also

$$\frac{\partial f(x,\theta)}{\partial\theta} = \dot{f}(x,\theta), \frac{\partial^2 f(x,\theta)}{\partial\theta^2} = \ddot{f}(x,\theta),$$
$$\frac{\partial l(x,\theta)}{\partial\theta} = \dot{l}(x,\theta) \text{ and } \frac{\partial^2 l(x,\theta)}{\partial\theta^2} = \ddot{l}(x,\theta) \tag{2a}$$

in the single-parameter case when $\Theta \subset \mathbb{R}$ and

Theory and Methods of Statistics. http://dx.doi.org/10.1016/B978-0-12-802440-9.00007-2

$$\frac{\partial f(x,\theta)}{\partial\theta_r} = \dot{f}_r(x,\theta), \quad \frac{\partial^2 f(x,\theta)}{\partial\theta_r\partial\theta_s} = \ddot{f}_{rs}(x,\theta),$$
$$\frac{\partial l(x,\theta)}{\partial\theta_r} = \dot{l}_r(x,\theta) \text{ and } \frac{\partial^2 l(x,\theta)}{\partial\theta_r\partial\theta_s} = \ddot{l}_{rs}(x,\theta) \tag{2b}$$

in the multiparameter case when $\Theta \subset \mathbb{R}^k$.

The following conditions were essentially introduced by Cramér [18, p. 500–501].

Regularity conditions (Cramér): single-parameter case.

In the notations of Eqs. (2a) and (2b):

1. $\dot{f}(x,t), \ddot{f}(x,t)$ exist for all (x,t) and there exists a nonnegative function $g(x)$ with $\int g < \infty$ such that $|\dot{f}(x,t)|$ and $|\ddot{f}(x,t)|$ are bounded above by $g(x)$ for all (x,t).
2. There exist nonnegative functions $H(x,\theta_0)$ and $\phi(\varepsilon)$ such that

$$\sup_{|t-\theta_0|\le\varepsilon} |\ddot{l}(x,t) - \ddot{l}(x,\theta_0)| \le H(x,\theta_0)\phi(\varepsilon),$$

   where $\lim_{\varepsilon\to 0}\phi(\varepsilon) = 0$ and $\mathrm{E}[H(X,\theta_0)] < \infty$.
3. $0 < -\mathrm{E}[\ddot{l}(X,\theta_0)] = I(\theta_0) < \infty$.

By dominated convergence, Condition 1 allows differentiation of $\int f(x,\theta)\,dx$ twice with respect to θ under the integral. Since $\int f(x,\theta) = 1$ for all θ, we have

$$\begin{aligned}
0 &= \frac{d}{d\theta}\int f(x,\theta)\,dx\Big|_{\theta=\theta_0} = \int \dot{f}(x,\theta_0)\,dx \\
&= \int \dot{l}(x,\theta_0) f(x,\theta_0)\,dx = \mathrm{E}[\dot{l}(X,\theta_0)], \text{ and} \\
0 &= \frac{d^2}{d\theta^2}\int f(x,\theta)\,dx\Big|_{\theta=\theta_0} = \int \ddot{f}(x,\theta_0)\,dx \\
&= \mathrm{E}[\ddot{l}(X,\theta_0)] + \mathrm{E}[\{\dot{l}(X,\theta_0)\}^2].
\end{aligned}$$

Thus

$$\mathrm{E}[\dot{l}(X,\theta_0)] = 0 \text{ and } I(\theta_0) = \mathrm{E}[-\ddot{l}(X,\theta_0)] = \mathrm{Var}(\dot{l}(X,\theta_0)). \tag{3a}$$

Regularity conditions in the multiparameter case: Condition 1 should hold for $\dot{f}_r(x,t)$ and $\ddot{f}_{rs}(x,t)$ for all (x,t) and for all r and s, Condition 2 should hold for $\ddot{l}_{rs}(x,t)$ for all r and s, and in Condition 3 we need all elements of the information matrix $I(\theta_0) = ((I_{rs}(\theta_0)))$ to exist and $I(\theta_0)$ to be positive definite.

By the same argument as in the single-parameter case, it follows from Condition 1 by dominated convergence, that

$$\begin{aligned}
&\mathrm{E}[\dot{l}_r(X,\theta_0)] = 0 \text{ and} \\
&I_{rs}(\theta_0) = \mathrm{E}[-\ddot{l}_{rs}(X,\theta_0)] = \mathrm{Cov}(\dot{l}_r(X,\theta_0), \dot{l}_s(X,\theta_0))
\end{aligned} \tag{3b}$$

for all r and s.

Theorem 7.1.1. *Under regularity Condition 1, the MLE $\hat{\theta}_n$ is consistent (ie, $\hat{\theta}_n \xrightarrow{P} \theta_0$ as $n \to \infty$).*

Proof. As shown in Section 5.5.1, the identifiably condition implies that $\mathrm{E}\,[l(X,t)]$ has a unique maximum at $t = \theta_0$. In the single-parameter case, we therefore have for any $\delta > 0$

$$\mathrm{E}[l(X,\theta_0)] > \max\{\mathrm{E}[l(X,\theta_0-\delta)], \mathrm{E}[l(X,\theta_0+\delta)]\}.$$

Hence for a given $\delta > 0$, there exists $\varepsilon > 0$ such that

$$\mathrm{E}[l(X,\theta_0)] - \mathrm{E}[l(X,t)] > 2\varepsilon \quad \text{for } t = \theta_0 \pm \delta.$$

Since $n^{-1}\sum l(X_i,t) \xrightarrow{P} \mathrm{E}[l(X,t)]$ for all t, it follows that

$$\lim_{n\to\infty} P\Big[n^{-1}\sum l(X_i,\theta_0) - n^{-1}\sum l(X_i,t) > \varepsilon\Big] = 1 \quad \text{for } t = \theta_0 \pm \delta.$$

But $\dot{l}(x,t)$ exist in a neighborhood of θ_0, so with probability tending to 1 as $n \to \infty$, the equation

$$g_n(t) = n^{-1}\sum_{i=1}^{n} \dot{l}(X_i,t) = 0$$

has a solution in the intervals $(\theta_0-\delta, \theta_0+\delta)$ for arbitrary $\delta > 0$; that is, there is a sequence of solutions $\{\hat{\theta}_n\}$ of the equation $g_n(t) = 0$ needed for (1) converges to θ_0 in probability as $n \to \infty$. The MLE is consistent in this sense.

In the multiparameter case, we modify the above proof by replacing $\theta_0 \pm \delta$ with $S(\theta_0,\delta) = \{t\colon \|t-\theta_0\| \le \delta\}$ and argue in the same way, assuming that the regularity conditions hold. □

Theorem 7.1.2 (Asymptotic Normality of MLEs: Single-Parameter Case). *Under regularity Conditions 1, 2, 3, the MLE $\hat{\theta}_n$ is asymptotically normal (ie, $\sqrt{n}(\hat{\theta}_n - \theta_0) \xrightarrow{\mathcal{L}} N(0, 1/I(\theta_0))$).*

Proof. Let $\{\hat{\theta}_n\}$ be the sequence described in the proof of Theorem 7.1.1 with $\hat{\theta}_n = \theta_0 + o_P(1)$ and $\sum_{i=1}^n \dot{l}(X_i,\hat{\theta}_n) = 0$. Then

$$0 = n^{-1/2}\sum_{i=1}^{n}\dot{l}(X_i,\hat{\theta}_n) = n^{-1/2}\sum_{i=1}^{n}\dot{l}(X_i,\theta_0) + \sqrt{n}(\hat{\theta}_n-\theta_0)n^{-1}\sum_{i=1}^{n}\ddot{l}\big(X_i,\tilde{\theta}_n\big),$$

where $\tilde{\theta}_n = \theta_0 + \lambda(\hat{\theta}_n - \theta_0)$, $0 \le \lambda \le 1$. Hence

$$\begin{aligned}\sqrt{n}\big(\hat{\theta}_n-\theta_0\big) &= \frac{n^{-1/2}\sum_{i=1}^n \dot{l}(X_i,\theta_0)}{-n^{-1}\sum_{i=1}^n \ddot{l}\big(X_i,\tilde{\theta}_n\big)} \\ &= \frac{n^{-1/2}\sum_{i=1}^n \dot{l}(X_i,\theta_0)}{-n^{-1}\sum_{i=1}^n \ddot{l}(X_i,\theta_0) + n^{-1}\sum_{i=1}^n R_n(X_i)}. \qquad (4)\end{aligned}$$

On the right-hand side of this expression, the numerator $n^{-1/2}\sum_{i=1}^{n}\dot{l}(X_i,\theta_0) \xrightarrow{\mathcal{L}} N((0,I(\theta_0))$ since $\dot{l}(X_i,\theta_0)$, $i = 1,2,\ldots$ are iid with mean 0 and variance $I(\theta_0)$ by Eq. (3a), while in the denominator,

$$-n^{-1}\sum_{i=1}^{n}\ddot{l}(X_i,\theta_0) \xrightarrow{P} \mathrm{E}[-\ddot{l}(X,\theta_0)] = I(\theta_0).$$

Finally,

$$\begin{aligned}|R_n(X_i)| &= |\ddot{l}(X_i,\theta_0+\lambda(\hat{\theta}_n-\theta_0)) - \ddot{l}(X_i,\theta_0)| \\ &\le \sup_{|\theta'-\theta_0|\le|\hat{\theta}_n-\theta_0|} |\ddot{l}(X_i,\theta') - \ddot{l}(X_i,\theta_0)| \le \varphi(|\hat{\theta}_n-\theta_0|)H(X_i,\theta_0),\end{aligned}$$

by Condition 3, so

$$\begin{aligned}\left|n^{-1}\sum_{i=1}^{n}R_n(X_i)\right| &\le \varphi(|\hat{\theta}_n-\theta_0|)n^{-1}\sum_{i=1}^{n}H(X_i,\theta_0) \\ &= o_P(1)\{O(1)+o_P(1)\} = o_P(1),\end{aligned}$$

since $|\hat{\theta}_n-\theta_0| = o_P(1)$ implies $\varphi(|\hat{\theta}_n-\theta_0|) = o_P(1)$ and $n^{-1}\sum_{i=1}^{n}H(X_i,\theta_0) = \mathrm{E}[H(X,\theta_0)] + o_P(1) = O(1)+o_P(1)$. Putting all this together in Eq. (4), we see that

$$\sqrt{n}(\hat{\theta}_n-\theta_0) = \frac{n^{-1/2}\sum_{i=1}^{n}\dot{l}(X_i,\theta_0)}{\{I(\theta_0)+o_P(1)\}+o_P(1)} \xrightarrow{\mathcal{L}} N(0,1/I(\theta_0)).$$

□

Suppose a sequence of unbiased estimators $T_n^* = T_n^*(X_1,\ldots,X_n)$ of θ_0 has $\mathrm{Var}[T_n^*] = 1/\{nI(\theta_0)\}$ which is the Cramér-Rao lower bound for unbiased estimators (under regularity conditions). If for a sequence of unbiased estimators, we define

$$e_n(T_n) = \frac{\text{C-R Lower Bound}}{\mathrm{Var}[T_n]} = \frac{1/\{nI(\theta_0)\}}{\mathrm{Var}[T_n]}$$

as the efficiency of T_n, then T_n^* described above has efficiency 1.

The large sample analog of this is to make comparison among all *consistent estimators* in terms of their *asymptotic variance.* Since the MLE $\hat{\theta}_n$ has variance $1/\{nI(\theta_0)\}$ in an asymptotic sense, we can say that the asymptotic efficiency of the MLE is 1, or simply that the MLE is asymptotically efficient. However, the justification of the last statement needs much deeper analysis as will be seen later.

Definition 7.1.1. Any estimator T_n with $\sqrt{n}(T_n-\theta_0) \xrightarrow{\mathcal{L}} N(0,1/I(\theta_0))$ is said to be a *best asymptotically normal* (BAN) estimator. (This is because of asymptotic normality and the property of asymptotic efficiency.)

Finding a BAN Estimator by the Newton-Raphson Method

In some situations, the likelihood equation does not have a closed form solution, so we cannot obtain the MLE $\hat{\theta}_n$ explicitly. However, in such a case, we can often find an estimator which is consistent. Starting with such an estimator $\tilde{\theta}_{n0} = \theta_0 + o_P(1)$ as an initial estimator we can use the Newton-Raphson method. After one iteration, we have

$$\tilde{\theta}_{n1} = \tilde{\theta}_{n0} + \frac{n^{-1}\sum_{i=1}^{n} \dot{l}(X_i, \tilde{\theta}_{n0})}{-n^{-1}\sum_{i=1}^{n} \ddot{l}(X_i, \tilde{\theta}_{n0})}.$$

By argument given in the course of the proof of Theorem 7.1.2, it can be shown that if $T_n = \theta_0 + o_P(1)$, then $\left|n^{-1}\sum_{i=1}^{n} \ddot{l}(X_i, T_n) - n^{-1}\sum_{i=1}^{n} \ddot{l}(X_i, \theta_0)\right| = o_P(1)$. Since $\tilde{\theta}_{n0} = \theta_0 + o_P(1)$, we then have

$$0 = n^{-1}\sum_{i=1}^{n} \dot{l}(X_i, \hat{\theta}_n) = n^{-1}\sum_{i=1}^{n} \dot{l}(X_i, \tilde{\theta}_{n0}) + (\hat{\theta}_n - \tilde{\theta}_{n0})n^{-1}\sum_{i=1}^{n} \ddot{l}(X_i, \tilde{\theta}_{n0})$$
$$+ o_P(1)(\hat{\theta}_n - \tilde{\theta}_{n0}),$$

so that

$$\hat{\theta}_n = \tilde{\theta}_{n0} + \frac{n^{-1}\sum_{i=1}^{n} \dot{l}(X_i, \tilde{\theta}_{n0})}{-n^{-1}\sum_{i=1}^{n} \ddot{l}(X_i, \tilde{\theta}_{n0})} + o_P(1)(\hat{\theta}_n - \tilde{\theta}_{n0}) = \tilde{\theta}_{n1} + o_P(1)(\hat{\theta}_n - \tilde{\theta}_{n0}).$$

Hence

$$\sqrt{n}(\tilde{\theta}_{n1} - \theta_0) = \sqrt{n}(\hat{\theta}_n - \theta_0) + o_P(1)\sqrt{n}(\hat{\theta}_n - \tilde{\theta}_{n0}) = \sqrt{n}(\hat{\theta}_n - \theta_0) + o_P(1)$$

under additional condition that $\sqrt{n}(\hat{\theta}_n - \tilde{\theta}_{n0}) = O_P(1)$ (ie, $\tilde{\theta}_{n0} - \theta_0 = O_P(n^{-1/2})$). In such a case, $\sqrt{n}(\tilde{\theta}_{n1} - \theta_0)$ has the same asymptotic distribution as that of $\sqrt{n}(\hat{\theta}_n - \theta_0)$, so $\tilde{\theta}_{n1}$ is a BAN estimator.

In the multiparameter case, $\boldsymbol{\theta}_0$ and $\boldsymbol{t}$ are k-dim column vectors with θ_{0r} and t_r as their rth coordinates, and

$$\dot{l}(x, \boldsymbol{t})^T = (\dot{l}_1(x, \boldsymbol{t}), \ldots, \dot{l}_k(x, \boldsymbol{t})), \quad \ddot{l}(x, \boldsymbol{t}) = ((\ddot{l}_{rs}(x, \boldsymbol{t}))).$$

By regularity conditions, $\mathrm{E}[\dot{l}_r(X, \boldsymbol{\theta}_0)] = 0$ and

$$\mathrm{E}[-\ddot{l}_{rs}(X, \boldsymbol{\theta}_0)] = \mathrm{Cov}[\dot{l}_r(X, \boldsymbol{\theta}_0), \dot{l}_s(X, \boldsymbol{\theta}_0)] = I_{rs}(\boldsymbol{\theta}_0)$$

as shown in Eq. (3b). In matrix notation,

$$\mathrm{E}[\dot{\boldsymbol{l}}(X, \boldsymbol{\theta}_0)] = \boldsymbol{0} \text{ and } \mathrm{E}[-((\ddot{l}_{rs}(X, \boldsymbol{\theta}_0)))] = ((I_{rs}(\boldsymbol{\theta}_0))) = \boldsymbol{I}(\boldsymbol{\theta}_0).$$

Let

$$A_{nr} = n^{-1/2}\sum_{i=1}^{n} \dot{l}_r(X_i, \boldsymbol{\theta}_0) \text{ and } \boldsymbol{A}_n^T = (A_{n1}, \ldots, A_{nk}).$$

Then $\boldsymbol{A}_n = n^{-1/2}\sum_{i=1}^{n} \dot{\boldsymbol{l}}(X_i, \boldsymbol{\theta}_0) \xrightarrow{\mathcal{L}} N_k(\boldsymbol{0}, \boldsymbol{I}(\boldsymbol{\theta}_0))$.

Next let $B_{nrs}(\boldsymbol{t}) = -n^{-1}\sum_{i=1}^{n}\ddot{l}_{rs}(X_i, \boldsymbol{t})$. Then

$$\boldsymbol{B}_n(\boldsymbol{t}) = ((B_{nrs}(\boldsymbol{t}))) = -n^{-1}\sum_{i=1}^{n}((\ddot{l}_{rs}(X_i, \boldsymbol{t}))) \xrightarrow{P} -\mathrm{E}[((\ddot{l}_{rs}(X, \boldsymbol{t})))].$$

By Condition 2, arguing in the same manner as in the single-parameter case, we see that $\boldsymbol{T}_n = \boldsymbol{\theta}_0 + o_P(1)$ implies $\boldsymbol{B}_n(\boldsymbol{T}_n) = \boldsymbol{B}_n(\boldsymbol{\theta}_0) + o_P(1)$. [For vectors and matrices, the $o_P(1)$ and $O_P(1)$ notations apply to each coordinate of the vectors and each entry of the matrices.]

By WLLN, we now have $\boldsymbol{B}_n(\boldsymbol{T}_n) = \boldsymbol{I}(\boldsymbol{\theta}_0) + o_P(1)$ if $\boldsymbol{T}_n = \boldsymbol{\theta}_0 + o_P(1)$.

Theorem 7.1.3 (Asymptotic Normality of MLEs: Multiparameter Case). *Under regularity Conditions 1, 2, 3, the MLE $\hat{\boldsymbol{\theta}}_n$ is asymptotically normal (ie, $\sqrt{n}(\hat{\boldsymbol{\theta}}_n - \boldsymbol{\theta}_0) \xrightarrow{\mathcal{L}} N_k(\boldsymbol{0}, \boldsymbol{I}(\boldsymbol{\theta}_0)^{-1})$).*

Proof. The MLE $\hat{\boldsymbol{\theta}}_n$ of $\boldsymbol{\theta}_0$ is the solution of the k equations

$$\sum_{i=1}^{n}\dot{l}_r(X_i, \hat{\boldsymbol{\theta}}_n) = 0, \quad r = 1, \ldots, k.$$

Expanding, as in the single-parameter case,

$$\begin{aligned}
0 &= n^{-1/2}\sum_{i=1}^{n}\dot{l}_r(X_i, \hat{\boldsymbol{\theta}}_n)\\
&= n^{-1/2}\sum_{i=1}^{n}\dot{l}_r(X_i, \boldsymbol{\theta}_0) + n^{-1/2}\sum_{i=1}^{n}\sum_{s=1}^{k}(\hat{\theta}_{ns} - \theta_{0s})\ddot{l}_{rs}(X_i, \boldsymbol{\theta}_0 + \lambda(\hat{\boldsymbol{\theta}}_n - \boldsymbol{\theta}_0))\\
&= n^{-1/2}\sum_{i=1}^{n}\dot{l}_r(X_i, \boldsymbol{\theta}_0)\\
&\quad - \sum_{s=1}^{k} n^{1/2}(\hat{\theta}_{ns} - \theta_{0s})n^{-1}\sum_{i=1}^{n}\{-\ddot{l}_{rs}(X_i, \boldsymbol{\theta}_0 + \lambda(\hat{\boldsymbol{\theta}}_n - \boldsymbol{\theta}_0))\}\\
&= A_{nr} - \sum_{s=1}^{k} n^{1/2}(\hat{\theta}_{ns} - \theta_{0s})B_{nrs}(\boldsymbol{\theta}_0 + \lambda(\hat{\boldsymbol{\theta}}_n - \boldsymbol{\theta}_0)), \quad 0 \le \lambda \le 1,
\end{aligned}$$

$$\text{ie, } \sum_{s=1}^{k} n^{1/2}(\hat{\theta}_{ns} - \theta_{0s})B_{nrs}(\boldsymbol{\theta}_0 + \lambda(\hat{\boldsymbol{\theta}}_n - \boldsymbol{\theta}_0)) = A_{nr}, \quad r = 1, \ldots, k.$$

Putting these k equations together in matrix notation, we get

$$\begin{aligned}
&\boldsymbol{B}_n(\boldsymbol{\theta}_0 + \lambda(\hat{\boldsymbol{\theta}}_n - \boldsymbol{\theta}_0))\sqrt{n}(\hat{\boldsymbol{\theta}}_n - \boldsymbol{\theta}_0) = \boldsymbol{A}_n, \text{ and hence}\\
&\sqrt{n}(\hat{\boldsymbol{\theta}}_n - \boldsymbol{\theta}_0) = \boldsymbol{B}_n(\boldsymbol{\theta}_0 + \lambda(\hat{\boldsymbol{\theta}}_n - \boldsymbol{\theta}_0))^{-1}\boldsymbol{A}_n\\
&\quad = [I(\boldsymbol{\theta}_0) + o_P(1)]^{-1}\boldsymbol{A}_n \xrightarrow{\mathcal{L}} I(\boldsymbol{\theta}_0)^{-1}\boldsymbol{W}, \text{ where } \boldsymbol{W} \sim N_k(\boldsymbol{0}, I(\boldsymbol{\theta}_0)).
\end{aligned}$$

□

Now consider the situation where the likelihood equations do not have a closed form solution. Here again, as in the single-parameter case, we use the Newton-Raphson method, starting with an initial estimator $\tilde{\boldsymbol{\theta}}_{n0}$ to calculate the next iterate $\tilde{\boldsymbol{\theta}}_{n1}$ by the formula

$$\tilde{\boldsymbol{\theta}}_{n1} = \tilde{\boldsymbol{\theta}}_{n0} + \boldsymbol{B}_n(\tilde{\boldsymbol{\theta}}_{n0})^{-1} n^{-1} \sum_{i=1}^{n} \dot{\boldsymbol{l}}(X_i, \tilde{\boldsymbol{\theta}}_{n0}).$$

If the initial estimator $\tilde{\boldsymbol{\theta}}_{n0}$ is $\sqrt{n}$ -consistent, then $\tilde{\boldsymbol{\theta}}_{n1}$ would be a BAN estimator.

Remark 7.1.1. We have seen that under some regularity conditions on $\{f(x, \theta), \theta \in \Theta\}$, the likelihood equation based on iid observations on $f(x, \theta_0)$ has a consistent sequence of solutions $\tilde{\theta}_n$ of θ_0 if θ_0 is an interior point of Θ and for such a sequence, $\sqrt{n}(\tilde{\theta}_n - \theta_0) \xrightarrow{\mathcal{L}} N(0, 1/I(\theta_0))$. However, these are local properties of the likelihood function, and such a sequence of solutions of the likelihood equation need not be the sequence of actual MLEs. On the other hand, globally, under another set of regularity conditions (due to Wald [27]), the actual MLE $\hat{\theta}_n$ converges almost surely to θ_0 (strong consistency). There is another aspect of the results about $\tilde{\theta}_n$ discussed above. Due to the fact that the asymptotic variance of $\tilde{\theta}_n$ is the same as the information lower bound of an unbiased estimator, some sort of asymptotic efficiency is suggested. Is it true that if $\sqrt{n}(T_n - \theta_0) \xrightarrow{\mathcal{L}} N(0, V(\theta_0))$ for some estimator T_n, then $V(\theta_0) \geq 1/I(\theta_0)$?

The following examples illustrate how each of these properties of the MLE can be violated. In each of these examples, we describe $f(x, \theta)$ from which iid observations $X_1, \ldots, X_n$ yield $\hat{\theta}_n$ or $\tilde{\theta}_n$. Example 7.1.2(a) is of a different nature.

Example 7.1.1. MLE is not asymptotically normal. Let $f(x, \theta) = I_{(0,\theta)}(x)$. Here the MLE $\hat{\theta}_n = X_{n:n} = \max(X_1, \ldots, X_n)$ and

$$P_\theta[n(\theta - X_{n:n}) \leq t] \rightarrow 1 - \exp(-t/\theta) \text{ as } n \rightarrow \infty.$$

(See Example 7.1.4)

Example 7.1.2 (MLE Is Not Consistent).

(a) Neyman and Scott [28]

Let (X_i, Y_i), $i = 1, 2, \ldots$ be independent with $(X_i, Y_i) \sim N_2((\mu_i, \mu_i), \sigma^2 I)$. The parameter of interest is σ^2 which is to be estimated, while $\mu_1, \mu_2 \ldots$ are nuisance parameters. The log likelihood based on $(X_1, Y_1), \ldots, (X_n, Y_n)$ is

$$\begin{aligned}\log L &= -\,n[\log 2\pi + \log \sigma^2] - (2\sigma^2)^{-1} \sum_{i=1}^{n} [(X_i - \mu_i)^2 + (Y_i - \mu_i)^2] \\ &= -\,n[\log 2\pi + \log \sigma^2] - (4\sigma^2)^{-1} \sum_{i=1}^{n} [Z_i^2 + 4\{(X_i + Y_i)/2 - \mu_i\}^2],\end{aligned}$$

where $Z_i = X_i - Y_i$. Obviously, the MLE for each μ_i is $\hat{\mu}_i = (X_i + Y_i)/2$ and it follows that the MLE of σ^2 is $\hat{\sigma}_n^2 = (4n)^{-1}\sum_{i=1}^n Z_i^2$. Now the Z_i's are iid $N(0, 2\sigma^2)$, so that $(2n)^{-1}\sum_{i=1}^n Z_i^2$ is a consistent estimator of σ^2. Hence $\hat{\sigma}_n^2 \xrightarrow{P} \sigma^2/2$.

(b) Basu [29]

Let A be the set of all rational numbers in $[0, 1]$ and B any (known) countable set of irrational numbers in $[0, 1]$. The iid observations $(X_1, \ldots, X_n)$ are 0–1 valued rv's which are *Bernoulli*(θ) for $\theta \in A$ and *Bernoulli*$(1-\theta)$ for $\theta \in B$. We therefore find

$$M_1 = \max_{\theta \in A} \theta^{\sum X_i}(1-\theta)^{n-\sum X_i} = \max_{\theta \in A} L(\theta) \text{ and}$$
$$M_2 = \max_{\theta \in B}(1-\theta)^{\sum X_i}\theta^{n-\sum X_i} = \max_{\theta \in B} L(\theta).$$

Thus the MLE of θ is

$$\hat{\theta}_n = \begin{cases} \arg\max_{\theta \in A} \theta^{\sum X_i}(1-\theta)^{n-\sum X_i} & \text{if } M_1 > M_2 \\ \arg\max_{\theta \in B}(1-\theta)^{\sum X_i}\theta^{n-\sum X_i} & \text{if } M_1 < M_2, \end{cases}$$

taking either one if $M_1 = M_2$.

Now $\max_{\theta \in [0,1]} \theta^{\sum X_i}(1-\theta)^{n-\sum X_i} = \left(\sum x_i/n\right)^{\sum Xi}\left(1 - \sum X_i/n\right)\right)^{n-\sum X_i}$ and the maximizer $\sum X_i/n \in A$ because it is rational. Hence

$$M_1 = \max_{\theta \in A} L(\theta) = \max_{\theta \in [0,1]} L(\theta) > \max_{\theta \in B} L(\theta) = M_2.$$

Thus $\hat{\theta}_n = \sum X_i/n$. But $\hat{\theta}_n \xrightarrow{P} \theta$ if $\theta \in A$ and $\hat{\theta}_n \xrightarrow{P} 1-\theta$ if $\theta \in B$, and is therefore inconsistent. So far we have not used the fact that B is countable. However, when B is countable, one can construct a consistent estimator.

(c) Ferguson [30]

Let

$$f(x, \theta) = (1-\theta)f_1(x, \theta) + \theta f_2(x, \theta), \quad \theta \in \Theta = [0, 1],$$

where

$$f_1(x, \theta) = \frac{1}{\delta(\theta)}\left[1 - \frac{|x-\theta|}{\delta(\theta)}\right] I_{[\theta-\delta(\theta), \theta+\delta(\theta)]}(x)$$

is a triangular density with base $[\theta - \delta(\theta), \theta + \delta(\theta)]$ and height $\delta(\theta)$, and $f_2(x, \theta) = (1/2)I_{[-1,1]}(x)$ is a uniform pdf on $[-1, 1]$. The function $\delta(\theta)$ is a continuous decreasing function of θ on $0 < \theta \le 1$ with $\delta(0) = 1$ and $0 < \delta(\theta) \le 1 - \theta$. As θ increases from 0 to 1, the triangle's base in $f_1(x, \theta)$ becomes shorter and shorter, its height becomes larger and larger, and it receives less and less weight. In this way, $f(x, \theta)$ continuously changes from the triangular to the rectangular density. Now suppose $\delta(\theta) \to 0$ as $\theta \to 1$ at a sufficiently fast rate so that $n^{-1}\log[(1 - X_{n:n})/\delta(X_{n:n})] \to \infty$ with probability 1 for $\theta = 0$ (triangular case) and hence for all θ. Then the MLE $\hat{\theta}_n \xrightarrow{a.s.} 1$, whatever the true value $\theta_0 \in [0, 1]$ may be.

Remark 7.1.2 (Super Efficiency). The MLE $\hat{\theta}_n$ has asymptotic variance $1/I(\theta_0)$. The possibility of estimators with asymptotic variance $V(\theta_0) \le 1/I(\theta_0)$ will be discussed in Section 7.1.2, Example 7.1.7.

Example 7.1.3. Let $(X_1, \ldots, X_n)$ be a random sample from *Cauchy*$(\theta, 1)$ with pdf

$$f(x,\theta) = \frac{1}{\pi[1+(x-\theta)^2]}.$$

Find a BAN estimator of θ and its asymptotic distribution.

Solution. Here

$$l(x,\theta) = \log f(x,\theta) = -\log\pi - \log[1+(x-\theta)^2],$$

$$\dot{l}(x,\theta) = \frac{2(x-\theta)}{1+(x-\theta)^2} \text{ and } \ddot{l}(x,\theta) = \frac{2[(x-\theta)^2-1]}{[1+(x-\theta)^2]^2}.$$

Then

$$I(\theta) = \mathrm{E}_\theta\left[\dot{l}^2(X,\theta)\right] = \int_{-\infty}^{\infty} \frac{4(x-\theta)^2}{[1+(x-\theta)^2]^2}\frac{dx}{\pi[1+(x-\theta)^2]}$$
$$= \frac{8}{\pi}\int_0^\infty \frac{y^2\,dy}{(1+y^2)^3} = 1/2.$$

Start with initial estimator $\tilde{\theta}_{n0}$ = Sample Median $= X_{n:[n/2]}$ which is $\sqrt{n}$-consistent (as will be shown in Chapter 8). Then a BAN estimator is

$$\tilde{\theta}_{n1} = \tilde{\theta}_{n0} + \frac{\sum_{i=1}^n \dot{l}(X_i,\tilde{\theta}_{n0})}{\sum_{i=1}^n(-\ddot{l}(X_i,\tilde{\theta}_{n0}))}$$
$$= \tilde{\theta}_{n0} + \sum_{i=1}^n\left[\frac{2(X_i-\tilde{\theta}_{n0})^2}{1+(X_i-\tilde{\theta}_{n0})^2}\right] \Big/ \sum_{i=1}^n\left[\frac{2\{(X_i-\tilde{\theta}_{n0})^2-1\}}{\{1+(X_i-\tilde{\theta}_{n0})^2\}^2}\right],$$

and $\sqrt{n}(\tilde{\theta}_{n1}-\theta_0) \xrightarrow{\mathcal{L}} N(0,2)$.

Example 7.1.4. Let $(X_1, \ldots, X_n)$ be a random sample from *Unif*$(0,\theta)$ with pdf $f(x,\theta) = \theta^{-1}I_{[0,\theta]}(x)$, $\theta > 0$. Find the MLE of θ and its asymptotic distribution.

Solution. Here

$$L(\theta) = \prod_{i=1}^n f(X_i,\theta) = \begin{cases} 0 & \theta < X_{n:n} = \max(X_1,\ldots,X_n) \\ \theta^{-n} & \theta > X_{n:n}. \end{cases}$$

It is easy to see that $L(\theta)$ has a unique maximum at $X_{n:n}$. Hence the MLE $\hat{\theta}_n = X_{n:n}$. To find the distribution of $\hat{\theta}_n$, note that

$$P[\hat{\theta}_n \le t] = P_\theta[X_i \le t, i=1,\ldots,n] = \begin{cases} (t/\theta)^n & 0 < t \le \theta \\ 1 & t > \theta \end{cases}, \text{ and}$$

$$P[n(\theta-\hat{\theta}_n) \le t] = P_\theta[\hat{\theta}_n > \theta - t/n] = 1 - P_\theta[\hat{\theta}_n \le \theta - t/n]$$
$$= 1 - \left(\frac{\theta - t/n}{\theta}\right)^n = 1 - \left(1 - \frac{t}{n\theta}\right)^n \to 1 - \mathrm{e}^{-t/\theta}.$$

Example 7.1.5. Let $\boldsymbol{X}_1, \ldots, \boldsymbol{X}_n$ be iid $N_k(\boldsymbol{\mu}, \boldsymbol{\Sigma})$ where $\boldsymbol{\Sigma}$ is positive definite. Find the MLE of $(\boldsymbol{\mu}, \boldsymbol{\Sigma})$.

Solution. The MLEs for $\boldsymbol{\mu}$ and $\boldsymbol{\Sigma}$ are $\bar{\boldsymbol{X}} = n^{-1}\sum_{i=1}^{n}\boldsymbol{X}_i$ and $\hat{\boldsymbol{\Sigma}} = n^{-1}\sum_{i=1}^{n}(\boldsymbol{X}_i - \bar{\boldsymbol{X}})(\boldsymbol{X}_i - \bar{\boldsymbol{X}})^T$. A proof is given later in the chapter on multivariate analysis.

Example 7.1.6. Let

$$f(x,\theta) = C(\theta)\exp[Q(\theta)T(x)]r(x), \quad \theta \in \Theta$$

be a pdf/pmf belonging to the single-parameter exponential family, where $\Theta = \{\theta\colon \int \exp[Q(\theta)T(x)]r(x)\,dx < \infty\}$. Find the MLE of θ from a random sample $(X_1, \ldots, X_n)$ from $f(x,\theta)$.

Solution. The log-likelihood and the likelihood equation are

$$l(\boldsymbol{X},\theta) = n\log C(\theta) + Q(\theta)\sum_{i=1}^{n}T(X_i) + \sum_{i=1}^{n}\log r(X_i), \text{ and}$$

$$\dot{l}(\boldsymbol{X},\theta) = n(C'(\theta)/C(\theta)) + Q'(\theta)\sum_{i=1}^{n}T(X_i) = 0.$$

The MLE $\hat{\theta}_n$ is the solution of

$$-\frac{C'(\theta)}{C(\theta)Q'(\theta)} = n^{-1}\sum_{i=1}^{n}T(X_i).$$

We can now find the MLEs of θ for some distributions as special cases of this result.

(a) *Bernoulli*(θ) with

$$f(x,\theta) = \theta^x(1-\theta)^{1-x} = (1-\theta)\exp\left[x\log\left(\frac{\theta}{1-\theta}\right)\right], \quad x = 0, 1.$$

Here

$$C(\theta) = 1-\theta,\ Q(\theta) = \log\left(\frac{\theta}{1-\theta}\right), \text{ and } T(x) = x, \text{ so } -\frac{C'(\theta)}{C(\theta)Q'(\theta)} = \theta,$$

and the likelihood equation is $\theta = n^{-1}\sum_{i=1}^{n}X_i = \bar{X}$. Thus the MLE of θ is $\hat{\theta}_n = \bar{X}$.

(b) *Poisson*(θ) with

$$f(x,\theta) = \exp(-\theta)\theta^x/x!, \quad x = 0, 1, \ldots.$$

Here

$$C(\theta) = e^{-\theta},\ Q(\theta) = \log\theta,\ T(x) = x, \text{ so } -\frac{C'(\theta)}{C(\theta)Q'(\theta)} = \theta,$$

and the likelihood equation is $\theta = n^{-1}\sum_{i=1}^{n}X_i = \bar{X}$. Thus the MLE of θ is $\hat{\theta}_n = \bar{X}$.

(c) *Gamma*(α,θ) with

$$f(x,\theta) = \frac{1}{\Gamma(\alpha)\theta^\alpha}\exp[-x/\theta]x^{\alpha-1}, \quad x > 0.$$

Here

$$C(\theta) = \frac{1}{\Gamma(\alpha)\theta^{\alpha}}, \; Q(\theta) = -1/\theta, \; T(x) = x, \text{ so } -\frac{C'(\theta)}{C(\theta)Q'(\theta)} = \alpha\theta,$$

and the likelihood equation is $\alpha\theta = n^{-1}\sum_{i=1}^{n} X_i = \bar{X}$. Thus the MLE of θ is $\hat{\theta}_n = \bar{X}/\alpha$.

7.1.1 Almost Sure Convergence (Strong Consistency) of MLEs

Let θ_0 be the true value of the unknown parameter θ and all probability statements are with respect to θ_0.

Notations

(i) For every $\theta \in \Theta$, let $N_{\theta,j} = \{\varphi\colon \|\varphi - \theta\| \leq j^{-1}\}$ be a decreasing sequence of neighborhoods of θ converging to $\{\theta\}$, and let

$$Z(\theta_0, N_{\theta,j}) = \inf_{\varphi \in N_{\theta,j}} \log[f(X,\theta_0)/f(X,\varphi)], \quad Z(\theta_0,\theta) = \log[f(X,\theta_0)/f(X,\theta)].$$

(ii) Let $N_{\infty,j} = \{\theta\colon \|\theta\| > j\}$, which can be thought of as a decreasing sequence of neighborhoods of ∞, converging to $\emptyset$, and let

$$Z(\theta_0, N_{\infty,j}) = \inf_{\varphi \in N_{\infty,j}} [f(X,\theta_0)/f(X,\varphi)].$$

We assume that the following *Regularity Conditions (Wald)* hold

1. The parameter space Θ is a closed subset of $\mathbb{R}^k$.
2. *Identifiability Condition.* For $\theta \neq \theta_0$, $\{x\colon f(x,\theta) \neq f(x,\theta_0)\}$ has positive probability under θ_0.
3. **(a)** For all θ, $\lim_{\theta_n \to \theta} f(x,\theta_n) = f(x,\theta)$ for all x.
   (b) $\lim_{\theta_n \to \infty} f(x,\theta_n) = 0$ for all x.
4. **(a)** For each θ, $I(\theta_0,\theta) = \mathrm{E}[Z(\theta_0,\theta)]$ exists.
   (b) For each θ, $I(\theta_0, N_{\theta,j}) = \mathrm{E}[Z(\theta_0, N_{\theta,j})] > -\infty$ for some $j = j_0$.
   (c) $I(\theta_0, N_{\infty,j}) = \mathrm{E}[Z(\theta_0, N_{\infty,j})] > -\infty$ for some $j = j_0$.

Theorem 7.1.4 (Wald). *Under Conditions 1–4,* $\hat{\theta}_n \xrightarrow{a.s.} \theta_0$.

The proof will be accomplished by first considering the case of finite $\Theta = \{\theta_0, \theta_1, \ldots, \theta_r\}$ and then extending to the case of Θ being an arbitrary closed subset of $\mathbb{R}^k$.

We start with the following lemma.

Lemma 7.1.1. $I(\theta_0,\theta) > 0$ *for all* $\theta \neq \theta_0$.

This result has already been proved in Section 5.5.1, using the Identifiability Condition.

In the case of finite Θ, we now have

$$\begin{aligned}
P[\hat{\theta}_n \neq \theta_0 \text{ i.o.}] &\leq \sum_{j=1}^{r} P[\hat{\theta}_n = \theta_j \text{ i.o.}] \\
&\leq \sum_{j=1}^{r} P\left[n^{-1}\sum_{i=1}^{n} \log f(X_i, \theta_0) \leq n^{-1}\sum_{i=1}^{n} \log f(X_i, \theta_j) \text{ i.o.}\right] \\
&= \sum_{j=1}^{r} P\left[n^{-1}\sum_{i=1}^{n} Z_i(\theta_0, \theta_j) \leq 0 \text{ i.o.}\right] = 0
\end{aligned}$$

by SLLN, because $\mathrm{E}[Z(\theta_0, \theta_j)] = I(\theta_0, \theta_j) > 0$ for all j by Lemma 7.1.1. This proves the theorem for finite Θ.

For the case of Θ being an arbitrary closed subset of $\mathbb{R}^k$, we need to show that for an arbitrary neighborhood N_0 of θ_0, $P[\hat{\theta}_n \notin N_0 \text{ i.o.}] = 0$. To extend the proof for the finite case to this generality, we shall cover $\Theta - N_0$ by a finite collection of sets $S_1, \ldots, S_r$ such that

$$I(\theta_0, S_j) = \mathrm{E}[Z(\theta_0, S_j)] = \mathrm{E}\left[\inf_{\varphi \in S_j} \log\{f(X, \theta_0)/f(X, \varphi)\}\right] > 0$$

for $j = 1, \ldots, r$. It then follows that

$$\begin{aligned}
P[\hat{\theta}_n \notin N_0 \text{ i.o.}] &\leq P[\hat{\theta}_n \in S_j \text{ i.o. for some } j = 1, \ldots, r] \\
&\leq \sum_{j=1}^{r} P[\hat{\theta}_n \in S_j \text{ i.o.}] \\
&\leq \sum_{j=1}^{r} P\left[n^{-1}\sum_{i=1}^{n} \log f(X_i, \theta_0) \leq \sup_{\varphi \in S_j} n^{-1}\sum_{i=1}^{n} \log f(X_i, \varphi) \text{ i.o.}\right] \\
&= \sum_{j=1}^{r} P\left[\inf_{\varphi \in S_j} n^{-1}\sum_{i=1}^{n} \log\{f(X_i, \theta_0)/f(X_i, \varphi)\} \leq 0 \text{ i.o.}\right] \\
&\leq \sum_{j=1}^{r} P\left[n^{-1}\sum_{i=1}^{n} \inf_{\varphi \in S_j} \log\{f(X_i, \theta_0)/f(X_i, \varphi)\} \leq 0 \text{ i.o.}\right] \\
&= \sum_{j=1}^{r} P\left[n^{-1}\sum_{i=1}^{n} Z_i(\theta_0, S_j) \leq 0 \text{ i.o.}\right] = 0
\end{aligned}$$

by SLLN because $\mathrm{E}[Z(\theta_0, S_j)] > 0$ for $j = 1, \ldots, r$.

We now construct $S_1, \ldots, S_r$ with the above-mentioned properties.

Lemma 7.1.2.

(a) *For each $\theta \neq \theta_0$, there is a neighborhood $N_{\theta,j}$ such that $I(\theta_0, N_{\theta,j}) = E[Z(\theta_0, N_{\theta,j})] > 0$.*

(b) *There exists a neighborhood $N_{\infty,j}$ of ∞ such that $I(\theta_0, N_{\infty,j}) = E[Z(\theta_0, N_{\infty,j})] > 0$.*

Proof. Fix $\theta \neq \theta_0$ and let

$$g_j(x) = \left\{ \inf_{\varphi \in N_{\theta,j}} \log(f(x,\theta_0)/f(x,\varphi)) \right\} f(x,\theta_0) - \left\{ \inf_{\varphi \in N_{\theta,j_0}} \log(f(x,\theta_0)/f(x,\varphi)) \right\} f(x,\theta_0)$$

for $j \geq j_0$, where j_0 is chosen by Condition 4(b). Since $\{N_{\theta,j}\}$ is decreasing, $\{g_j\}$ is increasing and are clearly nonnegative for $j \geq j_0$. Moreover, by Condition 3(a),

$$\lim_{j\to\infty} g_j(x) := g(x) = \left\{ \log(f(x,\theta_0)/f(x,\theta)) \right\} f(x,\theta_0) - \left\{ \inf_{\varphi \in N_{\theta,j_0}} \log(f(x,\theta_0)/f(x,\varphi)) \right\} f(x,\theta_0)$$

for all x. Hence by the Monotone Convergence Theorem,

$$\lim_{j\to\infty} \mathrm{E}[Z(\theta_0, N_{\theta,j})] - \mathrm{E}[Z(\theta_0, N_{\theta,j_0})] = \lim_{j\to\infty} \int g_j(x)\, dx = \int g(x)\, dx = \mathrm{E}[Z(\theta_0,\theta)] - \mathrm{E}[Z(\theta_0, N_{\theta,j_0})].$$

Since $\mathrm{E}[Z(\theta_0, N_{\theta,j_0})] > -\infty$, we cancel it from both sides of the above equality to obtain

$$\lim_{j\to\infty} \mathrm{E}[Z(\theta_0, N_{\theta,j})] = \mathrm{E}[Z(\theta_0,\theta)] > 0,$$

so that $\mathrm{E}[Z(\theta_0, N_{\theta,j})] > 0$ for sufficiently large j. This proves (a).

The proof of (b) is exactly the same, using Conditions 4(c) and 3(b) instead of Conditions 4(b) and 3(a). □

Proof of Theorem 7.1.4. First choose a set N_{∞,j^*} such that $\mathrm{E}[Z(\theta_0, N_{\infty,j^*})] > 0$. Now consider $\Theta \cap N_0^c \cap N_{\infty,j^*}^c$ (for an arbitrary neighborhood N_0 of θ_0), which is a closed and bounded set in $\mathbb{R}^k$ and therefore has the Heine-Borel property (compactness) of having a subcover for every open cover.

Now for every $\theta \in \Theta \cap N_0^c \cap N_{\infty,j^*}^c$, choose $N_{\theta,j}$ such that $\mathrm{E}[Z(\theta_0, N_{\theta,j})] > 0$. These sets $N_{\theta,j}$ for $\theta \in \Theta \cap N_0^c \cap N_{\infty,j^*}^c$ form an open cover of $\Theta \cap N_0^c \cap N_{\infty,j^*}^c$ and by the Heine-Borel property, there is a finite subcover $N_{\theta_1,j_1}, \ldots, N_{\theta_{r-1},j_{r-1}}$. Let

$$S_1 = N_{\theta_1,j_1}, \ldots, S_{r-1} = N_{\theta_{r-1},j_{r-1}} \text{ and } S_r = N_{\infty,j^*}.$$

Then $S_1, \ldots, S_r$ cover N_0^c and $\mathrm{E}[Z(\theta_0, S_j)] > 0$, $j = 1, \ldots, r$, and the theorem is proved by the argument given above. □

7.1.2 Asymptotic Efficiency of MLE

What is the connection between the information bound $1/I(\theta) = 1/\mathrm{E}_\theta[\dot{l}^2(X)]$ for the variance of an unbiased estimator of θ and the fact that the MLE of $\hat{\theta}_n$ of θ has the asymptotic distribution $\sqrt{n}(\hat{\theta}_n - \theta) \xrightarrow{\mathcal{L}} N(0, 1/I(\theta))$? [For simplicity, we are considering the case of a 1-dim parameter θ.]

We have seen that

(i) under regularity conditions on the pdf/pmf and restrictions on estimators, $n\text{Var}_\theta[T_n] \geq 1/I(\theta)$ for all unbiased estimators T_n of θ, and

(ii) the MLE $\hat{\theta}_n$ of θ has the property: $\sqrt{n}(\hat{\theta}_n - \theta) \xrightarrow{\mathcal{L}} N(0, 1/I(\theta))$.

From (i) and (ii), one may be tempted to hope that if $\sqrt{n}(T_n - \theta) \xrightarrow{\mathcal{L}} \xi_\theta$ under $f(\cdot, \theta)$, then ξ_θ must be at least as much dispersed as $N(0, 1/I(\theta))$ and in particular, $\sigma^2(\theta) = \text{E}_\theta[\xi_\theta^2] \geq 1/I(\theta)$ if ξ_θ is Gaussian. Can there be a situation in which $\text{E}_\theta[\xi_\theta^2] < 1/I(\theta)$? If so, then such an estimator would be called "superefficient." The following example shows that *superefficient estimators do exist.*

Example 7.1.7 (Hodges). Let $X_1, \ldots, X_n, \ldots$ be iid as $N(\theta, 1)$. Then the MLE $\hat{\theta}_n = \bar{X}_n = n^{-1}\sum_{i=1}^n X_i$ and $Z = \sqrt{n}(\bar{X}_n - \theta) \stackrel{\mathcal{D}}{=} N(0, 1)$. Also the information $I(\theta) = 1$. Now for $0 < a < 1$, let

$$T_n = \begin{cases} \bar{X}_n & \text{if } |\bar{X}_n| > n^{-1/4} \\ a\bar{X}_n & \text{if } |\bar{X}_n| \leq n^{-1/4}. \end{cases}$$

Then $\sqrt{n}(T_n - \theta)$ is distributed as Z if $|Z + \sqrt{n}\theta| > n^{1/4}$ and as $aZ + (a-1)\sqrt{n}\theta$ otherwise. From this it follows that $\sqrt{n}(T_n - \theta) \xrightarrow{\mathcal{L}} N(0, V(\theta))$, where $V(\theta) = 1$ if $\theta \neq 0$ and $V(\theta) = a^2 < 1$ if $\theta = 0$. This defines a superefficient estimator (see [31]).

Let us examine the behavior of $\{T_n\}$ in the above example under $\{P_{\theta_n}\}$ where $\theta_n = h/\sqrt{n}$. In the above proof, taking $\theta = \theta_n$, we have

$$\begin{aligned} \sqrt{n}(T_n - \theta_n) &\stackrel{\mathcal{D}}{=} ZI[|Z + h| > n^{1/4}] + \{aZ + (a-1)h\}I[|Z + h| \leq n^{1/4}] \\ &\xrightarrow{a.s.} aZ + (a-1)h. \end{aligned}$$

Hence under $\{P_{\theta_n}\}$, $\sqrt{n}(T_n - \theta_n) \xrightarrow{\mathcal{L}} N((a-1)h, a^2)$, although for $\theta = 0$, $\sqrt{n}(T_n - 0) \xrightarrow{\mathcal{L}} N(0, a^2)$. Thus we have a situation in which the asymptotic distribution of $\sqrt{n}(T_n - \theta_n)$ under $\{P_{\theta_n}\}$ with $\theta_n = h/\sqrt{n} \to 0$ as $n \to \infty$, depends not only on P_0 but actually on the particular sequence $\{P_{\theta_n}\} \to P_0$.

Hájek recognized this as the reason behind superefficiency and introduced the concept of *regular estimators* (or more precisely, locally regular estimators at a particular θ_0) to prevent superefficiency.

Definition 7.1.2. A sequence of estimators $\{T_n\}$ of θ is regular at $P = P_{\theta_0}$ if for every sequence $\{P_n\} = \{P_{\theta_n}\}$ with $\theta_n = \theta_0 + h/\sqrt{n} + o(1/\sqrt{n})$ for some $h \in \mathbb{R}$,

$$\mathcal{L}(\sqrt{n}(T_n - \theta_n)) \to \mathcal{L}(P) \text{ under } \{P_n\},$$

where $\mathcal{L}(P)$ depends on $P = P_{\theta_0}$, but not on the particular sequence $\{P_{\theta_n}\}$ (ie, not on h).

Theorem 7.1.5 (Hájek-Inagaki Decomposition Theorem) [See 67, 68]. *Suppose that* $\{T_n\}$ *is a regular estimator of* θ *at* $P = P_{\theta_0}$. *Then for every sequence* $\theta_n = \theta_0 + h/\sqrt{n} + o(1/\sqrt{n})$ *with* $h \in \mathbb{R}$,

$$\mathcal{L}(\sqrt{n}(T_n - \theta_n)) \to \mathcal{L}(P) = N(0, 1/I) * \mathcal{L}_1(P) \text{ under } P_n,$$

where $\mathcal{L}_1(P)$ *depends only on* P *(ie, on* θ_0*), but not on the particular sequence* $\{\theta_n\}$, *and* $*$ *denotes convolution.*

By this theorem, the asymptotic distribution of any regular estimator is more dispersed than $N(0, 1/I)$, which is the asymptotic distribution of the MLE. Thus *the MLE is asymptotically efficient among all regular estimators.*

7.1.3 MLE of Parameters in a Multinomial Distribution

Consider a multinomial distribution in m classes with probability $\pi_j(\boldsymbol{\theta})$, $j = 1, \ldots, m$ for the jth class, where $\pi_1(\cdot), \ldots, \pi_m(\cdot)$ are known functions of an unknown k-dim parameter vector $\boldsymbol{\theta}$. Let $n_1, \ldots, n_m$ denote the observed frequencies in the m classes in a random sample of size n. The following table summarizes this.

Class	**1**	**...**	***j***	**...**	***m***	**Total**
Probability	$\pi_1(\boldsymbol{\theta})$	...	$\pi_j(\boldsymbol{\theta})$	...	$\pi_m(\boldsymbol{\theta})$	1
Obs. frequency	n_1	...	n_j	...	n_m	n

Here the data consist of a random sample $\boldsymbol{X} = (\boldsymbol{X}_1, \ldots, \boldsymbol{X}_n)$ where the $\boldsymbol{X}_i$s are iid with

$$P[\boldsymbol{X}_i = \boldsymbol{e}_j] = f(\boldsymbol{e}_j, \boldsymbol{\theta}) = \pi_j(\boldsymbol{\theta}),\ j = 1, \ldots, m \text{ and } n_j = \sum_{i=1}^{n} I[\boldsymbol{X}_i = \boldsymbol{e}_j],$$

$\boldsymbol{e}_j$ being the m-dim jth coordinate vector. Hence

$$\dot{l}_r(\boldsymbol{e}_j, \boldsymbol{\theta}) = \frac{\partial}{\partial \theta_r} \log f(\boldsymbol{e}_j, \boldsymbol{\theta}) = \frac{\partial}{\partial \theta_r} \log \pi_j(\boldsymbol{\theta}) = \frac{1}{\pi_j(\boldsymbol{\theta})} \frac{\partial \pi_j(\boldsymbol{\theta})}{\partial \theta_r} := \frac{\dot{\pi}_{rj}(\boldsymbol{\theta})}{\pi_j(\boldsymbol{\theta})}, \text{ and}$$

$$S_{nr}(\boldsymbol{t}) = \sum_{i=1}^{n} \dot{l}_r(\boldsymbol{X}_i, \boldsymbol{t}) = \sum_{i=1}^{n} \sum_{j=1}^{m} I[\boldsymbol{X}_i = \boldsymbol{e}_j] \frac{\dot{\pi}_{rj}(\boldsymbol{t})}{\pi_j(\boldsymbol{t})} = \sum_{j=1}^{m} \frac{n_j}{\pi_j(\boldsymbol{t})} \dot{\pi}_{rj}(\boldsymbol{t}).$$

Thus the likelihood equations are

$$S_{nr}(\hat{\boldsymbol{\theta}}_n) = 0, \text{ ie, } \sum_{j=1}^{m} \frac{n_j}{\pi_j(\hat{\boldsymbol{\theta}}_n)} \dot{\pi}_{rj}(\hat{\boldsymbol{\theta}}_n) = 0, \quad r = 1, \ldots, k.$$

In a typical context, these equations would not have a closed form solution, so we obtain a BAN estimator by the Newton-Raphson method. First note that

$$P_{\boldsymbol{\theta}}\left[\dot{l}_r(\boldsymbol{X}, \boldsymbol{\theta}) = \frac{\dot{\pi}_{rj}(\boldsymbol{\theta})}{\pi_j(\boldsymbol{\theta})}\right] = P_{\boldsymbol{\theta}}[X = e_j] = \pi_j(\boldsymbol{\theta}), \quad j = 1, \ldots, m.$$

For $\boldsymbol{\theta} = \boldsymbol{\theta}_0$,

$$E_{\boldsymbol{\theta}_0}[\dot{l}_r(\mathbf{X}, \boldsymbol{\theta}_0)] = \sum_{j=1}^{m} \pi_j(\boldsymbol{\theta}_0) \frac{\dot{\pi}_{rj}(\boldsymbol{\theta}_0)}{\pi_j(\boldsymbol{\theta}_0)} = \sum_{j=1}^{m} \dot{\pi}_{rj}(\boldsymbol{\theta}_0) = \frac{\partial}{\partial \theta_r} \sum_{j=1}^{m} \pi_j(\boldsymbol{\theta}) \Bigg|_{\boldsymbol{\theta}=\boldsymbol{\theta}_0} = 0,$$

as it should be because $\sum_{j=1}^{m} \pi_j(\boldsymbol{\theta}) = 1$ for all $\boldsymbol{\theta}$. Next,

$$\begin{aligned} I_{rs}(\boldsymbol{\theta}_0) &= \text{Cov}_{\boldsymbol{\theta}_0}[\dot{l}_r(\mathbf{X}, \boldsymbol{\theta}_0), \dot{l}_s(\mathbf{X}, \boldsymbol{\theta}_0)] = E_{\boldsymbol{\theta}_0}[\dot{l}_r(\mathbf{X}, \boldsymbol{\theta}_0) \dot{l}_s(\mathbf{X}, \boldsymbol{\theta}_0)] \\ &= \sum_{j=1}^{m} \pi_j(\boldsymbol{\theta}_0) \{\dot{\pi}_{rj}(\boldsymbol{\theta}_0)/\pi_j(\boldsymbol{\theta}_0)\} \{\dot{\pi}_{sj}(\boldsymbol{\theta}_0)/\pi_j(\boldsymbol{\theta}_0)\} \\ &= \sum_{j=1}^{m} \dot{\pi}_{rj}(\boldsymbol{\theta}_0) \dot{\pi}_{sj}(\boldsymbol{\theta}_0)/\pi_j(\boldsymbol{\theta}_0). \end{aligned} \tag{5}$$

If $\boldsymbol{T}_n = \boldsymbol{\theta}_0 + o_P(1)$, then $I_{rs}(\boldsymbol{\theta}_0)$ is estimated by $I_{rs}(\boldsymbol{T}_n)$, which can be used in place of $B_{nrs}(\tilde{\boldsymbol{\theta}}_{n0})$ in the Newton-Raphson formula.

7.2 Likelihood Ratio Test

The theory of hypothesis testing, discussed in Chapter 6, based on N-P Lemma for simple H_0 vs simple H_1 and extended to some composite H_0 vs composite H_1 in exponential families, also in the presence of nuisance parameters, using the Generalized N-P Lemma, covered a limited range of problems. For testing a composite H_0: $\theta \in \Theta_0$ vs a composite H_1: $\theta \in \Theta_1$ in a general setting, we modify the likelihood ratio $\prod_{i=1}^{n} f(X_i, \theta_1)/\prod_{i=1}^{n} f(X_i, \theta_0)$ based on $X = (X_1, \ldots, X_n)$ in the simple vs simple problem by replacing θ_0 and θ_1 by $\hat{\theta}_i$ = MLE of θ restricted to Θ_i, $i = 0, 1$, respectively, or equivalently by

$$\Lambda_n = \frac{\sup_{\theta \in \Theta_0} \prod_{i=1}^{n} f(X_i, \theta)}{\sup_{\theta \in \Theta} \prod_{i=1}^{n} f(X_i, \theta)} = \frac{f(X_i, \tilde{\theta}_n)}{f(X_i, \hat{\theta}_n)}, \tag{6a}$$

where $\tilde{\theta}_n$ is the restricted MLE in Θ_0 and $\hat{\theta}_n$ is the unrestricted MLE in the entire parameter space $\Theta = \mathbb{R}^k$. The null hypothesis H_0 is rejected for small values of Λ_n, or equivalently, for large values of

$$-2 \log \Lambda_n = 2\Big[\sum \log f(X_i, \hat{\theta}_n) - \sum \log f(X_i, \tilde{\theta}_n)\Big]. \tag{6b}$$

This is known as the likelihood ratio test (LRT). The asymptotic properties of LRT will be discussed in this section.

Hypothesis testing was discussed in Chapter 6, mostly within the very restricted framework of exponential families of distributions. Here we take up the testing problem again in much broader context. Let $(X_1, \ldots, X_n)$ be a random sample from $f(x, \boldsymbol{\theta}), \boldsymbol{\theta} \in \mathbb{R}^k$.

We now consider the problem of testing H_0: $\boldsymbol{\theta} \in \Theta_0$ vs H_1: $\boldsymbol{\theta} \in \mathbb{R}^k - \Theta_0$, where Θ_0 is a d-dim hyperplane in $\mathbb{R}^k$ with $d < k$. Without loss of generality, let $\Theta_0 = \{\boldsymbol{\theta} = (\theta_1, \ldots, \theta_k)\colon \theta_{d+1} = \cdots = \theta_k = 0\}$.

The Neyman-Pearson simple-vs-simple likelihood ratio would generalize to the composite-vs-composite case as

$$\Lambda_n^* = \frac{\sup_{\boldsymbol{\theta}\in\Theta_0}\prod_{i=1}^n f(X_i,\boldsymbol{\theta})}{\sup_{\boldsymbol{\theta}\in\mathbb{R}^k-\Theta_0}\prod_{i=1}^n f(X_i,\boldsymbol{\theta})},$$

and reject H_0 if $\Lambda_n^* \leq$ critical value c. However, since determining the MLE of $\boldsymbol{\theta}$ restricted to $\mathbb{R}^k - \Theta_0$ causes complication, it is more convenient to modify the denominator and use the test statistic

$$\Lambda_n = \frac{\sup_{\boldsymbol{\theta}\in\Theta_0}\prod_{i=1}^n f(X_i,\boldsymbol{\theta})}{\sup_{\boldsymbol{\theta}\in\mathbb{R}^k}\prod_{i=1}^n f(X_i,\boldsymbol{\theta})} = \frac{\prod_{i=1}^n f(X_i,\tilde{\boldsymbol{\theta}}_n)}{\prod_{i=1}^n f(X_i,\hat{\boldsymbol{\theta}}_n)} \text{ as in } \textit{Eq.}\ (6a),$$

where $\tilde{\boldsymbol{\theta}}_n$ is the MLE of $\boldsymbol{\theta}$ restricted to Θ_0, $\hat{\boldsymbol{\theta}}_n$ is the unrestricted MLE of $\boldsymbol{\theta}$ based on $(X_1,\ldots,X_n)$ and H_0 is rejected if $\Lambda_n \leq$ critical value c. Clearly, $\Lambda_n = \Lambda_n^* < 1$ if $\boldsymbol{\theta}_n \in \mathbb{R}^k - \Theta_0$ and $\Lambda_n = 1 \leq \Lambda_n^*$ if $\hat{\boldsymbol{\theta}}_n \in \Theta_0$. Thus Λ_n is a nondecreasing function of Λ_n^* and therefore, rejecting H_0 for small values of Λ_n is equivalent to rejecting H_0 for small values of Λ_n^*.

We are thus led to the LRT: Reject H_0 for *small values of* Λ_n defined above, or equivalently, reject H_0 for *large values of*

$$-2\log\Lambda_n = 2\left[\sum_{i=1}^n l(X_i,\hat{\boldsymbol{\theta}}_n) - \sum_{i=1}^n l(X_i,\tilde{\boldsymbol{\theta}}_n)\right] \text{ as in } \textit{Eq.}\ (6b).$$

To construct a large-sample test of H_0 at a prescribed level of significance, we need the asymptotic distribution of $-2\log\Lambda_n$ under H_0. Without loss of generality, let the true parameter $\boldsymbol{\theta}_0 = 0$ (ie, $\theta_{01} = \cdots = \theta_{0k} = 0$). Since $\tilde{\boldsymbol{\theta}}_n$ is the MLE of $\boldsymbol{\theta}$ restricted to Θ_0, we must have $\tilde{\boldsymbol{\theta}}_n = (\tilde{\theta}_{n1},\ldots,\tilde{\theta}_{nd},0,\ldots,0)$.

Thus $\hat{\boldsymbol{\theta}}_n - \boldsymbol{\theta}_0 = (\hat{\theta}_{n1},\ldots,\hat{\theta}_{nd},\hat{\theta}_{n,d+1},\ldots,\hat{\theta}_{nk})$ and $\tilde{\boldsymbol{\theta}}_n - \boldsymbol{\theta}_0 = (\tilde{\theta}_{n1},\ldots,\tilde{\theta}_{nd},0,\ldots,0)$.

We use the notations of Section 5.2.3 and assume that the regularity conditions introduced for proving Theorem 7.1.2 hold. We also write

$$\dot{\boldsymbol{l}}_{(d)}(x,\boldsymbol{\theta})^T = (\dot{l}_1(x,\boldsymbol{\theta}),\ldots,\dot{l}_d(x,\boldsymbol{\theta})),\ \tilde{\boldsymbol{\theta}}_{n(d)}^T = (\tilde{\theta}_{n1},\ldots,\tilde{\theta}_{nd}),$$

$$\ddot{\boldsymbol{l}}_{(d)}(x,\boldsymbol{\theta}) = \begin{bmatrix} \ddot{l}_{11}(x,\boldsymbol{\theta}) & \cdots & \ddot{l}_{1d}(x,\boldsymbol{\theta}) \\ \vdots & & \vdots \\ \ddot{l}_{d1}(x,\theta) & \cdots & \ddot{l}_{dd}(x,\theta) \end{bmatrix} \text{ and } \boldsymbol{I}_{(d)} = \begin{bmatrix} I_{11} & \cdots & I_{1d} \\ \vdots & & \vdots \\ I_{d1} & \cdots & I_{dd} \end{bmatrix},$$

that is, $\dot{\boldsymbol{l}}_{(d)}(x,\boldsymbol{\theta})$ and $\tilde{\boldsymbol{\theta}}_{n(d)}$ are d-dim vectors consisting of the first d elements of $\dot{\boldsymbol{l}}(x,\boldsymbol{\theta})$ and $\tilde{\boldsymbol{\theta}}_n$, respectively, and $\ddot{\boldsymbol{l}}_{(d)}(x,\boldsymbol{\theta})$ and $\boldsymbol{I}_{(d)}$ are $d\times d$ matrices consisting of the upper left-hand elements of $\ddot{\boldsymbol{l}}(x,\boldsymbol{\theta})$ and $\boldsymbol{I}$, respectively.

Theorem 7.2.1. *Under regularity Conditions 1, 2, 3 (multiparameter),* $-2\log\Lambda_n \xrightarrow{\mathcal{L}} \chi^2_{k-d}$.

Proof. Expanding $\sum_{i=1}^{n} l(X_i, \hat{\boldsymbol{\theta}}_n)$ and $\sum_{i=1}^{n} l(X_i, \tilde{\boldsymbol{\theta}}_n)$ around $\boldsymbol{\theta}_0 = \mathbf{0}$, we have

$$\begin{aligned}
\sum_{i=1}^{n} l(X_i, \hat{\boldsymbol{\theta}}_n) &= \sum_{i=1}^{n} l(X_i, \mathbf{0}) + (\sqrt{n}\hat{\boldsymbol{\theta}}_n)^T n^{-1/2} \sum_{i=1}^{n} \dot{\boldsymbol{l}}(X_i, \mathbf{0}) \\
&\quad + (1/2)(\sqrt{n}\hat{\boldsymbol{\theta}}_n)^T \left\{ n^{-1} \sum_{i=1}^{n} \ddot{\boldsymbol{l}}(X_i, \mathbf{0}) + \boldsymbol{R}_n \right\} (\sqrt{n}\hat{\boldsymbol{\theta}}_n) \\
&= \sum_{i=1}^{n} l(X_i, \mathbf{0}) + (\sqrt{n}\hat{\boldsymbol{\theta}}_n)^T n^{-1/2} \sum_{i=1}^{n} \dot{\boldsymbol{l}}(X_i, \mathbf{0}) - (1/2)(\sqrt{n}\hat{\boldsymbol{\theta}}_n)^T \boldsymbol{I}(\mathbf{0})(\sqrt{n}\hat{\boldsymbol{\theta}}_n) + o_P(1),
\end{aligned}$$

treating the remainder term $\boldsymbol{R}_n$, using Condition 2 as in Section 7.1.1.

Similarly,

$$\begin{aligned}
\sum_{i=1}^{n} l(X_i, \tilde{\boldsymbol{\theta}}_n) &= \sum_{i=1}^{n} l(X_i, \mathbf{0}) + (\sqrt{n}\tilde{\boldsymbol{\theta}}_{n(d)})^T n^{-1/2} \sum_{i=1}^{n} \dot{\boldsymbol{l}}_{(d)}(X_i, \mathbf{0}) \\
&\quad + (1/2)(\sqrt{n}\tilde{\boldsymbol{\theta}}_{n(d)})^T \left\{ n^{-1} \sum_{i=1}^{n} \ddot{\boldsymbol{l}}_{(d)}(X_i, \mathbf{0}) + \boldsymbol{R}_n \right\} (\sqrt{n}\tilde{\boldsymbol{\theta}}_{n(d)}) \\
&= \sum_{i=1}^{n} l(X_i, \mathbf{0}) + (\sqrt{n}\tilde{\boldsymbol{\theta}}_{n(d)})^T n^{-1/2} \sum_{i=1}^{n} \dot{\boldsymbol{l}}_{(d)}(X_i, \mathbf{0}) \\
&\quad - (1/2)(\sqrt{n}\tilde{\boldsymbol{\theta}}_{n(d)})^T \boldsymbol{I}_{(d)}(\mathbf{0})(\sqrt{n}\tilde{\boldsymbol{\theta}}_{n(d)}) + o_P(1).
\end{aligned}$$

In the above expression, we now substitute (as seen in the proof of Theorem 7.1.3),

$$\begin{aligned}
\sqrt{n}\hat{\boldsymbol{\theta}}_n &= \sqrt{n}(\hat{\boldsymbol{\theta}}_n - \boldsymbol{\theta}_0) = [\boldsymbol{I}(\mathbf{0}) + o_P(1)]^{-1} n^{-1/2} \sum_{i=1}^{n} \dot{\boldsymbol{l}}(X_i, \mathbf{0}) + o_P(1), \text{ and similarly} \\
\sqrt{n}\tilde{\boldsymbol{\theta}}_{n(d)} &= \sqrt{n}(\tilde{\boldsymbol{\theta}}_{n(d)} - \boldsymbol{\theta}_0) = [\boldsymbol{I}_{(d)}(\mathbf{0}) + o_P(1)]^{-1} n^{-1/2} \sum_{i=1}^{n} \dot{\boldsymbol{l}}_{(d)}(X_i, \mathbf{0}) + o_P(1)
\end{aligned}$$

to obtain after simplification,

$$\begin{aligned}
-2\log \Lambda_n &= 2\left[\sum_{i=1}^{n} l(X_i, \hat{\boldsymbol{\theta}}_n) - \sum_{i=1}^{n} l(X_i, \tilde{\boldsymbol{\theta}}_n) \right] \\
&= \left\{ n^{-1/2} \sum_{i=1}^{n} \dot{\boldsymbol{l}}(X_i, \mathbf{0}) \right\}^T \boldsymbol{I}(\mathbf{0})^{-1} \left\{ n^{-1/2} \sum_{i=1}^{n} \dot{\boldsymbol{l}}(X_i, \mathbf{0}) \right\} \\
&\quad - \left\{ n^{-1/2} \sum_{i=1}^{n} \dot{\boldsymbol{l}}_{(d)}(X_i, \mathbf{0}) \right\}^T \boldsymbol{I}_{(d)}(\mathbf{0})^{-1} \left\{ n^{-1/2} \sum_{i=1}^{n} \dot{\boldsymbol{l}}_{(d)}(X_i, \mathbf{0}) \right\} + o_P(1).
\end{aligned}$$

By multivariate CLT,

$$n^{-1/2}\sum_{i=1}^{n}\dot{\boldsymbol{l}}(X_i,\mathbf{0}) \xrightarrow{\mathcal{L}} \boldsymbol{Y} \sim N_k(\mathbf{0},\boldsymbol{I}(\mathbf{0})) \text{ and } n^{-1/2}\sum_{i=1}^{n}\dot{\boldsymbol{l}}_{(d)}(X_i,\mathbf{0}) \xrightarrow{\mathcal{L}} \boldsymbol{Y}_{(d)},$$

where $\boldsymbol{Y}_{(d)}$ is the d-dim random vector consisting of the first d coordinates of $\boldsymbol{Y}$. Hence

$$-2\log\Lambda_n \xrightarrow{\mathcal{L}} \boldsymbol{Y}^T\boldsymbol{I}(\mathbf{0})^{-1}\boldsymbol{Y} - \boldsymbol{Y}_{(d)}^T\boldsymbol{I}_{(d)}(\mathbf{0})^{-1}\boldsymbol{Y}_{(d)}.$$

Using a suitable linear transformation, $\boldsymbol{Y}^T\boldsymbol{I}(\mathbf{0})^{-1}\boldsymbol{Y}$ and $\boldsymbol{Y}_{(d)}^T\boldsymbol{I}_{(d)}(\mathbf{0})^{-1}\boldsymbol{Y}_{(d)}$ can be simultaneously reduced to $\sum_{i=1}^{k}Z_i^2$ and $\sum_{i=1}^{d}Z_i^2$, respectively, where $Z_1,\ldots,Z_k$ are iid $N(0,1)$. Thus

$$-2\log\Lambda_n \xrightarrow{\mathcal{L}} \sum_{i=1}^{k}Z_i^2 - \sum_{i=1}^{d}Z_i^2 = \sum_{i=d+1}^{k}Z_i^2 \sim \chi^2_{k-d}.$$

□

An important property, essential for all desirable tests, is that the probability of type II error of the test at any level α tends to 0 (ie, the power tends to 1), as $n\to\infty$ for all departures from the null hypothesis. This property of a test is called *Consistency*.

Theorem 7.2.2. *The LRT is consistent, ie, for* $c>0$,

$$\lim_{n\to\infty}P_{\theta_0}[-2\log\Lambda_n < c] = 0 \textit{ if } \boldsymbol{\theta}_0 \notin \Theta_0,$$

under regularity conditions (Cramér and Wald).

Proof. Suppose that the true $\boldsymbol{\theta}_0 \notin \Theta_0$. Then arguing as in the proof of Theorem 7.2.1

$$\begin{aligned}
-2\log\Lambda_n &= -2\left[\sum_{i=1}^{n}l(X_i,\tilde{\boldsymbol{\theta}}_n) - \sum_{i=1}^{n}l(X_i,\boldsymbol{\theta}_0)\right] \\
&\quad + 2\left[\sum_{i=1}^{n}l(X_i,\hat{\boldsymbol{\theta}}_n) - \sum_{i=1}^{n}l(X_i,\boldsymbol{\theta}_0)\right] \\
&= -2\left[\sum_{i=1}^{n}l(X_i,\tilde{\boldsymbol{\theta}}_n) - \sum_{i=1}^{n}l(X_i,\boldsymbol{\theta}_0)\right] \\
&\quad + \left\{n^{-1/2}\sum_{i=1}^{n}\dot{\boldsymbol{l}}(X_i,\boldsymbol{\theta}_0)\right\}^T \boldsymbol{I}(\boldsymbol{\theta}_0)^{-1}\left\{n^{-1/2}\sum_{i=1}^{n}\dot{\boldsymbol{l}}(X_i,\boldsymbol{\theta}_0)\right\} + Y_n \\
&:= -2\left[\sum_{i=1}^{n}l(X_i,\tilde{\boldsymbol{\theta}}_n) - \sum_{i=1}^{n}l(X_i,\boldsymbol{\theta}_0)\right] + b_n^2 + Y_n,
\end{aligned}$$

where $Y_n = o_P(1)$.

Since

(i) $P[X_n+Y_n+b_n^2<c] \le P[X_n+Y_n<c] \le P[X_n<c+|Y_n|] \le P[X_n<2c]+P[|Y_n|\ge c]$,
(ii) $\lim_{n\to\infty}P[X_n<c]=0$ for all $c>0$, and
(iii) $Y_n = o_P(1)$

together imply $\lim_{n\to\infty} P[X_n + Y_n + b_n^2 < c] = 0$ for all $c > 0$, it is enough to show that for all $c > 0$,

$$\lim_{n\to\infty} P_{\boldsymbol{\theta}_0}\left[-\sum_{i=1}^{n} l(X_i, \tilde{\boldsymbol{\theta}}_n) + \sum_{i=1}^{n} l(X_i, \boldsymbol{\theta}_0) < c\right]$$
$$= \lim_{n\to\infty} P_{\boldsymbol{\theta}_0}\left[\inf_{\boldsymbol{\theta}\in\Theta_0} n^{-1}\sum_{i=1}^{n} \log\{f(X_i, \boldsymbol{\theta}_0)/f(X_i, \boldsymbol{\theta})\} < c/n\right] = 0$$

to prove the theorem.

Referring to the notations introduced in Section 7.1.1 and the proof of Theorem 7.1.4, define iid rv's

$$Z_i(\boldsymbol{\theta}_0, S) = \inf_{\boldsymbol{\theta}\in S} \log\{f(X_i, \boldsymbol{\theta}_0)/f(X_i, \boldsymbol{\theta})\}, \quad i = 1, 2, \ldots$$

for any set $S \subset \mathbb{R}^k$ so that $\boldsymbol{\theta}_0 \notin S$. Then under Wald's conditions, there exist a finite number of sets $S_0, S_1, \ldots, S_r$ such that

(i) $\Theta_0 \subset \bigcup_{j=0}^{r} S_j$ and (ii) for $\boldsymbol{\theta}_0 \notin \Theta_0$, $I(\boldsymbol{\theta}_0, S_j) = \mathrm{E}_{\boldsymbol{\theta}_0}[Z(\boldsymbol{\theta}_0, S_j)] > 0$, $j = 0, \ldots, r$.

We demonstrate the existence of such sets by the following steps

(a) Choose a neighborhood N_{∞,j^*} of ∞ such that

$$\mathrm{E}[Z(\boldsymbol{\theta}_0, N_{\infty,j^*} \cap \Theta_0)] \geq \mathrm{E}[Z(\boldsymbol{\theta}_0, N_{\infty,j^*})] > 0.$$

(b) $N^{c}_{\infty,j^*} \cap \Theta_0$ is a closed bounded set in $\mathbb{R}^k$ and therefore, has the Heine-Borel property of having a finite subcover for every open cover.

(c) For every $\boldsymbol{\theta} \in N^{c}_{\infty,j^*} \cap \Theta_0$, choose $N_{\boldsymbol{\theta},j}$ such that $\mathrm{E}[Z(\boldsymbol{\theta}_0, N_{\boldsymbol{\theta},j})] > 0$. These sets form an open cover of $N_{\infty,j^*} \cap \Theta_0$, from which we now choose a finite subcover $N_{\boldsymbol{\theta}_1,j_1}, \ldots, N_{\boldsymbol{\theta}_r,j_r}$.

(d) Call these sets $S_1 = N_{\boldsymbol{\theta}_1,j_1}, \ldots, S_r = N_{\boldsymbol{\theta}_r,j_r}$ and let $S_0 = N_{\infty,j^*} \cap \Theta_0$. Then $S_0, S_1, \ldots, S_r$ cover Θ_0 and satisfy (i) and (ii) above. The choice of N_{∞,j^*} and $N_{\boldsymbol{\theta},\boldsymbol{j}}$ for every $\theta \in N^{c}_{\infty,j} \cap \Theta_0$ with these properties is ensured by Wald's regularity conditions.

It now follows that

$$\inf_{\boldsymbol{\theta}\in\Theta_0} n^{-1}\sum_{i=1}^{n} \log\{f(X_i, \boldsymbol{\theta}_0)/f(X_i, \boldsymbol{\theta})\}$$
$$\geq \min_{0\leq j\leq r} \inf_{\boldsymbol{\theta}\in S_j} n^{-1}\sum_{i=1}^{n} \log\{f(X_i, \boldsymbol{\theta}_0)/f(X_i, \boldsymbol{\theta})\}$$
$$\geq \min_{0\leq j\leq r} n^{-1}\sum_{i=1}^{n} \inf_{\boldsymbol{\theta}\in S_j} \log\{f(X_i, \boldsymbol{\theta}_0)/f(X_i, \boldsymbol{\theta})\} = \min_{0\leq j\leq r} n^{-1}\sum_{i=1}^{n} Z_i(\boldsymbol{\theta}_0, S_j).$$

Thus

$$P_{\boldsymbol{\theta}_0}\left[\inf_{\boldsymbol{\theta}\in\Theta_0} n^{-1}\sum_{i=1}^{n}\log\{f(X_i,\boldsymbol{\theta}_0)/f(X_i,\boldsymbol{\theta})\} < c/n\right]$$

$$= P_{\boldsymbol{\theta}_0}\left[\min_{0\le j\le r} n^{-1}\sum_{i=1}^{n} Z_i(\boldsymbol{\theta}_0, S_j) < c/n\right] \le \sum_{j=0}^{r} P_{\boldsymbol{\theta}_0}\left[n^{-1}\sum_{i=1}^{n} Z_i(\boldsymbol{\theta}_0, S_j) < c/n\right],$$

which tends to 0 as $n \to \infty$, because

$$n^{-1}\sum_{i=1}^{n} Z_i(\boldsymbol{\theta}_0, S_j) \overset{a.s.}{\to} \mathrm{E}_{\boldsymbol{\theta}_0}[Z(\boldsymbol{\theta}_0, S_j)] = I(\boldsymbol{\theta}_0, S_j) > 0.$$

□

Example 7.2.1 (Homogeneity of Exponential Distributions). Let $(X_{i1}, \ldots, X_{in_i})$, $i = 1, \ldots, k$, be independent random samples from *Exponential*(θ_i), $i = 1, \ldots, k$, respectively. Find the LRT for H_0: $\theta_1 = \cdots = \theta_k$ against all possible departures from H_0.

Solution. Let $\hat{\theta}_{ni}$ and $\bar{\theta}_n$ be, respectively, the unrestricted MLE and the restricted MLE under H_0 of θ_i. Then

$$\hat{\theta}_{ni} = \bar{T}_i = T_i/n_i \text{ and } \bar{\theta}_N = \bar{T} = \sum_{i=1}^{k} T_i \Big/ \sum_{i=1}^{k} n_i, \text{ where}$$

$$T_i = \sum_{j=1}^{n_i} X_{ij}.$$

Now the LRT statistic is

$$\Lambda_N = \prod_{i=1}^{k}\prod_{j=1}^{n_i} f(X_{ij}, \bar{\theta}_N) \Big/ \prod_{i=1}^{k}\prod_{j=1}^{n_i} f(X_{ij}, \hat{\theta}_{ni})$$

$$= \left[\frac{1}{\bar{\theta}_N^N}\, \mathrm{e}^{-T/\bar{\theta}_N}\right] \Big/ \left[\prod_{i=1}^{k} \frac{1}{\hat{\theta}_{ni}^{n_i}}\, \mathrm{e}^{-T_i/\hat{\theta}_{ni}}\right] = \prod_{i=1}^{k}\left(\frac{\hat{\theta}_{ni}}{\bar{\theta}_N}\right)^{n_i}.$$

Since $\hat{\theta}_{ni}/\bar{\theta}_N = 1 + o_P(1)$ under H_0 and since $\log(1 + Y_N) = Y_N - (1/2)Y_N^2\{1 + o_P(1)\}$ if $Y_N = o_P(1)$, it follows that

$$-2\log\Lambda_N = -2\sum_{i=1}^{k} n_i \log(\hat{\theta}_{ni}/\bar{\theta}_N)$$

$$= -2\sum_{i=1}^{k} n_i[(\hat{\theta}_{ni}/\bar{\theta}_N - 1) - (1/2)(\hat{\theta}_{ni}/\bar{\theta}_N - 1)^2\{1 + o_P(1)\}]$$

under H_0. It is easy to check that $\sum_{i=1}^{k} n_i(\hat{\theta}_{ni}/\bar{\theta}_N - 1) = 0$. Hence

$$-2\log \Lambda_N = \sum_{i=1}^{k}\left[n_i(\hat{\theta}_{ni} - \bar{\theta}_N)^2/\bar{\theta}_N^2\right]\{1 + o_P(1)\} \xrightarrow{\mathcal{L}} \chi^2_{k-1}$$

under H_0 by Theorem 7.2.1. Moreover, since $\sqrt{n_i}(\hat{\theta}_{ni} - \theta)/\theta \xrightarrow{\mathcal{L}} N(0,1)$ independently for each i and since $\bar{\theta}_N = \theta + o_P(1)$, we can argue more directly, that

$$\sum_{i=1}^{k}\left[n_i(\hat{\theta}_{ni} - \bar{\theta}_N)^2/\bar{\theta}_N^2\right]\{1 + o_P(1)\}$$
$$= \sum_{i=1}^{k} n_i(\hat{\theta}_{ni} - \bar{\theta}_N)^2/\bar{\theta}^2 + o_P(1) \xrightarrow{\mathcal{L}} \chi^2_{k-1}.$$

Example 7.2.2 (Homogeneity of Multinomial Probabilities). Consider m mutually exclusive and exhaustive categories and let

$$\boldsymbol{\pi}_i = (\pi_{i1}, \ldots, \pi_{im}), \quad i = 1, \ldots, k, \; \pi_{ij} > 0 \text{ and } \sum_{j=1}^{m} \pi_{ij} = 1$$

be the probability distributions over these categories in k populations. These probability vectors are called multinomial probabilities. Let $(n_{i1}, \ldots, n_{im})$ be the frequencies in the m categories in independent random samples of sizes n_i, $i = 1, \ldots, k$, from these populations. Find the LRT for H_0: $\boldsymbol{\pi}_1 = \cdots = \boldsymbol{\pi}_k$ against all possible departures from H_0.

Solution. The unrestricted MLEs and restricted MLEs under H_0 of $\boldsymbol{\pi}_i$ are

$$\hat{\boldsymbol{\pi}}_i = (\hat{\pi}_{i1}, \ldots, \hat{\pi}_{im}) \text{ and } \hat{\boldsymbol{\pi}}_i^0 = (\hat{\pi}_{i1}^0, \ldots, \hat{\pi}_{im}^0),$$

respectively, where $\hat{\pi}_{ij} = n_{ij}/n_i$ and $\hat{\pi}_{ij}^0 = n_{0j}/N$, with $n_{0j} = \sum_{i=1}^{k} n_{ij}$ and $N = \sum_{i=1}^{k} n_i$. For notational convenience, we also let $n_i - \sum_{j=1}^{m} n_{ij} - n_{i0}$. The LRT statistic can now be written as

$$\Lambda_N = \prod_{i=1}^{k} f((n_{i1}, \ldots, n_{im}), \hat{\boldsymbol{\pi}}_i^0) \Big/ \prod_{i=1}^{k} f((n_{i1}, \ldots, n_{im}), \hat{\boldsymbol{\pi}}_i)$$
$$= \prod_{i=1}^{k}\prod_{j=1}^{m} (\hat{\pi}_{ij}^0/\hat{\pi}_{ij})^{n_{ij}},$$

canceling the multinomial coefficients from the numerator and denominator. Again, using $\hat{\pi}_{ij}^0/\hat{\pi}_{ij} = 1 + o_P(1)$ and the property of $\log(1 + Y_n)$ for $Y_n = o_P(1)$ as in Example 7.1.1, we have

$$
\begin{aligned}
-2\log \Lambda_N &= -2\sum_{i=1}^{k}\sum_{j=1}^{m} n_{ij}\log(\hat{\pi}_{ij}^{0}/\hat{\pi}_{ij}) \\
&= -2\sum_{i=1}^{k}\sum_{j=1}^{m} n_{i0}\hat{\pi}_{ij}\left[\frac{\hat{\pi}_{ij}^{0}-\hat{\pi}_{ij}}{\hat{\pi}_{ij}} - (1/2)\left(\frac{\hat{\pi}_{ij}^{0}-\hat{\pi}_{ij}}{\hat{\pi}_{ij}}\right)^2\{1+o_P(1)\}\right] \\
&= -2\sum_{i=1}^{k}\sum_{j=1}^{m} n_{i0}(\hat{\pi}_{ij}^{0}-\hat{\pi}_{ij}) + \sum_{i=1}^{k}\sum_{j=1}^{m} n_{i0}\frac{(\hat{\pi}_{ij}^{0}-\hat{\pi}_{ij})^2}{\hat{\pi}_{ij}}\{1+o_P(1)\}
\end{aligned}
$$

under H_0. Since $\sum_{i=1}^{k} n_{i0}(\hat{\pi}_{ij}^{0}-\hat{\pi}_{ij}) = 0$ for each j, and

$$
\begin{aligned}
\sum_{i=1}^{k}\sum_{j=1}^{m} n_{i0}\frac{(\hat{\pi}_{ij}^{0}-\hat{\pi}_{ij})^2}{\hat{\pi}_{ij}} &= \sum_{i=1}^{k}\sum_{j=1}^{m}\frac{(n_{ij}-(n_{i0}n_{0j}/N))^2}{n_{ij}} \\
&= \sum_{i=1}^{k}\sum_{j=1}^{m}\frac{(n_{ij}-(n_{i0}n_{0j}/N))^2}{(n_{i0}n_{0j}/N)}\{1+o_P(1)\}
\end{aligned}
$$

under H_0, we finally have

$$
-2\log \Lambda_N = \sum_{i=1}^{k}\sum_{j=1}^{m}\frac{(n_{ij}-(n_{i0}n_{0j}/N))^2}{(n_{i0}n_{0j}/N)} + o_P(1) \xrightarrow{\mathcal{L}} \chi^2_{(k-1)(m-1)}.
$$

7.3 Asymptotic Properties of MLE and LRT Based on Independent Nonidentically Distributed Data

In this section, we shall briefly indicate the behavior of MLE and LRT when the observations are independent but nonidentically distributed.

Suppose that $(X_{11},\ldots,X_{1n_1}), (X_{21},\ldots,X_{2n_2}),\ldots,(X_{m1},\ldots,X_{mn_m})$ are independent random samples from distributions with pdf/pmf $f_1(x,\boldsymbol{\theta}),\ldots,f_m(x,\boldsymbol{\theta})$, respectively, and let $n = n_1+\cdots+n_m$ be the total sample size.

We assume that the regularity conditions stated earlier, hold for each $\{f_j(x,\boldsymbol{\theta}), \boldsymbol{\theta}\in\Theta\subset\mathbb{R}^k\}$ and let

$$
I_{j,rs}(\boldsymbol{\theta}_0) = \mathrm{E}[-\ddot{l}_{j,rs}(X,\boldsymbol{\theta}_0)], \qquad I_j(\boldsymbol{\theta}_0) = ((I_{j,rs}(\boldsymbol{\theta}_0))), \quad j=1,\ldots,m.
$$

Let $\hat{\boldsymbol{\theta}}_n$ be the MLE of $\boldsymbol{\theta}$ based on the pooled data consisting of all X_{ji}, $i=1,\ldots,n_j$, $j=1,\ldots,m$.

If $n_1,\ldots,n_m\to\infty$ in such a way that $n_j/n\to\lambda_j>0$ for all j (of course $\lambda_1+\cdots+\lambda_m=1$), and if $\boldsymbol{I}(\boldsymbol{\theta}_0) = \sum_{j=1}^{m}\lambda_j\boldsymbol{I}_j(\boldsymbol{\theta}_0)$ is positive definite (even if some of the individual $\boldsymbol{I}_j(\boldsymbol{\theta}_0)$ is singular), then

$$
\sqrt{n}(\hat{\boldsymbol{\theta}}_n-\boldsymbol{\theta}_0) \xrightarrow{\mathcal{L}} \boldsymbol{W}\sim N_k(\boldsymbol{0},\boldsymbol{I}(\boldsymbol{\theta}_0)^{-1}).
$$

Asymptotic properties of the LRT discussed earlier for the iid case, also hold in the non-iid case described above.

7.4 Frequency χ^2

An important class of hypothesis testing problems arises in the context of the multinomial distribution *Multi*$(n; \pi_1, \ldots, \pi_m)$, where $\pi_j \geq 0$ with $\sum \pi_j = 1$ are the probabilities of a random sample belonging to the classes $(1, \ldots, m)$. The data consist of a random sample $(\boldsymbol{X}_1, \ldots, \boldsymbol{X}_n)$ where the $\boldsymbol{X}_i$'s are m-dim iid rv's with

$$f(\boldsymbol{e}_j, \boldsymbol{\pi}) = P(\boldsymbol{X}_i = \boldsymbol{e}_j) = \pi_j, \quad j = 1, \ldots, m,$$

where $\boldsymbol{e}_j$ is the m-dim vector with 1 at the jth coordinate and 0 for the rest of the coordinates, and $\boldsymbol{\pi}^T = (\pi_1, \ldots, \pi_m)$. As in Section 7.1.3, let

$$(n_1, \ldots, n_m)^T = \sum_{i=1}^{n} \boldsymbol{X}_i,$$

$$n_j = \sum_{i=1}^{n} I(\boldsymbol{X}_i = \boldsymbol{e}_j) = \text{observed frequency in the } j\text{th class.}$$

For testing H_0: $\pi_j = \pi_{j0}, j = 1, \ldots, m$, with $\pi_{j0} \geq 0$ with $\sum_{j=1}^{m} \pi_{j0} = 1$, against all possible alternatives satisfying these constraints, the test statistic

$$T_{0n} = \sum_{j=1}^{m} \frac{(n_j - n\pi_{j0})^2}{n\pi_{j0}}, \tag{7a}$$

known as the frequency χ^2, introduced by K. Pearson, is widely used. In many problems of practical importance such as testing for independence in $r \times s$ contingency tables, the null hypothesis value for the jth class is $\pi_{j0}(\theta_1, \ldots, \theta_k)$ where $\pi_{10}(\cdot), \ldots, \pi_{m0}(\cdot)$ are known functions of an unknown k-dim parameter $\boldsymbol{\theta} = (\theta_1, \ldots, \theta_k)^T$. The test statistic used to test such a composite hypothesis H_1: $\pi_j = \pi_{j0}(\boldsymbol{\theta}), j = 1, \ldots, m$ for given functions $\boldsymbol{\pi}_0(\cdot) = (\pi_{10}(\cdot), \ldots, \pi_{m0}(\cdot))^T$ of an unknown $\boldsymbol{\theta} \in \Theta \subset \mathbb{R}^k$ against all possible alternatives is

$$T_{1n} = \sum_{j=1}^{m} \frac{[n_j - n\pi_{j0}(\hat{\boldsymbol{\theta}}_n)]^2}{n\pi_{j0}(\hat{\boldsymbol{\theta}}_n)}, \tag{7b}$$

where $\hat{\boldsymbol{\theta}}_n = (\hat{\theta}_{n1}, \ldots, \hat{\theta}_{nk})^T$ is the MLE of $\boldsymbol{\theta} = (\theta_1, \ldots, \theta_k)^T$. In both the above problems, we reject the respective null hypotheses if T_{0n} or T_{1n} is too large. Our goal is to find the asymptotic distribution of T_{0n} under H_0 and of T_{1n} under H_1 so that the critical values $c_{0\alpha}$ and $c_{1\alpha}$ for level α tests can be determined so that

$$P_{H_0}[T_{0n} \geq c_{0\alpha}] \approx \alpha \text{ and } P_{H_1}[T_{1n} \geq c_{1\alpha}] \approx \alpha,$$

for large n. This will be accomplished in two steps:

1. Simplifying T_{0n} given by Eq. (7a) and T_{1n} given by Eq. (7b) to forms more convenient for asymptotics.
2. Deriving the asymptotic distributions of these simplified forms of T_{0n} and T_{1n}.

The statistic T_{0n} is called the frequency χ^2 because its null distribution follows the χ^2 distribution with $m-1$ df and the statistic T_{1n} also has a χ^2 distribution but with $m-k-1$ df. These asymptotic distributions will be derived in this section.

We first define some vectors and matrices which will be useful in handling T_{0n} and T_{1n}. In dealing with T_{0n}, let

$$q_j = \sqrt{\pi_{j0}}, \boldsymbol{q}^T = (q_1, \ldots, q_m), \boldsymbol{\Lambda}_q = \text{diag}((q_j)), \boldsymbol{\Lambda}_{1/q} = \boldsymbol{\Lambda}_q^{-1} = \text{diag}((1/q_j)).$$

Then $\text{diag}((\pi_{j0})) = \boldsymbol{\Lambda}_q^2, \boldsymbol{\pi}_0 = \boldsymbol{\Lambda}_q \boldsymbol{q}$, and under H_0,

$$\text{E}[\boldsymbol{X}] = \boldsymbol{\pi}_0 = \boldsymbol{\Lambda}_q \boldsymbol{q} \text{ and } \boldsymbol{\Sigma} = \text{E}[(\boldsymbol{X}-\boldsymbol{\pi}_0)(\boldsymbol{X}-\boldsymbol{\pi}_0)^T] = \boldsymbol{\Lambda}_q(\boldsymbol{I}-\boldsymbol{q}\boldsymbol{q}^T)\boldsymbol{\Lambda}_q \tag{8}$$

after some simplification. Next let

$$\boldsymbol{Z}_n = n^{-1/2}(n_1 - n\pi_{10}, \ldots, n_m - n\pi_{m0})^T = n^{-1/2}\sum_{i=1}^{n}(\boldsymbol{X}_i - \boldsymbol{\pi}_0).$$

Then

$$\boldsymbol{Z}_n = n^{-1/2}\sum_{i=1}^{n}(\boldsymbol{X}_i - \text{E}(\boldsymbol{X}_i)) \xrightarrow{\mathcal{L}} \boldsymbol{Z} \sim N_m(\boldsymbol{0}, \boldsymbol{\Sigma}) \text{ under } H_0. \tag{9}$$

Lemma 7.4.1. *Under H_0, $T_{0n} \xrightarrow{\mathcal{L}} \boldsymbol{U}^T\boldsymbol{U}$, where $\boldsymbol{U} \sim N_m(\boldsymbol{0}, \boldsymbol{I}-\boldsymbol{q}\boldsymbol{q}^T)$.*

Proof. Using Eq. (8) T_{0n} can be written as

$$\begin{aligned} T_{0n} &= \sum_{j=1}^{m}\left\{(\sqrt{n}q_j)^{-1}(n_j - n\pi_{j0})\right\}^2 = \sum_{j=1}^{m}(1/q_j^2)\left\{n^{-1/2}(n_j - n\pi_{j0})\right\}^2 \\ &= (\boldsymbol{\Lambda}_{1/q}\boldsymbol{Z}_n)^T(\boldsymbol{\Lambda}_{1/q}\boldsymbol{Z}_n) := \boldsymbol{U}_n^T\boldsymbol{U}_n \xrightarrow{\mathcal{L}} \boldsymbol{U}^T\boldsymbol{U}, \end{aligned}$$

where $\boldsymbol{U}_n = \boldsymbol{\Lambda}_{1/q}\boldsymbol{Z}_n \xrightarrow{\mathcal{L}} \boldsymbol{\Lambda}_{1/q}\boldsymbol{Z} = \boldsymbol{U} \sim N_m(\boldsymbol{0}, \boldsymbol{I}-\boldsymbol{q}\boldsymbol{q}^T)$, because the covariance matrix of $\boldsymbol{U}$ is $\boldsymbol{\Lambda}_{1/q}\boldsymbol{\Sigma}\boldsymbol{\Lambda}_{1/q} = \boldsymbol{I}-\boldsymbol{q}\boldsymbol{q}^T$ after some simplification. □

In dealing with T_{1n}, we continue to use the above notations with $\pi_{j0}(\boldsymbol{\theta}_0)$ instead of π_{j0}. The problem with T_{1n} is the presence of $\hat{\boldsymbol{\theta}}_n$ in both the numerator and the denominator. Replacing $\hat{\boldsymbol{\theta}}_n$ by $\boldsymbol{\theta}_0$ in the denominator is relatively easy and will be taken care at first. Handling $\hat{\boldsymbol{\theta}}_n$ in the numerator will take more work. We start with T_{1n} and after algebraic re-arrangements, write

$$
\begin{aligned}
T_{1n} &= \sum_{j=1}^{m} [n_j - n\pi_{j0}(\hat{\boldsymbol\theta}_n)]^2/(n\pi_{j0}(\hat{\boldsymbol\theta}_n)) \\
&= \sum_{j=1}^{m} [n_j - n\pi_{j0}(\hat{\boldsymbol\theta}_n)]^2/(n\pi_{j0}(\boldsymbol\theta_0))\{1 - R_{nj}\} := T^*_{1n} + R_n, \text{ where} \\
T^*_{1n} &= \sum_{j=1}^{m} [n_j - n\pi_{j0}(\hat{\boldsymbol\theta}_n)]^2/(n\pi_{j0}(\boldsymbol\theta_0)) \text{ and} \\
|R_n| &\le T^*_{1n} \max_{1\le j\le m} |R_{nj}| = T^*_{in} o_P(1),
\end{aligned}
$$

because under regularity conditions, $\pi_{j0}(\hat{\boldsymbol\theta}_n) - \pi_{j0}(\boldsymbol\theta_0) = o_P(1)$. We shall now show that $T^*_{1n} \xrightarrow{\mathcal{L}} T^*_1$ (to be determined), so that $T^*_{1n} = O_P(1)$ which would imply $R_n = O_P(1)o_P(1) = o_P(1)$, proving that $T_{1n} = T^*_{1n} + o_P(1) \xrightarrow{\mathcal{L}} T^*_1$.

We now work on

$$
\begin{aligned}
T^*_{1n} &= \sum_{j=1}^{m} [n_j - n\pi_{j0}(\hat{\boldsymbol\theta}_n)]^2/(n\pi_{j0}(\boldsymbol\theta_0)) \\
&= \sum_{j=1}^{m} \left[\frac{n_j - n\pi_{j0}(\boldsymbol\theta_0)}{\sqrt{n\pi_{j0}(\boldsymbol\theta_0)}} - \sqrt{n}\frac{\pi_{j0}(\hat{\boldsymbol\theta}_n) - \pi_{j0}(\boldsymbol\theta_0)}{\sqrt{\pi_{j0}(\boldsymbol\theta_0)}} \right]^2 \\
&= (\boldsymbol{U}_n - \boldsymbol{W}_n)^T(\boldsymbol{U}_n - \boldsymbol{W}_n), \qquad (10)
\end{aligned}
$$

where $\boldsymbol{U}_n = \boldsymbol{\Lambda}_{1/q}\boldsymbol{Z}_n$ as in Eq. (9) and the proof of Lemma 7.4.1 with $\boldsymbol\pi_0(\boldsymbol\theta_0)$ in place of $\boldsymbol\pi_0$, and

$$
\boldsymbol{W}_n^T = \sqrt{n}\left(\frac{\pi_{10}(\hat{\boldsymbol\theta}_n) - \pi_{10}(\boldsymbol\theta_0)}{\sqrt{\pi_{10}(\boldsymbol\theta_0)}}, \ldots, \frac{\pi_{m0}(\hat{\boldsymbol\theta}_n) - \pi_{m0}(\boldsymbol\theta_0)}{\sqrt{\pi_{m0}(\boldsymbol\theta_0)}} \right).
$$

Assuming regularity conditions including existence of $\partial\pi_{j0}/\partial\theta_r = \dot\pi_{rj}$ for $1 \le r \le k$, $1 \le j \le m$ and positive-definiteness of $\boldsymbol{I}(\boldsymbol\theta_0) = ((I_{rs}(\boldsymbol\theta_0)))$, let $d_{rj} = \dot\pi_{rj}/q_j$, $\boldsymbol{D} = ((d_{rj}))_{k\times m}$ and $I_{rs}(\boldsymbol\theta_0) = \sum_{j=1}^{m} d_{rj}d_{sj} = (r,s)$th element of $\boldsymbol{DD}^T$.

In these notations,

$$
\boldsymbol\pi_{j0}(\hat{\boldsymbol\theta}_n) - \boldsymbol\pi_{j0}(\boldsymbol\theta_0) = \sum_{r=1}^{k} \dot\pi_{rj}(\hat\theta_{nr} - \theta_{0r}) + o_P(1), \quad 1 \le j \le m,
$$

and now $\boldsymbol{W}_n^T$ can be expressed as

$$
\begin{aligned}
\boldsymbol{W}_n^T &= \sqrt{n}(\hat{\boldsymbol\theta}_n - \boldsymbol\theta_0)^T\boldsymbol{D} + o_P(1) = \left\{ \boldsymbol{I}(\boldsymbol\theta_0)^{-1} n^{-1/2} \sum_{i=1}^{n} \dot{\boldsymbol{l}}(\boldsymbol{X}_i, \boldsymbol\theta_0) \right\}^T \boldsymbol{D} + o_P(1) \\
&= \{I(\boldsymbol\theta_0)^{-1}\boldsymbol{DU}_n\}^T D + o_P(1), \qquad (11)
\end{aligned}
$$

using the expression for $\sqrt{n}(\hat{\boldsymbol{\theta}}_n - \boldsymbol{\theta}_0)$ obtained in the course of the proof of Theorem 7.1.3, because

$$\begin{aligned}
\sum_{i=1}^{n} \dot{\boldsymbol{l}}(\boldsymbol{X}_i, \boldsymbol{\theta}_0) &= \sum_{i=1}^{n}\sum_{j=1}^{m} I[\boldsymbol{X}_i = \boldsymbol{e}_j]\dot{\boldsymbol{l}}(\boldsymbol{e}_j, \boldsymbol{\theta}_0) = \sum_{j=1}^{m} n_j \dot{\boldsymbol{l}}(\boldsymbol{e}_j, \boldsymbol{\theta}_0) \\
&= \sum_{j=1}^{m} \frac{n_j}{\sqrt{\pi_j(\boldsymbol{\theta}_0)}} \left(\frac{\dot{\pi}_{1j}(\boldsymbol{\theta}_0)}{q_j}, \ldots, \frac{\dot{\pi}_{kj}(\boldsymbol{\theta}_0)}{q_j} \right)^T \\
&= \sqrt{n} \sum_{j=1}^{m} \frac{n_j - n\pi_{j0}(\boldsymbol{\theta}_0)}{\sqrt{n\pi_j(\boldsymbol{\theta}_0)}} \left(\frac{\dot{\pi}_{1j}(\boldsymbol{\theta}_0)}{q_j}, \ldots, \frac{\dot{\pi}_{kj}(\boldsymbol{\theta}_0)}{q_j} \right)^T \\
&= \sqrt{n} \sum_{j=1}^{m} U_{nj}(d_{1j}, \ldots, d_{kj}) = \sqrt{n}\boldsymbol{D}\boldsymbol{U}_n,
\end{aligned}$$

since $\sum_{j=1}^{m} \pi_{j0}(\boldsymbol{\theta}_0) = 1$ implies $\sum_{j=1}^{m} \dot{\pi}_{rj}(\boldsymbol{\theta}_0) = 0,\ r = 1, \ldots, k$.

From Eq. (11), it now follows that under H_1,

$$\begin{aligned}
&\boldsymbol{V}_n = \boldsymbol{U}_n - \boldsymbol{W}_n = \boldsymbol{U}_n - \boldsymbol{D}^T\{\boldsymbol{I}(\boldsymbol{\theta}_0)^{-1}\boldsymbol{D}\boldsymbol{U}_n\} = [\boldsymbol{I} - \boldsymbol{D}^T(\boldsymbol{D}\boldsymbol{D}^T)^{-1}\boldsymbol{D}]\boldsymbol{U}_n \\
&\xrightarrow{\mathcal{L}} \boldsymbol{V} = [\boldsymbol{I} - \boldsymbol{D}^T(\boldsymbol{D}\boldsymbol{D}^T)^{-1}\boldsymbol{D}]\boldsymbol{U} \sim N_m(\boldsymbol{0}, \boldsymbol{\Sigma}_V), \text{ where} \\
&\boldsymbol{\Sigma}_V = [\boldsymbol{I} - \boldsymbol{D}^T(\boldsymbol{D}\boldsymbol{D}^T)^{-1}\boldsymbol{D}](\boldsymbol{I} - \boldsymbol{q}\boldsymbol{q}^T)[\boldsymbol{I} - \boldsymbol{D}^T(\boldsymbol{D}\boldsymbol{D}^T)^{-1}\boldsymbol{D}] \\
&\quad = \boldsymbol{I} - \boldsymbol{q}\boldsymbol{q}^T - \boldsymbol{D}^T(\boldsymbol{D}\boldsymbol{D}^T)^{-1}\boldsymbol{D}.
\end{aligned}$$

Lemma 7.4.2. *Under* H_1, $T_{1n} \xrightarrow{\mathcal{L}} \boldsymbol{V}^T\boldsymbol{V}$ *where* $\boldsymbol{V} \sim N_m(\boldsymbol{0}, \boldsymbol{I} - \boldsymbol{q}\boldsymbol{q}^T - \boldsymbol{D}^T(\boldsymbol{D}\boldsymbol{D}^T)^{-1}\boldsymbol{D})$.

Proof. As shown in Eq. (10)

$$\begin{aligned}
T_{1n}^* &= (\boldsymbol{U}_n - \boldsymbol{W}_n)^T(\boldsymbol{U}_n - \boldsymbol{W}_n) = \boldsymbol{V}_n^T\boldsymbol{V}_n \\
&\xrightarrow{\mathcal{L}} T_1^* = \boldsymbol{V}^T\boldsymbol{V}.
\end{aligned}$$

Hence

$$T_{1n} = T_{1n}^* + R_n \xrightarrow{\mathcal{L}} T_1^* = \boldsymbol{V}^T\boldsymbol{V},$$

because $|R_n| \le T_n^* \max_{1\le j\le m} |R_{nj}| = o_P(1)$ as observed earlier. □

The distributions of $\boldsymbol{U}^T\boldsymbol{U}$ and $\boldsymbol{V}^T\boldsymbol{V}$ are obtained by using the following two lemmas.

Lemma 7.4.3. *If* $\boldsymbol{Y} \sim N_m(\boldsymbol{0}, \boldsymbol{C})$, *then* $\boldsymbol{Y}^T\boldsymbol{Y} \stackrel{\mathcal{D}}{=} \sum_{j=1}^{m} \lambda_j\xi_j^2$ *where* $\lambda_1, \ldots, \lambda_m$ *are the eigenvalues of* $\boldsymbol{C}$ *and* $\xi_1, \ldots, \xi_m$ *are iid* $N(0, 1)$.

Proof. Let $\boldsymbol{A} = [\boldsymbol{a}_1, \ldots, \boldsymbol{a}_m]$ and $\boldsymbol{\Lambda} = \text{diag}((\lambda_j))$, where $\boldsymbol{a}_1, \ldots, \boldsymbol{a}_m$ are orthonormal eigenvectors and $\lambda_1, \ldots, \lambda_m$ the corresponding eigenvalues of $\boldsymbol{C}$. Then $\boldsymbol{A}^T\boldsymbol{A} = \boldsymbol{A}\boldsymbol{A}^T = \boldsymbol{I}$ and $\boldsymbol{C} = \boldsymbol{A}\boldsymbol{\Lambda}\boldsymbol{A}^T$. Let $\boldsymbol{W} = \boldsymbol{A}^T\boldsymbol{Y}$. Then $\boldsymbol{W}^T\boldsymbol{W} = \boldsymbol{Y}^T\boldsymbol{Y}$ and $\boldsymbol{W} \sim N_m(\boldsymbol{0}, \boldsymbol{A}^T(\boldsymbol{A}\boldsymbol{\Lambda}\boldsymbol{A}^T)\boldsymbol{A}) = N_m(\boldsymbol{0}, \boldsymbol{\Lambda})$, so that

$$Y^TY = W^TW = \sum_{j=1}^{m} W_j^2 = \sum_{j=1}^{m} \lambda_j\left(W_j/\sqrt{\lambda_j}\right)^2 = \sum_{j=1}^{m} \lambda_j \xi_j^2,$$

where $\xi_1, \ldots, \xi_m$ are iid $N(0, 1)$. □

Lemma 7.4.4. *(i) The matrix $I - qq^T$ has $m-1$ eigenvalues equal to 1 and one eigenvalue equal to 0. (ii) The matrix $\boldsymbol{I} - \boldsymbol{qq}^T - \boldsymbol{D}^T(\boldsymbol{DD}^T)^{-1}\boldsymbol{D}$ has $m-k-1$ eigenvalues equal to 1 and $k+1$ eigenvalues equal to 0.*

Proof. First note that $\lambda_1, \ldots, \lambda_m$ are the eigenvalues of $\boldsymbol{C}$ iff $1-\lambda_1, \ldots, 1-\lambda_m$ are the eigenvalues of $\boldsymbol{I} - \boldsymbol{C}$, because $(\boldsymbol{I} - \boldsymbol{C})\boldsymbol{a} = (1-\lambda)\boldsymbol{a} \iff \boldsymbol{Ca} = \lambda\boldsymbol{a}$. Therefore, we need to consider eigenvalues of $\boldsymbol{qq}^T$ and $\boldsymbol{qq}^T + \boldsymbol{D}^T(\boldsymbol{DD}^T)^{-1}\boldsymbol{D}$. First, $(\boldsymbol{qq}^T)\boldsymbol{q} = \boldsymbol{q}(\boldsymbol{q}^T\boldsymbol{q}) = \boldsymbol{q}$ (since $\boldsymbol{q}^T\boldsymbol{q} = \sum_{j=1}^m \pi_{j0} = 1$), so 1 is an eigenvalue of $\boldsymbol{qq}^T$ with eigenvector $\boldsymbol{q}$. Now let $\boldsymbol{a}_1, \ldots, \boldsymbol{a}_{m-1}$ be the other $m-1$ eigenvectors of $\boldsymbol{qq}^T$. Then $\boldsymbol{q}^T\boldsymbol{a}_i = 0$, $i = 1, \ldots, m-1$, so $(\boldsymbol{qq}^T)\boldsymbol{a}_i = \boldsymbol{q}(\boldsymbol{q}^T\boldsymbol{a}_i) = 0$ for all i, showing that the other $m-1$ eigenvalues of $\boldsymbol{qq}^T$ are 0. This proves (i). Similarly, we can verify that $\boldsymbol{q}$ and the k row vectors of $\boldsymbol{D}$ are the eigenvectors of $\boldsymbol{qq}^T + \boldsymbol{D}^T(\boldsymbol{DD}^T)^{-1}\boldsymbol{D}$ with corresponding eigenvalues all equal to 1 and other eigenvalues all equal to 0. This proves (ii). □

Theorem 7.4.1. *Under H_0, $T_{0n} \xrightarrow{\mathcal{L}} \boldsymbol{U}^T\boldsymbol{U} \sim \chi^2_{m-1}$ and under H_1, $T_{1n} \xrightarrow{\mathcal{L}} \boldsymbol{V}^T\boldsymbol{V} \sim \chi^2_{m-k-1}$.*

Proof. Use Lemmas 7.4.1 and 7.4.4. □

Example 7.4.1. The number of flight cancelations between 6 am and noon at a certain airport was recorded for each day between the months of March and June. The data are summarized below with k = number of cancelations and n_k = number of days with k cancelations between 6 am and noon.

k	0	1	2	3	4	Total
n_k	35	55	22	8	2	$n = 122$

Test at level $\alpha = 0.05$ whether the number of flight cancelations follows a Poisson distribution.

Solution. Under the null hypothesis H_0 that the number of cancelations follows a Poison distribution with an unspecified mean θ, the MLE of θ is

$$\hat{\theta}_n = \sum_k kn_k/n = 1.074 \text{ and } e^{-\hat{\theta}_n} = 0.3417.$$

The expected frequencies under H_0 with $Poisson(\hat{\theta}_n)$ are given below, together with the observed frequencies. Since the frequency χ^2 test is a large sample test and in the data, the frequency $n_k = 2$ for $k = 4$ is too small, we have pooled the classes $k = 3$ and $k = 4$. (As a rule of thumb, we need the expected frequency in each class to be at least 5.)

k	0	1	2	≥ 3	Total
Observed frequency: n_k	35	55	22	10	122
Expected frequency: $nf(k, \hat{\theta}_n)$	41.69	44.77	24.04	11.52	122

From the observed and expected frequencies, the test statistic is obtained as

$$\chi^2 = \sum \frac{(\text{observed} - \text{expected})^2}{\text{expected}} = 1.07 + 2.34 + 0.18 + 0.19 = 3.78.$$

Under H_0, the test statistic is (asymptotically) distributed as χ^2 with

$$\text{df} = \text{number of classes} - \text{number of parameters estimated} - 1 = 4 - 1 - 1 = 2.$$

From the table of χ^2 distribution we have the critical value $\chi^2_{2,0.05} = 5.99$ (ie, $P[\chi^2_2 > 5.99] = 0.05$). Since the observed value of χ^2 is smaller than the critical value, we accept H_0 at level of significance $\alpha = 0.05$. In other words, at level $\alpha = 0.05$, the data indicate that the Poisson model is an acceptable fit.

Remark 7.4.1. Strictly speaking, having pooled some classes, one should calculate the MLE of the unspecified parameter(s) from the data after pooling. Since the calculation of MLE from pooled data is complicated, this issue is overlooked in practice. However, the use of MLE of θ from original data in calculating the Frequency χ^2 statistic from pooled data results in the asymptotic distribution of the test statistic to be stochastically larger than a χ^2 with prescribed distribution. This error is not serious in fitting a Poisson distribution, but may be so in fitting a normal distribution based on frequencies in class intervals and using MLEs of μ and σ^2 from raw data (see [32]).

Example 7.4.2 (Test for Independence in a Contingency Table). Let $(A_1, \ldots, A_k)$ and $(B_1, \ldots, B_m)$ be two classifications (into mutually exclusive and exhaustive categories) of a population with $P(A_i) = \pi_{i0} > 0$, $\sum_{i=1}^{k} \pi_{i0} = 1$ and $P(B_j) = \pi_{0j} > 0$, $\sum_{j=1}^{m} \pi_{0j} = 1$. Also let $P(A_i \cap B_j) = \pi_{ij}$. In a random sample of N observations from this population, the frequency distribution over such a cross-classification is called a contingency table having the following layout in which the frequency of A_iB_j is n_{ij}.

Contingency Table

	B_1		B_j		B_m	Subtotal
A_1	n_{11}	$\cdots$	n_{1j}	$\cdots$	n_{1m}	n_{10}
	$\vdots$		$\vdots$		$\vdots$	$\vdots$
A_i	n_{i1}	$\cdots$	n_{ij}	$\cdots$	n_{im}	n_{i0}
	$\vdots$		$\vdots$		$\vdots$	$\vdots$
A_k	n_{k1}	$\cdots$	n_{kj}	$\cdots$	n_{km}	n_{k0}
Subtotal	n_{01}	$\cdots$	n_{0j}	$\cdots$	n_{0m}	N

In this model, we need to test H_0: $\pi_{ij} = \pi_{i0}\pi_{0j}$ for all (i,j) (ie, the classifications $\{A_i\}$ and $\{B_j\}$ are independent), against all possible departures from H_0.

Solution. In this cross-classification, there are altogether km classes with observed frequencies n_{ij} and expected frequencies $N\pi_{i0}\pi_{0j}$ under H_0. Since H_0 involves unknown parameters π_{i0}, $i = 1,\ldots,k$, and π_{0j}, $j = 1,\ldots,m$, we use their MLEs $\hat{\pi}_{i0} = n_{i0}/N$ and $\hat{\pi}_{0j} = n_{0j}/N$. However, due to the constraints $\sum_{i=1}^{k}\pi_{i0} = 1 = \sum_{j=1}^{m}\pi_{0j}$, only $(k-1)+(m-1)$ parameters are actually estimated to find the expected frequencies $N\hat{\pi}_{i0}\hat{\pi}_{0j} = n_{i0}n_{0j}/N$. Thus the frequency χ^2 test statistic is

$$\chi^2 = \sum_{i=}^{k}\sum_{j=1}^{m}\frac{[n_{ij} - n_{i0}n_{0j}/N]^2}{n_{i0}n_{0j}/N},$$

which is asymptotically distributed under H_0 as a χ^2 with

$$\begin{aligned} \text{df} &= \text{number of classes} - \text{number of parameters estimated} - 1 \\ &= km - \{(k-1)+(m-1)\} - 1 = (k-1)(m-1). \end{aligned}$$

Note that conditionally, given $(n_{10},\ldots,n_{k0})$, the problem is the same as the one discussed in Example 7.2.2, where the LRT statistic is the same as the frequency χ^2 test statistics here, with the same asymptotic null distribution.

Exercises

7.1. Let $X_1,\ldots,X_n$ be iid with pdf $f(x,\theta) = \theta x^{-\theta-1}I_{(1,\infty)}(x), \theta > 2$.
   (a) Find the method of moments estimator $\tilde{\theta}_n$ of θ.
   (b) Find the MLE $\hat{\theta}_n$ of θ.
   (c) Find the asymptotic distributions of $\sqrt{n}(\tilde{\theta}_n - \theta)$ and $\sqrt{n}(\hat{\theta}_n - \theta)$.

7.2. Repeat Exercise 7.1 for the pdf $f(x,\theta) = \theta(\theta+1)x^{\theta-1}(1-x)I_{(0,1)}(x), \theta > 0$.

7.3. Let $X_1,\ldots,X_n$ be iid with pdf

$$f(x,\theta) = \theta I_{[0,1/3)}(x) + 2\theta I_{[1/3,2/3)}(x) + 3(1-\theta)I_{[2/3,1]}(x), \quad 0 < \theta < 1.$$

   (a) Find the MLE $\hat{\theta}_n$ of θ and its asymptotic distribution. Is $\hat{\theta}_n$ unbiased?
   (b) Is there a UMVUE of θ? If so, find it.

7.4. Let $X_1,\ldots,X_n$ be iid with pdf $f(x,\theta) = \exp[-(x-\theta)]I_{[\theta,\infty)}(x)$. Find the MLE $\hat{\theta}_n$ of θ and the asymptotic distribution of $\hat{\theta}_n$ after appropriate normalization.

7.5. Let $(X_1,\ldots,X_n)$ be a random sample from a log normal distribution with pdf

$$f(x,\theta) = \frac{1}{\sqrt{2\pi}x}\exp[-(\log x - \theta)^2/2]I_{(0,\infty)}(x).$$

   This means, $Y_i = \log X_i$, $i = 1,\ldots,n$, are iid $N(\theta,1)$ so that $(X_1,\ldots,X_n) \stackrel{\mathcal{D}}{=} (e^{Z_1+\theta},\ldots,e^{Z_n+\theta})$, where $Z_1,\ldots,Z_n$ are iid $N(0,1)$.
   (a) Find the method of moments estimator $\tilde{\theta}_n$ and the MLE $\hat{\theta}_n$ of θ based on $X_1,\ldots,X_n$.

(b) Find the asymptotic distributions of $\sqrt{n}(\tilde{\theta}_n - \theta)$ and $\sqrt{n}(\hat{\theta}_n - \theta)$.
(c) What is the asymptotic efficiency of $\tilde{\theta}_n$?

7.6. Suppose that we observe $U_{ij} = I_{(-\infty, a_i]}(X_{ij}), j = 1, \ldots, n_i$ and $i = 1, \ldots, k$, where $a_1 < \cdots < a_k$ are known and X_{ij} are independent $N(\mu, \sigma^2)$. Let $\boldsymbol{\theta}^{\mathsf{T}} = (\mu, \sigma^2)$.
(a) Let $\hat{\boldsymbol{\theta}}_n$ be the MLE $\boldsymbol{\theta}$ based on the observations $\{U_{ij}\}$. Find the asymptotic distribution of $\sqrt{n}(\hat{\boldsymbol{\theta}}_n - \boldsymbol{\theta})$ where $n = n_1 + \cdots + n_k \to \infty$ in such a way that $n_i/n \to \lambda_i \in (0, 1)$.
(b) How would you calculate $\hat{\boldsymbol{\theta}}_n$ or an estimator asymptotically equivalent to $\hat{\boldsymbol{\theta}}_n$? [Let $\hat{p}_i = \sum_{j=1}^{n_i} U_{ij}/n_i$. Then $\sqrt{n}(\hat{p}_i - \pi_i) = O_P(1)$, where $\pi_i = \Phi((a_i - \mu)/\sigma)$.]

7.7. A discrete distribution with pmf

$$f(x, \theta) = \frac{\exp(-\theta)}{1 - \exp(-\theta)} \frac{\theta^x}{x!}, \quad x = 1, 2, \ldots$$

is called a truncated Poisson distribution with parameter θ. Consider $\tilde{\theta}_n = 2\nu_2/\nu_1$, where ν_1 and ν_2 are observed frequencies of $X = 1$ and $X = 2$ in a random sample of n observations from $f(x, \theta)$ as an initial estimator θ and construct a BAN estimator $\hat{\theta}_n$ of θ. Find the asymptotic distribution of $\hat{\theta}_n$.

7.8. Consider the multinomial distribution $Multi(4, \boldsymbol{\pi}(\boldsymbol{\theta}))$, where $\boldsymbol{\theta} = (p, q)$, $\pi_O(\boldsymbol{\theta}) = r^2$, $\pi_A(\boldsymbol{\theta}) = p^2 + 2pr$, $\pi_B(\boldsymbol{\theta}) = q^2 + 2qr$, $\pi_{AB}(\boldsymbol{\theta}) = 2pq$, $p > 0$, $q > 0$, and $r = 1 - p - q > 0$. Suppose that in a random sample of size n, the cell frequencies are n_O, n_A, n_B, and n_{AB}. Set up the formulas for computing a BAN estimator of $\boldsymbol{\theta}$.

7.9. Let $X_1, \ldots, X_n$ be iid with pdf

$$f(x, \theta) = (1 - \mathrm{e}^{-c/\theta})^{-1}(1/\theta)\mathrm{e}^{-x/\theta} I_{(0,c]}(x), \quad \text{with a known } c,$$

and let $\hat{\theta}_n$ be the MLE of θ based on $X_1, \ldots, X_n$. Since the likelihood equation does not have a closed form solution here, we try to obtain a BAN estimator, starting with an initial estimator $\tilde{\theta}_{n0}$. For this let

$$\tilde{\theta}_{n0} = \frac{c}{2\log(p_n/q_n)}, \quad \text{where } p_n = n^{-1}\sum_{i=1}^{n} I_{(0,c/2]}(X_i) \text{ and } q_n = 1 - p_n.$$

Show that $\tilde{\theta}_{n0} = \theta + o_P(1)$ and find the BAN estimator using $\tilde{\theta}_{n0}$. [*Hint*: Let $\phi = \mathrm{e}^{-c/(2\theta)}$. Then p_n is a sample proportion, estimating $p = (1 - \phi)/(1 - \phi^2) = 1/(1 + \phi)$.]

7.10. (Mixture of distributions.) Suppose that U is a *Bernoulli*$(p(\theta))$ rv and conditionally, given U, X is distributed with pdf $f_U(x, \theta)$. Then the joint distribution of (U, X) is

$$g(u, x, \theta) = \{p(\theta)\}^u\{1 - p(\theta)\}^{1-u} f_u(x, \theta), \quad u = 0 \text{ or } 1,\ x \in R,\ \theta \in \Theta \subset R,$$

where $f_0(x, \theta)$ and $f_1(x, \theta)$ are pdf's/pmf's on R. Assume that the usual regularity conditions hold for f_0, f_1 and assume differentiability conditions on $p(\theta)$ as needed. Let $(U_1, X_1), \ldots, (U_n, X_n)$ be iid, as (U, X) and let $\hat{\theta}_n$ denote the MLE of θ

based on $(U_i, X_i), i = 1, \ldots, n$. Allow for the possibility for f_0 being discrete and f_1 continuous.

(a) Show that

$$\mathrm{E}_\theta\left[\frac{\partial \log g(U, X; \theta)}{\partial \theta}\right] = 0, \text{ and}$$

$$\begin{aligned} I(\theta) &= \mathrm{E}_\theta\left[-\frac{\partial^2 \log g(U, X, \theta)}{\partial \theta^2}\right] = \mathrm{E}_\theta\left[\left(\frac{\partial \log g(U, X, \theta)}{\partial \theta}\right)^2\right] \\ &= \mathrm{Var}_\theta\left(\frac{\partial \log g(U, X, \theta)}{\partial \theta}\right) \\ &= \frac{\{p'(\theta)\}^2}{p(\theta)(1 - p(\theta))} + p(\theta) I_0(\theta) + (1 - p(\theta)) I_1(\theta), \quad \text{where} \\ I_u(\theta) &= \mathrm{E}_\theta\left(\frac{\partial \log f_u(X, \theta)}{\partial \theta}\right)^2, \quad u = 0, 1. \end{aligned}$$

(b) How would you calculate the MLE $\hat{\theta}_n$ of θ and what is the asymptotic distribution of $\hat{\theta}_n$?

(c) Let $T_1, \ldots, T_n$ be iid exponential rv's which are right-censored at c, giving rise to observations $X_i = T_i I_{(0,c)}(T_i) + c I_{[c,\infty)}(T_i)$. Here $U_i = I_{(0,c)}(T_i)$ and conditionally on U_i, X_i is degenerate at $\{c\}$ if $U_i = 0$ and had pdf $(1/\theta)\mathrm{e}^{-x/\theta}(1 - \mathrm{e}^{-c/\theta})$ on $(0, c)$ if $U_i = 1$. Discuss the maximum likelihood estimation of θ from such data.

7.11. Let $X_1, \ldots, X_n$ be iid with pdf/pmf $f(x, \theta)$ and let $\hat{\theta}_n$ denote the MLE of θ based on $X_i, i = 1, \ldots, n$. Under regularity conditions in Section 7.1.1 show that the remainder terms in the expansion

$$0 = n^{-1/2} \sum_{i=1}^{n} \dot{l}(X_i, \theta) + \sqrt{n}(\hat{\theta}_n - \theta)\left[n^{-1} \sum_{i=1}^{n} \ddot{l}(X_i, \theta) + n^{-1} \sum_{i=1}^{n} R_n(X_i)\right]$$

satisfies $n^{-1} \sum_{i=1}^{n} R_n(X_i) = o_P(1)$. Extend this result to the case when $\boldsymbol{\theta}$ is k-dim.

7.12. Let $(X_1, \ldots, X_m)$ and $(Y_1, \ldots, Y_n)$ be independent samples from $N(\mu_1, \sigma^2)$ and $N(\mu_2, \sigma^2)$, respectively. Let $m, n \to \infty$ so that $m/(m + n) \to \alpha \in (0, 1)$.

(a) Find the MLE of (μ_1, μ_2, σ^2) and its asymptotic distribution.

(b) Find the MLE of (μ_1, μ_2, σ^2) under the restriction $\mu_1 = \mu_2$.

(c) Derive the LRT statistic $\Lambda_{m,n}$ for H_0: $\mu_1 = \mu_2$ vs H_1: $\mu_1 \neq \mu_2$, reduce it to a suitable form and find the asymptotic distribution of $-2 \log \Lambda_{m,n}$ under H_0,

(i) using the general properties of *LRT* statistics, and

(ii) from elementary considerations.

7.13. Suppose that in Section 7.1.1, the regularity condition

$$\sup_{|\theta - \theta_0| \leq \varepsilon} |\ddot{l}(x, \theta) - \ddot{l}(x, \theta_0)| \leq H(x, \theta_0)\phi(\varepsilon) \text{ with}$$

$$\lim_{\varepsilon \to 0} \phi(\varepsilon) = 0 \text{ and } \mathrm{E}[H(X, \theta_0)] < \infty$$

is strengthened by requiring

$$|\ddot{l}(x,\theta)| \le H(x) \text{ with } \mathrm{E}[H(X)] < \infty.$$

Show that the BAN property of

$$\tilde{\theta}_{n1} = \tilde{\theta}_{n0} - \sum \dot{l}(X_i, \tilde{\theta}_{n0}) / \sum \ddot{l}(X_i, \tilde{\theta}_{n0})$$

then holds if $n^{1/4}(\tilde{\theta}_{n0} - \theta_0) = o_P(1)$.

7.14. Let $(X_1, \ldots, X_n)$ be a random sample from a distribution with pdf/pmf $f(x, \boldsymbol{\theta})$ where $\boldsymbol{\theta} = (\theta_1, \ldots, \theta_k)^\mathsf{T} \in \Theta$ which is an open interval in $\mathbb{R}^k$, and $\boldsymbol{\theta}_0 = (\theta_{01}, \ldots, \theta_{k0})^\mathsf{T}$ is an interior point of Θ. Suppose that the family $\{f(x, \boldsymbol{\theta}) \colon \boldsymbol{\theta} \in \Theta\}$ satisfies the usual regularity conditions. We want to test $H_0 \colon \boldsymbol{\theta} = \boldsymbol{\theta}_0$ vs $H_1 \colon \boldsymbol{\theta} \in \Theta - \{\boldsymbol{\theta}_0\}$. Show that the following test statistics are equivalent (ie, differ from one another by $o_P(1)$) under H_0:

(a) Likelihood ratio statistic: $T_{1n} = -2 \log \Lambda_n$, where

$$\Lambda_n = \prod_{i=1}^{n} f(X_i, \boldsymbol{\theta}_0) \Bigg/ \left\{ \sup_{\boldsymbol{\theta} \in \Theta} \prod_{i=1}^{n} f(X_i, \boldsymbol{\theta}) \right\},$$

(b) Wald's statistics: $T_{2n} = \boldsymbol{D}_n^\mathsf{T} I(\hat{\boldsymbol{\theta}}_n) \boldsymbol{D}_n$, where $\boldsymbol{I}(\hat{\boldsymbol{\theta}}_n)$ is the Fisher-information matrix evaluated at the MLE $\hat{\boldsymbol{\theta}}_n$ of $\boldsymbol{\theta}$ and $\boldsymbol{D}_n = \sqrt{n}(\hat{\boldsymbol{\theta}}_n - \boldsymbol{\theta}_0)$,

(c) Rao's statistic: $T_{3n} = \boldsymbol{V}_n^\mathsf{T} \boldsymbol{I}(\boldsymbol{\theta}_0)^{-1} \boldsymbol{V}_n$, where $\boldsymbol{V}_n = n^{-1/2} \sum \dot{\boldsymbol{l}}(X_i, \boldsymbol{\theta}_0)$.

7.15. Let $(X_{i1}, \ldots, X_{in_i})$ be independent outcomes of Bernoulli trials in which X_{ij} takes values 1 or 0 with probabilities θ_i and $1 - \theta_i$, respectively, $i = 1, \ldots, 4$. We want to test $H_0 \colon \theta_{i+2}/\theta_{i+1} = \theta_{i+1}/\theta_i$, $i = 1, 2$, against the alternative $H_1 \colon \theta_{i+2}/\theta_{i+1} \neq \theta_{i+1}/\theta_i$ for at least one i. In this situation, computation of the LRT statistic $-2 \log \Lambda_n$ becomes messy. The following approach based on Wald's statistic involves only the unrestricted MLEs $\hat{\theta}_i = S_i/n_i$ of θ_i where $S_i = \sum_{1 \le j \le n_i} X_{ij}$, but leads to a test criterion which is asymptotically equivalent to $-2 \log \Lambda_n$ under H_0 (see Exercise 7.14 above). For the asymptotics, let $n_1, \ldots, n_4 \to \infty$ in such a way that $n_i / \sum n_k \to c_i > 0, i = 1, \ldots, 4$. Since $\theta_2^2 - \theta_1\theta_3 = \theta_3^2 - \theta_2\theta_4 = 0$ under H_0, large values of $|\hat{\theta}_2^2 - \hat{\theta}_1\hat{\theta}_3|$ and $|\hat{\theta}_3^2 - \hat{\theta}_2\hat{\theta}_4|$ would indicate departure from H_0.

(a) Show that under H_0, $h(\hat{\boldsymbol{\theta}})^T = \sqrt{n}(\hat{\theta}_2^2 - \hat{\theta}_1\hat{\theta}_3, \hat{\theta}_3^2 - \hat{\theta}_2\hat{\theta}_4)$ is asymptotically bivariate normal with mean vector **0** and find the covariance matrix $\Sigma(\boldsymbol{\theta})$ of this limiting distribution.

(b) Find the asymptotic distribution of $T_n = h(\hat{\boldsymbol{\theta}})^T \Sigma(\hat{\boldsymbol{\theta}})^{-1} h(\hat{\boldsymbol{\theta}})$ under H_0 and justify your answer.

(c) Explain how you would find the critical value of a test for H_0 based on T_n at level α.

7.16. Let $(X_{i1}, \ldots, X_{in_i})$ denote independent samples from $Poi(\theta_i), i = 1, \ldots, k$. We want to test $H_0 \colon \theta_1 = \cdots = \theta_k$ against all possible alternatives in $(0, \infty)^k$. Let $T_i = \sum_{1 \le j \le n_i} X_{ij}$, $\bar{T}_i = T_i/n_i$, $n = \sum n_i$, $T = \sum T_i$, and $\bar{T} = T/n$, then $(\bar{T}_1, \ldots, \bar{T}_k)$ and $(\bar{T}, \ldots, \bar{T})$ are, respectively, the unrestricted MLE and the restricted MLE of

$(\theta_1, \ldots, \theta_k)$ under H_0. [$T = 0$ causes some difficulties which we shall ignore for large sample purposes.]

(a) Show that the conditional distribution of $(T_1, \ldots, T_k)$ given T is $Multi(T, n_1/n, \ldots, n_k/n)$.

(b) Justify the use of $W_n = \sum[(T_i - n_i\bar{T})^2/(n_i\bar{T})]$ as a test statistic for testing H_0. What should be the rejection region of such a test at level α when n is large?

(c) Let Λ_n denote the LRT statistic for H_0. Show that

$$-2\log\Lambda_n = 2\sum n_i\bar{T}_i \log(\bar{T}_i/\bar{T}).$$

What is the asymptotic distribution of $-2\log\Lambda_n$ under H_0?

(d) Show that under H_0, $-2\log\Lambda_n = W_n + o_P(1)$. [Notice the similarity with Example 7.2.1 at the end of Section 7.2.]

7.17. The number of flies (X_1, X_2, X_3) in three categories, resulting from certain crossings is distributed as $Multi(n, \boldsymbol{\pi})$. According to the Hardy-Weinberg formula, the probabilities of this multinomial distribution are

$$\pi_1(\theta) = (1-\theta)^2,\ \pi_2(\theta) = 2\theta(1-\theta),\ \pi_3(\theta) = \theta^2,$$

for some $0 < \theta < 1$. In an experiment, the observed frequencies in the three categories are $x_1 = 45$, $x_2 = 58$, and $x_3 = 22$, in a random sample of size $n = 125$. Test whether the data support the above model at a level of significance $\alpha = 0.05$.

8

Distribution-Free Tests for Hypothesis Testing in Nonparametric Families

8.1 Ranks and Order Statistics

We start with the general case in which $\boldsymbol{X} = (X_1, \ldots, X_n)^T$ is a random vector with the joint pdf $f(\boldsymbol{x})$ on $\mathbb{R}^n$, and in this setting derive the joint distribution of the vector of order statistics $\boldsymbol{X}_{(n)} = (X_{n:1}, \ldots, X_{n:n})^T$ where $X_{n:1} < \cdots < X_{n:n}$ and the vector of ranks $\boldsymbol{R}_n = (R_{n:1}, \ldots, R_{n:n})^T$ where $R_{n:i} = 1 + \sum_{j \neq i=1}^{n} I_{(0,\infty)}(X_i - X_j)$ is the rank of X_i among $X_1, \ldots, X_n$.

Theorem 8.1.1. *If $(X_1, \ldots, X_n)$ has joint pdf f on $\mathbb{R}^n$, then*

(i) *the pdf of $\boldsymbol{X}_{(n)}$ is*

$$\bar{f}(y_1, \ldots, y_n) = \sum_{\boldsymbol{r}} f(y_{r_1}, \ldots, y_{r_n}), \quad y_1 < \cdots < y_n,$$

where the sum is over all $n!$ permutations $\boldsymbol{r} = (r_1, \ldots, r_n)^T$ of $(1, \ldots, n)$, and

(ii) $P[\boldsymbol{R} = \boldsymbol{r} | \boldsymbol{X}_{(n)} = \boldsymbol{y}] = f(y_{r_1}, \ldots, y_{r_n}) / \bar{f}(y_1, \ldots, y_n)$.

Proof. For $A \subset \{\boldsymbol{y}: y_1 < \cdots < y_n\}$,

$$P[\boldsymbol{X}_{(n)} \in A] = \int_{\boldsymbol{x}_{(n)} \in A} f(\boldsymbol{x})\, d\boldsymbol{x} = \sum_{\boldsymbol{r}} \int_{r(x) = \boldsymbol{r}, \boldsymbol{x}_{(n)} \in A} f(\boldsymbol{x})\, d\boldsymbol{x},$$

where it is understood that $\int$ in the above expressions refer to an n-dimensional integral and $d\boldsymbol{x} = dx_1 \cdots dx_n$. On $\{\boldsymbol{x}: r(\boldsymbol{x}) = \boldsymbol{r}\}$, transform $(y_1, \ldots, y_n) = (x_{n:1}, \ldots, x_{n:n})$ which is one-to-one with Jacobian equal to 1, and $r(\boldsymbol{x}) = \boldsymbol{r} \iff y_{r_i} = x_{n:r_i} = x_i$. Thus

$$P[\boldsymbol{X}_{(n)} \in A] = \sum_{r} \int_A f(y_{r_1}, \ldots, y_{r_n})\, d\boldsymbol{y} = \int_A \left\{ \sum_{r} f(y_{r_1}, \ldots, y_{r_n}) \right\} d\boldsymbol{y},$$

which proves (i). Moreover,

$$P[\boldsymbol{R} = \boldsymbol{r}, \boldsymbol{X}_{(n)} \in A] = \int_{r(x) = \boldsymbol{r}, \boldsymbol{x}_{(n)} \in A} f(\boldsymbol{x})\, d\boldsymbol{x} = \int_A f(y_{r_1}, \ldots, y_{r_n})\, d\boldsymbol{y}$$
$$= \int_A \frac{f(y_{r_1}, \ldots, y_{r_n})}{\bar{f}(y_1, \ldots, y_n)} \bar{f}(y_1, \ldots, y_n)\, d\boldsymbol{y},$$

proving (ii). □

Theory and Methods of Statistics. http://dx.doi.org/10.1016/B978-0-12-802440-9.00008-4

Corollary 8.1.1. *If* $f(y_{r_1},\ldots,y_{r_n}) = f(y_1,\ldots,y_r)$ *for all r, then*

(i) *the pdf of* $\mathbf{X}_{(n)}$ *equals* $n!f(y_1,\ldots,y_n)$, $y_1 < \cdots < y_n$,
(ii) $P[\mathbf{R} = \mathbf{r}|\mathbf{X}_{(n)} = \mathbf{y}] = 1/n!$ *for all* $\mathbf{r}$ *and* $\mathbf{y}$, *so* $P[\mathbf{R} = \mathbf{r}] = 1/n!$ *for all* $\mathbf{r}$, *and* $\mathbf{X}_{(n)}$ *and* $\mathbf{R}$ *are mutually independent.*

In particular, if $X_1,\ldots,X_n$ *are iid with common pdf* f, *then the pdf of* $\mathbf{X}_{(n)}$ *is* $n!\prod_{i=1}^n f(y_i)$, $y_1 < \cdots < y_n$ *as already proved in Section 1.11.*

For the next result, restrict to $X_1,\ldots,X_n$ which are iid with common pdf f. With slight abuse of notation, we shall write the joint pdf of $(X_1,\ldots,X_n)$ as $f(\mathbf{x}) = \prod_{i=1}^n f(x_i)$.

Suppose that the common pdf of f of $X_1,\ldots,X_n$ is symmetric (ie, $f(-x) = f(x)$ for all x). Consider

$$\mathbf{S} = (sign(X_1),\ldots,sign(X_n))^T,\ |\mathbf{X}| = (|X_1|,\ldots,|X_n|)^T, \text{ and let}$$
$$|\mathbf{X}|_{(n)} = (|X|_{n:1},\ldots,|X|_{n:n})^T$$

denote the vector of order statistics and $\mathbf{R}_n^+ = (R_{n:1}^+,\ldots,R_{n:n}^+)$ denote the ranks of $|X_i|$ among $|X_1|,\ldots,|X_n|$. Then

(i) for each i, $|X_i|$ and $sign(X_i)$ are independent,

$$P[S_i = 1] = P[S_i = -1] = 1/2,$$

and $|X_i|$ has pdf $2f(x)$, $x > 0$.
(ii) since $X_1,\ldots,X_n$ are independent, $S_1,\ldots,S_n$ and $|X_1|,\ldots,|X_n|$ are all mutually independent,
(iii) also, the ranks $\mathbf{R}_n^+$ and the order statistics $|\mathbf{X}|_{(n)}$ are independent, having distributions obtained in the Corollary of Theorem 8.1.1.

We thus have

Theorem 8.1.2. *If* $X_1,\ldots,X_n$ *are iid with common pdf* f *which is symmetric about* 0, *then*

(i) *the vectors* $\mathbf{S}$, $\mathbf{R}_n^+$, *and* $|\mathbf{X}|_{(n)}$ *are mutually independent,*
(ii) $P[\mathbf{R}_n^+ = \mathbf{r}] = 1/n!$ *for all* $\mathbf{r}$,
(iii) $|\mathbf{X}|_{(n)}$ *has joint pdf* $n!2^n\prod_{i=1}^n f(y_i)$, $0 < y_1 < \cdots < y_n$,
(iv) $P[\mathbf{S} = \mathbf{s}] = 1/2^n$ *for all* $\mathbf{s} = (\pm 1,\ldots,\pm 1)$.

The following rank-related lemma is for future use.

Lemma 8.1.1. *Let* $t(X_1,\ldots,X_n)$ *be a function of iid rv's* $X_1,\ldots,X_n$, *and suppose* $t(\mathbf{X})$ *has finite expectation. Then*

$$\begin{aligned} E[t(X_1,\ldots,X_n)|\mathbf{R} = \mathbf{r}] &= E[t(X_{n:r_1},\ldots,X_{n:r_n})|\mathbf{R}_n = \mathbf{r}] \\ &= E[t(X_{n:r_1},\ldots,X_{n:r_n})], \end{aligned}$$

because $\mathbf{X}_{(n)}$ *and* $\mathbf{R}_n$ *are independent.*

Application: Permutation Test

Let $\mathcal{P}_0$ be a family of pdf's on $\mathbb{R}^n$ defined by

$$\mathcal{P}_0 = \left\{ p\colon p(x_1, \ldots, x_N) = \prod_{i=1}^{N} f(x_i) \quad \text{where } f \text{ is an arbitrary pdf on } \mathbb{R} \right\}.$$

On the basis of a random sample from a distribution with pdf p, we want to test H_0: $p \in \mathcal{P}_0$ vs H_1: $p = q \notin \mathcal{P}_0$ for a specified q.

Since $(X_1, \ldots, X_N)$ is in one-to-one correspondence with $(\boldsymbol{X}_{(N)}, \boldsymbol{R}_N)$, the vectors of order statistics and ranks, we consider tests $\Psi(\boldsymbol{X}_{(N)}, \boldsymbol{R}_{(N)})$.

From Chapter 6, recall that a test Ψ is a similar region test of size α for H_0: $p \in \mathcal{P}_0$ if

$$\mathrm{E}_p[\Psi(\boldsymbol{X}_N, \boldsymbol{R}_N)] = \alpha \text{ for all } p \in \mathcal{P}_0.$$

Since the conditional distribution of $(X_1, \ldots, X_N)$ given $\boldsymbol{X}_{(N)}$ is uniformly distributed over the set of $N!$ permutations of $(X_1, \ldots, X_N)$ irrespective of $p \in \mathcal{P}_0, \boldsymbol{X}_{(N)}$ is sufficient for $p \in \mathcal{P}_0$ in $(X_1, \ldots, X_N)$. It can also be shown that $\boldsymbol{X}_{(N)}$ is complete, that is, $\mathrm{E}_p[g(\boldsymbol{X}_{(N)})] = 0$ for all $p \in \mathcal{P}_0$ implies $g(\boldsymbol{x}_{(N)}) = 0$, a.s. $\mathcal{P}_0$. By virtue of complete sufficiency of $\boldsymbol{X}_{(N)}$, the similar region property of Ψ holds iff Ψ has Neyman-structure with respect to $\boldsymbol{X}_{(N)}$ by Theorem 6.9.1, ie, for almost all $\boldsymbol{x}_{(N)}$,

$$\mathrm{E}_{H_0}[\Psi(\boldsymbol{X}_{(N)}, \boldsymbol{R}_N) | \boldsymbol{X}_{(N)} = \boldsymbol{x}_{(N)}] = \alpha.$$

Moreover, since $\boldsymbol{X}_{(N)}$ and $\boldsymbol{R}_N$ are independent under H_0, the last expression can be rewritten as

$$(1/N!) \sum_{\boldsymbol{r}} \Psi(\boldsymbol{x}_{(N)}, \boldsymbol{r}) = \alpha \text{ for all almost } \boldsymbol{x}_{(N)}, \tag{1}$$

and the problem of finding the most powerful test at level α for H_0: $p \in \mathcal{P}_0$ vs H_1: $p = q \notin \mathcal{P}_0$; that is, the problem of maximizing $\mathrm{E}_q[\Psi(\boldsymbol{X}_{(N)}, \boldsymbol{R}_N)]$ subject to Eq. (1) is solved by maximizing

$$\mathrm{E}_q[\Psi(\boldsymbol{X}_{(N)}, \boldsymbol{R}_N) | \boldsymbol{X}_{(N)} = \boldsymbol{x}_{(N)}] = \mathrm{E}_q[\Psi(\boldsymbol{x}_{(N)}, \boldsymbol{R}_N) | \boldsymbol{X}_{(N)} = \boldsymbol{x}_{(N)}]$$

subject to Eq. (1) for each $\boldsymbol{x}_{(N)}$.

The optimal Ψ is obtained by using the N-P Lemma conditionally, given $\boldsymbol{X}_{(N)} = \boldsymbol{x}_{(N)}$. By Theorem 8.1.1 and its Corollary, the conditional likelihood ratio

$$P_q[\boldsymbol{R}_N = \boldsymbol{r}_N | \boldsymbol{X}_{(N)} = \boldsymbol{x}_{(N)}] / P_p[\boldsymbol{R}_N = \boldsymbol{r}_N | \boldsymbol{X}_{(N)} = \boldsymbol{x}_{(N)}]$$

can be equivalently expressed as

$$N! q(x_{N:r_1}, \ldots, x_{N:r_N}) / q(\boldsymbol{x}_{(N)}).$$

The optimal Ψ is therefore given by

$$\begin{gathered} \Psi(\boldsymbol{x}_{(N)}, \boldsymbol{r}_N) = 0, \text{ or } \gamma(\boldsymbol{x}_{(N)}), \text{ or } 1, \text{ according as} \\ q(x_{N:r_1}, \ldots, x_{N:r_N}) <, \text{ or } =, \text{ or } > k(\boldsymbol{x}_{(N)}), \end{gathered}$$

where $k(\boldsymbol{x}_{(N)})$ and $0 \le \gamma(\boldsymbol{x}_{(N)}) \le 1$ are determined by the size α condition.

To find $k(\boldsymbol{x}_{(N)})$ and $\gamma(\boldsymbol{x}_{(N)})$ for a given $\boldsymbol{x}_{(N)}$, we arrange the $N!$ values of

$$q_0(\boldsymbol{x}_{(N)}, \boldsymbol{r}_N) = q(x_{N:r_1}, \ldots, x_{N:r_N})$$

for all permutations $\boldsymbol{r}_N$ of $(1, \ldots, N)$ in ascending order of magnitude and let $\boldsymbol{r}_N^*$ be such that

$$\nu_N = \#\{\boldsymbol{r}_N : q_0(\boldsymbol{x}_{(N)}, \boldsymbol{r}_N) \geq q_0(\boldsymbol{x}_{(N)}, \boldsymbol{r}_N^*)\} \geq N!\alpha, \text{ but}$$
$$\nu_N^- = \#\{\boldsymbol{r}_N : q_0(\boldsymbol{x}_{(N)}, \boldsymbol{r}_N) > q_0(\boldsymbol{x}_{(N)}, \boldsymbol{r}_N^*)\} < N!\alpha.$$

Then

$$k(\boldsymbol{x}_{(N)}) = q_0(\boldsymbol{x}_{(N)}, \boldsymbol{r}_N^*) \text{ and } \gamma(\boldsymbol{x}_{(N)}) = \left(N!\alpha - \nu_N^-\right)/\left(\nu_N - \nu_N^-\right).$$

Example 8.1.1. For $N = 4$, let $q(x_1, x_2, x_3, x_4) = f(x_1)f(x_2)f(x_3)g(x_4)$, where $f(x) = 1$, $0 \leq x \leq 1$ and $g(x) = 2x$, $0 \leq x \leq 1$, and let p denote the unknown joint pdf of $(X_1, \ldots, X_4)$. We want to test H_0: $p \in \mathcal{P}_0$ vs H_1: $p = q$ at level $\alpha = 0.1$ based on observations $(x_1, \ldots, x_4) = (0.2, 0.7, 0.4, 0.9)$.

Solution. The vectors of ranks and order statistics in the observed data are $\boldsymbol{r}_4 = (1, 3, 2, 4)$ and $\boldsymbol{x}_{(4)} = (0.2, 0.4, 0.7, 0.9)$. The values of $q(x_{4:r_1}, \ldots, x_{4:r_4})$ are

$$4x_3x_4 = 4(0.4)(0.9) = 1.44 \text{ for } \boldsymbol{r}_4 = (1,2,3,4), (1,2,4,3), (2,1,3,4), (2,1,4,3),$$
$$4x_1x_2 = 4(0.2)(0.7) = 0.56 \text{ for } \boldsymbol{r}_4 = (3,4,1,2), (3,4,2,1), (4,3,1,2), (4,3,2,1),$$
$$4x_2x_4 = 4(0.7)(0.9) = 2.52 \text{ for } \boldsymbol{r}_4 = (1,3,2,4), (1,3,4,2), (3,1,2,4), (3,1,4,2),$$
$$4x_1x_3 = 4(0.2)(0.4) = 0.32 \text{ for } \boldsymbol{r}_4 = (2,4,1,3), (2,4,3,1), (4,2,1,3), (4,2,3,1),$$
$$4x_2x_3 = 4(0.7)(0.4) = 1.12 \text{ for } \boldsymbol{r}_4 = (1,4,2,3), (1,4,3,2), (4,1,2,3), (4,1,3,2),$$
$$4x_1x_4 = 4(0.2)(0.9) = 0.72 \text{ for } \boldsymbol{r}_4 = (2,3,1,4), (2,3,4,1), (3,2,1,4), (3,2,4,1).$$

These values of $q(\cdot)$ are arranged as

$$0.32 < 0.56 < 0.72 < 1.12 < 1.44 < 2.52,$$

each value repeated four times, and $N!\alpha = (24)(0.1) = 2.40$.

Hence

$$q_0(\boldsymbol{x}_{(4)}, \boldsymbol{r}_4^*) = 2.52, \ \nu_4 = 4, \text{ and } \nu_4^- = 0, \text{ so}$$
$$k_n = 2.52 \text{ and } \gamma = 2.40/4 = 0.60.$$

For the observed data, $\boldsymbol{r}_4 = (1, 3, 2, 4)$ and $q_0 = 2.52$, so we randomize and reject H_0 with probability $\gamma = 0.60$.

Remark 8.1.1. The main problem in implementing a permutation test is that we cannot use standard tables. The critical value of $q_0(\boldsymbol{x}_{(N)}, \boldsymbol{r}_N)$ must be determined in every instance by the observed data.

8.1.1 Nonparametric Tests in Three Basic Problems

1. Test for symmetry in the one-sample problem.
   Let $X_1, \ldots, X_n$ be iid with common continuous cdf F (Unknown). We want to test H_0: $F(-x) = 1 - F(x)$ for all x (ie, the distribution is symmetric about zero), vs H_1 the

distribution is not symmetric about zero, or vs H_+: $F(-x) \le 1 - F(x)$ for all x with strict inequality for some $x\left(\text{or } \int_{-\infty}^{\infty}[1-F(-x)]\, dF(x) > 1/2\right)$.

Let $S_i = sign(X_i)$ and $R^+_{n:i}$ = rank of $|X_i|$ among $|X_1|, \ldots, |X_n|$.

Under H_0, high ranks and low ranks among $\{R^+_{n:i}\}$ will be equally associated with $S_i = 1$ or -1, but under H_+, the observations with $S_i = 1$ will tend to have higher $R^+_{n:i}$ and those with $S_i = -1$ will tend to have lower $R^+_{n:i}$. This leads to the consideration of the Wilcoxon signed-rank statistic:

$$T_n = \{n(n-1)\}^{-1}\sum_{i=1}^{n} S_i R^+_{n:i} \text{ or equivalently, } \sum_{\{i:\, X_i>0\}} R^+_{n:i} = (n+T_n)/2.$$

We reject H_0 in favor of H_+ (or H_1) if T_n (or $|T_n|$) $\ge$ critical value.

2. Test for homogeneity in the two-sample problem.
Let $X_1, \ldots, X_m$ be iid with common continuous cdf F and let $Y_1, \ldots, Y_n$ be iid with common continuous cdf G, the two samples being mutually independent. We want to test

$$H_0 : F = G \text{ vs } H_1: P_{F,G}\left[X_i > Y_j\right] = \int_{-\infty}^{\infty} G(x)\, dF(x) \ne 1/2, \text{ or}$$

$$H_+ : P_{F,G}\left[X_i > Y_j\right] > 1/2.$$

For notational convenience, write the combined sample as

$$X_1, \ldots, X_m, Y_1 = X_{m+1}, \ldots, Y_n = X_{m+n}.$$

The average ranks of the X-observations and the Y-observations in the combined sample are $m^{-1}\sum_{i=1}^{m} R_{m+n:i}$ and $n^{-1}\sum_{i=m+1}^{m+n} R_{m+n:i}$, respectively. Since the average ranks of the two samples would tend to be equal under H_0, the difference between the average ranks, known as the Wilcoxon two-sample rank-sum statistic defined as

$$W_{m,n} = m^{-1}\sum_{i=1}^{m} R_{m+n:i} - n^{-1}\sum_{i=m+1}^{m+n} R_{m+n:i}, \text{ or equivalently}$$

$$W'_{m,n} = \sum_{i=1}^{m} R_{m+n:i} \text{ (which is a linear function of } W_{m,n})$$

is an indicator of departure from H_0. Moreover, by algebraic rearrangement, we can write

$$W_{m,n} = (m+n)\left(T_{m,n} - 1/2\right), \text{ where } T_{m,n} = (mn)^{-1}\sum_{i=1}^{m}\sum_{j=1}^{n} I_{(0,\infty)}\left(X_i - Y_j\right)$$

which is called the Mann-Whitney statistic. Thus the three statistics $W_{m,n}$, $W'_{m,n}$, and $T_{m,n}$ are linear functions of one another and any of them can be used as a test statistic for testing H_0 vs H_1 or H_0 vs H_+. In particular, with the Mann-Whitney statistic, we can reject H_0 if favor of H_+ (or H_1) if $T_{m,n}$ (or $|T_{m,n}|$) $\ge$ critical value.

3. Test for independence in a bivariate distribution.

Let $(X_1, Y_1), \ldots, (X_n, Y_n)$ be iid as (X, Y). We want to test H_0: X and Y are independent vs H_1: X and Y are dependent, or H_+: X and Y are positively dependent. Let U, V, Z be independent rv's with unknown pdf's f, g, h. Then the alternative hypothesis can be formulated as $(X, Y) = (U + \Delta Z, V + \Delta Z)$, and then take H_1: $\Delta \neq 0$ and H_+: $\Delta > 0$.

Let $R_{n:i}$ = rank of X_i among $X_1, \ldots, X_n$ and $R'_{n:i}$ = rank of Y_i among $Y_1, \ldots, Y_n$. Then replacing $(X_1, Y_1), \ldots, (X_n, Y_n)$ by their ranks $\left(R_{n:1}, R'_{n:1}\right), \ldots, \left(R_{n:n}, R'_{n:n}\right)$, the Spearman's rank correlation defined as

$$\begin{aligned}\rho_S &= \text{correlation coefficient between } \left\{\left(R_{n:i}, R'_{n:i}\right),\ i = 1, \ldots, n\right\} \\ &= 12\left\{n\left(n^2 - 1\right)\right\}^{-1} \sum_{i=1}^{n} R_{n:i}R'_{n:i} - 3(n+1)/(n-1)\end{aligned}$$

(using $\bar{R}_n = \bar{R}'_n = (n+1)/2$ and $\sum_{i=1}^{n}\left(R_{n:i} - \bar{R}_n\right)^2 = \sum_{i=1}^{n}\left(R'_{n:i} - \bar{R}'_n\right)^2 = n\left(n^2 - 1\right)/12$) is an obvious candidate for testing H_0 vs H_1 or H_+. One can use either ρ_S or $\sum_{i=1}^{n} R_{n:i}R'_{n:i}$ as a test statistic.

Another test statistic for this problem is based on the simple idea that if we compare (X_i, Y_i) with (X_j, Y_j), then $X_i - X_j$ being positive or negative is independent of $Y_i - Y_j$ being positive or negative if X, Y are independent. An overall comparison between all pairs of data-points leads to

$$\tau_n = [n(n-1)]^{-1} \sum_{i=1}^{n} \sum_{j \neq i, j=1}^{n} sign\left(X_i - X_j\right) sign\left(Y_i - Y_j\right),$$

known as Kendall's tau statistic. We reject H_0 in favor of H_+ (or H_1) if τ_n (or $|\tau_n|$) $\geq$ critical value.

The statistic ρ_S differs from a linear function of τ_n by $o_P(1)$ as $n \to \infty$ under H_0, so a test based on ρ_S is carried out analogously (see [4]).

8.1.2 Exact Distribution of Rank Statistic Under H_0

Exact distributions of rank statistics under H_0 can be obtained by combinatorial arguments, using the properties of ranks under H_0. We illustrate this with the example of Mann-Whitney statistic in the two-sample problem.

Under H_0, the rv's $X_1, \ldots, X_m$, $Y_1, \ldots, Y_n$ are iid. Let $\nu_{m,n}(u)$ be the number of arrangements of m X's and n Y's in which exactly u pairs have $X > Y$ (ie, $I_{(0,\infty)}(X - Y) = u$). Since these arrangements are equally likely under H_0, we have

$$P_{H_0}\left[\sum_{i=1}^{m}\sum_{j=1}^{n} I_{(0,\infty)}\left(X_i - Y_j\right) = u\right] = \nu_{m,n}(u) \Big/ \binom{m+n}{m}.$$

The main thing is to find a formula for $\nu_{m,n}(u)$, which is obtained from the difference equation

$$\nu_{m,n}(u) = \nu_{m,n-1}(u) + \nu_{m-1,n}(u-n),$$

where $\nu_{m,n-1}(u)$ corresponds to sequences ending with Y and $\nu_{m-1,n}(u-n)$ corresponds to sequences ending in X. Using this difference equation with initial conditions

$$\nu_{m,0}(u) = \begin{cases} 1 & u=0 \\ 0 & u \neq 0 \end{cases} \text{ and } \nu_{0,n}(u) = \begin{cases} 1 & u=0 \\ 0 & u \neq 0 \end{cases},$$

we can compute $\nu_{m,n}(u)$ recursively.

Tables of exact tail probabilities are available for this and many other rank statistics for small to moderate sample sizes.

8.1.3 Asymptotic Distribution of Rank Statistics Under H_0 by U-Statistic Approach

One-sample U-statistic. Let $X_1, \ldots, X_n$ be iid with a common continuous cdf F. Many one-sample rank statistics are of the form

$$U_n = \left(1/n^{(r)}\right) \sum_{n,r} g(X_{i_1}, \ldots, X_{i_r}),$$

where $\sum_{n,r}$ denotes sum over all distinct $i_1, \ldots, i_r$ in $\{1, \ldots, n\}$ and $n^{(r)} = n(n-1) \cdots (n-r+1)$.

Assume without loss of generality that g is symmetric in its coordinates. [If not, replace g by g^* obtained by averaging over all permutations of its coordinates, which leaves U_n unchanged.] A statistic of this form is called U-statistic [33].

Suppose that $\mathrm{E}_F\left[g^2(X_1, \ldots, X_r)\right] < \infty$, and let

$$\begin{aligned} &\theta = \theta(F) = \mathrm{E}_F\left[g(X_1, \ldots, X_r)\right],\ h(X_1, \ldots, X_r) = g(X_1, \ldots, X_r) - \theta \text{ and} \\ &h_c(X_1, \ldots, X_c) = \mathrm{E}_F\left[h(X_1, \ldots, X_c, X_{c+1}, \ldots, X_r) | X_1, \ldots, X_c\right] \text{ for } c = 1, \ldots, r, \text{ and} \\ &\xi_c = \xi_c(F) = \mathrm{E}_F\left[h^2(X_1, \ldots, X_c)\right]. \end{aligned}$$

Then the mean and variance of U_n are

$$\begin{aligned} &\mathrm{E}[U_n] = \theta \text{ and } \mathrm{Var}[U_n] = \left(n^{(r)}\right)^{-2} \sum_{c=1}^{r} N_c \xi_c, \text{ where} \\ &N_n = \#\{(i_1, \ldots, i_r), (j_1, \ldots, j_r) \text{ with exactly } c \text{ elements in common}\} \\ &= \binom{n}{2r-c} \frac{(2r-c)!}{c!(r-c)!(r-c)!} (r!)^2 = \binom{n}{r}\binom{r}{c}\binom{n-r}{r-c}(r!)^2. \end{aligned}$$

Suppose that $\xi_1 > 0$. Since $N_{c+1}/N_c \to 0$ as $n \to \infty$, $\lim_{n\to\infty} n\mathrm{Var}[U_n] = r^2 \xi_1 > 0$.

Actually, U_n is asymptotically normal.

Theorem 8.1.3. *If* $\xi_1 > 0$, *then* $\sqrt{n}(U_n - \theta) \xrightarrow{\mathcal{L}} Z \sim N(0, r^2\xi_1)$.

Postponing the proof of Theorem 8.1.3, we look at *Two-sample U-statistics.* Let $\boldsymbol{X} = (X_1, \ldots, X_m)$ and $\boldsymbol{Y} = (Y_1, \ldots, Y_n)$ be mutually independent random samples from distributions with cdf's F and G, respectively, and define

$$U_{m,n} = \left\{m^{(r)}n^{(s)}\right\}^{-1}\sum_{m,r}\sum_{n,s} g\left(X_{i_1}, \ldots, X_{i_r}, Y_{j_1}, \ldots, Y_{j_s}\right), \text{ where}$$

$$g\left(x_{i_1}, \ldots, x_{i_r}, y_{j_1}, \ldots, y_{j_s}\right) = g(x_1, \ldots, x_r, y_1, \ldots, y_s)$$

for all permutations $(i_1, \ldots, i_r)$ of $(1, \ldots, r)$ and $(j_1, \ldots, j_s)$ of $(1, \ldots, s)$.

Suppose that $\mathrm{E}\left[g^2(X_1, \ldots, X_r, Y_1, \ldots, Y_s)\right] < \infty$ and define

$$\theta = \theta(F, G) = \mathrm{E}_{F,G}[g(\boldsymbol{X}, \boldsymbol{Y})], \; h(\boldsymbol{X}, \boldsymbol{Y}) = g(\boldsymbol{X}, \boldsymbol{Y}) - \theta$$

and for $c = 0, 1, \ldots, r$ and $d = 0, 1, \ldots, s$, let

$$\xi_{c,d} = \mathrm{E}\left[h^2_{c,d}(X_1, \ldots, X_c, Y_1, \ldots, Y_d)\right],$$

where

$$h_{c,d}(x_1, \ldots, x_c, y_1, \ldots, y_d) = \mathrm{E}[h(\boldsymbol{X}, \boldsymbol{Y})|X_1, \ldots, X_c, Y_1, \ldots, Y_d]$$

if $\min(c, d) \geq 1$ and $h_{0,0} = 0$. The mean and variance of $U_{m,n}$ are

$$\mathrm{E}[U_{m,n}] = \theta, \text{ and}$$

$$\begin{aligned}\mathrm{Var}[U_{m,n}] &= \left\{\binom{m}{r}\binom{n}{s}\right\}^{-2}\sum_{c=0}^{r}\sum_{d=0}^{s}\binom{r}{c}\binom{m-r}{r-c}\binom{s}{d}\binom{n-s}{s-d}\xi_{c,d}\\ &= \left(r^2/m\right)\xi_{1,0} + \left(s^2/n\right)\xi_{0,1} + o(1/\min(m, n)) \text{ as } m, n \to \infty.\end{aligned}$$

Thus if $m, n \to \infty$ in such a way that $m/(m + n) \to \lambda \in (0, 1)$, then

$$\lim_{m,n\to\infty}(m + n)\mathrm{Var}[U_{m,n}] = r^2\xi_{1,0}/\lambda + s^2\xi_{0,1}/(1 - \lambda),$$

and $U_{m,n}$ is asymptotically normal.

Theorem 8.1.4. *Under the above conditions,*

$$\sqrt{m+n}\left(U_{m,n} - \theta\right) \xrightarrow{\mathcal{L}} Z \sim N\left(0, r^2\xi_{1,0}/\lambda + s^2\xi_{0,1}/(1 - \lambda)\right).$$

The proofs of Theorems 8.1.3 and 8.1.4 are accomplished by the *Hájek Projection Method* of approximating an arbitrary function of independent rv's by a sum of functions of individual rv's.

Let $Z_1, \ldots, Z_n$ be independent rv's. Consider all rv's $\varphi(Z_1, \ldots, Z_n)$ with $\mathrm{E}\left[\varphi^2(\boldsymbol{Z})\right] < \infty$ as a vector space $\mathcal{J}$ with inner product $\langle T_1, T_2\rangle = \mathrm{Cov}[T_1, T_2]$. In $\mathcal{J}$, let $\mathcal{J}_0$ denote the subspace of all rv's of the form $S = \sum_{i=1}^{n}\psi_i(Z_i)$. We want to approximate an arbitrary $T \in \mathcal{J}$ with $\mathrm{E}[T] = 0$ by an rv $S \in \mathcal{J}_0$ with the smallest $\mathrm{E}\left[(T - S)^2\right] = \|T - S\|^2$ (ie, the projection of T on $\mathcal{J}_0$).

Theorem 8.1.5 (Hájek Projection Theorem). *If* $E[T] = 0$ *and* $E\left[T^2\right] < \infty$, *then the projection of* T *on* $\mathcal{J}_0$ *is* $T^* = \sum_{i=1}^{n} E[T|Z_i]$ *w.p. 1.*

Proof. Let $\psi_i^*(Z_i) = \mathrm{E}[T|Z_i]$ and consider an arbitrary $S = \sum_{i=1}^n \psi_i(Z_i) \in \mathcal{J}_0$. To prove the theorem, it is enough to show that $\mathrm{E}\big[(T-T^*)(T^*-S)\big] = 0$, because then we would have

$$\mathrm{E}\Big[(T-S)^2\Big] = \mathrm{E}\Big[(T-T^*)^2\Big] + \mathrm{E}\Big[(T^*-S)^2\Big] \geq \mathrm{E}\Big[(T-T^*)^2\Big], \tag{2}$$

with equality holding iff $S = T^*$ w.p. 1. Now

$$\begin{aligned}
\mathrm{E}[(S-T^*)(T-T^*)] &= \sum_{i=1}^n \mathrm{EE}\left[\{\psi_i(Z_i) - \psi_i^*(Z_i)\}\left\{T - \psi_i^*(Z_i) - \sum_{j\neq i, j=1}^n \psi_j^*(Z_j)\right\}\middle| Z_i\right] \\
&= \sum_{i=1}^n \mathrm{E}\left[\{\psi_i(Z_i) - \psi_i^*(Z_i)\}\left\{\mathrm{E}(T|Z_i) - \psi_i^*(Z_i) - \sum_{j\neq i, j=1}^n \mathrm{E}\big(\psi_j^*(Z_j)\big)\right\}\right] \\
&= \sum_{i=1}^n \mathrm{E}[\{\psi_i(Z_i) - \psi_i^*(Z_i)\}\{\psi_i^*(Z_i) - \psi_i^*(Z_i) - 0\}] = 0,
\end{aligned}$$

since Z_j is independent of Z_i for $j \neq i$, and

$$\mathrm{E}\Big[\psi_j^*(Z_j)\Big] = \mathrm{EE}\Big[T|Z_j\Big] = \mathrm{E}[T] = 0. \qquad \square$$

Corollary 8.1.2.

(a) *Taking* $S = 0$ *in Eq. (2), we have*

$E\Big[(T-T^*)^2\Big] = E\Big[T^2\Big] - E\Big[T^{*2}\Big]$, *ie,* $Var\big[T - T^*\big] = Var[T] - Var\big[T^*\big]$,

(b) *if* $E[T] = \mu$, *then* $T' = T - \mu$ *has mean 0 and the projection of* T' *on* $\mathcal{J}_0$ *is* $\sum_{i=1}^n E\big[T'|Z_i\big] = \sum_{i=1}^n E[T|Z_i] - n\mu$. *Hence the projection of* $T = T' + \mu$ *on* $\mathcal{J}_0$ *is* $T^* = \sum_{i=1}^n E[T|Z_i] - (n-1)E[T]$.

Proofs of Theorems 8.1.3 and 8.1.4.

We shall find the Hajek projections of the U-statistics and then apply the CLT. In the one-sample problem, the Hájek projection of $T_n = \sqrt{n}(U_n - \theta)$ is

$$\begin{aligned}
T_n^* &= \sum_{i=1}^n \mathrm{E}[T_n|X_i] = \sqrt{n}\Big\{n^{(r)}\Big\}^{-1} \sum_{i=1}^n \mathrm{E}\left[\sum_{n,r} h(X_{i_1}, \ldots, X_{i_r})|X_i\right] \\
&= rn^{-1/2} \sum_{i=1}^n h_1(X_i),
\end{aligned}$$

because for every $(i_1, \ldots, i_r)$ in the sum $\sum_{n,r}$,

$$\mathrm{E}[h(X_{i_1}, \ldots, X_{i_r})|X_i] = \begin{cases} h_1(X_i) & \text{if } i \in \{i_1, \ldots, i_r\} \\ \mathrm{E}[h(X_{i_1}, \ldots, X_{i_r})] = 0 & \text{otherwise,} \end{cases}$$

so that

$$\sum_{n,r} \mathrm{E}[h(X_{i_1},\ldots,X_{i_r})|X_i] = r(n-1)^{(r-1)} h_1(X_i) \text{ and}$$
$$\sqrt{n}(n-1)^{(r-1)}/n^{(r)} = n^{-1/2}.$$

Hence

(i) $T_n^* = rn^{-1/2}\sum_{i=1}^{n} h_1(X_i) \xrightarrow{\mathcal{L}} N\left(0, r^2\xi_1\right)$,

(ii) $\mathrm{Var}\left[T_n^*\right] = r^2\xi_1$,

(iii) $\mathrm{Var}[T_n] = n\mathrm{Var}[U_n] = r^2\xi_1 + o(1)$ as already shown, and

(iv) $\mathrm{E}\left[\left(T_n - T_n^*\right)^2\right] = \mathrm{Var}[T_n] - \mathrm{Var}\left[T_n^*\right] = o(1)$,

so that $T_n = T_n^* + o_P(1)$.

Thus $T_n = \sqrt{n}(U_n - \theta) \xrightarrow{\mathcal{L}} N(0, r^2\xi_1)$ by Slutsky's Theorem, completing the proof of Theorem 8.1.3.

Similarly, in the two-sample problem, the Hájek projection of $T_{m,n} = \sqrt{m+n}\left(U_{m,n} - \theta\right)$ is

$$\begin{aligned} T_{m,n}^* &= \sqrt{m+n}\left[\frac{r(m-1)^{(r-1)}}{m^{(r)}}\sum_{i=1}^{m} h_{1,0}(X_i) + \frac{s(n-1)^{(s-1)}}{n^{(s)}}\sum_{j=1}^{n} h_{0,1}(Y_j)\right] \\ &= \sqrt{m+n}\left[rm^{-1}\sum_{i=1}^{m} h_{1,0}(X_i) + sn^{-1}\sum_{j=1}^{n} h_{0,1}(Y_j)\right]. \end{aligned}$$

Now arguing as in one-sample problem, Theorem 8.1.4 follows. □

Examples

1. The Wilcoxon signed-rank statistic T_n can be written as

$$\begin{aligned} T_n &= \{n(n-1)\}^{-1}\sum_{i=1}^{n} S_i R_{n:i}^{+} = \left(1/n^{(2)}\right)\left[\sum_{i=1}^{n} S_i + \sum_{i \neq j i=1}^{n} S_i I_{(0,\infty)}\left(|X_i| - |X_j|\right)\right] \\ &= \left(1/n^{(2)}\right)\sum_{n,2}(1/2)\left[S_i I_{(0,\infty)}\left(|X_i| - |X_j|\right) + S_j I_{(0,\infty)}\left(|X_j| - |X_i|\right)\right] + O(1/n) \\ &= U_n + O(1/n), \text{ where} \\ U_n &= \left(1/n^{(2)}\right)\sum_{n,2} g(X_{i_1}, X_{i_2}),\ g(x_1, x_2) = I_{(0,\infty)}(x_1 + x_2) - 1/2. \end{aligned}$$

It is now easy to see that for this U statistic,

$$\theta(F) = \mathrm{E}_F\left[I_{(0,\infty)}(X_1 + X_2) - 1/2\right] = \int_{-\infty}^{\infty} [1 - F(-x)]\, dF(x) - 1/2$$

so under H_0: $F(-x) = 1 - F(x)$ for all x, $\theta = 0$ and $h(x_1, x_2) = g(x_1, x_2)$.
The variance of $h_1(X_1) = \mathrm{E}\big[h(X_1, X_2)|X_1\big] = (1/2) - \theta(F) - F(-X_1)$ is

$$\xi_1(F) = \int \left[1/2 - \theta(F) - F(-x)\right]^2 dF(x) > 0$$

iff $F(-X)$ is not constant with probability 1. For continuous F, this holds iff $0 < P_F[X < 0] < 1$. Under this condition

$$\sqrt{n}\left[T_n/\{n(n-1)\} - \theta(F)\right] \xrightarrow{\mathcal{L}} N(0, 4\xi_1(F)).$$

In particular, let the common cdf of $X_1, \ldots, X_n$ be $F(\cdot - \theta)$ where F has a pdf which is symmetric about 0. Then

$$\begin{aligned}\mu(\theta) = \theta(F) &= \int [1 - F(-x-\theta)]\, dF(x-\theta) - 1/2 \\ &= \int F(x + 2\theta)\, dF(x) - 1/2, \text{ and}\end{aligned}$$

$$\sigma^2(\theta) = 4\xi_1(F) = 4\left[\int F^2(x+2\theta)\, dF(x) - \left\{\int F(x+2\theta)\, dF(x)\right\}^2\right].$$

2. The Mann-Whitney statistic $U_{m,n} = (mn)^{-1}\sum_{i=1}^{m}\sum_{j=1}^{n} I_{(0,\infty)}(X_i - Y_j)$ based on independent rv's $(X_1, \ldots, X_m)$, $(Y_1, \ldots, Y_n)$, where the X_i's have a common cdf F and the Y_j's have a common cdf G is a two-sample U-statistic with

$$\theta = \theta(F, G) = P_{F,G}\big[X_i < Y_j\big] = \int F\, dG = 1/2 \text{ if } F = G.$$

It is also easy to check that

$$h_{10}(x) = 1 - G(x) - \mathrm{E}_F\big[1 - G(X)\big],\ h_{01}(y) = F(y) - \mathrm{E}_G[F(Y)],$$

so that $\xi_{10} = \mathrm{Var}_F\big[G(X)\big]$ and $\xi_{01} = \mathrm{Var}_G[F(Y)]$. If $F = G$, then $\xi_{10} = \xi_{01} = 1/12$. Thus under H_0: $F = G$,

$$\sqrt{m+n}\big(U_{m,n} - 1/2\big) \xrightarrow{\mathcal{L}} N\big(0, 1/(12\lambda(1-\lambda))\big),$$

if $m, n \to \infty$ so that $m/(m+n) \to \lambda \in (0, 1)$. On the other hand, if $G(x) = F(x - \theta)$, then $\mu(\theta) = \theta\big(F, G\big)$ and $\sigma^2(\theta) = \lambda^{-1}\xi_{10} + (1-\lambda)^{-1}\xi_{01}$ are given by

$$\mu(\theta) = \int F(x+\theta)\, dF(x) \text{ and}$$

$$\begin{aligned}\sigma^2(\theta) = {}& \lambda^{-1}\left[\int F^2(x-\theta)\, dF(x) - \left\{\int F(x-\theta)\, dF(x)\right\}^2\right] \\ &+ (1-\lambda)^{-1}\left[\int F^2(x+\theta)\, dF(x) - \left\{\int F(x+\theta)\, dF(x)\right\}^2\right].\end{aligned}$$

8.1.4 Asymptotic Comparison of Tests: Pitman's Approach

Let θ be a numerical characteristic of a population (or a combination of populations), for which we want to test H_0: $\theta = 0$ vs H_+: $\theta > 0$. Typically, a test statistic T_n based on a sample of size n would reject H_0 in favor of H_+ at level α if $T_n > c_n(\alpha)$. Here we look into the problem of comparing the asymptotic powers of two sequences of tests based on $\left\{T_n^{(1)}\right\}$ and $\left\{T_n^{(2)}\right\}$ with critical values $\left\{c_n^{(1)}(\alpha)\right\}$ and $\left\{c_n^{(2)}(\alpha)\right\}$, respectively, with same asymptotic Type I error probability at the same alternative. The following approach due to Pitman [See 69] attempts to make such a comparison by means of a measure of asymptotic relative efficiency (ARE) of one-test sequence with respect to the other. Since all reasonable tests are consistent (ie, having power at any $\theta > 0$ tending to 1 as $n \to \infty$ with Type I error probability fixed at $0 < \alpha < 1$), we shall consider asymptotic powers of two-test sequences at alternatives $\{\theta_n\}$ converging to the null hypothesis value $\theta = 0$ at the rate of $1/\sqrt{n}$. The following theorem provides a formula for computing such asymptotic powers.

Theorem 8.1.6 (Pitman). *Suppose that* $\{T_n\}$ *and* $\{c_n\}$ *are such that*

(i) $\lim_{n\to\infty} P_\theta\left[\sqrt{n}(T_n - \mu(\theta))/\sigma(\theta) \le t\right] = \Phi(t)$ *uniformly in a neighborhood of* $\theta = 0$, $\mu(\cdot)$ *is differentiable at 0 and* $\sigma(\cdot)$ *is continuous at 0 with* $\sigma(0) > 0$ *(here* Φ *denotes the cdf of the standard normal),*

(ii) $\lim_{n\to\infty} P_{\theta=0}[T_n \ge c_n] = \alpha$.

Then

$$\lim_{n\to\infty} P_{\theta=\delta/\sqrt{n}}[T_n \ge c_n] = \Phi\left(\Phi^{-1}(\alpha) + \delta\mu'(0)/\sigma(0)\right).$$

Proof. For fixed θ, the convergence in (i) is uniform in t by Polya's Theorem (Chapter 3, Theorem 3.2.5(VIII)), which will be used in the proof. Note that

$$\begin{aligned}\alpha &= \lim_{n\to\infty} P_0\left[\sqrt{n}(T_n - \mu(0))/\sigma(0) \ge \sqrt{n}(c_n - \mu(0))/\sigma(0)\right]\\ &= 1 - \Phi\left(\lim_{n\to\infty} \sqrt{n}(c_n - \mu(0))/\sigma(0)\right) = \Phi\left(-\lim_{n\to\infty} \sqrt{n}(c_n - \mu(0))/\sigma(0)\right).\end{aligned}$$

Hence

$$\lim_{n\to\infty} \sqrt{n}(c_n - \mu(0))/\sigma(0) = -\Phi^{-1}(\alpha).$$

Now

$$\begin{aligned}\lim_{n\to\infty} P_{\delta/\sqrt{n}}[T_n \ge c_n] &= \lim_{n\to\infty} P_{\delta/\sqrt{n}}\left[\sqrt{n}(T_n - \mu(\delta/\sqrt{n}))/\sigma(\delta/\sqrt{n}) \le \sqrt{n}(c_n - \mu(\delta/\sqrt{n}))/\sigma(\delta/\sqrt{n})\right]\\ &= 1 - \Phi\left(\lim_{n\to\infty} \sqrt{n}(c_n - \mu(\delta/\sqrt{n}))/\sigma(\delta/\sqrt{n})\right)\\ &= \Phi\left(-\lim_{n\to\infty} \sqrt{n}(c_n - \mu(\delta/\sqrt{n}))/\sigma(\delta/\sqrt{n})\right).\end{aligned}$$

Finally,

$$\begin{aligned}-\lim_{n\to\infty} \sqrt{n}(c_n - \mu(\delta/\sqrt{n}))/\sigma(\delta/\sqrt{n}) &= -\lim_{n\to\infty} \left\{\sqrt{n}(c_n - \mu(0))/\sigma(0)\right\}\left\{\sigma(0)/\sigma(\delta/\sqrt{n})\right\}\\ &\quad + \lim_{n\to\infty} \left\{(\mu(\delta/\sqrt{n}) - \mu(0))/(\delta/\sqrt{n})\right\}\left\{\delta/\sigma(\delta/\sqrt{n})\right\}\\ &= \Phi^{-1}(\alpha) + \delta\mu'(0)/\sigma(0).\end{aligned}$$

□

In view of this theorem, if $T_{n_i}^{(i)}$, $i = 1, 2$ have the same asymptotic power at $\delta_i/\sqrt{n_i}$, $i = 1, 2$, and if $\delta_1/\sqrt{n_1} = \delta_2/\sqrt{n_2} = \theta_n$ (say), then

$$\lim_{n_i\to\infty} \beta_{T_{n_i}^{(i)}}(\theta_n) = \Phi\left(\Phi^{-1}(\alpha) + \delta_i\mu_i'(0)/\sigma_i(0)\right), \quad i = 1, 2,$$

must be equal, that is

$$\sqrt{n_2/n_1} = \delta_2/\delta_1 = \frac{\mu_1'(0)/\sigma_1(0)}{\mu_2'(0)/\sigma_2(0)}.$$

The ratio n_2/n_1 is a measure of relative efficiency of $\left\{T_n^{(1)}\right\}$ in comparison with $\left\{T_n^{(2)}\right\}$. For this reason, the ratio

$$e_{1,2} = \frac{\{\mu_1'(0)/\sigma_1(0)\}^2}{\{\mu_2'(0)/\sigma_2(0)\}^2}$$

is called the *asymptotic relative efficiency (ARE)* of $\left\{T_n^{(1)}\right\}$ with respect to $\left\{T_n^{(2)}\right\}$. The quantity $\{\mu_i'(0)/\sigma_i(0)\}^2$ is called the *asymptotic efficacy* of the sequence $\left\{T_n^{(i)}\right\}$.

Example 8.1.2. Let F be an unknown cdf with pdf $f = F'$ which is symmetric about 0 and let $G(x, \theta) = F(x - \theta)$. Based on a random sample $(X_1, \ldots, X_n)$ from $G(x, \theta)$, we want to test H_0: $\theta = 0$ vs H_+: $\theta > 0$. Consider three test statistics

$$T_n^{(1)} = \bar{X}_n/s_n,\ T_n^{(2)} = \bar{S}_n, \text{ and } T_n^{(3)} = \{n(n-1)\}^{-1}\sum_{i=1}^{n} S_i R_{n:i}^+,$$

for the t-test, the sign test, and the Wilcoxon singed-rank test. Then all three sequences $\left\{T_n^{(j)} : j = 1, 2, 3\right\}$ satisfy the conditions of Theorem 8.1.6 with

$$\mu_1(\theta) = \theta/\sigma(F),\ \sigma_1^2(\theta) = 1 + \theta^2 h(\theta) \text{ where } h(\theta) = \left\{\mu_4(F) - \sigma^4(F)\right\}/\left\{4\sigma^6(F)\right\} \text{ for } T_n^{(1)},$$

$$\mu_2(\theta) = 1 - 2F(-\theta),\ \sigma_2^2(\theta) = 4F(\theta)\{1 - F(\theta)\} \text{ for } T_n^{(2)}, \text{ and}$$

$$\mu_3(\theta) = \int F(x + 2\theta)\, dF(x) - 1/2,\ \sigma_3^2(\theta) = 4\mathrm{Var}_F[F(X + 2\theta)], \text{ where}$$

$$\sigma^2(F) = \mathrm{E}_F\left[(X - \theta)^2\right] \text{ and } \mu_4(F) = \mathrm{E}_F\left[(X - \theta)^4\right].$$

Thus the asymptotic efficacies of $\left\{T_n^{(1)}\right\}$, $\left\{T_n^{(2)}\right\}$, and $\left\{T_n^{(3)}\right\}$ are

$$\{\mu_1'(0)/\sigma_1(0)\}^2 = 1/\sigma^2(F),\ \{\mu_2'(0)/\sigma_2(0)\}^2 = 4f^2(0), \text{ and}$$

$$\{\mu_3'(0)/\sigma_3(0)\}^2 = 12\left\{\int_{-\infty}^{\infty} f^2(x)\, dx\right\}^2.$$

If f is the pdf of $N(0,1)$, then the AREs of $\left\{T_n^{(2)}\right\}$ and $\left\{T_n^{(3)}\right\}$ with respect to $\left\{T_n^{(1)}\right\}$ are

$$e_{2,1} = 4f^2(0)\sigma^2(F) = 2/\pi \text{ and } e_{3,1} = 12\left\{\int_{-\infty}^{\infty} f^2(x)\,dx\right\}^2 \sigma^2(F) = 3/\pi.$$

An Outline of Contiguity Theory

Let S_N be a test statistic based on observations $X_1,\ldots,X_N$, iid as X distributed with pdf/pmf $f(x,\theta)$ to test H_0: $\theta = 0$ vs H_1: $\theta > 0$. In the next example, we shall illustrate the use of Contiguity Theory in deriving the asymptotic distribution of S_N under $f\left(x,\delta/\sqrt{N}\right)$ from its asymptotic distribution under $f(x,0)$ (see [4, 34]).

For $N = 1,2,\ldots$, let P_N and Q_N be probabilities on $(\mathcal{X}_N,\mathcal{A}_N)$. The sequence $\{Q_N\}$ is said to be contiguous to $\{P_N\}$ if for any sequence $A_N \in \mathcal{A}_N$, $\lim_{N\to\infty} P_N(A_N) = 0$ implies $\lim_{N\to\infty} Q_N(A_N) = 0$. Let L_N denote the likelihood ratio of Q_N to P_N.

For our purpose, $\mathcal{X}_N$ is the sample space of a dataset $(X_1,\ldots,X_N)$, $\mathcal{A}_N$ is a family of events in $\mathcal{X}_N$, and P_N, Q_N are joint distributions of $(X_1,\ldots,X_N)$ under two models such as a null hypothesis and a sequence of alternatives. For each N, let S_N be a test statistic based on $(X_1,\ldots,X_N)$ and suppose that we know the asymptotic distribution of $\{S_N\}$ under $\{P_N\}$. The aim is to find the asymptotic distribution of $\{S_N\}$ under $\{Q_N\}$ from this by using $\{L_N\}$ as a link between $\{P_N\}$ and $\{Q_N\}$. Operationally, this is achieved by finding the asymptotic joint distribution of $\{(S_N,\log L_N)\}$ under $\{P_N\}$ when $\{Q_N\}$ is contiguous to $\{P_N\}$.

The following results due to LeCam (LeCam's First and Third Lemma) provide the main tools:

I. If $\log L_N \overset{\mathcal{L}}{\to} N(-\sigma^2/2,\sigma^2)$ for some $\sigma^2 > 0$ under $\{P_N\}$, then $\{Q_N\}$ is contiguous to $\{P_N\}$.

II. If $\begin{pmatrix} S_N \\ \log L_N \end{pmatrix} \overset{\mathcal{L}}{\to} N_2\left(\begin{bmatrix}\mu_1\\ \mu_2\end{bmatrix},\begin{bmatrix}\sigma_1^2 & \sigma_{12}\\ \sigma_{12} & \sigma_2^2\end{bmatrix}\right)$ with $\mu_2 = -\sigma_2^2/2$ under $\{P_N\}$, then $S_N \overset{\mathcal{L}}{\to} N(\mu_1+\sigma_{12},\sigma_1^2)$ under $\{Q_N\}$.

Example 8.1.3. Let $X_1,\ldots,X_N$ be iid as X distributed with pdf/pmf $f(x;\theta)$ and let P_N, Q_N denote, respectively, the joint distribution of $(X_1,\ldots,X_N)$ under $f(x;0)$ and $f(x;\delta/\sqrt{n})$. The log-likelihood ratio of Q_N to P_N is

$$\log L_N = \sum_{i=1}^{N}\left[\log f\left(X_i;\delta/\sqrt{N}\right) - \log f(X_i;0)\right].$$

Under P_N, subject to regularity conditions,

$$\log L_N = \left(\delta/\sqrt{N}\right)\sum_{i=1}^{N} \dot{l}(X_i;0) - \left(\delta^2/2\right)I(f) + o_P(1), \text{ as } N\to\infty, \text{ where}$$

$$\dot{l}(x;0) = \frac{\partial}{\partial\theta}\log f(x;\theta)\bigg|_{\theta=0} \text{ with } \mathrm{E}_{\theta=0}\left[\dot{l}(X;0)\right] = 0, \text{ and}$$

$$\mathrm{E}_{\theta=0}\left[\dot{l}(X;0)^2\right] = I(f) \text{ (Fisher-information) [see Chapter 5, Section 5.2].}$$

Suppose that $S_N^{(1)}$ and $S_N^{(2)}$ are two statistics based on $(X_1, \ldots, X_N)$, which under P_N, are asymptotically of the form

$$S_N^{(j)} = N^{-1/2}\sum_{i=1}^{N}\psi_j(X_i) + o_P(1),\ j = 1, 2, \text{ as } N \to \infty, \text{ where}$$

$$\mathrm{E}_{\theta=0}\left[\psi_j(X)\right] = 0,\ \mathrm{E}_{\theta=0}\left[\psi_j^2(X)\right] = \sigma_j^2, \text{ and } \delta\mathrm{E}_{\theta=0}\left[\dot{l}(X;0)\psi_j(X)\right] = \mu_j.$$

(a) Find the asymptotic joint distribution of $\left(S_N^{(j)}, \log L_N\right), j = 1, 2$, under P_N. What do these asymptotic distributions imply?

(b) Find the asymptotic distribution of $S_N^{(j)}$ under Q_N, $j = 1, 2$.

(c) Find the Pitman ARE of $S_N^{(1)}$ with respect to $S_N^{(2)}$.

Solution. We shall use the Contiguity Theory.

(a) By the bivariate CLT and Slutsky's Theorem, for $j = 1, 2$, and denoting $\mu_j = \delta\mathrm{E}_{\theta=0}\left[\dot{l}(X;0)\psi_j(X)\right]$,

$$\begin{bmatrix} S_N^{(j)} \\ \log L_N \end{bmatrix} \xrightarrow{\mathcal{L}} N_2\left(\begin{bmatrix} 0 \\ -\left(\delta^2/2\right)I(f) \end{bmatrix}, \begin{bmatrix} \sigma_j^2 & \mu_j \\ \mu_j & \delta^2 I(f) \end{bmatrix}\right), \text{ where}$$

$$\mu_j = \delta\mathrm{E}_{\theta_0}\left[\dot{l}\left(X_j;0\right)\psi_j(X)\right].$$

(b) Since $\log L_N \xrightarrow{\mathcal{L}} N(-(\delta/2)I(f), \delta^2 I(f))$, $\{Q_N\}$ is contiguous to $\{P_N\}$ by LeCam's First Lemma, so LeCam's Third Lemma applies, by which

$$S_N^{(j)} \xrightarrow{\mathcal{L}} N\left(0 + \mu_j, \sigma_j^2\right) = N\left(\delta\mathrm{E}_{\theta=0}\left[\dot{l}(X;0)\psi_j(X)\right], \sigma_j^2\right), \quad j = 1, 2.$$

(c) The Pitman ARE of $S_N^{(2)}$ with respect to $S_N^{(1)}$ is

$$e_{2,1} = \frac{\mathrm{E}_0\left[\dot{l}(X;0)\psi_2(X)\right]/\sigma_2^2}{\mathrm{E}_0\left[\dot{l}(X;0)\psi_1(X)\right]/\sigma_1^2}.$$

8.2 Locally Most Powerful Rank Tests

Tests discussed in the last section were proposed on an ad hoc basis, on intuitive grounds, without any kind of optimality criterion in mind. Although some of these tests have good asymptotic properties under normality in terms of Pitman's ARE, their performances under other models will vary.

It should be mentioned that for data which one suspects to be normally distributed without being sure about it, Fisher and Yates [35] proposed the test statistic

$$\sum_{i=m+1}^{m+n} \mathrm{E}\left[\Phi^{-1}\left(U_{N:R_{N:i}}\right)\right], \quad N = m+n$$

for the two-sample problem, where $U_{N:1} < \cdots < U_{N:N}$ are the order statistics in a random sample of size N from *Unif* $(0, 1)$ and Φ is the cdf of $N(0, 1)$. Further investigations on this and related problems were carried out by Hoeffding [36] and Terry [37].

The material presented in this section is mainly based on the development by Hájek and Šidák [4] aiming at construction of rank tests with a local optimality property for a general class of nonparametric hypotheses.

Definition 8.2.1. Let $\mathcal{P}$ and $\{q_\Delta, \Delta \geq 0\}$ be families of pdf's on $\mathbb{R}^N$, so that $q_0 \in \mathcal{P}$ but $\{q_\Delta, \Delta > 0\}$ is distinct from $\mathcal{P}$, and suppose that p is the joint pdf of rv's $(X_1, \ldots, X_N)$. A rank test Ψ^* is a locally most powerful (LMP) rank test at level α for H_0: $p \in \mathcal{P}$ vs H_1: $p = q_\Delta$, $\Delta > 0$ based on $(X_1, \ldots, X_N)$ if

(i) $\mathrm{E}_p[\Psi^*(\boldsymbol{R}_N)] \leq \alpha$ for all $P \in \mathcal{P}$ where $\boldsymbol{R}_N$ is the vector of ranks of $(X_1, \ldots, X_N)$ and
(ii) there exists $\varepsilon > 0$ such that $\mathrm{E}_p[\Psi^*(\boldsymbol{R}_N)] \geq \mathrm{E}_p[\Psi(\boldsymbol{R}_N)]$ for all Ψ satisfying (i) and for all $p \in \{q_\Delta, 0 < \Delta \leq \varepsilon\}$.

In this section, we shall construct LMP rank tests for several nonparametric hypothesis testing problems. These problems will involve the following families of probabilities

$$\mathcal{P}_0 = \left\{ p\text{: } p(\boldsymbol{x}) = \prod_{i=1}^{N} f(x_i),\ f \text{ is an unknown pdf on } \mathbb{R} \right\},$$

$$\mathcal{P}_1 = \left\{ p\text{: } p(\boldsymbol{x}) = \prod_{i=1}^{N} f(x_i),\ f \text{ is an unknown } \textit{symmetric} \text{ pdf on } \mathbb{R} \right\},$$

$$\mathcal{P}_2 = \left\{ p\text{: } p(\boldsymbol{x}, \boldsymbol{y}) = \prod_{i=1}^{N} f(x_i)g(y_i),\ f \text{ and } g \text{ are unknown pdf's on } \mathbb{R} \right\}.$$

The problems of testing $p \in \mathcal{P}_0$ or $\mathcal{P}_1$ or $\mathcal{P}_2$ vs various alternatives $\{q_\Delta, \Delta > 0\}$ will be taken up in the following.

8.2.1 Testing H_0: $p \in \mathcal{P}_0$ Against a General Alternative

Regularity Conditions A. Let $\{f(x, \theta), \theta \in J\}$ be a family of pdf's on $\mathbb{R}$ where J is an open interval in $\mathbb{R}$ containing 0 and the following hold:

(i) $f(x, \theta)$ is absolutely continuous in θ for almost all x (see Section 2.1 of Chapter I of Hájek and Šidák [4]),

(ii) for each θ in a neighborhood of 0, the partial derivative

$$\dot{f}(x,\theta) = \lim_{h\to 0}\{f(x,\theta+h) - f(x,\theta)\}/h$$

exists for almost all x,

(iii) $\lim_{\theta\to 0}\int_{-\infty}^{\infty}|\dot{f}(x,\theta)|\,dx = \int_{-\infty}^{\infty}|\dot{f}(x,0)|\,dx < \infty$.

Theorem 8.2.1. *Let p denote an unknown joint pdf of* $(X_1,\ldots,X_N)$ *and let* $q_\Delta(x_1,\ldots,x_N) = \prod_{i=1}^{N} f(x_i, c_i\Delta)$, *where the family* $\{f(x,\theta), \theta\in J\}$ *is specified and satisfies the regularity conditions A. Then the LMP rank test at level* α *for* H_0: $p \in \mathcal{P}_0$ *vs* H_1: $p = q_\Delta$, $\Delta > 0$ *is given by*

$$\Psi(\mathbf{r}) = 0, \text{ or } \gamma, \text{ or } 1 \text{ according as}$$

$$\sum_{i=1}^{N} c_i a_N(r_i, f) <, \text{ or } =, \text{ or } > k, \text{ where}$$

$$a_N(i,f) = E_{q_0}\left[\frac{\dot{f}}{f}(X_{N:i}, 0)\right],$$

and $0 \le \gamma \le 1$ *and k are determined by* $E_p[\Psi(\mathbf{R})] = \alpha$ *for all* $p \in \mathcal{P}_0$.

Before going into the proof, we shall look into some special cases of the family $\{f(x,\theta)\}$ which determines the *score function* $a_N(i,f)$.

Location and Scale Families

Definition 8.2.2. The family $\{f(x,\theta) = f(x-\theta), -\infty < \theta < \infty\}$ is called a location family and the family $\{f(x,\theta) = \mathrm{e}^{-\theta} f((x-\mu)\mathrm{e}^{-\theta}), -\infty < \theta < \infty\}$ is called a scale family.

Lemma 8.2.1. *Let f be an absolutely continuous pdf. Then*

(a) *the location family satisfies Condition A if* $\int_{-\infty}^{\infty}|f'(x)|\,dx < \infty$, *and*

(b) *the scale family satisfies Condition A if* $\int_{-\infty}^{\infty}|xf'(x)|\,dx < \infty$.

In the location family with $f(x,\theta) = f(x-\theta)$, we have

$$\dot{f}(x,\theta) = -f'(x-\theta) \text{ and } \frac{\dot{f}}{f}(X_{n:i}, 0) = -\frac{f'}{f}(X_{N:i}).$$

Let $F(x,\theta) = F(x-\theta)$ where F is the cdf corresponding to f. Then

$$F^{-1}(u,\theta) = \theta + F^{-1}(u), \quad 0 < u < 1, \text{ and}$$

$$\frac{\dot{f}}{f}\left(F^{-1}(u,0), 0\right) = -\frac{f'}{f}\left(F^{-1}(u)\right) := \varphi(u,f).$$

Thus $a_N(i,f) = \mathrm{E}[\varphi(U_{N:i}, f)]$ where $U_{N:1} < \cdots < U_{N:N}$ are the order statistics in a random sample $(U_1,\ldots,U_N)$ from $\mathit{Unif}(0,1)$.

In the scale family,

$$\dot{f}(x,\theta) = e^{-\theta} f\left((x-\mu)e^{-\theta}\right) - (x-\mu)e^{-2\theta} f'\left((x-\mu)e^{-\theta}\right)$$
$$= f(x,\theta)\left[-1 - (x-\mu)e^{-\theta}\frac{f'}{f}\left((x-\mu)e^{-\theta}\right)\right], \text{ and}$$
$$F(x,\theta) = \int_{-\infty}^{x} e^{-\theta} f\left((y-\mu)e^{-\theta}\right) dy = \int_{-\infty}^{(x-\mu)e^{-\theta}} f(y)\, dy = F\left((x-\mu)e^{-\theta}\right).$$

Then $F^{-1}(u,\theta) = \mu + e^{\theta}F^{-1}(u)$, $0 < u < 1$, and

$$\frac{\dot{f}}{f}\left(F^{-1}(u,0),0\right) = -1 - F^{-1}(u)\frac{f'}{f}(u) := \varphi_1(u,f).$$

The score function $a_N(i,f)$ in this case is now denoted by

$$a_{1N}(i,f) = \mathrm{E}[\varphi_1(U_{N:i},f)].$$

The above results are summarized below.

Let F denote the cdf corresponding to the pdf f by means of which the location and scale families are defined and let $U_{N:i}$ denote the ith-order statistic in a random sample of sine N from $Unif(0,1)$. Then the score functions in the location and the scale family are, respectively,

$$a_N(i,f) = \mathrm{E}[\varphi(U_{N:i},f)] \text{ where } \varphi(u,f) = -\frac{f'}{f}\left(F^{-1}(u)\right), \tag{3}$$
$$a_{1N}(i,f) = \mathrm{E}[\varphi_1(U_{N:i},f)] \text{ where } \varphi_1(u,f) = -1 - F^{-1}(u)\frac{f'}{f}\left(F^{-1}(u)\right). \tag{4}$$

Next consider the important case when

$$c_1 = \cdots = c_m = 0 \text{ and } c_{m+1} = \cdots = c_N = 1,$$

which corresponds to the *two-sample problem* in which $(X_1,\ldots,X_m)$ and $(X_{m+1},\ldots,X_N)$ are independent random samples from two populations and we want to test where they have the same pdf, or whether their pdf's differ in location (or scale). The test criterion for the LMP rank tests in the two-sample location problem and the two-sample scale problem are, respectively,

$$\sum_{i=m+1}^{N} a_N(r_i,f) \text{ and } \sum_{i=m+1}^{N} a_{1N}(r_i,f),$$

where the score functions are given by Eqs. (3) and (4).

We now give an outline of the proof of Theorem 8.2.1. The proofs of three subsequent theorems dealing with LMP rank tests in which $\mathcal{P}_0$ is replaced by $\mathcal{P}_1$ or $\mathcal{P}_2$ in H_0 and q_Δ have other forms, will be concerned with the likelihood ratio of $\boldsymbol{R}_N$ or $(\boldsymbol{R}_n^+, \boldsymbol{S})$, etc., namely,

$$P_{q_\Delta}[\boldsymbol{R}_N = \boldsymbol{r}]/P_{q_0}[\boldsymbol{R}_N = \boldsymbol{r}], \text{ etc.},$$

which will be written as

$$1 + N![P_{q_\Delta}(\boldsymbol{R}_N = \boldsymbol{r}) - P_{q_0}(\boldsymbol{R}_N = \boldsymbol{r})]$$
$$= 1 + N!\Delta \int_{R_N = r_N} \Delta^{-1} \left\{ \prod_{i=1}^{N} f(x_i, \Delta c_i) - \prod_{i=1}^{N} f(x_i, 0) \right\} d\boldsymbol{x}$$

in the proof of Theorem 8.2.1 and similarly in the other proofs. The integral in the expression above is an N-dimensional integral. The integrand is a difference of two products, which will be simplified by the identity

$$\prod_{i=1}^{N} a_i - \prod_{i=1}^{N} b_i = \sum_{k=1}^{k} (a_k - b_k) \prod_{j=1}^{k-1} a_j \prod_{j=k+1}^{N} b_j \tag{5}$$

resulting in a sum of k integrals. The crucial step is taking $\lim_{\Delta \downarrow 0}$ under the integrals, where the regularity conditions come into play. The proof is then completed by some routine simplifications, using Lemma 8.2.1.

Proof of Theorem 8.2.1. By N-P Lemma, it is enough to show that there exists $\varepsilon > 0$ such that for all $\Delta \in (0, \varepsilon]$, the likelihood ratios

$$P_{q_\Delta}[\boldsymbol{R}_N = \boldsymbol{r}]/P_{q_0}[\boldsymbol{R}_N = \boldsymbol{r}] \quad \text{for all } \boldsymbol{r}$$

are in the same ascending order as the numbers $\sum_{i=1}^{N} c_i a_N(r_i, f)$. Since $P_{q_0}[\boldsymbol{R}_N = \boldsymbol{r}] = 1/N!$ for all $\boldsymbol{r}$, the likelihood ratio equals

$$1 + N! \int_{\boldsymbol{R}_N = \boldsymbol{r}} \left\{ \prod_{i=1}^{N} f(x_i, \Delta c_i) - \prod_{i=1}^{N} f(x_i, 0) \right\} d\boldsymbol{x} = 1 + N!\Delta \sum_{k=1}^{N} c_k I_{Nk}(\boldsymbol{r}, f; \Delta), \text{ where}$$
$$I_{Nk} = \int_{\boldsymbol{R}_N = \boldsymbol{r}} \left\{ \frac{f(x_k, \Delta c_k) - f(x_k, 0)}{\Delta c_k} \right\} \prod_{j=1}^{k-1} f\left(x_j, \Delta c_j\right) \prod_{j=k+1}^{N} f\left(x_j, 0\right) d\boldsymbol{x}, \tag{6}$$

by Eq. (5). Hence the theorem will be proved by showing that

$$\lim_{\Delta \downarrow 0} I_{Nk}(\boldsymbol{r}, f; \Delta) = a_N(r_k, f) = \mathrm{E}_{q_0}\left[\frac{\dot{f}}{f}(X_{N:k}, 0) | \boldsymbol{R}_N = \boldsymbol{r} \right]$$
$$= \mathrm{E}_{q_0}\left[\frac{\dot{f}}{f}(X_{N:k}, 0) \right],$$

using Lemma 8.1.1, and because $N! \prod_{j=1}^{N} f(x_j, 0)$ is the conditional pdf of $\boldsymbol{X}$ given $\boldsymbol{R}_N = \boldsymbol{r}$ under q_0.

The justification for taking $\lim_{\Delta \downarrow 0}$ under the integral is provided by the regularity conditions on $\{f(x, \theta), \theta \in J\}$. We omit the technicalities of this demonstration, referring to Hájek and Šidák [4]. □

8.2.2 One-Sample Location Problem, Assuming Symmetry

Here we shall consider rank-sign tests based on $(\boldsymbol{R}_N^+, \boldsymbol{S})$, the properties of which under symmetry are given in Theorem 8.1.2, and a rank-sign test has been discussed in Section 8.1.1.

Theorem 8.2.2. *Suppose that* $(X_1, \ldots, X_N)$ *have joint pdf* $p(\boldsymbol{x})$ *on* R^N. *Then the LMP rank-sign test at level* α *for* H_0: $p \in \mathcal{P}_1$ *vs* H_1: $p(\boldsymbol{x}) = q_\Delta(\boldsymbol{x}) = \prod_{i=1}^N f(x_i - \Delta)$, $\Delta > 0$, *where* f *is a specified symmetric absolutely continuous pdf on R with* $\int_{-\infty}^{\infty} |f'(x)|\, dx < \infty$, *is given by*

$$\Psi(\boldsymbol{r}, \boldsymbol{s}) = 0, \text{ or } \gamma, \text{ or } 1 \text{ according as}$$

$$\sum_{i=1}^N s_i a_N^+(r_i, f) <, \text{ or } = \text{ or } > k,$$

the constant k *and* $0 \leq \gamma \leq 1$ *being determined by* $E_{H_0}[\Psi(\boldsymbol{R}_N^+, \boldsymbol{S})] = \alpha$ *and the score function is*

$$a_N^+(i, f) = E_{q_0}\left[-\frac{f'}{f}(|X|_{N:i})\right] = E_{q_0}\left[-\frac{f'}{f}\left(F^{-1}(1/2 + (1/2)U_{N:i})\right)\right]$$
$$:= E_{q_0}[\varphi^+(U_{N:i}, f)].$$

Proof. We need to show that the likelihood ratios

$$\left(2^N N!\right) P_{q_\Delta}\left[\left(\boldsymbol{R}_N^+, \boldsymbol{S}\right) = (\boldsymbol{r}, \boldsymbol{s})\right]$$

for the $2^N N!$ different $(\boldsymbol{r}, \boldsymbol{s})$ are ordered in the same way as $\sum_{i=1}^N s_i a_N^+(r_i, f)$ for all sufficiently small Δ. As indicated earlier, we write these likelihood ratios as

$$1 + \left(2^N N!\right) \Delta \sum_{i=1}^N I_{Nk}(\boldsymbol{r}, \boldsymbol{s}, f; \Delta)$$

using Eq. (5) where the integrals I_{Nk} have the same form as in Eq. (6) except that the integration is over the set $\{(\boldsymbol{R}_N^+, \boldsymbol{S}) = (\boldsymbol{r}, \boldsymbol{s})\}$. By Lemma 8.2.1(a), the conditions on f allow taking $\lim_{\Delta \downarrow 0}$ under these integrals, showing that

$$\lim_{\Delta \downarrow 0} I_{Nk}(\boldsymbol{r}, \boldsymbol{s}, f, \Delta) = \int_{(\boldsymbol{R}_N^+, \boldsymbol{S}) = (\boldsymbol{r}, \boldsymbol{s})} \left\{-\frac{f'}{f}(x_k)\right\} f(\boldsymbol{x}) d\boldsymbol{x}.$$

Finally, since f is symmetric,

$$\frac{f'}{f}(x) = sign(x) \frac{f'}{f}(|x|).$$

Hence

$$\lim_{\Delta \downarrow 0} I_{Nk}(\boldsymbol{r}, \boldsymbol{s}, f; \Delta) = s_k \int_{(\boldsymbol{R}_N^+, \boldsymbol{S}) = (\boldsymbol{r}, \boldsymbol{s})} \left\{-\frac{f'}{f}(|x_k|)\right\} f(\boldsymbol{x})\, d\boldsymbol{x}$$
$$= \frac{s_k}{2^N N!} \mathrm{E}_{q_0}\left[-\frac{f'}{f}(|X_k|) | \boldsymbol{R}_N^+ = \boldsymbol{r}, S = s\right]$$

$$= \frac{s_k}{2^N N!} \mathrm{E}_{q_0}\left[-\frac{f'}{f}(|X_k|)|\boldsymbol{R}_N^+ = \boldsymbol{r}\right]$$
$$= \frac{s_k}{2^N N!} \mathrm{E}_{q_0}\left[-\frac{f'}{f}(|X|_{N:r_k})\right] \propto s_k a_N^+(r_k, f)$$

by Lemma 8.1.1 and since $|X_k|$ is independent of $\boldsymbol{S}$ under q_0. This concludes the proof, leaving the verification of $a_N^+(i,f) = \mathrm{E}\left[\varphi^+(U_{N:i}, f)\right]$ as an exercise. □

8.2.3 Two-Sample Scale Problem, Assuming Symmetry

Theorem 8.2.3. *Let $p(\boldsymbol{x})$ be the joint pdf of $(X_1, \ldots, X_N)$. Then the LMP rank test for H_0: $p \in \mathcal{P}_1$ vs H_1: $p(\boldsymbol{x}) = q_\Delta(\boldsymbol{x}) = \prod_{i=1}^m f(x_i, 0) \prod_{i=m+1}^N f(x_i, \Delta)$, where $\{f(x,\theta) = e^{-\theta} f(xe^{-\theta})\}$ is a scale family with a specified symmetric absolutely continuous pdf f on $\mathbb{R}$ having $\int_{-\infty}^{\infty} |xf'(x)|\, dx < \infty$, is given by*

$$\Psi(\boldsymbol{r}) = 0, \text{ or } \gamma, \text{ or } 1 \text{ according as}$$

$$\sum_{i=m+1}^N a_{1N}^+(r_i, f) <, \text{ or } =, \text{ or } > k,$$

the constants k and $0 \leq \gamma \leq 1$ being determined by $E_{H_0}\left[\Psi(\boldsymbol{R}_N^+)\right] = \alpha$ and the score function is

$$\begin{aligned} a_{1N}^+(i,f) &= E_{q_0}\left[-1 - |X|_{N:i}\frac{f'}{f}(|X|_{N:i})\right] \\ &= E\left[-1 - F^{-1}(1/2 + 1/2U_{N:i})\frac{f'}{f}\left(F^{-1}(1/2 + 1/2U_{N:i})\right)\right] \\ &:= E\left[\varphi_1^+(U_{N:i}, f)\right]. \end{aligned}$$

Proof. This is a special case of Theorem 8.2.1 with $c_1 = \cdots = c_m = 0$, $c_{m+1} = \cdots = c_N = 1$, and $f(x,\theta) = \mathrm{e}^{-\theta} f(x\mathrm{e}^{-\theta})$. Lemma 8.2.1(b) and the conditions on f imply Condition A for the validity of Theorem 8.2.1. We have already seen that for this $f(x,\theta)$,

$$\dot{f}(x,\theta)/f(x,\theta) = -1 - x\mathrm{e}^{-\theta}\frac{f'}{f}\left(x\mathrm{e}^{-\theta}\right), \text{ so}$$
$$\frac{\dot{f}}{f}(x,0) = -1 - x\frac{f'}{f}(x).$$

Also, since f is symmetric,

$$x = sign(x)|x| \text{ and } \frac{f'}{f}(x) = sign(x)\frac{f'}{f}(|x|).$$

Hence the score function is

$$a_{1N}^+(i,f) = \mathrm{E}_{q_0}\left[-1 - |X|_{N:i}\frac{f'}{f}(|X|_{N:i})\right].$$

The verification of $a^+_{1N}(i,f) = \mathrm{E}\left[\varphi_1^+(U_{N:i},f)\right]$ is left as an exercise. □

8.2.4 Test for Independence in a Bivariate Population

Theorem 8.2.4. *Let $p(\boldsymbol{x},\boldsymbol{y})$ be the joint pdf of $\{(X_i, Y_i)\colon i = 1,\dots,N\}$ and $\boldsymbol{R}_N, \boldsymbol{R}'_N$ the vectors of ranks of $(X_1,\dots,X_N)$ and $(Y_1,\dots,Y_N)$, respectively. Then the LMP rank test at level α based on $\boldsymbol{R}_N$ and $\boldsymbol{R}'_N$ for $H_0\colon p \in \mathcal{P}_2$ vs $H_1\colon p(\boldsymbol{x},\boldsymbol{y}) = q_\Delta(\boldsymbol{x},\boldsymbol{y}) = \prod_{i=1}^N h_\Delta(x_i,y_i)$, $\Delta > 0$, where $h_\Delta(x,y) = \int f(x-\Delta z)g(y-\Delta z)\, dM(z)$ with specified pdf's f and g on $\mathbb{R}$ having continuous f', g' and specified cdf M on $\mathbb{R}$ with finite variance σ_M^2 is*

$$\Psi(\boldsymbol{r},\boldsymbol{r}') = 0, \text{ or } \gamma, \text{ or } 1 \text{ according as}$$

$$\sum_{i=1}^N a_N(r_i,f)a_N(r'_i,g) <, \text{ or } =, \text{ or } > k,$$

the constants k and $0 \le \gamma \le 1$ being determined by $E_{H_0}\left[\Psi(\boldsymbol{R}_N,\boldsymbol{R}'_n)\right] = \alpha$ and the score functions $a_N(i,f), a_N(i,g)$ are as in Eq. (3).

Remark 8.2.1. Under q_Δ, the (X_i, Y_i)'s are iid as $(X,Y) = (X^* + \Delta Z, Y^* + \Delta Z)$ where X^* with pdf f, Y^* with pdf g, and Z with cdf M are mutually independent.

In the proof of this theorem, we shall use

$$p(\boldsymbol{x},\boldsymbol{y}) = \prod_{i=1}^N \{f_\Delta(x_i)g_\Delta(y_i)\} \text{ with}$$

$$f_\Delta(x) = \int f(x-\Delta z)\, dM(z) \text{ and } g_\Delta(y) = \int g(y-\Delta z)\, dM(z).$$

Note that this $p(\boldsymbol{x},\boldsymbol{y}) \in \mathcal{P}_2$ and call this distribution $Q_{0,\Delta}$.

Lemma 8.2.2. $\lim_{\Delta\downarrow 0} \Delta^{-2}\left[h_\Delta(x,y) - f_\Delta(x)g_\Delta(y)\right] = f'(x)g'(y)\sigma_M^2$.

Proof. By algebraic rearrangements, we can write

$$\begin{aligned} &h_\Delta(x,y) - f_\Delta(x)g_\Delta(y) \\ &= \iint \left[f(x-\Delta z)g(y-\Delta z) - f(x-\Delta z)g(y-\Delta z')\right] dM(z)\, dM(z') \\ &= (1/2)\iint \{f(x-\Delta z) - f(x-\Delta z')\}\{g(y-\Delta z) - g(y-\Delta z')\}\, dM(z)\, dM(z'). \end{aligned}$$

Hence

$$\begin{aligned} &\Delta^{-2}\left[h_\Delta(x,y) - f_\Delta(x)g_\Delta(y)\right] \\ &= \iint \left\{\frac{f(x-\Delta z) - f(x-\Delta z')}{\Delta(z-z')}\right\}\left\{\frac{g(y-\Delta z) - g(y-\Delta z')}{\Delta(z-z')}\right\}(1/2)(z-z')^2\, dM(z)\, dM(z'). \end{aligned}$$

Let $A(\Delta,\delta)$ be the part of this integral over $\left\{|z|, |z'| \le \delta/\Delta\right\}$ and let $R(\Delta,\delta)$ be the remainder. Then $A(\Delta,\delta)$ can be made arbitrarily close to

$$f'(x)g'(y)(1/2)\int_{|z|\le\delta/\Delta}\int_{|z'|\le\delta/\Delta}(z-z')^2\, dM(z)\, dM(z')$$

by making $\delta > 0$ sufficiently small, and then this tends to $f'(x)g'(y)\sigma_M^2$ as $\Delta \to 0$. Finally, with $\|f\|, \|g\|$ as sup norms of f, g and $C = \{|z|, |z'| \le \delta/\Delta\}^c$,

$$\begin{aligned} R(\Delta,\delta) &\le \left[2\|f\|\,\|g\|/\Delta^2\right]\iint_C dM(z)\,dM(z') \\ &\le \left[4\|f\|\,\|g\|/\delta^2\right](\delta/\Delta)^2 P_M[|Z| > \delta/\Delta] \\ &\le \left[4\|f\|\,\|g\|/\delta^2\right]\int_{|z|>\delta/\Delta} z^2\,dM(z) \to 0 \end{aligned}$$

for arbitrary δ as $\Delta \to 0$. □

Proof. As in the proofs of Theorems 8.2.1–8.2.3, we consider the likelihood ratio

$$\begin{aligned} L(\boldsymbol{r},\boldsymbol{r}') &= \frac{Q_\Delta(\boldsymbol{R}=\boldsymbol{r},\boldsymbol{R}'=\boldsymbol{r}')}{Q_{0\Delta}(\boldsymbol{R}=\boldsymbol{r},\boldsymbol{R}'=\boldsymbol{r}')} \\ &= 1 + (N!)^2\Delta^2\int_{\boldsymbol{R}=\boldsymbol{r}}\int_{\boldsymbol{R}'=\boldsymbol{r}'}\Delta^{-2}\left[\prod_{i=1}^N h_\Delta(x_i,y_i) - \prod_{i=1}^N\{f_\Delta(x_i)g_\Delta(y_i)\}\right] dx\,dy \\ &= 1 + N!\Delta^2\sum_{k=1}^N I_{Nk}(\boldsymbol{r},\boldsymbol{r}',f,g;\Delta) \end{aligned}$$

using Eq. (5), where

$$I_{Nk} = \int_{\boldsymbol{R}=\boldsymbol{r}}\int_{\boldsymbol{R}'=\boldsymbol{r}'}\frac{h_\Delta(x_k,y_k) - f_\Delta(x_k)g_\Delta(y_k)}{\Delta^2}\prod_{j=1}^{k-1} h_\Delta\left(x_j,y_j\right)\prod_{j=k+1}^N\left\{f_\Delta(x_j)g_\Delta(y_j)\right\} dxdy.$$

Since the likelihood ratios $L(\boldsymbol{r},\boldsymbol{r}')$ are ordered as $\sum_{k=1}^N I_{Nk}(\boldsymbol{r},\boldsymbol{r}',f,g;\Delta)$, the theorem will be proved by showing that for each (N,K),

$$(N!)^2\lim_{\Delta\downarrow 0} I_{Nk}(\boldsymbol{r},\boldsymbol{r}',f,g;\Delta) = a_N(r_i,f)a_N(r_i',g).$$

By Lemma 8.2.2, for each (N,k) the integrand of $I_{Nk}(\boldsymbol{r},\boldsymbol{r}',f,g;\Delta)$ converges to (as $\Delta \to 0$)

$$\frac{f'}{f}(x_k)\frac{g'}{g}(y_k)\prod_{j=1}^N\left\{f(x_j)g(y_j)\right\}.$$

Now taking $\lim_{\Delta\downarrow 0}$ under the integral in I_{Nk}, for the justification of which we refer to Hájek and Šidák [4], we have

$$
\begin{aligned}
&(N!)^2 \lim_{\Delta \downarrow 0} I_{Nk}(\boldsymbol{r}, \boldsymbol{r}', f, g; \Delta) \\
&= \sigma_M^2 \int_{\boldsymbol{R}=\boldsymbol{r}} \int_{\boldsymbol{R}'=\boldsymbol{r}'} \frac{f'}{f}(x_k) \frac{g'}{g}(y_k) \frac{\prod_{j=1}^N \{f(x_j) g(y_j)\}}{(1/N!)^2} dx\, dy \\
&= \sigma_M^2 \mathrm{E}_{q_0}\left[\frac{f'}{f}(X_k) \frac{g'}{g}(Y_k) | \boldsymbol{R}_N = \boldsymbol{r}, \boldsymbol{R}_N' = \boldsymbol{r}'\right] \\
&= \sigma_M^2 \mathrm{E}_{q_0}\left[\frac{f'}{f}(X_k) | \boldsymbol{R}_N = \boldsymbol{r}\right] \mathrm{E}_{q_0}\left[\frac{g'}{g}(Y_k) | \boldsymbol{R}_N' = \boldsymbol{r}'\right] \\
&= \sigma_M^2 \mathrm{E}_{q_0}\left[\frac{f'}{f}(X_{N:r_k})\right] \mathrm{E}_{q_0}\left[\frac{g'}{g}\left(Y_{N:r_k'}\right)\right] = \sigma_M^2 a_N(r_k, f) a_N\left(r_k', g\right),
\end{aligned}
$$

using Lemma 8.1.1. □

8.2.5 Specific Rank Tests Using Approximate Scores

First, note that the scores $a_N(i,f), a_{1N}(i,f), \ldots$ can only be obtained from tables (if available) due to their complicated expressions. For example, there are tables for $\mathrm{E}[\Phi^{-1}(U_{N:i})]$. However, since $\mathrm{E}[U_{N:i}] = i/(N+1)$ and $\mathrm{Var}[U_{N:i}] = i(N-i+1)/\{(N+1)^2(N+2)\} \to 0$ as $N \to \infty$, the distribution of $U_{N:i}$ is concentrated near $i/(N+1)$; so

$$
a_N(i,f) = \mathrm{E}[\varphi(U_{N:i}, f)] \approx \varphi(\mathrm{E}[U_{N:i}], f) = \varphi(i/(N+1), f)
$$

if φ is sufficiently smooth near $u = i/(N+1)$.

We now look at a number of specific problems.

I. Two-sample location
- **(a)** Normal: $\varphi(u,f) = \Phi^{-1}(u)$, $a_N(i,f) = \mathrm{E}[\Phi^{-1}(U_{N:i})] \approx \Phi^{-1}(i/(N+1))$.
- **(b)** Logistic: $\varphi(u,f) = 2u-1$, $a_N(i,f) = \mathrm{E}[2U_{N:i}-1] = (2/(N+1))i - 1$, or equivalently, i (Wilcoxon test).
- **(c)** Double exponential: $\varphi(u,f) = sign(2u-1)$, $a_N(i,f) = \mathrm{E}[sign(2U_{N:i}-1)] \approx sign(2i/(N+1)-1) = sign(i-(N+1)/2)$. The test statistic $\sum_{i=m+1}^N sign(r_i - (N+1)/2)$, or equivalently, $\sum_{i=m+1}^N (1/2)[sign(r_i-(N+1)/2)+1]$ can be expressed as
  # observations in the second sample exceeding the pooled median (Median test).

II. Two-sample scale
- **(a)** Normal: $1 + a_{1N}(i,f) = \mathrm{E}\left[\{\Phi^{-1}(U_{N:i})\}^2\right] \approx \{\Phi^{-1}(i/(N+1))\}^2$.
- **(b)** Cauchy-type tail: $f(x) = \frac{1}{2(1+|x|)^2}$; $1 + \varphi_1(u,f) = 2|2u-1|$, $1 + a_{1N}(i,f) = 2\mathrm{E}[|2U_{N:i} - 1|]$ has led to several statistics.

III. One-sample location, assuming symmetry.
- **(a)** Normal: $a_N^+(i,f) = \mathrm{E}[\Phi^{-1}(1/2 + (1/2)U_{N:i})] \approx \Phi^{-1}(1/2 + (1/2)i/(N+1))$.

(b) Logistic: $\varphi(u,f) = 2u - 1$, $\varphi^{+}(u,f) = \varphi(1/2 + (1/2)u,f) = u$, $a_N^{+}(i,f) = \mathrm{E}[U_{N:i}] = i/(N+1)$. Test statistic is $\sum_{X_i>0} R_{N:i}^{+}/(N+1)$, or equivalently, $\sum_{X_i>0} R_{N:i}^{+}$ (Wilcoxon signed-rank test).

IV. Bivariate independence: For f and g both logistic, test statistic is Spearman's rank correlation ρ_S.

The verification of these scores and resulting test statistics are left as exercises.

8.2.6 Asymptotic Distribution of Test Statistics for LMP Rank Tests

We need the distribution under H_0 of the test statistic of an LMP rank test in order to find the critical value at a given level α. In some cases, these statistics have special forms (at least approximately), for which exact distributions may be manageable (see Section 8.1.2), while in some other cases, asymptotic distributions can be derived by the U-statistic approach (see Section 8.1.3).

We first state without proof the following theorem from Hájek and Šidák [4], providing the asymptotic null distributions of a large class of rank statistics such as those constructed in Theorem 8.2.1.

Theorem 8.2.5. *Suppose that φ is a square-integrable function on $[0,1]$ with $\bar{\varphi} = \int_0^1 \varphi(u)\,du$ and that the sequence $\{c_N\}$ satisfies*

$$\lim_{N\to\infty} \frac{\sum_{i=1}^{N}(c_{Ni} - \bar{c}_N)^2}{\max_{1\le i\le N}(c_{Ni} - \bar{c}_N)^2} = \infty, \text{ where } \bar{c}_N = N^{-1}\sum_{i=1}^{N} c_{Ni}.$$

Let $S_{cN} = \sum_{i=1}^{N} c_{Ni} a_N^{\varphi}(R_{N:i})$, where $\mathbf{R}_N$ is the rank vector of $(X_1,\ldots,X_N)$ having joint pdf $p(x) \in \mathcal{P}_0$ and $a_N^{\varphi}(i) = E[\varphi(U_{N:i})]$. Then $(S_{cN} - \mu_{cN})/\sigma_{cN} \xrightarrow{\mathcal{L}} N(0,1)$, where

$$\mu_{cN} = c_N \sum_{i=1}^{N} a_N^{\varphi}(i) \text{ and } \sigma_{cN}^2 = \sum_{i=1}^{N}(c_{Ni} - \bar{c}_N)^2 \int_0^1 \{\varphi(u) - \bar{\varphi}\}^2\,du,$$

which is assumed to be positive.

In the two-sample case with $c_{N1} = \cdots = c_{Nm_N} = 0$, $c_{N,m_N+1} = \cdots = c_{N,m_N+n_N} = c_{NN} = 1$, $\bar{c}_N = n_N/N = 1 - \lambda_N$, the condition on $\{c_N\}$ holds if $m_N/N \to \lambda \in (0,1)$. Here $S_{cN} = \sum_{i=m_N+1}^{N} a_N^{\varphi}(R_{N:i})$, $\mu_{cN} = (1-\lambda_N)\sum_{i=1}^{N} a_N^{\varphi}(i)$, and $\sigma_{cN}^2 = N\lambda_N(1-\lambda_N)\int_0^1\{\varphi(u) - \bar{\varphi}\}^2\,du$.

We now discuss a different approach by Chenoff and Savage [38] for deriving the asymptotic distribution of the LMP rank statistic in the two-sample location problem under the null hypothesis as well as under location-shift.

Let $(X_1,\ldots,X_m)$ and $(Y_1,\ldots,Y_n)$ be independent random samples from populations with cdf's F and G, respectively, on $\mathbb{R}$ and let the empirical cdf's of the two samples be denoted by

$$F_m(x) = m^{-1}\sum_{i=1}^{m} I_{(-\infty,x]}(X_i) \text{ and } G_n(y) = n^{-1}\sum_{i=1}^{n} I_{(-\infty,y]}(Y_i).$$

Next let

$$H(x) = \lambda_N F(x) + (1-\lambda_N)G(x) \text{ and } H_N(x) = \lambda_N F_m(x) + (1-\lambda_N)G_n(x),$$

where $N = m + n$ and $m/N = \lambda_N$. Finally, let $\{J_N\}$ be a sequence of functions on $[0, 1]$ by suitably extending $J_N(i/N) = a_N(i)$, $i = 1, \ldots, N$ (eg, as a step function), which converges to a function J obeying certain conditions to be stated later.

The main idea of the Chernoff-Savage approach is to recognize that

$H_N(X_i) =$ empirical cdf of the combined sample evaluated at $X_i = R_{N:i}/N$ is the rank statistic of interest,

where $R_{N:i} =$ rank of X_i in the combined sample, so if $J_N(i/N) = a_N(i)$, then

$$T_N = \int J_N\left[H_N(x)\right] dF_m(x) = m^{-1}\sum_{i=1}^{m} a_N(R_{N:i}).$$

To find the asymptotic distribution of T_N, write

$$\begin{aligned} T_N &- \int J[H(x)]\, dF(x) \\ &= \int \{J[H_N(x)] - J[H(x)]\}\, dF_m(x) + \int J[H(x)] d[F_m(x) - F(x)] \\ &+ \int \{J_N[H_N(x)] - J[H_N(x)]\}\, dF_m(x) \end{aligned} \tag{7}$$

On the right-hand side of Eq. (7), expand the integrand of the first term as

$$\{H_N(x) - H(x)\}J'[H(x)] + (1/2)\{H_N(x) - H(x)\}^2 J''\left[\alpha_N H_N(x) + (1-\alpha_N)H(x)\right],$$

$0 \le \alpha_N \le 1$ and make rearrangement of all the terms. Thus

$$\begin{aligned} T_N - \int J[H]\, dF &= \int \left[\lambda_N(F_m - F) + (1-\lambda_N)(G_n - G)\right]J'(H)\, dF \\ &+ \int J(H) d(F_m - F) + R_{N1} + R_{N2} + R_{n3}, \end{aligned} \tag{8}$$

where

$$R_{N1} = \int (H_N - H)J'(H) d(F_m - F),$$

$$R_{N2} = (1/2)\int (H_N - H)^2 J'\left(\alpha_N H_N + (1-\alpha_N)H\right) dF_m, \text{ and}$$

$$R_{N3} = \int \{J_N(H_N) - J(H_N)\}\, dF_m.$$

Integrating by parts, the main terms of Eq. (8) become

$$\begin{aligned} &\int \left[\lambda_N(F_m - F) + (1-\lambda_N)G_n - G\right]J'(H)\, dF \\ &\quad - \int (F_m - F)J'(H) d\left[\lambda_N F + (1-\lambda_N)G\right] \end{aligned}$$

$$= -(1-\lambda_N)\int (F_m - F)J'(H)\, dG + (1-\lambda_N)\int (G_n - G)J'(H)\, dF.$$

Now letting

$$J'[H(x)]\, dG(x) = dB(x),\ J'[H(x)]\, dF(x) = dB^*(x),$$

and integrating by parts, we are led to

$$\begin{aligned} T_N &- \int J[H(x)]\, dF(x) \\ &= -(1-\lambda_N)\int [F_m(x) - F(x)]\, dB(x) + (1-\lambda_N)\int \big[G_n(x) - G(x)\big]\, dB^*(x) + R_N \\ &= (1-\lambda_N)\int B(x) d[F_m(x) - F(x)] - (1-\lambda_N)\int B^*(x) d\big[G_n(x) - G(x)\big] + R_N \\ &= (1-\lambda_N)\left[m^{-1}\sum_{i=1}^{m}\{B(X_i) - \mathrm{E}B(X_i)\} - n^{-1}\sum_{i=1}^{n}\{B^*(Y_i) - \mathrm{E}B^*(Y_i)\}\right] + R_N. \end{aligned}$$

The remainder term $R_N = R_{N1} + R_{N2} + R_{N3}$ can be shown to be $o_P\big(N^{-1/2}\big)$ under the following conditions:

1. $J(u) = \lim_{N\to\infty} J_N(u)$ exists for $0 < u < 1$ and is not a constant.
2. $\int_{\{x:\, 0<H_N(x)<1\}} \{J_N[H_N(x)] - J[H_N(x)]\}\, dF_m(x) = o_P\big(N^{-1/2}\big)$.
3. $J_N(1) = O\big(N^{-1/2}\big)$.
4. $|J^{(i)}(u)| = \left|\frac{d^iJ}{du^i}\right| \le \text{constant}|u(1-u)|^{-i-1/2+\delta}$, $i = 0, 1, 2$ for some $\delta > 0$.

Also, since $H = \lambda_N F + (1-\lambda_N)G$ depends on N, so do the iid sequences $\{B(X_i), i = 1, \ldots, m\}$ and $\{B^*(Y_i), i = 1, \ldots, n\}$. The asymptotic normality of $\sqrt{N}[T_N - \int J[H(x)]\, dF(x)\big]$ would, therefore, have to be justified for this triangular array situation. This justification is also provided by the above conditions.

We now arrive at the following theorem.

Theorem 8.2.6 (Chernoff-Savage). *Under Conditions 1–4, if* $0 < \lambda_0 \le \lambda_N \le 1 - \lambda_0 < 1$ *for all N, then*

$$\frac{\sqrt{N}\Big[T_N - \int_{-\infty}^{\infty} J[H(x)]\, dF(x)\Big]}{\sqrt{(1-\lambda_N)\Big\{\Big(\frac{1-\lambda_N}{\lambda_N}\Big) Var[B(X_1)] + Var\big[B^*(Y_1)\big]\Big\}}} \xrightarrow{\mathcal{L}} N(0,1).$$

To make the statement of the above theorem explicit, we need $\mathrm{Var}[B(X_1)]$ and $\mathrm{Var}\big[B^*(Y_1)\big]$. Let $F_1(x) = I(X_1 \le x)$ and $G_1(y) = I\big(Y_1 \le y\big)$. Then

$$\mathrm{E}\big[F_1(x)\big] = F(x),\ \mathrm{Cov}\big[F_1(x), F_1(y)\big] = \min\big\{F(x), F(y)\big\} - F(x)F(y)$$

and likewise for $\mathrm{E}\big[G_1(x)\big]$ and $\mathrm{Cov}\big[G_1(x), G_1(y)\big]$. Using these, we have

$$\mathrm{Var}\big[B(X_1)\big] = 2\iint_{-\infty<x<y<\infty} F(x)\big\{1 - F(y)\big\}J'[H(x)]J'\big[H(y)\big]\, dG(x)\, dG(y) \tag{9}$$

and $\text{Var}\big[B^*(Y_1)\big]$ is obtained by interchanging F and G in the above formula for $\text{Var}[B(X_1)]$. The asymptotic variance

$$V_N = (1-\lambda_N)\left\{\left(\frac{1-\lambda_N}{\lambda_N}\right)\text{Var}\big[B(X_1)\big] + \text{Var}\big[B^*(Y_1)\big]\right\}$$

can now be written in an explicit form.

In the two-sample problem, with $G(x) = F(x-\theta_N)$ and $F' = f$, the asymptotic normality holds uniformly in $0 < \lambda_0 \le \lambda_N \le 1-\lambda_0 < 1$ and θ_N in some neighborhood of 0. If $\theta_N \to 0$, then

$$\begin{aligned}
\left(\frac{\lambda_N}{1-\lambda_N}\right)V_N &\to 2\iint_{x<y} F(x)\{1-F(y)\}J'[F(x)]J'\big[F(y)\big]\,dF(x)\,dF(y)\\
&= 2\iint_{0<u<v<1} u(1-v)J'(u)J'(v)\,du\,dv\\
&= 2\iiiint_{0<s<u<v<t<1} J'(u)J'(v)\,du\,dv\,ds\,dt\\
&= \iint_{0<s<t<1}\Big[J^2(s) + J^2(t) - 2J(s)J(t)\Big]\,ds\,dt\\
&= \int_0^1 J^2(t)\,dt - \left\{\int_0^1 J(t)\,dt\right\}^2. \qquad (10)
\end{aligned}$$

Verifying the details of the derivations of Eqs. (9) and (10) are left as exercises.

Going back to the original problem with

$$J_N(i/N) = a_N(i) = \text{E}\Big[K^{-1}(U_{N:i})\Big],$$

K being a strictly increasing absolutely continuous cdf, we need $J = K^{-1}$ to satisfy Condition 4 of Theorem 8.2.6 and $\lambda_N \to \lambda \in (0,1)$. Then $\sqrt{N}[T_N - \mu(\theta)]/\sigma(\theta) \xrightarrow{\mathcal{L}} N(0,1)$, where $\mu(\theta)$ and $\sigma^2(\theta)$ are obtained by letting $J = K^{-1}$ in $\int J[H(x)]\,dF(x)$ and the formula for V_N using Eq. (9) and its counterpart for $\text{Var}\big[B^*(Y_1)\big]$. From these, we get

$$\mu'(0) = -(1-\lambda)\int_{-\infty}^{\infty}\left(K^{-1}\right)'[F(x)]f(x)\,dF(x), \text{ and}$$

$$\sigma^2(0) = \left(\frac{1-\lambda}{\lambda}\right)\left[\int_0^1\left\{K^{-1}(u)\right\}^2 du - \left\{\int_0^1 K^{-1}(u)\,du\right\}^2\right] = \left(\frac{1-\lambda}{\lambda}\right)\sigma_K^2,$$

if $\sigma_K^2 = \text{Var}_K[X] < \infty$. Hence the Pitman asymptotic efficacy of T_N is

$$\{\mu'(0)\}^2/\sigma^2(0) = \frac{\lambda(1-\lambda)}{\sigma_K^2}\left\{\int_{-\infty}^{\infty} K^{-1\prime}[F(x)]f(x)\,dF(x)\right\}^2.$$

Now take $J = K^{-1} = \Phi^{-1}$. If $F(x) = \Phi\big((x-a)/b\big)$ and we take $K = \Phi$ in T_N, then

$$\int_{-\infty}^{\infty} K^{-1\prime}[F(x)]f(x)\, dF(x) = b^{-1}\int_{-\infty}^{\infty} \Phi^{-1\prime}[\Phi(t)]\phi(t)\, d\Phi(t)$$
$$= b^{-1}\int_{-\infty}^{\infty} d\Phi(t) = b^{-1},$$

so $\{\mu'(0)\}^2/\sigma^2(0) = \lambda(1-\lambda)/b^2$.

For $F(x) = \Phi\big((x-a)/b\big)$, the two-sample t-test has the same asymptotic efficacy. Thus the Pitman ARE of the LMP rank test with score function Φ^{-1} with respect to the t-test $= 1$.

8.3 Tests Based on Empirical Distribution Function

Let $X_1, \ldots, X_n, \ldots$ be a sequence of iid rv's with common cdf F which we assume to be continuous and strictly increasing. The random function

$$F_n(x) = n^{-1}\sum_{i=1}^{n} I_{(-\infty,x]}(X_i), \quad -\infty < x < \infty,$$

called the empirical distribution function (edf) based on $X_1, \ldots, X_n$ has already been defined in Chapter 3.

Theorem 8.3.1 (Glivenko-Cantelli). $\sup_x |F_n(x) - F(x)| \to 0$ *a.s., as* $n \to \infty$.

Proof.

(i) If $|F_n(a) - F(a)| < \varepsilon/2$, $|F_n(b) - F(b)| < \varepsilon/2$ and $F(b) - F(a) < \varepsilon/2$ for $a < b$, then $|F_n(x) - F(x)| < \varepsilon$ for all $a \le x \le b$ and

(ii) there are only a finite number of points at which F has a jump bigger than $\varepsilon/2$. From these facts it follows that for every $\varepsilon > 0$, there exists a constant C and another constant α such that

$$P\Big[\sup_x |F_n(x) - F(x)| > \varepsilon\Big] \le C\exp\big(-\alpha n\varepsilon^2\big) \quad \text{for all } n.$$

[By a result due to Dvoretzky et al. [39], this probability inequality holds with $\alpha = 2$.] The theorem now follows from the Borel-Cantelli Lemma. The details of the proof are left as an exercise. □

8.3.1 Test Statistics

There are some well-known tests for the nonparametric hypothesis H_0: $F = F_0$ (where F_0 is specified continuous cdf) against H_1: $F \ne F_0$ which are based on the random function $\{\sqrt{n}[F_n(x) - F_0(x)]: x \in \mathbb{R}\}$. Test statistics for two such tests are

Kolmogorov-Smirnov statistic: $D_n = \sup_x \sqrt{n}|F_n(x) - F_0(x)|$, and
Cramér-von Mises statistic: $W_n^2 = n\int_{-\infty}^{\infty}[F_n(x) - F_0(x)]^2\, dF_0(x)$.

An important fact is the *distribution-free property* of these test statistics. To see this, suppose that F_0 is *strictly* increasing (a condition which can be removed if we argue a little more carefully), and let $U_i = F_0(X_i)$. In Section 1.11 we have seen that $X_1, \ldots, X_n$ are iid with cdf $F_0 \iff U_1, \ldots, U_n$ are iid *Unif* $(0, 1)$, so that

$$\{\sqrt{n}[F_n(x) - F_0(x)]:\ -\infty < x < \infty\} \overset{D}{=} \{\sqrt{n}[G_n \circ F_0(x) - F_0(x)], -\infty < x < \infty\}, \tag{11}$$

where $G_n(t) = n^{-1}\sum_{i=1}^{n} I_{(0,t]}(U_i)$ is the edf of $(U_1, \ldots, U_n)$.

Now letting $t = F_0(x)$, the statistics D_n and W_n^2 can be expressed as

$$D_n = \sup_{0\le t\le 1} \sqrt{n}|G_n(t) - t| \text{ and } W_n^2 = n\int_0^1 [G_n(t) - t]^2\, dt. \tag{12}$$

This shows that the distribution of D_n and W_n^2 under H_0: $F = F_0$ is the same for all F_0.

The statistic D_n can also be used to construct a confidence band for an unknown continuous cdf. Since D_n is distribution-free, we can find a constant c_α for a given $0 < \alpha < 1$ such that for all F we have

$$\begin{aligned}1 - \alpha &= P_F\Big[\sup_x \sqrt{n}|F_n(x) - F_0(x)| \le c_\alpha\Big]\\ &= P_F\Big[F_n(x) - n^{-1/2}c_\alpha \le F(x) \le F_n(x) + n^{-1/2}c_\alpha \text{ for all } x\Big].\end{aligned}$$

Consequently, $\{F_n(x) \pm n^{-1/2}c_\alpha, -\infty < x < \infty\}$ is a confidence band for F with confidence coefficient $1 - \alpha$.

One-sided versions of the test statistic D_n, namely,

$$D_n^+ = \sup_x \sqrt{n}[F_n(x) - F_0(x)] \text{ and } D_n^- = \sup_x \sqrt{n}[F_0(x) - F_n(x)] \tag{13}$$

can be used to test H_0: $F = F_0$ against one-sided alternatives.

The above ideas extend to two-sample problems as follows.

Let $X_{11}, \ldots, X_{1m}, \ldots$ and $X_{21}, \ldots, X_{2n}, \ldots$ be two independent iid sequences with common cdf's F_1, F_2, respectively, and let

$$F_{1m}(x) = m^{-1}\sum_{i=1}^{m} I_{(-\infty,x]}(X_{1i}) \text{ and } F_{2n}(x) = n^{-1}\sum_{i=1}^{n} I_{(-\infty,x]}(X_{2i})$$

be the edf based on $X_{11}, \ldots, X_{1m}$ and $X_{21}, \ldots, X_{2n}$, respectively. Then the two-sample extensions of the Kolmogorov-Smirnov and the Cramér-von Mises statistics are, respectively,

$$\begin{aligned}D_{mn} &= \sqrt{\frac{mn}{m+n}}\sup_x |F_{1m}(x) - F_{2n}(x)| \text{ and}\\ W_{mn}^2 &= \frac{mn}{m+n}\int_{-\infty}^{\infty}[F_{1m}(x) - F_{2n}(x)]^2\, d\Big(\frac{mF_{1m}(x) + nF_{2n}(x)}{m+n}\Big),\end{aligned} \tag{14}$$

which can be used to test H_0: $F_1 = F_2$ vs H_1: $F_1 \ne F_2$.

We now examine the distribution of the stochastic process $\{\sqrt{n}[F_n(x) - F_0(x)], -\infty < x < \infty\}$ under H_0: $F = F_0$ or equivalently, the process $\{\sqrt{n}[G_n(t) - t], 0 \le t \le 1\}$ where $G_n(t)$ is the edf of a random sample $U_1, \ldots, U_n$ from *Unif*$(0, 1)$ given in Eq. (11). Although exact results for small sample are available for the statistics D_n^+ and D_n, we shall only give an outline of the asymptotic property of $\{\sqrt{n}[G_n(t) - t], 0 \le t \le 1\}$ as $n \to \infty$.

First note that

$$\sqrt{n}[G_n(t) - t] = n^{-1/2} \sum_{i=1}^{n} [I_{[0,t]}(U_i) - t], \text{ and}$$

$$E[I_{[0,t]}(U_i)] = t, \ \mathrm{Var}[I_{[0,t]}(U_i)] = t(1-t),$$

$$\mathrm{Cov}[I_{[0,s]}(U_i), I_{[0,t]}(U_i)] = \min(s,t) - st = s(1-t) \quad \text{for } 0 \le s \le t \le 1.$$

By multivariate CLT, it follows that for all sets of finite points $0 < t_1 < \cdots < t_k < 1$,

$$\left(\sqrt{n}[G_n(t_1) - t_1], \ldots, \sqrt{n}[G_n(t_k) - t_k]\right)^T \xrightarrow{\mathcal{L}} N_k(\mathbf{0}, \Sigma_{t_1,\ldots,t_k}), \text{ where}$$

$$\Sigma_{t_1,\ldots,t_k} = \begin{bmatrix} t_1(1-t_1) & t_1(1-t_2) & \cdots & t_1(1-t_k) \\ t_1(1-t_2) & t_2(1-t_2) & \cdots & t_2(1-t_k) \\ \vdots & \vdots & \vdots & \vdots \\ t_1(1-t_k) & t_2(1-t_k) & \cdots & t_k(1-t_k) \end{bmatrix}. \tag{15}$$

From this, it seems plausible that the entire stochastic process $\{\sqrt{n}[G_n(t) - t], 0 \le t \le 1\}$ should converge (in some sense) to a *Gaussian process* with mean value function 0 and covariance function $\rho(s,t) = s(1-t)$, $0 \le s \le t \le 1$, that is, a stochastic process $\{X(t)\colon 0 \le t \le 1\}$ such that for any $0 \le t_1 < \cdots < t_k \le 1$, $(X(t_1), \ldots, X(t_k))$ follows a k-dim normal distribution with mean vector $\mathbf{0}$ and covariance matrix $\Sigma_{t_1,\ldots,t_k}$ given by Eq. (15). This was conjectured by Doob [40]. For a formal description of this phenomenon, we shall first review some basic facts about Brownian Motion and Weak Convergence.

8.3.2 Brownian Motion: Some Basic Facts

A standard Brownian Motion (B.M.) is a stochastic process $\{X(t), t \ge 0\}$ (ie, a collection of rv's on some probability space) with the following properties:

(i) $X(0) = 0$ w.p. 1.

(ii) For all k and $0 = t_0 < t_1 < \cdots < t_k$, $X(t_i) - X(t_{i-1})$, $1 \le i \le k$, are mutually independent and $X(t_i) - X(t_{i-1}) \sim N(0, t_i - t_{i-1})$. Equivalently, $(X(t_1), \ldots, X(t_k))^T \sim N_n(0, ((\min(t_i, t_j))))$. [This is the property of independent and stationary increments in a Gaussian Process.]

Wiener showed that it is possible to construct a probability distribution on a suitable σ-field of continuous functions on $\mathbb{R}$ such that $\{X(t) = X(t, \omega), t \ge 0\}$ defined on continuous functions ω would have properties (i) and (ii). For this reason, the probability distribution of a standard B.M. is called the "Wiener Measure." From now on, we assume that the sample paths of a B.M. are continuous w.p. 1.

In general a B.M. $\{Y(t), t \geq 0\}$ with mean μ and variance σ^2 per unit time is $Y(t) = \mu t + \sigma X(t)$, where $\{X(t), t \geq 0\}$ is a standard B.M.

We now state some properties of a standard B.M.

(I) If $\{X(t), t \geq 0\}$ is a standard B.M., then so are

(a) $\{-X(t), t \geq 0\}$ (Symmetry).

(b) $\{X(s+t) - X(s), t \geq 0\}$ for fixed s. Moreover, the process is independent of $\{X(\tau), 0 \leq \tau \leq s\}$ (Markov Property).

(c) $\{tX(1/t), t \geq 0\}$ (Inversion).

(d) $\left\{\alpha^{-1/2}X(\alpha t), t \geq 0\right\}$ for $\alpha > 0$ (Scale Change).

(e) $\{X(t_0) - X(t_0 - t), 0 \leq t \leq t_0\}$ (Time Reversal).

For proofs of the above, it is enough to check that the covariance function of the processes in (a)–(e) is $\min(t_1, t_2)$.

The next three properties involve the concept of martingales. In Section 3.3, we have mentioned the martingale property of the stochastic process $\left\{S_k = \sum_{i=1}^{k} X_i, k = 1, 2, \ldots\right\}$ where $\{X_i\}$ is a sequence of independent rv's with mean zero. More generally, a stochastic process $\{X(t), t \geq 0\}$ on a probability space $(\Omega, \mathcal{A}, P)$ is a martingale if $\mathrm{E}[|X(t)|] < \infty$ for all t and $\mathrm{E}[X(t)|\{X(\tau), 0 \leq \tau \leq s\}] = X(s)$ for all $s \leq t$.

(II) **(a)** From the Markov Property, it follows that for $s < t$,

$$\begin{aligned}\mathrm{E}[X(t)|X(\tau), \tau \leq s] &= \mathrm{E}[X(t) - X(s) + X(s)|X(\tau), \tau \leq s] \\ &= \mathrm{E}[X(t) - X(s)] + X(s) = X(s),\end{aligned}$$

that is, $\{X(t), t \geq 0\}$ is a martingale.

(b) It is also easy to verify that $\left\{X(t)^2 - t, t \geq 0\right\}$ is a martingale.

(c) Let $\xi(t) = \mathrm{e}^{\theta X(t)}/\mathrm{E}\left[\mathrm{e}^{\theta X(t)}\right] = \mathrm{e}^{\theta X(t) - \theta^2 t/2}$, since

$$\mathrm{E}\left[\mathrm{e}^{\theta X(t)}\right] = \text{mgf of } N(0, t) \text{ at } \theta = \mathrm{e}^{\theta^2 t/2}.$$

For $s \leq t$

$$\begin{aligned}\mathrm{E}[\xi(t)|\xi(\tau), \tau \leq s] &= \mathrm{E}\left[\exp\left(\theta X(s) + \theta(X(t) - X(s)) - \theta^2 t/2\right)|\xi(\tau), \tau \leq s\right] \\ &= \exp\left(\theta X(s) - \theta^2 t/2\right)\mathrm{E}\left[\exp(\theta(X(t) - X(s)))|\xi(\tau), \tau \leq s\right] \\ &= \exp\left(\theta X(s) - \theta^2 t/2\right)\exp\left(\theta^2(t-s)/2\right) \\ &= \exp\left(\theta X(s) - \theta^2 s/2\right) = \xi(s),\end{aligned}$$

showing that $\{\xi(t), t \geq 0\}$ is a martingale.

(III) Stopping Time.

For a continuous time stochastic process $\{X(t), t \geq 0\}$, let $T \geq 0$ be an rv defined in such a manner that the event $\{T \leq t\}$ depends only on $\{X(\tau), \tau \leq t\}$. Such an rv is called a stopping time. Examples are $T_a = \inf\{t\colon X(t) = a\}$, $T_{ab} = \inf\{t\colon X(t) \notin (a, b)\}$ for $a < 0 < b$.

If T is a stopping time, then $\min(T, t)$ is also a stopping time. Suppose that $\{X(t), t \geq 0\}$ is a martingale with continuous sample paths and let $T < \infty$ be a stopping time with $E[|X(T)|] < \infty$. Then under integrability conditions, $E[X(T)] = E[X(0)]$. See Freedman [41] Brownian Motion and Diffusion, p. 193 and Breiman [42] Probability, Generalization of Corollary 5.31, p. 98 and 274 for details.

The martingale $\{\xi(t), t \geq 0\}$ defined in II(c) above satisfies these conditions, so

$$E\big[\xi(\min(T, t))\big] = E[\xi(0)]. \tag{16}$$

8.3.3 Weak Convergence of $\{Y_n(t) = \sqrt{n}(G_n(t) - t), 0 \leq t \leq 1\}$

The random functions $Y_n(\cdot)$ are not in $C[0, 1]$ due to jumps in $G_n(t)$ given in Eq. (11). However, we can take care of this difficulty by a minor adjustment of $G_n(t)$. Let $U_{n:1} < \cdots < U_{n:n}$ denote the order statistics of the random sample $(U_1, \ldots, U_n)$ from *Unif* $(0, 1)$. The edf G_n has jumps of $1/n$ at each of the order statistics. Now define $\tilde{G}_n$ by linear interpolation between the points:

$$(0,0), (U_{n:1}, 1/(n+1)), (U_{n:2}, 2/(n+1)), \ldots, (U_{n:n}, n/(n+1)), (1,1), \quad \text{ie,}$$

$$\tilde{G}_n(t) = (n+1)^{-1}\big[(i-1) + (t - U_{n:i-1})/(U_{n:i} - U_{n:i-1})\big] \text{ for } U_{n:i-1} \leq t \leq U_{n:i}$$

letting $U_{n:0} = 0$ and $U_{n:n+1} = 1$. Then $|\tilde{G}_n(t) - G_n(t)| \leq 1/n$ for all $t \in [0, 1]$, and $\left\{\tilde{Y}_n(t) = \sqrt{n}\left(\tilde{G}_n(t) - t\right), 0 \leq t \leq 1\right\}$ is in $C[0, 1]$. We can, therefore, use the theory of weak convergence in $C[0, 1]$ outlined in Section A.5 to find the weak limit of $\{Y_n(\cdot)\}$ which is the same as the weak limit of $\left\{\tilde{Y}_n(\cdot)\right\}$, because $\sup_{0 \leq t \leq 1} |Y_n(t) - \tilde{Y}_n(t)| \leq n^{-1/2}$.

As mentioned at the end of Section 8.3.1, the natural candidate for the weak limit of $\{Y_n(t) = \sqrt{n}(G_n(t) - t), 0 \leq t \leq 1\}$ is a Gaussian process on $[0, 1]$ with mean value function 0 and covariance function $\{\rho(s, t) = s(1 - t),\ 0 \leq s \leq t \leq 1\}$. Now the process $\{Y(t) = X(t) - tX(1), 0 \leq t \leq 1\}$ where $\{X(t)\}$ is a standard B.M. fits this description, because $\{Y(t)\}$ is a Gaussian process with $E[Y(t)] = 0$ for all t and for $s < t$,

$$\begin{aligned}\text{Cov}[Y(s), Y(t)] &= \text{Cov}[X(s), X(t)] - s\text{Cov}[X(t), X(1)] - t\text{Cov}[X(s), X(1)] + st\text{Var}[X(1)] \\ &= s - st - st + st = s(1 - t).\end{aligned}$$

To accomplish the actual proof of $Y_n(\cdot) \xrightarrow{w} Y(\cdot)$ via weak convergence of $\left\{\tilde{Y}_n(\cdot)\right\}$, we invoke Theorem A.5.2 in Section A.5, of which Condition (i) regarding convergence of fdd's is already seen, Condition (iia) that $Y_n(0) = O_P(1)$ is trivial and Condition (iib) is verified by a lengthy analysis (see [43, p. 105–108]).

The process $\{Y(t) = X(t) - tX(1), 0 \leq t \leq 1\}$ where $\{X(t)\}$ is a standard B.M. is called a *Brownian Bridge* because it connects the points $(0, Y(0)) = (0, 0)$ and $(1, Y(1)) = (1, 0)$ by a continuous sample path.

8.3.4 Asymptotic Distributions of D_n^+, D_n, and W_n^2

All three statistics

$$D_n^+ = \sup_{0\le t\le 1} \sqrt{n}\left[G_n(t) - t\right] = \sup_{0\le t\le 1} Y_n(t)$$

$$D_n = \sup_{0\le t\le 1} |Y_n(t)| \text{ and } W_n^2 = \int_0^1 Y_n(t)^2\, dt$$

given by Eqs. (12) and (13) are continuous functions of $\{Y_n(\cdot)\}$. It therefore follows from the Continuous Mapping Theorem A.5.1, Section A.5 that

Theorem 8.3.2. *$D_n^+ \xrightarrow{\mathcal{L}} \sup_{0\le t\le 1} Y(t)$, $D_n \xrightarrow{\mathcal{L}} \sup_{0\le t\le 1} |Y(t)|$, and $W_n^2 \xrightarrow{\mathcal{L}} \int_0^1 Y(t)^2\, dt$, where $\{Y(t)\}$ is the Brownian Bridge on $[0, 1]$.*

We first find the asymptotic distribution of D_n^+.

Theorem 8.3.3. *Let $\{Y(t) = X(t) - tX(1), 0 \le t \le 1\}$, where $\{X(t)\}$ is a standard B.M. Then*

$$P\left[\sup_{0\le t\le 1} Y(t) \ge y\right] = e^{-2y^2}, \quad y > 0.$$

Proof. Note that

$$\sup_{0\le t\le 1} Y(t) \ge y \Longleftrightarrow \sup_{s\ge 0} Y\big(s/(1+s)\big) \ge y$$
$$\Longleftrightarrow X^*(s) = (1+s)Y\big(s/(1+s)\big) \ge (1+s)y$$

for some $s \ge 0$. Now $\left\{X^*(t) = (1+t)Y\big(t/(1+t)\big), t \ge 0\right\}$ is a standard B.M., because it is a Gaussian process with $\mathrm{E}\left[X^*(t)\right] = 0$ and

$$\mathrm{Cov}\left[X^*(s), X^*(t)\right] = (1+s)(1+t)\left[s/(1+s)\right]\left[1 - t/(1-t)\right] = s, \quad \text{for } s \le t.$$

Hence

$$P\left[\sup_{0\le t\le 1} Y(t) \ge y\right] = P\left[X(s) \ge ys + y \quad \text{for some } s \ge 0\right], \tag{17}$$

where $\{X(t)\}$ is a standard B.M.

We shall now find $P\left[X(t) \ge at + b \text{ for some } t \ge 0\right]$ by a martingale approach.

Consider the stopping time

$$T = \min\left\{t\colon X(t) = at + b\right\} \text{ if } X(t) = at + b \text{ for some } t \ge 0 \text{ and}$$
$$T = \infty \text{ if } X(t) < at + b \text{ for all } t > 0.$$

Then

$$P\left[X(t) \ge at + b \text{ for some } t \ge 0\right] = P[T < \infty]. \tag{18}$$

Recall that for each t, $T \wedge t = \min(T, t)$ is a stopping time, and $\left\{\xi(t) = e^{\theta X(t)}/\mathrm{E}\left[e^{\theta X(t)}\right] = e^{\theta X(t) - \theta^2 t/2}\right\}$ is a martingale (property II(c) of a standard B.M.). Consequently, Eq. (16) holds.

For $\theta = 2a > 0$,

$$\begin{aligned}
\xi(T \wedge t) &= \exp\Big[\theta X(T \wedge t) - \theta^2 (T \wedge t)/2\Big] \\
&= \begin{cases} \exp\Big[\theta X(t) - \theta^2 t/2\Big] < \exp\Big[\theta(at+b) - \theta^2 t/2\Big], & t < T \\ \exp\Big[\theta X(T) - \theta^2 T/2\Big] = \exp\Big[\theta(aT+b) - \theta^2 T/2\Big], & t \geq T \end{cases} \\
&\leq \exp\Big[\theta\{a(T \wedge t) + b\} - \theta^2 (T \wedge t)/2\Big] \\
&= \exp\Big[2a\{a(T \wedge t) + b\} - (2a)^2 (T \wedge t)/2\Big] = \mathrm{e}^{2ab}. \qquad (19)
\end{aligned}$$

Next note that

$$\xi(t) = \exp\Big[\theta X(t) - \theta^2 t/2\Big] \to 0 \text{ a.s. as } t \to \infty, \qquad (20)$$

because if $\{Z_i\}$ are iid $N(0,1)$ and $X(n) = \sum_{i=1}^n Z_i$, then $\theta X(n) - \theta^2 n/2 = \theta n(\bar{Z}_n - \theta/2) \to -\infty$ a.s.

Now consider

$$\lim_{t\to\infty} \xi(T \wedge t) = \lim_{t\to\infty} [\xi(t) I(t \leq T) + \xi(T) I(T < t)],$$

of which the first term is 0 a.s. by Eq. (20) and the second term is $\xi(T) \lim_{t\to\infty} I(T < t) = \xi(T) I(T < \infty)$. Thus

$$\lim_{t\to\infty} \xi(T \wedge t) = \xi(T) I(T < \infty), \text{ a.s.}$$

Hence by dominated convergence, using Eq. (19),

$$\begin{aligned}
\lim_{t\to\infty} \mathrm{E}[\xi(T \wedge t)] &= \mathrm{E}[\xi(T) I(T < \infty)] \\
&= \mathrm{E}\Big[\exp[\theta(aT+b) - \theta^2 T/2] I(T < \infty)\Big] \\
&= \mathrm{E}\Big[\exp[2a(aT+b) - (2a)^2 T/2] I(T < \infty)\Big] \\
&= \mathrm{e}^{2ab} P[T < \infty]
\end{aligned}$$

for $\theta = 2a$. But by Eq. (16), $\mathrm{E}[\xi(T \wedge t)] = 1$ for all t, so

$$\lim_{t\to\infty} \mathrm{E}[\xi(T \wedge t)] = 1 = \mathrm{e}^{2ab} P[T < \infty].$$

Thus $P[T < \infty] = \mathrm{e}^{-2ab}$ for $a > 0$ and $b \geq 0$. Taking $a = b = y > 0$ in this formula and using Eqs. (17) and (18), the theorem follows. □

The distribution of $\sup_{0 \leq t \leq 1} |Y(t)|$ is obtained by lengthy analysis of the joint behavior of the maximum and the minimum of a standard B.M. We state the result in the following theorem, referring to Billingsley [43, p. 77–80 and 83–85] for a proof.

Theorem 8.3.4. *Let* $\{Y(t), 0 \leq t \leq 1\}$ *be a Brownian Bridge on* $[0,1]$. *Then*

$$P\Big[\sup_{0 \leq t} |Y(t)| \leq y\Big] = 1 + 2\sum_{k=1}^{\infty} (-1)^k e^{-2k^2 y^2}, \quad y > 0.$$

The distribution of $\int_0^1 Y(t)^2\, dt$ is obtained by using an orthogonal expansion of a standard B.M. in terms of an infinite sequence of iid $N(0,1)$ rv's $\{Y_0, Y_1, \ldots\}$ due to Wiener [44]. We sketch the derivation of this distribution, referring to Breiman [42], Section 12.7, and Hájek and Šidák [4], V3.3 Theorem C for details.

Let $\{X(t)\}$ be a standard B.M. on $[0,\pi]$. Then $\mathrm{Cov}[X(s), X(t)]$ can be expressed by the following identity:

$$st/\pi + (2/\pi)\sum_{k=1}^{\infty}(\sin ks)(\sin kt)/k^2 = \min(s,t) \quad \text{for } 0 \le s,t \le \pi. \tag{21}$$

Hence $\{X(t)\}$ has the representation:

$$X(t) = (1/\sqrt{\pi})tY_0 + \sqrt{2/\pi}\sum_{k=1}^{\infty}\{(\sin kt)/k\}Y_k, \quad 0 \le t \le \pi, \tag{22}$$

where $Y_0, Y_1, \ldots$ are iid $N(0,1)$ rv's, because the right-hand side of Eq. (22) is a Gaussian process with $\mathrm{E}[X(t)] = 0$ and $\mathrm{Cov}[X(s), X(t)]$ is the same as in Eq. (21).

Rewrite Eq. (21), replacing s/π, t/π by s^*, t^*, respectively, and rewrite Eq. (22) by scale change to obtain

$$X^*(t) = (1/\sqrt{\pi})X(\pi t) = tY_0 + \sqrt{2}\sum_{k=1}^{\infty}\{(\sin k\pi t)/(k\pi)\}Y_k, \quad 0 \le t \le 1$$

and

$$s^*t^* + 2\sum_{k=1}^{\infty}(\sin k\pi s^*)(\sin k\pi t^*)/(k\pi)^2 = \min(s^*, t^*), \quad 0 \le s^*, t^* \le 1. \tag{23}$$

Hence $\{X^*(t)\}$ is a standard B.M. on $[0,1]$, because $\mathrm{Cov}\big[X^*(s), X^*(t)\big] = \min(s,t)$ by Eq. (23).

Since $X^*(1) = Y_0$, it follows that

$$Y(t) = X^*(t) - tX^*(1) = \sqrt{2}\sum_{k=1}^{\infty}\{(\sin k\pi t)/(k\pi)\}Y_k \tag{24}$$

is a Brownian Bridge on $[0,1]$. We thus arrive at the following theorem.

Theorem 8.3.5. *$\int_0^1 Y(t)^2\, dt = \sum_{k=1}^{\infty}(k\pi)^{-2}Y_k^2$, which is a mixture of χ^2's.*

Proof. By Eq. (24),

$$\int_0^1 Y(t)^2\, dt = 2\sum_{j=1}^{\infty}\sum_{k=1}^{\infty}\left(jk\pi^2\right)^{-1}Y_jY_k\int_0^1 \sin j\pi t \sin k\pi t\, dt = \sum_{k=1}^{\infty}(k\pi)^{-2}Y_k^2,$$

since $2\int_0^1(\sin j\pi t)(\sin k\pi t)\, dt = 1$ when $j = k$ and $= 0$ when $j \ne k$. □

Finally, we consider the two-sample statistics D_{mn} and W^2_{mn} given by Eq. (14) under the null hypothesis $F_1 = F_2 = F$ (unknown).

For $m, n \to \infty$ in such a way that $m/(m+n) = m/N \to \lambda$ and $n/(m+n) = n/N \to 1-\lambda$ for $0 < \lambda < 1$, write

$$\begin{aligned}
D_{mn} &= \sqrt{\frac{mn}{m+n}} \sup_x |\{F_{1m}(x) - F(x)\} - \{F_{2n}(x) - F(x)\}| \\
&= \sqrt{\frac{mn}{m+n}} \sup_{0\le t\le 1} |m^{-1/2}[\sqrt{m}\{G_{1m}(t) - t\}] - n^{-1/2}[\sqrt{n}\{G_{2n}(t) - t\}]| \\
&\xrightarrow{\mathcal{L}} \sqrt{\lambda(1-\lambda)} \sup_{0\le t\le 1} |\lambda^{-1/2} Y_1(t) - (1-\lambda)^{-1/2} Y_2(t)| \\
&= \sup_{0\le t\le 1} |\sqrt{1-\lambda} Y_1(t) - \sqrt{\lambda} Y_2(t)|,
\end{aligned}$$

where $\{Y_1(t)\}$ and $\{Y_2(t)\}$ are independent Brownian Bridges on $[0, 1]$. Since $\left\{Y(t) = \sqrt{1-\lambda} Y_1(t) - \sqrt{\lambda} Y_2(t)\right\}$ is a Gaussian process with mean value function 0 and

$$\text{Cov}[Y(s), Y(t)] = (1-\lambda)s(1-t) + \lambda s(1-t) = s(1-t) \quad \text{for } 0 \le s, t \le 1,$$

$D_{mn} \xrightarrow{\mathcal{L}} \sup_{0\le t\le 1} |Y(t)|$ and $\{Y(t)\}$ is a Brownian Bridge on $[0, 1]$. Thus the asymptotic distribution of D_{mn} is the same as that of the one-sample statistic D_n under null hypothesis.

For W^2_{mn}, analogous argument holds with the additional observation that $(m+n)^{-1}\left[mG_{1m}(t) + nG_{2n}(t)\right] \to t$, a.s. uniformly by Glivenko-Cantelli Lemma. Hence $W^2_{mn} \xrightarrow{\mathcal{L}} \int_0^1 Y(t)^2\, dt$.

Exercises

8.1. Let p be the joint pdf of $X_1, \ldots, X_m, X_{m+1}, \ldots, X_{m+n}$. We want a UMP similar region test at level α for H_0: $p \in P_0$ vs H_1: $p = q_\Delta$, $\Delta > 1$, where

$$q_\Delta(x_1, \ldots, x_{n+m}) = (2\pi)^{(m+n)/2} \Delta^{-n/2} \exp\left[-(1/2)\sum_{i=1}^{m} x_i^2 - (2\Delta)^{-1} \sum_{i=m+1}^{n+m} x_i^2\right].$$

(a) Show that the UMP similar region test at level α for H_0 vs H_1 is of the form: $\varphi(\boldsymbol{x}_{(m+n)}, \boldsymbol{r}_{(m+n)}) = 0$, or $\gamma(\boldsymbol{x}_{(m+n)})$, or 1 according as $\sum_{i=m+1}^{m+n} x_i^2 <$, or $=$, or $> k(\boldsymbol{x}_{(m+n)})$.

(b) For $m = 6$, $n = 9$, and $\Delta = 2$, generate $(x_1, \ldots, x_{m+n})$ from q_Δ and apply the above test at level $\alpha = 0.05$ by determining $k(x_{m+n})$ and $\gamma(x_{m+n})$ from

(i) the permutation distribution of $\sum_{i=m+1}^{m+n} x_i^2$,

(ii) a random sample of 100 combinations.

8.2. The Mann-Whitney statistic, properly normalized, is

$$T_n = \frac{\sqrt{n}}{n_1 n_2}\sum_{i=1}^{n_1}\sum_{j=1}^{n_2}\{I_{[0,\infty)}(X_i - Y_j) - \theta\},$$

$$n_1 = [n\lambda], n_2 = [n(1-\lambda)], \quad 0 < \lambda < 1,$$

where $\{X_i\}$ and $\{Y_j\}$ are independent iid sequences from F and G, respectively, and $\theta = \int G(x)\, dF(x)$. Find

(a) the Hájek projection T_n^* of T_n when $G = F$,

(b) the asymptotic distribution of T_n^*, and

(c) the asymptotic distribution of T_n with proper justification.

8.3. The Kendall's tau statistic based on a random sample $(X_1, Y_1), \ldots, (X_N, Y_N)$ from a continuous bivariate distribution is

$$\tau_n = \frac{1}{N(N-1)}\sum_{i=1}^{N}\sum_{j\neq i=1}^{N} sign\left(X_i - X_j\right) sign\left(Y_i - Y_j\right).$$

(a) Find the Hájek projection τ_N^* of τ_N and the asymptotic distribution of $\sqrt{N}\tau_N^*$ under H_0: X and Y are independent.

(b) Use (a) and the property of $\text{Var}\left[\tau_N - \tau_N^*\right]$ to derive the asymptotic distribution of $\sqrt{N}\tau_N$ under H_0.

8.4. Let $X_1, \ldots, X_m$, $Y_1 = X_{m+1}, \ldots, Y_n = X_{m+n}$ be independent rv's, $N = m + n$.

(a) Under H_0: $X_i \sim N(0,1)$, $i = 1, \ldots, N$ and under contiguous alternative H_N: $X_i \sim N(0,1)$, $i = 1, \ldots, m$ and $X_i \sim N\left(\delta/\sqrt{N}, 1\right)$, $i = m+1, \ldots, N$, find the asymptotic distributions of the following:

S_{N1} = Two-sample t-statistic $= (\bar{X} - \bar{Y}) \Big/ \left[\left(\frac{1}{m} + \frac{1}{n}\right)\frac{(m-1)S_X^2 + (n-1)S_Y^2}{m+n-2}\right]^{1/2}$,

S_{N2} = Wilcoxon statistic $= m^{-1}\sum_{i=1}^{m} R_{N:i} - n^{-1}\sum_{i=m+1}^{m+n} R_{N:i}$, or equivalently, $S_{N2} = \sum_{i=1}^{m} R_{N:i}$,

S_{N3} = Fisher-Yates normal scores rank statistic $= \sum_{i=1}^{m} \text{E}\left[\Phi^{-1}\left(U_{N:R_{N:i}}\right)\right]$, using usual notations.

(b) Find the Pitman AREs of S_{N2} and S_{N3} with respect to S_{N1}.

8.5. Let $X_1, \ldots, X_n$ be a random sample from a distribution with mean μ, variance σ^2, and finite fourth central moment $\beta = \text{E}\left[(X-\mu)^4\right]$. We want the asymptotic distribution of the sample variance $s_n^2 = (n-1)^{-1}\sum_{i=1}^{n}\left(X_i - \bar{X}_n\right)^2$.

(a) Show that $s_n^2 = \{n(n-1)\}^{-1}\sum_{i=1}^{n}\sum_{j=1}^{n}\left(X_i - X_j\right)^2$, which is a U-statistic.

(b) Use this U-statistic form to find the Hájek projection of $2s_n^2$ and then the asymptotic distribution of $\sqrt{n}\left(s_n^2 - \sigma^2\right)$.

8.6. Let $(X_1, Y_1), \ldots, (X_N, Y_N)$ be iid as (X, Y) following the bivariate normal distribution with $\text{E}[X] = \text{E}[Y] = 0$, $\text{E}\left[X^2\right] = \text{E}\left[Y^2\right] = 1$, and $\text{E}[XY] = \theta$. For testing H_0: $\theta = 0$ vs H_1: $\theta > 0$ (ie, independence vs positive dependence), find the Pitman ARE of

$S_{N1} = \sqrt{N}\tau_N$ with respect to $S_{N2} = \sqrt{N}r_N$, where τ_N is Kendall's tau and r_N is the product-moment correlation based on $(X_1, Y_1), \ldots, (X_N, Y_N)$ as in Example 8.1.3. Use the Hájek projection of τ_N and its asymptotic property obtained in Exercise 8.3 and show that $\sqrt{N}r_N = N^{-1/2}\sum_{i=1}^{N} X_iY_i + o_P(1)$. Now find $\dot{l}(x, y; 0)$ and I, express $\log L_N$, S_{N1}, and S_{N2} in the desired form. [*Hint*: If Φ and ϕ are the cdf and pdf of $N(0, 1)$, then $\int_{-\infty}^{\infty} x\Phi(x)\, d\Phi(x) = \int_{-\infty}^{\infty}\left[\int_{-\infty}^{x}\phi(t)\, dt\right]x\phi(x)\, dx = \int_{-\infty}^{\infty}\left[\int_t^{\infty} x\phi(x)\, dx\right]\phi(t)\, dt$. Evaluate this.]

8.7. In this problem, Kendall's tau and Spearman's rank correlation are used in a different context. Let $X_i = \theta d_i + Z_i$, $i = 1, \ldots, n$, where $Z_1, \ldots, Z_n$ are iid with a continuous cdf and $d_1, \ldots, d_n$ are equally spaced constants in increasing order, which can be taken to be $1, \ldots, n$, without loss of generality. The following statistics can be used to test $H_0\colon \theta = 0$ vs $H_1\colon \theta > 0$.

(i) Moore-Wallis Difference-Sign Statistic: $D_n = \sum_{i=2}^{n} I_{(0,\infty)}(X_i - X_{i-1})$,

(ii) Difference-Sign Correlation Coefficient (Kendall):
$\tau_N = [4n(n-1)]^{-1}\sum_{i=1}^{n-1}\sum_{j=i+1}^{n} I_{(0,\infty)}(X_i - X_j)$,

(iii) Rank Correlation Coefficient (Spearman):
$\rho_n = 1 - \left[6n(n^2-1)\right]^{-1}\sum_{i=1}^{n-1}\sum_{j=i+1}^{n}(j-i)\left[1 + sign(X_i - X_j)\right]$,

(iv) The t-statistic: $T_n = \sum_{i=1}^{n}\left(d_i - \bar{d}_n\right)(X_i - \bar{X}_n) \Big/ \sqrt{\sum_{i=1}^{n}(X_i - \bar{X}_n)^2}$.

We now ask the following questions, regrading the asymptotic properties of these statistics:

(a) Express Kendall's tau and Spearman's rank correlation coefficient in the form given in (ii) and (iii).

(b) Express the t-statistic in the form given in (iv). [This is the test statistic for $H_0\colon \theta = 0$ in the linear model $X_i = \mu + d_i\theta + Z_i$, where $Z_1, \ldots, Z_n$ are iid $N(0, \sigma^2)$, obtained from the least squares estimator and the residual sum of squares. See Example 6.9.10.]

(c) Find the asymptotic distribution of D_n, τ_n, ρ_n, and T_n (properly normalized) under H_0 and under H_1, assuming that the Z_i's are iid as $N(0, \sigma^2)$. For H_1, consider contiguous alternatives.

(d) Calculate the Pitman AREs of the tests based on D_n, τ_n, and ρ_n with respect to the test based on T_n.

[*Hint*: The statistic τ_n is a U-statistic, but ρ_n is not; so find the Hájek projection (adjusting for the mean) and check that the Hájek projection differs from the original statistic by $o_P(1)$. Express T_n in terms of $b_n = \sqrt{\sum_{i=1}^{n}\left(d_i - \bar{d}_n\right)^2}$, $Y_n = \sum_{i=1}^{n}\left(d_i - \bar{d}_n\right)Z_i/b_n \sim N(0, 1)$, and $W_n = \sum_{i=1}^{n}(Z_i - \bar{Z}_n)^2 - Y_n^2 \sim \chi^2_{n-2}$, and in the nonnull case use $\theta = \theta_n = \delta/b_n$. For D_n, use Theorem 3.3.3 of Chapter 3 for m-dependent processes.]

8.8. Let $U_{n1} = \left(1/n^{(r)}\right)\sum_{n,r} g_1\left(X_{i_1}, \ldots, X_{i_r}\right)$ and $U_{n2} = \left(1/n^{(s)}\right)\sum_{n,s} g_2\left(X_{i_1}, \ldots, X_{i_s}\right)$ be two U-statistics based on iid rv's $X_1, \ldots, X_n$, both g_1 and g_2 being symmetric in their arguments. Let $\theta_1 = \mathrm{E}\left[g_1(X_1, \ldots, X_r)\right]$, $\theta_2 = \mathrm{E}\left[g_2(X_1, \ldots, X_s)\right]$ and assume that

$\mathrm{E}\left[g_1^2(X_1,\ldots,X_r)\right]$, and $\mathrm{E}\left[g_2^2(X_1,\ldots,X_s)\right]$ are finite. We want the asymptotic distribution of $\sqrt{n}[\varphi(U_{n1},U_{n2})-\varphi(\theta_1,\theta_2)]$, where φ has continuous first partial derivatives at (θ_1,θ_2). Let

$$h_1(x_1,\ldots,x_r)=g_1(x_1,\ldots,x_r)-\theta_1,\; h_2(x_1,\ldots,x_s)=g_1(x_1,\ldots,x_s)-\theta_2,$$
$$h_1^*(X_1)=\mathrm{E}\left[h_1(X_1,\ldots,X_r)|X_1\right],\; h_2^*(X_1)=\mathrm{E}\left[h_2(X_1,\ldots,X_s)|X_1\right],$$
$$\sigma_{11}=\mathrm{E}\left[h_1^{*^2}(X_1)\right],\; \sigma_{22}=\mathrm{E}\left[h_2^{*^2}(X_1)\right],\; \sigma_{12}=\mathrm{E}\left[h_1^*(X_1)h_2^*(X_1)\right],\text{ and}$$

suppose that $\Sigma=\begin{bmatrix}\sigma_{11} & \sigma_{12}\\ \sigma_{12} & \sigma_{22}\end{bmatrix}$ is positive-definite.

(a) Find two-dimensional Hájek projection $\left(V_{n1}^*,V_{n2}^*\right)$ of $(V_{n1},V_{n2})=\left(\sqrt{n}(U_{n1}-\theta_1),\sqrt{n}(U_{n2}-\theta_2)\right)$.

(b) Show that $\left(V_{n1}^*,V_{n2}^*\right)-(V_{n1},V_{n2})=o_P(1)$.

(c) Find the asymptotic distribution of (V_{n1},V_{n2}).

(d) From (c), find the asymptotic distribution of $\sqrt{n}[\varphi(U_{n1},U_{n2})-\varphi(\theta_1,\theta_2)]$, using the delta method.

8.9. Let $X_1,\ldots,X_n$ be iid positive-valued rv's with cdf F having mean μ and variance σ^2. Let

$$\Delta=\mathrm{E}_F\left[|X_1-X_2|\right]=\iint |x_1-x_2|\,dF(x_1)\,dF(x_2)\text{ and}$$
$$\zeta=\mathrm{E}_F\left[X_1|X_1-X_2|\right]=\iint x_1|x_1-x_2|\,dF(x_1)\,dF(x_2).$$

The statistic $G_n=D_n/\left(2\bar{X}_n\right)$, where $\bar{X}_n=n^{-1}\sum_{i=1}^n X_i$ and $D_n=[n(n-1)]^{-1}\sum_{i=1}^n\sum_{j\neq i=1}^n |X_i-X_j|$ is known as Gini's coefficient of concentration.

Use the result of Exercise 8.8 to find the asymptotic distribution of $\sqrt{n}\left(G_n-\Delta/(2\mu)\right)$.

8.10. Verify that the score functions

$$a_N^+(i,f)=\mathrm{E}_{q_0}\left[-\frac{f'}{f}(|X|_{N:i})\right]=\mathrm{E}\left[\varphi^+(U_{N:i},f)\right]\text{ and}$$
$$a_{1N}^+(i,f)=\mathrm{E}_{q_0}\left[-1-|X|_{N:i}\frac{f'}{f}(|X|_{N:i})\right]=\mathrm{E}\left[\varphi_1^+(U_{N:i},f)\right]$$

are as stated in Theorems 8.2.2 and 8.2.3.

8.11. Verify that the scores and the resulting test statistics for the specific problems in Section 8.2.5 are as stated in the text.

8.12. Let p be the joint pdf of $(X_1,\ldots,X_{m+n})$. We want to test H_0: $p\in\mathcal{P}_0$ vs H_1: $p=q_\Delta$, $\Delta>0$. Show that if the joint cdf Q_Δ corresponding to q_Δ is given by (Lehmann Alternative)

$$Q_\Delta[X_i \le x_i, 1 \le i \le m+n] = \prod_{i=1}^{m}\left[(1-\Delta)F(x_i) + \Delta F^2(x_i)\right] \prod_{i=m+1}^{m+n} F(x_i),$$

where F is an arbitrary absolutely continuous cdf, then the LMP rank test of H_0: $p \in \mathcal{P}_0$ vs H_1: $p = q_\Delta$, $\Delta > 0$, has a critical region of the form $\sum_{i=1}^{m} R_{m+n:i} \ge$ constant.

8.13. In the set-up of Exercise 8.12, suppose that the joint cdf Q_Δ is given by

$$Q_\Delta[X_i \le x_i, 1 \le i \le m+n] = \prod_{i=1}^{m} F(x_i)^{1+\Delta} \prod_{i=m+1}^{m+n} \left[1 - \{1 - F(x_i)\}^{1+\Delta}\right],$$

where F is an arbitrary absolutely continuous cdf. Show that the LMP rank test of H_0: $\mathcal{P} \in P_0$ vs H_1: $p = q_\Delta$, $\Delta > 0$, has a critical region of the form $\sum_{i=1}^{m} a_N(R_{m+n:i}) \ge$ constant, with $a_N(i) = \sum_{j=0}^{i-1} 1/(N-j) - \sum_{j=0}^{N-i} 1/(N-j)$, $N = m+n$. [*Hint*: First show that $\mathrm{E}[-\log(1-U_{N:i})] = \sum_{j=0}^{i-1} 1/(N-j)$ and $\mathrm{E}[-\log U_{N:i}] = \sum_{j=0}^{N-i} 1/(N-j)$, where $U_{N:1} < \cdots < U_{N:N}$ is an ordered random sample from *Uniform*$(0,1)$.]

8.14. Verify the formula for $\mathrm{Var}[B(X)]$ given in Eq. (9) and the limit of $\lambda_N(1-\lambda_N)^{-1}V_N$ as $\theta_N \to 0$ given by Eq. (10).

8.15. Suppose that $X_1, \ldots, X_N$ are iid rv's and let $R_{N:i}$ be the rank of X_i among $X_1, \ldots, X_N$. Consider the scores

$$a_N(i) = \begin{cases} i & 1 \le i \le N-1 \\ N^2/2 & i = N \end{cases},$$

and let $S_{mn} = \sum_{i=1}^{m} a_N(R_{N:i})$, $m+n = N$. Show that if $N \to \infty$, $m \to \infty$, and $m/N \to 0$, then

(a) $\{\mathrm{E}[S_{mn}] - mN/2\}^2/(mnN/12) \to \infty$,

(b) $\mathrm{Var}[S_{mn}]/(mnN/12) \to \infty$, and

(c) $(S_{mn} - mN/2) \xrightarrow{\mathcal{L}} N(0,1)$.

8.16. Give a detailed proof of the Glivenko-Cantelli Theorem using the outline in the text.

8.17. Let $\{X(t), 0 \le t \le 1\}$ be a collection of rv's such that

(i) $X(0) = 0$ with probability 1,

(ii) for any $0 = t_0 < t_1 < \cdots < t_k \le 1$, the increments $X(t_i) - X(t_{i-1})$ are independent $N(0, t_i - t_{i-1})$.

Show that for any $0 < t_1 < \cdots < t_k \le 1$ $(X(t_1), \ldots, X(t_k))$ follows the k-dim normal distribution with mean vector $\mathbf{0}$ and covariance matrix $((\sigma(t_i, t_j))) = ((\min(t_i, t_j)))$.

8.18. Let $U_1 U_2, \ldots$ be iid *Uniform*$(0,1)$ rv's and $F_n(t) = n^{-1}\sum_{i=1}^{n} I_{[0,t]}(U_i)$. Let $Y_n(t) = \sqrt{n}[F_n(t) - t]$. Show that for any $0 \le t_1 < \cdots < t_k \le 1$, $(Y_n(t_1), \ldots, Y_n(t_k)) \xrightarrow{\mathcal{L}} (Y(t_1), \ldots, Y(t_k))$, where $Y(t) \stackrel{\mathcal{D}}{=} X(t) - tX(1)$ and $\{X(t)\colon 0 \le t \le 1\}$ is an in Exercise 8.17.

9

Curve Estimation

9.1 Introduction

This chapter is concerned with three problems:

Problem 1. Let $X_1, \ldots, X_n$ be a random sample from a cdf F with pdf $f = F'$. We want to estimate the function f.

Problem 2. Let $(X_1, Y_1), \ldots, (X_n, Y_n)$ be a random sample from a bivariate distribution with regression function

$$m(x) = \mathrm{E}[Y|X = x] = \int_{-\infty}^{\infty} y f_{Y|X}(y|x)\, dy = \varphi(x)/f(x), \text{ where}$$
$$\varphi(x) = \int_{-\infty}^{\infty} y f_{XY}(x, y)\, dy$$

and f is the pdf of X. We want to estimate the function m.

Problem 3. Let (T, C) be independent positive-valued rv's, where T is the survival time of sample unit with cdf F and pdf f, observations on which may be stopped at time C. We want to estimate the survival function $S(t) = 1 - F(t)$ and the hazard function $\lambda(t) = f(t)/S(t)$ based on iid observations on (T, C).

We shall develop methods for estimating the functions f and m, and look at the asymptotic properties of these estimators as $n \to \infty$. Methods for estimating the survival function and integrated hazard function will also be constructed.

9.2 Density Estimation

Since the empirical cdf $F_n(x) = n^{-1} \sum_{i=1}^{n} I_{(-\infty,x]}(X_i)$ is a natural estimator of F, we can attempt to estimate $f = F'$ via the estimator F_n of F, using the relation

$$\begin{aligned} f(x) &= \lim_{h \downarrow 0} h^{-1}\left[F(x + h/2) - F(x - h/2)\right] \\ &\approx h^{-1}\left[F(x + h/2) - F(x - h/2)\right] \end{aligned}$$

Theory and Methods of Statistics. http://dx.doi.org/10.1016/B978-0-12-802440-9.00009-6

for small $h > 0$. This leads to the estimator

$$\begin{aligned} f_n(x) &= h_n^{-1}\left[F_n(x + h_n/2) - F_n(x - h_n/2)\right] \\ &= (nh_n)^{-1}\sum_{i=1}^{n} I_{(x-h_n/2,x+h_n/2]}(X_i) \\ &= (nh_n)^{-1}\sum_{i=1}^{n} I_{(-1/2,1/2]}\left(\frac{x - X_i}{h_n}\right). \end{aligned} \quad (1)$$

The data distribution is a discrete distribution with mass $1/n$ at each X_i. If we spread the discrete mass uniformly over an interval of length h_n centered around X_i, then the mass $1/n$ at X_i is replaced by a histogram of height $(nh_n)^{-1}$ on the interval $(X_i - h_n/2, X_i + h_n/2]$. Putting all these histograms around $X_1, \ldots, X_n$ together, we obtain $f_n(x)$ defined by Eq. (1). In this intuitive description, we could spread the mass $1/n$ at each X_i by an arbitrary pdf $h_n^{-1}K\left((\cdot - X_i)/h_n\right)$ instead of the uniform pdf $h_n^{-1}I_{(-1/2,1/2]}\left((\cdot - X_i)/h_n\right)$. This would define a general class of estimators of the form

$$f_n(x) = (nh_n)^{-1}\sum_{i=1}^{n} K((x - X_i)/h_n), \quad (2)$$

where K is a pdf (ie, $K(u) \geq 0$ and $\int_{-\infty}^{\infty} K(u)\, du = 1$). Such an estimator is called a *kernel estimator* with *kernel K* and *bandwidth* h_n. Since h_n serves the purpose of spreading discrete masses over the support of K, it is called a *smoothing parameter* of the estimator $f_n(x)$.

Note that f_n is itself a pdf, because $f_n(x) \geq 0$ for all x and it is easy to check that $\int_{-\infty}^{\infty} f_n(x)\, dx = 1$.

The kernel K and the bandwidth h_n are chosen by the user. The following kernels are often used:

$$\begin{aligned} \text{Uniform: } K(u) &= I_{[-1/2,1/2]}(u), \\ \text{Logistic: } K(u) &= e^u/(1 + e^u)^2, \\ \text{Epanechnikov: } K(u) &= (3/4)\left(1 - u^2\right)I_{[-1,1]}(u). \end{aligned}$$

Actually, the estimators are not very sensitive to the choice of K, but they are very sensitive to the choice of h_n.

For early work on kernel estimators of density functions and regression functions (discussed in the next section), we refer to Rosenblatt [45], Parzen [46], and Nadaraya [47].

Properties of Kernel Estimators of Density Functions

We first look at the effect of h_n on the estimator f_n in terms of its mean and variance.

It should be intuitively clear that the bias increases as h_n increases (eg, with uniform kernel, $f_n(x)$ estimates $h_n^{-1}\left[F(x + h_n/2) - F(x - h_n/2)\right]$ instead of $\lim_{h\downarrow 0} h^{-1}\left[F(x + h/2) - F(x - h/2)\right]$), and the variance increases as h_n decreases (because fewer observations make appreciable contributions as the bandwidth shrinks). Therefore, we should choose h_n appropriately to strike a balance between bias and variance in order to achieve the best rate at which the mean-square error $(MSE) = Bias^2 + Variance$ tends to 0 as $n \to \infty$, remembering that $h_n \downarrow 0$ as $n \to \infty$ anyway.

$$\begin{aligned} \mathrm{E}[f_n(x)] &= \mathrm{E}\left[h_n^{-1}K((x-X)/h_n)\right] = \int_{-\infty}^{\infty} h_n^{-1}K((x-y)/h_n)f(y)\,dy \\ &= \int_{-\infty}^{\infty} K(u)f(x-h_n u)\,du, \end{aligned} \tag{3}$$

letting $u = (x-y)/h_n$, and similarly,

$$\begin{aligned} \mathrm{Var}[f_n(x)] = n^{-1}\Big[h_n^{-1}\int_{-\infty}^{\infty} K^2(u)f(x-h_n u)\,du \\ -\left\{\int_{-\infty}^{\infty} K(u)f(x-h_n u)\,du\right\}^2\Big]. \end{aligned} \tag{4}$$

We make the following assumptions:

1. (a) K is a symmetric pdf (ie, $K(-u) = K(u)$ for all u), (b) $\sigma_K^2 = \int_{-\infty}^{\infty} u^2K(u)\,du < \infty$, and (c) $\|K\|^2 = \int_{-\infty}^{\infty} K^2(u)\,du < \infty$.
2. f'' is bounded and continuous.

Assumption 1 poses no problem, because K is chosen by the user. If Assumption 2 is replaced by other smoothness conditions of F, then the results we now derive, will change accordingly.

Expand $f(x-h_n u)$ in the expressions (3) and (4) of $\mathrm{E}[f_n(x)]$ and $\mathrm{Var}[f_n(x)]$:

$$\begin{aligned} f(x-h_n u) =& f(x) - h_n u f'(x) + (1/2)h_n^2u^2f''(x) \\ &+ (1/2)h_n^2u^2\{f''(-\lambda h_n u) - f''(x)\}, \quad 0 \le \lambda \le 1. \end{aligned} \tag{5}$$

Since $\int_{-\infty}^{\infty} uK(u)\,du = 0$ by Assumption 1(a) and

$$\lim_{n\to\infty}\int_{-\infty}^{\infty}\{f''(x-h_n u) - f''(x)\}u^2K(u)\,du = 0$$

by Assumption 2 and Dominated Convergence, using Eq. (5) we have

$$\begin{aligned} &Bias[f_n(x)] = \mathrm{E}[f_n(x)] - f(x) = (1/2)h_n^2\left[\sigma_K^2 f''(x) + o(1)\right] \text{ and} \\ &\mathrm{Var}[f_n(x)] = (nh_n)^{-1}\left[\|K\|^2 f(x) + o(1)\right]. \end{aligned}$$

Combining the above two formulas, we have

$$MSE[f_n(x)] = \left(h_n^4\right)\left[\sigma_K^4 f''(x)^2/4 + o(1)\right] + (nh_n)^{-1}\left[\|K\|^2 f(x) + o(1)\right],$$

so that

$$n^{4/5}MSE[f_n(x)] = \left(n^{1/5}h_n\right)^4\left[\sigma_K^4 f''(x)^2/4 + o(1)\right] + \left(n^{1/5}h_n\right)^{-1}\left[\|K\|^2 f(x) + o(1)\right]. \tag{6}$$

On the right-hand side of Eq. (6), the first term $\to\infty$ if $n^{1/5}h_n \to \infty$ and the second term $\to\infty$ if $n^{1/5}h_n \to 0$. Thus $n^{4/5}MSE[f_n(x)] \to \infty$ if $n^{1/5}h_n$ either tends to 0 or to ∞, and remains bounded if h_n is of the order of magnitude $n^{-1/5}$. Taking $h_n = tn^{-1/5}$, we have

$$n^{4/5}MSE[f_n(x)] = t^4\sigma_K^4 f''(x)^2/4 + t^{-1}\|K\|^2 f(x) + o(1).$$

Finally, $t^4a + t^{-1}b$ with $a, b > 0$ is minimized at $t_0 = \left(b/(4a)\right)^{1/5}$.

Using this, we see that $n^{4/5}MSE[f_n(x)]$ is minimized with $h_n = n^{-1/5}t_0$ where $t_0 = \left[\{\|K\|^2 f(x)\}/\{\sigma_K^4 f''(x)^2\}\right]^{1/5}$.

For example, if K is the uniform kernel on $[-1/2, 1/2]$, then with $\|K\|^2 = 1$ and $\sigma_K^2 = 1/12$, we have $t_0 = \left[144f(x)/f''(x)^2\right]^{1/5}$.

Since t_0 involves the unknown $f(x)$ and $f''(x)$, one way to implement this in practice would be to obtain initial (consistent) estimates of $f(x)$ and $f''(x)$ from the data and then "plug in" these estimates in the formula for t_0.

So far we have considered the estimation of $f(x)$ at a specific x. However, we are often interested in estimating the entire function f, in which case, one would like to use the same bandwidth for all x. For this, the integrated mean-square error $IMSE(f_n) = \int MSE[f_n(x)]\ dx$ or integrated square error $\int [f_n(x) - f(x)]^2\ dx$ would be a reasonable criterion to minimize. There is a huge literature on the issue of bandwidth choice in density estimation and regression estimation, some of which will be discussed at the end of this chapter.

We shall now establish the following asymptotic properties of f_n:

Theorem 9.2.1 (Strong Uniform Consistency). *In addition to Assumptions 1(a, b) and 2, suppose that the kernel K is of bounded variation on $(-\infty, \infty)$. Then $\sup_x |f_n(x) - f(x)| \to 0$ a.s., provided that $nh_n^2/\log n \to \infty$ as $n \to \infty$. [In particular, this holds for $h_n = O(n^{-1/5})$.]*

Theorem 9.2.2 (Asymptotic Normality). *Under Assumptions 1(a, b, c) and 2, with $h_n = n^{-1/5}t$,*

$$n^{2/5}[f_n(x) - f(x)] \xrightarrow{\mathcal{L}} N\left((1/2)t^2\sigma_K^2 f''(x), t^{-1}\|K\|^2 f(x)\right).$$

We now prove the strong uniform consistency property. The asymptotic normality of $f_n(x)$ will be proved together with that of the regression estimator $m_n(x)$ of $m(x)$ to be discussed later.

We first state a result from Real Analysis and a probability inequality to be used in the proof of Theorem 9.2.1 before proving the theorem.

Theorem 9.2.3. *If K is of bounded variation on $(-\infty, \infty)$ and $\int |K(u)|\ du < \infty$, then* $\lim_{u\to\pm\infty} K(u) = 0$.

Theorem 9.2.4 (Dvoretzky-Kiefer-Wolfowitz). *If F_n is the empirical cdf of a random sample of size n from F, then there exists a constant C so that*

$$P\left[\sup_x |F_n(x) - F(x)| > a\right] \le C\exp\left(-2na^2\right).$$

Proof of Theorem 9.2.1. Note that $\sup_x |f_n(x) - f(x)| \le A_n + B_n$, where

$$\begin{aligned}
B_n &= \sup_x |\mathrm{E}[f_n(x)] - f(x)| = \sup_x \left| \int_{-\infty}^{\infty} K(u)\{f(x - h_n u) - f(x)\}\, du \right| \\
&= \sup_x \left| \int_{-\infty}^{\infty} \left[-h_n u f'(x) + (1/2) h_n^2 u^2 f''(x - \lambda h_n u) \right] K(u)\, du \right| \\
&\le (1/2) h_n^2 \sigma_K^2 \sup_x |f''(x)| = O\left(h_n^2\right) = o(1),
\end{aligned}$$

using Assumptions 1(a, b) and 2, and

$$\begin{aligned}
A_n &= \sup_x |f_n(x) - \mathrm{E}[f_n(x)]| \\
&= \sup_x \left| (nh_n)^{-1} \sum_{i=1}^{n} K((x - X_i)/h_n) - \mathrm{E}\left[(nh_n)^{-1} \sum_{i=1}^{n} K((x - X_i)/h_n) \right] \right| \\
&= h_n^{-1} \sup_x \left| \int_{-\infty}^{\infty} K((x - y)/h_n) d[F_n(y) - F(y)] \right| \\
&= h_n^{-1} \sup_x \left| \int_{-\infty}^{\infty} -[F_n(y) - F(y)] K((x - y)/h_n)\, dK((x - y)/h_n) \right| \\
&\le (\mu/h_n) \sup_x |F_n(x) - F(x)|,
\end{aligned}$$

where $\mu = \bigvee_{-\infty}^{\infty} K$ is the total variation of K on $(-\infty, \infty)$, having used integration by parts and observing that $\lim_{y \to \pm\infty} h_n^{-1} K((x - y)/h_n)\{F_n(y) - F(y)\} = 0$, all this by using Theorem 9.2.3. We shall now show that for arbitrary $\varepsilon > 0$, $\sum_{n=1}^{\infty} P[\sup_x |f_n(x) - \mathrm{E}[f_n(x)]| > \varepsilon] < \infty$ and then $A_n \to 0$ a.s., will follow by the Borel-Cantelli Lemma. To this end, we use Theorem 9.2.4 to see that

$$\begin{aligned}
P\left[\sup_x |f_n(x) - \mathrm{E}[f_n(x)]| > \varepsilon \right] &\le P\left[(\mu/h_n) \sup_x |F_n(x) - F(x)| > \varepsilon \right] \\
&= P\left[\sup_x |F_n(x) - F(x)| > h_n \varepsilon/\mu \right] \\
&\le C \exp\left[-2n (h_n \varepsilon/\mu)^2 \right] = C \exp\left[-2 (\varepsilon/\mu)^2 n h_n^2 \right].
\end{aligned}$$

Since $nh_n^2/\log n \to \infty$ as $n \to \infty$, $(\varepsilon/\mu)^2 nh_n^2 > \log n$ (ie, $\exp\left[-2(\varepsilon/\mu)^2 nh_n^2\right] < 1/n^2$ for large n) and $\sum_{n=1}^{\infty} 1/n^2 < \infty$. □

Functions of bounded variation, Stieltjes integral and integration by parts are discussed in Appendix A.3.

Remark 9.2.1. Strong uniform consistency of f_n holds under milder conditions than stated above. It is enough to assume:

1*. K is a pdf of bounded variation.
2*. f is uniformly continuous (and is therefore bounded).
3*. $h_n \downarrow 0$ and $nh_n^2/\log n \to \infty$ as $n \to \infty$.

To prove Theorem 9.2.1 under these conditions, note that only 1* and 3* are used in proving $A_n \to 0$ a.s. in the above proof, so we only need to modify the proof of the fact that $B_n = \sup_x | \mathrm{E}[f_n(x)] - f(x)| \to 0$ as $n \to \infty$ using $\int_{-\infty}^{\infty} K(u)\, du < \infty$ and 2*. The main points to note are that under these conditions, for every $\varepsilon > 0$,

(i) we can choose M such that $\int_{|u|>M} K(u) < \varepsilon/\{4 \sup_x f(x)\}$, so that

$$\int_{|u|>M} |f(x - h_n u) - f(x)| K(u)\, du < \varepsilon/2 \quad \text{for all } x, \text{ and}$$

(ii) we can choose $\delta_\varepsilon > 0$ such that

$$|h| < \delta_\varepsilon \Longrightarrow \sup_x |f(x-h) - f(x) < \varepsilon/2.$$

The details of the proof is left as an *exercise*.

Remark 9.2.2. For rates of convergence of kernel estimators of density functions and their derivatives, see Bhattacharya [48]. An estimator of Fisher-information of a location family of unknown form has also been constructed in this paper.

9.3 Regression Estimation

The regression function Y on X is $m(\cdot)$, where

$$m(x) = \varphi(x)/f_X(x), \; \varphi(x) = \int_{-\infty}^{\infty} y f_{XY}(x, y)\, dy.$$

Since $f_X(x)$ is estimated by $f_n(x)$ given in Eq. (2), the main thing is to estimate $\varphi(x)$. Analogous to $f(x) = F'(x)$, $\varphi(x)$ can be expressed as

$$\begin{aligned}
\varphi(x) &= \frac{d}{dx} \int_{t=-\infty}^{x} \left[\int_{y=-\infty}^{\infty} y f_{XY}(t, y)\, dy \right] dt \\
&= \lim_{h \downarrow 0} h^{-1} \int_{x-h/2}^{x+h/2} \int_{y=-\infty}^{\infty} y f_{XY}(t, y)\, dy\, dt \\
&= \lim_{h \downarrow 0} h^{-1} \int_{-\infty}^{\infty} \int_{-\infty}^{\infty} I_{[x-h/2, x+h/2]}(t) y f_{XY}(t, y)\, dy\, dt \\
&= \lim_{h \downarrow 0} \mathrm{E}\left[h^{-1} I_{[x-h/2, x+h/2]}(X) Y\right] = \lim_{h \downarrow 0} \mathrm{E}\left[h^{-1} I_{[-1/2, 1/2]}((x - X)/h) Y\right],
\end{aligned}$$

so a natural estimator of $\varphi(x)$ is

$$\varphi_n(x) = (nh_n)^{-1} \sum_{i=1}^{n} I_{[-1/2, 1/2]}((x - X_i)/h_n) Y_i$$

with small $h_n > 0$, or more generally,

$$\varphi_n(x) = (nh_n)^{-1} \sum_{i=1}^{n} K((x - X_i)/h_n)Y_i, \tag{7}$$

using a pdf K as the kernel.

This leads to the kernel regression estimator

$$m_n(x) = \frac{(nh_n)^{-1} \sum_{i=1}^{n} K((x - X_i)/h_n)Y_i}{(nh_n)^{-1} \sum_{i=1}^{n} K((x - X_i)/h_n)} = \frac{\varphi_n(x)}{f_n(x)}, \tag{8}$$

where $h_n \downarrow 0$ as $n \to \infty$.

The estimator $m_n(\cdot)$ also has the strong uniform convergence property, which we state below without proof.

Theorem 9.3.1. *Suppose that the following conditions hold:*

1. **(a)** $P[a \le X \le b, c \le Y \le d] = 1$ *for some* $a < b$ *and* $c < d$,
   (b) $f_X(x)$ *is bounded away from 0 on* $[a, b]$,
   (c) $|f_{XY}(x_1, y) - f_{XY}(x_2, y)| \le M|x_1 - x_2|$ *for some* M *and for all* $x_1, x_2 \in [a, b]$ *and* $y \in [c, d]$.
2. K *is a bounded symmetric pdf on* $[-1, 1]$.
3. $h_n \to 0$ *and* $nh_n/\log n \to \infty$ *as* $n \to \infty$.

Then $\sup_{a \le x \le b} |m_n(x) - m(x)| \to 0$ *a.s. at the rate of* $r_n = h_n + \sqrt{\log n/(nh_n)}$ *(ie,* $r_n^{-1} \sup_{a \le x \le b} |m_n(x) - m(x)|$ *is bounded with probability 1).*

We now state a theorem on the asymptotic normality of the bivariate sequence $(f_n(x), \varphi_n(x))$ given by Eqs. (2) and (7). In particular, this will establish the asymptotic normality of $f_n(x)$, proving Theorem 9.2.2, and the asymptotic normality of $m_n(x) = \varphi_n(x)/f_n(x)$ given in Eq. (8), will follow by the delta method.

We shall make the following assumptions:

1. The second derivative f_X'', m'', and v'', where $v(x) = \mathrm{Var}[Y|X = x]$ exist, and are bounded and continuous.
2. The kernel K is a symmetric pdf with $\sigma_K^2 = \int u^2 K(u)\, du < \infty$ and $\|K\|^2 = \int K^2(u)\, du < \infty$.

Theorem 9.3.2. *Under Assumptions 1 and 2, with* $h_n = n^{-1/5} t$,

$$n^{2/5} \begin{bmatrix} f_n(x) - f(x) \\ \varphi_n(x) - \varphi(x) \end{bmatrix}$$

$$\xrightarrow{\mathcal{L}} N_2\left(\frac{1}{2} t^2 \sigma_K^2 \begin{bmatrix} f''(x) \\ \varphi''(x) \end{bmatrix}, t^{-1} \|K\|^2 f(x) \begin{bmatrix} 1 & m(x) \\ m(x) & v(x) + m^2(x) \end{bmatrix}\right).$$

Corollary 9.3.1. *Convergence of the first coordinate gives us Theorem 9.2.2, and using the delta method with $g(u, w) = w/u$, we have*

$$n^{2/5}[m_n(x) - m(x)] \xrightarrow{\mathcal{L}} N(\beta(x), \Psi(x)), \text{ where}$$
$$\beta(x) = (1/2)t^2\sigma_K^2 \frac{\varphi''(x) - f''(x)m(x)}{f(x)} \text{ and } \Psi(x) = t^{-1}\|K\|^2 \frac{v(x)}{f(x)}.$$

Verification of the formulas for $\beta(x)$ and $\Psi(x)$ is left as an *exercise*.

Proof of Theorem 9.3.2. We shall use the Cramér-Wold device, namely,

$$a_1U_n + a_2W_n \xrightarrow{\mathcal{L}} a_1U + a_2W \text{ for all } (a_1, a_2) \Rightarrow (U_n, W_n) \xrightarrow{\mathcal{L}} (U, W).$$

Let us, therefore, consider

$$\begin{aligned}\xi_{ni} &= a_1h_n^{-1}K((x - X_i)/h_n) + a_2h_n^{-1}K((x - X_i)/h_n)Y_i \\ &= h_n^{-1}K((x - X_i)/h_n)(a_1 + a_2Y_i).\end{aligned}$$

Then for each n, $\{\xi_{ni}, i = 1, \ldots, n\}$ are iid. Let

$$\mu_{ni} = \mathrm{E}[\xi_{ni}],\ \sigma_{ni}^2 = \mathrm{Var}[\xi_{ni}],\ A_n = \sum_{i=1}^{n}\mu_{ni} = n\mu_{n1}, \text{ and } B_n^2 = \sum_{i=1}^{n}\sigma_{ni}^2 = n\sigma_{n1}^2.$$

Letting $m(x) = \mathrm{E}[Y|X = x]$ and $v(x) = \mathrm{Var}[Y|X = x]$, we have

$$\begin{aligned}\mu_{ni} &= \iint h_n^{-1}K((x - t)/h_n)(a_1 + a_2y)f_{Y|X}(y|t)f_X(t)\, dy\, dt \\ &= \iint K(u)(a_1 + a_2y)f_{Y|X}(y|x - h_nu)f_X(x - h_nu)\, dy\, du \\ &= \int K(u)\{a_1 + a_2m(x - h_nu)\}f_X(x - h_nu)\, du,\end{aligned}$$

and similarly,

$$\sigma_{ni}^2 = h_n^{-1}\int K^2(u)\left[\{a_1 + a_2m(x - h_nu)\}^2 + a_2^2v(x - h_nu)\right]f(x - h_nu)\, du - \mu_{ni}^2.$$

Assumptions 1 and 2 allow us to expand $f(x - h_nu)$, $m(x - h_nu)$, and $v(x - h_nu)$ to second-order terms about x in the above expressions. After algebraic manipulations, this leads to

$$\begin{aligned}&\mu_{ni} = \{a_1 + a_2m(x)\}f(x) + (1/2)h_n^2[\alpha(x) + o(1)] \text{ and} \\ &\sigma_{ni}^2 = h_n^{-1}[\gamma(x) + o(1)],\ B_n = n\sigma_{n1}^2 = nh_n^{-1}[\gamma(x) + o(1)]\end{aligned} \tag{9}$$

where

$$\begin{aligned}&\alpha(x) = \sigma_K^2[a_1f''(x) + a_2\{f(x)m''(x) + f''(x)m(x) + 2f'(x)m'(x)\}], \\ &\gamma(x) = \|K\|^2\left[\{a_1 + a_2m(x)\}^2 + a_2^2v(x)\right]f(x).\end{aligned} \tag{9a}$$

To establish asymptotic normality of $\sum_{i=1}^n(\xi_{ni}-\mu_{ni})/B_n$, we now check the Lindeberg Condition:

$$\lim_{n\to\infty} B_n^{-2}\sum_{i=1}^n \mathrm{E}\Big[I_{(\varepsilon B_n,\infty)}\big(|\xi_{ni}-\mu_{ni}|\big)(\xi_{ni}-\mu_{ni})^2\Big]=0,$$

the verification of which is left as an *exercise.*

We thus have $\sum_{i=1}^n(\xi_{ni}-\mu_{ni})/B_n \xrightarrow{\mathcal{L}} N(0,1)$.

Using Eqs. (9) and (9a), $\sum_{i=1}^n(\xi_{ni}-\mu_{ni})/B_n$ simplifies to

$$\sqrt{\frac{nh_n}{\gamma(x)+o(1)}}\Big[a_1\Big\{f_n(x)-f(x)-(1/2)h_n^2\sigma_K^2 f''(x)(1+o(1))\Big\}$$
$$+a_2\Big\{\varphi_n(x)-\varphi(x)-(1/2)h_n^2\sigma_K^2\varphi''(x)(1+o(1))\Big\}\Big],$$

which $\xrightarrow{\mathcal{L}} N(0,1)$ for all (a_1,a_2).

Writing

$$\gamma(x)=\begin{pmatrix}a_1 & a_2\end{pmatrix}\begin{bmatrix}1 & m(x)\\ m(x) & v(x)+m^2(x)\end{bmatrix}\begin{pmatrix}a_1\\ a_2\end{pmatrix}\|K\|^2 f(x),$$

this implies

$$\begin{bmatrix}\sqrt{nh_n}\{f_n(x)-f(x)\}-(1/2)\sqrt{nh_n^5}\sigma_K^2 f''(x)(1+o(1))\\ \sqrt{nh_n}\{\varphi_n(x)-\varphi(x)\}-(1/2)\sqrt{nh_n^5}\sigma_K^2\varphi''(x)(1+o(1))\end{bmatrix}$$
$$\xrightarrow{\mathcal{L}} N_2\left(\begin{bmatrix}0\\0\end{bmatrix}, \|K\|^2 f(x)\begin{bmatrix}1 & m(x)\\ m(x) & v(x)+m^2(x)\end{bmatrix}\right).$$

Taking $h_n=n^{-1/5}t$, we arrive at the desired result. □

9.4 Nearest Neighbor Approach

Consider the uniform kernel with bandwidth h_n. In both density estimation and regression estimation, the methods discussed will suffer from the drawback that for those x where $f(x)$ is small, very few datapoints will have their X-values falling in the window $x \pm h_n/2$. As a result, very few datapoints will contribute to the construction of $f_n(x)$ or $m_n(x)$ where $f(x)$ is small.

To overcome this difficulty, one may decide to let the data determine the bandwidth h_n. The idea is to choose h_n large enough so that *exactly* $k=k_n$ of the X_i's are included in the window $x \pm h_n/2$. Clearly, this would put the burden of MSE more on the bias when $f(x)$ is small (causing h_n to be large) and more on the variance when $f(x)$ is large (causing h_n to be small).

Let $d_n = \inf\left\{h\colon \sum_{i=1}^{n} I_{[x-h,x+h]}(X_i) \geq k_n\right\}$. Then $\sum_{i=1}^{n} I_{[x-d_n,x+d_n]}(X_i) = k_n$. Now if we let d_n play the role of $h_n/2$ in the kernel estimation procedure with uniform kernel, then

$$\begin{aligned} f_n(x) &= (nh_n)^{-1}\sum_{i=1}^{n} I_{[x-h_n/2,x+h_n/2]}(X_i) \\ &= \frac{1}{n(2d_n)}\sum_{i=1}^{n} I_{[x-d_n,x+d_n]}(X_i) = \frac{k_n}{2nd_n}. \end{aligned}$$

This d_n is called the k_n-nearest neighbor (k_n-NN) distance from x and $f_n(x) = k_n/(2nd_n)$ is called the k_n-NN estimator of $f(x)$.

Here k_n is the smoothing parameter and the optimal rate at which $k_n \to \infty$ as $n \to \infty$ is $O(n^{4/5})$ (ie, we choose $k_n = [n^{4/5}t]$), where $[\cdot]$ is the greatest integer function. This is under the second-order smoothness condition assumed earlier.

Analogous argument in regression estimation leads to the k_n-NN regression estimator

$$\begin{aligned} m_n(x) &= \frac{(n2d_n)^{-1}\sum_{i=1}^{n} I_{[x-d_n,x+d_n]}(X_i)Y_i}{(n2d_n)^{-1}\sum_{i=1}^{n} I_{[x-d_n,x+d_n]}(X_i)} \\ &= k_n^{-1}\sum_{i=1}^{n} I_{[x-d_n,x+d_n]}(X_i)Y_i = k_n^{-1}\sum_{\{i\colon |X_i-x|\leq d_n\}} Y_i, \end{aligned}$$

that is, $m_n(x)$ is the mean of those k_n values of Y_i corresponding to the X_i's which are k_n nearest neighbors of x.

Another way to express this is to

(i) replace (X_i, Y_i) by $(|X_i - x|, Y_i) := (Z_i, Y_i), i = 1, \ldots, n$,
(ii) rank $Z_i = |X_i - x|, i = 1, \ldots, n$ as $0 < Z_{n:1} < \cdots < Z_{n:n}$ (strict inequality w.p. 1),
(iii) let $Y_{n:i}$ be the Y-value associated with $Z_{n:i}$, ie, $Y_{n:i} = Y_j \iff Z_{n:i} = Z_j$.

Then the k_n-NN estimator of $m(x)$ is $m_n(x) = k_n^{-1}\sum_{i=1}^{k_n} Y_{n:i}$.

9.5 Curve Estimation in Higher Dimension

Let $\boldsymbol{X}$ be a d-dim rv with pdf f and Y be a real-valued rv whose regression on $\boldsymbol{X}$ is $m(\boldsymbol{x}) = \mathrm{E}[Y|\boldsymbol{X} = \boldsymbol{x}]$. So far, we have discussed the problems of estimating the pdf f from iid observations on X and of estimating the regression function from iid observation $(X_1, Y_1), \ldots, (X_n, Y_n)$ when $d = 1$. We now consider these problems for $d \geq 2$. For this, the kernel and NN methods and the theoretical results for their asymptotics described so far, extend in a straightforward manner to higher dimensions. However, the actual sample size n needed for these estimators to perform reasonably well, increases rapidly with d.

For d-dim $\boldsymbol{X}_1, \ldots, \boldsymbol{X}_n$, we work with d -dim kernel K_d which is symmetric in each coordinate and define

$$f_n(\boldsymbol{x}) = \left(nh_n^d\right)^{-1} \sum_{i=1}^{n} K_d((\boldsymbol{x} - \boldsymbol{X}_i)/h_n) \text{ and}$$

$$m_n(\boldsymbol{x}) = \left(nh_n^d\right)^{-1} \sum_{i=1}^{n} K_d((\boldsymbol{x} - \boldsymbol{X}_i)/h_n) Y_i / f_n(\boldsymbol{x}).$$

In particular, we may choose $K_d(u_1, \ldots, u_d) = \prod_{j=1}^{d} K(u_j)$ where K is a pdf on $\mathbb{R}$ which is symmetric about 0.

To understand the difficulty of high dimensionality, let K_d be the uniform kernel on $\left[-1/2, 1/2\right]^d$. Then the only datapoints contributing to the construction of $f_n(\boldsymbol{x})$ and $m_n(\boldsymbol{x})$ are those with $X_i - x \in \left[-h_n/2, h_n/2\right]^d$, the number of which is of the order of nh_n^d. Without going into detailed calculations, it is easy to see that the bias of $f_n(\boldsymbol{x})$ or $m_n(\boldsymbol{x})$ will still be of the order of h_n^2 (as in the case of $d = 1$), but their variances will be of the order of $1/\left(nh_n^d\right)$. This will result in mean square error of the form

$$A(\boldsymbol{x})h_n^4 + B(\boldsymbol{x})/\left(nh_n^d\right) + R_n(\boldsymbol{x}).$$

Neglecting the remainder term, we see that

$$n^{4/(d+4)} MSE \approx A(\boldsymbol{x})\left(n^{1/(d+4)} h_n\right)^4 + B(\boldsymbol{x})\left(n^{1/(d+4)} h_n\right)^{-d},$$

which blows up if $n^{1/(d+4)} h_n$ either $\to 0$ or ∞. Thus the optimal $h_n = n^{-1/(d+4)} t$ with a suitable t, resulting in $MSE = O\left(n^{-4/(d+4)}\right)$. This shows how the rate at which the MSE converges to 0 slows down as d increases. In the Curve Estimation literature, this phenomenon is called "Curse of Dimensionality."

For example, if the X_i's are sampled from $N_d(\boldsymbol{0}, \boldsymbol{I})$, then in order to estimate this density at $\boldsymbol{x} = \boldsymbol{0}$ by the kernel method using normal kernel and optimal bandwidth, the sample size n needed for

$$\text{Relative } MSE[f_n(\boldsymbol{x})] = \mathrm{E}\left[\{f_n(\boldsymbol{x}) - f(\boldsymbol{x})\}^2\right] / f^2(\boldsymbol{x})$$

to be <0.1 increases from $n = 4$ for $d = 1$ to $n = 67$ for $d = 3$ to $n = 10{,}700$ for $d = 7$ (see [49]).

9.6 Curve Estimation Using Local Polynomials

Let us first briefly review the result on the kernel estimator $m_n(x)$ of the unknown regression function $m(x)$ as given in Corollary 9.3.1, where asymptotic normality of the estimator $m_n(x)$ is established after suitable renormalization. Asymptotic bias is

$$\text{asymptotic bias: } \left(h_n^2/2\right)\sigma_K^2 \frac{\varphi''(x) - f''(x)m(x)}{f(x)},$$

where $h_n = n^{-1/5}t$, $t > 0$. Since $\varphi(x) = m(x)f(x)$, we get $\varphi''(x) - f''(x)m(x) = m''(x)f(x) + 2m'(x)f'(x)$. From Corollary 9.3.1 we get the asymptotic bias and asymptotic variance of $m_n(x)$ as

$$\text{asymptotic bias: } \left(h_n^2/2\right)\sigma_K^2\left[m''(x) + 2m'(x)f'(x)/f(x)\right], \text{ and}$$
$$\text{asymptotic variance: } \left(nh_n\right)^{-1}\|K\|^2 v(x)/f(x),$$

where $v(x) = \text{Var}[Y|X = x]$. Note that the bias term involves $f'(x)$ (ie, it depends on the smoothness of the marginal density of X). In other words, the design of the X_i's enters the picture in regression estimation. Moreover, the estimator $m_n(x)$ has another aspect which makes it difficult to use in practice. Suppose that f is supported on a compact interval which we take to be $[0, 1]$ without loss of generality. The bias of the estimator $m_n(x)$ at or near the boundary points 0 or 1 is of order h_n and not h_n^2. Thus the regression estimate is less reliable at or near the boundary points. The local polynomial method seeks to remove these negative aspects of the kernel estimator m_n.

In order to simplify notations let us denote

$$w_i(x) = \frac{K((x - X_i)/h_n)}{\sum_{j=1}^n K\left(\left(x - X_j\right)/h_n\right)}.$$

Thus we may rewrite the kernel estimator as $m_n(x) = \sum_{i=1}^n w_i(x)Y_i$. Since $w_i(x) \geq 0$ for all i, and $\sum_{i=1}^n w_i(x) = 1$, $m_n(x)$ is a weighted average of Y_i's with weights $w_i(x)$. If the kernel K is compactly supported, say on $\left[-1/2, 1/2\right]$, then $w_i(x) = 0$ whenever $|X_i - x| > h_n/2$. Note that a regression model for (X_i, Y_i) is of the form $Y_i = m(X_i) + \varepsilon_i$, where $\{\varepsilon_i\}$ are independent with $E[\varepsilon_i|X_i] = 0$ for all i. If X_i is in the neighborhood $N_n(x) = \left\{u: |u - x| \leq h_n/2\right\}$ (ie, $|X_i - x| \leq h_n/2$) and we approximate $m(X_i)$ by a constant $\beta_0 = m(x)$, then the regression model takes the approximate form $Y_i = \beta_0 + \varepsilon_i$ when X_i is in $N_n(x)$. If we obtain an estimate of β_0 from a weighted least squares criterion of the form $\sum_{i=1}^n (Y_i - \beta_0)^2 w_i(x)$, which is minimized with respect to β_0, then $m_n(x)$ is that value of β_0 which minimizes this local weighted least squares.

Now if we approximate $m(X_i)$ by a straight line instead of a constant when X_i is in $N_n(x)$, then a simple Taylor series expansion yields

$$m(X_i) = \beta_0 + \beta_1(X_i - x) + O\left(\left(X_i - x\right)^2\right),$$

where $\beta_0 = m(x)$ and $\beta_1 = m'(x)$. Thus the regression model described in the last paragraph can be approximately expressed as $Y_i = \beta_0 + \beta_1(X_i - x) + \varepsilon_i$ when X_i is in $N_n(x)$. The parameters β_0 and β_1 can be estimated by using the method of weighted least squares. If the estimate of β_0 is $\hat{\beta}_0$, then our estimate of $m(x)$ is $\hat{\beta}_0$. More formally, we seek to minimize $Q = \sum_{i=1}^n \left[Y_i - \beta_0 - \beta_1(X_i - x)\right]^2 w_i(x)$ with respect to β_0 and β_1. If we differentiate Q with respect to β_0 and β_1 and equate the derivatives to 0, we are led to the equations

$$\beta_0 + \sum w_i(x)(X_i - x)\beta_1 = \sum w_i(x)Y_i,$$
$$\sum w_i(x)(X_i - x)\beta_0 + \sum w_i(x)(X_i - x)^2\beta_1 = \sum w_i(x)(X_i - x)Y_i.$$

We can get an explicit expression for the estimate of β_0 from these equations after some tedious algebra. Let us simplify some notations by denoting

$$c_s(x) = \sum w_i(x)(X_i - x)^s, \quad s = 1, 2, 3.$$

Then the estimate of β_0 is

$$\hat{\beta}_0 = \frac{\sum w_i(x)\left[c_2(x) - c_1(x)(X_i - x)\right]Y_i}{\sum w_i(x)\left[c_2(x) - c_1(x)(X_i - x)\right]}.$$

Thus the local linear estimate of $m(x)$ is given by

$$m_n^{(LL)}(x) = \frac{\sum w_i(x)\left[c_2(x) - c_1(x)(X_i - x)\right]Y_i}{\sum w_i(x)\left[c_2(x) - c_1(x)(X_i - x)\right]} = \sum l_i(x)Y_i, \text{ where}$$

$$l_i(x) = \frac{w_i(x)\left[c_2(x) - c_1(x)(X_i - x)\right]}{\sum w_j(x)\left[c_2(x) - c_1(x)\left(X_j - x\right)\right]} = \frac{w_i(x)\left[c_2(x) - c_1(x)(X_i - x)\right]}{c_2(x) - c_1(x)^2}.$$

The local linear estimate of $m(x)$ is also a linear combination of Y_i 's with weights $l_i(x)$, and the weights sum to 1 (ie, $\sum l_i(x) = 1$). These weights also have the property $\sum l_i(x)(X_i - x) = 0$. This is unlike the regular kernel estimate $m_n(x) = \sum w_i(x)Y_i$ where the weights $w_i(x)$ do not necessarily satisfy the equation $\sum w_i(x)(X_i - x) = 0$. Why is this property important? It provides a correction term in the bias part. In order to see this, let us write

$$m_n^{(LL)}(x) - m(x) = \sum l_i(x)\left[Y_i - m(X_i)\right] + \sum l_i(x)\left[m(X_i) - m(x)\right].$$

The first sum in the last expression contributes toward the asymptotic variance of $m_n^{(LL)}(x)$, whereas the second sum contributes toward the asymptotic bias. We now examine the second sum in some detail. Let us assume that K is supported on a compact interval, say $\left[-1/2, 1/2\right]$, and it is symmetric about 0, m is twice differentiable, m'' is continuous on the compact interval $\left[a, b\right]$ on which m is being estimated and f is continuous on $\left[a, b\right]$. Under these conditions it can be shown that $\inf_i l_i(x) \geq 0$ with probability converging to 1. Since m'' is continuous on a compact interval, it is also uniformly continuous. Since $w_i(x) = 0$ whenever $|X_i - x| > h_n/2$, using a two-term Taylor expansion of $m(X_i)$ around $m(x)$, we have

$$\begin{aligned}\sum l_i(x)\left[m(X_i) - m(x)\right] &= \sum l_i(x)\left[(X_i - x)m'(x) + (1/2)(X_i - x)^2 m''(x)\right] + o_P\left(h_n^2\right) \\ &= (1/2)\sum l_i(x)(X_i - x)^2 m''(x) + o_P\left(h_n^2\right),\end{aligned}$$

because the sum involving the linear term with $X_i - x$ is exactly equal to zero (ie, $\sum l_i(x)(X_i - x) = 0$). A similar argument for the bias part for $m_n(x)$ (ie, $\sum w_i(x)\left[m(X_i) - m(x)\right]$) would involve a term of the form $\sum w_i(x)(X_i - x)$. In order for this term to be of order h_n^2, it is required that f is differentiable and f' is continuous. Such a condition is not needed for the local linear estimate.

Let us now examine the bias term for the local linear estimator $m_n^{(LL)}(x)$ a bit more closely and establish that $\sum l_i(x)(X_i - x)^2 = h_n^2[\sigma_K^2 + o_P(1)]$. Then the bias term would be $(1/2)h_n^2\sigma_K^2 m''(x) + o_P(h_n^2)$.

Using the expression for $l_i(x)$ we get

$$\sum l_i(x)(X_i - x)^2 = \sum \frac{w_i(x)[c_2(x) - c_1(x)(X_i - x)](X_i - x)^2}{c_2(x) - c_1(x)^2}$$
$$= \frac{c_2(x)^2 - c_1(x)c_3(x)}{c_2(x) - c_1(x)^2}.$$

For any nonnegative integer r, using Chebychev's inequality, it can be established that

$$\left(nh_n^{s+1}\right)^{-1} \sum K((x - X_i)/h_n)(X_i - x)^s$$
$$= (nh_n)^{-1} \sum K((x - X_i)/h_n)\{(X_i - x)/h_n\}^s \xrightarrow{P} \int u^s K(u)\, du\, f(x).$$

Noting that $\int u^s K(u)\, du = 0$ for $s = 1$ and $s = 3$ (since K is symmetric about 0), we have

$$c_1(x) = h_n\left[\int uK(u)\, du + o_P(1)\right] = o_P(h_n),$$
$$c_2(x) = h_n^2\left[\int u^2 K(u)\, du + o_P(1)\right] = h_n^2\left[\sigma_K^2 + o_P(1)\right], \text{ and}$$
$$c_3(x) = h_n^3\left[\int u^3 K(u)\, du + o_P(1)\right] = o_P\left(h_n^3\right).$$

Plugging in these approximations for $c_1(x)$, $c_2(x)$, and $c_3(x)$ in the expression for $\sum l_i(x)(X_i - x)^2$, we have

$$\sum l_i(x)(X_i - x)^2 = \frac{c_2(x)^2 - c_1(x)c_3(x)}{c_2(x) - c_1(x)^2}$$
$$= \frac{h_n^4\left[\sigma_K^2 + o_P(1)\right]^2 - o_P\left(h_n^4\right)}{h_n^2\left[\sigma_K^2 + o_P(1)\right] - o_P\left(h_n^2\right)} = h_n^2\left[\sigma_K^2 + o_P(1)\right].$$

Using arguments similar to the ones in proving Corollary 9.3.1, we can also get a similar result for the local linear estimate $m_n^{(LL)}(x)$.

Theorem 9.6.1. *Assume that*

(a) *on a compact interval $[a, b]$, m is twice differentiable, and m'', v, and f are continuous,*
(b) *the kernel K is a bounded pdf on a compact interval $[\ \ c, c]$ and is symmetric about zero, and*
(c) $\sup_{x \in [a,b]} E[Y^4 | X = x] < \infty$.

Taking $h_n = n^{-1/5}t$, *it can be shown that for* $x \in [a, b]$,

$$n^{2/5}\left[m_n^{(LL)}(x) - m(x)\right] \xrightarrow{\mathcal{L}} N\left(\beta^{(LL)}(x), \Psi^{(LL)}(x)\right), \text{ where}$$
$$\beta^{(LL)}(x) = (1/2)t^2\sigma_K^2 m''(x) \text{ and } \Psi(x) = t^{-1}\|K\|^2\frac{v(x)}{f(x)}.$$

Note that the asymptotic variances of $m_n(x)$ and $m_n^{(LL)}(x)$ are the same, but their asymptotic biases are different.

Remark 9.6.1. In this section, a few aspects of local polynomial method have been highlighted. More applications and details can be found in the book by Fan and Gijbels [50].

1. Instead of the local linear method for estimating $m(x)$, one may consider local polynomial estimation by minimizing $Q = \sum[Y_i - \beta_0 - \beta_1(X_i - x) - \cdots - \beta_p(X_i - x)^p]^2 w_i(x)$ with respect to $\beta_0, \beta_1, \ldots, \beta_p$. Then the local polynomial estimate of $m(x)$ is $m_n^{(LP)}(x) = \hat{\beta}_0$.
2. If the independent variable X is vector valued, one can obtain a local linear (or more generally a local polynomial) estimate of $m(\boldsymbol{x})$. In this case, one minimizes $Q = \sum[Y_i - \beta_0 - \boldsymbol{\beta}_1^T(\boldsymbol{X}_i - \boldsymbol{x})]^2 w_i(\boldsymbol{x})$ with respect to β_0 and $\boldsymbol{\beta}_1$, and as before the estimate of $m(\boldsymbol{x})$ is given by $\hat{\beta}_0$.
3. It is also possible to carry out density estimation using the local linear or local polynomial method.
4. The choice of bandwidth h_n is crucial as in any other curve estimation problem. A cross-validation method can also be employed to obtain an appropriate value of h_n. The method of cross-validation is described in the next subsection in the context of kernel density and regression estimation problems.

9.6.1 Choice of h_n in Density Estimation

Ideally, we would choose h_n which minimizes the Integrated Square-Error

$$\int (f_n - f)^2 = \int f_n^2 - 2\int f_n f + \int f^2,$$

in which the last term does not depend on f_n. So the aim is to minimize $R(f_n) = \int f_n^2 - 2\int f_n f$. The first term $\int f_n^2$ is calculated directly from f_n. So the main thing is to estimate the second term from the data. The idea of "leave-one-out" is used for this purpose. Let

$$f_{n,-i}(x) = \left[\{(n-1)h_n\}^{-1}\sum_{j\neq i=1}^{n} K\left(\left(x - X_j\right)/h_n\right)\right] \quad \text{for each } i = 1, \ldots, n. \tag{10}$$

Then

$$\mathrm{E}\left[n^{-1}\sum_{i=1}^{n} f_{n,-i}(X_i)\right] = \mathrm{E}[f_{n,-n}(X_n)] = \mathrm{E}\mathrm{E}[f_{n,-n}(X_n)|X_1,\ldots,X_{n-1}]$$
$$= \mathrm{E}\int f_{n,-n}(x)f(x)\,dx = \mathrm{E}\int f_n(x)f(x)\,dx,$$

because

$$\mathrm{E}[f_{n,-n}(x)] = \mathrm{E}\left[\{(n-1)h_n\}^{-1}\sum_{j=1}^{n-1} K\left(\left(x-X_j\right)/h_n\right)\right]$$
$$= \mathrm{E}\left[(nh_n]^{-1}\sum_{j=1}^{n} K\left(\left(x-X_j\right)/h_n\right)\right] = \mathrm{E}[f_n(x)].$$

Thus the estimator

$$\hat{R}(f_n) = \int f_n^2 - (2/n)\sum_{i=1}^{n} f_{n,-i}(X_i) \tag{11}$$

has the property: $\mathrm{E}\left[\hat{R}(f_n)\right] = \mathrm{E}\left[R(f_n)\right]$.

Hence minimizing $\mathrm{E}\left[\hat{R}(f_n)\right]$ is equivalent to minimizing $\mathrm{E}\left[R(f_n)\right]$, so a choice of h_n which minimizes $\hat{R}(f_n)$ will, hopefully, approximate the optimal h_n which minimizes $R(f_n)$ itself.

We now describe the actual cross-validation procedure. First, for computational facility, we express

$$\int f_n^2 = (nh_n)^{-2}\sum_{i=1}^{n}\sum_{j=1}^{n}\int K((X_i-x)/h_n)K\left(\left(x-X_j\right)/h_n\right)dx$$
$$= \left(n^2h_n\right)^{-1}\sum_{i=1}^{n}\sum_{j=1}^{n}\int K(X_i/h_n-u)K\left(u-X_j/h_n\right)du$$
$$= \left(n^2h_n\right)^{-1}\sum_{i=1}^{n}\sum_{j=1}^{n}\int K\left(\left(X_i-X_j\right)/h_n-\left(u-X_j/h_n\right)\right)K\left(u-X_j/h_n\right)du$$
$$= \left(n^2h_n\right)^{-1}\sum_{i=1}^{n}\sum_{j=1}^{n}K^{(2)}\left(\left(X_i-X_j\right)/h_n\right),$$

where $K^{(2)}$ is the convolution of K with itself, that is

$$K^{(2)}(z) = \int K(z-u)K(u)\,du = \text{ pdf of } Z_1 + Z_2,$$

where Z_1 and Z_2 are iid with pdf K. Next rewrite the term $n^{-1}\sum_{i=1}^{n} f_{n,-i}(X_i)$ with a minor modification to obtain

$$
\begin{aligned}
n^{-1}\sum_{i=1}^{n} f_{n,-i}(X_i) &= n^{-1}\sum_{i=1}^{n}[(n-1)h_n]^{-1}\sum_{j\neq i=1}^{n} K\left(\left(X_i - X_j\right)/h_n\right) \\
&\approx \left(n^2 h_n\right)^{-1}\sum_{i=1}^{n}\sum_{j=1}^{n} K\left(\left(X_i - X_j\right)/h_n\right) - (nh_n)^{-1}K(0).
\end{aligned}
$$

Putting the two terms together in Eq. (11), we have

$$
\begin{aligned}
\hat{R}(f_n) \approx& \hat{R}_1(f_n) \\
=& \left(n^2 h_n\right)^{-1}\sum_{i=1}^{n}\sum_{j=1}^{n} K^{(2)}\left(\left(X_i - X_j\right)/h_n\right) \\
& - 2\left\{\left(n^2 h_n\right)^{-1}\sum_{i=1}^{n}\sum_{j=1}^{n} K\left(\left(X_i - X_j\right)/h_n\right) - (nh_n)^{-1}K(0)\right\} \\
=& \left(n^2 h_n\right)^{-1}\sum_{i=1}^{n}\sum_{j=1}^{n} K^*\left((X_i - X_j\right)/h_n) + 2(nh_n)^{-1}K(0),
\end{aligned}
$$

where $K^*(\cdot) = K^{(2)}(\cdot) - 2K(\cdot)$.

The least squares cross-validation procedure is to choose the bandwidth as the value of h_n which minimizes $\hat{R}_1(f_n)$ (see [49, 51]).

Let $I(f_n; h_n) = \int (f_n - f)^2$, where f_n is given by Eq. (2) and let $h_n(CV)$, $h_n(opt)$ denote, respectively, the h_n obtained by minimizing $\hat{R}_1(f_n)$ and the unknown optimal h_n which minimizes $I(f_n; h_n)$ for the given data. Then the following optimality property holds for $h_n(CV)$ whenever f is bounded and K satisfies some mild conditions:

$$
\lim_{n\to\infty} \frac{I(f_n; h_n(CV))}{I(f_n; h_n(opt))} = 1 \quad \text{with probability 1.}
$$

The above discussion was for $d = 1$. All of this goes through for $d \geq 2$, by replacing h_n by h_n^d in the formula for $\hat{R}_1(f_n)$.

9.6.2 Regression Estimation

In regression estimation, analogous to density estimation, our goal may be to minimize

$$
\begin{aligned}
d(m_n; h_n) &= \int (m_n - m)^2 wf \\
&= \int m_n^2 wf - 2\int m_n m wf + \int m^2 wf,
\end{aligned}
$$

where f is the pdf of X and $w(x) > 0$ is a weight function. Again, the last terms does not depend on m_n, while the first two terms involve the unknown m and f and therefore, need to be estimated.

To motivate the proposed procedure, let (X_0, Y_0) be an observation on (X, Y) which is independent of the data $(X_1, Y_1), \ldots, (X_n, Y_n)$ and then rewrite the two terms as

$$\begin{aligned}\int m_n^2 wf - 2\int m_n mwf &= \mathrm{E}\left[\left\{m_n^2(X_0) - 2m_n(X_0)\mathrm{E}(Y_0|X_0)\right\}w(X_0)|Data\right] \\ &= \mathrm{E}\left[\mathrm{E}\left(\left\{m_n^2(X_0) - 2m_n(X_0)Y_0\right\}w(X_0)|X_0, Data\right)|Data\right] \\ &= \mathrm{E}\left[\left\{m_n^2(X_0) - 2m_n(X_0)Y_0\right\}w(X_0)|Data\right] \\ &= \mathrm{E}\left[\{Y_0 - m_n(X_0)\}^2 w(X_0)|Data\right] - \mathrm{E}\left[Y_0^2 w(X_0)\right].\end{aligned}$$

Since $\mathrm{E}\left[Y_0^2 w(X_0)\right]$ does not depend on the choice of h_n, we only need to estimate $\mathrm{E}\left[\{Y_0 - m_n(X_0)\}^2 w(X_0)|Data\right]$ and then choose h_n to minimize this estimate.

Again using the "leave-one-out" method, we construct the estimate

$$\hat{d}_1(m_n; h_n) = n^{-1}\sum_{i=1}^{n}\{Y_i - m_{n,-i}(X_i)\}^2 w(X_i)$$

for $\mathrm{E}\left[\{Y_0 - m_n(X_0)\}^2 w(X_0)|Data\right]$, where

$$m_{n,-i}(x) = \left[(n-1)h_n\right]^{-1}\sum_{j\neq i=1}^{n} K\left(\left(x - X_j\right)/h_n\right)Y_j/f_{n,-i}(x)$$

with $f_{n,-i}(x)$ as in Eq. (10).

Then the cross-validated choice $h_n(CV)$ and the optimal choice $h_n(opt)$ of h_n are, respectively, the minimizer of $\hat{d}_1(m_n; h_n)$ and $d(m_n; h_n)$. However, for technical reason, the choice of h_n is restricted to $\left[C_1 n^{-\delta}, C_2\right]$ for some constants C_1, C_2 and $\delta > 0$. This technical condition is also in conformity with the practice. To see this, assume that the kernel K is supported on $\left[-1/2, 1/2\right]$. Then the nonparametric regression estimate at x is simply a weighted average of Y_i's for which X_i is in the interval $\left[x - h_n/2, x + h_n/2\right]$. Now, if h_n is too small and x is not one of the X_i's, then this interval may not have X-observations, thus there are no observations to average on and one cannot get a nonparametric estimate of m at x.

Suppose that f is supported on a compact set in $\mathbb{R}^d$, on which it is bounded away from 0, and that f and m are continuous on this set. Then under assumption of boundedness of conditional moments of all orders of Y given X, the following property holds for $h_n(CV)$:

$$\frac{d(m_n; h_n(CV))}{d(m_n; h_n(opt))} \xrightarrow{P} 1.$$

9.7 Estimation of Survival Function and Hazard Rates Under Random Right-Censoring

Let $(T_1, C_1), \ldots, (T_n, C_n)$ be iid pairs of positive-valued rv's where for each i, T_i and C_i are independent. For each i, T_i is the *survival time* (ie, the time until death of a sample subject or the failure time of a sample equipment) and C_i is the *censoring time* (ie, time at which observation is stopped for this sample unit). Thus the observed data consist of $(Y_1, \delta_1), \ldots, (Y_n, \delta_n)$ where $Y_i = T_i \wedge C_i$ and $\delta_i = I(T_i \leq C_i)$, so $(Y_i, \delta_i) = (y, 0)$ means that the observation on the ith unit was censored at time y and $(Y_i, \delta_i) = (y, 1)$ means that the observation continued until death or failure which occurred at time y. This is *random right censoring* of survival time. Foe simplicity, we shall assume that there are no ties.

Let F denote the common cdf of the T_i's and let f denote the corresponding pdf.

Definition 9.7.1. The function $S(t) = 1 - F(t) = P[T > t]$ is called the survival function and $\lambda(t) = f(t)/[1 - F(t)]$ is called the hazard function or the hazard rate of the rv T.

Since $\lambda(t)\,dt \approx P\big[t < T < t + dt|T > t\big]$, it is also called the instantaneous failure rate.

The survival function $S(t)$ and the hazard function $\lambda(t)$ are related by the formula

$$S(t) = \exp\left[-\int_0^t \lambda(u)\,du\right], \tag{12}$$

the proof of which is left as an *exercise*.

Estimation of the Survival Function

We now consider the problem of estimating the survival function $S(t)$ from randomly right-censored data $\{(Y_i, \delta_i),\ i = 1, \ldots, n\}$. Let $Y_{n:1} < \cdots < Y_{n:n}$ denote the order statistics of $Y_1, \ldots, Y_n$ and let $\delta_{n:1}, \ldots, \delta_{n:n}$ be defined by $\delta_{n:i} = \delta_j \iff Y_{n:i} = Y_j$. At each $Y_{n:i}$, either a death or a censoring occurs, depending on whether $\delta_{n:i} = 1$ (death) or $\delta_{n:i} = 0$ (censoring).

Consider the intervals $I_i = (Y_{n:i-1}, Y_{n:i}]$, taking $Y_{n:0} = 0$, and let

$\mathcal{R}(t) =$ Risk set at time t consisting of those who are still alive at time $t-$,

$n_i = \#\mathcal{R}(Y_{n:i}) =$ number alive at time $Y_{n:i}-$,

$d_i =$ number dying at time $Y_{n:i} = \delta_{n:i}$,

$p_i = P\big[T > Y_{n:i}|T \geq Y_{n:i-1}\big]$ and $q_i = 1 - p_i$.

The natural estimates of q_i and p_i are

$$\hat{q}_i = d_i/n_i \text{ and } \hat{p}_i = 1 - \hat{q}_i = \begin{Bmatrix} 1 - 1/n_i & \text{if } \delta_{n:i} = 1 \\ 1 & \text{if } \delta_{n:i} = 0 \end{Bmatrix} = (1 - 1/n_i)^{\delta_{n:i}}.$$

The Product-Limit (PL) estimator of $S(t)$ due to Kaplan and Meier [52] is

$$\hat{S}(t) = \prod_{i:Y_{n:i} \leq t} \hat{p}_i = \prod_{i:Y_{n:i} \leq t} (1 - 1/n_i)^{\delta_{n:i}}, \tag{13}$$

because $n_i = \#\mathcal{R}(Y_{n:i-1}) = n - \#\{\text{dead or censored before } Y_{n:i-1}\} = n - i + 1$.

Remark 9.7.1. On the interval $(Y_{n:i-1}, Y_{n:i}]$, $\hat{S}(t)$ remains unchanged if the observation at $Y_{n:i}$ is censored and it is reduced by a factor of $(n-i)/(n-i+1)$ if $Y_{n:i}$ is uncensored (death).

Variance of $\hat{S}(t)$

Since $n_i\hat{p}_i \sim Bin(n_i, p_i)$,

$$\text{Var}[\log \hat{p}_i] \approx \text{Var}[\hat{p}_i](d\log p_i/dp_i)^2 = (p_iq_i/n_i)\left(1/p_i^2\right) = q_i/(n_ip_i)$$

by the delta method. Hence

$$\text{Var}\left[\log \hat{S}(Y_{n:i})\right] = \text{Var}\left[\sum_{j=1}^{i} \log \hat{p}_j\right] = \sum_{j=1}^{i} \text{Var}\left[\log \hat{p}_j\right] \approx \sum_{j=1}^{i} \hat{q}_j/\left(n_j\hat{p}_j\right),$$

assuming $\log \hat{p}_1, \log \hat{p}_2, \ldots$ are independent. Using the delta method again, we have

$$\begin{aligned}\text{Var}\left[\hat{S}(Y_{n:i})\right] &= \text{Var}\left[\exp\left(\log \hat{S}_{n:i}\right)\right] \approx \exp\left(2\log \hat{S}_{n:i}\right)\text{Var}\left[\log \hat{S}_{n:i}\right] \\ &\approx \hat{S}^2(Y_{n:i})\sum_{j=1}^{i} \hat{q}_j/\left(n_j\hat{p}_j\right).\end{aligned}$$

Thus

$$\text{Var}\left[\hat{S}(t)\right] \approx \hat{S}^2(t) \sum_{i:Y_{n:i}\le t} \hat{q}_i/(n_i\hat{p}_i) = \hat{S}^2(t) \sum_{i:Y_{n:i}\le t} \delta_{n:i}/\{n_i(n_i-1)\}.$$

Redistribute-to-the-Right Algorithm

This is another method, due to Efron [53], of calculating the PL estimator $\hat{S}(t)$ given by Eq. (13). If we start with a sample of size n,

- **(i)** first put a probability of $1/n$ at each $Y_{n:i}$;
- **(ii)** if $Y_{n:i_1}$ is the first censored observation, redistribute the probability $1/n$ assigned to $Y_{n:i_1}$ equally to $Y_{n:i_1+1}, \ldots, Y_{n:n}$, so that each of these observations now carries a probability of $(1/n)(1+1/(n-i_1))$;
- **(iii)** if $Y_{n:i_2}$ is the second censored observation, distribute the probability on $Y_{n:i_2}$ equally to $Y_{n:i_2+1}, \ldots, Y_{n:n}$, so that each of these observations now carries a probability of $(1/n)(1+1/(n-i_1))(1+1/(n-i_2))$, and so on.

A proof that this method leads to the same estimator as the one given by Eq. (13) is left as an *exercise*.

Estimation of the Integrated Hazard Function

Estimation of the hazard function from censored data would involve estimation of the pdf of T from censored data, which is difficult. On the other hand, estimating the integrated hazard function $\Lambda(t) = \int_0^t \lambda(u)\, du$ is straightforward, since $S(t) = \exp[-\Lambda(t)]$ by Eq. (12). We therefore estimate $\Lambda(t)$ by the estimator of $-\log S(t)$, that is,

$$\hat{\Lambda}_1(t) = -\log \hat{S}(t) = -\sum_{Y_{n:i}\le y} \log(1-\delta_{n:i}/(n-i+1))$$

because $\left(1-1/(n-i+1)\right)^{\delta_{n:i}} = 1-\delta_{n:i}/\left(n-i+1\right)$. Also since $-\log(1-x) \approx x$ for small x, we have another estimator

$$\hat{\Lambda}_2(t) = \sum_{Y_{n:i}\le t} \delta_{n:i}/(n-i+1)$$

which is approximately the same as $\hat{\Lambda}_1$.

Exercises

9.1. Show that a sufficient condition for a kernel estimator $f_n(x)$ of $f(x)$ with bandwidth h_n to be a consistent estimator is that $h_n \downarrow 0$ and $nh_n \to \infty$.

9.2. Suppose that f is m times differentiable and $f^{(m)}$ is bounded. Drop the condition that the kernel K is a pdf, but satisfies the conditions:
$\int K(u)\, du = 1$, $\int u^r K(u)\, du = 0$, $r = 1, \ldots, m-1$, $\int |u|^m K(u)\, du < \infty$, and $\int K^2(u)\, du < \infty$.

(a) Find the asymptotic bias and variance of the estimator $f_n(x) = \left(nh_n\right)^{-1} \sum_{i=1}^n K\left((x-X_i)/h_n\right)$ using such a kernel K.

(b) Determine the optimal rate at which h_n should tend to 0 and the corresponding rate of convergence of the MSE of $f_n(x)$.

9.3. Let $m_n(x)$ be a kernel estimator of the regression function $m(x)$ of Y on X at $X = x$ based on a random sample of size n. Verify the formula for the mean $\beta(x)$ and the variance $\Psi(x)$ of the asymptotic distribution of $n^{2/5}[m_n(x) - m(x)]$ given in the text.

9.4. Give a detailed proof of Theorem 9.2.1 under the milder conditions 1*, 2*, and 3*.

9.5. Find the formulas of bias and variance of the k_n-NN estimators of a pdf $f(x)$ and a regression function $m(x)$, and verify that the optimal rate at which $k_n \to \infty$ as $n \to \infty$ is $O\left(n^{4/5}\right)$.

9.6. Prove formula (12) on the relation between the survival function $S(t)$ and the hazard function $\lambda(t)$.

9.7. Prove that the redistribute-to-the-right algorithm leads to the same estimator as the one given by Eq. (13).

10

Statistical Functionals and Their Use in Robust Estimation

10.1 Introduction

Let $\mathcal{F}_0$ be a family of cdf's in $\mathbb{R}^d$. Then $T\colon \mathcal{F}_0 \to \mathbb{R}$ is called a statistical functional. Most statistical problems involve inference about such a $T(F)$ for an unknown $F \in \mathcal{F}_0$ based on $T(F_n)$, where F_n is the empirical distribution function of a random sample $X_1, \ldots, X_n$ from F. The behavior of $T(F_n) - T(F)$ is, therefore, of interest. Study of statistical functionals was introduced by von Mises [54].

Examples

In these examples, for simplicity, $d = 1$.

1. $T(F) = \mathrm{E}_F\left[g(X)\right] = \int g(x)\, dF(x)$ on $\mathcal{F}_0 = \left\{F\colon \mathrm{E}_F\left[|g(X)|\right] < \infty\right\}$, $T(F_n) = n^{-1}\sum_{i=1}^n g(X_i)$.
2. $T(F) = \mathrm{E}_F\left[(X - \xi(F))^k\right]$ on $\mathcal{F}_0 = \left\{F\colon \mathrm{E}_F|X|^k < \infty\right\}$, where $\xi(F) = \mathrm{E}_F[X]$,
   $T(F_n) = n^{-1}\sum_{i=1}^n \left(X_i - \bar{X}_n\right)^k$ where $\bar{X}_n = n^{-1}\sum_{i=1}^n X_i$.
3. $T(F) = F^{-1}(p) = \inf\left\{x\colon F(x) \geq p\right\}$, $0 < p < 1$, on $\mathcal{F}_0 = \left\{\text{all cdf's on } \mathbb{R}\right\}$, $T(F_n) = X_{n:\,[np]}$.
4. $T(F) = \mathrm{E}_F\left[g(X_1, \ldots, X_r)\right]$ where g is symmetric in its coordinates,
   $\mathcal{F}_0 = \left\{F\colon \mathrm{E}_F\left[|g(X_1, \ldots, X_r)|\right] < \infty\right\}$, $T(F_n) = n^{-r}\sum_{i_1=1}^n \cdots \sum_{i_r=1}^n g\left(X_{i_1}, \ldots, X_{i_r}\right)$.

In the last example, $T(F_n)$ is called a V-statistic which differs from a U-statistic by its inclusion of all $\left(i_1, \ldots, i_r\right)$ rather than only those for which $i_1 \neq \cdots \neq i_r$ and then dividing the sum by n^r instead of $n^{(r)} = n(n-1)\cdots(n-r+1)$.

In Section 10.2, we shall introduce an expansion of $\sqrt{n}[T(F_n) - T(F)]$ by means of "differentials" of $T(F)$, analogous to the expansion of $\sqrt{n}\left[T\left(\hat{\theta}_n\right) - T(\theta)\right]$. The leading term of this expansion will provide a functional delta method subject to the remainder term being $o_P(1)$. Postponing the issue of remainder terms, Sections 10.3–10.5 will be devoted to the L- and M-estimators. These estimators are called "robust" because their properties hold for a wide class of distributions, unlike estimators focused on squared-error loss or the maximum likelihood estimators which are susceptible to distributions with heavy tails. Finally, the issue of remainder terms is taken up in Section 10.6.

Theory and Methods of Statistics. http://dx.doi.org/10.1016/B978-0-12-802440-9.00010-2

10.2 Functional Delta Method

We shall assume that $\mathcal{F}_0$ is a collection of cdf's such that

(i) $F \in \mathcal{F}_0 \Rightarrow F_n \in \mathcal{F}_0$ and
(ii) $\mathcal{F}_0$ is convex (ie, $F, G \in \mathcal{F}_0$ implies $F + \lambda(G - F) \in \mathcal{F}_0$ for any $\lambda \in [0, 1]$).

First consider a k-dim parameter $\theta = \theta(F)$ being estimated by $\hat{\theta}_n$ and a real-valued function $T(\theta)$ of θ being estimated by $T\left(\hat{\theta}_n\right)$. For large n, taking the leading term in the expansion of $\sqrt{n}\left[T\left(\hat{\theta}_n\right) - T(\theta)\right]$, we have

$$\sqrt{n}\left[T\left(\hat{\theta}_n\right) - T(\theta)\right] = \langle \nabla T(\theta), \sqrt{n}\left(\hat{\theta}_n - \theta\right)\rangle + R_n,$$

where $R_n = o_P(1)$ if $\sqrt{n}\left(\hat{\theta}_n - \theta\right) \xrightarrow{\mathcal{L}} W$, a random vector, and T is continuously differentiable. In that case, the asymptotic distribution of $\sqrt{n}\left[T\left(\hat{\theta}_n\right) - T(\theta)\right]$ is the same as that of the leading term in the expansion (ie, $\sqrt{n}\left[T\left(\hat{\theta}_n\right) - T(\theta)\right] \xrightarrow{\mathcal{L}} \langle \nabla T(\theta), W\rangle$). This is commonly called the delta-method (see Theorem 3.2.6, Chapter 3). For a similar analysis of $\sqrt{n}[T(F_n) - T(F)]$, we need an analogous expansion which involves *differentiation of a functional T at F*.

To understand the meaning of differentiation of T at F, let us again look at $T: \mathbb{R}^k \to \mathbb{R}$ and examine the one-term Taylor expansion of $T(\theta + \Delta) - T(\theta)$ for small Δ:

$$T(\theta + \Delta) - T(\theta) = \sum_{r=1}^{k} T'(r; \theta)\Delta_r + o(\|\Delta\|), \tag{1}$$

where $\|\Delta\| = \left(\sum_{r=1}^k \Delta_r^2\right)^{1/2}$ or $\max_{1\le r\le k} |\Delta_r|$, and $T'(r;\theta) = \partial T(u)/\partial u_r|_{u=\theta}$.

The leading term is the differential $L_\theta(\Delta) := \sum_{r=1}^k T'(r;\theta)\Delta_r$, which is linear in Δ.

Replacing θ by F, $\theta + \Delta$ by G, r by x, $\sum_{r=1}^k$ by $\int_{x\in\mathbb{R}^d}$, and Δ_r by $d\Delta(x) = d\left[G(x) - F(x)\right]$, the expansion (1) would take the form

$$\begin{aligned} T(G) - T(F) &= L_F(G - F) + o(\|\Delta\|_\rho) \\ &= \int T'(x; F)\, d\left[G(x) - F(x)\right] + o(\|\Delta\|_\rho). \end{aligned}$$

Such an expansion would be valid if there exists a linear functional L_F on $\mathcal{D} = \left\{c(G - F)\colon c \in \mathbb{R} \text{ and } F,\ G \in \mathcal{F}_0\right\}$ which is not identically 0 and satisfies

$$\lim_{j\to\infty} \frac{T\left(G_j\right) - T(F) - L_F\left(G_j - F\right)}{\|G_j - F\|_\rho} = 0,$$

whenever $\{G_j\}$ is a sequence in $\mathcal{F}_0$ with $\|G_j - F\|_\rho = \rho(G_j, F) \to 0$ as $j \to \infty$. Here ρ is a distance such that

$$\rho(F, F + t(G - F)) = |t|\rho(F, G) \iff \|t(G_j - F)\| = |t| \, \|G_j - F\|_\rho$$

holds for all t in $\mathbb{R}$ and F, $G \in \mathcal{F}_0$.

In general, existence of L_F with this property depends on how the sequence $\{G_j\}$ is allowed to approach F in the metric ρ. Three such schemes are described below. In each scheme, the linear functional L_F, if it exists, is called the *differential*.

A. *Gâteaux differentiability.* T is Gâteaux differentiable if the differential exists for every sequence $G_j = F + t_j \Delta \in \mathcal{F}_0$, where $\Delta = G - F \in \mathcal{D}$ is fixed and $t_j \to 0$. Here $\rho(G_j, F) = \|G_j - F\|_\rho = \|t_j \Delta\|_\rho = |t_j| \, \|\Delta\|_\rho \to 0$ irrespective of the metric ρ.

B. *ρ-Hadamard differentiability.* T is ρ -Hadamard differentiable if the differential exists for every sequence $G_j = F + t_j \Delta_j \in \mathcal{F}_0$, where $\|\Delta_j - \Delta\|_\rho \to 0$ for a fixed $\Delta \in \mathcal{D}$ and $t_j \to 0$.

C. *ρ-Fréchet differentiability.* T is ρ-Fréchet differentiable if the differential exists for every $G_j \in \mathcal{F}_0$ with $\rho(G_j, F) \to 0$.

Since these differentiability conditions are increasingly stringent, ρ-Fréchet differentiability $\Rightarrow$ ρ-Hadamard differentiability $\Rightarrow$ Gâteaux differentiability, but the reverse implications do not hold. Moreover, in the expansion

$$T(G) - T(F) = L_F(G - F) + o(\rho(G, F)) \quad \text{as } \rho(G, F) \to 0,$$

if the differential L_F in C exists, it is the same as L_F in A and B, and if L_F in B exists, it is the same as in L_F in A.

We now examine the nature of the Gâteaux differential L_F of T by the following heuristics. Due to linearity of L_F

$$\lim_{t \to 0} \frac{T(F + t\Delta) - T(F) - L_F(t\Delta)}{\|t\Delta\|} = 0$$

$$\iff \lim_{t \to 0} \frac{T(F + t\Delta) - T(F) - tL_F(\Delta)}{t} = 0$$

$$\iff \lim_{t \to 0} \frac{T(F + t\Delta) - T(F)}{t} = L_F(\Delta), \text{ ie,}$$

$$L_F(G - F) = \frac{d}{dt} T((1 - t)F + tG)|_{t=0}.$$

Let δ_x be the cdf with its entire mass at x (ie, $\delta_x(u) = I_{[x,\infty)}(u)$). Then for every u,

$$\int \{\delta_x(u) - F(u)\} \, dG(x) = \int_{x \le u} dG(x) - F(u) = G(u) - F(u) \text{ and}$$

$$\int \{\delta_x(u) - F(u)\} \, dF(x) = 0, \text{ ie,}$$

$$\Delta = G - F = \int (\delta_x - F) \, d[G(x) - F(x)]. \tag{2}$$

Thus, if we let

$$T'(x, F) = \frac{d}{dt} T((1-t)F + t\delta_x)|_{t=0} = L_F(\delta_x - F),$$

then by linearity of L_F, using Eq. (2), we should have

$$\begin{aligned} L_F(G-F) &= L_F \int (\delta_x - F)\, d\big[G(x) - F(x)\big] \\ &= \int L_F(\delta_x - F)\, d\big[G(x) - F(x)\big] = \int T'(x; F)\, d\big[G(x) - F(x)\big]. \end{aligned} \tag{3}$$

The first-order differential expansion of $T(G)$ can now be written as

$$\begin{aligned} T(G) - T(F) &= \int T'(x; F)\, dG(x) + \text{Rem} \\ &= \int T'(x; F)\, d\big[G(x) - F(x)\big] + \text{Rem}, \end{aligned}$$

where the remainder term $\to 0$ as $G \to F$.

Letting F_n play the role of G, we now have

$$\begin{aligned} \sqrt{n}[T(F_n) - T(F)] &= \sqrt{n} \int T'(x; F)\, dF_n(x) + R_n \\ &= n^{-1/2} \sum_{i=1}^{n} T'(X_i; F) + R_n. \end{aligned} \tag{4}$$

Since $\sqrt{n}\|F_n - F\|_\infty = O_P(1)$, following the analogy of the parametric delta method, *we would expect* R_n to be $o_P(1)$ and then the asymptotic distribution of $\sqrt{n}[T(F_n) - T(F)]$ would be the same as that of $n^{-1/2}\sum_{i=1}^{n} T'(X_i; F)$. Interchanging the order of operations L_F and the integration as in Eq. (3) (which needs justification), *we would expect*

$$\begin{aligned} E_F\big[T'(X; F)\big] &= \int T'(x; F)\, dF(x) = \int L_F(\delta_x - F)\, dF(x) \\ &= L_F \int (\delta_x - F)\, dF(x) = L_F(0) = 0, \end{aligned}$$

and if we let

$$\sigma^2(F) = \text{Var}_F\big[T'(X; F)\big] = \int \{T'(x; F)\}^2\, dF(x),$$

assuming that it exists, then we would have

$$\sqrt{n}[T(F_n) - T(F)] = n^{-1/2} \sum_{i=1}^{n} T'(X_i; F) + o_P(1) \xrightarrow{\mathcal{L}} N\big(0, \sigma^2(F)\big).$$

This is known as the functional delta method.

Remark 10.2.1.

1. In the robustness literature [55, 56], $T'(x; F)$ is called the *influence function* and it is denoted by

$$IF(x; F, T) = T'(x; F) = \frac{d}{dt} T((1-t)F + t\delta_x)|_{t=0}.$$

The function $IF(x; F, T)$ measures the rate at which $T(F)$ changes when F is contaminated by δ_x with a small probability. The contamination δ_x is called *gross-error* and

$$\lambda^* = \sup_x |IF(x; F, T)|$$

is called the *gross-error sensitivity* of T at F.

2. Since Gâteaux differentiability is too weak, there is no guarantee that $R_n = o_P(1)$, as seen in the following example [57]:
Define $T(F) = \sum_{x\in[0,1]} [F(x) - F(x-)]^\alpha$, $\alpha > 1$, as a measure of jumps of F on $\mathbb{R}$. For $F = U$ (uniform distribution on $[0, 1]$) which has no jumps,

$$T'(x; U) = \frac{d}{dt} T((1-t)U + t\delta_x)|_{t=0} = \frac{d}{dt} \sum_{y\in[0,1]} t^\alpha [\delta_x(y) - \delta_x(y-)]^\alpha |_{t=0}$$
$$= \frac{d}{dt} t^\alpha |_{t=0} = \alpha t^{\alpha-1}|_{t=0} = 0 \text{ for } \alpha > 1,$$

because U has no jump and $\delta_x(y)$ has exactly one jump of magnitude 1 at $y = x$. Now F_n has n jumps of $1/n$ each with probability 1, so $T(F_n) = n(1/n)^\alpha = n^{1-\alpha}$. Hence

$$\sqrt{n}[T(F_n) - T(U)] = \sqrt{n}\left[n^{1-\alpha} - 0\right] = n^{3/2-\alpha}$$

with probability 1. Thus the expansion

$$\sqrt{n}[T(F_n) - T(F)] = n^{-1/2} \sum_{i=1}^{n} T'(X_i; F) + R_n$$

becomes $n^{3/2-\alpha} = 0 + R_n$ and for $1 < \alpha < 3/2$, $R_n = o_P(1)$ is false.

We now summarize the "potential" of the Gâteaux differentiability approach based on the above heuristics:
Let $T'(x; F) = \frac{d}{dt} T((1-t)F + t\delta_x)|_{t=0}$. Then

$$\sqrt{n}[T(F_n) - T(F)] = n^{-1/2} \sum_{i=1}^{n} T'(X_i; F) + R_n.$$

If $\mathrm{E}_F\left[T'(X; F)\right] = 0$, $0 < \mathrm{Var}_F\left[T'(X; F)\right] = \sigma^2(F) < \infty$ and if $R_n = o_P(1)$, then $\sqrt{n}[T(F_n) - T(F)] \xrightarrow{\mathcal{L}} N\left(0, \sigma^2(F)\right)$ as $n \to \infty$.

To put the above heuristics to work, the main thing is to demonstrate that $R_n = o_P(1)$ which can be attempted in one of the two ways.

I. Use the expansion given in Eq. (4) as a working formula and then carry out the following steps:
(i) Calculate $T'(x; F) = \frac{d}{dt} T((1-t)F + t\delta_x)|_{t=0}$ which involves a simple one-variable differentiation.
(ii) Check the condition $\mathrm{E}_F\left[T'(X; F)\right] = \int T'(x; F)\, dF(x) = 0$.
(iii) Check the condition $R_n = o_P(1)$ by examining the particular case.

(iv) Calculate $\sigma^2(F) = \int \{T'(x; F)\}^2 \, dF(x)$.

II. Obtain conditions on T in terms of ρ-Hadamard or ρ -Fréchet differentiability, so that $R_n = o_P(1)$.

We first illustrate the first approach by the examples listed in Section 10.1 of this chapter and then go into some theoretical considerations needed to pursue the second approach.

Example 10.2.1 (The Mean). Here $T(F) = \int u \, dF(u)$, $T(F_n) = n^{-1}\sum_{i=1}^n X_i = \bar{X}_n$. The influence function is

$$T'(x; F) = \frac{d}{dt}\int u \, d[(1-t)F(u) + t\delta_x(u)]|_{t=0} = x - \int u \, dF(u)$$

$$= x - \mathrm{E}_F(X) := x - \xi(F).$$

In this case, the first-order approximation is an identity, because the function $T(F)$ is already linear. If the sample space is $\mathbb{R}$, then the influence function is unbounded, so $\bar{X}_n$ is not robust.

Example 10.2.2 (The kth Central Moment). Let $\mu_k = T(F) = \int [u - \xi(F)]^k \, dF(u)$, where $\xi(F) = \mathrm{E}_F(X)$. The influence function is

$$\begin{aligned}
T'(x; F) &= \frac{d}{dt}T((1-t)F + t\delta_x)|_{t=0} \\
&= \frac{d}{dt}\int [u - \xi(F) - t(x - \xi(F))]^k d[F(u) + t(\delta_x(u) - F(u))]|_{t=0} \\
&= -k(x - \xi(F))\mu_{k-1} + (x - \xi(F))^k - \mu_k,
\end{aligned}$$

which is unbounded. By routine calculations, we have

$$\begin{aligned}
&\mathrm{E}_F\big[T'(X; F)\big] = 0 \text{ and} \\
&\sigma^2(F) = \mu_{2k} - \mu_k^2 - 2k\mu_{k-1}\mu_{k+1} + k^2\mu_{k-1}\mu_2.
\end{aligned}$$

Finally, the remainder term of this one-term expansion of $\sqrt{n}[T(F_n) - T(F)]$ is

$$\begin{aligned}
R_n &= \sqrt{n}[T(F_n) - T(F)] - n^{-1/2}\sum_{i=1}^n T'(X_i; F) \\
&= \sqrt{n}\left[n^{-1}\sum_{i=1}^n (X_i - \bar{X}_n)^k - \mu_k\right] \\
&\quad - n^{-1/2}\sum_{i=1}^n \big[(X_i - \xi(F))^k - \mu_k - k\mu_{k-1}(X_i - \xi(F))\big] \\
&= n^{-1/2}\sum_{i=1}^n \left[(X_i - \bar{X}_n)^k - (X_i - \xi(F))^k\right] + k\mu_{k-1}\sqrt{n}(\bar{X}_n - \xi(F)) \\
&= n^{-1/2}\sum_{i=1}^n \left[(Y_i - \bar{Y}_n)^k - Y_i^k\right] + k\mu_{k-1}\sqrt{n}\bar{Y}_n \qquad (5)
\end{aligned}$$

with $Y_i = X_i - \xi(F)$. Then $E_F[Y_i] = 0$ and $E_F\left[Y_i^{k-j}\right] = \mu_{k-j}$. Now expand the last expression in Eq. (5), make some algebraic rearrangements and note that $\sqrt{n}\bar{Y}_n = O_P(1)$, $n^{-1}\sum_{i=1}^n Y_i^{k-j} = \mu_{k-j} + o_P(1)$ for $j = 2, \ldots, k$. This shows that $R_n = o_P(1)$.

Example 10.2.3 (The p-Quantile). For $0 < p < 1$, the p-quantile of F is

$$T(F) = F^{-1}(p) = \inf\{x\colon F(x) \geq p\}, \quad \text{so that } F\left(F^{-1}(p)\right) = p.$$

Let $F_t = (1-t)F + t\delta_x$ for a give x. Now differentiating both sides of the identity

$$p = F_t\left(F_t^{-1}(p)\right) = (1-t)F\left(F_t^{-1}(p)\right) + t\delta_x\left(F_t^{-1}(p)\right)$$

with respect to t, evaluated at 0, we have.

$$0 = \left[\left\{-F\left(F_t^{-1}(p)\right) + (1-t)f\left(F_t^{-1}(p)\right)\frac{d}{dt}F_t^{-1}(p)\right\} + \left\{\delta_x\left(F_t^{-1}(p)\right) + t\frac{d}{dt}\delta_x\left(F_t^{-1}(p)\right)\right\}\right]_{t=0}$$

$$= -F\left(F^{-1}(p)\right) + f\left(F^{-1}(p)\right)\frac{d}{dt}F_t^{-1}(p)\Big|_{t=0} + I_{[x,\infty)}\left(F^{-1}(p)\right) + 0$$

$$= f\left(F^{-1}(p)\right)\frac{d}{dt}F_t^{-1}(p)\Big|_{t=0} - \left\{p - I_{[-\infty,F^{-1}(p))}(x)\right\}.$$

Hence

$$T'(x;F) = \frac{d}{dt}F_t^{-1}(p)|_{t=0} = \frac{p - I_{(-\infty,F^{-1}(p)]}(x)}{f(F^{-1}(p))}, \tag{6}$$

and the leading term of the expression of $\sqrt{n}[T(F_n) - T(F)]$ is

$$n^{-1/2}\sum_{i=1}^n T'(X_i;F) = \frac{1}{\sqrt{n}}\sum_{i=1}^n \frac{p - I_{(-\infty,F^{-1}(p)]}(X_i)}{f(F^{-1}(p))}$$

$$\xrightarrow{\mathcal{L}} N\left(0, \frac{p(1-p)}{f^2(F^{-1}(p))}\right),$$

provided that the remainder term $R_n = o_P(1)$. Indeed, $R_n = O_{a.s.}(n^{-1/4}\log n)$, ie, there exists C such that $P\left[|R_n| > Cn^{-1/4}\log n \text{ i.o.}\right] = 0$. For a proof, Bahadur [58].

Remark 10.2.2.

1. The expression (6), rewritten as

$$X_{n:[np]} = F^{-1}(p) + \frac{1}{n}\sum_{i=1}^n \frac{p - I_{(-\infty,F^{-1}(p)]}(X_i)}{f(F^{-1}(p))} + R_n$$

is known as Bahadur representation of a sample quantile.

2. Since $\left[p - I_{(-\infty,F^{-1}(p)]}(X_i)\right]/f(F^{-1}(p))$, $i = 1, \ldots, n$ are iid with mean zero and variance $p(1-p)/f^2(F^{-1}(p))$, $X_{n:[np]} \overset{a.s.}{\to} F^{-1}(p)$.
3. The Bahadur representation can be obtained by a simple heuristic argument. Let F_n be the edf of $X_1, \ldots, X_n$. Then $F_n(X_{n:[np]}) = [np]/n$. Now note that if $\xi \le X_{n:[np]}$, then

$$\begin{aligned}(X_{n:[np]} - \xi)f(\xi) &\approx F(X_{n:[np]}) - F(\xi) \approx F_n(X_{n:[np]}) - F_n(\xi)\\ &= [np] - n^{-1}\sum_{i=1}^{n} I_{(-\infty,\xi]}(X_i) \approx n^{-1}\sum_{i=1}^{n}[p - I_{(-\infty,\xi]}(X_i)]\end{aligned}$$

and similarly, for $X_{n:[np]} \le \xi$,

$$(\xi - X_{n:[np]})f(\xi) \approx n^{-1}\sum_{i=1}^{n}[I_{(-\infty,\xi]}(X_i) - p].$$

In both cases,

$$X_{n:[np]} \approx \xi + n^{-1}\sum_{i=1}^{n}[p - I_{(-\infty,\xi]}(X_i)]/f(\xi),$$

where $\xi = F^{-1}(p)$.
4. Asymptotic joint distributions of several sample quantiles can be obtained similarly.

Example 10.2.4 (The V-Statistic). Let $g\colon \mathbb{R}^r \to \mathbb{R}$ (or more generally, $g\colon \mathbb{R}^d \times \cdots \times \mathbb{R}^d \to \mathbb{R}$) be such that $g(y_{i_1}, \ldots, y_{i_r}) = g(y_1, \ldots, y_r)$ for all $(y_1, \ldots, y_r)$ and all permutations $(i_1, \ldots, i_r)$ of $(1, \ldots, r)$. Then $T(F) = \int \cdots \int g(y_1, \ldots, y_r) \prod_{i=1}^{r} dF(y_i)$ for $F \in \mathcal{F}_0 = \{F\colon \mathrm{E}_F[|g(Y_1, \ldots, Y_r)|] < \infty\}$ is called a V-functional and its corresponding V-statistic is

$$T(F_n) = n^{-r}\sum_{i_1=1}^{n}\cdots\sum_{i_r=1}^{n} g(X_{i_1}, \ldots, X_{i_r})$$

based on a random sample $X_1, \ldots, X_n$ from F. Since g is symmetric in its coordinates,

$$\begin{aligned}T((1-t)F + t\delta_x) &= \int \cdots \int g(y_1, \ldots, y_r)\prod_{i=1}^{r} d[(1-t)F(y_i) + t\delta_x(y_i)]\\ &= (1-t)^r T(F) + \sum_{c=1}^{r} t^c(1-t)^{r-c} r^{(c)}\\ &\quad \times \int \cdots \int g(x, \ldots, x, y_{c+1}, \ldots, y_r)\prod_{i=c+1}^{r} dF(y_i)\end{aligned}$$

which is a polynomial in t, so its derivative with respect to t, evaluated at 0 is the coefficient of t. Thus

$$\begin{aligned}
T'(x;F) &= \frac{d}{dt}T((1-t)F+t\delta_x)|_{t=0} \\
&= r\int\cdots\int g(x,y_2,\ldots,y_r)\prod_{i=2}^{r} dF(y_i) - rT(F) \\
&= r\{\mathrm{E}_F[g(X_1,X_2,\ldots,X_r)|X_1=x] - T(F)\} = rh_1(x), \text{ where} \\
h_1(x) &= \mathrm{E}_F[g(X_1,X_2,\ldots,X_r)|X_1=x] - T(F).
\end{aligned}$$

Clearly,

$$r^{-1}\mathrm{E}_F[T'(X;F)] = 0 \text{ and } \sigma^2(F) = r^2\mathrm{E}_F\left[h_1^2(X)\right].$$

Hence subject to the verification of the remainder term

$$R_n = \sqrt{n}[T(F_n) - T(F)] - rn^{-1/2}\sum_{i=1}^{n} h_1(X_i)$$

being $o_P(1)$, we have

$$\sqrt{n}[T(F_n) - T(F)] \xrightarrow{\mathcal{L}} N\left(0,\sigma^2(F)\right).$$

Examine the discrepancy between the V-statistic and the corresponding U -statistic to see that this result is what one would expect.

10.3 The L-Estimators

Let $F(\cdot-\theta)$ be a cdf with pdf $f(\cdot-\theta)$ where f is symmetric about 0 and $\theta\in\mathbb{R}$ is unknown. Then θ is a location parameter which is the point of symmetry of the unknown distribution, which is the median of $F(\cdot-\theta)$ and also the mean if it exists. If $X_1,\ldots,X_n$ is a random sample from $F(\cdot-\theta)$, then the sample mean $\bar{X}_n$, being the UMVUE and the MLE of θ if $F(\cdot-\theta)$ is normal with mean θ, is a very good estimator, but it is not so good if $F(\cdot-\theta)$ is Cauchy with median θ. The median $N_{n:[n/2]}$ is no good in case of normal distributions, but does not break down like $\bar{X}_n$ for Cauchy distributions.

The reason for $\bar{X}_n$ performing so poorly for the Cauchy distribution is due to its heavy tails. The Cauchy pdf tends to 0 at the rate of $1/x^2$ as $x\to\pm\infty$ as opposed to the $\mathrm{e}^{-x^2/2}$ rate for the normal pdf. This makes the extreme-order statistics unstable. The sample mean can also be viewed as $n^{-1}\sum_{i=1}^{n} X_{n:i}$, where $X_{n:1} < \cdots < X_{n:n}$ are the order statistics in $(X_1,\ldots,X_n)$. To protect $\bar{X}_n$ from being drastically affected by possibly heavy tails of the underlying distribution, it would seem reasonable to redistribute the weights on the order statistics so that the extreme ones are de-emphasized. This leads to the consideration of estimators which are linear functions of order statistics. These estimators include the sample mean, the sample median, and also a class of estimators called α -trimmed means with $0<\alpha<1/2$ defined by $\bar{X}_{n(\alpha)} = (n-2[\alpha n])^{-1}\sum_{i=[\alpha n]+1}^{n-[\alpha n]} X_{n:i}$, which fall between the sample mean and the sample median.

Definition 10.3.1. An L-functional $T\colon \mathcal{F}_0 \to \mathbb{R}$ is defined as

$$T(F) = \int_{-\infty}^{\infty} xJ[F(x)]\, dF(x) = \int_0^1 F^{-1}(u)J(u)\, du,$$
$$F \in \mathcal{F}_0 \text{ and } J\colon [0,1] \to R,$$

and if $X_1, \ldots, X_n$ is a random sample from $F \in \mathcal{F}_0$ with edf F_n, then

$$T(F_n) = \int_{-\infty}^{\infty} xJ[F_n(x)]\, dF_n(x) = n^{-1}\sum_{i=1}^{n} J(i/n)X_{n:i}$$

is called the L-estimator of $T(F)$. The function J is called the score function.

We now calculate the influence function and the remainder $R(G,F)$ in the expansion

$$T(G) - T(F) = \int T'(x;F)\, d\big[G(x) - F(x)\big] + R(G,F).$$

Differentiating

$$T(F + t(\delta_x - F)) = \int uJ[F(u) + t(\delta_x(u) - F(u))]d[F(u) + t(\delta_x(u) - F(u))]$$

with respect to t at $t = 0$, we have

$$T'(x;F) = \int u(\delta_x(u) - F(u))J'[F(u)]\, dF(u) + \int uJ[F(u)]\, d(\delta_x(u) - F(u)).$$

Integrate the second integral by parts and make some algebraic rearrangements to get

$$\begin{aligned} T'(x;F) &= -\int (\delta_x(u) - F(u))J[F(u)]\, du \\ &= \int_{-\infty}^{x} J[F(u)]\, du - \int_{-\infty}^{\infty} (1 - F(u))J[F(u)]\, du. \end{aligned}$$

The remainder term is

$$R(G,F) = T(G) - T(F) - \int T'(x;F)\, d\big[G(x) - F(x)\big],$$

where

$$\begin{aligned} T(G) - T(F) &= \int_0^1 \big[G^{-1}(u) - F^{-1}(u)\big]J(u)\, du = \int_0^1 \left[\int_{F^{-1}(u)}^{G^{-1}(u)} dx\right] J(u)\, du \\ &= \int_0^1 \int_{-\infty}^{\infty} I_{[F^{-1}(u),G^{-1}(u)]}(x)J(u)\, dx\, du \\ &= \int_{-\infty}^{\infty} \int_0^1 I_{[G(x),F(x)]}(u)J(u)\, du\, dx \\ &= \int_{-\infty}^{\infty} \left[\int_{G(x)}^{F(x)} J(u)\, du\right] dx = -\int_{-\infty}^{\infty} \left[\int_{F(x)}^{G(x)} J(u)\, du\right] dx, \text{ and} \end{aligned}$$

$$\begin{aligned} \int T'(x;F)\, d\big[G(x) - F(x)\big] &= \int_{-\infty}^{\infty} \left[\int_{-\infty}^{x} J[F(u)]\, du\right] d\big[G(x) - F(x)\big] - 0 \\ &= -\int_{-\infty}^{\infty} \big[G(x) - F(x)\big]J[F(x)]\, dx. \end{aligned}$$

Thus

$$R(G,F) = -\int_{-\infty}^{\infty}\left[\int_{F(x)}^{G(x)} J(u)\,du - \big(G(x)-F(x)\big)J(F(x))\right]dx$$
$$= -\int_{-\infty}^{\infty} W_G(x)\big[G(x)-F(x)\big]\,dx, \text{ where} \tag{7}$$
$$W_G(x) = \big(G(x)-F(x)\big)^{-1}\int_{F(x)}^{G(x)} J(u)\,du - J(F(x)).$$

Now letting F_n play the role of G, we have

$$\sqrt{n}[T(F_n)-T(F)] = n^{-1/2}\sum_{i=1}^{n} T'(X_i;F) + R_n, \text{ where}$$
$$T'(x;F) = -\int (\delta_x(u)-F(u))J[F(u)]\,du \text{ and } R_n = R(F_n,F). \tag{8}$$

We now verify that $\mathrm{E}_F\big[T'(X;F)\big]=0$ and obtain a formula for $\sigma^2(F) = \mathrm{Var}_F\big[T'(X;F)\big]$. This will then give us the asymptotic distribution of the L-estimator, namely,

$$\sqrt{n}[T(F_n)-T(F)] \xrightarrow{\mathcal{L}} N\Big(0,\sigma^2(F)\Big)$$

subject to $R_n = R(F_n,F) = o_P(1)$.

First,

$$\mathrm{E}_F\big[T'(X;F)\big] = -\int\left[\int (\delta_x(u)-F(u))J[F(u)]\,du\right]dF(x)$$
$$= -\int\left[\int (\delta_x(u)-F(u))\,dF(x)\right]J(F(u))\,du = 0,$$

because $\int(\delta_x(u)-F(u))\,dF(x) = 0$ for all u.

Next,

$$\sigma^2(F) = \mathrm{Var}_F\big[T'(X;F)\big]$$
$$= \int\left[\int (\delta_x(u)-F(u))J[F(u)]\,du\right]^2 dF(x)$$
$$= 2\iint_{u<v}\left[\int (\delta_x(u)-F(u))(\delta_x(v)-F(v))\,dF(x)\right]J(F(u))J(F(v))\,du\,dv$$
$$= 2\iint_{u<v} F(u)(1-F(v))J(F(u))J(F(v))\,du\,dv$$
$$= \iint\big[F(\min(u,v)) - F(u)F(v)\big]J(F(u))J(F(v))\,du\,dv, \tag{9}$$

because

$$\int (\delta_x(u)-F(u))(\delta_x(v)-F(v))\,dF(x) = (1-F(u))(1-F(v))F(u)$$
$$- F(u)(1-F(v))(F(v)-F(u)) + F(u)F(v)(1-F(v))$$
$$= F(u)(1-F(v)).$$

An alternate formula for $\sigma^2(F)$ can be obtained by first replacing (u, v) by (x, y) and then letting $F(x) = u$ and $F(y) = v$ in Eq. (9). Thus

$$\begin{aligned}\sigma^2(F) &= 2\iint_{x<y} F(x)(1-F(y))J(F(x))J(Fny))\,dx\,dy\\ &= 2\iint_{u<v} u(1-v)\left\{J(u)/f\left(F^{-1}(u)\right)\right\}\left\{J(v)/f(F^{-1}(v))\right\}\,du\,dv\\ &= 2\iint_{u<v} u(1-v)A'(u)A'(v)\,du\,dv, \text{ where}\\ A(u) &= \int_0^u \frac{J(t)}{f(F^{-1}(t))}\,dt + c\end{aligned}$$

where c is such that $A(1) = 0$, making $A'(u) = J(u)/f\left(F^{-1}(u)\right)$. Thus

$$\begin{aligned}\sigma^2(F) &= 2\int_{u=0}^1\left[\int_{v=u}^1 A'(v)\,dv\right]uA'(u)\,du - \int_0^1\int_0^1 uvA'(u)A'(v)\,du\,dv\\ &= 2\int_0^1 [0-A(u)]uA'(u)\,du - \left(\int_0^1 aA'(u)\,du\right)^2\\ &= -\int_0^1 u\,dA^2(u) - \left(\int_0^1 u\,dA(u)\right)^2 = \int_0^1 A^2(u)\,du - \left(\int_0^1 A(u)\,du\right)^2,\end{aligned} \tag{10}$$

using integration by parts in both integrals.

10.3.1 Asymptotic Distribution of α-Trimmed Mean When f Is Symmetric About θ

The L-functional for α-trimmed mean is: $T(F) = (1-2\alpha)^{-1}\int_{F^{-1}(\alpha)}^{F^{-1}(1-\alpha)} x\,dF(x)$ (ie, the score function is $J(t) = (1-2\alpha)^{-1}I_{[\alpha,1-\alpha]}(t))$. Since $f = F'$ is symmetric about θ, $F^{-1}(1-\alpha) - \theta = \theta - F^{-1}(\alpha)$, so that

$$\int_{F^{-1}(\alpha)}^{F^{-1}(1-\alpha)} (x-\theta)\,dF(x) = 0 \text{ and } T(F) = (1-2\alpha)^{-1}\int_{F^{-1}(\alpha)}^{F^{-1}(1-\alpha)} x\,dF(x) = \theta.$$

Next,

$$\begin{aligned}A(u) &= \int_0^u \frac{(1-2\alpha)^{-1}I_{[\alpha,1-\alpha]}(t)}{f(F^{-1}(t))}\,dt\\ &= \begin{cases}0 & \text{if } 0 < u < \alpha\\ (1-2\alpha)^{-1}\left[F^{-1}(u) - F^{-1}(\alpha)\right] & \text{if } \alpha \le u \le 1-\alpha\\ (1-2\alpha)^{-1}[F^{-1}(1-\alpha) - F^{-1}(\alpha)] & \text{if } 1-\alpha < u < 1.\end{cases}\end{aligned}$$

To calculate $\int_0^1 A(t)\,dt$ and $\int_0^1 A^2(t)\,dt$, use that due to symmetry of $f = F'$ about θ, $F^{-1}(1-\alpha) - \theta = \theta - F^{-1}(\alpha)$. Using these facts, we obtain, after algebraic simplifications,

$$\sigma^2(F) = \int_0^1 A^2(u)\,du - \left(\int_0^1 A(u)\,du\right)^2$$
$$= \frac{1}{(1-2\alpha)^2}\left[2\alpha\left(\theta - F^{-1}(\alpha)\right)^2 + \int_{F^{-1}(\alpha)}^{F^{-1}(1-\alpha)} (x-\theta)^2\,dF(x)\right],$$

by Eq. (10).

10.4 The M-Estimators

The method of maximum likelihood is based on the fact that subject to identifiability of the family $\{f(\cdot, t)\}$, if X has pdf $f(x, \theta)$, then the function $E_\theta[\log f(X, t)]$ has $t = \theta$ as its unique maximizer. The MLE of θ based on a random sample $X_1, \dots, X_n$ is the maximizer of $n^{-1}\sum_{i=1}^n \log f(X_i; t)$ which is a natural estimate of $E_\theta[\log f(X, t)]$. Equivalently, the MLE of θ is a solution of $\sum_{i=1}^n \Psi(X_i, t) = 0$, where $\Psi(x, t) = \partial \log f(x, t)/\partial t$, assuming that f is smooth. In the location problem,

$$\Psi(x, t) = \partial \log f(x-t)/\partial t = -\frac{f'}{f}(x-t),$$

so the MLE of a location parameter is the solution of $\sum_{i=1}^n \Psi(X_i - t) = 0$ where $\Psi = -f'/f$.

The MLEs have good properties under the correct model, but if not, then $\Psi(x, t)$ may be very unstable for some x, as in the location problem in which $-(f'/f)(x) = x$ for $N(0, 1)$ and $X_i - t$ for extreme observations are unstable if the true distribution is heavy-tailed.

The M-estimators attempt to overcome this weakness of MLEs by using a function $\rho(x-t)$ in the location problem instead of, but somewhat similar to $\log f(x-t)$, and then solving for t in the equation $\sum_{i=1}^n \Psi(X_i - t) = 0$, where $\Psi = \rho'$ instead of $-f'/f$. In general, we replace $-\log f(x, t)$ by $\rho(x, t)$ and solve for t in the equation $\sum_{i=1}^n \Psi(X_i, t) = 0$, where $\Psi(x, t) = \partial\rho(x, t)/\partial t$.

We now formally define an M-functional $T(F)$, of which $T(F_n)$ will be an M-estimator.

Definition 10.4.1. Let ρ: $\mathbb{R}^d \times \mathbb{R} \to \mathbb{R}$ and let Θ be an open subset of $\mathbb{R}$, Then

$$T(F) = \arg\min_{t\in\Theta} \int \rho(x, t)\,dF(x), \quad F \in \mathcal{F}_0$$

is an M-functional, and if F_n is the edf of a random sample $X_1, \dots, X_n$ from F, then

$$T(F_n) = \arg\min_{t\in\Theta} \sum_{i=1}^n \rho(X_i, t)$$

is the M-estimator of $T(F)$.

If $\Psi(x, t) = \partial\rho(x, t)/\partial t$ exists and if

$$\lambda_F(t) = \int \Psi(x, t)\,dF(x) = \frac{\partial}{\partial t}\int \rho(x, t)\,dF(x), \tag{11}$$

then $T(F)$ is a solution of $\lambda_F(t) = 0$, and so

$$\lambda_F(T(F)) = \int \Psi(x, T(F))\, dF(x) = 0 \quad \text{for all } F \in \mathcal{F}_0. \tag{12}$$

The M-estimator $T(F_n)$ of $T(F)$ can be equivalently be expressed as a solution of the equation $\sum_{i=1}^{n} \Psi(X_i, t) = 0$.

To calculate $T'(x; F) = \frac{d}{dt} T((1-t)F + t\delta_x)|_{t=0}$, start with

$$\begin{aligned} 0 &= \lambda_{(1-t)F+t\delta_x}((1-t)F + t\delta_x) \\ &= \int \Psi(u, T((1-t)F + t\delta_x))\, d[(1-t)F(u) + t\delta_x(x)] \\ &= (1-t)\int \Psi(u, T((1-t)F + t\delta_x))\, dF(u) + t\Psi(x, T((1-t)F + t\delta_x)), \end{aligned}$$

and differentiate both sides with respect to t at $t = 0$ to obtain

$$\begin{aligned} 0 &= -\int \Psi(u, T(F))\, dF(u) + \frac{d}{dt}\int \Psi(u, T((1-t)F + t\delta_x))\, dF(u)|_{t=0} \\ &\quad + \Psi(x, T(F)) + 0 \\ &= -\lambda_F(T(F)) + \frac{d}{dt}\lambda_F(T((1-t)F + t\delta_x))|_{t=0} + \Psi(T(F)) \\ &= 0 + \lambda_F'(T(F))T'(x; F) + \Psi(x, T(F)), \end{aligned}$$

using Eqs. (11) and (12).

Assuming $\lambda_F'(T(F)) \neq 0$, we thus have

$$T'(x; F) = -\Psi(x, T(F))/\lambda_F'(T(F)). \tag{13}$$

We now check that

$$\mathrm{E}_F[T'(X; F)] = -\mathrm{E}_F[\Psi(X, T(F))]/\lambda_F'(T(F)) = -\lambda_F(T(F))/\lambda_F'(T(F)) = 0$$

by Eqs. (11) and (12), and

$$\sigma^2(F) = \sigma^2(F, \Psi) = \mathrm{Var}_F[T'(X; F)] = \frac{\int \Psi^2(x, T(F))\, dF(x)}{\{\lambda_F'(T(F))\}^2}. \tag{14}$$

We can now conclude that

$$\sqrt{n}[T(F_n) - T(F)] = n^{-1/2}\sum_{i=1}^{n} T'(X_i; F) + R_n \xrightarrow{\mathcal{L}} N\left(0, \sigma^2(F, \Psi)\right),$$

subject to the verification $R_n = o_P(1)$.

The MLE is also an M-estimator with $\Psi(x, t) = -\partial \log f(x, t)/\partial t$, and letting $F_\theta = F(\cdot, \theta)$, $T(F_\theta) = \theta$. Now

$$\lambda_{F_\theta}'(T(F_\theta)) = \lambda_{F_\theta}'(\theta) = \int \left(-\frac{\partial^2 \log f(x, t)}{\partial t^2}\right)_{t=\theta} f(x, \theta)\, dx = I_f(\theta), \text{ and}$$

$$\int \Psi^2(x, T(F_\theta))\, dF_\theta(x) = \int \left(\frac{\partial \log f(x, t)}{\partial t}\right)^2_{t=\theta} f(x, \theta)\, dx = I_f(\theta),$$

where $I_f(\theta)$ is the Fisher-information of the family $\{f(x,t), t \in \Theta\}$ at $t = \theta$. Thus $\sigma^2(F_\theta, \Psi) = 1/I_f(\theta)$ by Eq. (14).

In the location problem, $f(x,\theta) = f(x-\theta)$ where f is a symmetric (about 0) pdf. Here

$$\left.\frac{\partial \log f(x-t)}{\partial t}\right|_{t=\theta} = -\frac{f'(x-\theta)}{f(x-\theta)} \text{ and}$$

$$I_f(\theta) = \int \left\{\frac{f'(x-\theta)}{f(x-\theta)}\right\}^2 f(x-\theta)\, dx = \int \left\{\frac{f'(x)}{f(x)}\right\}^2 f(x)\, dx \text{ for all } \theta.$$

10.4.1 A Minimax Approach to the Choice of Ψ

For robust estimation of a location parameter θ, instead of assuming F to be a known cdf, we work within the model

$$\mathcal{F} = \mathcal{F}(G,\varepsilon) = \left\{F = (1-\varepsilon')G + \varepsilon' H \colon H \text{ is a cdf symmetric about } 0,\ \varepsilon' \in [0,\varepsilon]\right\},$$

where G is a specified cdf which is symmetric about 0 and ε is a specified positive number (ie, we assume F to be a symmetric cdf lying "within ε distance" of a specified symmetric cdf G). For given G and ε, we now look for $\Psi_0 = \Psi_0(G,\varepsilon)$ such that

$$\sup_{F \in \mathcal{F}(G,\varepsilon)} \sigma^2(F, \Psi_0(G,\varepsilon)) \le \sup_{F \in \mathcal{F}(G,\varepsilon)} \sigma^2(F, \Psi) \quad \text{for all } \psi.$$

Then the M-estimator with $\Psi = \Psi_0(G,\varepsilon)$ is minimax for the family $\mathcal{F}(G,\varepsilon)$ in the sense of minimizing the maximum possible asymptotic variance.

For $G = \Phi$, the cdf of $N(0,1)$, the solution of this minimax problem lies in the class of "Huber Functions":

$$\Psi_0(x) = \begin{cases} -k & x \le -k \\ x & |x| < k \\ k & x \ge k, \end{cases}$$

where k is given in terms of ε by the formula

$$\int_{-k}^{k} \phi(x)\, dx + (2/k)\phi(k) = 1/(1-\varepsilon), \quad \phi = \Phi' = \text{pdf of } N(0,1).$$

See Huber [56].

An Alternative Derivation of the Asymptotic Distribution of M-Estimators When the Score Function Ψ Is Monotone

Let $\hat{\theta}_n$ be an M-estimator of the location parameter θ of a family of pdf's $\{f(x,\theta) = f(x-\theta),\ \theta \in \mathbb{R}\}$ where f is symmetric about 0 and suppose that the score function is antisymmetric and monotone increasing.

Since $\hat{\theta}_n$ is the solution of $\sum_{i=1}^n \Psi(X_i - t) = 0$ and $\sum_{i=1}^n \Psi(X_i - t)$ is monotone decreasing in t, $\sum_{i=1}^n \Psi(X_i - t) \le 0$ for all $t \ge \hat{\theta}_n$ and $\sum_{i=1}^n \Psi(X_i - t) \ge 0$ for all $t \le \hat{\theta}_n$. Hence

$$
\begin{aligned}
P_\theta\left[\sqrt{n}\left(\hat{\theta}_n-\theta\right)\le a\right] &= P_\theta\left[\theta+a/\sqrt{n}\ge\hat{\theta}_n\right]\\
&= P_\theta\left[n^{-1/2}\sum_{i=1}^{n}\Psi(X_i-\theta-a/\sqrt{n})\le 0\right]\\
&= P_0\left[n^{-1/2}\sum_{i=1}^{n}\Psi(X_i-a/\sqrt{n})\le 0\right].
\end{aligned}
$$

Now write

$$
n^{-1/2}\sum_{i=1}^{n}\Psi(X_i-a/\sqrt{n}) = n^{-1/2}\sum_{i=1}^{n}\Psi(X_i) - an^{-1}\sum_{i=1}^{n}\Psi'(X_i)+R_n.
$$

Since $\mathrm{E}_0[\Psi(X_i)]=0$ due to Ψ being antisymmetric,

$$
n^{-1/2}\sum_{i=1}^{n}\Psi(X_i)\xrightarrow{\mathcal{L}} N(0,\mathrm{Var}_0[\Psi(X_i)])
$$

provided that $\mathrm{Var}_0[\Psi(X_i)]=\int\Psi^2(x)f(x)\,dx<\infty$. Also,

$$
n^{-1}\sum_{i=1}^{n}\Psi'(X_i)=\mathrm{E}_0\left[\Psi'(X_i)\right]+o_P(1)
$$

provided that $\mathrm{E}_0\left[\Psi'(X_i)\right]=\int\Psi'(x)f(x)\,dx$ exists.

Thus if $R_n=o_P(1)$ by regularity conditions on F and Ψ, then

$$
n^{-1/2}\sum_{i=1}^{n}\Psi(X_i-a/\sqrt{n})\xrightarrow{\mathcal{L}} N(-a\mathrm{E}_0\left[\Psi'(X)\right],\mathrm{Var}_0[\Psi(X)]).
$$

Consequently,

$$
\begin{aligned}
\lim_{n\to\infty}P_\theta\left[\sqrt{n}\left(\hat{\theta}_n-\theta\right)\le a\right] &= \lim_{n\to\infty}P_0\left[n^{-1/2}\sum_{i=1}^{n}\Psi(X_i-a/\sqrt{n})\le 0\right]\\
&= \Phi\left(\frac{a\mathrm{E}_0\left[\Psi'(X)\right]}{\sqrt{\mathrm{Var}_0[\Psi(X)]}}\right)=\Phi\left(\frac{a}{\sqrt{\mathrm{Var}_0[\Psi(X)]/\mathrm{E}_0^2\left[\Psi'(X)\right]}}\right),
\end{aligned}
$$

provided that $\mathrm{E}_0\left[\Psi'(X)\right]\ne 0$, that is,

$$
\sqrt{n}\left(\hat{\theta}_n-\theta\right)\xrightarrow{\mathcal{L}} N\left(0,\sigma^2(F,\Psi)\right),\text{ where}
$$

$$
\sigma^2(F,\Psi)=\frac{\mathrm{Var}_0[\Psi(X)]}{\mathrm{E}_0^2\left[\Psi'(X)\right]}=\frac{\int\Psi^2(x)f(x)\,dx}{\left[\int\Psi'(x)f(x)\,dx\right]^2}.
$$

In the above, Ψ is assumed to be strictly increasing.

If Ψ is merely nondecreasing, then the equation $\sum_{i=1}^{n}\Psi(X_i-t)=0$ may be satisfied on an entire interval. In that case, the above argument would need minor modification.

10.5 A Relation Between L-Estimators and M-Estimators

Let F be a cdf, symmetric about 0 and let $f = F'$ be the pdf. For simplicity, assume F to be strictly increasing.

Consider an M-estimator of θ based on iid observations $X_1, \ldots, X_n$ from $F(x - \theta)$, with score function Ψ having $\Psi(-x) = -\Psi(x)$ for all x. Note that nothing changes if Ψ is multiplied by a constant, so we normalize Ψ to make $\int \Psi'(x)f(x)\ dx = 1$ (assuming that $\int \Psi'(x)f(x)\ dx \neq 0$). Then the asymptotic variance of the M-estimator with score function Ψ is

$$\sigma^2(F,\Psi) = \frac{\int \Psi^2(x)f(x)\,dx}{\left[\int \Psi'(x)f(x)\,dx\right]^2} = \int \Psi^2(x)f(x)\,dx.$$

Now consider an L-estimator with score function

$$J_\Psi(u) = \Psi'\left(F^{-1}(u)\right), \quad 0 < u < 1.$$

Then the asymptotic variance of this L-estimator is the same as $\sigma^2(F,\Psi)$. To see this, note that here

$$A'_\Psi(u) = J_\Psi(u)/f(F^{-1}(u)) = \Psi'\left(F^{-1}(u)\right)/f\left(F^{-1}(u)\right),$$

so that

$$\begin{aligned} A_\Psi(t) &= \int_{1/2}^{t} A'_\Psi(u)\,du = \int_{1/2}^{t} \left\{\Psi'\left(F^{-1}(u)\right)/f\left(F^{-1}(u)\right)\right\} du \\ &= \int_{F^{-1}(1/2)}^{F^{-1}(t)} \left\{\Psi'(x)/f(x)\right\} dF(x) = \int_0^{F^{-1}(t)} \Psi'(x)\,dx \\ &= \Psi\left(F^{-1}(t)\right) - \Psi(0) = \Psi\left(F^{-1}(t)\right). \end{aligned}$$

Hence

$$\int_0^1 A_\Psi(t)\,dt = \int_0^1 \Psi\left(F^{-1}(t)\right) dt = \int_{-\infty}^{\infty} \Psi(x)f(x)\,dx = 0 \text{ and}$$

$$\int_0^1 A_\Psi^2(t)\,dt = \int_0^1 \left\{\Psi\left(F^{-1}(t)\right)\right\}^2 dt = \int_{-\infty}^{\infty} \Psi\left({}^2x\right)f(x)\,dx.$$

Thus the asymptotic variance of the L-estimator with score function J_Ψ is

$$\int_0^1 A_\Psi^2(t)\,dt - \left(\int_0^1 A_\Psi(t)\,dt\right)^2 = \int_{-\infty}^{\infty} \Psi^2(x)f(x)\,dx = \sigma^2(F,\Psi).$$

10.6 The Remainder Term R_n

Throughout, we consider the case of $d = 1$, that is, $X_1, \ldots, X_n$ are real-valued iid rv's with cdf F and F_n is the edf of $X_1, \ldots, X_n$.

We shall consider the following distances in

$\mathcal{F}$ = collection of all cdf's on $\mathbb{R}$, and
$\mathcal{F}_1$ = collection of all cdf's on $\mathbb{R}$ with finite mean:

Sup-norm distance: $\rho_\infty(F_1, F_2) = \sup_x |F_1(x) - F_2(x)|$ for all F_1 and $F_2 \in \mathcal{F}$.

L_p-distance: $\rho_{L_p}(F_1, F_2) = \left[\int_{\mathbb{R}} |F_1(x) - F_2(x)|^p \, dx\right]^{1/p}$, $p \geq 1$ for all F_1 and $F_2 \in \mathcal{F}_1$.

Lemma 10.6.1. *(i)* $\rho_\infty(F_n, F) = O_P(n^{-1/2})$, *(ii)* $\rho_{L_p}(F_n, F) = O_P(n^{-1/2})$ *if either* $1 \leq p < 2$ *and* $\int [F(x)(1 - F(x))]^{p/2} \, dx < \infty$ *or* $p \geq 2$.

Proof. By the DKW Theorem (See Theorem 9.2.4),

$$\begin{aligned} \mathrm{E}\left[\sqrt{n}\rho_\infty(F_n, F)\right] &= \int_0^\infty P\left[\sqrt{n}\rho_\infty(F_n, F) > y\right] dy \\ &\leq \int_0^\infty C \exp\left[-2n(y/\sqrt{n})^2\right] dy \\ &= C \int_0^\infty \exp\left[-2y^2\right] dy \quad \text{for all } n. \end{aligned}$$

Now by Markov inequality,

$$\begin{aligned} P\left[\sqrt{n}\rho_\infty(F_n, F) > M\right] &\leq M^{-1}\mathrm{E}\left[\sqrt{n}\rho_\infty(F_n, F)\right] \\ &\leq CM^{-1} \int_0^\infty \exp\left[-2y^2\right] dy \to 0 \quad \text{as } M \to \infty, \end{aligned}$$

proving that $\sqrt{n}\rho_\infty(F_n, F) = O_P(1)$ (ie, $\rho_\infty(F_n, F) = O_P(n^{-1/2})$).

The proof of (ii) is longer and we omit it. □

The following theorem provides conditions under which the remainder term

$$R_n = \sqrt{n}[T(F_n) - T(F)] - n^{-1/2} \sum_{i=1}^n T'(X_i; F) = o_P(1).$$

Theorem 10.6.1.

(i) *If T is ρ_∞-Hadamard differentiable at F, then $R_n = o_P(1)$.*
(ii) *If T is ρ_∞-Fréchet differentiable at F and if $\sqrt{n}\rho(F_n, F) = O_P(1)$, then $R_n = o_P(1)$.*

Proof. Part (i) is proved by advanced techniques, for which we refer to Fernholtz [57].

To prove part (ii), note the following:

By definition of ρ-Fréchet differentiability, $\rho(F_n, F) \to 0$ implies

$$\begin{aligned} \frac{T(F_n) - T(F) - L_F(F_n - F)}{\rho(F_n, F)} &= \frac{\sqrt{n}[T(F_n) - T(F)] - \sqrt{n}L_F(F_n - F)}{\sqrt{n}\rho(F_n, F)} \\ &= \frac{R_n}{\sqrt{n}\rho(F_n, F)} \to 0, \end{aligned}$$

because if $L_F(F_n - F)$ exists as a ρ-Fréchet differential, then it is the same as the Gâteaux differential which also exists, so that $\sqrt{n}L_F(F_n - F) = n^{-1/2}\sum_{i=1}^{n} T'(X_i; F)$. Thus for every $\varepsilon > 0$, there exists $\delta > 0$ such that

$$\rho(F_n, F) < \delta \text{ implies } |R_n| < \varepsilon\sqrt{n}\rho(F_n, F) \quad \text{for all large } n.$$

Hence for arbitrary $\eta > 0$,

$$P[|R_n| > \eta] \leq P[\rho(F_n, F) \geq \delta] + P\left[\sqrt{n}\rho(F_n, F) > \eta/\varepsilon\right]$$

for large n, so

$$\limsup_n P[|R_n| > \eta] \leq \limsup_n P[\rho(F_n, F) \geq \delta] + \limsup_n P\left[\sqrt{n}\rho(F_n, F) > \eta/\varepsilon\right].$$

Now (i) $\sqrt{n}\rho(F_n, F) = O_P(1)$ implies $\rho(F_n, F) = o_P(1)$, $\lim_{n\to\infty} P[\rho(F_n, F) \geq \delta] = 0$ and (ii) $\sqrt{n}\rho(F_n, F) = O_P(1)$ also implies $P\left[\sqrt{n}\rho(F_n, F) > \eta/\varepsilon\right]$ can be made arbitrarily small by choosing η/ε sufficiently large (ie, by making ε sufficiently small).

This completes the proof. □

We now examine the remainder terms of the L- and M-estimators.

Theorem 10.6.2. *Let $T(G) = \int xJ(G(x))\, dG(x)$ be an L-functional.*

(i) *If J is bounded, $J(u) = 0$ for all $u \notin (\alpha, \beta)$ for some $0 < \alpha < \beta < 1$, and J is continuous at $F(x)$ for almost all x, then T is ρ_∞-Fréchet differentiable at F with $T'(x; F)$ bounded and continuous.*

(ii) *If $|J(u) - J(v)| \leq C|u - v|^{p-1}$ for some $C > 0$ and $p > 1$, then T is ρ_{L_p}-Fréchet differentiable at F.*

Proof.

(i) Let $\{G_j\}$ satisfy $\lim_{j\to\infty} \rho(G_j, F) = 0$. Choose $c < d$ such that $F(c) < \alpha < \beta < F(d)$. Then for $x \notin [c, d]$, $F(x) \notin (\alpha, \beta)$, so $J(F(x)) = 0$ for all $x \notin [c, d]$. Since $\rho_\infty(G_j, F) = 0$, for sufficiently large j, $G_j(x)$ is also $\notin (\alpha, \beta)$ for $x \notin [c, d]$, so $J(G_j(x)) = 0$ for all $x \notin [c, d]$. Hence for sufficiently large j and for $x \notin [c, d]$, $J(u) = 0$ for $u = F(x)$ and for all u lying between $F(x)$ and $G_j(x)$, and therefore, $\int_{F(x)}^{G_j(x)} [J(u) - J(F(x))]\, du = 0$ and $W_{G_j}(x) = 0$.

Hence for large j,

$$|R(G_j, F)| = \left|\int_c^d W_{G_j(x)}\left[G_j(x) - F(x)\right] dx\right| \leq \rho_\infty(G_j, F)\int_c^d |W_{G_j}(x)|\, dx.$$

Now use the continuity of J at $F(x)$ for almost all x to see that $W_{G_j}(x) \to 0$ a.e., and is bounded since J is bounded. Hence $\int_c^d |W_{G_j}(x)|\, dx \to 0$ by dominated convergence. Thus $\lim_{j\to\infty} |R(G_j, F)|/\rho_\infty(G_j, F) = 0$ showing that T is ρ_∞-Fréchet differentiable at

F. Again, for $y \notin [c, d]$, $F(y) \notin [\alpha, \beta]$ so $J(F(y)) = 0$. Thus

$$T'(x; F) = -\int_c^d [\delta_x(y) - F(y)]J(F(y))\, dy$$

is bounded and continuous. This proves (i).

(ii) Let $\{G_j\}$ satisfy $\lim_{j\to\infty} \rho_{L_p}(G_j, F) = 0$ and note that

$$\begin{aligned}
|R(G_j, F)| &= \left| \int \left[\int_{F(x)}^{G_j(x)} \{J(u) - J(F(x))\}\, du \right] dx \right| \\
&\le \int \left[\int_{F(x)}^{G_j(x)} |J(u) - J(F(x))|\, du \right] dx \\
&\le \int \left[\int_{F(x)}^{G_j(x)} C|u - F(x)^{p-1}|\, du \right] dx \\
&\le C \int |G_j(x) - F(x)|^p\, dx = C\left\{\rho_{L_p}(G_j, F)\right\}^p.
\end{aligned}$$

Hence

$$\lim_{j\to\infty} \frac{|R(G_j, F)|}{\rho_{L_p}(G_j, F)} \le C \lim_{j\to\infty} \left\{\rho_{L_p}(G_j, F)\right\}^{p-1} = 0 \quad \text{for } p > 1,$$

showing that T is ρ_{L_p}-Fréchet differentiable at F. □

Corollary 10.6.1. *In the expansion*

$$\sqrt{n}[T(F_n) - T(F)] = n^{-1/2} \sum_{i=1}^{n} T'(X_i; F) + R_n$$

of the L-estimator $T(F_n)$ of $T(f) = \int xJ(F(x))\, dx$, the remainder term $R_n = o_P(1)$ if the function J satisfies the conditions of Theorem 10.6.2(i) or (ii).

Proof. Under the condition of Theorem 10.6.2(i), T is ρ_∞-Fréchet differentiable at F and by Lemma 10.6.1(i), $\rho_\infty(F_n, F) = O_P(n^{-1/2})$. Therefore, Theorem 10.6.1(ii) applies, proving $R_n = o_P(1)$.

Under the condition of Theorem 10.6.2(ii) with $p > 1$, T is ρ_{L_p}-Fréchet differentiable and by Lemma 10.6.1(ii), $\rho_{L_p}(F_n, F) = O_P(n^{-1/2})$ provided that $\int [F(x)(1 - F(x))^{p/2}\, dx < \infty$ also holds for the case $1 < p < 2$. Hence Theorem 10.6.1(ii) applies, proving $R_n = o_P(1)$. □

Theorem 10.6.3. *Let T be an M-functional with score function $\Psi(x, t)$ which is bounded and continuous, and suppose that the function $\lambda_F(t) = \int \Psi(x, t)\, dF(x)$ is continuously differentiable at $T(F)$ while $\lambda_F'(T(F)) \neq 0$. Then T is ρ_∞-Hadamard differentiable at F.*

Proof. Let $G_j = F + t_j\Delta_j$ where $\|\Delta_j - \Delta\|_\infty \to 0$ on $\mathcal{D} = \{c(F_1 - F_2), c > 0\}$ and $t_j \to 0$. We argue through the following steps:

(i) Since the ρ_∞-Hadamard differential, if it exists, is the same as the Gâteaux differential,

$$L_F\left(G_j - F\right) = \int T'(x;F)\, d\left[G_j(x) - F(x)\right] = -\int \frac{\Psi(x, T(F))}{\lambda'_F(T(F))} d\left[G_j(x) - F(x)\right],$$

using the formula for $T'(x;F)$ given in Eq. (11). It will, therefore, be enough to show that

$$\lim_{j\to\infty} \frac{T\left(G_j\right) - T(F) + \{\lambda'_F(T(F))\}^{-1} \int \Psi(x, T(F))\, d\left[G_j(x) - F(x)\right]}{\|t_j\Delta_j\|_\infty} \tag{15}$$
$$= \lim_{j\to\infty} R_j/\|t_j\Delta_j\|_\infty,$$

where R_j is the numerator of Eq. (15).

(ii) Since $\lambda_F(T(F)) = \lambda_{G_j}\left(T(G_j)\right) = 0$ by (12),

$$\begin{aligned}\lambda_F\left(T(G_j)\right) - \lambda_F(T(F)) &= \lambda_F\left(T(G_j)\right) - \lambda_{G_j}\left(T(G_j)\right)\\ &= -\int \Psi\left(x, T(G_j)\right) d\left[G_j(x) - F(x)\right]\\ &= -\int \Psi\left(x, T(G_j)\right) d\left[t_j\Delta_j(x)\right] \to 0,\end{aligned}$$

because $t_j \to 0$, $\|\Delta_j - \Delta\|_\infty \to 0$ and Ψ is bounded.

(iii) Since $\lambda'_F(T(F)) \neq 0$, the inverse function $\lambda_F^{-1}(\cdot)$ of $\lambda_F(\cdot)$ exists and is continuous in a neighborhood of $\lambda_F(T(F)) = 0$. Hence $\lambda_F\left(T(G_j)\right) - \lambda_F(T(F)) \to 0$ implies

$$T\left(G_j\right) - T(F) = \lambda_F^{-1}\left(\lambda_F\left(T\left(G_j\right)\right)\right) - \lambda_F^{-1}\left(\lambda_F(T(F))\right) \to 0.$$

(iv) Use (ii) to write

$$T\left(G_j\right) - T(F) = -\frac{T\left(G_j\right) - T(F)}{\lambda_F\left(T(G_j)\right) - \lambda_F(T(F))} \int \Psi\left(x, T(G_j)\right) d\left[t_j\Delta_j(x)\right].$$

(v) From (i) and (iv),

$$\begin{aligned}R_j &= \frac{1}{\lambda'_F(T(F))} \int \Psi(x, T(F))\, d\left[t_j\Delta_j(x)\right]\\ &\quad - \frac{T\left(G_j\right) - T(F)}{\lambda_F\left(T(G_j)\right) - \lambda_F(T(F))} \int \Psi\left(x, T(G_j)\right) d\left[t_j\Delta_j(x)\right]\\ &= t_j\left[\{\lambda'_F(T(F))\}^{-1} - \left\{\frac{\lambda_F\left(T(G_j)\right) - \lambda_F(T(F))}{T\left(G_j\right) - T(F)}\right\}^{-1} \int \Psi(x, T(F))\, d\Delta_j(x)\right]\end{aligned}$$

$$- t_j \frac{T\left(G_j\right) - T(F)}{\lambda_F\left(T(G_j)\right) - \lambda_F(T(F))} \int \left[\Psi\left(x, T(G_j)\right) - \Psi(x, T(F))\right] d\Delta_j(x)$$

$$:= t_j\left(R_{1j} + R_{2j}\right).$$

(vi) By (iii), $T\left(G_j\right) - T(F) \to 0$, so

$$\left\{\frac{\lambda_F\left(T(G_j)\right) - \lambda_F(T(F))}{T\left(G_j\right) - T(F)}\right\}^{-1} - \{\lambda'_F(T(F))\}^{-1} \to 0,$$

which together with boundedness of Ψ and $\|\Delta_j - \Delta\|_\infty \to 0$ implies $R_{1j} \to 0$. Also $T\left(G_j\right) - T(F) \to 0$ implies

$$\frac{T\left(G_j\right) - T(F)}{\lambda_F\left(T(G_j)\right) - \lambda_F(T(F))} \to \{\lambda'_F(T(F))\}^{-1} \text{ and}$$

$$\int [\Psi\left(x, T(G_j)\right) - \Psi(x, T(F))]\, d\Delta_j(x) \to 0,$$

because Ψ is bounded and continuous, and $\|\Delta_j - \Delta\|_\infty \to 0$. Thus $R_{2j} \to 0$. Putting all this together, we have

$$\lim_{j\to\infty} \frac{R_j}{\|t_j\Delta_j\|_\infty} = \lim_{j\to\infty} \frac{t_j\left(R_{1j} + R_{2j}\right)}{\|t_j\Delta_j\|_\infty} = \lim_{j\to\infty} \frac{R_{1j} + R_{2j}}{\|\Delta_j\|_\infty} = 0,$$

showing that Eq. (15) holds. □

Corollary 10.6.2. *If $T(F)$ is an M-functional with score function $\Psi(x, t)$ satisfying the conditions of Theorem 10.6.3 and if $T(F_n)$ is the corresponding M-estimator, then in the expansion*

$$\sqrt{n}[T(F_n) - T(F)] = n^{-1/2} \sum_{i=1}^{n} T'(X_i; F) + R_n,$$

the remainder term R_n is $o_P(1)$.

Proof. Under the conditions of Theorem 10.6.3, T is ρ_∞-Hadamard differentiable at F, so Theorem 10.6.1(i) applies. □

10.7 The Jackknife and the Bootstrap

In this section we will briefly deal with two well-known resampling methods — the jackknife and the bootstrap. These methods can be used to obtain approximate bias, variance, and distribution of estimates without having to obtain their analytic expressions which may be quite complicated in many cases (see Efron and Tibshirani [59]). At the beginning of this chapter, some examples of statistical functional were given, and in

Examples 10.2.1–10.2.3, the first derivatives (influence functions) were explicitly obtained for three functionals. It is important to add that there are many more useful functionals for the univariate case as in Exercises 10.5–10.11 or in the multivariate case such as the correlation coefficient, multiple correlation, etc. In this section, the discussion will be informal, starting with the concept of second derivative of a functional. Extending the definition given in Section 10.2 (for the first-order expansion of a statistical functional), $T''(x_1, x_2; F)$ is the second derivative if

$$T(F+t\Delta) - T(F) - t\int T'(x;F)\,d\Delta(x) - t^2\int T''(x_1,x_2;F)\,d\Delta(x_1)\,d\Delta(x_2)$$
$$= o\left(t^2\right),$$

as $t \downarrow 0$. We ignore the issue of whether the convergence is uniform in Δ. This has been discussed in the previous sections. We now examine the bias and the variance of the estimate $T(F_n)$ of $T(F)$, where F_n is the empirical cdf on the basis of iid sample $X_1, \ldots, X_n$. If we write $\Delta_n = \sqrt{n}(F_n - F)$ and $t = n^{-1/2}$, under appropriate conditions we have

$$\begin{aligned} T(F_n) &= T(F) + n^{-1/2}\int T'(x;F)\,d\Delta_n(x) \\ &\quad + n^{-1}\int T''(x_1,x_2;F)\,d\Delta_n(x_1)\,d\Delta_n(x_2) + R_n \\ &:= T(F) + L_n(F) + Q_n(F) + R_n, \end{aligned} \tag{16}$$

where we assume that the remainder term is $o_P(1/n)$. However, we should point out that, usually for many functionals, $R_n = O_P(n^{-3/2})$.

We now discuss the concepts of asymptotic bias (ABias) and asymptotic variance (AVar) of a statistical functional. In all our subsequent discussions in this section, *we ignore the remainder term R_n and assume that $T(F_n) = T(F) + L_n(F) + Q_n(F)$*, and that $\mathrm{E}\left[\{T'(X;F)\}^2\right]$, $\mathrm{E}\left[\{T''(X,X;F)\}^2\right]$, and $\mathrm{E}\left[\{T''(X_1,X_2;F)\}^2\right]$ are finite.

Note that $L_n(F)$ has mean zero and variance

$$\mathrm{Var}[L_n(F)] = n^{-1}\mathrm{Var}\big[T'(X;F)\big].$$

A tedious calculation, whose justification will be given later, shows

$$\mathrm{Var}\big[L_n(F) + Q_n(F)\big] = \mathrm{Var}[L_n(F)] + O(n^{-2}).$$

The expected value of $Q_n(F)$ is

$$\begin{aligned} \mathrm{E}\big[Q_n(F)\big] &= n^{-1}\big[\mathrm{E}\{T''(X,X;F)\} - \mathrm{E}\{T''(X_1,X_2;F)\}\big], \text{ and hence} \\ \mathrm{E}\big[L_n(F) + Q_n(F)\big] &= \mathrm{E}\big[Q_n(F)\big] = n^{-1}\big[\mathrm{E}\{T''(X,X;F)\} - \mathrm{E}\{T''(X_1,X_2;F)\}\big]. \end{aligned}$$

We define the asymptotic bias and asymptotic variance of $T(F_n)$ as

$$\begin{aligned} \mathrm{ABias}(T(F_n)) &= \mathrm{E}\big[Q_n(F)\big] \\ &= n^{-1}\big[\mathrm{E}\{T''(X,X;F)\} - \mathrm{E}\,\{T''(X_1,X_2;F)\}\big] := n^{-1}b(F), \end{aligned} \tag{17}$$

$$\text{AVar}[T(F_n)] = \text{Var}[L_n(F)] = n^{-1}\text{Var}\big[T'(X;F)\big] := n^{-1}v(F). \tag{18}$$

In some cases, it is possible to find explicit expressions for ABias and AVar, and in such cases, we may simply replace $b(F)$ and $v(F)$ by $b(F_n)$ and $v(F_n)$. However, in many others, explicit expressions for ABias and AVar are quite complicated, and it is useful to have simple sample-based methods for estimating these quantities. We now discuss estimation of these two quantities: asymptotic bias and asymptotic variance.

Remark 10.7.1. It is important to note that $T(F_n)$ may not have finite mean and variance in many cases. For instance, let $X_1, \ldots, X_n$ be iid discrete random variables taking values in $\mathbb{N} = \{0, 1, 2, \ldots\}$. Let $\mu(F) = \text{E}[X]$ and $\sigma^2(F) = \text{Var}[X]$, and we are interested in estimating $T(F) = \log(\mu(F))$ or the coefficient-of-variation $T(F) = \sigma(F)/\mu(F)$. Note that if $P[X = 0] > 0$, then $T(F_n)$ in each of these cases does not have finite mean and the variance is not defined. However, in statistical applications, the issues of interest are estimation of $T(F)$ and construction of its confidence interval. Hence if $P[T(F_n) = \infty] \to 0$ as $n \to \infty$, we may bypass the problem of estimating the actual bias and variance (may not exist) by assuming that the expression for $T(F_n)$ is given by the expression on the right-hand side of Eq. (16) without the remainder term R_n.

10.7.1 Estimation of Asymptotic Bias and Asymptotic Variance

Let F_{ni} be the empirical cdf of F on the basis of $X_1, \ldots, X_{i-1}, X_{i+1}, \ldots, X_n$ (ie, F_{ni} is based on $n-1$ observations, deleting X_i). Let $T(F_{ni})$ be the estimate of $T(F)$ based on F_{ni}. Tukey's pseudo values are defined to be

$$V_i = nT(F_n) - (n-1)T(F_{ni}), \quad i = 1, \ldots, n.$$

These pseudo values are *approximately iid* with mean $T(F)$ and variance $v(F)/n$. As a matter of fact, the following turn out to be true

$$\text{E}\big[\bar{V}\big] = T(F),$$
$$\text{E}\left[(n-1)^{-1}\sum_{i=1}^{n}(V_i - \bar{V})^2\right] = v(F) + o\big(n^{-1}\big),$$

where $\bar{V} = n^{-1}\sum_{i=1}^{n} V_i$. Since $\text{E}\big[T(F_n) - \bar{V}\big] = \text{E}[T(F_n)] - T(F)$, the Jackknife estimates of ABias$(T(F_n))$ is given by

$$\text{ABias}^{(J)}(T(F_n)) = T(F_n) - \bar{V} = (n-1)\big[\bar{T}_n - T(F_n)\big],$$

where $\bar{T}_n = n^{-1}\sum_{i=1}^{n} T(F_{ni})$.

The Jackknife estimate of AVar$(T(F_n))$ is

$$\text{AVar}^{(J)}(T(F_n)) = [n(n-1)]^{-1}\sum_{i-1}^{n}(V_i - \bar{V})^2 = \frac{n-1}{n}\sum_{i-1}^{n}(T(F_{ni}) - \bar{T}_n)^2.$$

The bootstrap method is conceptually simple: it basically seeks to replace $b_n(F)$ by its empirical estimate $b_n(F_n)$. However, it does so without having to obtain any analytic expression for $b_n(F)$. The method described here is known as the "nonparametric" bootstrap. Let $X_1^*, \ldots, X_n^*$ be iid with cdf F_n, then the bootstrap estimate of ABias is given by

$$\text{ABias}^{(B)}(T(F_n)) = \mathrm{E}\big[\{T(F_n^*) - T(F_n)\}|X_1, \ldots, X_n\big],$$

that is, the (conditional) expectation is taken over the random sample $X_1^*, \ldots, X_n^*$. In practice, it is calculated as follows. Draw a random sample of size n with replacement from the data $\{X_1, \ldots, X_n\}$ and repeat this N times. Let $F_{n(t)}^*$ be the empirical cdf on the basis of the tth sample $\left\{X_{1(t)}^*, \ldots, X_{n(t)}^*\right\}$, and let $T\left(F_{n(t)}^*\right)$ be the estimate of $T(F)$ based on $F_{n(t)}^*$. Then, one calculates the quantity

$$N^{-1}\sum_{t=1}^{N}\left[T\left(F_{n(t)}^*\right) - T(F_n)\right],$$

which, by the weak law of large numbers, converges to $\text{ABias}^{(B)}(T(F_n))$ as $N \to \infty$.

The bootstrap estimate of the asymptotic variance is

$$\text{AVar}^{(B)}(T(F_n)) = \text{Var}\big[T(F_n^*)|X_1, \ldots, X_n\big],$$

where the conditional variance is over the bootstrap sample $(X_1^*, \ldots, X_n^*)$.

The bootstrap procedure may also be used to obtain an estimate of the sampling distribution $Q_n(z) = P\big[\sqrt{n}\{T(F_n) - T(F)\} \le z\big]$, z real. Even though such sampling distributions are approximately normal under appropriate conditions when the sample size n is large, one may nonetheless use the bootstrap method in such cases. A bootstrap estimate of $Q_n(z)$ is given by $Q_n^{(B)}(z) = P\big[\sqrt{n}\{T(F_n^*) - T(F_n)\} \le z|X_1, \ldots, X_n\big]$, where the (conditional) probability is over the bootstrap sample $(X_1^*, \ldots, X_n^*)$. In general if there is a functional of the form $T(F_n, F)$ and one wishes to estimate $\mathrm{E}[T(F_n, F)]$, then its bootstrap estimate is $\mathrm{E}\big[T(F_n^*, F_n)|X_1, \ldots, X_n\big]$.

Remark 10.7.2. It is important to point out that the jackknife and bootstrap procedures may not always work. For instance, the jackknife estimate of bias cannot provide consistent estimates for the ABias and AVar when estimating a quantile $F^{-1}(p)$. Success of the jackknife method depends on the smoothness of the functional (ie, on the validity of the expansion given in Eq. (16)). Bootstrap works well for quantile estimation as long as p is away from 0 or 1. However, it cannot provide consistent estimates for ABias and AVar when estimating extreme quantiles (ie, when p is close to 0 or 1).

10.7.2 Heuristic Justification for the Jackknife and the Bootstrap

Let us first justify the validity of the bootstrap estimate. In the arguments given here we denote $n^{-1/2}\Delta_n = F_n - F$ by D_n and $F_n^* - F_n$ by D_n^*. We can simplify notations by writing $\int T'(x; F)\, dD_n(x)$ by $L_F(D_n)$ and $\int T''(x_1, x_2; F)\, dD_n(x_1)\, dD_n(x_2)$ by $Q_F(D_n)$. If we expand $T(F_n^*)$ about F_n, then we have

$$T(F_n^*) - T(F_n) = L_{F_n}(D_n^*) + Q_{F_n}(D_n^*).$$

Conditional expectation (given $X_1, \ldots, X_n$) of the first term on the right-hand side of the last expression equals zero, and the conditional expectation of the second term is $n^{-1}b_n(F_n)$. The result follows once we note that $|b_n(F_n) - b_n(F)| \xrightarrow{P} 0$ as $n \to \infty$. The justification for the bootstrap estimate of the variance is similar.

Let us now look at the jackknife estimate. The validity of the bias estimate is fairly easy to establish since

$$\mathrm{E}[V_i] = \mathrm{E}[nT(F_n) - (n-1)T(F_{ni})] = b(F) - b(F) = 0.$$

Thus $\mathrm{E}[\bar{V}] = 0$.

Justification for the validity of the jackknife estimate of the variance requires a bit of work. Simple algebra tells us

$$\begin{aligned}\sum_{i=1}^{n}(V_i - \bar{V})^2 &= (2n)^{-1} \sum_{1 \le i \ne i' \le n} (V_i - V_{i'})^2 \\ &= (2n)^{-1}(n-1)^2 \sum_{1 \le i \ne i' \le n} [T(F_{ni}) - T(F_{ni'})]^2. \end{aligned} \tag{19}$$

We will work with $T(F_{ni}) - T(F_{ni'})$ for the proof.

For notational simplicity, we suppress F in the notations of T' and T''. We also assume that T'' is symmetric in its arguments. Since if it is not, we can symmetrize it by using $[T''(x_1, x_2) + T''(x_2, x_1)]/2$, and such symmetrization does not alter the value of $\int T''(x_1, x_2)\, dH(x_1)\, dH(x_2)$, where H is a cdf or a difference of two cdfs. For any x, x_1, and x_2, define

$$\begin{aligned} T_1(x) &= T'(x) - \mathrm{E}[T'(X)], \\ T_2(x_1, x_2) &= T''(x_1, x_2) - \mathrm{E}[T''(x_1, X_2)] - \mathrm{E}[T''(X_1, x_2)] + \mathrm{E}[T''(X_1, X_2)]. \end{aligned}$$

Then

$$\begin{aligned} L_n(F) &= \int T'(x)\, dD_n(x) = \int T_1(x)\, dF_n(x) = n^{-1}\sum_{j=1}^{n} T_1(X_j), \\ Q_n(F) &= \int T''(x_1, x_2)\, dD_n(x_1)\, dD_n(x_2) \\ &= \int T_2(x_1, x_2)\, dF_n(x_1)\, dF_n(x_2) = n^{-2} \sum_{1 \le j,k \le n} T_2(X_j, X_k). \end{aligned}$$

Thus

$$T(F_n) = T(F) + n^{-1}\sum_{j=1}^{n} T_1(X_j) + n^{-2} \sum_{1 \le j,k \le n} T_2(X_j, X_k).$$

The two sums in the last expression are not uncorrelated. A simple argument can be used to create two uncorrelated sums. Note that $b(F) = \mathrm{E}[T_2(X, X)]$ and let

$$\begin{aligned} U_i &= T_1(X_i) + n^{-1}[T_2(X_i, X_i) - b(F)], \\ W_{j,k} &= T_2(X_j, X_k). \end{aligned}$$

Then $T(F_n)$ can be rewritten as

$$T(F_n) = T(F) + b(F)/n + n^{-1}\sum_{j=1}^{n} U_i + n^{-2}\sum_{1\le j\ne k\le n} W_{j,k}. \tag{20}$$

It is easy to see that U_i and $W_{j,k}$, $j \ne k$, have zero means, $\mathrm{Cov}[U_i, W_{j,k}] = 0$ whenever $j \ne k$, and $\mathrm{Cov}[W_{j_1,k_1}, W_{j_2,k_2}] = 0$ except when $(j_1, k_1) = (j_2, k_2)$ or $(j_1, k_1) = (k_2, j_2)$. We therefore have

$$\begin{aligned}
\mathrm{E}[T(F_n)] &= T(F) + b(F)/n,\\
\mathrm{Var}[T(F_n)] &= \mathrm{Var}\left[n^{-1}\sum_{j=1}^{n} U_i\right] + \mathrm{Var}\left[n^{-2}\sum_{1\le j\ne k\le n} W_{j,k}\right]\\
&= n^{-1}\mathrm{Var}[U_1] + O\left(n^{-2}\right)\\
&= n^{-1}\left\{\mathrm{Var}[T_1(X)] + O\left(n^{-1}\right)\right\} + O\left(n^{-2}\right)\\
&= n^{-1}v(F) + O\left(n^{-2}\right).
\end{aligned}$$

Applying the equality in Eq. (20) for $T(F_{ni})$ and $T(F_{ni'})$ for $i \ne i'$, we have

$$T(F_{ni}) - T(F_{ni'}) = (n-1)^{-1}[U_{i'} - U_i] + 2(n-1)^{-2}\sum_{j\ne i, j\ne i'}\left[W_{i',j} - W_{i,j}\right].$$

Thus

$$\begin{aligned}
\mathrm{E}\left\{\sum_{1\le i\ne i'\le n}[T(F_{ni}) - T(F_{ni'})]^2\right\} &= \sum_{i\ne i'}\mathrm{E}\left\{(n-1)^{-1}[U_{i'} - U_i]\right\}^2\\
&\quad + \sum_{i\ne i'}\mathrm{E}\left\{2(n-1)^{-2}\sum_{j\ne i, j\ne i'}\left[W_{i',j} - W_{i,j}\right]\right\}^2\\
&= 2n(n-1)^{-1}\mathrm{Var}[U_1] + O\left(n^{-1}\right)\\
&= 2n(n-1)^{-1}\left\{\mathrm{Var}[T_1(X)] + O\left(n^{-1}\right)\right\} + O\left(n^{-1}\right)\\
&= 2n(n-1)^{-1}\mathrm{Var}[T_1(X)] + O\left(n^{-1}\right).
\end{aligned}$$

From Eq. (19), we thus have

$$\mathrm{E}\left[\sum_{i=1}^{n}(V_i - \bar{V})^2\right] = (n-1)\mathrm{Var}[T_1(X)] + O(1) = (n-1)v(F) + O(1).$$

This shows that $\mathrm{E}\left[AVar^{(J)}(T(F_n))\right] = n^{-1}v(F) + O(n^{-2})$.

Exercises

In the following problems, $X_1, \ldots, X_n$ is a random sample from a population with cdf F and pdf f, $\bar{X}_n$ and s_n^2 are the sample mean and sample variance, $X_{n:1} < \cdots < X_{n:n}$ are the order statistics, and $Q_{1n} = X_{n:[n/4]}$, $M_n = X_{n:[n/2]}$, and $Q_{3n} = X_{n:[3n/4]}$ are, respectively, the sample first quartile, the sample median, and the sample third quartile.

10.1. Let F be the cdf of $N(\mu, \sigma^2)$. Consider two tests with critical regions $T_{1n} \geq c_{1n}$ and $T_{2n} \geq c_{2n}$ for testing H_0: $\mu = 0$ vs H_1: $\mu > 0$, where $T_{1n} = \sqrt{n}\bar{X}_n/s_n$, $T_{2n} = Q_{1n} + Q_{3n}$, and c_{1n}, c_{2n} are chosen so as to control the Type I error probability at a given level α, approximately for large n. For alternatives of the order $1/\sqrt{n}$, calculate the Pitman ARE of the test based on T_{2n} with respect to the one based on T_{1n}.

10.2. If F is the cdf of $N(\mu, \sigma^2)$, find the asymptotic joint distribution of $(\bar{X}_n, M_n)$, properly normalized.

10.3. Let ξ_1 and ξ_3 denote the first and the third quartiles of F (ie, $\xi_1 = F^{-1}(1/4)$ and $\xi_3 = F^{-1}(3/4)$).

(a) Write down the Bahadur representation of Q_{1n} and Q_{3n}, stating conditions on F for their validity.

(b) Let F be the cdf of $N(\mu, \sigma^2)$. Find the asymptotic distribution of the inter-quartile range $D_n = Q_{3n} - Q_{1n}$.

(c) Find a constant c such that $T_n = cD_n$ is a consistent estimator of σ. What can you say about the asymptotic efficiency of T_n? [If ϕ is the pdf and Φ is cdf of $N(0, 1)$, then $\Phi^{-1}(3/4) = 0.6745$ and $\phi(\Phi^{-1}(3/4)) = 0.3178$.]

10.4. Suppose the pdf involves a parameter θ and

$$f(x, \theta) = (1/2 - \theta)e^{x} I_{(-\infty,0)}(x) + (1/2)e^{-x/(1+2\theta)}$$
$$= \begin{cases} (1/2 - \theta)e^{-|x|} & x < 0 \\ (1/2 + \theta)\frac{1}{1+2\theta}e^{-|x|/(1+2\theta)} & x \geq 0. \end{cases}$$

For $\theta = 0$, the pdf $f(x, 0) = (1/2)e^{-|x|}$, $-\infty < x < \infty$ is symmetric with mean = median = 0. For $\theta \neq 0$, none of this is true. To test H_0: $\theta = 0$ vs H_1: $\theta \neq 0$ we can, therefore, test for symmetry on the basis of the mean or the median. Another possibility is to test whether the MLE $\hat{\theta}_n$ of θ exhibits distributional property that it should have under H_0. With this in mind, consider the following test statistics:

$$T_{0n} = \hat{\theta}_n,\ T_{1n} = \bar{X}_n,\ T_{2n} = M_n,$$
$$T_{3n} = Q_{3n} - 2M_n + Q_{1n} \text{ (a measure of asymmetry)}.$$

(a) Find the asymptotic distributions of T_{0n}, T_{1n}, T_{2n}, and T_{3n} (properly normalized).

(b) Consider tests with critical regions $T_{jn} \geq c_{jn}$, $j = 0, 1, 2, 3$, each c_{jn} being chosen to control the Type I error probabilities at a given level α approximately, for large n. For alternative $\theta > 0$ of the order $1/\sqrt{n}$, calculate the Pitman AREs of the tests based on T_{1n}, T_{2n}, T_{3n} with respect to the one based on T_{0n}.

10.5. **(a)** Let T_1 and T_2 be statistical functionals of a family of cdf's $\mathcal{F}_0$ having finite third moments. Express the influence function $T'(x; F)$ of $T = T_1/T_2$ in terms of $T_1(F)$, $T_2(F)$ and their influence functions $T_1'(x; F)$ and $T_2'(x; F)$.

(b) Let $T(F) = \mathrm{E}_F\big[(X - \xi(F))^3\big]/\big\{\mathrm{E}_F\big[(X - \xi(F))^2\big]\big\}^{3/2}$, where $\xi(F) = \mathrm{E}_F[X]$. Use the result of (a) and the formula for the influence function of $\mathrm{E}_F\Big[(X - \xi(F))^k\Big]$ given in Section 10.2 to derive the influence function of this $T(F)$.

(c) Verify for this $T(F)$ that $\mathrm{E}_F\big[T'(X; F)\big] = 0$, $R_n = \sqrt{n}[T(F_n) - T(F)] - n^{-1/2} \sum_{i=1}^{n} T'(X_i; F) = o_P(1)$, and then find the asymptotic distribution of $\sqrt{n}[T(F_n) - T(F)]$ for arbitrary $F \in \mathcal{F}_0$ and when F is the cdf of $N\big(\mu, \sigma^2\big)$.

10.6. Let $\mathcal{F}_0$ be the set of all cdf's on $\mathbb{R}$ with finite mean.

(a) Find $T(F)$ on $\mathcal{F}_0$ such that $T(F_n) = n^{-1} \sum_{i=1}^{n} |X_i - \bar{X}_n|$.

(b) Find the influence function of $T(F)$.

(c) Specializing to the case of symmetric F, indicate how you would show that $R_n = \sqrt{n}[T(F_n) - T(F)] - n^{-1/2} \sum_{i=1}^{n} T'(X_i; F) = o_P(1)$.

10.7. Let $T(F) = \mathrm{E}_F\big[|X - X'|\big]$, where X, X' are iid with cdf F.

(a) Find the influence function of $T(F)$.

(b) Write down the expression for $T(F_n)$ and derive the asymptotic distribution of $\sqrt{n}[T(F_n) - T(F)]$.

10.8. Let $T(F) = \big(\int x \, dF(x)\big)^{1/2} = \xi(F)^{1/2}$ for F having finite mean $\xi(F) > 0$.

(a) Find the influence function of $T'(x; F)$.

(b) Verify that $\mathrm{E}_F\big[T'(X; F)\big] = 0$,
$R_n = \sqrt{n}[T(F_n) - T(F)] - n^{-1/2} \sum_{i=1}^{n} T'(X_i; F) = o_P(1)$, and then find the asymptotic distribution of $\sqrt{n}[T(F_n) - T(F)]$.

10.9. Let $T(F) = \big[\int (x - \xi(F))^2 \, dF(x)\big]^{1/2}$ for a cdf with mean $\mathrm{E}_F[X] = \xi(F)$ and finite second moment. Find the influence function of $T(F)$ and the asymptotic distribution of $\sqrt{n}[T(F_n) - T(F)]$.

10.10. For cdf's on $\mathbb{R}$ with nonzero mean $\xi(F)$, define $T(F) = 1/\xi(F)$. Find the influence function of $T(F)$ and the asymptotic distribution of $\sqrt{n}[T(F_n) - T(F)]$.

10.11. For cdf's on $\mathbb{R}$ with positive mean $\xi(F)$ and finite second moment, let $T(F) = \big[\int (x - \xi(F))^2 \, dF(x)\big]^{1/2}/\xi(F)$ denote the coefficient of variation. Find the influence function of $T(F)$ using the results of Exercises 10.5(a), 10.9, and 10.10, and then find the asymptotic distribution of $\sqrt{n}[T(F_n) - T(F)]$.

10.12. The pdf $f(x, \theta)$ of X is of the form

$$f(x, \theta) = \begin{cases} 1/4 & |x - \theta| \le 1 \\ (1/4)\exp[-|x - \theta| + 1] & |x - \theta| > 1, \end{cases}$$

which is uniform in the middle and double-exponential in the tails.

(a) Find formulas for the α-quantile and the $(1 - \alpha)$-quantile of this distribution for $0 < \alpha < 1/4$ and for $1/4 < \alpha < 1/2$.

(b) Let $\bar{X}_{n(\alpha)} = (n-2[\alpha n])^{-1}\sum_{i=[\alpha n]+1}^{n-[\alpha n]} X_{n:i}$ denote the α-trimmed mean based on $X_1,\ldots,X_n$. Find the asymptotic distribution of $\sqrt{n}\left[\bar{X}_{n(\alpha)}-\theta\right]$.

(c) Plot the asymptotic variance $\sigma_1^2(\alpha)$ obtained in (b) for $\alpha = 0.1, 0.2, 0.3, 0.4$ to determine the best choice of trimming among these.

(d) Let $M_n(k)$ denote the M-estimator of θ obtained by solving $\sum_{i=1}^{n}\Psi(X_i-t)=0$, using the Huber function $\Psi(x) = xI(|x|\le k) + k\, sign(x) I(|x|>k)$. Find the asymptotic distribution of $\sqrt{n}\left[M_n(k)-\theta\right]$.

(e) Plot the asymptotic variance $\sigma_2^2(k)$ obtained in (d) for suitably chosen values of k (starting with $k = 0.5, 1.0, 1.5, 2.0$) to determine as good a choice of k you can.

(f) Compare the performances of $\bar{X}_{n(\alpha)}$ with α chosen in (c) and $M_n(k)$ with k chosen in (e).

In the following two problems, the population cdf and pdf are, respectively, $F_\theta(x) = F(x-\theta)$ and $f_\theta(x) = f(x-\theta)$, where f is symmetric about 0.

10.13. **(a)** Suppose that T is an L-functional with score function J. Show that if J is symmetricabout 1/2 with $\int_0^1 J(u)\,du = 1$, then $T(F_\theta)=\theta$.

(b) Let $\bar{X}_{n(\alpha)}$ denote the α-trimmed mean. Find the asymptotic variance of $\sqrt{n}\left[\bar{X}_{n(\alpha)}-\theta\right]$ when $\alpha = 1/4$ and $f(x)$ is the pdf of the standard Cauchy distribution.

10.14. Let $T(F)$ be the solution of the equation $\int \Psi(x-t)\,dF(x)=0$ and let $\hat{\theta}_n$ be the corresponding M-estimator. Make the following assumptions as needed:

(i) differentiation under the integral sign is valid,

(ii) $xf(x)$ is of bounded variation,

(iii) $\int |\Psi(x)||f'(x)|\,dx < \infty$ and $\neq 0$,

(iv) $\int \Psi^2(x) f(x)\,dx < \infty$.

(a) Show that if $\Psi(-x) = -\Psi(x)$ for all x, then $T(F_\theta)=\theta$.

(b) Let $0 < I_f = \int \{f'(x)/f(x)\}^2 f(x)\,dx < \infty$ denote the Fisher-information of the location family $\{f(x-\theta), \theta\in\mathbb{R}\}$ and let $\sigma^2(F,\Psi)$ denote the asymptotic variance of $\sqrt{n}\left(\hat{\theta}_n-\theta\right)$. Show that $\sigma^2(F,\Psi)\ge I_f^{-1}$. [*Hint*: Use integration by parts to rewrite the denominator of $\sigma^2(F,\Psi)$ and then apply Cauchy-Schwarz inequality.]

(c) Calculate $\sigma^2(F,\Psi)$ when f is the pdf of the standard Cauchy distribution and $\Psi(x) = -xI_{[-c,c]}(x)$. [*Hint*: Show that $\int_0^c (1+x^2)^{-1}\,dx = \arctan c$ and $\int_0^c x^2(1+x^2)^{-1}\,dx = c - \arctan c$.]

10.15. The sample mean in a random sample from a Cauchy distribution can be stabilized by excluding just one- or two-order statistics at each end. Let $\bar{X}^*_{n,r} = (n-2r)^{-1}\sum_{i=r+1}^{n-r} X_{n:i}$. Show that $\bar{X}^*_{n,r}$ is an unbiased estimator of the Cauchy median θ if $r\ge 1$ and that $\bar{X}^*_{n,r}$ has finite variance if $r\ge 2$.

10.16. Calculate the ARE of the M-estimator with score function $\Psi(x) = -xI_{[-c,c]}(x)$ of the median θ of a Cauchy distribution, comparing $\sigma^2(F, \Psi)$ with I_f^{-1}. Use the results of Exercise 10.14(c) with $c = 1$.

10.17. Let $F_\theta(x) = F(x - \theta)$ be a cdf with pdf $f_\theta(x) = f(x - \theta)$, where $f(x) = (1/2)\mathrm{e}^{-|x|}$, θ is unknown and we want to estimate θ. Let $T(F_\theta)$ be a solution to the equation

$$\lambda_{F_\theta}(t) = \int \Psi(x - t)\, dF_\theta(x) = 0 \text{ where } \Psi(x) = \begin{cases} |x|^{1/2} & x \le 0 \\ -|x|^{1/2} & x > 1. \end{cases}$$

Then the influence function of $T(F_\theta)$ is $T'(x; F_\theta) = -\Psi(x - T(F_\theta))/\lambda'_{F_\theta}(T(F_\theta))$.

(a) Show that $T(F_\theta) = \theta$.

(b) Express the corresponding M-estimator $\hat{\theta}_n = T(F_n)$ in terms of $(X_1, \ldots, X_n)$ of which F_n is the edf.

(c) Write down one-term Taylor expansion of $\sqrt{n}\left(\hat{\theta}_n - \theta\right)$. Assume that the remainder term is $o_P(1)$.

(d) Show that $\mathrm{E}_\theta\left[T'(X; F_\theta)\right] = 0$ and evaluate $\sigma^2(F_\theta, \Psi) = \mathrm{Var}\left[T'(X; F_\theta)\right]$.

(e) Find the asymptotic distribution of $\sqrt{n}\left(\hat{\theta}_n - \theta\right)$. [*Hint*: To find $\lambda'_{F_\theta}(T(F_\theta)) = d\lambda_{F_\theta}(t)/\, dt|_{t=\theta}$, differentiate under the integral and use $\Psi'(x) = -(1/2)|x|^{-1/2}$ for $x \ne 0$. Although Ψ' does not exist at $x = 0$, this calculation is still valid. Also note that $\int_0^\infty x^{-1/2}\mathrm{e}^{-x}\, dx = \Gamma(1/2) = \sqrt{\pi}$.]

10.18. Let $f_\theta(x) = f(x - \theta)$ be the pdf of a double exponential distribution with mean θ (ie, $f_\theta(x) = (1/2)\mathrm{e}^{-|x-\theta|}$) having cdf F_θ and let F_n the empirical cdf of a random sample from F_θ. Let $T(F_n)$ be the L-estimator of θ with score function $J(u) = 4uI\left(0 \le u \le 1/2\right) + 4(1 - u)I\left(1/2 < u \le 1\right)$.

(a) Express $T(F_n)$ as a linear function of the order statistics $X_{n:1} < \cdots < X_{n:n}$, using coefficients $J\left(i/(n+1)\right)$ instead of $J\left(i/n\right)$, with n even.

(b) Show that $T(F_\theta) = \theta$ for the corresponding L-functional.

(c) Find the asymptotic distribution of $\sqrt{n}[T(F_n) - \theta]$. [*Hint*: The asymptotic variance $\sigma^2(F, J)$ of $\sqrt{n}[T(F_n) - T(F)]$ involves F^{-1}. If F is the cdf corresponding to $f(x) = (1/2)\mathrm{e}^{-|x|}$, then $F^{-1}(u) = \log(2u)$ if $0 \le u \le 1/2$ and $= -\log(2(1 - u))$ if $1/2 < u \le 1$.]

11

Linear Models

11.1 Introduction

Linear models are widely used in statistical data analysis when the dependent or the response variable is quantitative, whereas the independent variables may be quantitative, qualitative, or both. It can also be used for some types of nonlinear modeling as an example given below will show. A few well-known classes of linear models are

(i) regression: all the variables are quantitative,
(ii) analysis of variance (ANOVA): all the independent variables are qualitative, and
(iii) analysis of covariance (ANCOVA): some of the independent variables are quantitative and some qualitative.

An obvious example of a linear model is simple linear regression with one independent variable. If the observations are (Y_i, X_i), $i = 1, \ldots, n$, where $\{Y_i\}$ are the values of the dependent variable and $\{X_i\}$ are the values of the independent variable, then the simple linear regression model is

$$Y_i = \beta_0 + \beta_1 X_i + \varepsilon_i, \quad i = 1, \ldots, n,$$

where β_0 is the intercept, β_1 is the slope, and $\{\varepsilon_i\}$ are mutually uncorrelated random errors with mean 0 and common variance σ^2. This model may also be written as $\boldsymbol{Y} = \boldsymbol{X\beta} + \boldsymbol{\varepsilon}$, where

$$\boldsymbol{Y} = \begin{pmatrix} Y_1 \\ \cdot \\ \cdot \\ \cdot \\ Y_n \end{pmatrix}, \quad \boldsymbol{X} = \begin{pmatrix} 1 & X_1 \\ \cdot & \cdot \\ \cdot & \cdot \\ \cdot & \cdot \\ 1 & X_n \end{pmatrix}, \quad \boldsymbol{\beta} = \begin{pmatrix} \beta_0 \\ \beta_1 \end{pmatrix}, \quad \boldsymbol{\varepsilon} = \begin{pmatrix} \varepsilon_1 \\ \cdot \\ \cdot \\ \cdot \\ \varepsilon_n \end{pmatrix},$$

where $\mathrm{E}[\boldsymbol{\varepsilon}] = \boldsymbol{0}$ and $\mathrm{Cov}\,[\boldsymbol{\varepsilon}] = \sigma^2 \boldsymbol{I}$. In the linear model terminology, $\boldsymbol{X}$ is called the design matrix and the goal is to obtain estimates of the unknown parameters $\boldsymbol{\beta}$ and σ^2, and carry out inferences on them.

It turns out that all the models mentioned in (i)–(iii) can be rewritten in the framework of a Gauss-Markov model which is

$$\boldsymbol{Y} = \boldsymbol{X\beta} + \boldsymbol{\varepsilon}, \quad \mathrm{E}[\boldsymbol{\varepsilon}] = \boldsymbol{0}, \quad \mathrm{Cov}[\boldsymbol{\varepsilon}] = \sigma^2 \boldsymbol{I}, \tag{1}$$

Theory and Methods of Statistics. http://dx.doi.org/10.1016/B978-0-12-802440-9.00011-4

where $\boldsymbol{Y}$ is $n \times 1$ vector of observed response values, the design matrix $\boldsymbol{X}$ is of order $n \times p$ and $\boldsymbol{\beta}$ is a $p \times 1$ vector of unknown parameters. In standard applications, the errors are often taken to be iid $N(0, \sigma^2)$. In this chapter, the columns of $\boldsymbol{X}$ will sometimes be referred to as independent variables.

Important Assumptions

Throughout this chapter, we assume that the design matrix $\boldsymbol{X}$ is nonrandom, or if it is random all the calculations such as $\mathrm{E}[\cdot]$, $\mathrm{Cov}[\cdot]$, etc., are carried out conditionally on $\boldsymbol{X}$. We also assume that $n > p$ and $\boldsymbol{X}$ has full rank (ie, $rank(\boldsymbol{X}) = p$). For the Gauss-Markov model,

$$\mathrm{E}[\boldsymbol{Y}] = \boldsymbol{X\beta} \text{ and } \mathrm{Cov}[\boldsymbol{Y}] = \sigma^2 \boldsymbol{I}.$$

The problems of inference involve estimation of $\boldsymbol{\beta}$ and σ^2, construction of confidence intervals, and hypotheses tests for $\boldsymbol{\beta}$ and its linear functions, deciding if some columns of $\boldsymbol{X}$ can be deleted from the model, prediction of Y at a future set of values of the independent variables, etc.

11.2 Examples of Gauss-Markov Models

Even though the description of the Gauss-Markov in Eq. (1) requires only the mutual uncorrelatedness of the random errors, this assumption is too general to be useful in applications. Therefore, following the standard practice, we assume that these random errors are iid with mean 0 and variance σ^2 in all the examples below.

Example 11.2.1 (Linear Regression). Suppose that we have n observation vectors $(Y_i, X_{i,1}, \ldots, X_{i,p-1})$, $i = 1, \ldots, n$, where the response for the ith case is Y_i and values of the independent variables are $X_{i,1}, \ldots, X_{i,p-1}$. Then a linear regression model is

$$Y_i = \beta_0 + \beta_1 X_{i,1} + \cdots + \beta_{p-1} X_{i,p-1} + \varepsilon_i,$$

where $\{\varepsilon_i\}$ are iid with mean 0 and variance σ^2. The statistical analysis involves estimation of the unknown constants $\beta_0, \beta_1, \ldots, \beta_{p-1}$ and σ^2 from the data $(Y_i, X_{i,1}, \ldots, X_{i,p-1})$, $i = 1, \ldots, n$. This model can be expressed in the Gauss-Markov framework given in Eq. (1) with

$$\boldsymbol{Y} = \begin{pmatrix} Y_1 \\ \cdot \\ \cdot \\ \cdot \\ Y_n \end{pmatrix}, \quad \boldsymbol{X} = \begin{pmatrix} 1 & X_{1,1} & \cdot & \cdot & \cdot & X_{1,p-1} \\ \cdot & \cdot & \cdot & \cdot & \cdot & \cdot \\ \cdot & \cdot & \cdot & \cdot & \cdot & \cdot \\ \cdot & \cdot & \cdot & \cdot & \cdot & \cdot \\ 1 & X_{n,1} & \cdot & \cdot & \cdot & X_{n,p-1} \end{pmatrix},$$

$$\boldsymbol{\beta} = \begin{pmatrix} \beta_0 \\ \cdot \\ \cdot \\ \cdot \\ \beta_{p-1} \end{pmatrix}, \quad \boldsymbol{\varepsilon} = \begin{pmatrix} \varepsilon_1 \\ \cdot \\ \cdot \\ \cdot \\ \varepsilon_n \end{pmatrix}.$$

Example 11.2.2 (Nonlinear Regression). The structure given in the last example is quite general as it can accommodate nonlinear cases. For instance, if we have only one independent variable but it is believed that the relation between the independent variable and the dependent variable is nonlinear, then we may fit a polynomial model to account for the nonlinearity. If a polynomial of degree $p-1$ is considered, when the observations are $\{Y_i, X_i\}$, $i = 1, \ldots, n$, then we may consider the model

$$Y_i = \beta_0 + \beta_1 X_i + \beta_2 X_i^2 + \cdots + \beta_{p-1} X_i^{p-1} + \varepsilon_i.$$

This is clearly of the form given in the last example if we take $X_{i,1} = X_i, X_{i,2} = X_i^2, \ldots, X_{i,p-1} = X_i^{p-1}$. It is worthwhile to point out that in actual data analysis one may not use the powers of X_i to create independent variables since it may lead to the problem of very strong correlation among independent variables (also called the problem of high multicollinearity), which leads to instability (high variance) in the parameter estimates, especially when p is not small. In such cases, one may employ orthogonal polynomials to create the independent variables instead of using the powers of the independent variable to form the columns of the design matrix $\boldsymbol{X}$.

Remark 11.2.1. The last two examples show that any linear regression model can be reexpressed in the Gauss-Markov framework. In order to show that the ANOVA and ANCOVA models are in the Gauss-Markov setup, it will be enough to show that they can be written as linear regression models and this is the approach taken in the examples below.

Example 11.2.3 (One-Factor Analysis of Variance (ANOVA)). The superintendant of a school district may be interested in comparing the mathematical aptitudes of the students in k different schools in a city. In order to achieve this, n_i students are selected at random from the ith school, $i = 1, \ldots, k$, and the score of each of the $n = n_1 + \cdots + n_k$ students on a standardized mathematics test is recorded. This is an example of a one-factor study where "school" is called a factor with k levels. Thus there are k populations and n_i iid observations are available from the ith population. A typical assumption is that the populations may have different means $\{\mu_i\}$ but the variances are the same. If Y_{ij} is the score of the jth student in the ith school, then the one-factor ANOVA model can be written as

$$Y_{ij} = \mu_i + \varepsilon_{ij}, \quad j = 1, \ldots, n_i, \quad i = 1, \ldots, k,$$

where $\{\varepsilon_{ij}\}$ are iid with mean 0 and variance σ^2. This model is called *balanced* if n_i's are the same (ie, $n_i = n_0$ for all i).

If μ is an overall (weighted) average of $\{\mu_i\}$ (ie, $\mu = \sum_{i=1}^k w_i \mu_i$ where $w_i \geq 0$ and $\sum w_i = 1$), then the factor effect for the ith school is defined to be $\alpha_i = \mu_i - \mu$ and $\{\alpha_i\}$ satisfy the constraint $\sum w_i \alpha_i = 0$. In practice, the weights are often taken to be $w_i \equiv 1/k$ or $w_i = n_i/n$, $i = 1, \ldots, k$, though other choices are also possible. The one-factor ANOVA model may also be written as a factor-effects model

$$Y_{ij} = \mu + \alpha_i + \varepsilon_{ij}, \quad j = 1, \ldots, n_i, \quad i = 1, \ldots, k.$$

Note that $\mathrm{E}[Y_{ij}] = \mu + \alpha_i = \mu_i$, irrespective of how μ and α_i are defined.

We now examine how this model may be recast in the regression setup. If the main effects satisfy the constraint $\sum \alpha_i = 0$ (ie, $w_i \equiv 1/k$), and if $\alpha_1, \ldots, \alpha_{k-1}$ are known, then $\alpha_k = -\alpha_1 - \cdots - \alpha_{k-1}$ is known. Thus there are really k free parameters in the factor-effects model and they are $\mu, \alpha_1, \ldots, \alpha_{k-1}$. Define $k-1$ variables as follows

$$X_{ij,1} = \begin{cases} 1 & j=1 \\ -1 & j=k \\ 0 & \text{otherwise} \end{cases}, \qquad X_{ij,2} = \begin{cases} 1 & j=2 \\ -1 & j=k \\ 0 & \text{otherwise} \end{cases}, \ldots,$$

$$X_{ij,k-1} = \begin{cases} 1 & j=k-1 \\ -1 & j=k \\ 0 & \text{otherwise.} \end{cases}$$

Then the ANOVA model can be expressed in the regression framework

$$Y_{ij} = \beta_0 + \beta_1 X_{ij,1} + \cdots + \beta_{k-1} X_{ij,k-1} + \varepsilon_{ij}, \text{ with}$$
$$\beta_0 = \mu, \quad \beta_1 = \alpha_1, \ldots, \beta_{k-1} = \alpha_{k-1}.$$

Note that when $j = k$,

$$\mu + \alpha_1 X_{ij,1} + \cdots + \alpha_{k-1} X_{ij,k-1} = \mu - (\alpha_1 + \cdots + \alpha_{k-1}) = \mu + \alpha_k.$$

Example 11.2.4 (One-Factor ANOVA Continued). As mentioned in the last example, there are many ways to define the overall mean μ and the factor effects $\{\alpha_i\}$. If the overall mean μ is defined to be $\mu = \sum w_i \mu_i$ with $w_i = n_i/n$ and $\alpha_i = \mu_i - \mu$, then the constraint on the factor effects is $\sum (n_i/n)\alpha_i = 0$ (ie, $\sum n_i \alpha_i = 0$). In this case one may define the X-variables as

$$X_{ij,1} = \begin{cases} 1 & i=1 \\ -n_1/n_k & i=k \\ 0 & \text{otherwise} \end{cases}, \qquad X_{ij,2} = \begin{cases} 1 & i=2 \\ -n_2/n_k & i=k \\ 0 & \text{otherwise} \end{cases}, \ldots,$$

$$X_{ij,k-1} = \begin{cases} 1 & i=k-1 \\ -n_{k-1}/n_k & i=k \\ 0 & \text{otherwise.} \end{cases}$$

Then the ANOVA model can be rewritten as

$$Y_{ij} = \beta_0 + \beta_1 X_{ij,1} + \cdots + \beta_{k-1} X_{ij,k-1} + \varepsilon_{ij}, \text{ with}$$
$$\beta_0 = \mu, \quad \beta_1 = \alpha_1, \ldots, \beta_{k-1} = \alpha_{k-1}.$$

A one-factor ANOVA model may also be written as a regression of Y on the following $k-1$ indicator variables as defined below

$$X_{ij,1} = \begin{cases} 1 & i=1 \\ 0 & i \neq 1 \end{cases}, \ldots, X_{ij,k-1} = \begin{cases} 1 & i=k-1 \\ 0 & i \neq k-1. \end{cases}$$

Example 11.2.5 (Two-Factor ANOVA). If in Example 11.2.3, ethnicity/race of each student is recorded, then we have a two-factor study with factors "school" and "ethnicity." Changing the notations a bit, suppose that there a levels of factor A (school) and b levels of factor B (ethnic group). Assume that a random sample of n_{ij} students is taken from the ith school with the jth ethnic background and the observed scores are $\{Y_{ijk}\colon k = 1, \ldots, n_{ij}\}$. If

μ_{ij} is the mean score of the students in the ith school with the jth ethnic background, then the cell means model is

$$Y_{ijk} = \mu_{ij} + \varepsilon_{ijk}, \quad k = 1, \ldots, n_{ij}, \; j = 1, \ldots, b, \; i = 1, \ldots, a,$$

where $\{\varepsilon_{ijk}\}$ are usually assumed to be iid random errors with mean 0 and variance σ^2. A two-factor study is called balanced if $n_{ij} = n_0$ for all i and j.

In order to rewrite the two-factor model as a factor-effects model, let us define

$$\begin{aligned}
&\mu = (ab)^{-1} \sum_{j=1}^{a} \sum_{i=1}^{a} \mu_{ij}, \; \mu_{i\cdot} = b^{-1} \sum_{j=1}^{b} \mu_{ij}, \; \mu_{\cdot j} = a^{-1} \sum_{i=1}^{a} \mu_{ij}, \\
&\alpha_i = \mu_{i\cdot} - \mu, \; \beta_j = \mu_{\cdot j} - \mu, \text{ and} \\
&(\alpha\beta)_{ij} = \mu_{ij} - \mu_{i\cdot} - \mu_{\cdot j} + \mu = \mu_{ij} - \left(\mu + \alpha_i + \beta_j\right), \text{ and hence} \\
&\mu_{ij} = \mu + \alpha_i + \beta_j + (\alpha\beta)_{ij}.
\end{aligned}$$

Here μ is the overall mean, $\{\alpha_i\}$ are called the main effects of factor A, $\{\beta_j\}$ the main effects of factor B, and $\left\{(\alpha\beta)_{ij}\right\}$ the interaction effects. The main effects and the interactions satisfy the constraints

$$\begin{aligned}
&\sum \alpha_i = 0, \quad \sum \beta_j = 0, \\
&\sum_j (\alpha\beta)_{ij} = 0 \text{ for any } i, \quad \text{and} \quad \sum_i (\alpha\beta)_{ij} = 0 \text{ for any } j.
\end{aligned}$$

Thus the two-factor ANOVA model may be written as a factor-effects model

$$Y_{ijk} = \mu + \alpha_i + \beta_j + (\alpha\beta)_{ij} + \varepsilon_{ijk}.$$

If it turns out that the interaction effects are zero, then we end up with an additive model (ie, additive in factor effects)

$$Y_{ijk} = \mu + \alpha_i + \beta_j + \varepsilon_{ijk}.$$

In order to express the two-factor ANOVA model in the regression framework, we need to define variables for factors A and B. We can create $a - 1$ factor A variables as

$$X^{(A)}_{ijk,1} = \begin{cases} 1 & i = 1 \\ -1 & i = a \\ 0 & \text{otherwise} \end{cases}, \quad X^{(A)}_{ijk,2} = \begin{cases} 1 & i = 2 \\ -1 & i = a \\ 0 & \text{otherwise} \end{cases}, \ldots,$$

$$X^{(A)}_{ijk,a-1} = \begin{cases} 1 & i = a - 1 \\ -1 & i = a \\ 0 & \text{otherwise.} \end{cases}$$

Similarly, the $b - 1$ factor B variables are

$$X^{(B)}_{ijk,1} = \begin{cases} 1 & j = 1 \\ -1 & j = b \\ 0 & \text{otherwise} \end{cases}, \quad X^{(B)}_{ijk,2} = \begin{cases} 1 & j = 2 \\ -1 & j = b \\ 0 & \text{otherwise} \end{cases}, \ldots,$$

$$X^{(B)}_{ijk,b-1} = \begin{cases} 1 & j = b-1 \\ -1 & j = b \\ 0 & \text{otherwise.} \end{cases}$$

Then the two-factor ANOVA model can be expressed as

$$\begin{aligned} Y_{ijk} &= \mu + \sum_{l=1}^{a-1} \alpha_l X^{(A)}_{ijk,l} + \sum_{m=1}^{b-1} \beta_m X^{(B)}_{ijk,m} \\ &\quad + \sum_{l=1}^{a-1} \sum_{m=1}^{b-1} (\alpha\beta)_{lm} X^{(A)}_{ijk,l} X^{(B)}_{ijk,m} + \varepsilon_{ijk}. \end{aligned}$$

It is of interest to note that the interaction effects are the coefficients associated with the product terms of $X^{(A)}$ and $X^{(B)}$.

The additive model (ie, when the interactions are not present) can be written as

$$Y_{ijk} = \mu + \sum_{l=1}^{a-1} \alpha_l X^{(A)}_{ijk,l} + \sum_{m=1}^{b-1} \beta_m X^{(B)}_{ijk,m} + \varepsilon_{ijk}.$$

Example 11.2.6 (Analysis of Covariance). Analysis of covariance models come up when some of the independent variables are quantitative and others are qualitative. Let us first discuss a case with one qualitative variable (factor) and one quantitative variable. If in Example 11.2.3, family income level of each student in the sample is recorded, then any modeling should take into account the school effect (factor) and the income levels $\{Z_{ij}\}$. A simple model in this case is

$$\begin{aligned} &Y_{ij} = \mu + \alpha_i + \gamma Z_{ij} + \varepsilon_{ij}, \text{ or more generally} \\ &Y_{ij} = \mu + \alpha_i + \gamma_i Z_{ij} + \varepsilon_{ij}, \quad j = 1, \ldots, n_i, \quad i = 1, \ldots, k, \end{aligned}$$

where $\{\varepsilon_{ij}\}$ are iid with mean 0 and variance σ^2. The first is an additive model, whereas the second model contains an interaction between the qualitative and the quantitative variables. In the first model, there are k parallel regression lines, whereas the second model allows for k separate regression lines with possibly different intercepts and different slopes. In both cases, it is assumed that $\sum \alpha_i = 0$. In order to write these two models in the Gauss-Markov framework, we may define the indicator variables $X_{ij,1}, \ldots, X_{ij,k}$ for k schools as in Example 11.2.4, that is,

$$X_{ij,1} = \begin{cases} 1 & \text{if } i = 1 \\ 0 & \text{otherwise} \end{cases}, \ldots, X_{ij,k} = \begin{cases} 1 & \text{if } i = k \\ 0 & \text{otherwise.} \end{cases}$$

Then the first model can be written as

$$Y_{ij} = \sum_{i=1}^{k} \beta_l X_{ij,l} + \gamma Z_{ij} + \varepsilon_{ij},$$

where $\beta_l = \mu + \alpha_l$. The second model (ie, the model with interactions) can be written as

$$Y_{ij} = \sum_{l=1}^{k} \beta_l X_{ij,l} + \sum_{l=1}^{k} \gamma_l X_{ij,l} Z_{ij} + \varepsilon_{ij},$$

with $\beta_l = \mu + \alpha_l$.

Example 11.2.7 (Analysis of Covariance). A researcher wishes to investigate the effects of k different diets on the growth (weight) of animals. Let Y_{ij} be the growth of the jth subject on the ith diet and let Z_{ij1} and Z_{ij2} be the initial weight and age of the subject. Analysis of covariance model may be written as

$$Y_{ij} = \mu + \alpha_i + \gamma_1 Z_{ij1} + \gamma_2 Z_{ij2} + \varepsilon_{ij}, \quad j = 1, \ldots, n_i, \quad i = 1, \ldots, k,$$

where $\{\varepsilon_{ij}\}$ are iid with mean 0 and variance σ^2. If $\{X_{ij,l}\colon l = 1, \ldots, k\}$ are created as in the last example, then we may write this model as

$$Y_{ij} = \sum_{l=1}^{k} \beta_l X_{ij,l} + \sum_{l=1}^{2} \gamma_l Z_{ijl} + \varepsilon_{ij}.$$

If the researcher wishes to consider, in addition to the diet (factor A, a levels), the effect of gender (factor B), then she may have n_{ij} subjects of gender j assigned to diet i. If Y_{ijk} is the growth rate of the kth subject with gender j and diet i, we may consider the model

$$Y_{ijk} = \mu + \alpha_i + \beta_j + (\alpha\beta)_{ij} + \gamma_1 z_{ijk1} + \gamma_2 Z_{ijk2} + \varepsilon_{ijk}, \quad k = 1, \ldots, n_{ij}, \quad j = 1, 2, \quad i = 1, k,$$

where $\{\alpha_i\}$, $\{\beta_j\}$ are the main effects of the factors and $\left\{(\alpha\beta)_{ij}\right\}$ the interaction effects, and they satisfy the constraints stated in Example 11.2.5.

Remark 11.2.2. In the last example, the first model with diet as the factor, and initial weight and age as the covariates, may be written as

$$\boldsymbol{Y} = \boldsymbol{X\beta} + \boldsymbol{Z\gamma} + \boldsymbol{\varepsilon},$$

where $\boldsymbol{Z}$ is an $n \times 2$ matrix whose columns consist of $\{Z_{ij1}\}$ and $\{Z_{ij2}\}$. The first component on the right-hand side is $\boldsymbol{X\beta}$ and this consists of information on the factor levels. The second component $\boldsymbol{Z\gamma}$ consists of information on the covariates. Note that this follows the Gauss-Markov framework with the $n \times (k+2)$ design matrix $[\boldsymbol{X}\ \boldsymbol{Z}]$ and the unknown vector of parameters $\boldsymbol{\beta} = (\beta_1, \ldots, \beta_k, \gamma_1, \gamma_2)^T$. The same representation holds for the model with interaction in Example 11.2.6 (which allows for regression lines with different slopes and intercepts), except that $\boldsymbol{Z}$ is now an $n \times k$ matrix whose lth column has $\{X_{ij,l} Z_{ij}\}$, $l = 1, \ldots, k$.

11.3 Gauss-Markov Models: Estimation

This section is devoted to estimation of the unknown parameters of a Gauss-Markov model as given in Eq. (1) (ie, to estimate $\boldsymbol{\beta}$ and σ^2 when $\boldsymbol{Y}$ and $\boldsymbol{X}$ are observed). The method of least squares is a standard procedure for obtaining an estimate of $\boldsymbol{\beta}$ and it is done by minimizing the quantity $G(\boldsymbol{b}) = \|\boldsymbol{Y} - \boldsymbol{Xb}\|^2$ with respect to $\boldsymbol{b}$ in $\mathbb{R}^p$. It turns out that G has a unique minimum and if the minimum is attained at $\boldsymbol{b} = \hat{\boldsymbol{\beta}}$, then $\hat{\boldsymbol{\beta}}$ is taken to be an estimate of $\boldsymbol{\beta}$ and an estimate of the unknown mean vector $\boldsymbol{X\beta}$ is $\hat{\boldsymbol{Y}} = \boldsymbol{X}\hat{\boldsymbol{\beta}}$. The vector of residuals $\hat{\boldsymbol{\varepsilon}} = \boldsymbol{Y} - \hat{\boldsymbol{Y}}$, which is an estimate of the error vector $\boldsymbol{\varepsilon}$, can be used to estimate σ^2.

The column space of $\boldsymbol{X}$ is $\mathcal{M}(\boldsymbol{X}) = \{\boldsymbol{Xb}: \boldsymbol{b} \in \mathbb{R}^p\}$ and the (orthogonal) projection on it is given by $\boldsymbol{Q}_{\boldsymbol{X}} = \boldsymbol{X}(\boldsymbol{X}^T\boldsymbol{X})^{-1}\boldsymbol{X}^T$ (Section B.6). Note that $Q_{\boldsymbol{X}}$ is symmetric and idempotent (ie, $\boldsymbol{Q}_{\boldsymbol{X}}^2 = \boldsymbol{Q}_{\boldsymbol{X}}$), and $\boldsymbol{I} - \boldsymbol{Q}_{\boldsymbol{X}}$ is the projection on $\mathcal{M}(\boldsymbol{X})^{\perp}$, the orthogonal complement of $\mathcal{M}(\boldsymbol{X})$.

11.3.1 Estimation of $\boldsymbol{\beta}$ and σ^2

We begin with the discussion on estimation of $\boldsymbol{\beta}$. Note that

$$G(\boldsymbol{b}) = \boldsymbol{Y}^T\boldsymbol{Y} - 2\boldsymbol{b}^T\boldsymbol{X}^T\boldsymbol{Y} + \boldsymbol{b}^T\boldsymbol{X}^T\boldsymbol{Xb}.$$

The gradient and Hessian of G are

$$\partial G/\partial \boldsymbol{b} = -2\boldsymbol{X}^T\boldsymbol{Y} + 2\boldsymbol{X}^T\boldsymbol{Xb}, \quad \partial^2 G/\partial \boldsymbol{b}\partial \boldsymbol{b} = 2\boldsymbol{X}^T\boldsymbol{X}.$$

If $\hat{\boldsymbol{\beta}}$ is a solution of $\partial G/\partial \boldsymbol{b} = \boldsymbol{0}$, then clearly $\boldsymbol{X}^T\boldsymbol{X}\hat{\boldsymbol{\beta}} = \boldsymbol{X}^T\boldsymbol{Y}$. Since $rank(\boldsymbol{X}) = rank(\boldsymbol{X}^T\boldsymbol{X}) = p$, the Hessian $\partial^2 G/(\partial \boldsymbol{b}\partial \boldsymbol{b}) = 2\boldsymbol{X}^T\boldsymbol{X}$ is positive definite and the function G is strictly convex. Therefore, G has a unique minimum at $\boldsymbol{b} = \hat{\boldsymbol{\beta}}$.

The estimated mean vector and the vector of the residuals are

$$\hat{\boldsymbol{Y}} = \boldsymbol{X}\hat{\boldsymbol{\beta}} = \boldsymbol{X}\left(\boldsymbol{X}^T\boldsymbol{X}\right)^{-1}\boldsymbol{X}^T\boldsymbol{Y} = \boldsymbol{Q}_{\boldsymbol{X}}\boldsymbol{Y}, \text{ and}$$
$$\hat{\boldsymbol{\varepsilon}} = \boldsymbol{Y} - \hat{\boldsymbol{Y}} = \boldsymbol{Y} - \boldsymbol{X}\hat{\boldsymbol{\beta}} = (\boldsymbol{I} - \boldsymbol{Q}_{\boldsymbol{X}})\boldsymbol{Y}.$$

Since $\hat{\boldsymbol{Y}} = \boldsymbol{X}\hat{\boldsymbol{\beta}}$ is in $\mathcal{M}(\boldsymbol{X})$ and $\hat{\boldsymbol{\varepsilon}}$ is in $\mathcal{M}(\boldsymbol{X})^{\perp}$, $\hat{\boldsymbol{Y}}$ is orthogonal to the vector of residuals. Thus we are led to the following important result on the least squares method for estimating $\boldsymbol{\beta}$.

Theorem 11.3.1. *Consider the function* $G(\boldsymbol{b}) = \|\boldsymbol{Y} - \boldsymbol{Xb}\|^2$, $\boldsymbol{b} \in \mathbb{R}^p$.

(a) *The function G has a unique minimum and denote by* $\hat{\boldsymbol{\beta}}$ *the vector at which G achieves its minimum. Then we have* $\boldsymbol{X}^T\boldsymbol{X}\hat{\boldsymbol{\beta}} = \boldsymbol{X}^T\boldsymbol{Y}$ *(ie,* $\hat{\boldsymbol{\beta}} = (\boldsymbol{X}^T\boldsymbol{X})^{-1}\boldsymbol{X}^T\boldsymbol{Y}$*).*

(b) *Let* $\hat{\boldsymbol{\varepsilon}} = \boldsymbol{Y} - \boldsymbol{X}\hat{\boldsymbol{\beta}}$ *be the vector of residuals. Then* $\boldsymbol{X}^T\hat{\boldsymbol{\varepsilon}} = 0$.

(c) *The vector of residuals is orthogonal to the estimated mean vector* $\hat{\boldsymbol{Y}} = \boldsymbol{X}\hat{\boldsymbol{\beta}}$ *(ie,* $\hat{\boldsymbol{Y}}^T\hat{\boldsymbol{\varepsilon}} = 0$*).*

A few important properties of the least squares estimate $\hat{\boldsymbol{\beta}}$ of $\boldsymbol{\beta}$, the estimated mean vector $\hat{\boldsymbol{Y}} = \boldsymbol{X}\hat{\boldsymbol{\beta}}$, and the residual vector $\hat{\boldsymbol{\varepsilon}}$ can be derived rather easily using some basic algebra. Since $\hat{\boldsymbol{\beta}} = (\boldsymbol{X}^T\boldsymbol{X})^{-1}\boldsymbol{X}^T\boldsymbol{Y}$, we have

$$\mathrm{E}\left[\hat{\boldsymbol{\beta}}\right] = \left(\boldsymbol{X}^T\boldsymbol{X}\right)^{-1}\boldsymbol{X}^T\mathrm{E}[\boldsymbol{Y}] = \left(\boldsymbol{X}^T\boldsymbol{X}\right)^{-1}\boldsymbol{X}^T\boldsymbol{X}\boldsymbol{\beta} = \boldsymbol{\beta},$$
$$\mathrm{Cov}\left[\hat{\boldsymbol{\beta}}\right] = \left(\boldsymbol{X}^T\boldsymbol{X}\right)^{-1}\boldsymbol{X}^T\mathrm{Cov}[\boldsymbol{Y}]\boldsymbol{X}\left(\boldsymbol{X}^T\boldsymbol{X}\right)^{-1} = \sigma^2\left(\boldsymbol{X}^T\boldsymbol{X}\right)^{-1},$$
$$\mathrm{E}\left[\boldsymbol{X}\hat{\boldsymbol{\beta}}\right] = \boldsymbol{X}\mathrm{E}\left[\hat{\boldsymbol{\beta}}\right] = \boldsymbol{X}\boldsymbol{\beta},$$
$$\mathrm{Cov}\left[\boldsymbol{X}\hat{\boldsymbol{\beta}}\right] = \boldsymbol{X}\mathrm{Cov}\left[\hat{\boldsymbol{\beta}}\right]\boldsymbol{X}^T = \sigma^2\boldsymbol{X}\left(\boldsymbol{X}^T\boldsymbol{X}\right)^{-1}\boldsymbol{X}^T = \sigma^2\boldsymbol{Q}_{\boldsymbol{X}}.$$

Since $\boldsymbol{I} - \boldsymbol{Q_X}$ is symmetric and idempotent, we have

$$\begin{aligned}
\mathrm{E}[\hat{\boldsymbol{\varepsilon}}] &= (\boldsymbol{I} - \boldsymbol{Q_X})\mathrm{E}[\boldsymbol{Y}] = (\boldsymbol{I} - \boldsymbol{Q_X})\boldsymbol{X\beta} = \boldsymbol{0}, \\
\mathrm{Cov}[\hat{\boldsymbol{\varepsilon}}] &= (\boldsymbol{I} - \boldsymbol{Q_X})\mathrm{Cov}[\boldsymbol{Y}](\boldsymbol{I} - \boldsymbol{Q_X})^T \\
&= \sigma^2(\boldsymbol{I} - \boldsymbol{Q_X})(\boldsymbol{I} - \boldsymbol{Q_X})^T = \sigma^2(\boldsymbol{I} - \boldsymbol{Q_X})^2 = \sigma^2(\boldsymbol{I} - \boldsymbol{Q_X}), \\
\mathrm{E}\left[\|\hat{\boldsymbol{\varepsilon}}\|^2\right] &= \mathit{trace}(\mathrm{Cov}[\hat{\boldsymbol{\varepsilon}}]) = \sigma^2\, \mathit{trace}(\boldsymbol{I} - \boldsymbol{Q_X}) \\
&= \sigma^2[n - \mathit{trace}(\boldsymbol{Q_X})] = \sigma^2(n-p),
\end{aligned}$$

the last step is justified since the rank of the projection matrix $\boldsymbol{Q_X}$ equals its trace. The last result indicates that an unbiased estimator of σ^2 is given by $\hat{\sigma}^2 = \|\hat{\boldsymbol{\varepsilon}}\|^2/(n-p)$. In the literature,

(i) $\|\hat{\boldsymbol{\varepsilon}}\|^2$ is usually called the *residual sum of squares* and is denoted by *SSE*,
(ii) degrees of freedom (df) of the SSE is defined to be $n - \mathit{rank}(\boldsymbol{X}) = n - p$,
(iii) mean square error (denoted by MSE) is defined to be $\mathit{MSE} = \|\hat{\boldsymbol{\varepsilon}}\|^2/(n-p)$.

For any linear function $\boldsymbol{l}^T\boldsymbol{Y}$ of $\boldsymbol{Y}$, $\boldsymbol{l} \in \mathbb{R}^n$, we have

$$\mathrm{Cov}\left[\boldsymbol{l}^T\hat{\boldsymbol{Y}}, \hat{\boldsymbol{\varepsilon}}\right] = \mathrm{Cov}\left[\boldsymbol{l}^T\boldsymbol{Q_X}\boldsymbol{Y}, (\boldsymbol{I} - \boldsymbol{Q_X})\boldsymbol{Y}\right] = \sigma^2\boldsymbol{l}^T\boldsymbol{Q_X}(\boldsymbol{I} - \boldsymbol{Q_X}) = \boldsymbol{0}.$$

Thus any linear function of the estimated mean vector $\hat{\boldsymbol{Y}}$ is uncorrelated with the vector of residuals $\hat{\boldsymbol{\varepsilon}}$. This observation is crucial in inference since under the assumption of normality of $\boldsymbol{\varepsilon}$, uncorrelatedness implies independence and thus $\boldsymbol{X}\hat{\boldsymbol{\beta}}$ is independent of $\hat{\boldsymbol{\varepsilon}}$. The discussion above leads to the following result.

Theorem 11.3.2. *Let $\hat{\boldsymbol{Y}} = \boldsymbol{X}\hat{\boldsymbol{\beta}}$ be the fitted mean vector and $\hat{\boldsymbol{\varepsilon}}$ be the vector of residuals as in Theorem 11.3.1. Then the following hold:*

(a) $E\left[\hat{\boldsymbol{\beta}}\right] = \boldsymbol{\beta}$, $Cov\left[\hat{\boldsymbol{\beta}}\right] = \sigma^2(\boldsymbol{X}^T\boldsymbol{X})^{-1}$.
(b) $E\left[\hat{\boldsymbol{Y}}\right] = \boldsymbol{X\beta}$, $Cov\left[\hat{\boldsymbol{Y}}\right] = \sigma^2\boldsymbol{Q_X}$.
(c) $E[\hat{\boldsymbol{\varepsilon}}] = 0$, $Cov[\hat{\boldsymbol{\varepsilon}}] = \sigma^2(\boldsymbol{I} - \boldsymbol{Q_X})$.
(d) *The residual vector $\hat{\boldsymbol{\varepsilon}}$ is uncorrelated to any linear function of the estimated mean vector $\hat{\boldsymbol{Y}} = \boldsymbol{X}\hat{\boldsymbol{\beta}}$.*
(e) $E[\mathit{MSE}] = \sigma^2$, *where* $\mathit{MSE} = \|\hat{\boldsymbol{\varepsilon}}\|^2/(n-p)$.

11.3.2 Estimation of Linear Functions of $\boldsymbol{\beta}$

Often it is of interest to estimate linear functions of the unknown parameter $\boldsymbol{\beta}$. If $\tilde{\boldsymbol{\beta}}$ is an unbiased for $\boldsymbol{\beta}$, then a linear function $\boldsymbol{a}^T\tilde{\boldsymbol{\beta}}$, $\boldsymbol{a} \in \mathbb{R}^p$, is also unbiased for $\boldsymbol{a}^T\boldsymbol{\beta}$. Let $\boldsymbol{L}$ be a known $p \times m$ matrix of rank $m \le p$, and consider the problem of estimating the linear function $\boldsymbol{\theta} = \boldsymbol{L}^T\boldsymbol{\beta}$. Least squares estimate of $\boldsymbol{\theta}$ is defined to be equal to $\hat{\boldsymbol{\theta}} = \boldsymbol{L}^T\hat{\boldsymbol{\beta}}$, where $\hat{\boldsymbol{\beta}}$ is the least squares estimate of $\boldsymbol{\beta}$. The following simple result shows that $\hat{\boldsymbol{\theta}}$ is an unbiased estimate of $\boldsymbol{\theta}$.

Lemma 11.3.1. $E\left[\hat{\boldsymbol{\theta}}\right] = \boldsymbol{\theta}$, $Cov\left[\hat{\boldsymbol{\theta}}\right] = \sigma^2 \boldsymbol{L}^T (\boldsymbol{X}^T\boldsymbol{X})^{-1}\boldsymbol{L}$.

11.3.3 Best Linear Unbiased Estimation

We begin with a definition.

Definition 11.3.1. A linear function $\tilde{\boldsymbol{\beta}}$ of $\boldsymbol{Y}$ is called a best linear unbiased estimator (BLUE) of $\boldsymbol{\beta}$ if

(i) $\tilde{\boldsymbol{\beta}}$ is an unbiased estimator of $\boldsymbol{\beta}$, and

(ii) for any $\boldsymbol{a} \in \mathbb{R}^p$, $\text{Var}\left[\boldsymbol{a}^T\tilde{\boldsymbol{\beta}}\right] \le \text{Var}\left[\boldsymbol{l}^T\boldsymbol{Y}\right]$ for all linear unbiased estimators $\boldsymbol{l}^T\boldsymbol{Y}$ of $\boldsymbol{a}^T\boldsymbol{\beta}$, $\boldsymbol{l} \in \mathbb{R}^n$.

It is clear from this definition that $\tilde{\boldsymbol{\beta}}$ is a BLUE of $\boldsymbol{\beta}$ if $\boldsymbol{a}^T\tilde{\boldsymbol{\beta}}$ is BLUE of $\boldsymbol{a}^T\boldsymbol{\beta}$ for any $\boldsymbol{a} \in \mathbb{R}^p$. The following argument will show that the BLUE exists, it is unique and it is equal to the least squares estimate $\hat{\boldsymbol{\beta}}$. It will be enough to show that, if for any $\boldsymbol{a} \in \mathbb{R}^p$, $\boldsymbol{l}^T\boldsymbol{Y}$ is a BLUE for $\boldsymbol{a}^T\boldsymbol{\beta}$, then $\boldsymbol{l}^T\boldsymbol{Y} = \boldsymbol{a}^T\hat{\boldsymbol{\beta}}$.

If $\boldsymbol{l}^T\boldsymbol{Y}$ is an unbiased estimator of $\boldsymbol{a}^T\boldsymbol{\beta}$, then $\boldsymbol{a}^T\boldsymbol{\beta} = \text{E}\left[\boldsymbol{l}^T\boldsymbol{Y}\right] = \boldsymbol{l}^T\boldsymbol{X}\boldsymbol{\beta}$ for all $\boldsymbol{\beta}$ and hence $\boldsymbol{X}^T\boldsymbol{l} = \boldsymbol{a}$. If $\boldsymbol{l}^T\boldsymbol{Y}$ is a BLUE of $\boldsymbol{a}^T\boldsymbol{\beta}$, then for any linear unbiased estimator $\boldsymbol{m}^T\boldsymbol{Y}$ of 0 (ie, $\text{E}[\boldsymbol{m}^T\boldsymbol{Y}] = 0$ for all $\boldsymbol{\beta}$), $(\boldsymbol{l} + t\boldsymbol{m})^T\boldsymbol{Y}$ is also unbiased for $\boldsymbol{a}^T\boldsymbol{\beta}$, where t is a real number. Let

$$h(t) = \text{Var}\left[(\boldsymbol{l} + t\boldsymbol{m})^T\boldsymbol{Y}\right] = \sigma^2\|\boldsymbol{l} + t\boldsymbol{m}\|^2.$$

Since $\boldsymbol{l}^T\boldsymbol{Y}$ is a BLUE, the function h achieves a minimum at $t = 0$, thus $0 = h'(0) = 2\sigma^2\boldsymbol{l}^T\boldsymbol{m}$ (ie, $\boldsymbol{l}^T\boldsymbol{m} = 0$). Since $\boldsymbol{m}^T\boldsymbol{Y}$ is an unbiased estimator of 0, we have $\boldsymbol{m}^T\boldsymbol{X}\boldsymbol{\beta} = 0$ for all $\boldsymbol{\beta}$ and thus $\boldsymbol{X}^T\boldsymbol{m} = \boldsymbol{0}$. Since $\boldsymbol{l}^T\boldsymbol{m} = 0$ for all $\boldsymbol{m}$ satisfying the condition $\boldsymbol{X}^T\boldsymbol{m} = \boldsymbol{0}$ (ie, for all $\boldsymbol{m} \in \mathcal{M}(\boldsymbol{X})^T$), it follows that $\boldsymbol{l}$ must be in $\mathcal{M}(\boldsymbol{X})$. Thus $\boldsymbol{l} = \boldsymbol{X}\boldsymbol{c}$ for some $\boldsymbol{c} \in \mathbb{R}^p$. Since $\boldsymbol{l}^T\boldsymbol{Y}$ is unbiased for $\boldsymbol{a}^T\boldsymbol{\beta}$, we have $\boldsymbol{a}^T\boldsymbol{\beta} = \boldsymbol{l}^T\boldsymbol{X}\boldsymbol{\beta} = \boldsymbol{c}^T\boldsymbol{X}^T\boldsymbol{X}\boldsymbol{\beta}$ for all $\boldsymbol{\beta}$. This implies that $\boldsymbol{c} = (\boldsymbol{X}^T\boldsymbol{X})^{-1}\boldsymbol{a}$ and hence $\boldsymbol{l} = \boldsymbol{X}(\boldsymbol{X}^T\boldsymbol{X})^{-1}\boldsymbol{a}$. Thus if $\boldsymbol{l}^T\boldsymbol{Y}$ is a BLUE of $\boldsymbol{a}^T\boldsymbol{\beta}$, then

$$\boldsymbol{l}^T\boldsymbol{Y} = \boldsymbol{a}^T\left(\boldsymbol{X}^T\boldsymbol{X}\right)^{-1}\boldsymbol{X}^T\boldsymbol{Y} = \boldsymbol{a}^T\hat{\boldsymbol{\beta}}.$$

Uniqueness is clear since any BLUE must have this form.

Thus we are led to the following important result.

Lemma 11.3.2. *The BLUE of $\boldsymbol{\beta}$ is unique and it is equal to the least squares estimate $\hat{\boldsymbol{\beta}}$.*

11.4 Decomposition of Total Sum of Squares

In each of the examples given in Section 11.2 of this chapter, the design matrix $\boldsymbol{X}$ has a column consisting of 1's. In regression, it corresponds to the intercept term and in ANOVA models, it corresponds to the overall mean. Let us assume without loss of generality that the first column of $\boldsymbol{X}$ consists of 1's and the vector of parameters is $\boldsymbol{\beta} = (\beta_0, \beta_1, \ldots, \beta_{p-1})^T$. A model which only has the first column (and ignores the last $p-1$ columns) is $Y_i = \beta_0 + \varepsilon_i$, $i = 1, \ldots, n$. Clearly the least squares estimate of β_0 is $\bar{Y} = n^{-1}\sum_{i=1}^n Y_i$ and the residual sum of squares of this simple model $\sum\left(Y_i - \hat{\beta}_0\right)^2 = \sum(Y_i - \bar{Y})^2$ is usually called the total sum of squares (SSTO). Let $\hat{\boldsymbol{Y}} = \boldsymbol{X}\hat{\boldsymbol{\beta}}$ be the least squares estimate of the mean $\boldsymbol{X\beta}$ for the full model $\boldsymbol{Y} = \boldsymbol{X\beta} + \boldsymbol{\varepsilon}$. We already know that the residual vector $\hat{\boldsymbol{\varepsilon}} = \boldsymbol{Y} - \boldsymbol{X}\hat{\boldsymbol{\beta}} = \boldsymbol{Y} - \hat{\boldsymbol{Y}}$ is orthogonal to any vector in $\mathcal{M}(\boldsymbol{X})$ and hence to $\bar{Y}\mathbf{1}$, where $\mathbf{1}$ is an $n \times 1$ vector of 1's. Then

$$
\begin{aligned}
SSTO &= \sum(Y_i - \bar{Y})^2 = \|\boldsymbol{Y} - \bar{Y}\mathbf{1}\|^2 = \|\boldsymbol{X}\hat{\boldsymbol{\beta}} - \bar{Y}\mathbf{1}\|^2 + \|\boldsymbol{Y} - \boldsymbol{X}\hat{\boldsymbol{\beta}}\|^2 \\
&= \sum\left(\hat{Y}_i - \bar{Y}\right)^2 + \sum\left(Y_i - \hat{Y}_i\right)^2 \\
&= \sum\left(\hat{Y}_i - \bar{Y}\right)^2 + SSE.
\end{aligned}
$$

Since SSTO is the residual sum of squares for the model $Y_i = \beta_0 + \varepsilon_i$, $i = 1, \ldots, n$, the quantity $\sum\left(\hat{Y}_i - \bar{Y}\right)^2$ is the reduction in the residual sum of squares when we go from the simple model $\boldsymbol{Y} = \beta_0\mathbf{1} + \boldsymbol{\varepsilon}$ to the full model $\boldsymbol{Y} = \boldsymbol{X\beta} + \boldsymbol{\varepsilon}$ and

$$
R^2 = \sum\left(\hat{Y}_i - \bar{Y}\right)^2 / SSTO = 1 - SSE/SSTO
$$

is the proportional reduction in the residual sum of squares. It is also called the *coefficient of determination* and is widely used in practice. Clearly, R^2 is between 0 and 1, and if R^2 is close 1, it is taken as an indication that the full model $\boldsymbol{Y} = \boldsymbol{X\beta} + \boldsymbol{\varepsilon}$ explains the data well. The reduction $\sum\left(\hat{Y}_i - \bar{Y}\right)^2$ is called the "regression sum of squares" in the regression model and "treatment sum of squares" in the one-factor ANOVA model.

Even though R^2 is popular as a descriptive measure it has some flaws. If there is a true model $\boldsymbol{Y} = \boldsymbol{X}_*\boldsymbol{\beta}_* + \boldsymbol{\varepsilon}$ with $rank(\boldsymbol{X}_*) = p_*$ and it is nested in the model under consideration $\boldsymbol{Y} = \boldsymbol{X\beta} + \boldsymbol{\varepsilon}$, where $\mathcal{M}(\boldsymbol{X}_*) \subset \mathcal{M}(\boldsymbol{X})$ and $rank(\boldsymbol{X}) = p > p_*$, then clearly the SSE for the latter model is smaller than that of the former (ie, the true lower dimensional model), and thus the latter model has a higher R^2. As a matter of fact, for any class of nested models, the value of R^2 will always increase as we consider higher dimensional models, and this is true regardless of what the true model is. In order to remedy this, we first need to identify a parameter ρ^2 that R^2 is trying to estimate. It turns out that a descriptive measure called the adjusted R^2, denoted by R^2_{adj}, is a better estimate of ρ^2 than R^2 is. Before we go any further, let us first state a simple lemma and then define the concept of degrees of freedom associated with residual sums of squares for any linear model.

We consider here a general structure for the $n \times 1$ observation vector $\boldsymbol{Y}$ with mean $\boldsymbol{\mu}$ such that

$$
\boldsymbol{Y} = \boldsymbol{\mu} + \boldsymbol{\varepsilon}, \text{ with } \mathrm{E}[\boldsymbol{\varepsilon}] = \mathbf{0} \quad \text{and} \quad \mathrm{Cov}[\boldsymbol{\varepsilon}] = \sigma^2\boldsymbol{I}.
$$

If $\boldsymbol{X}$ is $n \times p$ of rank p, then the minimum of $\|\boldsymbol{\mu} - \boldsymbol{X\beta}\|^2$, $\boldsymbol{\beta} \in \mathbb{R}^p$, is attained at $\boldsymbol{\beta}^* = \left(\boldsymbol{X}^T\boldsymbol{X}\right)^{-1}\boldsymbol{X}^T\boldsymbol{\mu}$, and the *projected mean* is $\boldsymbol{X\beta}^* = \boldsymbol{X}\left(\boldsymbol{X}^T\boldsymbol{X}\right)^{-1}\boldsymbol{X}^T\boldsymbol{\mu} = \boldsymbol{Q_X\mu}$. So when a model $\boldsymbol{Y} = \boldsymbol{X\beta} + \boldsymbol{\varepsilon}$ is fitted to the data, then $\boldsymbol{Q_XY}$ is estimating the projected mean $\boldsymbol{Q_X\mu}$ and the model is a true description of the data if $\boldsymbol{\mu} = \boldsymbol{Q_X\mu}$.

Lemma 11.4.1. *Assume that $\boldsymbol{Y} = \boldsymbol{\mu} + \boldsymbol{\varepsilon}$, where $\boldsymbol{Y}$ is $n \times 1$ observation vector, $E[\boldsymbol{\varepsilon}] = \mathbf{0}$, $Cov[\boldsymbol{\varepsilon}] = \sigma^2\boldsymbol{I}$, and a model of the form $\boldsymbol{Y} = \boldsymbol{X\beta} + \boldsymbol{\varepsilon}$, where $\boldsymbol{X}$ is $n \times p$ of rank p, is fitted to the* *data. A measure of deviation of* $\boldsymbol{Y}$ *from the projected mean* $\boldsymbol{Q_X\mu}$ *is* $D = \mathrm{E}\left[\|\boldsymbol{Y} - \boldsymbol{Q_X\mu}\|^2\right]$ *and an estimate of* D *is* $SSE = \|\boldsymbol{Y} - \boldsymbol{Q_X Y}\|^2$. *The following hold:*

(a) $D = n\sigma^2 + \|(\boldsymbol{I} - \boldsymbol{Q_X})\boldsymbol{\mu}\|^2$,
(b) $\mathrm{E}[SSE] = (n - p) + \|(\boldsymbol{I} - \boldsymbol{Q_X})\boldsymbol{\mu}\|^2$.

Definition 11.4.1. If a model of the form $\boldsymbol{Y} = \boldsymbol{X\beta} + \boldsymbol{\varepsilon}$, where $\boldsymbol{X}$ is $n \times p$ matrix of rank p, is fitted to the observation vector $\boldsymbol{Y}$ where $\boldsymbol{Y} = \boldsymbol{\mu} + \boldsymbol{\varepsilon}$, $\mathrm{E}[\boldsymbol{\varepsilon}] = \boldsymbol{0}$, $\mathrm{Cov}[\boldsymbol{\varepsilon}] = \sigma^2\boldsymbol{I}$, then $\boldsymbol{Q_X\mu}$ is called the projected mean of the model. The degrees of freedom (df) of the residual sum of squares $SSE = \|\boldsymbol{Y} - \boldsymbol{Q_X Y}\|^2$ is $n - p$, and the mean square error is $MSE = SSE/(n - p)$.

For the simple model $Y_i = \beta_0 + \varepsilon_i$, $i = 1, \ldots, n$, the residual sum of squares is $SSTO$ and its mean square error is $MSTO/(n - 1)$. The adjusted R^2 is defined as

$$R^2_{\text{adj}} = 1 - MSE/MSTO.$$

For the model $\boldsymbol{Y} = \beta_0\boldsymbol{1} + \boldsymbol{\varepsilon}$, let $\boldsymbol{Q_0\mu}$ be the projection of $\boldsymbol{\mu}$ on $\boldsymbol{1}$ (ie, $\boldsymbol{Q_0\mu} = \beta_0\boldsymbol{1}$ with $\beta_0 = \boldsymbol{1}^T\boldsymbol{\mu}/n$). Noting that $\bar{Y}$ estimates β_0 and SSTO estimates $\mathrm{E}\left[\|\boldsymbol{Y} - \boldsymbol{Q_0\mu}\|^2\right]$, the deviation of $\boldsymbol{Y}$ from $\boldsymbol{Q_0X\beta} = \beta_0\boldsymbol{1}$. Similarly, SSE is an estimate of $\mathrm{E}\left[\|\boldsymbol{Y} - \boldsymbol{Q_X\mu}\|^2\right]$, when the model $\boldsymbol{Y} = \boldsymbol{X\beta} + \boldsymbol{\varepsilon}$ (where the first column of $\boldsymbol{X}$ consists of 1's) is fitted to the data. We may thus define the proportional reduction in the true deviation of $\boldsymbol{Y}$ from the projected mean when we go from the model $\boldsymbol{Y} = \beta_0\boldsymbol{1} + \boldsymbol{\varepsilon}$ to the model $\boldsymbol{Y} = \boldsymbol{X\beta} + \boldsymbol{\varepsilon}$ as

$$\begin{aligned}
\rho^2 &= 1 - \mathrm{E}\left[\|\boldsymbol{Y} - \boldsymbol{Q_X\mu}\|^2\right] \Big/ \mathrm{E}\left[\|\boldsymbol{Y} - \boldsymbol{Q_0\mu}\|^2\right] \\
&= 1 - \frac{n\sigma^2 + \|(\boldsymbol{I} - \boldsymbol{Q_X})\boldsymbol{\mu}\|^2}{n\sigma^2 + \|(\boldsymbol{I} - \boldsymbol{Q_0})\boldsymbol{\mu}\|^2} \\
&= \frac{\|(\boldsymbol{Q_X} - \boldsymbol{Q_0})\boldsymbol{\mu}\|^2}{n\sigma^2 + \|(\boldsymbol{I} - \boldsymbol{Q_0})\boldsymbol{\mu}\|^2},
\end{aligned}$$

using the results in the lemma above. Note that R^2 is an estimate of ρ^2, but it has a serious flaw as discussed above.

Under the assumption of normality of the error terms, $\|\boldsymbol{Y} - \boldsymbol{Q_0\mu}\|^2/\sigma^2 \sim \chi^2_{n-1}(\delta_0^2)$, where the noncentrality parameter is $\delta_0^2 = (1/2)\|(\boldsymbol{I} - \boldsymbol{Q_0})\boldsymbol{\mu}\|^2/\sigma^2$. Similarly, $\|\boldsymbol{Y} - \boldsymbol{Q_X Y}\|^2/\sigma^2 \sim \chi^2_{n-p}(\delta^2)$ with $\delta^2 = (1/2)\|(\boldsymbol{I} - \boldsymbol{Q_X})\boldsymbol{\mu}\|^2/\sigma^2$. If n is large, it can be shown that

$$\begin{aligned}
MSE &= \|\boldsymbol{Y} - \boldsymbol{Q_X Y}\|^2/(n - p) \\
&= \sigma^2 + n^{-1}\|(\boldsymbol{I} - \boldsymbol{Q_X})\boldsymbol{\mu}\|^2 + O_P\left(n^{-1/2}\right), \text{ and} \\
MSTO &= \|\boldsymbol{Y} - \boldsymbol{Q_0 Y}\|^2/(n - 1) \\
&= \sigma^2 + n^{-1}\|(\boldsymbol{I} - \boldsymbol{Q_0})\boldsymbol{\mu}\|^2 + O_P\left(n^{-1/2}\right),
\end{aligned}$$

assuming that the quantity $n^{-1}\|(\boldsymbol{I} - \boldsymbol{Q_0})\boldsymbol{\mu}\|^2$ stays bounded as $n \to \infty$. These results are generally true under reasonable technical conditions even if the error terms are not normally distributed. Thus

$$
\begin{aligned}
R^2_{\text{adj}} &= 1 - MSE/MSTO = \frac{MSTO - MSE}{MSTO} \\
&= \frac{n^{-1}\|(\boldsymbol{Q_X} - \boldsymbol{Q}_0)\boldsymbol{\mu}\|^2}{\sigma^2 + n^{-1}\|(\boldsymbol{I} - \boldsymbol{Q}_0)\boldsymbol{\mu}\|^2} + O_P\left(n^{-1/2}\right) \\
&= \rho^2 + O_P\left(n^{-1/2}\right).
\end{aligned}
$$

Clearly, R^2_{adj} is a $\sqrt{n}$ consistent estimate of ρ^2.

Example 11.4.1. In the multiple regression case with $p-1$ independent variables, the normal equations $\boldsymbol{X}^T\boldsymbol{X}\boldsymbol{\beta} = \boldsymbol{X}^T\boldsymbol{Y}$ are

$$
\begin{bmatrix}
n & \sum X_{i,1} & \cdots & \sum X_{i,p-1} \\
\sum X_{i,1} & \sum X_{i,1}^2 & \cdots & \sum X_{i,1}X_{i,p-1} \\
\vdots & \vdots & \cdots & \vdots \\
\sum X_{i,p-1} & \sum X_{i,p-1}X_{i,1} & \cdots & \sum X_{i,p-1}^2
\end{bmatrix}
\begin{bmatrix}
\beta_0 \\ \beta_1 \\ \vdots \\ \beta_{p-1}
\end{bmatrix}
=
\begin{bmatrix}
\sum X_{i,1}Y_i \\ \sum X_{i,2}Y_i \\ \vdots \\ \sum X_{i,p-1}Y_i
\end{bmatrix},
$$

where all the sums are over i from 1 through n. When $p = 2$ (ie, there is only one independent variable), the solutions are

$$
\hat{\beta}_1 = \sum (X_{i,1} - \bar{X}_1)(Y_i - \bar{Y}) \Big/ \sum (X_{i,1} - \bar{X}_1)^2, \quad \hat{\beta}_0 = \bar{Y} - \hat{\beta}_1 \bar{X}_1,
$$

where $\bar{X}_1 = n^{-1}\sum_{i=1}^n X_{i,1}$ and $\bar{Y} = n^{-1}\sum_{i=1}^n Y_i$. When $p > 2$, there are no such simple expressions for the estimates of β_0, β_1, etc. Typically, solving these equations require computing packages, which are widely available. If $\hat{\boldsymbol{Y}} = \boldsymbol{X}\hat{\boldsymbol{\beta}}$ (ie, $\hat{Y}_i = \hat{\beta}_0 + \hat{\beta}_1 X_{i,1} + \cdots + \hat{\beta}_{p-1}X_{i,p-1}$, $i = 1, \ldots, n$), then the quantity $\sum\left(\hat{Y}_i - \bar{Y}\right)$ is called the "regression sum of squares" and thus $R^2 = SSR/SSTO$.

Example 11.4.2. In the one-factor ANOVA case as in Example 11.2.3, it is fairly easy to get the least squares estimate of μ_i

$$
\hat{\mu}_i = n_i^{-1}\sum_{j=1}^{n_i} Y_{ij} := \bar{Y}_{i\cdot}.
$$

The residual sum of squares is $SSE = \sum_i \sum_j (Y_{ij} - \bar{Y}_{i\cdot})^2$ and $MSE = SSE/(n-k)$ is an unbiased estimate of σ^2, where $n = n_1 + \cdots + n_k$ is the total number of observations. The estimates of the overall mean $\mu = \sum w_i\mu_i$ and $\alpha_i = \mu_i - \mu$ are obtained by substituting $\{\mu_i\}$ by $\{\hat{\mu}_i\}$. For instance, if $w_i = n_i/n$, then

$$
\hat{\mu} = \sum (n_i/n)\bar{Y}_{i\cdot} = \sum_i \sum_j Y_{ij}/n := \bar{Y}_{\cdot\cdot} \quad \text{and} \quad \hat{\alpha}_i = \bar{Y}_{i\cdot} - \bar{Y}_{\cdot\cdot}.
$$

As before if we denote $\hat{\boldsymbol{Y}} = \boldsymbol{X}\hat{\boldsymbol{\beta}}$, then $\hat{Y}_{ij} = \bar{Y}_{i\cdot}$ and $\sum_i \sum_j \left(\hat{Y}_{ij} - \bar{Y}_{\cdot\cdot}\right)^2 = \sum n_i(\bar{Y}_{i\cdot} - \bar{Y}_{\cdot\cdot})^2$ is called the "treatment sum of squares" (SSTR) or the "between group sum of squares." For this case, the *SSE* is sometimes called the "within group sum of squares." The decomposition of the total sum of squares is

$$SSTO = \sum_i \sum_j \left(Y_{ij} - \bar{Y}_{..}\right)^2 = \sum n_i \hat{\alpha}_i^2 + SSE, \text{ ie,}$$
$$SSTO = SSTR + SSE.$$

In this case $R^2 = SSTR/SSTO$. The quantity $SSTR/(k-1)$ is usually called $MSTR$, mean square for the treatment. It can be shown that (left as an exercise)

$$\mathrm{E}[MSTR]/\sigma^2 = 1 + (k-1)^{-1} \sum n_i(\mu_i - \mu)^2/\sigma^2.$$

The quantity $(k-1)^{-1} \sum n_i(\mu_i - \mu)^2/\sigma^2$ is a unit-free measure of the variability of $\{\mu_i\}$. This measure equals 0 if and only if μ_i's are all the same. Thus the ratio $F = MSTR/MSE$, which fluctuates about 1 if and only if $\mu_1 = \cdots = \mu_k$, is used for testing the hypothesis that the means are the same. Under the assumption of normality (ie, $\{\varepsilon_{ij}\}$ are iid $N(0, \sigma^2)$),

$$SSE/\sigma^2 \sim \chi^2_{n-k},\ SSTR/\sigma^2 \sim \chi^2_{k-1}\left(\delta^2\right), \text{ where}$$
$$\delta^2 = (1/2) \sum n_i \alpha_i^2/\sigma^2,$$

and $F = MSTR/MSE \sim F_{k-1,n-k}(\delta^2)$. Hence the F-statistic can be used to test H_0: $\alpha_1 = \cdots = \alpha_k = 0$, since $F \sim F_{k-1,n-k}$ under H_0.

If the overall mean μ is defined as $k^{-1} \sum \mu_i$, then the estimates of μ and α_i are $\hat{\mu} = k^{-1} \sum \bar{Y}_{i.}$ and $\hat{\alpha}_i = \bar{Y}_{i.} - \hat{\mu}$. Note that $\mathrm{E}[Y_{ij}]$ is always equal to μ_i irrespective of how μ and $\{\alpha_i\}$ are defined, $\hat{Y}_{ij} = \bar{Y}_{i.}$ and $\hat{\varepsilon}_{ij} = Y_{ij} - \bar{Y}_{i.}$, and hence the residual sum of squares also remains the same.

Example 11.4.3. In general ANOVA models (one- or multifactor), one may first obtain the estimates of the means of all the factor combinations and then use them to estimate the overall mean, factor effects, interactions, etc. For instance, in the two-factor ANOVA model, the estimate of μ_{ij} is $\hat{\mu}_{ij} = \bar{Y}_{ij.}$, where $\bar{Y}_{ij.} = n_{ij}^{-1} \sum_{k=1}^{n_{ij}} Y_{ijk}$, and this can be used to estimate μ, $\mu_{i.}$, $\mu_{.j}$, the factor effects $\{\alpha_i\}$, $\{\beta_j\}$, and the interactions $\left\{(\alpha\beta)_{ij}\right\}$, which are all linear functions of $\{\mu_{ij}\}$. The residual sum of squares is $SSE = \sum_i \sum_j \sum_k \left(Y_{ijk} - \bar{Y}_{ij.}\right)^2$, and $MSE = SSE/(n - ab)$, where $n = \sum_i \sum_j n_{ij}$ is the total number of observations, is an unbiased estimate of σ^2. Thus

$$\hat{\mu} = (ab)^{-1} \sum_i \sum_j \hat{\mu}_{ij} = (ab)^{-1} \sum_i \sum_j \bar{Y}_{ij.}$$
$$\hat{\mu}_{i.} = b^{-1} \sum_j \bar{Y}_{ij.},\ \hat{\mu}_{.j} = a^{-1} \sum_i \bar{Y}_{ij.},$$
$$\hat{\alpha}_i = \hat{\mu}_{i.} - \hat{\mu},\ \hat{\beta}_j = \hat{\mu}_{.j} - \hat{\mu},\ \widehat{(\alpha\beta)}_{ij} = \hat{\mu}_{ij} - \hat{\mu}_{i.} - \hat{\mu}_{.j} + \hat{\mu}, \text{ and}$$
$$\hat{Y}_{ijk} = \bar{Y}_{ij.} = \hat{\mu} + \hat{\alpha}_\iota + \hat{\beta}_j + \widehat{(\alpha\beta)}_{ij}.$$

Example 11.4.4. In the balanced two-factor ANOVA model (ie, $n_{ij} \equiv n_0$), the estimates are much simpler when one uses the following notations

$$\bar{Y}_{...} = (n_0 ab)^{-1} \sum_i \sum_j \sum_k Y_{ijk},$$
$$\bar{Y}_{i..} = (n_0 b)^{-1} \sum_j \sum_k Y_{ijk},\ \bar{Y}_{.j.} = (n_0 a)^{-1} \sum_i \sum_k Y_{ijk}, \text{ and}$$
$$\bar{Y}_{ij.} = n_0^{-1} \sum_k Y_{ijk}.$$

With these notations

$$\hat{\mu} = \bar{Y}_{...},\ \hat{\alpha}_i = \bar{Y}_{i..} - \bar{Y}_{...},\ \hat{\beta}_j = \bar{Y}_{.j.} - \bar{Y}_{...},$$
$$\widehat{(\alpha\beta)}_{ij} = \bar{Y}_{ij.} - \bar{Y}_{i..} - \bar{Y}_{.j.} + \bar{Y}_{...}, \text{ and}$$
$$Y_{ijk} - \hat{\mu} = \hat{\alpha}_i + \hat{\beta}_j + \widehat{(\alpha\beta)}_{ij} + \hat{\varepsilon}_{ijk},$$

where $\{\hat{\varepsilon}_{ijk} = Y_{ijk} - \bar{Y}_{ij.}\}$ are the residuals. If both sides are squared and summed over i, j, and k, all the cross-product terms vanish to yield the following decomposition of the total sum of squares

$$\begin{aligned} SSTO &= \sum_i \sum_j \sum_k \left(Y_{ijk} - \bar{Y}_{...}\right)^2 \\ &= \sum_i \sum_j \sum_k \left(\bar{Y}_{ij.} - \bar{Y}_{...}\right)^2 + \sum_i \sum_j \sum_k \left(Y_{ijk} - \bar{Y}_{ij.}\right)^2 \\ &= (n_0 b) \sum \hat{\alpha}_i^2 + (n_0 a) \sum \hat{\beta}_j^2 + n_0 \sum_i \sum_j \widehat{(\alpha\beta)}_{ij}^2 + SSE \\ &:= SSA + SSB + SSAB + SSE. \end{aligned}$$

SSA, *SSB*, and *SSAB* are called the sums of squares due to the main effects of factor *A*, main effects of factor *B*, and the interactions, respectively. It is important to point out that this decomposition of *SSTO* is no longer valid for the unbalanced case.

It can also be shown that (left as an exercise)

$$\mathrm{E}[SSA] = (a-1)\sigma^2 + (n_0 b) \sum \alpha_i^2,\quad \mathrm{E}[SSB] = (b-1)\sigma^2 + (n_0 a) \sum \beta_j^2,$$
$$\mathrm{E}[SSAB] = (a-1)(b-1)\sigma^2 + n_0 \sum_i \sum_j (\alpha\beta)_{ij}^2,$$
$$\mathrm{E}[SSE] = (n - ab)\sigma^2.$$

How do we view these results in matrix terms? Consider the following (sub)models

$$Y_{ijk} = \mu + \varepsilon_{ijk} \text{ (model 0)},$$
$$Y_{ijk} = \mu + \alpha_i + \varepsilon_{ijk} \text{ (model 1)},$$
$$Y_{ijk} = \mu + \beta_j + \varepsilon_{ijk} \text{ (model 2)},$$
$$Y_{ijk} = \mu + \alpha_i + \beta_j + \varepsilon_{ijk} \text{ (model 3), and}$$
$$Y_{ijk} = \mu + \alpha_i + \beta_j + (\alpha\beta)_{ij} + \varepsilon_{ijk} \text{ (model 4)},$$

where model 4 is the full (true) model and the rest are submodels of the full model. The fitted values (or estimated means) of these models are

$$\hat{Y}^{(0)}_{ijk} = \hat{\mu} = \bar{Y}_{...},\ \hat{Y}^{(1)}_{ijk} = \bar{Y}_{i..} = \hat{\mu} + \hat{\alpha}_i,$$
$$\hat{Y}^{(2)}_{ijk} = \bar{Y}_{.j.} = \hat{\mu} + \hat{\beta}_j,\ \hat{Y}^{(3)}_{ijk} = \bar{Y}_{i..} + \bar{Y}_{.j.} - \bar{Y}_{...} = \hat{\mu} + \hat{\alpha}_i + \hat{\beta}_j, \text{ and}$$
$$\hat{Y}^{(4)}_{ijk} = \bar{Y}_{ij.} = \hat{\mu} + \hat{\alpha}_i + \hat{\beta}_j + \widehat{(\alpha\beta)}_{ij}.$$

For each of the five linear models above, the postulated mean belongs to a linear space and the vector of fitted values is a projection of $\boldsymbol{Y}$ on this column space. If $\boldsymbol{Q}_j\boldsymbol{Y}$ denotes the projection of $\boldsymbol{Y}$ for model $j, j = 0, \ldots, 4$, then the decomposition $Y_{ijk} - \hat{\mu} = \hat{\alpha}_i + \hat{\beta}_j + \widehat{(\alpha\beta)}_{ij} + \hat{\varepsilon}_{ijk}$ can be written in the matrix form as

$$(\boldsymbol{I} - \boldsymbol{Q}_0)\boldsymbol{Y} = (\boldsymbol{Q}_1 - \boldsymbol{Q}_0)\boldsymbol{Y} + (\boldsymbol{Q}_2 - \boldsymbol{Q}_0)\boldsymbol{Y} + (\boldsymbol{Q}_4 - \boldsymbol{Q}_3)\boldsymbol{Y} + (\boldsymbol{I} - \boldsymbol{Q}_4)\boldsymbol{Y}, \text{ or}$$
$$(\boldsymbol{I} - \boldsymbol{Q}_0)\boldsymbol{Y} = \boldsymbol{M}_1\boldsymbol{Y} + \boldsymbol{M}_2\boldsymbol{Y} + \boldsymbol{M}_3\boldsymbol{Y} + \boldsymbol{M}_4\boldsymbol{Y},$$

where $\boldsymbol{M}_1 = \boldsymbol{Q}_1 - \boldsymbol{Q}_0$, $\boldsymbol{M}_2 = \boldsymbol{Q}_2 - \boldsymbol{Q}_0$, $\boldsymbol{M}_3 = \boldsymbol{Q}_4 - \boldsymbol{Q}_3$, and $\boldsymbol{M}_4 = \boldsymbol{I} - \boldsymbol{Q}_4$. It is fairly easy to check that $\boldsymbol{M}_j, j = 1, 2, 3, 4$, are projection matrices. Moreover, for the balanced two-factor ANOVA model, $\boldsymbol{M}_i\boldsymbol{M}_j = 0$, $1 \leq i \neq j \leq 4$. When $\{\varepsilon_{ijk}\}$ are iid $N(0, \sigma^2)$, using the results in Section B.7 we have

$$SSA/\sigma^2 = \|\boldsymbol{M}_1\boldsymbol{Y}\|^2/\sigma^2 \sim \chi^2_{a-1}\left(\delta_1^2\right),\ \delta_1^2 = (1/2)(n_0 b)\sum \alpha_i^2/\sigma^2,$$
$$SSB/\sigma^2 = \|\boldsymbol{M}_2\boldsymbol{Y}\|^2/\sigma^2 \sim \chi^2_{b-1}\left(\delta_2^2\right),\ \delta_2^2 = (1/2)(n_0 a)\sum \beta_j^2/\sigma^2,$$
$$SSAB/\sigma^2 = \|\boldsymbol{M}_3\boldsymbol{Y}\|^2/\sigma^2 \sim \chi^2_{(a-1)(b-1)}\left(\delta_3^2\right),\ \delta_3^3 = (1/2)n_0 \sum_i \sum_j (\alpha\beta)^2_{ij}/\sigma^2,$$
$$SSE/\sigma^2 = \|\boldsymbol{M}_4\boldsymbol{Y}\|^2/\sigma^2 \sim \chi^2_{n-ab},$$

and *SSA*, *SSB*, *SSAB*, and *SSE* are independent.

Example 11.4.5 (Estimation in One-Factor ANOVA). In one-factor ANOVA or multifactor models, one is often interested in comparing the means or comparing the factor effects. For instance, in a one-factor model, it is of interest to estimate the pairwise differences of the means $\mu_i - \mu_{i'} = \alpha_i - \alpha_{i'}$, $i \neq i'$. In general, one may be interested in estimating a linear combination of the means $\theta = \sum c_i\mu_i$, where $\{c_i\}$ are known constants. A linear combination $\theta = \sum c_i\mu_i$ is called a contrast if $\sum c_i = 0$. Thus, any pairwise difference of the means is a contrast.

The least squares estimate of $\theta = \sum c_i\mu_i$ is $\hat{\theta} = \sum c_i\bar{Y}_{i.}$. It is fairly easy to see that its mean, variance, and the estimated variance are

$$\mathrm{E}\left[\hat{\theta}\right] = \theta,\ \mathrm{Var}\left[\hat{\theta}\right] = \sigma^2 \sum c_i^2/n_i, \text{ and } s^2\left(\hat{\theta}\right) = \widehat{\mathrm{Var}\left[\hat{\theta}\right]} = MSE \sum c_i^2/n_i.$$

Since $\hat{\theta}$ is a function of the estimated mean vector, it is independent of *SSE* and hence of *MSE*. Under the assumption of normality $\left(\hat{\theta} - \theta\right)/s\left(\hat{\theta}\right) \sim t_{n-k}$, and this result can be used for constructing confidence intervals or for testing hypotheses.

Example 11.4.6 (ANCOVA With One Factor and One Covariate). Consider the following model with one factor and one covariate

$$Y_{ij} = \mu + \alpha_i + \gamma Z_{ij} + \varepsilon_{ij}, \quad j = 1, \dots, n_i, \quad i = 1, \dots, a.$$

If γ were known then we could rewrite the model as $Y_{ij}^* = \mu + \alpha_i + \varepsilon_{ij}$, where $Y_{ij}^* = Y_{ij} - \gamma Z_{ij}$, and the estimates of μ and α_i would be

$$\tilde{\mu} = \bar{Y}_{..}^* = \bar{Y}_{..} - \gamma \bar{Z}_{..}, \quad \text{and} \quad \tilde{\alpha}_i = \bar{Y}_{i.}^* - \bar{Y}_{..}^* = \bar{Y}_{i.} - \bar{Y}_{..} - \gamma(\bar{Z}_{i.} - \bar{Z}_{..}).$$

When γ is unknown (which is usually the case in practice) and it is estimated by $\hat{\gamma}$, then we may plug in the estimate of γ in the above expressions in order to obtain the estimates of μ and α_i. It turns out that this reasoning is valid and it will be discussed in a separate section later.

We now outline a simple strategy for obtaining the least squares estimate of γ. Rewriting the ANCOVA model as

$$Y_{ij} = \mu + \alpha_i^* + \gamma \tilde{Z}_{ij} + \varepsilon_{ij},$$

where $\tilde{Z}_{ij} = Z_{ij} - \bar{Z}_{i.}$ and $\alpha_i^* = \alpha_i + \gamma \bar{Z}_{i.}$, the least squares criterion is

$$\sum_i \sum_j \left(Y_{ij} - \mu - \alpha_i^* - \gamma \tilde{Z}_{ij}\right)^2 = \sum_i \sum_j Y_{ij}^2 + \sum_i \sum_j (\mu + \alpha_i^*)^2 + \sum_i \sum_j \left(\gamma \tilde{Z}_{ij}\right)^2$$
$$- 2\sum_i \sum_j Y_{ij}(\mu + \alpha_i^*) - 2\sum_i \sum_j Y_{ij}\left(\gamma \tilde{Z}_{ij}\right),$$

since the cross-product term involving $\mu + \alpha_i^*$ and $\gamma \tilde{Z}_{ij}$ equals 0. This allows for estimation of γ and $\mu + \alpha_i^*$ separately. Thus

$$\hat{\gamma} = \sum_i \sum_j Y_{ij}\tilde{Z}_{ij} \Big/ \sum_i \sum_j \tilde{Z}_{ij}^2,$$
$$\hat{\mu} = \bar{Y}_{..} - \hat{\gamma}\bar{Z}_{..}, \; \hat{\alpha}_i = \bar{Y}_{i.} - \bar{Y}_{..} - \hat{\gamma}(\bar{Z}_{i.} - \bar{Z}_{i.}), \quad \text{and} \quad \hat{Y}_{ij} = \bar{Y}_{i.} + \hat{\gamma}\tilde{Z}_{ij}.$$

Details on ANCOVA models appear in a later section.

11.5 Estimation Under Linear Restrictions on β

In Section 11.4 of this chapter, a simple submodel $\boldsymbol{Y} = \beta_0 \mathbf{1} + \boldsymbol{\varepsilon}$ of $\boldsymbol{Y} = \boldsymbol{X\beta} + \boldsymbol{\varepsilon}$ was considered when discussing the concept of R^2. However, more general submodels can also be considered, and an analogous result on the decomposition of the residual sum of squares for the submodel can be obtained. The details will be given later, but the result is as follows. If $\hat{\boldsymbol{Y}} = \boldsymbol{X}\hat{\boldsymbol{\beta}}$ is the estimated mean vector for the full model $\boldsymbol{Y} = \boldsymbol{X\beta} + \boldsymbol{\varepsilon}$ and $\tilde{\boldsymbol{Y}} = \boldsymbol{X}\tilde{\boldsymbol{\beta}}$ is the estimated mean vector under a submodel (reduced model) of the full model, where $\tilde{\boldsymbol{\beta}}$ is the least squares estimate of $\boldsymbol{\beta}$ in the submodel, then it turns out that

$$\|\boldsymbol{Y} - \tilde{\boldsymbol{Y}}\|^2 = \|\hat{\boldsymbol{Y}} - \tilde{\boldsymbol{Y}}\|^2 + \|\boldsymbol{Y} - \hat{\boldsymbol{Y}}\|^2.$$

Thus if we write $SSE_F = \|\boldsymbol{Y} - \hat{\boldsymbol{Y}}\|^2$ and $SSE_R = \|\boldsymbol{Y} - \tilde{\boldsymbol{Y}}\|^2$ as the residual sum of squares under the full and reduced models, then

$$SSE_R = \|\hat{\boldsymbol{Y}} - \tilde{\boldsymbol{Y}}\|^2 + SSE_F.$$

Analogous to R^2, we can obtain the proportional reduction (also known as the *coefficient of partial determination*) in the residual sum of squares when going from the submodel to the full model

$$\frac{SSE_R - SSE_F}{SSE_R}.$$

Let us now see when and how the above results hold. Suppose we wish to estimate the mean vector $\boldsymbol{X\beta}$ in the Gauss-Markov model under the restriction $\boldsymbol{L}^T\boldsymbol{\beta} = \boldsymbol{\theta}_0$, $\boldsymbol{\theta}_0$ known, where L is a $p \times m$ matrix of rank $m \leq p$. Such a problem usually comes up in hypothesis testing where the null is H_0: $\boldsymbol{L}^T\boldsymbol{\beta} = \boldsymbol{\theta}_0$ against the alternative H_1: $\boldsymbol{L}^T\boldsymbol{\beta} \neq \boldsymbol{\theta}_0$. For such a testing problem, we need to obtain the residual sums of squares for the full model and reduced model (ie, the submodel model with the constraint $\boldsymbol{L}^T\boldsymbol{\beta} = \boldsymbol{\theta}_0$), and then use them to carry out the test which will be described later.

The restricted least squares estimate of $\boldsymbol{\beta}$ has a complicated expression and some notational simplifications make the arguments clearer. Since the design matrix $\boldsymbol{X}$ may not have orthogonal columns, it helps to reexpress $\boldsymbol{X\beta}$ as $\boldsymbol{X}_0\boldsymbol{\gamma}$ so that the columns of $\boldsymbol{X}_0$ are orthonormal. Let $(\boldsymbol{X}^T\boldsymbol{X})^{1/2}$ be a symmetric square root of $\boldsymbol{X}^T\boldsymbol{X}$ and define

$$\boldsymbol{X}_0 = \boldsymbol{X}\left(\boldsymbol{X}^T\boldsymbol{X}\right)^{-1/2}, \boldsymbol{\gamma} = \left(\boldsymbol{X}^T\boldsymbol{X}\right)^{1/2}\boldsymbol{\beta}, \boldsymbol{L}_0 = \left(\boldsymbol{X}^T\boldsymbol{X}\right)^{-1/2}\boldsymbol{L}, \text{ so that}$$
$$\boldsymbol{X\beta} = \boldsymbol{X}_0\boldsymbol{\gamma} \text{ and } \boldsymbol{\theta} = \boldsymbol{L}^T\boldsymbol{\beta} = \boldsymbol{L}_0^T\boldsymbol{\gamma}.$$

With these notations, let us note the following:

(i) $\boldsymbol{X}_0^T\boldsymbol{X}_0 = \boldsymbol{I}$,
(ii) the least squares estimate of $\boldsymbol{\gamma}$ in the full model is $\hat{\boldsymbol{\gamma}} = (\boldsymbol{X}^T\boldsymbol{X})^{1/2}\hat{\boldsymbol{\beta}} = \boldsymbol{X}_0^T\boldsymbol{Y}$,
(iii) $\mathrm{E}[\hat{\boldsymbol{\gamma}}] = \boldsymbol{\gamma}$, $\mathrm{Cov}[\hat{\boldsymbol{\gamma}}] = \sigma^2\boldsymbol{I}$,
(iv) $\mathrm{E}\left[\hat{\boldsymbol{\theta}}\right] = \boldsymbol{\theta}$ and $\mathrm{Cov}\left[\hat{\boldsymbol{\theta}}\right] = \sigma^2\boldsymbol{L}_0^T\boldsymbol{L}_0$, where $\hat{\boldsymbol{\theta}} = \boldsymbol{L}^T\hat{\boldsymbol{\beta}} = \boldsymbol{L}_0^T\hat{\boldsymbol{\gamma}}$, and
(v) the least squares estimate of the mean vector $\boldsymbol{\mu} = \boldsymbol{X\beta}$ in the full model is $\hat{\boldsymbol{\mu}} = \boldsymbol{X}\hat{\boldsymbol{\beta}} = \boldsymbol{X}_0\hat{\boldsymbol{\gamma}}$.

For the restricted least squares case, we now find the estimate of $\boldsymbol{X\beta}$. We minimize $\|\boldsymbol{Y} - \boldsymbol{X\beta}\|^2 = \|\boldsymbol{Y} - \boldsymbol{X}_0\boldsymbol{\gamma}\|^2$ with respect to $\boldsymbol{\gamma}$ subject to the constraint $\boldsymbol{L}_0^T\boldsymbol{\gamma} = \boldsymbol{\theta}_0$. The method of Lagrangian multiplier is useful for a constrained optimization problem, and we minimize

$$\|\boldsymbol{Y} - \boldsymbol{X}_0\boldsymbol{\gamma}\|^2 + \boldsymbol{\lambda}^T\left(\boldsymbol{L}_0^T\boldsymbol{\gamma} - \boldsymbol{\theta}_0\right)$$

with respect to $\boldsymbol{\gamma}$ where $\boldsymbol{\lambda}$ is the Lagrangian multiplier vector. Differentiating the last expression with respect to $\boldsymbol{\gamma}$ and $\boldsymbol{\lambda}$, and equating the derivatives to $\mathbf{0}$, we have

$$\tilde{\boldsymbol{\gamma}} = \boldsymbol{X}_0^T \boldsymbol{Y} - (1/2)\boldsymbol{L}_0\boldsymbol{\lambda} \quad \text{and} \quad \boldsymbol{L}_0^T\tilde{\boldsymbol{\gamma}} = \boldsymbol{\theta}_0.$$

Premultiplying the first equation with $\boldsymbol{L}_0^T$ we get $\boldsymbol{L}_0^T\tilde{\boldsymbol{\gamma}} = \boldsymbol{L}_0^T\boldsymbol{X}_0^T\boldsymbol{Y} - (1/2)\boldsymbol{L}_0^T\boldsymbol{L}_0\boldsymbol{\lambda}$. Since $\boldsymbol{L}_0^T\tilde{\boldsymbol{\gamma}} = \boldsymbol{\theta}_0$, then $\boldsymbol{\lambda} = 2(\boldsymbol{L}_0^T\boldsymbol{L}_0)^{-1}[\boldsymbol{L}_0^T\boldsymbol{X}_0^T\boldsymbol{Y} - \boldsymbol{\theta}_0]$. Thus we get a solution to the restricted least squares problem as

$$\begin{aligned}
\tilde{\boldsymbol{\gamma}} &= \boldsymbol{X}_0^T\boldsymbol{Y} - \boldsymbol{L}_0\left(\boldsymbol{L}_0^T\boldsymbol{L}_0\right)^{-1}\left(\boldsymbol{L}_0^T\boldsymbol{X}_0^T\boldsymbol{Y} - \boldsymbol{\theta}_0\right) \\
&= \hat{\boldsymbol{\gamma}} - \boldsymbol{L}_0\left(\boldsymbol{L}_0^T\boldsymbol{L}_0\right)^{-1}\left(\boldsymbol{L}_0^T\hat{\boldsymbol{\gamma}} - \boldsymbol{\theta}_0\right), \text{ and} \\
\tilde{\boldsymbol{\beta}} &= \left(\boldsymbol{X}^T\boldsymbol{X}\right)^{-1/2}\tilde{\boldsymbol{\gamma}}.
\end{aligned}$$

The estimated mean vector and the vector of residuals $\tilde{\boldsymbol{\varepsilon}}$ for the restricted least squares are

$$\begin{aligned}
\boldsymbol{X}\tilde{\boldsymbol{\beta}} &= \boldsymbol{X}_0\tilde{\boldsymbol{\gamma}} = \boldsymbol{X}_0\hat{\boldsymbol{\gamma}} - \boldsymbol{X}_0\boldsymbol{L}_0\left(\boldsymbol{L}_0^T\boldsymbol{L}_0\right)^{-1}\left(\boldsymbol{L}_0^T\hat{\boldsymbol{\gamma}} - \boldsymbol{\theta}_0\right) \\
&= \boldsymbol{X}\hat{\boldsymbol{\beta}} - \boldsymbol{X}_0\boldsymbol{L}_0\left(\boldsymbol{L}_0^T\boldsymbol{L}_0\right)^{-1}\left(\hat{\boldsymbol{\theta}} - \boldsymbol{\theta}_0\right), \\
\tilde{\boldsymbol{\varepsilon}} &= \boldsymbol{Y} - \boldsymbol{X}\tilde{\boldsymbol{\beta}} = \hat{\boldsymbol{\varepsilon}} + \boldsymbol{X}_0\boldsymbol{L}_0\left(\boldsymbol{L}_0^T\boldsymbol{L}_0\right)^{-1}\left(\hat{\boldsymbol{\theta}} - \boldsymbol{\theta}_0\right).
\end{aligned}$$

Since $\boldsymbol{X}_0\boldsymbol{L}_0(\boldsymbol{L}_0^T\boldsymbol{L}_0)^{-1}\left(\hat{\boldsymbol{\theta}} - \boldsymbol{\theta}_0\right)$ is in $\mathcal{M}(\boldsymbol{X})$ and $\hat{\boldsymbol{\varepsilon}}$ is orthogonal to $\mathcal{M}(\boldsymbol{X})$, we have

$$\begin{aligned}
\|\tilde{\boldsymbol{\varepsilon}}\|^2 &= \|\hat{\boldsymbol{\varepsilon}}\|^2 + \|\boldsymbol{X}_0\boldsymbol{L}_0\left(\boldsymbol{L}_0^T\boldsymbol{L}_0\right)^{-1}\left(\hat{\boldsymbol{\theta}} - \boldsymbol{\theta}_0\right)\|^2 \\
&= \|\hat{\boldsymbol{\varepsilon}}\|^2 + \left(\hat{\boldsymbol{\theta}} - \boldsymbol{\theta}_0\right)^T\left(\boldsymbol{L}_0^T\boldsymbol{L}_0\right)^{-1}\left(\hat{\boldsymbol{\theta}} - \boldsymbol{\theta}_0\right).
\end{aligned}$$

Therefore,

$$\begin{aligned}
SSE_R &= SSE_F + \left(\hat{\boldsymbol{\theta}} - \boldsymbol{\theta}_0\right)^T\left(\boldsymbol{L}_0^T\boldsymbol{L}_0\right)^{-1}\left(\hat{\boldsymbol{\theta}} - \boldsymbol{\theta}_0\right), \text{ or} \\
SSE_R - SSE_F &= \left(\hat{\boldsymbol{\theta}} - \boldsymbol{\theta}_0\right)^T\left(\boldsymbol{L}_0^T\boldsymbol{L}_0\right)^{-1}\left(\hat{\boldsymbol{\theta}} - \boldsymbol{\theta}_0\right). \qquad (2)
\end{aligned}$$

The discussion above leads to the following important result.

Theorem 11.5.1. *Let $\tilde{\boldsymbol{\beta}}$ be the least squares estimate of $\boldsymbol{\beta}$ in the Gauss-Markov model in Eq. (1) under the restriction $\boldsymbol{L}^T\boldsymbol{\beta} = \boldsymbol{\theta}_0$, where $\boldsymbol{\theta}_0$ is known, and let $\boldsymbol{X}\tilde{\boldsymbol{\beta}}$ be the estimated mean vector under the restriction. Let $\boldsymbol{\theta} = \boldsymbol{L}^T\boldsymbol{\beta}$, $\boldsymbol{L}_0 = (\boldsymbol{X}^T\boldsymbol{X})^{-1/2}\boldsymbol{L}$, and $\hat{\boldsymbol{\theta}} = \boldsymbol{L}^T\hat{\boldsymbol{\beta}}$, where $\hat{\boldsymbol{\beta}}$ is the unrestricted least squares estimate of $\boldsymbol{\beta}$. Denote the residual vectors $\hat{\boldsymbol{\varepsilon}} = \boldsymbol{Y} - \boldsymbol{X}\hat{\boldsymbol{\beta}}$ and $\tilde{\boldsymbol{\varepsilon}} = \boldsymbol{Y} - \boldsymbol{X}\tilde{\boldsymbol{\beta}}$. Then:*

(a) $\boldsymbol{X}\tilde{\boldsymbol{\beta}} = \boldsymbol{X}\hat{\boldsymbol{\beta}} - \boldsymbol{X}_0\boldsymbol{L}_0(\boldsymbol{L}_0^T\boldsymbol{L}_0)^{-1}\left(\hat{\boldsymbol{\theta}} - \boldsymbol{\theta}_0\right)$,

(b) $\tilde{\boldsymbol{\varepsilon}} = \hat{\boldsymbol{\varepsilon}} + \boldsymbol{X}_0\boldsymbol{L}_0(\boldsymbol{L}_0^T\boldsymbol{L}_0)^{-1}\left(\hat{\boldsymbol{\theta}} - \boldsymbol{\theta}_0\right)$,

(c) $\|\tilde{\boldsymbol{\varepsilon}}\|^2 = \|\hat{\boldsymbol{\varepsilon}}\|^2 + \left(\hat{\boldsymbol{\theta}} - \boldsymbol{\theta}_0\right)^T(\boldsymbol{L}_0^T\boldsymbol{L}_0)^{-1}\left(\hat{\boldsymbol{\theta}} - \boldsymbol{\theta}_0\right)$.

11.6 Gauss-Markov Models: Inference

Throughout this section we assume a Gauss-Markov model with normal errors (ie, $\boldsymbol{\varepsilon} \sim N_n(\mathbf{0}, \sigma^2 \boldsymbol{I})$). Theorem 11.5.1 can be used for constructing confidence regions and hypotheses testing. If $\boldsymbol{\theta} = \boldsymbol{L}^T\boldsymbol{\beta}$, where $\boldsymbol{L}$ is a $p \times m$ matrix of rank $m \leq p$, then its least square estimate is $\hat{\boldsymbol{\theta}} = \boldsymbol{L}^T\hat{\boldsymbol{\beta}}$, where $\hat{\boldsymbol{\beta}}$ is the least square estimate of $\boldsymbol{\beta}$. Since $\hat{\boldsymbol{\theta}}$ is a linear function of the estimated mean vector $\hat{\boldsymbol{Y}} = \boldsymbol{X}\hat{\boldsymbol{\beta}}$, it is uncorrelated with the residual vector $\hat{\boldsymbol{\varepsilon}} = \hat{\boldsymbol{Y}} - \boldsymbol{X}\hat{\boldsymbol{\beta}}$ (Theorem 11.3.2). Thus, under the assumption of normality, $\hat{\boldsymbol{\theta}}$ is independent of $\hat{\boldsymbol{\varepsilon}}$ and hence of $MSE = \|\hat{\boldsymbol{Y}} - \boldsymbol{X}\hat{\boldsymbol{\beta}}\|^2/(n-p)$. Note that

(i) $\hat{\boldsymbol{\theta}} \sim N_m(\boldsymbol{\theta}, \sigma^2\boldsymbol{L}_0^T\boldsymbol{L}_0)$, where $\boldsymbol{L}_0 = (\boldsymbol{X}^T\boldsymbol{X})^{-1/2}\boldsymbol{L}$, and
$\left(\hat{\boldsymbol{\theta}} - \boldsymbol{\theta}\right)^T\left(\boldsymbol{L}_0^T\boldsymbol{L}_0\right)^{-1}\left(\hat{\boldsymbol{\theta}} - \boldsymbol{\theta}\right)^T/\sigma^2 \sim \chi^2_m$,

(ii) $SSE/\sigma^2 = \|\hat{\boldsymbol{Y}} - \boldsymbol{X}\hat{\boldsymbol{\beta}}\|^2/\sigma^2 \sim \chi^2_{n-p}$,

(iii) $\hat{\boldsymbol{\theta}}$ and SSE are independent.

The F-ratio

$$F = \frac{\left(\hat{\boldsymbol{\theta}} - \boldsymbol{\theta}\right)^T\left(\boldsymbol{L}_0^T\boldsymbol{L}_0\right)^{-1}\left(\hat{\boldsymbol{\theta}} - \boldsymbol{\theta}\right)^T/\left(\sigma^2 m\right)}{\|\hat{\boldsymbol{Y}} - \boldsymbol{X}\hat{\boldsymbol{\beta}}\|^2/(\sigma^2(n-p))}$$

$$= \frac{\left(\hat{\boldsymbol{\theta}} - \boldsymbol{\theta}\right)^T\left(\boldsymbol{L}_0^T\boldsymbol{L}_0\right)^{-1}\left(\hat{\boldsymbol{\theta}} - \boldsymbol{\theta}\right)^T/m}{MSE}$$

$$= \left(\hat{\boldsymbol{\theta}} - \boldsymbol{\theta}\right)^T\left[\boldsymbol{s}^2\left(\hat{\boldsymbol{\theta}}\right)\right]^{-1}\left(\hat{\boldsymbol{\theta}} - \boldsymbol{\theta}\right)/m, \text{ where}$$

$$\boldsymbol{s}^2\left(\hat{\boldsymbol{\theta}}\right) = MSE\left[\boldsymbol{L}_0^T\boldsymbol{L}_0\right]^{-1},$$

has an F-distribution with $df = (m, n-p)$.

If $\boldsymbol{L}$ is a vector (ie, $m = 1$), then $\theta = \boldsymbol{L}^T\boldsymbol{\beta}$ is a real number and $\left(\hat{\theta} - \theta\right)/s\left(\hat{\theta}\right) \sim t_{n-p}$. Then a confidence interval for θ with confidence coefficient $1-\alpha$ is $\hat{\theta} \pm t_{n-p,\alpha/2}s\left(\hat{\theta}\right)$. Similarly, if we want to test H_0: $\theta = \theta_0$ against the alternative H_1: $\theta \neq \theta_0$, then one may reject the null hypothesis if $\left|\left(\hat{\theta} - \theta_0\right)/s\left(\hat{\theta}\right)\right| > t_{n-p,\alpha/2}$.

If $\boldsymbol{L}$ is a matrix of order $p \times m$ with rank $m \leq p$ and $\boldsymbol{\theta} = \boldsymbol{L}^T\boldsymbol{\beta}$, and it is desired to carry out a test H_0: $\boldsymbol{\theta} = \boldsymbol{\theta}_0$ against H_1: $\boldsymbol{\theta} \neq \boldsymbol{\theta}_0$, then there are two equivalent ways to express the F-statistic for this test. One expression involves SSE_F and SSE_R, whereas the other involves $\hat{\boldsymbol{\theta}} = \boldsymbol{L}^T\hat{\boldsymbol{\beta}}$ (where $\hat{\boldsymbol{\beta}}$ is the least squares estimate under the full model) and $\boldsymbol{s}^2\left(\hat{\boldsymbol{\theta}}\right)$, the estimate of $\mathrm{Cov}\left[\hat{\boldsymbol{\theta}}\right]$. For a particular application, one may use the form that is more convenient to obtain the F-statistic. If MSE_F is the mean square error under the full model, then the F-statistic can be written in the two equivalent forms

$$F = \frac{(SSE_R - SSE_F)/m}{MSE_F} \tag{3a}$$

$$= \left(\hat{\theta} - \theta_0\right)^T \left[s^2\left(\hat{\theta}\right)\right]^{-1} \left(\hat{\theta} - \theta_0\right)/m. \tag{3b}$$

Since $\hat{\theta} \sim N_m\left(\theta, \sigma^2 L^T (X^T X)^{-1} L\right)$, using Theorem 11.5.1, we have

$$[SSE_R - SSE_F]/\sigma^2 = \left(\hat{\theta} - \theta_0\right)^T \left[L_0^T L_0\right]^{-1} \left(\hat{\theta} - \theta_0\right)/\sigma^2 \sim \chi_m^2\left(\delta^2\right),$$

where $\delta^2 = (1/2)(\theta - \theta_0)^T [L_0^T L_0]^{-1} (\theta - \theta_0)/\sigma^2$. Since $\hat{\beta}$ is independent of the residual vector $Y - X\hat{\beta}$, $SSE_R - SSE_F$, which is a function of $\hat{\beta}$, is independent of SSE_F, a function of $Y - X\hat{\beta}$. Thus $SSE_F/\sigma^2 \sim \chi_{n-p}^2$, and under H_0, $(SSE_R - SSE_F)/\sigma^2 \sim \chi_m^2(\delta^2)$, where δ^2 is given above. Therefore

$$F = \frac{(SSE_R - SSE_F)/m}{SSE_F/(n-p)} = \frac{(SSE_R - SSE_F)/m}{MSE_F}$$

$$= \left(\hat{\theta} - \theta_0\right)^T \left[s^2\left(\hat{\theta}\right)\right]^{-1} \left(\hat{\theta} - \theta_0\right)/m \sim F_{m,n-p}\left(\delta^2\right).$$

Thus we reject $H_0\colon \theta = \theta_0$ in favor of $H_1\colon \theta \neq \theta_0$ if the value of the F-statistic given above is higher than the critical value obtained from the F-distribution with $df = (m, n-p)$.

A few important facts come out from the above discussions:

(a) Any linear hypothesis about β induces a reduced model.
(b) There are two alternate ways to derive the F-statistic for testing $H_0\colon \theta = \theta_0$ vs $H_1\colon \theta \neq \theta_0$.
(c) $SSE_R - SSE_F$ depends on Y only through $\hat{\theta} = L^T\hat{\beta}$. Since $X\hat{\beta}$ is uncorrelated with $\hat{\varepsilon} = Y - X\hat{\beta}$, $\hat{\beta}$ is independent of $\hat{\varepsilon}$. Consequently, $SSE_R - SSE_F$ (a function of $\hat{\beta}$) is independent of SSE_F (a function of $\hat{\varepsilon}$).
(d) $df(SSE_F) = n - p$, $df(SSE_R) = n - p + m$, $df(SSE_R) - df(SSE_F) = m$.
(e) Under $H_0\colon \theta = \theta_0$, the F-statistic as given in Eqs. (3a) and (3b) has an F-distribution with $df = (m, n-p)$.

We will summarize the above discussion in the following result.

Theorem 11.6.1. *Assume the Gauss-Markov setup as in Eq. (1) with $\varepsilon \sim N_n(\mathbf{0}, \sigma^2 I)$. Let L be a known matrix of order $p \times m$ with rank $m \leq p$, and denote $\theta = L^T\beta$. We wish to test $H_0\colon \theta = \theta_0$ against $H_1\colon \theta \neq \theta_0$. Let $\hat{\beta}$ be the least squares estimate of β and $\hat{\theta} = L^T\hat{\beta}$. Let $SSE_F = \|Y - X\hat{\beta}\|^2$ and $SSE_R = \|Y - X\tilde{\beta}\|^2$, where $\tilde{\beta}$ is the least squares estimate of β under the restriction $L^T\beta = \theta_0$. Then*

(a) *$\hat{\theta}$ is independent of SSE_F,*
(b) *$SSE_F/\sigma^2 \sim \chi_{n-p}^2$ and $SSE_R - SSE_R$ is independent of SSE_F,*
(c) *$(SSE_R - SSE_F)/\sigma^2 \sim \chi_m^2(\delta^2)$ where the noncentrality parameter is $\delta^2 = (1/2)(\theta - \theta_0)^T [L_0^T L_0]^{-1} (\theta - \theta_0)/\sigma^2$ with $L_0 = L(X^T X)^{-1/2}$, and*

(d) *for the F-statistic in Eqs. (3a) and* (3b), $F \sim F_{m,n-p}(\delta^2)$, *where* δ^2 *is as given in part (c), and* $F \sim F_{m,n-p}$ *under* H_0.

Example 11.6.1 (Deleting a Variable From the Regression). Let us consider a regression model with $p - 1 = 4$ independent variables. Suppose we wish to find out if variable X_1 should be dropped from the model. This is equivalent to testing H_0: $\beta_1 = 0$ against H_1: $\beta_1 \neq 0$. In this case, we can write $\boldsymbol{L}$ as a row vector of length $p = 5$ whose second element is 1 and the rest are zeros. In such a case, clearly, $\theta = \boldsymbol{L}^T\boldsymbol{\beta} = \beta_1$. We can carry out either a t-test or an F-test for this purpose. The reduced model (ie, the model under H_0) has $p-2$ independent variables instead of $p-1$. If we call the original model the full model, then we have

$$Y = \beta_0 + \beta_1 X_1 + \beta_2 X_2 + \beta_3 X_3 + \beta_4 X_4 + \varepsilon \text{ (full)}$$
$$Y = \beta_0 + \beta_2 X_2 + \beta_3 X_3 + \beta_4 X_4 + \varepsilon \text{ (reduced)}.$$

In this case, we may obtain the residual sums of squares for the full and the reduced models and $df(SSE_F) = n - 5$, $df(SSE_R) = n - 4$. The F-statistic for this test

$$F = \frac{SSE_R - SSE_F}{MSE_F}$$

is exactly equal to $\left[\hat{\beta}_1/s\left(\hat{\beta}_1\right)\right]^2$. Thus we may simply use the t-statistic $t = \hat{\beta}_1/s\left(\hat{\beta}_1\right)$ for testing H_0: $\beta_1 = 0$ against H_1: $\beta_1 \neq 0$ and avoid calculating SSE_R.

Example 11.6.2 (Deleting More Than One Variable From the Regression Model). Suppose the set-up is the same as the last example, but now we wish to know if we can delete the first two independent variables from the model. This is equivalent to testing H_0: $\beta_1 = \beta_2 = 0$ against H_1: at least one β_1, β_2 is nonzero. In this case, $\boldsymbol{L}$ can be written as a $p \times 2$ matrix which has all zeros except for 1's at the second element of the first column and at the third element of the second column. We need to test H_0: $\boldsymbol{\theta} = \boldsymbol{\theta}_0$ against H_1: $\boldsymbol{\theta} \neq \boldsymbol{\theta}_0$, where $\boldsymbol{\theta}_0 = (0, 0)^T$. The F-statistic for this test is $F = \left(\hat{\boldsymbol{\theta}} - \boldsymbol{\theta}_0\right)^T \left[\boldsymbol{s}^2\left(\hat{\boldsymbol{\theta}}\right)\right]^{-1} \left(\hat{\boldsymbol{\theta}} - \boldsymbol{\theta}_0\right)/2$. However, we can derive this same test statistic in a different way. The full and the reduced (under the null) models are

$$Y = \beta_0 + \beta_1 X_1 + \beta_2 X_2 + \beta_3 X_3 + \beta_4 X_4 + \varepsilon \text{ (full)},$$
$$Y = \beta_0 + \beta_3 X_3 + \beta_4 X_4 + \varepsilon \text{ (reduced)}.$$

Here, $df(SSE_F) = n - 5$, $df(SSE_R) = n - 3$, and we can write

$$F = \frac{(SSE_R - SSE_F)/2}{MSE_F},$$

and $F \sim F_{2,n-5}$ under H_0.

Example 11.6.3 (One-Factor ANOVA Model). Consider a one-factor model with k levels. If it is desired to test H_0: $\alpha_1 = \cdots = \alpha_k = 0$ vs H_1: not all α_i's are 0, then the full and the reduced models are

$$Y_{ij} = \mu + \alpha_i + \varepsilon_{ij} \text{ (full)}, \quad Y_{ij} = \mu + \varepsilon_{ij} \text{ (reduced)}.$$

In this case,

$$SSE_F = \sum_i \sum_j \left(Y_{ij} - \bar{Y}_{i\cdot}\right)^2, \quad df(SSE_F) = n - k,$$
$$SSE_R = \sum_i \sum_j \left(Y_{ij} - \bar{Y}_{\cdot\cdot}\right)^2, \quad df(SSE_R) = n - 1, \text{ and}$$
$$SSR_R - SSE_F = \sum_i \sum_j (\bar{Y}_{i\cdot} - \bar{Y}_{\cdot\cdot})^2 = \sum n_i \hat{\alpha}_i^2 := SSTR.$$

The F-statistic is

$$F = \frac{[SSR_R - SSE_F]/(k-1)}{MSE_F} = \frac{SSTR/(k-1)}{MSE_F} = \frac{MSTR}{MSE},$$

where $MSE = MSE_F$. Denoting $\delta^2 = (1/2)\sum n_i \alpha_i^2/\sigma^2$, we get the result $F \sim F_{k-1,n-k}(\delta^2)$ and $F \sim F_{k-1,n-k}$ under H_0.

Example 11.6.4 (Balanced Two-Factor ANOVA). In the two-factor balanced ANOVA model, we have seen in Example 11.4.4 that *SST*0, the total sum of squares, can be decomposed as the sum of *SSA*, *SSB*, and *SSAB*. If we want to test H_0: $(\alpha\beta)_{ij} = 0$ for all i and j, vs H_1: at least one $(\alpha\beta)_{ij}$ is not zero, the full and reduced models are

$$Y_{ijk} = \mu + \alpha_i + \beta_j + (\alpha\beta)_{ij} + \varepsilon_{ijk} \text{ (full)},$$
$$Y_{ijk} = \mu + \alpha_i + \beta_j + \varepsilon_{ijk} \text{ (reduced)}.$$

Estimated mean values for the full and reduced models are

$$\hat{Y}_{ijk} = \bar{Y}_{ij\cdot} = \hat{\mu} + \hat{\alpha}_i + \hat{\beta}_j + \widehat{(\alpha\beta)}_{ij}$$
$$\tilde{Y}_{ijk} = \hat{\mu} + \hat{\alpha}_i + \hat{\beta}_j,$$

where $\hat{\mu}$, $\hat{\alpha}_i$, and $\hat{\beta}_j$ are as given in Example 11.4.4. The sums of squares for the full and the reduced models are

$$SSE_F = \sum_i \sum_j \sum_k \left(Y_{ijk} - \bar{Y}_{ij\cdot}\right)^2, \quad df(SSE_F) = n - ab,$$
$$SSE_R = \sum_i \sum_j \sum_k \left(Y_{ijk} - \hat{\mu} - \hat{\alpha}_i - \hat{\beta}_j\right)^2, \quad df(SSE_R) = n - (a + b - 1),$$
$$SSE_R - SSE_F = \sum_i \sum_j \sum_k \left(\bar{Y}_{ij\cdot} - \hat{\mu} - \hat{\alpha}_i - \hat{\beta}_j\right)^2$$
$$= n_0 \sum_i \sum_j \widehat{(\alpha\beta)}_{ij}^2 = SSAB.$$

Thus the F-statistic for this testing problem is

$$F = \frac{SSAB/(ab - a - b + 1)}{MSE_F} = \frac{MSAB}{MSE},$$

where *MSE* denotes MSE_F. Under H_0, $F \sim F_{(a-1)(b-1),n-ab}$.

Even though it may not be a standard practice to test for the main effects in the presence of interactions, we can still formulate the statistical problem and describe the test statistic. Let H_0: $\alpha_1 = \cdots = \alpha_a = 0$ vs H_1: not all α_i's are zero. Then the full and reduced models are

$$Y_{ijk} = \mu + \alpha_i + \beta_j + (\alpha\beta)_{ij} + \varepsilon_{ijk} \text{ (full)},$$
$$Y_{ijk} = \mu + \beta_j + (\alpha\beta)_{ij} + \varepsilon_{ijk} \text{ (reduced)}.$$

Estimated mean values for the full and reduced models, and their sums of squares are

$$\hat{Y}_{ijk} = \hat{\mu} + \hat{\alpha}_i + \hat{\beta}_j + \widehat{(\alpha\beta)}_{ij}, \quad \tilde{Y}_{ijk} = \hat{\mu} + \hat{\beta}_j + \widehat{(\alpha\beta)}_{ij},$$
$$SSE_F = \sum_i \sum_j \sum_k \left(Y_{ijk} - \bar{Y}_{ij.}\right)^2, \quad df(SSE_F) = n - ab,$$
$$SSE_R = \sum_i \sum_j \sum_k \left(Y_{ijk} - \hat{\mu} - \hat{\beta}_j - \widehat{(\alpha\beta)}_{ij}\right)^2, \quad df(SSE_R) = n - [1 + a(b-1)],$$
$$SSE_R - SSE_F = SSA,$$

and the F-statistic is

$$F = \frac{SSA/(a-1)}{MSE_F} = \frac{MSA}{MSE},$$

where $MSE = MSE_F$. Under H_0, $F \sim F_{a-1,n-ab}$.

Example 11.6.5 (Two-Factor ANOVA: One Observation per Cell). Consider a two-factor ANOVA model as in the previous example with $n_{ij} = 1$ for all i and j. If Y_{ij} is the response when factor A is at level i and factor B is at level j, then a model without interactions is

$$Y_{ij} = \mu + \alpha_i + \beta_j + \varepsilon_{ij}, \quad j = 1, \ldots, b, \quad i = 1, \ldots, a,$$

where $\sum \alpha_i = 0$, $\sum \beta_j = 0$, and $\{\varepsilon_{ij}\}$ are iid $N(0, \sigma^2)$. Estimates of μ, α_i, and β_j are exactly the same as before (ie, $\hat{\mu} = \bar{Y}_{..}$, $\hat{\alpha}_i = \bar{Y}_{i.} - \bar{Y}_{..}$, and $\hat{\beta}_j = \bar{Y}_{.j} - \bar{Y}_{..}$). In this case, the fitted Y values, residuals, and the residual sum of squares are

$$\hat{Y}_{ij} = \hat{\mu} + \hat{\alpha}_i + \hat{\beta}_j, \quad \hat{\varepsilon}_{ij} = Y_{ij} - \left(\hat{\mu} + \hat{\alpha}_i + \hat{\beta}_j\right),$$
$$SSE = \sum_i \sum_j \left(Y_{ij} - \hat{\mu} - \hat{\alpha}_i - \hat{\beta}_j\right)^2, \text{ and}$$
$$df(SSE) = ab - (a + b - 1) = (a-1)(b-1), \text{ and}$$
$$MSE = SSE/[(a-1)(b-1)].$$

One can then carry out inferences on $\{\alpha_i\}$ and $\{\beta_j\}$ such as construction of simultaneous confidence intervals or tests such as H_0: $\alpha_1 = \cdots = \alpha_a = 0$ vs H_1: not all α_i are 0.

Now if it is desired to investigate if the interaction effects are present, one may think of the usual model in the two-factor case

$$Y_{ij} = \mu + \alpha_i + \beta_j + (\alpha\beta)_{ij} + \varepsilon_{ij}, \quad j = 1, \ldots, b, \quad i = 1, \ldots, a,$$

with the usual constraints on $\{\alpha_i\}$, $\{\beta_j\}$, and $\left\{(\alpha\beta)_{ij}\right\}$. For this model, the number of unknown parameters (excluding σ^2) is $n = ab$, the estimated mean is $\hat{Y}_{ij} = Y_{ij}$ and the residuals are $Y_{ij} - \hat{Y}_{ij} = 0$. Thus there is no way to estimate σ^2 as there are too many interaction parameters, and we cannot use this model to determine if the interaction effects are present. One way to approach this issue is to consider a more restrictive type of the interaction of the form $\theta\alpha_i\beta_j$, θ real. This leads to the consideration of Tukey's interaction model

$$Y_{ij} = \mu + \alpha_i + \beta_j + \theta\alpha_i\beta_j + \varepsilon_{ij}, \quad j = 1, \ldots, b, \quad i = 1, \ldots, a,$$

where $\sum \alpha_i = 0$, $\sum \beta_j = 0$, and $\{\varepsilon_{ij}\}$ are iid $N(0, \sigma^2)$. For this model, absence or presence of interaction effects can be judged by testing H_0: $\theta = 0$ against H_1: $\theta \neq 0$. This is known as "Tukey's one degree of freedom test for nonadditivity", details of which are described below.

Now Tukey's model is no longer a linear model, but if one estimates $\hat{\mu}$, $\hat{\alpha}_i$, and $\hat{\beta}_j$ as in the additive model, then minimizing $\sum_i \sum_j \left(Y_{ij} - \tilde{Y}_{ij} - \theta\hat{\alpha}_i\hat{\beta}_j\right)^2$, where $\tilde{Y}_{ij} = \hat{\mu} + \hat{\alpha}_i + \hat{\beta}_j$, with respect to θ leads to an estimate

$$\hat{\theta} = \frac{\sum_i \sum_j \tilde{\varepsilon}_{ij}\hat{\alpha}_i\hat{\beta}_j}{S_\alpha S_\beta}, \text{ where}$$

$$\tilde{\varepsilon}_{ij} = Y_{ij} - \tilde{Y}_{ij},\ S_\alpha = \sum \hat{\alpha}_i^2, \text{ and } S_\beta = \sum \hat{\beta}_j^2.$$

Since $\{\tilde{\varepsilon}_{ij}\}$ are independent of $\left\{\tilde{Y}_{ij}\right\}$, $\mathrm{E}\left[\tilde{\varepsilon}_{ij}|\tilde{\boldsymbol{Y}}\right] = \mathrm{E}\left[\tilde{\varepsilon}_{ij}\right] = \theta\alpha_i\beta_j$, where $\tilde{\boldsymbol{Y}}$ is the vector of $\tilde{Y}_{ij}$'s, and

$$\mathrm{E}\left[\hat{\theta}|\tilde{\boldsymbol{Y}}\right] = \frac{\sum_i \sum_j \mathrm{E}\left[\tilde{\varepsilon}_{ij}|\tilde{\boldsymbol{Y}}\right]\hat{\alpha}_i\hat{\beta}_j}{S_\alpha S_\beta} = \frac{\sum_i \sum_j \left(\theta\alpha_i\beta_j\right)\hat{\alpha}_i\hat{\beta}_j}{S_\alpha S_\beta}$$

$$= \theta \frac{\sum_i \sum_j \alpha_i\beta_j\hat{\alpha}_i\hat{\beta}_j}{S_\alpha S_\beta} := \theta A.$$

For two sequences of constants $\{e_i, i = 1, \ldots, a\}$ and $\{f_j, j = 1, \ldots, b\}$ satisfying the constraints $\sum e_i = 0$ and $\sum f_j = 0$,

$$\sum_i \sum_j \tilde{\varepsilon}_{ij} e_i f_j = \sum_i \sum_j Y_{ij} e_i f_j, \text{ and}$$

$$\mathrm{Var}\left[\sum_i \sum_j \tilde{\varepsilon}_{ij} e_i f_j \middle| \tilde{\boldsymbol{Y}}\right] = \mathrm{Var}\left[\sum_i \sum_j \tilde{\varepsilon}_{ij} e_i f_j\right]$$

$$= \mathrm{Var}\left[\sum_i \sum_j Y_{ij} e_i f_j\right] = \sigma^2 \sum_i \sum_j e_i^2 f_j^2.$$

Thus

$$\text{Var}\left[\hat{\theta}|\tilde{\boldsymbol{Y}}\right] = \sigma^2/[S_\alpha S_\beta].$$

In order to carry out a test on θ, we need an estimate of σ^2. If we write $\hat{Y}_{ij} = \tilde{Y}_{ij} + \hat{\theta}\hat{\alpha}_i\hat{\beta}_j$ and $\hat{\varepsilon}_{ij} = Y_{ij} - \hat{Y}_{ij} = \tilde{\varepsilon}_{ij} - \hat{\theta}\hat{\alpha}_i\hat{\beta}_j$, then a simple calculation will show that

$$SSE_R = \sum_i\sum_j \tilde{\varepsilon}_{ij}^2 = \sum_i\sum_j \hat{\varepsilon}_{ij}^2 + \hat{\theta}^2 S_\alpha S_\beta = SSE_F + \hat{\theta}^2 S_\alpha S_\beta.$$

Conditionally on $\tilde{\boldsymbol{Y}}$,

(i) $\left[SSE_R - SSE_F\right]/\sigma^2 = \hat{\theta}^2 S_\alpha S_\beta/\sigma^2 \sim \chi_1^2(\delta_1^2)$,
where $\delta_1^2 = (1/2)\left\{\text{E}\left[\tilde{\theta}|\tilde{\boldsymbol{Y}}\right]\right\}^2/\sigma^2 = (1/2)\theta^2 A^2/\sigma^2$,
(ii) $SSE_R/\sigma^2 \sim \chi^2_{(a-1)(b-1)}(\delta^2)$, where $\delta^2 = \theta^2\sum_j\sum_i \alpha_i^2\beta_j^2/\sigma^2$, and
(iii) SSE_F is nonnegative.

Hence by an application of Lemma B.7.2 in Appendix B, we can conclude that, conditionally on $\tilde{\boldsymbol{Y}}$, $SSE_F/\sigma^2 \sim \chi^2_{(a-1)(b-1)-1}(\delta^2 - \delta_1^2)$ and SSE_F/σ^2 is independent of $\left[SSE_R - SSE_F\right]/\sigma^2$. This allows us to construct a test for H_0: $\theta = 0$ vs H_1: $\theta \neq 0$ using the F-statistic

$$F = \frac{SSE_R - SSE_F}{MSE_F} = \frac{\hat{\theta}^2 S_\alpha S_\beta}{MSE_F}, \text{ where}$$

$$MSE_F = \left[\sum_i\sum_j \tilde{\varepsilon}_{ij}^2 - \hat{\theta}^2 S_\alpha S_\beta\right]\Big/[(a-1)(b-1)-1],$$

and $F \sim F_{1,(a-1)(b-1)-1}$ under H_0.

Example 11.6.6 (Nested ANOVA). Let us begin with a simple example. The school superintendent has asked every school in a town to try a pilot training program in order to improve the quantitative aptitude of the students. The town has a schools and each school has its own b designated teachers for this training program. A random sample of n_{ij} students is chosen in the ith school (factor A), $i = 1, \ldots, a$, and assigned to the jth teacher in that school, $j = 1, \ldots, b$, and after 6 months of training, the students are given a standardized test to evaluate their performances. Note that the teachers in different schools are entirely different and thus the teacher effect (factor B) is nested in the school (factor A). If Y_{ijk} is the score of the kth student assigned to teacher j in the ith school, then a reasonable model is

$$Y_{ijk} = \mu + \alpha_i + \beta_{j(i)} + \varepsilon_{ijk}, \quad k = 1, \ldots, n_{ij}, \; j = 1, \ldots, b, \; i = 1, \ldots, a,$$

where $\{\alpha_i\}$ are the factor A (school) effects, $\left\{\beta_{j(i)}\right\}$ are the factor B (teacher) effects nested in factor A, and $\{\varepsilon_{ijk}\}$ are iid $N(0, \sigma^2)$ errors. This is an example of a simple *nested ANOVA model*. For this model, it is assumed that

$$\text{(a) } \sum \alpha_i = 0, \text{ (b) } \sum_{j=1}^{b} \beta_{j(i)} = 0 \quad \text{for each } i = 1, \ldots, a.$$

Let $\bar{Y}_{\ldots}$, $\bar{Y}_{i\ldots}$, etc., be as in the two-factor ANOVA model. Then estimates of μ, α_i, $\beta_{j(i)}$, fitted Y values, and the residuals are

$$\hat{\mu} = (ab)^{-1} \sum_i \sum_j \bar{Y}_{ij\cdot}, \ \hat{\alpha}_i = b^{-1} \sum_j \bar{Y}_{ij\cdot} - \hat{\mu}, \ \hat{\beta}_{j(i)} = \bar{Y}_{ij\cdot} - \hat{\mu} - \hat{\alpha}_i,$$
$$\hat{Y}_{ijk} = \bar{Y}_{ij\cdot}, \ \hat{\varepsilon}_{ijk} = Y_{ijk} - \bar{Y}_{ij\cdot}, \text{ and}$$
$$SSE = \sum_i \sum_j \sum_k \hat{\varepsilon}^2_{ijk}.$$

As usual, $SSE/\sigma^2 \sim \chi^2_{n-ab}$, and $MSE = SSE/(n - ab)$ is an unbiased estimate of σ^2, where n is the total number of observations. One can obtain variances of $\hat{\mu}$, $\{\hat{\alpha}_i\}$, and $\left\{\hat{\beta}_{j(i)}\right\}$, but they are a bit cumbersome in the unbalanced case. In the balanced case (ie, $n_{ij} = n_0$ for all i and j), the expressions of $\hat{\mu}$, $\{\hat{\alpha}_i\}$, and $\left\{\hat{\beta}_{j(i)}\right\}$ are simpler

$$\hat{\mu} = \bar{Y}_{\ldots}, \ \hat{\alpha}_i = \bar{Y}_{i\cdot\cdot} - \bar{Y}_{\ldots}, \ \hat{\beta}_{j(i)} = \bar{Y}_{ij\cdot} - \hat{\mu} - \hat{\alpha}_i = \bar{Y}_{ij\cdot} - \bar{Y}_{i\cdot\cdot},$$

and

$$\begin{aligned} \mathrm{Var}[\hat{\mu}] &= \mathrm{Var}[\bar{Y}_{\ldots}] = \sigma^2/n, \\ \mathrm{Var}[\hat{\alpha}_i] &= \mathrm{Var}[\bar{Y}_{i\cdot\cdot} - \bar{Y}_{\ldots}] = \sigma^2(a-1)/n, \\ \mathrm{Var}\left[\hat{\beta}_{j(i)}\right] &= \mathrm{Var}\left[\bar{Y}_{ij\cdot} - \bar{Y}_{i\cdot\cdot}\right] = \sigma^2(b-1)/(n_0 b). \end{aligned}$$

Since $\hat{\mu}$, $\{\hat{\alpha}_i\}$, and $\left\{\hat{\beta}_{j(i)}\right\}$ are functions of the fitted mean vector $\hat{\boldsymbol{Y}}$, they are independent of SSE and hence of MSE, and this fact can be used to carry out inferences such as hypotheses testing and construction of confidence intervals.

If we wish to test the hypothesis of no teacher effect (ie, test H_0: $\beta_{j(i)} = 0$ for all j and i) against the alternative H_1: not all $\beta_{j(i)}$ are zero, then the full and reduced models and fitted Y values are

$$Y_{ijk} = \mu + \alpha_i + \beta_{j(i)} + \varepsilon_{ijk}, \ \hat{Y}_{ijk} = \bar{Y}_{ij\cdot} = \hat{\mu} + \hat{\alpha}_i + \hat{\beta}_{j(i)} \text{ (full)},$$
$$Y_{ijk} = \mu + \alpha_i + \varepsilon_{ijk}, \ \tilde{Y}_{ijk} = \hat{\mu} + \hat{\alpha}_i \text{ (reduced)}.$$

The residual sums of squares are

$$SSE_F = \sum_i \sum_j \sum_k \left(Y_{ijk} - \bar{Y}_{ij\cdot}\right)^2, \ df(SSE_F) = n - ab,$$

$$SSE_R = \sum_i \sum_j \sum_k \left(Y_{ijk} - \hat{\mu} - \hat{\alpha}_i\right)^2, df(SSE_R) = n - a,$$

$$SSE_R - SSE_F = \sum_i \sum_j n_{ij} \hat{\beta}^2_{j(i)} := SSB(A).$$

The quantity $SSB(A)$ is the sum of squares due to teachers (factor B) nested in school (factor A). The test statistic is

$$F = \frac{SSB(A)/(ab-a)}{MSE_F} = \frac{MSB(A)}{MSE},$$

where $MSB(A) = SSB(A)/(ab-a)$ and $MSE = MSE_F$. Under H_0, $F \sim F_{ab-a,n-ab}$.

In order to test the hypothesis of no school effect (ie, H_0: $\alpha_1 = \cdots = \alpha_a = 0$ vs H_1: at least one α_i is not zero), the full and reduced models along with the fitted values are

$$Y_{ijk} = \mu + \alpha_i + \beta_{j(i)} + \varepsilon_{ijk},\ \hat{Y}_{ijk} = \bar{Y}_{ij\cdot} = \hat{\mu} + \hat{\alpha}_i + \hat{\beta}_{j(i)} \text{ (full)},$$
$$Y_{ijk} = \mu + \beta_{j(i)} + \varepsilon_{ijk},\ \tilde{Y}_{ijk} = \hat{\mu} + \hat{\beta}_{j(i)} \text{ (reduced)}.$$

The residual sum of squares are

$$SSE_F = \sum_i \sum_j \sum_k \left(Y_{ijk} - \bar{Y}_{ij\cdot}\right)^2,\ df(SSE_F) = n - ab,$$
$$SSE_R = \sum_i \sum_j \sum_k \left(Y_{ijk} - \hat{\mu} - \hat{\beta}_{j(i)}\right)^2,$$
$$df(SSE_R) = n - 1 - a(b-1), \text{ and}$$
$$SSE_R - SSE_F = \sum\sum n_{ij}\hat{\alpha}_i^2 := SSA.$$

Thus the F-statistic

$$F = \frac{SSA/(a-1)}{MSE_F} = \frac{MSA}{MSE},$$

where $MSE = MSE_F$, follows an F-distribution with df $(a-1, n-ab)$ under H_0.

For the balanced case, the total sum of squares admits the following decomposition

$$SSTO = SSA + SSB(A) + SSE.$$

11.6.1 Simultaneous Inference

We now address the issue of constructing simultaneous confidence intervals for $\boldsymbol{\theta} = \boldsymbol{L}^T\boldsymbol{\beta}$ and its linear functions, where $\boldsymbol{L}$ is a matrix of order $p \times m$ of rank $m \leq p$. The main ingredients are the basic distributional results on the least squares estimate $\hat{\boldsymbol{\beta}}$ of $\boldsymbol{\beta}$ and $\hat{\boldsymbol{\theta}}$ of $\boldsymbol{\theta}$, as outlined in this section. Recall that the estimate of the error variance σ^2 in the Gauss-Markov model $\boldsymbol{Y} = \boldsymbol{X\beta} + \boldsymbol{\varepsilon}$ is given by $MSE = \|\boldsymbol{Y} - \boldsymbol{X}\hat{\boldsymbol{\beta}}\|^2/(n-p)$. Thus for a real-valued parameter $\theta = \boldsymbol{L}^T\boldsymbol{\beta}$, $\left(\hat{\theta} - \theta\right)/s\left(\hat{\theta}\right) \sim t_{n-p}$ where $\hat{\theta} = \boldsymbol{L}^T\hat{\boldsymbol{\beta}}$ and $s^2\left(\hat{\theta}\right) = MSE\left\{\boldsymbol{L}^T(\boldsymbol{X}^T\boldsymbol{X})^{-1}\boldsymbol{L}\right\}^{-1}$. Hence $\hat{\theta} \pm t_{n-p,\alpha/2}s\left(\hat{\theta}\right)$ is a confidence interval for θ with confidence coefficient $1-\alpha$. Now if $\boldsymbol{\theta}$ is vector valued (ie, $\boldsymbol{L}$ is a $p \times m$ matrix), then there are different methods for constructing simultaneous confidence intervals for $\theta_1, \ldots, \theta_m$, the components of $\boldsymbol{\theta}$, and for the linear combinations of $\boldsymbol{\theta}$. In the literature, there are many methods for simultaneous inference, but the discussion below will be for only three well-known methods: Bonferroni, Scheffé, and Tukey. Even though we only discuss the problem of constructing simultaneous confidence intervals, these can also be employed for simultaneous hypotheses testing.

Bonferroni Method

According to the Bonferroni method, simultaneous confidence intervals with a family confidence coefficient of at least $1-\alpha$ are given by

$$\theta_j\text{: } \hat{\theta}_j \pm Bs\left(\hat{\theta}_j\right), \quad j=1,\ldots,m, \quad \text{with } B=t_{n-p,\alpha/(2m)},$$

where $s^2\left(\hat{\theta}_j\right)$ is the jth diagonal element of the matrix $s^2\left(\hat{\boldsymbol{\theta}}\right)$. Even though this method is valid for any m, its usefulness is questionable when m is not small since the multiplier $t_{n-p,\alpha/(2m)}$ associated with the confidence intervals increases as m increases. Mathematically, the multiplier converges to ∞ as $m \to \infty$. In reality, the Bonferroni method is an inefficient method for constructing simultaneous confidence intervals when m is larger than 3 or 4. The Scheffé procedure is more appropriate when m is large. Before describing the Scheffé method, let us briefly see why the Bonferroni method leads to a simultaneous confidence of at least $1-\alpha$. Let A_j denote the random event $\left\{\theta_j\text{: } \theta_j \in \left[\hat{\theta}_j - Bs\left(\hat{\theta}_j\right), \hat{\theta}_j + Bs\left(\hat{\theta}_j\right)\right]\right\}$, $j=1,\ldots,m$. Now $P\left(A_j^c\right)=\alpha/m$ and

$$P\left(\bigcap_{j=1}^{m} A_j\right) = 1 - P\left(\bigcup_{j=1}^{m} A_j^c\right) \geq 1 - \sum_{j=1}^{m} P\left(A_j^c\right) = 1-\alpha.$$

It shows that the probability that θ_j is inside $\hat{\theta}_j \pm Bs\left(\hat{\theta}_j\right)$ for all $j=1,\ldots,m$, is at least $1-\alpha$, which justifies the Bonferroni approach. This leads to the following lemma.

Lemma 11.6.1. *The confidence intervals* $\hat{\theta}_j \pm Bs\left(\hat{\theta}_j\right)$ *for* θ_j, $j=1,\ldots,m$, *have a simultaneous confidence of at least* $1-\alpha$, *where* $B=t_{n-p,\alpha/(2m)}$.

Scheffé Method

This method obtains simultaneous confidence intervals for all linear combinations $\boldsymbol{a}^T\boldsymbol{\theta}$, $\boldsymbol{a} \in \mathbb{R}^m$. Note that the least squares estimate of $\boldsymbol{a}^T\boldsymbol{\theta}$ is $\boldsymbol{a}^T\hat{\boldsymbol{\theta}} = \boldsymbol{a}^T\boldsymbol{L}^T\hat{\boldsymbol{\beta}}$, where $\hat{\boldsymbol{\beta}}$ is the least squares estimate of $\boldsymbol{\beta}$. It is also clear that $\mathrm{E}\left[\boldsymbol{a}^T\hat{\boldsymbol{\theta}}\right] = \boldsymbol{a}^T\boldsymbol{\theta}$, $\mathrm{Var}\left[\boldsymbol{a}^T\hat{\boldsymbol{\theta}}\right] = \boldsymbol{a}^T\mathrm{Cov}\left[\hat{\boldsymbol{\theta}}\right]\boldsymbol{a}$, and an estimate of $\mathrm{Var}\left[\boldsymbol{a}^T\hat{\boldsymbol{\theta}}\right]$ is given by $s^2\left(\boldsymbol{a}^T\hat{\boldsymbol{\theta}}\right) = \boldsymbol{a}^T s^2\left(\hat{\boldsymbol{\theta}}\right)\boldsymbol{a}$.

This method states that simultaneous confidence intervals for all linear combinations of $\boldsymbol{\theta}$ with an overall confidence level of $1-\alpha$ are given by

$$\boldsymbol{a}^T\boldsymbol{\theta}\text{: } \boldsymbol{a}^T\hat{\boldsymbol{\theta}} \pm Ss\left(\boldsymbol{a}^T\hat{\boldsymbol{\theta}}\right), \quad \boldsymbol{a} \in \mathbb{R}^m, \text{ where } S=\sqrt{mF_{m,n-p,\alpha}}.$$

The Scheffé method is closely related to a procedure for obtaining a confidence region for $\boldsymbol{\theta}$, known as the confidence ellipsoid, which is given by

$$A = \left\{\boldsymbol{\theta} \in \mathbb{R}^m: \left(\boldsymbol{\theta} - \hat{\boldsymbol{\theta}}\right)^T \left[\boldsymbol{s}^2\left(\hat{\boldsymbol{\theta}}\right)\right]^{-1} \left(\boldsymbol{\theta} - \hat{\boldsymbol{\theta}}\right)/m \le F_{m,n-p,\alpha}\right\}.$$

Note that $P_\theta[A] = 1 - \alpha$ and thus the confidence ellipsoid provides a confidence region for $\boldsymbol{\theta}$ with confidence $1-\alpha$. However, this is not a useful method in practice as it is not possible to visualize the region when $m > 3$.

There is a connection between Scheffé's simultaneous confidence intervals and the confidence ellipsoid, and it is through the following equality

$$\sup_{\boldsymbol{a}\in\mathbb{R}^m} \left(\boldsymbol{a}^T\hat{\boldsymbol{\theta}} - \boldsymbol{a}^T\boldsymbol{\theta}\right)^2 / s^2\left(\boldsymbol{a}^T\hat{\boldsymbol{\theta}}\right) = \left(\boldsymbol{\theta} - \hat{\boldsymbol{\theta}}\right)^T \left[\boldsymbol{s}^2\left(\hat{\boldsymbol{\theta}}\right)\right]^{-1} \left(\boldsymbol{\theta} - \hat{\boldsymbol{\theta}}\right),$$

which basically follows from the Cauchy-Schwarz inequality (Section B.1) since

$$\left(\boldsymbol{a}^T\hat{\boldsymbol{\theta}} - \boldsymbol{a}^T\boldsymbol{\theta}\right)^2 / s^2\left(\boldsymbol{a}^T\hat{\boldsymbol{\theta}}\right) = \left[\boldsymbol{a}^T\left(\hat{\boldsymbol{\theta}} - \boldsymbol{\theta}\right)\right]^2 / \left[\boldsymbol{a}^T\boldsymbol{s}^2\left(\hat{\boldsymbol{\theta}}\right)\boldsymbol{a}\right].$$

Thus we have

$$\begin{aligned}
1 - \alpha &= P_\theta\left[\left(\hat{\boldsymbol{\theta}} - \boldsymbol{\theta}\right)^T \left[\boldsymbol{s}^2\left(\hat{\boldsymbol{\theta}}\right)\right]^{-1} \left(\hat{\boldsymbol{\theta}} - \boldsymbol{\theta}\right)/m \le F_{m,n-p,\alpha}\right] \\
&= P_\theta\left[\sup_{\boldsymbol{a}\in\mathbb{R}^m} \left(\boldsymbol{a}^T\hat{\boldsymbol{\theta}} - \boldsymbol{a}^T\boldsymbol{\theta}\right)^2 / \left[\boldsymbol{a}^T\boldsymbol{s}^2\left(\hat{\boldsymbol{\theta}}\right)\boldsymbol{a}\right] \le S^2\right] \\
&= P_\theta\left[\left(\boldsymbol{a}^T\hat{\boldsymbol{\theta}} - \boldsymbol{a}^T\boldsymbol{\theta}\right)^2 / \left[\boldsymbol{a}^T\boldsymbol{s}^2\left(\hat{\boldsymbol{\theta}}\right)\boldsymbol{a}\right] \le S^2 \text{ for all } \boldsymbol{a} \in \mathbb{R}^m\right] \\
&= P_\theta\left[\boldsymbol{a}^T\hat{\boldsymbol{\theta}} - Ss\left(\boldsymbol{a}^T\hat{\boldsymbol{\theta}}\right) \le \boldsymbol{a}^T\boldsymbol{\theta} \le \boldsymbol{a}^T\hat{\boldsymbol{\theta}} + Ss\left(\boldsymbol{a}^T\hat{\boldsymbol{\theta}}\right) \text{ for all } \boldsymbol{a} \in \mathbb{R}^m\right].
\end{aligned}$$

This justifies the validity of the Scheffé method and we have the following result.

Theorem 11.6.2. *Let $\boldsymbol{\theta}$ be a linear function of $\boldsymbol{\beta}$ of the form $\boldsymbol{L}^T\boldsymbol{\beta}$ where $\boldsymbol{L}$ is a $p \times m$ matrix of rank $m \le p$. Then simultaneous confidence intervals for all linear combinations of $\boldsymbol{\theta}$ with an overall confidence coefficient of $1 - \alpha$ are*

$$\boldsymbol{a}^T\boldsymbol{\theta}: \boldsymbol{a}^T\hat{\boldsymbol{\theta}} \pm Ss\left(\boldsymbol{a}^T\hat{\boldsymbol{\theta}}\right), \quad \boldsymbol{a} \in \mathbb{R}^m, \text{ where } S = \sqrt{mF_{m,n-p,\alpha}}.$$

Tukey Method

The basic argument behind Tukey's method is given below and it will be clear from an example (given below) how the method can be used. Let $W_i \sim N(\theta_i, \tau^2)$, $i = 1, \ldots, t$, be independent random variables and let S^2 be an unbiased estimate of τ^2 such that

$$\text{(i) } S^2 \text{ is independent of } W_1, \ldots, W_t, \text{ (ii) } \nu S^2/\tau^2 \sim \chi^2_\nu.$$

Consider the following rv, called the studentized range variable,

$$Q = \left\{ \max_i (W_i - \theta_i) - \min_i (W_i - \theta_i) \right\} \Big/ S.$$

The distribution of Q is known as the studentized range distribution (denoted by $Q(t, \nu)$) and it is available in statistical packages. Many statistics textbooks also have the table of this distribution. Tukey simultaneous confidence intervals with family confidence coefficient $1 - \alpha$ for all pairwise differences $\theta_i - \theta_j$ is

$$W_i - W_j \pm Q_{t,\nu,\alpha} S,$$

where $Q_{t,\nu,\alpha}$ is the $(1 - \alpha)$-quantile of the studentized range distribution with degrees of freedom (t, ν).

Application of Tukey's Method to One-Factor Balanced ANOVA Models

Let $Y_{ij} = \mu + \alpha_i + \varepsilon_{ij}$, $j = 1, \ldots, n_0$, $i = 1, \ldots, k$, where $\{\varepsilon_{ij}\}$ are iid $N(0, \sigma^2)$ and let $\mu_i = \mu + \alpha_i$, $i = 1, \ldots, k$. Suppose that we want to construct simultaneous confidence intervals for $\mu_i - \mu_{i'} = \alpha_i - \alpha_{i'}$, $i \neq i'$, with a family confidence coefficient of $1 - \alpha$. Take $W_i = \sqrt{n_0}\bar{Y}_{i\cdot}$, $\theta_i = \sqrt{n_0}\mu_i$, $S^2 = MSE$, $\tau^2 = \sigma^2$, $t = k$, and $\nu = n - k$, where $n = kn_0$ is the total number of observations. Then simultaneous confidence intervals for $\theta_i - \theta_{i'} = \sqrt{n_0}(\alpha_i - \alpha_{i'})$, $i \neq i'$, with family confidence coefficient $1 - \alpha$ are given by

$$\sqrt{n_0}(\bar{Y}_{i\cdot} - \bar{Y}_{i'\cdot}) \pm Q_{k,n-k,\alpha} MSE.$$

Consequently, simultaneous confidence intervals for $\theta_i - \theta_{i'}$, $i \neq i'$, are

$$(\bar{Y}_{i\cdot} - \bar{Y}_{i'\cdot}) \pm Ts(\bar{Y}_{i\cdot} - \bar{Y}_{i'\cdot}),$$

where $T = Q_{k,n-k,\alpha}/\sqrt{2}$.

11.6.2 Prediction Intervals

If we rewrite the Gauss-Markov model as

$$Y_i = \boldsymbol{\beta}^T \boldsymbol{x}_i + \varepsilon_i, \quad i = 1, \ldots, n,$$

where $\boldsymbol{x}_i^T$ is the ith row of the design matrix, it is sometimes of interest to predict Y_0 at a vector $\boldsymbol{x}_0$ of values in $\mathbb{R}^p$ where $Y_0 = \boldsymbol{\beta}^T \boldsymbol{x}_0 + \varepsilon_0$ and $\varepsilon_0 \sim N(0, \sigma^2)$ is independent of the observation vector $\boldsymbol{Y}$. If $\boldsymbol{\beta}$ were known, the best predictor of Y_0 would be $\boldsymbol{\beta}^T \boldsymbol{x}_0$. Usually, $\boldsymbol{\beta}$ is unknown and the predicted value of Y_0 is $\hat{Y}_0 = \hat{\boldsymbol{\beta}}^T \boldsymbol{x}_0$, where $\hat{\boldsymbol{\beta}}$ is the least squares estimate of $\boldsymbol{\beta}$. Noting that ε_0 is independent of $\hat{\boldsymbol{\beta}}$ and hence of $\hat{Y}_0$, we have

$$\mathrm{E}\left[\hat{Y}_0 - Y_0\right] = \mathrm{E}\left[\hat{\boldsymbol{\beta}}^T \boldsymbol{x}_0 - \boldsymbol{\beta}^T \boldsymbol{x}_0 - \varepsilon_0\right] = 0, \text{ and}$$

$$\begin{aligned}\mathrm{Var}\left[\hat{Y}_0 - Y_0\right] &= \mathrm{Var}\left[\hat{\boldsymbol{\beta}}^T \boldsymbol{x}_0 - \boldsymbol{\beta}^T \boldsymbol{x}_0 - \varepsilon_0\right] = \mathrm{Var}\left[\hat{\boldsymbol{\beta}}^T \boldsymbol{x}_0 - \boldsymbol{\beta}^T \boldsymbol{x}_0\right] + \mathrm{Var}[\varepsilon_0] \\ &= \boldsymbol{x}_0^T \mathrm{Cov}\left[\hat{\boldsymbol{\beta}}\right] \boldsymbol{x}_0 + \sigma^2 = \sigma^2 \boldsymbol{x}_0^T \left(\boldsymbol{X}^T \boldsymbol{X}\right)^{-1} \boldsymbol{x}_0 + \sigma^2.\end{aligned}$$

Thus an estimate of $\text{Var}\left[\hat{Y}_0 - Y_0\right]$ is

$$s^2\left(\hat{Y}_0 - Y_0\right) = MSE\, \boldsymbol{x}_0^T \left(\boldsymbol{X}^T\boldsymbol{X}\right)^{-1} \boldsymbol{x}_0 + MSE := s^2 \text{ (pred)}.$$

Since $\hat{Y}_0 - Y_0$ is normally distributed with mean 0 and variance given above, and is independent of $s^2\left(\hat{Y}_0 - Y_0\right)$, the random variable $\left(\hat{Y}_0 - Y_0\right)/s(\text{pred}) \sim t_{n-p}$. Thus a prediction interval for Y_0 with confidence coefficient $1-\alpha$ is given by $\hat{Y}_0 \pm t_{n-p,\alpha/2}s(\text{pred})$.

Simultaneous prediction intervals for m different Y values at k different $\boldsymbol{x}$ vectors can be constructed using the Bonferroni method by taking the multiplier associated with the prediction intervals to be equal to $B = t_{n-p,\alpha/(2m)}$.

11.7 Analysis of Covariance

As we have discussed in Remark 11.2.2, the ANCOVA model can be written as

$$\boldsymbol{Y} = \boldsymbol{X\beta} + \boldsymbol{Z\gamma} + \boldsymbol{\varepsilon}, \tag{4}$$

where $\boldsymbol{Y}$ is $n \times 1$ vector of observations, $\boldsymbol{X}$ is a known $n \times p$ matrix of rank p, $\boldsymbol{Z}$ is a known $n \times q$ matrix of rank q, and $\boldsymbol{\varepsilon}$ is $N_n(\boldsymbol{0}, \sigma^2 I)$. We further assume that $\mathcal{M}(\boldsymbol{X}) \cap \mathcal{M}(\boldsymbol{Z}) = \{\boldsymbol{0}\}$, and hence the rank of the augmented matrix $[\boldsymbol{X}\,\boldsymbol{Z}]$ is $p+q$. Here $\boldsymbol{\beta}$ and $\boldsymbol{\gamma}$ are vectors of unknown parameters to be estimated.

Here we will be concerned with estimation of $\boldsymbol{\beta}$ and $\boldsymbol{\gamma}$, and inference on them. For instance, in Example 11.2.7 we may be interested in testing that diet has no effect on the growth rate (ie, H_0: $\alpha_1 = \cdots = \alpha_k = 0$ vs H_1: not all α_i's equal 0). We may be interested in testing that age has no effect on growth (ie, H_0: $\gamma_2 = 0$ vs H_1: $\gamma_2 \neq 0$). Or we may be interested in finding out neither initial weight nor age has any effect on the growth rate, H_0: $\gamma_1 = \gamma_2 = 0$ vs H_1: not both of γ_1 and γ_2 are zero.

Here we discuss the following issues:

(i) estimation of $\boldsymbol{\beta}$ and $\boldsymbol{\gamma}$, and their linear functions $\boldsymbol{\theta} = \boldsymbol{L}^T\boldsymbol{\beta}$ and $\eta = \boldsymbol{M}^T\boldsymbol{\gamma}$,
(ii) test for H_0: $\boldsymbol{\theta} = \boldsymbol{\theta}_0$ vs H_1: $\boldsymbol{\theta} \neq \boldsymbol{\theta}_0$, $\boldsymbol{\theta}_0$ known, and
(iii) test for H_0: $\eta = \eta_0$ vs H_1: $\eta \neq \eta_0$, η_0 known.

11.7.1 Estimation of $\boldsymbol{\beta}$ and $\boldsymbol{\gamma}$

If $\boldsymbol{W} = [\boldsymbol{X}\,\boldsymbol{Z}]$ and $\boldsymbol{\theta}$ is the $(p+q)$-dim vector obtained by stacking $\boldsymbol{\beta}$ and $\boldsymbol{\gamma}$ vertically, then the normal equations would be $\boldsymbol{W}^T\boldsymbol{W\theta} = \boldsymbol{W}^T\boldsymbol{Y}$. The matrix $\boldsymbol{W}^T\boldsymbol{W} = \begin{pmatrix} \boldsymbol{X}^T\boldsymbol{X} & \boldsymbol{X}^T\boldsymbol{Z} \\ \boldsymbol{Z}^T\boldsymbol{X} & \boldsymbol{Z}^T\boldsymbol{Z} \end{pmatrix}$ is not necessarily block diagonal. It would help in obtaining a simple estimate of $\boldsymbol{\gamma}$ if we could rewrite the model in order to get a block diagonal matrix on the left-hand side of the normal equations. In order to achieve this, we can argue as follows. Let $\boldsymbol{Q_X}$

be the orthogonal projection on the column space of $\boldsymbol{X}$ (ie, $\boldsymbol{Q_X} = \boldsymbol{X}(\boldsymbol{X}^T\boldsymbol{X})^{-1}\boldsymbol{X}^T$ and $\tilde{\boldsymbol{Z}} = (\boldsymbol{I} - \boldsymbol{Q_X})\boldsymbol{Z})$. Then we can rewrite the ANCOVA model as

$$\boldsymbol{Y} = \boldsymbol{X}\boldsymbol{\delta} + \tilde{\boldsymbol{Z}}\boldsymbol{\gamma} + \boldsymbol{\varepsilon},$$

where $\boldsymbol{\delta} = \boldsymbol{\beta} + (\boldsymbol{X}^T\boldsymbol{X})^{-1}\boldsymbol{X}^T\boldsymbol{Z}\boldsymbol{\gamma}$. Since $\boldsymbol{X}^T\tilde{\boldsymbol{Z}} = \boldsymbol{0}$, the least squares method produces the normal equations

$$\begin{pmatrix} \boldsymbol{X}^T\boldsymbol{X} & \boldsymbol{0} \\ \boldsymbol{0} & \tilde{\boldsymbol{Z}}^T\tilde{\boldsymbol{Z}} \end{pmatrix}\begin{pmatrix} \boldsymbol{\delta} \\ \boldsymbol{\gamma} \end{pmatrix} = \begin{pmatrix} \boldsymbol{X}^T\boldsymbol{Y} \\ \tilde{\boldsymbol{Z}}^T\boldsymbol{Y} \end{pmatrix}.$$

From the second set of equations we get

$$\hat{\boldsymbol{\gamma}} = \left(\tilde{\boldsymbol{Z}}^T\tilde{\boldsymbol{Z}}\right)^{-1}\tilde{\boldsymbol{Z}}^T\boldsymbol{Y}.$$

From the first set of equations we get

$$\boldsymbol{X}^T\boldsymbol{Y} = \boldsymbol{X}^T\boldsymbol{X}\hat{\boldsymbol{\delta}} = \boldsymbol{X}^T\boldsymbol{X}\left(\hat{\boldsymbol{\beta}} + \left(\boldsymbol{X}^T\boldsymbol{X}\right)^{-1}\boldsymbol{X}^T\boldsymbol{Z}\hat{\boldsymbol{\gamma}}\right) = \boldsymbol{X}^T\boldsymbol{X}\hat{\boldsymbol{\beta}} + \boldsymbol{X}^T\boldsymbol{Z}\hat{\boldsymbol{\gamma}}, \text{ and}$$
$$\boldsymbol{X}^T\boldsymbol{X}\hat{\boldsymbol{\beta}} = \boldsymbol{X}^T(\boldsymbol{Y} - \boldsymbol{Z}\hat{\boldsymbol{\gamma}}).$$

The discussion above leads to the following simple result.

Lemma 11.7.1. *The least squares estimates of $\boldsymbol{\beta}$ and $\boldsymbol{\gamma}$ for the ANCOVA model are*

$$\text{(a) } \hat{\boldsymbol{\gamma}} = \left(\tilde{\boldsymbol{Z}}'\tilde{\boldsymbol{Z}}\right)^{-1}\tilde{\boldsymbol{Z}}^T\boldsymbol{Y}, \text{ (b) } \hat{\boldsymbol{\beta}} = \left(\boldsymbol{X}^T\boldsymbol{X}\right)^{-1}\boldsymbol{X}^T(\boldsymbol{Y} - \boldsymbol{Z}\hat{\boldsymbol{\gamma}}).$$

The following lemma states a few basic results on the least squares estimates and its proof is left to the reader.

Lemma 11.7.2. *If $\hat{\boldsymbol{\beta}}$ and $\hat{\boldsymbol{\gamma}}$ are least squares estimates of $\boldsymbol{\beta}$ and $\boldsymbol{\gamma}$, respectively, then we have*

(a) $E[\hat{\boldsymbol{\gamma}}] = \boldsymbol{\gamma}$, $E\left[\hat{\boldsymbol{\beta}}\right] = \boldsymbol{\beta}$,

(b) $Cov[\hat{\boldsymbol{\gamma}}] = \sigma^2\left(\tilde{\boldsymbol{Z}}^T\tilde{\boldsymbol{Z}}\right)^{-1}$,

(c) $Cov\left[\hat{\boldsymbol{\beta}}\right] = \sigma^2\left[(\boldsymbol{X}^T\boldsymbol{X})^{-1} + (\boldsymbol{X}^T\boldsymbol{X})^{-1}\boldsymbol{X}^T\boldsymbol{D}\boldsymbol{X}(\boldsymbol{X}^T\boldsymbol{X})^{-1}\right]$, *where* $\boldsymbol{D} = \boldsymbol{Z}\left(\tilde{\boldsymbol{Z}}^T\tilde{\boldsymbol{Z}}\right)^{-1}\boldsymbol{Z}^T$,

(d) $Cov\left[\boldsymbol{X}\hat{\boldsymbol{\beta}}\right] = \sigma^2(\boldsymbol{Q_X} + \boldsymbol{Q_X}\boldsymbol{D}\boldsymbol{Q_X})$,

(e) $Cov\left[\hat{\boldsymbol{\gamma}}, \boldsymbol{X}\hat{\boldsymbol{\beta}}\right] = -\sigma^2\left(\tilde{\boldsymbol{Z}}^T\tilde{\boldsymbol{Z}}\right)^{-1}\boldsymbol{Z}^T\boldsymbol{Q_X}$.

Remark 11.7.1. Here is an intuitive way to view the matrix $\tilde{\boldsymbol{Z}}$. Suppose that the columns of $\boldsymbol{Z}$ are $\boldsymbol{Z}_1, \ldots, \boldsymbol{Z}_q$. Then the ith column vector for the matrix $\tilde{\boldsymbol{Z}}$ is $\tilde{\boldsymbol{Z}}_i = (\boldsymbol{I} - \boldsymbol{Q_X})\boldsymbol{Z}_i$. Now for any column of $\boldsymbol{Z}$, say $\boldsymbol{Z}_1$, we can view $\boldsymbol{Q_X}\boldsymbol{Z}_1$ as the vector of fitted values when fitting the model $\boldsymbol{Z}_1 = \boldsymbol{X}\boldsymbol{\beta} + error$, and hence the vector of residuals is $(\boldsymbol{I} - \boldsymbol{Q_X})\boldsymbol{Z}_1$. Similar interpretation holds for all the columns of $\tilde{\boldsymbol{Z}}$.

11.7.2 Residual Sum of Squares

The vectors of fitted values and the residuals for the ANCOVA model are

$$\hat{Y} = X\hat{\beta} + Z\hat{\gamma} = Q_X(Y - Z\hat{\gamma}) + Z\hat{\gamma} = Q_X Y + \tilde{Z}\hat{\gamma},$$
$$\hat{\varepsilon} = Y - \hat{Y} = (I - Q_X)Y - \tilde{Z}\hat{\gamma}.$$

Since $\hat{\varepsilon}$ is orthogonal to any linear combination of X and Z, and hence to $\tilde{Z}\hat{\gamma}$, we have

$$\|(I - Q_X)Y\|^2 = \|\hat{\varepsilon} + \tilde{Z}\hat{\gamma}\|^2 = \|\hat{\varepsilon}\|^2 + \|\tilde{Z}\hat{\gamma}\|^2, \text{ and hence}$$
$$SSE = \|\hat{\varepsilon}\|^2 = \|(I - Q_X)Y\|^2 - \|\tilde{Z}\hat{\gamma}\|^2.$$

Since the rank of the matrix $[X\ Z]$ is $p + q$ and $df(SSE) = n - p - q$, an unbiased estimate of σ^2 is

$$MSE = SSE/(n - p - q).$$

11.7.3 Inference for γ and β

We first discuss inference on γ and its linear functions.

Inference for γ

If we want to estimate $\eta = M^T\gamma$, where M is a $q \times s$ matrix of rank $s \le q$, then the least squares estimate of η is given by $\hat{\eta} = M^T\hat{\gamma}$, where $\hat{\gamma}$ is the least squares estimate of γ. If $\varepsilon \sim N_n(\mathbf{0}, \sigma^2 I)$, then $\hat{\eta} \sim N_s(\eta, \sigma^2 M^T \text{Cov}[\hat{\gamma}]M)$, where $\text{Cov}[\hat{\gamma}]$ is as in Lemma 11.7.2 and it can be estimated by $s^2(\hat{\eta}) = MSE\left[M^T\left(\tilde{Z}^T\tilde{Z}\right)^{-1}M\right]$. We can now easily handle all the inferential issues regarding η such as construction of confidence intervals and tests of hypotheses. In particular, suppose we want to test H_0: $\eta = \eta_0$ vs H_1: $\eta \ne \eta_0$, where η_0 is known. A test statistic is

$$F = (\hat{\eta} - \eta_0)^T\left[s^2(\hat{\eta})\right]^{-1}(\hat{\eta} - \eta_0)/s.$$

Degrees of freedom associated with this F-test are $(s, n - p - q)$.

Inference for β

If we want to estimate $\theta = L^T\beta$, where L is a $p \times r$ matrix of rank $r \le p$, then its least squares estimate is given by $\hat{\theta} = L^T\hat{\beta}$ and the distribution of $\hat{\theta}$ is an r-dim normal with mean θ and covariance matrix $L^T\text{Cov}\left[\hat{\beta}\right]L$, where the expression for $\text{Cov}\left[\hat{\beta}\right]$ is given in Lemma 11.7.2. An estimate of $\text{Cov}\left[\hat{\theta}\right]$ is

$$s^2\left(\hat{\theta}\right) = MSE\, L^T\left[\left(X^TX\right)^{-1} + \left(X^TX\right)^{-1}X^TDX\left(X^TX\right)^{-1}\right].$$

In order to carry out a test H_0: $\boldsymbol{\theta} = \boldsymbol{\theta}_0$ vs H_1: $\boldsymbol{\theta} \neq \boldsymbol{\theta}_0$, $\boldsymbol{\theta}_0$ known, we can use the following test statistic

$$F = \left(\hat{\theta} - \theta_0\right)^T \left[s^2\left(\hat{\theta}\right)\right]^{-1} \left(\hat{\theta} - \theta_0\right)/r.$$

The degrees of freedom for this F-test are $(r, n - p - q)$.

We should point that we can also carry out tests for $\boldsymbol{\theta}$ or $\boldsymbol{\eta}$ using the alternate expression for the F-statistic which involves obtaining SSE_R and SSE_F, the residual sums of squares for the reduced and full models, as described in Section 11.6.

Example 11.7.1. Consider the following model with one factor and one covariate

$$Y_{ij} = \mu + \alpha_i + \gamma Z_{ij} + \varepsilon_{ij}, \quad j = 1, \ldots, n_i, \quad i = 1, \ldots, k.$$

In Example 11.4.6, estimates of μ, α_i, and γ are given. Here $\boldsymbol{Z}$ has only one column, $\boldsymbol{Q_X Z}$ is the vector of fitted values when fitting the model $Z_{ij} = \mu + \alpha_i + error$, and $\tilde{\boldsymbol{Z}} = (\boldsymbol{I} - \boldsymbol{Q_X})\boldsymbol{Z}$ is the vector of residuals. Hence $\tilde{Z}_{ij} = Z_{ij} - \bar{Z}_{i\cdot}$, where $\bar{Z}_{i\cdot} = \sum_{j=1}^{n_i} Z_{ij}/n_i$. Similarly, let $\tilde{Y}_{ij} = Y_{ij} - \bar{Y}_{i\cdot}$, where $\bar{Y}_{i\cdot} = \sum_{j=1}^{n_i} Y_{ij}/n_i$. Consequently,

$$\begin{aligned}
\hat{\gamma} &= \left(\tilde{\boldsymbol{Z}}^T\tilde{\boldsymbol{Z}}\right)^{-1}\tilde{\boldsymbol{Z}}^T\boldsymbol{Y} = \left(\sum_i\sum_j \tilde{Z}_{ij}Y_{ij}\right) \Big/ \left(\sum_i\sum_j \tilde{Z}_{ij}^2\right) \\
&= \left(\sum_i\sum_j \tilde{Z}_{ij}\tilde{Y}_{ij}\right) \Big/ \left(\sum_i\sum_j \tilde{Z}_{ij}^2\right).
\end{aligned}$$

The residual sum of squares is

$$\begin{aligned}
SSE &= \tilde{\boldsymbol{Y}}^T\tilde{\boldsymbol{Y}} - \left(\tilde{\boldsymbol{Y}}^T\tilde{\boldsymbol{Z}}\right)\left(\tilde{\boldsymbol{Z}}^T\tilde{\boldsymbol{Z}}\right)^{-1}\left(\tilde{\boldsymbol{Z}}^T\tilde{\boldsymbol{Y}}\right) \\
&= \sum_i\sum_j \tilde{Y}_{ij}^2 - \left(\sum_i\sum_j \tilde{Z}_{ij}\tilde{Y}_{ij}\right)^2 \Big/ \left(\sum_i\sum_j \tilde{Z}_{ij}^2\right).
\end{aligned}$$

Note that $df(SSE) = n - k - 1$, where $n = \sum n_i$ is the total number of observations. So an estimate of σ^2 is $MSE = SSE/(n - k - 1)$.

If we want to test H_0: $\alpha_1 = \cdots = \alpha_k = 0$ vs H_1: not all α_i's are equal to zero, then we need to obtain the residual sum of squares SSE_R of the reduced model.

The reduced model (under H_0) is $Y_{ij} = \mu + \gamma Z_{ij} + \varepsilon_{ij}$ and

$$\begin{aligned}
SSE_R &= \inf_{\mu,\gamma} \sum_i\sum_j \left[Y_{ij} - \mu - \gamma Z_{ij}\right]^2 \\
&= \sum_i\sum_j \left[\left(Y_{ij} - \bar{Y}_{\cdot\cdot}\right) - \gamma^*\left(Z_{ij} - \bar{Z}_{\cdot\cdot}\right)\right]^2, \text{ where}
\end{aligned}$$

$$\gamma^* = \sum_i \sum_j \left(Z_{ij} - \bar{Z}_{..}\right) Y_{ij} \Big/ \sum_i \sum_j \left(Z_{ij} - \bar{Z}_{..}\right)^2,$$
$$df(SSE_R) = n - 2.$$

So the test statistic is

$$F = \frac{(SSE_R - SSE_F)/(k-1)}{MSE},$$

where $MSE = MSE_F$ and the degrees of freedom for this test are $(k-1, n-k-1)$.

If we wish to construct a confidence interval for $\theta = \alpha_i - \alpha_{i'}$, $i \neq i'$, then its estimate is $\hat{\theta} = \bar{Y}_{i.} - \bar{Y}_{i'.} - \hat{\gamma}(\bar{Z}_{i.} - \bar{Z}_{i'.})$. It is left to the reader to verify that

$$\mathrm{Cov}\left[\bar{Y}_{i.}, \hat{\gamma}\right] = 0 \quad \text{for any } i = 1, \ldots, k.$$

Fairly simple calculations will show that

$$\mathrm{E}\left[\hat{\theta}\right] = \theta \text{ and}$$
$$\mathrm{Var}\left[\hat{\theta}\right] = \sigma^2\left[1/n_i + 1/n_{i'} + (\bar{Z}_{i.} - \bar{Z}_{i'.})^2/S_{\tilde{Z}}\right], \text{ where}$$
$$S_{\tilde{Z}} = \sum_i \sum_j \tilde{Z}_{ij}^2.$$

Thus an estimate of $\mathrm{Var}\left[\hat{\theta}\right]$ is given by

$$s^2\left(\hat{\theta}\right) = MSE\left[1/n_i + 1/n_{i'} + (\bar{Z}_{i.} - \bar{Z}_{i'.})^2/S_{\tilde{Z}}\right].$$

Since $\left(\hat{\theta} - \theta\right)/s\left(\hat{\theta}\right) \sim t_{n-k-1}$, we can construct a confidence interval for θ. As a matter of fact, we can carry out many pairwise comparisons $\alpha_i - \alpha_{i'}$, $i \neq i'$, using the Bonferroni or Scheffé methods.

11.7.4 Application of ANCOVA to Missing Data

Suppose that we have the usual linear model $\boldsymbol{Y} = \boldsymbol{X}\boldsymbol{\beta} + \boldsymbol{\varepsilon}$, where $\boldsymbol{Y}$ is $n \times 1$ and $\boldsymbol{X}$ is $n \times p$. However, the last r observations on the response (ie, Y values) are missing. Let $\boldsymbol{Y}_*$ and $\boldsymbol{Y}_r$ denote the vectors of $n-r$ available observations and the vectors of r missing observations, respectively. Similarly, let $\boldsymbol{X}_*$ and $\boldsymbol{X}_r$ denote the submatrices of $\boldsymbol{X}$ corresponding to the available and missing cases. Note that $\boldsymbol{X}_*$ is $(n-r) \times p$ and $\boldsymbol{X}_r$ is $r \times p$. The basic idea here is to formulate an ANCOVA by introducing one unknown parameter for every missing observation. Consider the ANCOVA model

$$\begin{pmatrix} Y_* \\ Y_r \end{pmatrix} = \begin{pmatrix} X_* \\ X_r \end{pmatrix}\beta + \begin{pmatrix} \mathbf{0} \\ I_r \end{pmatrix}\gamma + \varepsilon,$$

where I_r is the $r \times r$ identity matrix, and the $r \times 1$ vector of unknown parameters γ is introduced here for the missing cases. If we carry out a straightforward least squares with the design matrix $\begin{pmatrix} X_* & \mathbf{0} \\ X_r & I_r \end{pmatrix}$ in order to estimate the unknown parameters $\begin{pmatrix} \beta \\ \gamma \end{pmatrix}$, then the normal equations are

$$\begin{pmatrix} X_*^T X_* + X_r^T X_r & X_r^T \\ X_r & I_r \end{pmatrix}\begin{pmatrix} \hat{\beta} \\ \hat{\gamma} \end{pmatrix} = \begin{pmatrix} X_*^T Y_* + X_r^T Y_r \\ Y_r \end{pmatrix}.$$

The two sets of equations (obtainable from the normal equations above) are

$$X_*^T X_* \hat{\beta} + X_r^T X_r \hat{\beta} + X_r^T \hat{\gamma} = X_*^T Y_* + X_r^T Y_r,$$
$$X_r \hat{\beta} + \hat{\gamma} = Y_r.$$

Premultiply the second equation above by X_r^T and then subtract it from the first equation to obtain

$$X_*^T X_* \hat{\beta} = X_*^T Y_*, \text{ and hence}$$
$$\hat{\gamma} = Y_r - X_r \hat{\beta}.$$

We now have explicit expressions for $\hat{\beta}$ and $\hat{\gamma}$. The vectors of fitted values and residuals are

$$\hat{Y} = \begin{pmatrix} X_* \\ X_r \end{pmatrix}\hat{\beta} + \begin{pmatrix} \mathbf{0} \\ I_r \end{pmatrix}\hat{\gamma} = \begin{pmatrix} X_* \hat{\beta} \\ X_r \hat{\beta} + \hat{\gamma} \end{pmatrix} = \begin{pmatrix} X_* \hat{\beta} \\ Y_r \end{pmatrix},$$
$$Y - \hat{Y} = \begin{pmatrix} Y_* \\ Y_r \end{pmatrix} - \begin{pmatrix} X_* \hat{\beta} \\ Y_r \end{pmatrix} = \begin{pmatrix} Y_* - X_* \hat{\beta} \\ \mathbf{0} \end{pmatrix}.$$

Consequently, the residual sum of squares is given by

$$SSE = \|Y_* - X_* \hat{\beta}\|^2.$$

Note that this residual sum of squares turns out to be identical to that for the model $(Y_* = X_* \beta + \varepsilon_*)$ with only available observations. The degrees of freedom for the *SSE* is $n-p-r$. Thus, the degrees of freedom of the SSE is the same as the one that can be obtained from an analysis of the available observations assuming that $rank(X) = rank(X_*)$.

11.8 Model Selection

In the Gauss-Markov setup one is often concerned with selecting a model from an appropriate class of candidate models. There are a number of methods for model selection and properties of these methods have been investigated by many authors, but we focus on a few of them instead of describing all. Consider the general framework $Y = \mu + \varepsilon$, where Y

is the $n \times 1$ response vector, $\boldsymbol{\mu}$ is the mean vector, and $\boldsymbol{\varepsilon}$ is the vector of iid errors with mean 0 and variance σ^2. We assume that there is a class of candidate models $\{\boldsymbol{X}_k\boldsymbol{\beta}_k: k = 1, \ldots, K\}$ for describing $\boldsymbol{\mu}$, where $\boldsymbol{X}_k$ is a known $n \times p_k$ matrix of rank p_k and $\boldsymbol{\beta}_k$ is unknown.

11.8.1 An Overview of Various Model Selection Criteria

For the kth model, if $\hat{\boldsymbol{\mu}}_k = \boldsymbol{X}_k\hat{\boldsymbol{\beta}}_k$ is the estimated mean vector, where $\hat{\boldsymbol{\beta}}_k$ is the least squares estimate of $\boldsymbol{\beta}_k$, then the residual sum of squares is $SSE_k = \|\boldsymbol{Y} - \hat{\boldsymbol{\mu}}_k\|^2$. Note that the residual sum of squares $SSE_k = \|\boldsymbol{Y} - \boldsymbol{X}_k\hat{\boldsymbol{\beta}}_k\|^2$ decreases for a nested class of models (where for each k, the kth model is contained in the $k+1$-st model) and the minimum of $\{SSE_k\}$ is attained for the model with the largest number of parameters. Thus one cannot hope to select an appropriate model by minimizing SSE_k over k. A reasonable measure of how well $\hat{\boldsymbol{\mu}}_k$ estimates $\boldsymbol{\mu}$ is given by $D_k = \mathrm{E}\left[\|\boldsymbol{\mu} - \hat{\boldsymbol{\mu}}_k\|^2\right]$, the expected squared distance between $\boldsymbol{\mu}$ and $\hat{\boldsymbol{\mu}}_k$. Ideally, one would choose the model for which D_k is the smallest, but D_k is unknown as it depends on the unknown population parameters. Therefore, one then first obtains a good estimate $\hat{D}_k$ of D_k, then minimizes $\hat{D}_k$ over $k = 1, \ldots, K$. If the minimum is attained at $k = \hat{k}$, then $\left\{\boldsymbol{X}_{\hat{k}}\hat{\boldsymbol{\beta}}_{\hat{k}}\right\}$ is considered the most appropriate estimate of the mean vector $\boldsymbol{\mu}$. Akaike's FPE and Mallows' criteria seek to obtain an unbiased (or nearly unbiased) estimate of D_k.

Cross-validation (CV) and generalized cross-validation (GCV) seek to estimate the prediction error of the kth fitted model. In this setup, it is assumed that $(Y_i, \boldsymbol{x}_{k,i})$, $i = 1, \ldots, n$, where the observed values of the covariates $\left\{\boldsymbol{x}_{k,i}^T\right\}$ are the rows of $\boldsymbol{X}_k$, are iid and $(Y_{n+1}, \boldsymbol{x}_{k,n+1})$ is an independent copy of $(Y_i, \boldsymbol{x}_{k,i})$. If $\hat{\boldsymbol{\beta}}_k$ is the least squares estimate of $\boldsymbol{\beta}_k$, then the error for predicting Y_{n+1} using the estimated kth model is $PE(k) = \mathrm{E}\left[Y_{n+1} - \hat{\boldsymbol{\beta}}_k^T\boldsymbol{x}_{k,n+1}\right]^2$. Both CV and GCV obtain nearly unbiased estimators of this prediction error. Thus one chooses the model for which $\widehat{PE}(k)$ (as given by the CV or GCV criterion) is the smallest.

Both Akaike Information Criterion (AIC) and Bayesian Information Criterion (BIC) are likelihood-based methods, which seek to model the joint pdf f of $\boldsymbol{Y} = (Y_1, \ldots, Y_n)^T$. Suppose there is a class of candidate probability models $\{f_k(\cdot, \boldsymbol{\theta}_k): k = 1, \ldots, K\}$ for f under consideration. In the case of linear models, f_k may be the pdf of $N_n(\boldsymbol{X}_k\boldsymbol{\beta}_k, \sigma^2\boldsymbol{I})$ and $\boldsymbol{\theta}_k = \begin{pmatrix}\boldsymbol{\beta}_k \\ \sigma_k\end{pmatrix}$. As in the case of residual sum of squares, $-2\log f\left(\boldsymbol{Y}, \hat{\boldsymbol{\theta}}_k\right)$, where $\hat{\boldsymbol{\theta}}_k$ is the MLE for $\boldsymbol{\theta}_k$, will always decrease if we take a sequence of nested models and thus it is not possible to select an appropriate model by trying to minimize $-2\log f\left(\boldsymbol{Y}, \hat{\boldsymbol{\theta}}_k\right)$ over k, since the minimum is always attained at the model with the largest number of parameters. A true measure of how good the kth fitted model is can be judged by how well it performs for a dataset that is independent of the data $\boldsymbol{Y}$ and this is what is done in the arguments involving the AIC. The AIC seeks to obtain an estimate of 2 times the Kullback-Leibler divergence $KL(k) = \mathrm{E}\left[\log f\left(\tilde{\boldsymbol{Y}}\right)/f_k\left(\tilde{\boldsymbol{Y}},\hat{\boldsymbol{\theta}}_k\right)\right]$, where $\tilde{\boldsymbol{Y}}$ has the same distribution as $\boldsymbol{Y}$, but is independent of it. The model corresponding to the smallest value of AIC is considered the most appropriate one.

As the name suggests, the BIC is based on Bayesian considerations. In this setting, there is an "index of model" variable J which takes values in $\{1,\ldots,K\}$ so that the prior probability of choosing model k is π_k (ie, $P[J=k] = \pi_k$). Given $J = k$, the pdf of $\boldsymbol{Y}$ is $f_k(\cdot,\boldsymbol{\theta}_k)$, where $\boldsymbol{\theta}_k$ has a prior g_k. The goal to find the model for which the conditional probability $P[J=k|\boldsymbol{Y}]$ is maximized or $-2\log P[J=k|\boldsymbol{Y}]$ is minimized. Since these conditional probabilities are unknown, the BIC obtains an estimate of $-2\log P[J=k|\boldsymbol{Y}]$ for each k.

For the kth linear model, let $SSE_k = \|\boldsymbol{Y} - \boldsymbol{X}_k\hat{\boldsymbol{\beta}}_k\|^2$ and $\hat{\sigma}_k^2 = SSE_k/n$, where $\hat{\boldsymbol{\beta}}_k$ is the least squares estimate of $\boldsymbol{\beta}_k$. Under normality (ie, $\boldsymbol{\varepsilon} \sim N_n(\boldsymbol{0},\sigma^2\boldsymbol{I})$), the MLE of σ^2 is $\hat{\sigma}_k^2$. The various model selection criteria mentioned above are

$$\begin{aligned}
FPE(k) &= \frac{n+p_k}{n-p_k}SSE_k,\\
MAL(k) &= SSE_k + 2p_k\hat{\sigma}^2,\\
AIC(k) &= n\log\hat{\sigma}_k^2 + 2p_k,\\
BIC(k) &= n\log\hat{\sigma}_k^2 + \log(n)p_k,\\
\widehat{PE}^{(cv)}(k) &= n^{-1}\sum_{i=1}^{n}\left(Y_i - \hat{\boldsymbol{\beta}}_{k,(-i)}^T\boldsymbol{x}_{k,i}\right)^2, \text{ and}\\
\widehat{PE}^{(gcv)}(k) &= n^{-1}SSE_k/(1-p_k/n)^2.
\end{aligned}$$

For Mallows' criterion, it is assumed that $\hat{\sigma}^2$ is a consistent estimate of σ^2. In the cross-validation criterion, it is understood that, for the kth model, the rows of $\boldsymbol{X}_k$ are $\left\{\boldsymbol{x}_{k,i}^T, i=1,\ldots,n\right\}$ and $\hat{\boldsymbol{\beta}}_{k,(-i)}$ is the least squares estimate of $\boldsymbol{\beta}_k$ based on $n-1$ observations, deleting $(Y_i,\boldsymbol{x}_{k,i})$.

Remark 11.8.1. Both AIC and BIC have forms which are more general than what are written above. More general versions are given below and they are described in detail. It can be shown that FPE, Mallows', CV, GCV, and AIC criteria are equivalent in an asymptotic sense as $n \to \infty$. The BIC is different from the others as its use may lead to models with fewer parameters. If the correct model is in the candidate class, then mathematical arguments, under appropriate conditions, show that BIC selects the correct model with probability converging to 1. The other model selection criteria given above tend to choose the "best" predictive model. It is important to keep in mind that the correct model (if it exists) is not necessarily the best predictive model.

11.8.2 Akaike's FPE and Mallows' Criteria

Suppose $\hat{\boldsymbol{\beta}}_k$ is the least squares estimate of $\boldsymbol{\beta}_k$ and $\hat{\boldsymbol{\mu}}_k = \boldsymbol{X}_k\hat{\boldsymbol{\beta}}_k$. A measure of divergence between $\boldsymbol{\mu}$ and $\hat{\boldsymbol{\mu}}_k$ is given by $D_k = \mathrm{E}[\|\boldsymbol{\mu} - \hat{\boldsymbol{\mu}}_k\|^2]$, and the goal is to obtain a reasonable estimate of D_k. Now we can write $\hat{\boldsymbol{\mu}}_k = \boldsymbol{Q}_k\boldsymbol{Y}$, where $\boldsymbol{Q}_k = \boldsymbol{X}_k(\boldsymbol{X}_k^T\boldsymbol{X}_k)^{-1}\boldsymbol{X}_k^T$ is the projection on the column space of $\boldsymbol{X}_k$ and $rank(\boldsymbol{Q}_k) = trace(\boldsymbol{Q}_k) = p_k$. Moreover,

$$\boldsymbol{\mu} - \hat{\boldsymbol{\mu}}_k = \boldsymbol{\mu} - \boldsymbol{Q}_k\boldsymbol{Y} = (\boldsymbol{I} - \boldsymbol{Q}_k)\boldsymbol{\mu} - \boldsymbol{Q}_k\boldsymbol{\varepsilon} = \bar{\boldsymbol{Q}}_k\boldsymbol{\mu} - \boldsymbol{Q}_k\boldsymbol{\varepsilon},$$

where $\bar{\boldsymbol{Q}}_k = \boldsymbol{I} - \boldsymbol{Q}_k$. Since $\bar{\boldsymbol{Q}}_k\boldsymbol{\mu}$ and $\boldsymbol{Q}_k\boldsymbol{\varepsilon}$ are orthogonal, $\|\boldsymbol{\mu} - \hat{\boldsymbol{\mu}}_k\|^2 = \|\bar{\boldsymbol{Q}}_k\boldsymbol{\mu}\|^2 + \|\boldsymbol{Q}_k\boldsymbol{\varepsilon}\|^2$ and taking expectation on both sides, we have

$$D_k = \mathrm{E}\|\boldsymbol{\mu} - \hat{\boldsymbol{\mu}}_k\|^2 = \|\bar{\boldsymbol{Q}}_k\boldsymbol{\mu}\|^2 + p_k\sigma^2.$$

Let us now examine the residual sum of squares $SSE_k = \|\boldsymbol{Y} - \hat{\boldsymbol{\mu}}_k\|^2$. We can write $\boldsymbol{Y} - \hat{\boldsymbol{\mu}}_k = \bar{\boldsymbol{Q}}_k\boldsymbol{Y} = \bar{\boldsymbol{Q}}_k\boldsymbol{\mu} + \bar{\boldsymbol{Q}}_k\boldsymbol{\varepsilon}$. Hence

$$\begin{aligned} SSE_k &= \|\bar{\boldsymbol{Q}}_k\boldsymbol{\mu}\|^2 + \|\bar{\boldsymbol{Q}}_k\boldsymbol{\varepsilon}\|^2 + 2\left(\bar{\boldsymbol{Q}}_k\boldsymbol{\mu}\right)^T\left(\bar{\boldsymbol{Q}}_k\boldsymbol{\varepsilon}\right), \text{ and} \\ \mathrm{E}\left[SSE_k\right] &= \|\bar{\boldsymbol{Q}}_k\boldsymbol{\mu}\|^2 + (n - p_k)\sigma^2. \end{aligned}$$

This shows that

$$D_k = \mathrm{E}\left[SSE_k\right] - (n - 2p_k)\sigma^2 = \mathrm{E}\left[SSE_k\right] + 2p_k\sigma^2 - n\sigma^2.$$

The last term $n\sigma^2$ does not depend on k and hence there is no need to estimate it. So if we have an estimate $\hat{\sigma}_k^2$ of σ^2, then we can get a criterion

$$\hat{D}_k = SSE_k + 2p_k\hat{\sigma}_k^2 - n\sigma^2.$$

The criteria given in Section 11.8.1 ignore the $n\sigma^2$ term. How one estimates σ^2 depends on the type of problem at hand. If σ^2 is estimated by $MSE_k = SSE_k/(n - p_k)$, then one ends up with Akaike's FPE criterion given above. Another possibility is to take a large enough model (maybe the largest model in the candidate class) so that the model bias is low and use the mean square error of that model to estimate σ^2. In such a case, the estimate of σ^2 does not depend on k and the resulting $\hat{D}_k$ is a special case of Mallows' criterion.

11.8.3 AIC and BIC

Let f be the joint pdf of $\boldsymbol{Y} = (Y_1, \ldots, Y_n)^T$ and the goal is to find a suitable approximation for f among a class of candidate probability models. Let the kth model be $\{f_k(y, \boldsymbol{\theta}_k), \boldsymbol{\theta}_k \in \Omega_k\}$, $k = 1, \ldots, K$, where Ω_k is a nonempty open subset of $\mathbb{R}^{p_k}$. If $\hat{\boldsymbol{\theta}}_k$ is the MLE of $\boldsymbol{\theta}_k$ for the kth model, then the AIC and BIC are

$$\begin{aligned} AIC(k) &= -2\log f_k\left(\boldsymbol{Y}, \hat{\boldsymbol{\theta}}_k\right) + 2p_k, \\ BIC(k) &= -2\log f_k\left(\boldsymbol{Y}, \hat{\boldsymbol{\theta}}_k\right) + \log(n)p_k. \end{aligned}$$

If a criterion is minimized at $k = \hat{k}$, then $f_{\hat{k}}$ is declared to be the most appropriate model according to that criterion. Note that in both the criteria, $-2\log f_k\left(\boldsymbol{Y}, \hat{\boldsymbol{\theta}}_k\right)$ are penalized by a constant times the number of parameters estimated. However, the BIC has a higher penalty than the AIC and thus it may select a model with fewer parameters than the AIC.

In the case of linear models with normal errors, if one has a sequence of models $\{\boldsymbol{X}_k\boldsymbol{\beta}_k: \boldsymbol{\beta}_k \in \mathbb{R}^{p_k}\}$, then there are simple expressions for these criteria. Let $\boldsymbol{\theta}_k = \begin{pmatrix}\boldsymbol{\beta}_k \\ \sigma_k\end{pmatrix}$ be the vector of parameters for the kth model, then the MLE are

$$\hat{\boldsymbol{\beta}}_k = \left(\boldsymbol{X}_k^T\boldsymbol{X}_k\right)^{-1}\boldsymbol{X}_k^T\boldsymbol{Y}, \quad \hat{\sigma}_k^2 = \|\boldsymbol{Y} - \boldsymbol{X}_k\hat{\boldsymbol{\beta}}_k\|^2/n,$$

and thus we have

$$\begin{aligned} f_k\left(\boldsymbol{Y}, \hat{\boldsymbol{\theta}}_k\right) &= \left(\frac{1}{\sqrt{2\pi}\,\hat{\sigma}_k}\right)^n \exp\left[-\|\boldsymbol{Y} - \boldsymbol{X}_k\hat{\beta}_k\|^2 \big/ \left(2\hat{\sigma}_k^2\right)\right] \\ &= \left(\frac{1}{\sqrt{2\pi}\,\hat{\sigma}_k}\right)^n \exp[-n/2], \text{ and} \\ -2\log f_k\left(\boldsymbol{Y}, \hat{\boldsymbol{\theta}}_k\right) &= n\log\hat{\sigma}_k^2 + n + n\log(2\pi). \end{aligned}$$

The last two terms in the last expression do not involve k, and after ignoring them we arrive at

$$AIC(k) = n\log\hat{\sigma}_k^2 + 2p_k \quad \text{and} \quad BIC(k) = n\log\hat{\sigma}_k^2 + \log(n)p_k.$$

Theoretical arguments used in deriving these criteria are different. The arguments given here are heuristic, but they can be made rigorous at the expense of fairly cumbersome calculations. The AIC is obtained by trying to approximate a predictive likelihood, whereas the derivation of the BIC involves approximating the probability of choosing a model given the data $\boldsymbol{Y}$.

Mathematical Settings

I. Suppose $\tilde{\boldsymbol{Y}}$ has the same distribution as $\boldsymbol{Y}$ but is independent of it. AIC is obtained by an approximation of $\mathrm{E}\left[-2\log f\left(\tilde{\boldsymbol{Y}}, \hat{\boldsymbol{\theta}}_k\right)\right]$, where $\hat{\boldsymbol{\theta}}_k$ is the MLE based on the available data.

II. Suppose J is a discrete rv taking values $1, \ldots, K$, and $P[J = k] = \pi_k$. Also assume that given $J = k$, the pdf of the model is $f(\cdot, \boldsymbol{\theta}_k)$, where $\boldsymbol{\theta}_k$ has a prior $g_k(\cdot)$. Then the goal is to find the value of k which maximizes $P[J = k|\boldsymbol{Y}]$ or minimizes $-2\log P[J = k|\boldsymbol{Y}]$. The BIC is obtained by finding an approximation to $-2\log P[J = k|\boldsymbol{Y}]$. It turns out that as long as the priors $\{\pi_k\}$ and $\{g_k(\cdot)\}$ are reasonably "flat" on the parameter spaces (ie, they are not degenerate or close to being degenerate), the dominant terms in the asymptotic expansion of $-2\log P[J = k|\boldsymbol{Y}]$ do not depend on the priors.

Heuristic Derivation of AIC

Let us assume that there exists a unique $\boldsymbol{\theta}_{k0} \in \Omega_k$ when $\mathrm{E}[\log f(\boldsymbol{Y}, \boldsymbol{\theta}_k)]$ is maximized over $\boldsymbol{\theta}_k \in \Omega_k$, where the expectation is over the true pdf f of $\boldsymbol{Y}$. Let f_i be the pdf of Y_i and $f_{k,i}(\cdot, \boldsymbol{\theta}_k)$ be the pdf of Y_i under the kth model and denote

$$\xi_{k,i}(\cdot,\boldsymbol{\theta}_k) = -\log f_{k,i}(\cdot,\boldsymbol{\theta}_k),\ \xi_k(\boldsymbol{y},\boldsymbol{\theta}_k) = \sum_{i=1}^{n} \xi_{k,i}(y_i,\boldsymbol{\theta}_k),$$

$$\dot{\xi}_{k,i}(\cdot,\boldsymbol{\theta}_k) = \frac{\partial}{\partial \boldsymbol{\theta}_k}\xi_{k,i}(\cdot,\boldsymbol{\theta}_k),\ \ddot{\xi}_{k,i}(\cdot,\boldsymbol{\theta}_k) = \frac{\partial^2}{\partial \boldsymbol{\theta}_k \partial \boldsymbol{\theta}_k}\xi_{k,i}(\cdot,\boldsymbol{\theta}_k),$$

$$\dot{\xi}_k(\boldsymbol{y},\boldsymbol{\theta}_k) = \sum_{i=1}^{n} \dot{\xi}_{k,i}(y_i,\boldsymbol{\theta}_k), \quad \text{and} \quad \ddot{\xi}(\boldsymbol{y},\boldsymbol{\theta}_k) = \sum_{i=1}^{n} \ddot{\xi}_{k,i}(y_i,\boldsymbol{\theta}_k).$$

In these notations, $f(\boldsymbol{y},\boldsymbol{\theta}_k) = \exp[-\xi_k(\boldsymbol{y},\boldsymbol{\theta}_k)]$. Let

$$\boldsymbol{I}_k(f) = \mathrm{E}\left[n^{-1}\sum_{i=1}^{n} \dot{\xi}_{k,i}(Y_i,\boldsymbol{\theta}_{k0})\dot{\xi}_{k,i}(Y_i,\boldsymbol{\theta}_{k0})^T\right],\ \tilde{\boldsymbol{I}}_k(f) = \mathrm{E}\left[n^{-1}\ddot{\xi}_k(\boldsymbol{Y},\boldsymbol{\theta}_{k0})\right],$$

where the expectations are taken over the true pdf $f(\cdot)$ of $\boldsymbol{Y}$. Suppose that $\hat{\boldsymbol{\theta}}_k$ is the MLE of $\boldsymbol{\theta}_k$ based on the data $\boldsymbol{Y}$, $\tilde{\boldsymbol{Y}}$ is an independent copy of $\boldsymbol{Y}$, and the following hold (heuristic justifications given below)

$$\mathrm{E}\left[\xi_k\left(\tilde{\boldsymbol{Y}},\hat{\boldsymbol{\theta}}_k\right)\right] = \mathrm{E}\left[\xi_k\left(\boldsymbol{Y},\hat{\boldsymbol{\theta}}_k\right)\right] + (1/2)trace\left(\tilde{\boldsymbol{I}}_k(f)^{-1}\boldsymbol{I}_k(f)\right)[1+o(1)], \tag{5a}$$

$$\mathrm{E}\left[\xi_k\left(\boldsymbol{Y},\hat{\boldsymbol{\theta}}_k\right)\right] = \mathrm{E}\left[\xi_k\left(\boldsymbol{Y},\hat{\boldsymbol{\theta}}_k\right)\right] - (1/2)trace\left(\tilde{\boldsymbol{I}}_k(f)^{-1}\boldsymbol{I}_k(f)\right)[1+o(1)]. \tag{5b}$$

Then a reasonable estimate of $\mathrm{E}\left[\xi_k\left(\tilde{\boldsymbol{Y}},\hat{\boldsymbol{\theta}}_k\right)\right]$ is given by

$$\xi_k\left(\boldsymbol{Y},\hat{\boldsymbol{\theta}}_k\right) + trace\left(\tilde{\boldsymbol{I}}_k(f)^{-1}\boldsymbol{I}_k(f)\right).$$

Since $trace\left(\tilde{\boldsymbol{I}}_k(f)^{-1}\boldsymbol{I}_k(f)\right)$ is unknown, it needs to be estimated. If it is assumed that $f_k(\cdot,\boldsymbol{\theta}_{k0})$ is reasonably close to f, then we may replace $\tilde{\boldsymbol{I}}_k(f)$ and $\boldsymbol{I}_k(f)$ by $\tilde{\boldsymbol{I}}_k(f_k(\cdot,\boldsymbol{\theta}_{k0}))$ and $\boldsymbol{I}(f_k(\cdot,\boldsymbol{\theta}_{k0}))$, respectively. Nothing that $\tilde{\boldsymbol{I}}_k(f_k(\cdot,\boldsymbol{\theta}_{k0})) = \boldsymbol{I}_k(f_k(\cdot,\boldsymbol{\theta}_{k0}))$ (using Eqs. (3a) and (3b) in Section 7.1, and ξ_k is the negative of the log-likelihood), we may estimate $trace\left(\tilde{\boldsymbol{I}}_k(f)^{-1}\boldsymbol{I}_k(f)\right)$ by

$$trace\left(\tilde{\boldsymbol{I}}_k(f_k(\cdot,\boldsymbol{\theta}_{k0}))^{-1}\boldsymbol{I}_k(f_k(\cdot,\boldsymbol{\theta}_{k0}))\right) = trace(\boldsymbol{I}) = p_k.$$

Thus an estimate of $\mathrm{E}\left[\xi_k\left(\tilde{\boldsymbol{Y}},\hat{\boldsymbol{\theta}}_k\right)\right]$ is given by

$$\xi_k\left(\boldsymbol{Y},\hat{\boldsymbol{\theta}}_k\right) + p_k = -\log f_k\left(\boldsymbol{Y},\hat{\boldsymbol{\theta}}_k\right) + p_k.$$

The AIC is two times the quantity given above.

What remains to be shown is that the approximate expansions given in Eqs. (5a) and (5b) are valid. The MLE $\ddot{\boldsymbol{\theta}}_k$ satisfies the likelihood equation $\dot{\xi}_k\left(\boldsymbol{Y},\ddot{\boldsymbol{\theta}}_k\right) = \boldsymbol{0}$. Since $\boldsymbol{\theta}_{k0}$ maximizes $\mathrm{E}[\log f(\boldsymbol{Y},\boldsymbol{\theta}_k)]$, $\mathrm{E}\left[\dot{\xi}_k(\boldsymbol{Y},\boldsymbol{\theta}_{k0})\right] = 0$. For notational convenience, we denote $\boldsymbol{I}_k(f)$

and $\tilde{I}_k(f)$ by I_k and $\tilde{I}_k$. Arguments used in obtaining the asymptotic properties of the MLE (Chapter 7) can be employed to conclude the following

$$\begin{aligned}
n^{-1}\ddot{\xi}_k(Y,\theta_{k0}) - \tilde{I}_k &\xrightarrow{P} \mathbf{0}, \\
Z_{n,k} = n^{-1/2}I_k^{-1/2}\dot{\xi}_k(Y,\theta_{k0}) &\xrightarrow{\mathcal{D}} Z_k \sim N_{p_k}(\mathbf{0},I), \text{ and} \\
\hat{\theta}_k - \theta_{k0} &= -\ddot{\xi}_k(Y,\theta_{k0})^{-1}\dot{\xi}_k(Y,\theta_{k0})[1+o_P(1)] \\
&= -n^{-1/2}\tilde{I}_k^{-1}I_k^{1/2}Z_{n,k}[1+o_P(1)].
\end{aligned}$$

Now

$$\begin{aligned}
\xi_k\left(\tilde{Y},\hat{\theta}_k\right) &= \xi_k\left(\tilde{Y},\theta_{k0}\right) + \left(\hat{\theta}_k - \theta_{k0}\right)^T\dot{\xi}_k\left(\tilde{Y},\theta_{k0}\right) \\
&+ (1/2)\left(\hat{\theta}_k - \theta_{k0}\right)^T\ddot{\xi}_k\left(\tilde{Y},\theta_{k0}\right)\left(\hat{\theta}_k - \theta_{k0}\right)[1+o_P(1)].
\end{aligned}$$

Since $\tilde{Y}$ is independent of Y and hence of $\hat{\theta}_k$,

$$\mathrm{E}\left[\xi_k\left(\tilde{Y},\theta_{k0}\right)\right] = \mathrm{E}[\xi_k(Y,\theta_{k0})],\ \mathrm{E}\left[\dot{\xi}_k\left(\tilde{Y},\theta_{k0}\right)\right] = \mathrm{E}\left[\dot{\xi}(Y,\theta_{k0})\right] = \mathbf{0}.$$

Noting that $n^{-1}\ddot{\xi}_k\left(\tilde{Y},\theta_{k0}\right) - \tilde{I}_k \xrightarrow{P} \mathbf{0}$, we have

$$\begin{aligned}
&\left(\hat{\theta}_k - \theta_{k0}\right)^T\ddot{\xi}_k\left(\tilde{Y},\theta_{k0}\right)\left(\hat{\theta}_k - \theta_{k0}\right) \\
&= Z_{n,k}^T I_k^{1/2}\tilde{I}_k^{-1}\left[n^{-1}\ddot{\xi}_k\left(\tilde{Y},\theta_{k0}\right)\right]\tilde{I}_k^{-1}I_k^{1/2}Z_{n,k}[1+o_P(1)] \\
&= Z_{n,k}^T I_k^{1/2}\tilde{I}_k^{-1}I_k^{1/2}Z_{n,k}[1+o_P(1)].
\end{aligned}$$

Since $Z_{n,k} \xrightarrow{\mathcal{D}} Z_k$ and $\mathrm{E}\left[Z_k Z_k^T\right] = I$, ignoring the $o_P(1)$ term we have

$$\begin{aligned}
\mathrm{E}\left[\left(\hat{\theta}_k - \theta_{k0}\right)^T\ddot{\xi}_k\left(\tilde{Y},\theta_{k0}\right)\left(\hat{\theta}_k - \theta_{k0}\right)\right] &= \mathrm{E}\left[Z_k^T I_k^{1/2}\tilde{I}_k^{-1}I_k^{1/2}Z_k\right][1+o(1)] \\
&= trace\left(I_k^{1/2}\tilde{I}_k^{-1}I_k^{1/2}\mathrm{E}\left[Z_k Z_k^T\right]\right)[1+o(1)] \\
&= trace\left(\tilde{I}_k^{-1}I_k\right)[1+o(1)], \text{ and} \\
\mathrm{E}\left[\xi_k\left(\tilde{Y},\hat{\theta}_k\right)\right] &= \mathrm{E}[\xi_k(Y,\theta_k)] + (1/2)trace\left(\tilde{I}_k^{-1}I_k\right)[1+o(1)].
\end{aligned}$$

This shows that Eq. (5a) holds. The above argument assumes that the $o_P(1)$ term can be turned into an $o(1)$ term when taking the expectation. Such a step requires the concept of *uniform integrability* which we ignore here for the sake of simplicity of discussion.

In order to verify Eq. (5b), expand $\xi_k\left(Y,\hat{\theta}_k\right)$ about $\xi_k(Y,\theta_{k0})$ to get

$$\begin{aligned}
\xi_k\left(Y,\hat{\theta}_k\right) &= \xi_k(Y,\theta_{k0}) + \left(\hat{\theta}_k - \theta_{k0}\right)^T\dot{\xi}_k(Y,\theta_{k0}) \\
&+ (1/2)\left(\hat{\theta}_k - \theta_{k0}\right)^T\ddot{\xi}_k(Y,\theta_{k0})\left(\hat{\theta}_k - \theta_{k0}\right)[1+o_P(1)].
\end{aligned}$$

Since $\hat{\boldsymbol{\theta}}_k - \boldsymbol{\theta}_{k0} = -\ddot{\boldsymbol{\xi}}_k(\boldsymbol{Y}, \boldsymbol{\theta}_{k0})^{-1}\dot{\boldsymbol{\xi}}_k(\boldsymbol{Y}, \boldsymbol{\theta}_{k0})[1 + o_P(1)]$, we have

$$\xi_k\left(\boldsymbol{Y}, \hat{\boldsymbol{\theta}}_k\right) = \xi_k(\boldsymbol{Y}, \boldsymbol{\theta}_{k0}) - (1/2)\left(\hat{\boldsymbol{\theta}}_k - \boldsymbol{\theta}_{k0}\right)^T \ddot{\boldsymbol{\xi}}_k(\boldsymbol{Y}, \boldsymbol{\theta}_{k0})\left(\hat{\boldsymbol{\theta}}_k - \boldsymbol{\theta}_{k0}\right)[1 + o_P(1)].$$

Using the same argument as in the expansion of $\xi_k\left(\tilde{\boldsymbol{Y}}, \hat{\boldsymbol{\theta}}_k\right)$, we get

$$\mathrm{E}\left[\xi_k\left(\boldsymbol{Y}, \hat{\boldsymbol{\theta}}_k\right)\right] = \mathrm{E}[\xi_k(\boldsymbol{Y}, \boldsymbol{\theta}_k)] - (1/2)trace\left(\tilde{\boldsymbol{I}}_k^{-1}\boldsymbol{I}_k\right)[1 + o(1)].$$

Heuristic Derivation of BIC

Recall that in the Bayesian setting, the goal is to find k for which $P[J = k|Y]$ is maximized where the discrete variable J can take values $1, \ldots, K$ with probabilities $\pi_1, \ldots, \pi_K$. Now

$$P[J = k|\boldsymbol{Y} = \boldsymbol{y}] = \frac{\int_{\Omega_k} f_k(\boldsymbol{y}, \boldsymbol{\theta}_k) g_k(\boldsymbol{\theta}_k)\pi_k\, d\boldsymbol{\theta}_k}{f(\boldsymbol{y})},$$

where $f(\cdot)$ is the marginal pdf of $\boldsymbol{Y}$. Since the denominator of $f(\boldsymbol{y})$ does not depend on k, we may simply ignore it as it plays no role in optimization of $P[J = k|\boldsymbol{Y}]$ over k. We outline the basic arguments in approximating the integral in the numerator when g_k is taken as a noninformative prior (ie, $g_k(\boldsymbol{\theta}) \equiv 1$). When Ω_k is not compact, the use of such a prior can be justified by taking a uniform prior on a compact region $A_{k,j} \subset \Omega_k$ and letting $A_{k,j} \to \Omega_k$ as $j \to \infty$ in an appropriate manner.

Let ξ_k, $\xi_{k,i}$, etc., be the same as in the derivation of the AIC. Then $f_k(\boldsymbol{Y}, \boldsymbol{\theta}_k) = \exp[-\xi_k(\boldsymbol{Y}, \boldsymbol{\theta}_k)]$. The MLE $\hat{\boldsymbol{\theta}}_k$ of $\boldsymbol{\theta}_k$ is a solution to the equation $\dot{\boldsymbol{\xi}}_k(\boldsymbol{Y}, \boldsymbol{\theta}_k) = 0$ and assume that $n^{-1}\ddot{\boldsymbol{\xi}}_k\left(\boldsymbol{Y}, \hat{\boldsymbol{\theta}}_k\right) \xrightarrow{P} \tilde{\boldsymbol{I}}_k$, a positive definite matrix. Reexpress $\boldsymbol{\theta}_k$ as $\hat{\boldsymbol{\theta}}_k + n^{-1/2}\boldsymbol{u}$ where $\boldsymbol{u} = \sqrt{n}\left(\boldsymbol{\theta}_k - \hat{\boldsymbol{\theta}}_k\right)$ and hence

$$\begin{aligned}\int_{\Omega_k} f_k(\boldsymbol{Y}, \boldsymbol{\theta}_k)\, d\boldsymbol{\theta}_k &= n^{-p_k/2}\int_{\Omega_k'} f_k\left(\boldsymbol{Y}, \hat{\boldsymbol{\theta}}_k + n^{-1/2}\boldsymbol{u}\right) d\boldsymbol{u} \\ &= n^{-p_k/2}\int_{\Omega_k'} \exp\left[-\xi_k\left(\boldsymbol{Y}, \hat{\boldsymbol{\theta}}_k + n^{-1/2}\boldsymbol{u}\right)\right] d\boldsymbol{u},\end{aligned}$$

where $\Omega_k' = \left\{\sqrt{n}\left(\boldsymbol{z} - \hat{\boldsymbol{\theta}}_k\right) : \boldsymbol{z} \in \Omega_k\right\}$. If it can be shown that (heuristic details given below)

$$\int_{\Omega_k'} \exp\left[-\xi_k\left(\boldsymbol{Y}, \hat{\boldsymbol{\theta}}_k + n^{-1/2}\boldsymbol{u}\right)\right] d\boldsymbol{u} = \exp\left[-\xi_k\left(\boldsymbol{Y}, \hat{\boldsymbol{\theta}}_k\right)\right]\left(\sqrt{2\pi}\right)^{p_k}\left|\tilde{\boldsymbol{I}}_k\right|^{-1/2}[1 + o_P(1)],$$

then we have

$$\int_{\Omega_k} f_k(\boldsymbol{Y}, \boldsymbol{\theta}_k)\, d\boldsymbol{\theta}_k = n^{-p_k/2}\exp\left[-\xi_k\left(\boldsymbol{Y}, \hat{\boldsymbol{\theta}}_k\right)\right]\left(\sqrt{2\pi}\right)^{p_k}\left|\tilde{\boldsymbol{I}}_k\right|^{-1/2}[1 + o_P(1)].$$

Since $f_k(\boldsymbol{Y}, \hat{\boldsymbol{\theta}}_k) = \exp\left[-\xi_k\left(\boldsymbol{Y}, \hat{\boldsymbol{\theta}}_k\right)\right]$, keeping only the dominant terms, -2 times the logarithm of $\int_{\Omega_k} f_k(\boldsymbol{Y}, \boldsymbol{\theta}_k)\, d\boldsymbol{\theta}_k$ is approximately equal to

$$2\xi_k\left(\boldsymbol{Y}, \hat{\boldsymbol{\theta}}_k\right) + \log(n)p_k = -2\log f_k\left(\boldsymbol{Y}, \hat{\boldsymbol{\theta}}_k\right) + \log(n)p_k,$$

which is the BIC.

Let us now outline the heuristic arguments involved in the approximation of the integral $\int_{\Omega'_k} \exp\left[-\xi_k\left(\boldsymbol{Y}, \hat{\boldsymbol{\theta}}_k + n^{-1/2}\boldsymbol{u}\right)\right] d\boldsymbol{u}$. Assuming that $\hat{\boldsymbol{\theta}}_k$ is in a compact set with probability converging to 1, $\Omega'_k \rightarrow \mathbb{R}^{p_k}$ as $n \rightarrow \infty$. Since $\hat{\boldsymbol{\theta}}_k$ is the MLE, we have $\dot{\xi}_k\left(\boldsymbol{Y}, \hat{\boldsymbol{\theta}}_k\right) = 0$. Expanding $\xi_k\left(\boldsymbol{Y}, \hat{\boldsymbol{\theta}}_k + n^{-1/2}\boldsymbol{u}\right)$ about $\xi_k\left(\boldsymbol{Y}, \hat{\boldsymbol{\theta}}_k\right)$, we have (under reasonable regularity conditions on f_k)

$$\xi_k\left(\boldsymbol{Y}, \hat{\boldsymbol{\theta}}_k + n^{-1/2}\boldsymbol{u}\right) = \xi_k\left(\boldsymbol{Y}, \hat{\boldsymbol{\theta}}_k\right) + (1/2)\boldsymbol{u}^T\tilde{\boldsymbol{I}}_k\boldsymbol{u}[1 + o_P(1)].$$

Thus

$$\begin{aligned}
&\int_{\Omega'_k} \exp\left[-\xi_k\left(\boldsymbol{Y}, \hat{\boldsymbol{\theta}}_k + n^{-1/2}\boldsymbol{u}\right)\right] d\boldsymbol{u}\\
&= \int_{\Omega'_k} \exp\left[-\xi_k\left(\boldsymbol{Y}, \hat{\boldsymbol{\theta}}_k\right) - (1/2)\boldsymbol{u}^T\tilde{\boldsymbol{I}}_k\boldsymbol{u}(1 + o_P(1))\right] d\boldsymbol{u}\\
&= \exp\left[-\xi_k\left(\boldsymbol{Y}, \hat{\boldsymbol{\theta}}_k\right)\right] \int_{\Omega'_k} \exp\left[-(1/2)\boldsymbol{u}^T\tilde{\boldsymbol{I}}_k\boldsymbol{u}(1 + o_P(1))\right] d\boldsymbol{u}\\
&= \exp\left[-\xi_k\left(\boldsymbol{Y}, \hat{\boldsymbol{\theta}}_k\right)\right] \int_{\mathbb{R}^{p_k}} \exp\left[-(1/2)\boldsymbol{u}^T\tilde{\boldsymbol{I}}_k\boldsymbol{u}\right] d\boldsymbol{u}[1 + o_P(1)].
\end{aligned}$$

This kind of approximation of an integral is known as the Laplace approximation. We have omitted a lot of details in the above approximation including how the integral over Ω'_k can be approximated by the integral over $\mathbb{R}^{p_k}$, how the $o_P(1)$ terms can be brought out of the exponential term, etc. However, all the calculations can be justified using appropriate mathematical conditions.

The integrand in the last integral is proportional to the pdf of a p_k -dim normal random vector with mean $\boldsymbol{0}$ and covariance matrix $\tilde{\boldsymbol{I}}_k{}^{-1}$. Thus

$$\int_{R^k} \exp\left[-(1/2)\boldsymbol{u}^T\tilde{\boldsymbol{I}}_k\boldsymbol{u}\right] d\boldsymbol{u} = \left(\sqrt{2\pi}\right)^{p_k}\left|\tilde{\boldsymbol{I}}_k\right|^{-1/2}.$$

Thus an approximation to $\int_{\Omega_k} f_k(\boldsymbol{Y}, \boldsymbol{\theta}_k)\, d\boldsymbol{\theta}_k$ is given by

$$\exp\left[-\xi_k\left(\boldsymbol{Y}, \hat{\boldsymbol{\theta}}_k\right)\right]\left(\sqrt{2\pi}\right)^{p_k}\left|\tilde{\boldsymbol{I}}_k\right|^{-1/2}.$$

11.8.4 Cross-Validation and Generalized Cross-Validation

Recall that the kth model under consideration is $\boldsymbol{Y} = \boldsymbol{X}_k\boldsymbol{\beta}_k + \boldsymbol{\varepsilon}$, where $\boldsymbol{X}_k$ is an $n \times p_k$ matrix of rank p_k, and assume that the rows of $\boldsymbol{X}_k$ are $\boldsymbol{x}_{k,i}^T$, $i = 1, \ldots, n$, and that $(Y_i, \boldsymbol{x}_{k,i})$,

$i = 1, \dots, n$, are iid. If $(Y_{n+1}, \boldsymbol{x}_{k,n+1})$ is a future copy of $(Y_i, \boldsymbol{x}_{k,i})$, which is independent of the data $\{(Y_i, \boldsymbol{x}_{k,i}), i = 1, \dots, n\}$, then the mean square error in predicting Y_{n+1} by $\hat{\boldsymbol{\beta}}_k^T \boldsymbol{x}_{k,n+1}$ is $PE(k) = \mathrm{E}\left[Y_{n+1} - \hat{\boldsymbol{\beta}}_k^T \boldsymbol{x}_{k,n+1}\right]^2$, where $\hat{\boldsymbol{\beta}}_k = (\boldsymbol{X}_k^T \boldsymbol{X}_k)^{-1} \boldsymbol{X}_k^T \boldsymbol{Y}$ is the least squares estimate of $\boldsymbol{\beta}_k$. Cross-validation method seeks to estimate this prediction error $PE(k)$. The method is as follows. Obtain the least squares estimate $\hat{\boldsymbol{\beta}}_{k,(-i)}$ of $\boldsymbol{\beta}_k$ based on $n-1$ observations deleting the ith case. Then an estimate of the prediction error is given by

$$\widehat{PE}^{(cv)}(k) = n^{-1} \sum_{i=1}^{n} \left(Y_i - \hat{\boldsymbol{\beta}}_{k,(-i)}^T \boldsymbol{x}_{k,i} \right)^2.$$

As will be shown below, there is no need to calculate all the n different estimates of $\boldsymbol{\beta}_k$. Let $\boldsymbol{Q}_k = \boldsymbol{X}_k (\boldsymbol{X}_k^T \boldsymbol{X}_k)^{-1} \boldsymbol{X}_k^T$ be the projection matrix which projects on the columns of $\boldsymbol{X}_k$. Let $\hat{Y}_i = \hat{\boldsymbol{\beta}}_k^T \boldsymbol{x}_i$ and $\hat{\varepsilon}_i = Y_i - \hat{Y}_i$. It will be shown that

$$\hat{\boldsymbol{\beta}}_{k,(-i)} = \hat{\boldsymbol{\beta}}_k - (1 - q_{k,ii})^{-1} \left(\boldsymbol{X}_k^T \boldsymbol{X}_k \right)^{-1} \boldsymbol{x}_{k,i} \hat{\varepsilon}_i, \tag{6a}$$

$$Y_i - \hat{\boldsymbol{\beta}}_{k,(-i)}^T \boldsymbol{x}_{k,i} = \left(Y_i - \hat{\boldsymbol{\beta}}_k^T \boldsymbol{x}_{k,i} \right) / (1 - q_{k,ii}), \tag{6b}$$

where $q_{k,ii}$ is the ith diagonal element of $\boldsymbol{Q}_k$. So $PE^{(cv)}$ can be written as

$$\widehat{PE}^{(cv)}(k) = n^{-1} \sum_i \left(Y_i - \hat{\boldsymbol{\beta}}_k^T \boldsymbol{x}_i \right)^2 / (1 - q_{k,ii})^2.$$

If any or some of the values of $q_{k,ii}$ is/are close to 1, then the cross-validated estimate of $PE(k)$ may become unstable. For this reason, it has been proposed to replace $q_{k,ii}$ by its average

$$\bar{q}_k = \sum_i q_{k,ii}/n = trace(\boldsymbol{Q}_k)/n = p_k/n.$$

This leads to what is known as the generalized cross-validation estimate

$$\widehat{PE}^{(gcv)}(k) = n^{-1} \sum_i \left(Y_i - \hat{\boldsymbol{\beta}}_k^T \boldsymbol{x}_i \right)^2 / (1 - p_k/n)^2.$$

It should be noted that the expression of the $\widehat{PE}^{(gcv)}(k)$ is almost proportional to Akaike' FPE criterion. More precisely,

$$\widehat{PE}^{(gcv)}(k) = n^{-1} \left[1 + O(p_k/n)^2 \right] FPE_k,$$

assuming that $p_k/n \to 0$ as $n \to \infty$.

We now verify the identities given in Eqs. (6a) and (6b). For notational convenience we suppress k in the expressions for $\boldsymbol{X}_k$, $\hat{\boldsymbol{\beta}}_k$, $\boldsymbol{x}_{k,i}$, etc. Let $\boldsymbol{X}_{(-i)}$ be the matrix obtained from $\boldsymbol{X}$ by deleting its ith row. Thus $\boldsymbol{X}_{(-i)}$ has $n-1$ rows. Similarly let $\boldsymbol{Y}_{(-i)}$ be the vector with

$n-1$ rows after deleting the ith row from $\boldsymbol{Y}$. Note that the normal equations for the least squares estimate of $\boldsymbol{\beta}$ after deleting the ith observation $(Y_i, \boldsymbol{x}_i)$ is given by the solution of $\boldsymbol{X}_{(-i)}^T\boldsymbol{X}_{(-i)}\boldsymbol{\beta} = \boldsymbol{X}_{(-i)}^T\boldsymbol{Y}_{(-i)}$. Since

$$\boldsymbol{X}_{(-i)}^T\boldsymbol{X}_{(-i)} = \boldsymbol{X}^T\boldsymbol{X} - \boldsymbol{x}_i\boldsymbol{x}_i^T \quad \text{and} \quad \boldsymbol{X}_{(-i)}^T\boldsymbol{Y}_{(-i)} = \boldsymbol{X}^T\boldsymbol{Y} - \boldsymbol{x}_iY_i,$$

using the Sherman-Morrison formula (Section B.1), and denoting $(\boldsymbol{X}^T\boldsymbol{X})^{-1}$ by $\boldsymbol{D}$, we have

$$\begin{aligned}
\left(\boldsymbol{X}_{(-i)}^T\boldsymbol{X}_{(-i)}\right)^{-1} &= \left(\boldsymbol{X}^T\boldsymbol{X} - \boldsymbol{x}_i\boldsymbol{x}_i^T\right)^{-1} \\
&= \left(\boldsymbol{X}^T\boldsymbol{X}\right)^{-1} + \left[1 - \boldsymbol{x}_i^T\left(\boldsymbol{X}^T\boldsymbol{X}\right)^{-1}\boldsymbol{x}_i\right]^{-1}\left(\boldsymbol{X}^T\boldsymbol{X}\right)^{-1}\boldsymbol{x}_i\boldsymbol{x}_i^T\left(\boldsymbol{X}^T\boldsymbol{X}\right)^{-1} \\
&= \left(\boldsymbol{X}^T\boldsymbol{X}\right)^{-1} + (1 - q_{ii})^{-1}\left(\boldsymbol{X}^T\boldsymbol{X}\right)^{-1}\boldsymbol{x}_i\boldsymbol{x}_i^T\left(\boldsymbol{X}^T\boldsymbol{X}\right)^{-1} \\
&= \boldsymbol{D} + (1 - q_{ii})^{-1}\boldsymbol{D}\boldsymbol{x}_i\boldsymbol{x}_i^T\boldsymbol{D},
\end{aligned}$$

where the ith diagonal element of $\boldsymbol{Q}$ is $q_{ii} = \boldsymbol{x}_i^T(\boldsymbol{X}^T\boldsymbol{X})^{-1}\boldsymbol{x}_i = \boldsymbol{x}_i^T\boldsymbol{D}\boldsymbol{x}_i$. Thus

$$\begin{aligned}
\hat{\boldsymbol{\beta}}_{(-i)} &= \left(\boldsymbol{X}_{(-i)}^T\boldsymbol{X}_{(-i)}\right)^{-1}\boldsymbol{X}_{(-i)}^T\boldsymbol{Y}_{(-i)} \\
&= \left[\boldsymbol{D} + (1 - q_{ii})^{-1}\boldsymbol{D}\boldsymbol{x}_i\boldsymbol{x}_i^T\boldsymbol{D}\right]\left[\boldsymbol{X}^T\boldsymbol{Y} - \boldsymbol{x}_iY_i\right] \\
&= \left[\boldsymbol{I} + (1 - q_{ii})^{-1}\boldsymbol{D}\boldsymbol{x}_i\boldsymbol{x}_i^T\right]\left[\boldsymbol{D}\boldsymbol{X}^T\boldsymbol{Y} - \boldsymbol{D}\boldsymbol{x}_iY_i\right] \\
&= \left[\boldsymbol{I} + (1 - q_{ii})^{-1}\boldsymbol{D}\boldsymbol{x}_i\boldsymbol{x}_i^T\right]\left[\hat{\boldsymbol{\beta}} - \boldsymbol{D}\boldsymbol{x}_iY_i\right] \\
&= \hat{\boldsymbol{\beta}} - \boldsymbol{D}\boldsymbol{x}_iY_i + (1 - q_{ii})^{-1}\boldsymbol{D}\boldsymbol{x}_i\left[\boldsymbol{x}_i^T\hat{\boldsymbol{\beta}} - \boldsymbol{x}_i^T\boldsymbol{D}\boldsymbol{x}_iY_i\right] \\
&= \hat{\boldsymbol{\beta}} - \boldsymbol{D}\boldsymbol{x}_iY_i + (1 - q_{ii})^{-1}\boldsymbol{D}\boldsymbol{x}_i\left[\hat{Y}_i - q_{ii}Y_i\right] \\
&= \hat{\boldsymbol{\beta}} - \boldsymbol{D}\boldsymbol{x}_i\left[Y_i - (1 - q_{ii})^{-1}\left(\hat{Y}_i - q_{ii}Y_i\right)\right] \\
&= \hat{\boldsymbol{\beta}} - \boldsymbol{D}\boldsymbol{x}_i\left[(1 - q_{ii})^{-1}\left(Y_i - \hat{Y}_i\right)\right] = \hat{\boldsymbol{\beta}} - (1 - q_{ii})^{-1}\boldsymbol{D}\boldsymbol{x}_i\hat{\varepsilon}_i.
\end{aligned}$$

This show that equality (6a) holds. In order to verify Eq. (6b), note that

$$\begin{aligned}
\hat{\boldsymbol{\beta}}_{(-i)}^T\boldsymbol{x}_i &= \hat{\boldsymbol{\beta}}^T\boldsymbol{x}_i - (1 - q_{ii})^{-1}\boldsymbol{x}_i^T\boldsymbol{D}\boldsymbol{x}_i\hat{\varepsilon}_i \\
&= \hat{\boldsymbol{\beta}}^T\boldsymbol{x}_i - (1 - q_{ii})^{-1}q_{ii}\hat{\varepsilon}_i, \text{ and} \\
Y_i - \hat{\boldsymbol{\beta}}_{(-i)}^T\boldsymbol{x}_i &= \left(Y_i - \hat{\boldsymbol{\beta}}^T\boldsymbol{x}_i\right) + (1 - q_{ii})^{-1}q_{ii}\hat{\varepsilon}_i \\
&= \hat{\varepsilon}_i + (1 - q_{ii})^{-1}q_{ii}\hat{\varepsilon}_i = (1 - q_{ii})^{-1}\hat{\varepsilon}_i.
\end{aligned}$$

11.9 Some Alternate Methods for Regression

In a regression setting with $p - 1$ independent variables, in some cases, Y is related to only $k < p - 1$ independent variables, and thus it makes sense to remove some of the independent variables from the model. Stepwise regression, all subsets regression, or a penalty method such as lasso may be used for this purpose. In some cases, another situation may be true where Y is related to all the independent variables and the independent variables are well correlated among themselves. In such cases, it may be reasonable to regress Y on a few (say $k < p - 1$) appropriately created linear combinations of the independent variables. Partial least squares (PLS) or principal components regression (PCR) may be used for this purpose. The phenomenon of all or some of the independent variables being well-correlated among themselves is called *multicollinearity*. Another method that has been proposed in the literature for multicollinear cases is known as the ridge regression.

Theoretical properties of some of the procedures described above are either nontrivial or not well-understood. For this reason, only outlines of these procedures will be given. It is also important to point out there are other regression procedures in addition to the ones mentioned above.

For all the methods to be discussed here, we assume the Gauss-Markov setup given in Eq. (1) is true with $\boldsymbol{\varepsilon}$ as the vector of iid variables with mean 0 and variance σ^2, and the goal is to estimate the mean vector $\boldsymbol{X\beta}$ on the basis of the observation vector $\boldsymbol{Y}$. Even though the typical assumption is that $\{\varepsilon_i\}$ are iid, in some cases, such as stepwise regression, the typical assumption is $\boldsymbol{\varepsilon} \sim N_n(\boldsymbol{0}, \sigma^2\boldsymbol{I})$.

11.9.1 All Subsets Regression

All subsets regression consists of fitting all possible submodels and choosing an appropriate (sub)model using a criterion such as AIC, Mallows' or BIC (ie, choose the model with the smallest value of the criterion). If there are $p - 1$ independent variables then there are 2^{p-1} possible submodels. When p is not small, then the number of submodels 2^{p-1} is quite large and, in such cases, carrying out all subsets regression may not be feasible. For this reason, one may use a computationally feasible method such as stepwise regression which is described below.

11.9.2 Stepwise Regression

Let us first discuss the simplest versions of this method: forward selection and backward elimination.

In the forward selection method, one builds up a model beginning with no independent variable and adding one variable at a time till it is not possible to add any more. It is typical to start with the model $Y = \beta_0 + \varepsilon$ and then try each of the $p - 1$ regression models, $Y = \beta_0 + \beta_j X_j + \varepsilon$, $j = 1, \ldots, p - 1$. When comparing the models $Y = \beta_0 + \varepsilon$ (reduced) to $Y = \beta_0 + \beta_j X_j + \varepsilon$ (full), one can test H_0: $\beta_j = 0$ against H_1: $\beta_j \neq 0$. This leads to an F-statistic and, since there are $p-1$ models, there are $p-1$ such F-values. The variable with the largest F-value is the best candidate for inclusion. If this largest F-statistic is larger than a preselected threshold (called F-to-enter), then the corresponding independent variable is entered in the model. If the value of the largest F-statistic is smaller than the threshold, then no variable can be added and it is declared that $Y = \beta_0 + \varepsilon$ is the most appropriate model.

Suppose that variable X_2 has been selected in step 1, then one considers $p-2$ models, $Y = \beta_0 + \beta_2 X_2 + \beta_j X_j + \varepsilon$, $1 \leq j \neq 2 \leq p-1$. Now there are $p-2$ testing problems H_0: $\beta_j = 0$ against H_1: $\beta_j \neq 0$, $1 \leq j \neq 2 \leq p-1$. Thus there are $p-2$ F-values and the variable with the largest F-value is entered if it is larger than the threshold. Otherwise $Y = \beta_0 + \beta_2 X_2 + \varepsilon$ is considered the most appropriate model. In this manner, independent variables can be added one at a time till it is not possible to enter any more.

Remark 11.9.1. In each step, instead of using the F-values, one may use p-values and enter a variable if the corresponding p-value is the smallest, and it is smaller than a threshold p-value (called p-to-enter).

The selected model depends on the threshold F-value (often taken to be equal to 3.5 or 4) or the threshold p-value (often taken to be equal to 0.05 or 0.10). Changing the threshold value may lead to a different model. For this reason, sometimes it is considered appropriate to look at the best candidate's AIC or BIC value at each step. The model with the smallest value AIC or BIC is considered the most appropriate.

In the backward elimination method, one starts with the full model and then deletes variables one at a time using the F-values and a threshold called F-to-delete till no variable can be dropped. As in the forward selection method, one may use p-values instead of F-values or use a criterion such as AIC to carry out this procedure.

Stepwise regression can be done in a forward or in a backward manner. In the forward stepwise regression, variables are added as in the forward stepwise regression. However, in each step it also has the option of deleting a variable that is already in the model. So forward stepwise regression is basically a forward selection method which includes a backward elimination step. For this reason, a forward stepwise method needs two threshold values: one for forward selection and the other for backward elimination.

Similarly, backward stepwise regression is basically a backward elimination method with the option of reentering a variable that is outside the model. As in the forward stepwise method, it needs two threshold values.

11.9.3 Penalty Methods

Penalty-based approaches to regression seek to minimize the least squares criterion $G(\boldsymbol{b}) = \|\boldsymbol{Y} - \boldsymbol{X}\boldsymbol{b}\|^2$ with restrictions on $\boldsymbol{b}$. This can usually be achieved by minimizing $G(\boldsymbol{b})$ plus a penalty term as given below

$$G(\boldsymbol{b}, \lambda) = G(\boldsymbol{b}) + \lambda p(\boldsymbol{b}),$$

where the penalty parameter λ is nonnegative and the penalty function $p(\boldsymbol{b}) \geq 0$ is equal to 0 at $\boldsymbol{b} = \boldsymbol{0}$. There are many possible choices for the penalty function, but we will mention only two: $p(\boldsymbol{b}) = \sum c_j|b_j|$ and $p(\boldsymbol{b}) = \sum c_j^2 b_j^2$, where c_j^2 is the jth diagonal element of $\boldsymbol{X}^T\boldsymbol{X}$. For the first case, also known as lasso, there is no explicit expression for the minimizer. However, for the second case, known as the ridge regression, there is an explicit expression for the minimizer $\hat{\boldsymbol{\beta}}(\lambda) = \left(\boldsymbol{X}^T\boldsymbol{X} + \lambda\boldsymbol{C}\right)^{-1}\boldsymbol{X}^T\boldsymbol{Y}$, where $\boldsymbol{C}$ is a diagonal matrix with diagonal elements $\left\{c_j^2\right\}$. When the penalty term $\lambda = 0$, both methods are the same as the ordinary least squares. In general, the penalty parameter can be chosen by cross-validation or its modified versions.

Ridge regression works well in estimating the mean response when the independent variables are well correlated among themselves. However, if the independent variables are mutually uncorrelated, its performance may not be satisfactory. With appropriate choice of λ and under regularity conditions, lasso can recover the regression model when some of the beta parameters are nonzero and the rest are zeros. However, the ridge estimates will typically result in nonzero estimates for all the beta parameters.

The penalty methods mentioned above can be motivated by a Bayesian consideration where the true parameters are assumed to randomly distributed. If all the variables have been standardized, then there is no need for an intercept term and the penalty term can be taken to be equal to $\lambda \sum \beta_j^2 = \lambda\|\boldsymbol{\beta}\|^2$. If the true regression coefficients $\{\beta_j\}$ are assumed to be iid $N(0, \tau^2)$, then ignoring the terms that do not depend on $\boldsymbol{Y}$ or $\boldsymbol{\beta}$, -2 times the logarithm of the joint pdf of $\boldsymbol{Y}$ and $\boldsymbol{\beta}$ is

$$\|\boldsymbol{Y} - \boldsymbol{X}\boldsymbol{\beta}\|^2/\sigma^2 + \|\boldsymbol{\beta}\|^2/\tau^2 = \sigma^{-2}\left[\|\boldsymbol{Y} - \boldsymbol{X}\boldsymbol{\beta}\|^2 + \lambda\|\boldsymbol{\beta}\|^2\right], \quad \lambda = \sigma^2/\tau^2.$$

Minimizing the above with respect to $\boldsymbol{\beta}$ leads to the ridge estimate of $\boldsymbol{\beta}$. If instead of assuming normal distribution of $\{\beta_j\}$, one assumes $\{\beta_j\}$ to be iid double exponential with the pdf $(2\tau)^{-1}\exp\left[-|u|/\tau\right]$, $-\infty < u < \infty$, $\tau > 0$, then the argument above leads to a lasso estimate of $\boldsymbol{\beta}$.

It can be shown that there exists $\lambda > 0$ such that the ridge estimate $\boldsymbol{X}\hat{\boldsymbol{\beta}}(\lambda)$ is a superior estimate of $\boldsymbol{\mu} = \mathrm{E}[\boldsymbol{Y}] = \boldsymbol{X}\boldsymbol{\beta}$ than the least squares estimate $\boldsymbol{X}\hat{\boldsymbol{\beta}}$ of $\boldsymbol{\mu}$. We write down the result for the case when all the variables have been standardized so that the penalty term is $\lambda\|\boldsymbol{\beta}\|^2$ and the ridge estimate of $\boldsymbol{\beta}$ is $\hat{\boldsymbol{\beta}}(\lambda) = \left(\boldsymbol{X}^T\boldsymbol{X} + \lambda\boldsymbol{I}\right)^{-1}\boldsymbol{X}^T\boldsymbol{Y}$.

Lemma 11.9.1. *Let us assume that all the variables have been standardized in a regression model and the ridge estimate of* $\boldsymbol{\beta}$ *is* $\hat{\boldsymbol{\beta}}(\lambda) = \left(\boldsymbol{X}^T\boldsymbol{X} + \lambda\boldsymbol{I}\right)^{-1}\boldsymbol{X}^T\boldsymbol{Y}$, $\lambda > 0$. *Let* $D(\lambda) = E\left[\|\boldsymbol{X}\hat{\boldsymbol{\beta}}(\lambda) - \boldsymbol{X}\boldsymbol{\beta}\|^2\right]$ *be the expected value of the squared distance between* $\boldsymbol{X}\hat{\boldsymbol{\beta}}(\lambda)$ *and* $\boldsymbol{X}\boldsymbol{\beta}$. *There exists* $\lambda > 0$ *such that* $D(\lambda) < D(0)$, *ie, there exists a ridge estimate* $\boldsymbol{X}\hat{\boldsymbol{\beta}}(\lambda)$, $\lambda > 0$, *which is a superior estimate of* $\boldsymbol{X}\boldsymbol{\beta}$ *than the least squares estimate* $\boldsymbol{X}\hat{\boldsymbol{\beta}} = \boldsymbol{X}\hat{\boldsymbol{\beta}}(0)$.

Proof of Lemma 11.9.1. Clearly,

$$\boldsymbol{\beta}(\lambda) = \mathrm{E}\left[\hat{\boldsymbol{\beta}}(\lambda)\right] = \left(\boldsymbol{X}^T\boldsymbol{X} + \lambda\boldsymbol{I}\right)^{-1}\boldsymbol{X}^T\boldsymbol{X}\boldsymbol{\beta} = \boldsymbol{\beta} - \lambda\left(\boldsymbol{X}^T\boldsymbol{X} + \lambda\boldsymbol{I}\right)^{-1}\boldsymbol{\beta}.$$

Hence, writing $\boldsymbol{A}_\lambda = \boldsymbol{X}^T\boldsymbol{X} + \lambda\boldsymbol{I}$ for notational convenience, we have

$$\begin{aligned}
\boldsymbol{X}\hat{\boldsymbol{\beta}}(\lambda) - \boldsymbol{X}\boldsymbol{\beta} &= \boldsymbol{X}\hat{\boldsymbol{\beta}}(\lambda) - \boldsymbol{X}\boldsymbol{\beta}(\lambda) + \boldsymbol{X}\boldsymbol{\beta}(\lambda) - \boldsymbol{X}\boldsymbol{\beta} \\
&= \boldsymbol{X}\left(\boldsymbol{X}^T\boldsymbol{X} + \lambda\boldsymbol{I}\right)^{-1}\boldsymbol{X}^T\boldsymbol{\varepsilon} - \lambda\boldsymbol{X}\left(\boldsymbol{X}^T\boldsymbol{X} + \lambda\boldsymbol{I}\right)^{-1}\boldsymbol{\beta} \\
&= \boldsymbol{X}\boldsymbol{A}_\lambda^{-1}\boldsymbol{X}^T\boldsymbol{\varepsilon} - \lambda\boldsymbol{X}\boldsymbol{A}_\lambda^{-1}\boldsymbol{\beta}, \text{ and} \\
D(\lambda) &= \mathrm{E}\left[\|\boldsymbol{X}\hat{\boldsymbol{\beta}}(\lambda) - \boldsymbol{X}\boldsymbol{\beta}\|^2\right] = \mathrm{E}\left[\|\boldsymbol{X}\boldsymbol{A}_\lambda^{-1}\boldsymbol{X}^T\boldsymbol{\varepsilon}\|^2\right] + \lambda^2\|\boldsymbol{X}\boldsymbol{A}_\lambda^{-1}\boldsymbol{\beta}\|^2 \\
&:= D_V(\lambda) + D_B(\lambda).
\end{aligned}$$

We may regard $D_V(\lambda)$ and $D_B(\lambda)$ as the variance and the bias-squared terms. Since $\mathrm{Cov}[\boldsymbol{\varepsilon}] = \sigma^2\boldsymbol{I}$, we have

$$\begin{aligned}
D_V(\lambda) &= \mathrm{E}\left[\|\boldsymbol{X}\boldsymbol{A}_\lambda^{-1}\boldsymbol{X}^T\boldsymbol{\varepsilon}\|^2\right] = trace\left(\boldsymbol{X}\boldsymbol{A}_\lambda^{-1}\boldsymbol{X}^T\mathrm{Cov}[\boldsymbol{\varepsilon}]\boldsymbol{X}\boldsymbol{A}_\lambda^{-1}\boldsymbol{X}^T\right) \\
&= \sigma^2\, trace\left(\boldsymbol{X}\boldsymbol{A}_\lambda^{-1}\boldsymbol{X}^T\boldsymbol{X}\boldsymbol{A}_\lambda^{-1}\boldsymbol{X}^T\right) \\
&= \sigma^2\, trace\left(\boldsymbol{X}^T\boldsymbol{X}\boldsymbol{A}_\lambda^{-1}\boldsymbol{X}^T\boldsymbol{X}\boldsymbol{A}_\lambda^{-1}\right) = \sigma^2\, trace\left(\left[\boldsymbol{X}^T\boldsymbol{X}\boldsymbol{A}_\lambda^{-1}\right]^2\right).
\end{aligned}$$

Let the spectral decomposition of $\boldsymbol{X}^T\boldsymbol{X}$ be $\sum_{j=1}^{p}\lambda_j\boldsymbol{u}_j\boldsymbol{u}_j^T$ with $\lambda_1 \geq \cdots \geq \lambda_p$, and denote $\gamma_j = \boldsymbol{u}_j^T\boldsymbol{\beta}$. Then

$$\begin{aligned}
\boldsymbol{A}_\lambda^{-1} &= \sum\left(\lambda_j + \lambda\right)^{-1}\boldsymbol{u}_j\boldsymbol{u}_j^T, \\
\boldsymbol{X}^T\boldsymbol{X}\boldsymbol{A}_\lambda^{-1} &= \sum\lambda_j\left(\lambda_j + \lambda\right)^{-1}\boldsymbol{u}_j\boldsymbol{u}_j^T, \text{ and} \\
\boldsymbol{A}_\lambda^{-1}\boldsymbol{X}^T\boldsymbol{X}\boldsymbol{A}_\lambda^{-1} &= \sum\lambda_j\left(\lambda_j + \lambda\right)^{-2}\boldsymbol{u}_j\boldsymbol{u}_j^T.
\end{aligned}$$

Therefore

$$\begin{aligned}
D_V(\lambda) &= \sigma^2\, trace\left(\left[\boldsymbol{X}^T\boldsymbol{X}\boldsymbol{A}_\lambda^{-1}\right]^2\right) = \sum\left[\lambda_j/\left(\lambda_j + \lambda\right)\right]^2, \\
D_B(\lambda) &= \lambda^2\|\boldsymbol{X}\boldsymbol{A}_\lambda^{-1}\boldsymbol{\beta}\|^2 = \lambda^2\sum\lambda_j\left(\lambda_j + \lambda\right)^{-2}\gamma_j^2.
\end{aligned}$$

It is fairly easy to see that $D_V(\lambda)$ is a decreasing function of λ, but $D_B(\lambda)$ is increasing in λ. As a matter of fact

$$\begin{aligned}
D_V'(\lambda) &= -2\sigma^2\sum\lambda_j^2\left(\lambda_j + \lambda\right)^{-3}, \quad D_B'(\lambda) = 2\lambda\sum\lambda_j^{-2}\left(\lambda_j + \lambda\right)^{-3}\gamma_j^2, \\
D'(\lambda) &= D_V'(\lambda) + D_B'(\lambda) \\
&= -2\sigma^2\sum\lambda_j^2\left(\lambda_j + \lambda\right)^{-3} + 2\lambda\sum\lambda_j^{-2}\left(\lambda_j + \lambda\right)^{-3}\gamma_j^2, \text{ and} \\
D'(0) &= -2\sigma^2\sum\lambda_j^{-1} < 0.
\end{aligned}$$

Thus $D(\lambda)$ is decreasing in a neighborhood of 0 and hence $D(\lambda) < D(0)$ for some $\lambda > 0$. □

11.9.4 Partial Least Squares and Principal Components Regression

The main idea behind PCR and PLS is to create a design matrix $\boldsymbol{Z}$ from $\boldsymbol{X}$ whose columns are mutually orthogonal. Thus if an $n \times p$ matrix $\boldsymbol{Z} = [\boldsymbol{X u}_1, \ldots, \boldsymbol{X u}_p]$ is created so that

$$\boldsymbol{Z}_j^T \boldsymbol{Z}_k = (\boldsymbol{X u}_j)^T (\boldsymbol{X u}_k) = \boldsymbol{u}_j^T \boldsymbol{X}^T \boldsymbol{X u}_k = 0, \quad j \neq k,$$

then $\boldsymbol{Z}^T\boldsymbol{Z}$ is a diagonal matrix. If we write $\boldsymbol{U} = [\boldsymbol{u}_1, \ldots, \boldsymbol{u}_p]$, then $\boldsymbol{X\beta} = \boldsymbol{XUU}^{-1}\boldsymbol{\beta} = \boldsymbol{Z\alpha}$, with $\boldsymbol{\alpha} = \boldsymbol{U}^{-1}\boldsymbol{\beta}$. So we have a new Gauss-Markov model $\boldsymbol{Y} = \boldsymbol{Z\alpha} + \boldsymbol{\varepsilon}$ and the least squares estimate of $\boldsymbol{\alpha}$ is $\hat{\boldsymbol{\alpha}} = (\boldsymbol{Z}^T\boldsymbol{Z})^{-1}\boldsymbol{Z}^T\boldsymbol{Y}$, and $\hat{\alpha}_j = \boldsymbol{Z}_j^T\boldsymbol{Y}/\boldsymbol{Z}_j^T\boldsymbol{Z}_j$ is the estimate of α_j, the jth component of $\boldsymbol{\alpha}$. It is important to note that in this formulation, estimate of α_j depends only on $\boldsymbol{Z}_j$ and $\boldsymbol{Y}$, $\text{Cov}[\hat{\boldsymbol{\alpha}}] = \sigma^2(\boldsymbol{Z}^T\boldsymbol{Z})^{-1}$, and whenever $j \neq k$, $\text{Cov}[\hat{\alpha}_j, \hat{\alpha}_k] = 0$ since $\boldsymbol{Z}^T\boldsymbol{Z}$ is a diagonal matrix. Model selection now involves choosing a few of the newly created independent variables $\boldsymbol{Z}_1, \ldots, \boldsymbol{Z}_p$ and this can be carried out by using stepwise regression with an appropriate model selection criterion described in the last section. Since $\boldsymbol{\beta} = \boldsymbol{U\alpha}$, once we have an estimate $\hat{\boldsymbol{\alpha}}$ of $\boldsymbol{\alpha}$ (and this includes the case when some α_j are set be zero if we decide to delete the corresponding $\boldsymbol{Z}_j$'s), we can get the estimate of $\boldsymbol{\beta}$ as $\hat{\boldsymbol{\beta}} = \boldsymbol{U}\hat{\boldsymbol{\alpha}} = \sum \hat{\alpha}_j \boldsymbol{u}_j$.

PCR and PLS differ in the way the vectors $\boldsymbol{u}_1, \ldots, \boldsymbol{u}_p$ are created or the new mutually orthogonal vectors $\boldsymbol{Z}_1, \ldots, \boldsymbol{Z}_p$ are created. In PCR, $\{\boldsymbol{u}_j\}$ are taken to be the orthonormal eigenvectors obtained from the spectral decomposition of $\boldsymbol{X}^T\boldsymbol{X}$. Thus if $\lambda_1, \ldots, \lambda_p$ are the eigenvalues of $\boldsymbol{X}^T\boldsymbol{X}$ with the corresponding orthonormal eigenvectors $\boldsymbol{u}_1, \ldots, \boldsymbol{u}_p$, then $\boldsymbol{Z}_j$ is taken to be $\boldsymbol{Z}_j = \boldsymbol{X u}_j$, $j = 1, \ldots, p$, then clearly $\{\boldsymbol{Z}_j\}$ are mutually orthogonal.

In the presence of multicollinearity, it is of interest to create the mutually uncorrelated variables economically so that a linear combination of a few of them can model the response variable Y well. It is thus desirable to have a method which creates the vectors $\{\boldsymbol{u}_j\}$ using both $\boldsymbol{Y}$ and $\boldsymbol{X}$. PCR uses only information from $\boldsymbol{X}$, whereas PLS uses both $\boldsymbol{X}$ and $\boldsymbol{Y}$. It should be pointed out that PLS is basically what is known as the Conjugate Gradient Method in Numerical Analysis. It is an iterative method for obtaining the minimum of a quadratic form $\boldsymbol{\beta}^T\boldsymbol{S\beta} - 2\boldsymbol{b}^T\boldsymbol{\beta}$ over $\boldsymbol{\beta} \in \mathbb{R}^p$, where $\boldsymbol{S}$ is a $p \times p$ positive definite matrix.

Remark 11.9.2. It is generally advisable to standardize all the variables for the PCR and the PLS so that there is no need to include an intercept term in the model and all the variables are on the same scale.

There are many equivalent ways of describing PLS, including a version with statistical interpretation. This procedure generates $\boldsymbol{u}_1, \boldsymbol{u}_2, \ldots$ in a recursive manner along with the estimated coefficients $\hat{\alpha}_1, \hat{\alpha}_2, \ldots$ and the corresponding estimates of $\boldsymbol{\beta}$. Here is a version which follows the numerical analyst's description. For notational convenience, let $\boldsymbol{S} = n^{-1}\boldsymbol{X}^T\boldsymbol{X}$, $\boldsymbol{b} = n^{-1}\boldsymbol{X}^T\boldsymbol{Y}$ so that the normal equations can be written as $\boldsymbol{S\beta} = \boldsymbol{b}$. The following iterative scheme describes how the columns of the matrix $\boldsymbol{Z}$, and $\hat{\alpha}_1, \hat{\alpha}_2, \ldots$etc., are generated recursively.

Step 1. Take $\boldsymbol{u}_1 = \boldsymbol{b}$. Then $\boldsymbol{Z}_1 = \boldsymbol{X u}_1$, $\hat{\alpha}_1 = \boldsymbol{Z}_1^T Y/\boldsymbol{Z}_1^T\boldsymbol{Z}_1 = \boldsymbol{u}_1^T\boldsymbol{b}/\boldsymbol{u}_1^T\boldsymbol{S u}_1$, and $\hat{\boldsymbol{\beta}}^{(1)} = \hat{\alpha}_1\boldsymbol{u}_1$. Denote the remainder $\boldsymbol{b} - \boldsymbol{S}\hat{\boldsymbol{\beta}}^{(1)}$ by $\boldsymbol{r}_1$.

We will describe how to carry out this recursion. Suppose that $\boldsymbol{u}_1, \ldots, \boldsymbol{u}_k$ has been generated with the corresponding estimates $\hat{\boldsymbol{\beta}}^{(1)}, \ldots, \hat{\boldsymbol{\beta}}^{(k)}$ and the remainders $\boldsymbol{r}_1 = \boldsymbol{b} - \boldsymbol{S}\hat{\boldsymbol{\beta}}^{(1)}, \ldots, \boldsymbol{r}_k = \boldsymbol{b} - \boldsymbol{S}\hat{\boldsymbol{\beta}}^{(k)}$, and we now want to create $\boldsymbol{u}_{k+1}$.

Step $k+1$. Let $\boldsymbol{u}_{k+1} = \boldsymbol{r}_k - c_k\boldsymbol{u}_k$, where c_k is such that $\boldsymbol{Z}_{k+1} = \boldsymbol{X}\boldsymbol{u}_{k+1}$ is orthogonal to $\boldsymbol{Z}_k = \boldsymbol{X}\boldsymbol{u}_k$. This leads to $c_k = \boldsymbol{u}_k^T\boldsymbol{S}\boldsymbol{r}_k/\boldsymbol{u}_k^T\boldsymbol{S}\boldsymbol{u}_k$. So we have

$$\begin{aligned}
\boldsymbol{u}_{k+1} &= \boldsymbol{r}_k - c_k\boldsymbol{u}_k,\ \boldsymbol{Z}_{k+1} = \boldsymbol{X}\boldsymbol{u}_{k+1} \text{ with } c_k = \boldsymbol{u}_k^T\boldsymbol{S}\boldsymbol{r}_k/\boldsymbol{r}_k^T\boldsymbol{S}\boldsymbol{r}_k,\\
\hat{\alpha}_{k+1} &= \boldsymbol{Z}_{k+1}^T\boldsymbol{Y}/\boldsymbol{Z}_{k+1}^T\boldsymbol{Z}_{k+1} = \boldsymbol{u}_{k+1}^T\boldsymbol{b}/\boldsymbol{u}_{k+1}^T\boldsymbol{S}\boldsymbol{u}_{k+1}, \text{ and}\\
\hat{\boldsymbol{\beta}}^{(k+1)} &= \hat{\alpha}_1\boldsymbol{u}_1 + \cdots + \hat{\alpha}_k\boldsymbol{u}_k + \hat{\alpha}_{k+1}\boldsymbol{u}_{k+1} = \hat{\boldsymbol{\beta}}^{(k)} + \hat{\alpha}_{k+1}\boldsymbol{u}_{k+1},
\end{aligned}$$

and the remainder is $\boldsymbol{r}_{k+1} = \boldsymbol{b} - \boldsymbol{X}\hat{\boldsymbol{\beta}}^{(k+1)}$.

Remark 11.9.3.

(a) It can be shown that for the procedure described above $\boldsymbol{Z}_1 = \boldsymbol{X}\boldsymbol{u}_1, \boldsymbol{Z}_2 = \boldsymbol{X}\boldsymbol{u}_2, \ldots$ are orthogonal.

(b) If $\boldsymbol{S}$ is a diagonal matrix, the iteration stops after the first iteration. If $\boldsymbol{X}$ is $n \times p$, then the procedure stops after p iterations and $\hat{\boldsymbol{\beta}}^{(p)}$ equals the least squares estimate $\hat{\boldsymbol{\beta}}$ of $\boldsymbol{\beta}$.

(c) How do we decide when to stop the iterations? One may use a criterion such as AIC or cross-validation and terminate the iteration when the value of the AIC stops decreasing.

11.10 Random- and Mixed-Effects Models

Random-effects models come up in many situations of practical interest. Consider a simple example. Suppose in a state, it is desired to know the average performance of children in some standardized mathematics test. Since the state has many schools and the student performance may vary from school to school, it may be desirable to choose k schools at random and, for each selected school, the standardized test is given to some randomly selected children. Thus if Y_{ij} is the performance of the jth child in the ith school, one may write a one-factor ANOVA model as $Y_{ij} = \mu + \alpha_i + \varepsilon_{ij}$, where μ is the overall mean, α_i is the school effect, and ε_{ij} is the random error. Since the schools are randomly selected it is reasonable to assume that, for this one-factor model, $\{\alpha_i\}$ are iid $N(0, \sigma_1^2)$. This is perhaps the simplest of all random-effect models.

Example 11.10.1 (One random factor). Suppose we have observations $\{Y_{ij} : j = 1, \ldots, n_i, i = 1, \ldots, k\}$ where the random factor has k levels and the model is

$$Y_{ij} = \mu + \alpha_i + \varepsilon_{ij}, \quad j = 1, \ldots, n_i, \quad i = 1, \ldots, k,$$

where μ is a constant,

(i) $\{\varepsilon_{ij}\}$ are iid $N(0,\sigma^2)$, $\{\alpha_i\}$ are iid $N(0,\sigma_1^2)$, and
(ii) $\{\alpha_i\}$ and $\{\varepsilon_{ij}\}$ are mutually independent.

Thus, we have

$$E\left[Y_{ij}\right] = \mu, \ \mathrm{Var}\left[Y_{ij}\right] = \sigma_1^2 + \sigma^2,$$
$$\mathrm{Cov}\left[Y_{ij}, Y_{ij'}\right] = \sigma_1^2, \quad j \neq j', \text{ and}$$
$$\mathrm{Cov}\left[Y_{ij}, Y_{i'j'}\right] = 0, \quad i \neq i'.$$

Now the variability of Y_{ij} has two components: variability among schools and the error variance. For this reason, σ_1^2 and σ^2 are also called the components of variance of Y_{ij}. The observations $\{Y_{i1}, \ldots, Y_{in_i}\}$ in a particular school are correlated, unlike in the fixed factor case. The correlation between Y_{ij} and $Y_{ij'}, j \neq j'$, which equals $\sigma_1^2/(\sigma_1^2 + \sigma^2)$, is called the intraclass correlation. The goal is often to estimate μ and the intraclass correlation coefficient.

Example 11.10.2 (Two Random Factors). Suppose we have observations $\{Y_{ijk}: k = 1, \ldots, n_{ij}, j = 1, \ldots, b, i = 1, , a\}$ where factor A has a levels, factor B has b levels, and both factors are random. Thus it can be written as

$$Y_{ijk} = \mu + \alpha_i + \beta_j + (\alpha\beta)_{ij} + \varepsilon_{ijk},$$

where it is assumed that μ is a constant; $\{\alpha_i\}$ are iid $N(0,\sigma_1^2)$; $\{\beta_j\}$ are iid $N(0,\sigma_2^2)$; $\left\{(\alpha\beta)_{ij}\right\}$ are iid $N(0,\sigma_3^2)$; $\{\varepsilon_{ijk}\}$ are iid $N(0,\sigma^2)$; and $\{\alpha_i\}$, $\{\beta_j\}$, $\left\{(\alpha\beta)_{ij}\right\}$, and $\{\varepsilon_{ij}\}$, are mutually independent. Here

$$E\left[Y_{ijk}\right] = \mu \text{ and } \mathrm{Var}\left[Y_{ijk}\right] = \sigma_1^2 + \sigma_2^2 + \sigma_3^2 + \sigma^2.$$

So the components of variance are σ_1^2, σ_2^2, σ_3^2, and σ^2. The goal is often to estimate these variance components along with the overall mean μ.

Example 11.10.3 (Mixed-Effects Model: One Factor Fixed and One Factor Random). In the last example, if factor A is fixed but factor B is random, then the model is the same except that μ and α_i are constants with $\sum \alpha_i = 0$, but $\{\beta_j\}$, $\left\{(\alpha\beta)_{ij}\right\}$ are $\{\varepsilon_{ijk}\}$ are iid $N(0,\sigma_1^2), N(0,\sigma_2^2)$ and $N(0,\sigma^2)$, respectively. It is also assumed that $\{\beta_j\}$, $\left\{(\alpha\beta)_{ij}\right\}$ and $\{\varepsilon_{ijk}\}$ are mutually independent. Thus

$$E\left[Y_{ij}\right] = \mu + \alpha_i \quad \text{and} \quad \mathrm{Var}\left[Y_{ijk}\right] = \sigma_1^2 + \sigma_2^2 + \sigma^2.$$

This is a mixed-effects model with the variance components σ_1^2 , σ_2^2, and σ^2.

Example 11.10.4 (Mixed-Effects Model: One Random Factor With a Covariate). We can also have a mixed model in which some factors are random, some fixed, and there are one or more covariates. Consider a case with one random factor and one covariate. Thus if $(Y_{ij}, X_{ij}), j = 1, \ldots, n_i$, are the observed values of the response and the covariate for the factor at level i, then a model can be written as

$$Y_{ij} = \beta_0 + \beta_1 X_{ij} + \alpha_i + \varepsilon_{ij}, \quad j = 1, \ldots, n_i, \quad i = 1, \ldots, k,$$

where β_0, β_1 are unknown constants, $\{\alpha_i\}$ are iid $N(0, \sigma_1^2)$, $\{\varepsilon_{ij}\}$ are iid $N(0, \sigma^2)$, and $\{\alpha_i\}$ and $\{\varepsilon_{ij}\}$ are mutually independent. In this case, we have

$$\mathrm{E}\left[Y_{ij}\right] = \beta_0 + \beta_1 X_{ij} \quad \text{and} \quad \mathrm{Var}\left[Y_{ij}\right] = \sigma_1^2 + \sigma^2.$$

Example 11.10.5 (Mixed Model: Growth Model With Random Slopes). Suppose we have k children whose heights (growths) are measured at various ages. Let Y_{ij} be the height of the ith child at age $t_j, j = 1, \ldots, n_0$, and let X_{ij} be the vector of covariates (available nutrition, parents' heights, family income, etc.). If we model the height as a polynomial of degree r in age with random slopes, then a reasonable model is

$$Y_{ij} = \beta_0 + \boldsymbol{\beta}_1^T X_{ij} + \sum_{l=1}^{r} \gamma_{il} t_j^l + \varepsilon_{ij}, \quad j = 1, \ldots, n_0, \quad i = 1, \ldots, k,$$

where β_0, β_1 are nonrandom, $\{\gamma_{il}: i = 1, \ldots, k\}$ are iid $N(0, \sigma_l^2)$ for each l, $\{\varepsilon_{ij}\}$ are iid $N(0, \sigma^2)$, and $\{\gamma_{il}\}$ and $\{\varepsilon_{ij}\}$ are mutually independent. This is an example of a growth model which allows different rates of growths for different children.

In Example 11.10.4 we can create k indicator variables $\{Z_{ij1}, \ldots Z_{ijk}\}$ as follows. Let $Z_{ijl} = 1$ if $i = l$ and 0 otherwise. If we denote the random effects by $\{\gamma_i\}$ instead of $\{\alpha_i\}$, then

$$Y_{ij} = \beta_0 + \beta_1 X_{ij} + Z_{ij1}\gamma_1 + \cdots Z_{ijk}\gamma_k + \varepsilon_{ij}.$$

We can express this model in a matrix form. Let $n = n_1 + \cdots + n_k$ be the total number of observations. Let $\boldsymbol{X}$ be the $n \times 2$ matrix whose first column has only ones and its second column consists of values of the covariate. Let $\boldsymbol{Z}$ be an $n \times k$ matrix whose first column consists of the values of Z_{ij1}, second column consists of the values Z_{ij2}, etc. Then we may rewrite the last model as

$$\boldsymbol{Y} = \boldsymbol{X\beta} + \boldsymbol{Z\gamma} + \boldsymbol{\varepsilon},$$

where $\boldsymbol{\gamma}$ is the $k \times 1$ vector of $\gamma_1, \ldots, \gamma_k$. Here $\boldsymbol{X}$ and $\boldsymbol{Z}$ are known matrices, $\boldsymbol{\beta}$ is the vector of unknown parameters, $\boldsymbol{\gamma} \sim N_k(\boldsymbol{0}, \sigma_1^2 \boldsymbol{I})$, $\boldsymbol{\varepsilon} \sim N_n(\boldsymbol{0}, \sigma^2 \boldsymbol{I})$, and $\boldsymbol{\gamma}$ and $\boldsymbol{\varepsilon}$ are independent.

One-factor random-effects model can also be written in this form. In order to accommodate the two-factor random- and mixed-effects models, which have more than two variance components, we consider a more general model

$$\boldsymbol{Y} = \boldsymbol{X\beta} + \boldsymbol{Z}_1\boldsymbol{\gamma}_1 + \cdots + \boldsymbol{Z}_r\boldsymbol{\gamma}_r + \boldsymbol{\varepsilon}, \tag{7}$$

where $X, Z_1, \ldots, Z_r$ are known matrices of orders $n \times p$, $n \times q_1, \ldots, n \times q_r$, respectively, β is a vector of unknown parameters, $\gamma_j \sim N_{q_j}\left(0, \sigma_j^2 I\right)$, $j = 1, \ldots, r$, $\varepsilon \sim N_n(0, \sigma^2 I)$, and $\gamma_1, \ldots, \gamma_k$ and ε are all independent. In this framework

$$\mathrm{E}[Y] = X\beta \text{ and } \mathrm{Cov}[Y] = \sigma_1^2 Z_1 Z_1^T + \cdots + \sigma_k^2 Z_r Z_r^T + \sigma^2 I.$$

Two-factor mixed- and random-effects models (Examples 11.10.2 and 11.10.5) can also be expressed in this form and it is left as an exercise. It is often useful to express the above model by taking $Z = [Z_1, \ldots, Z_k]$ and γ as a column vector with $\gamma_1, \ldots, \gamma_k$ stacked vertically. Thus Z is $n \times q$ and γ is $q \times 1$, where $q = q_1 + \cdots + q_k$. Thus

$$Y = X\beta + Z\gamma + \varepsilon, \tag{8}$$

where $\gamma \sim N_m(0, D)$, where D is a block diagonal matrix with the diagonal blocks $\mathrm{Cov}[\gamma_1], \ldots, \mathrm{Cov}[\gamma_r]$.

Assumption

For the model in Eq. (8) we assume that $rank(X) = p$, $rank(Z_j) = q_j$, $j = 1, \ldots, r$.

11.10.1 Estimation of β and Prediction of γ

We now discuss estimation of β and prediction of γ assuming all the variance components to be known. Since γ is random, its estimation (or linear function of it) is called prediction.

Definition 11.10.1.

(i) A linear function $\hat{\beta}$ of Y is called a BLUE of β if it is an unbiased estimator of β, that is, $\mathrm{E}\left[\hat{\beta}\right] = \beta$, and, for any $a \in \mathbb{R}^p$, $\mathrm{Var}\left[a^T\hat{\beta}\right] \le \mathrm{Var}\left[l^T Y\right]$ for all linear unbiased estimators $l^T Y$ of $a^T\beta$.

(ii) A linear function $\hat{\gamma}$ of Y is called a best linear unbiased predictor (BLUP) of γ if it is an unbiased predictor of γ, that is, $\mathrm{E}[\hat{\gamma} - \gamma] = 0$, and for any $b \in \mathbb{R}^q$, $\mathrm{E}\left[b^T\hat{\gamma} - b^T\gamma\right]^2 \le \mathrm{E}\left[l^T Y - b^T\gamma\right]^2$ for all linear unbiased predictors $l^T Y$ of $b^T\gamma$.

According to the definition given above, $\hat{\beta}$ is a BLUE of β if $a^T\hat{\beta}$ is the BLUE of $a^T\beta$ for any $a \in \mathbb{R}^p$. Similarly, $\hat{\gamma}$ is BLUP for γ if $b^T\hat{\gamma}$ is the BLUP of $b^T\gamma$ for any $b \in \mathbb{R}^q$.

For the mixed model, the BLUE of β is no longer equal to $(X^T X)^{-1} X^T Y$. As a matter of fact, it now depends on the unknown matrix D and σ^2. Let $\Sigma = \mathrm{Cov}[Y] = ZDZ^T + \sigma^2 I$. Then, if we multiply both sides of the mixed model by $\Sigma^{-1/2}$, where $\Sigma^{1/2}$ is a symmetric square root of Σ, it leads to a reexpression of the mixed model as $\tilde{Y} = \tilde{X}\beta + \tilde{\varepsilon}$, where $\tilde{Y} = \Sigma^{-1/2} Y$, $\tilde{X} = \Sigma^{-1/2} X$, and $\tilde{\varepsilon} = \Sigma^{-1/2}(Z\gamma + \varepsilon)$. Thus the reexpressed mixed model is in the standard Gauss-Markov setup since $\mathrm{E}[\tilde{\varepsilon}] = 0$ and $\mathrm{Cov}[\tilde{\varepsilon}] = I$, and the BLUE of β is $\hat{\beta} = \left(\tilde{X}^T\tilde{X}\right)^{-1}\tilde{X}^T\tilde{Y} = (X^T\Sigma^{-1}X)^{-1} X^T\Sigma^{-1} Y$. We leave it to the reader to verify that the BLUE of β is unique.

Let us first find out what the BLUP of $\boldsymbol{b}^T\boldsymbol{\gamma}$ would be if $\boldsymbol{\beta}$ were known. The best linear predictor of $\boldsymbol{b}^T\boldsymbol{\gamma}$ can be obtained by a linear regression of $\boldsymbol{b}^T\boldsymbol{\gamma}$ on $\boldsymbol{Y}$. So the best linear predictor $h_0 + \boldsymbol{h}_1^T\boldsymbol{Y}$ has the form

$$\begin{aligned} \boldsymbol{h}_1 &= [\mathrm{Cov}(\boldsymbol{Y})]^{-1}\mathrm{Cov}\left[\boldsymbol{Y}, \boldsymbol{b}^T\boldsymbol{\gamma}\right] = \boldsymbol{\Sigma}^{-1}\boldsymbol{ZDb}, \\ h_0 &= \mathrm{E}\left[\boldsymbol{b}^T\boldsymbol{\gamma}\right] - \boldsymbol{h}_1^T\mathrm{E}[\boldsymbol{Y}] = -\boldsymbol{h}_1^T\boldsymbol{X\beta}, \text{ and thus} \\ h_0 + \boldsymbol{h}_1^T\boldsymbol{Y} &= \boldsymbol{h}_1^T(\boldsymbol{Y} - \boldsymbol{X\beta}). \end{aligned}$$

If $\boldsymbol{\beta}$ were known, the BLUP of $\boldsymbol{b}^T\boldsymbol{\gamma}$ would be

$$\boldsymbol{h}_1^T(\boldsymbol{Y} - \boldsymbol{X\beta}) = \boldsymbol{b}^T\boldsymbol{DZ}^T\boldsymbol{\Sigma}^{-1}(\boldsymbol{Y} - \boldsymbol{X\beta}).$$

Since $\boldsymbol{\beta}$ is unknown, it therefore makes sense to replace it by its BLUE. Thus the BLUP of $\boldsymbol{b}^T\boldsymbol{\gamma}$ should be $\boldsymbol{b}^T\boldsymbol{DZ}^T\boldsymbol{\Sigma}^{-1}\left(\boldsymbol{Y} - \boldsymbol{X}\hat{\boldsymbol{\beta}}\right)$ and the BLUP of $\boldsymbol{\gamma}$ should be $\hat{\boldsymbol{\gamma}} = \boldsymbol{DZ}^T\boldsymbol{\Sigma}^{-1}\left(\boldsymbol{Y} - \boldsymbol{X}\hat{\boldsymbol{\beta}}\right)$ (proof given below). We leave it to the reader to show that the BLUE of $\boldsymbol{\gamma}$ is unique.

We now write down expressions for the BLUE and BLUP of $\boldsymbol{\beta}$ and $\boldsymbol{\gamma}$ in the following result.

Theorem 11.10.1. *The BLUE of $\boldsymbol{\beta}$ and the BLUP for $\boldsymbol{\gamma}$ in the mixed linear model are given by*

$$\hat{\boldsymbol{\beta}} = \left(\boldsymbol{X}^T\boldsymbol{\Sigma}^{-1}\boldsymbol{X}\right)^{-1}\boldsymbol{X}^T\boldsymbol{\Sigma}^{-1}\boldsymbol{Y}, \quad \hat{\boldsymbol{\gamma}} = \boldsymbol{D}^{-1}\boldsymbol{Z}^T\boldsymbol{\Sigma}^{-1}\left(\boldsymbol{Y} - \boldsymbol{X}\hat{\boldsymbol{\beta}}\right).$$

Remark 11.10.1. Unlike in the Gauss-Markov model discussed in Section 11.3 of this chapter, the normal equations for estimating $\boldsymbol{\beta}$ are no longer of the form $\boldsymbol{X}^T\boldsymbol{X\beta} = \boldsymbol{X}^T\boldsymbol{Y}$. Except in some balanced random- and mixed-effects models, as an example given below will show, the BLUE of $\boldsymbol{\beta}$ now depends on $\boldsymbol{\Sigma}$, the unknown covariance matrix of $\boldsymbol{Y}$. Similarly prediction of $\boldsymbol{\gamma}$ also requires the knowledge of $\boldsymbol{\Sigma}$. In practice, however, $\boldsymbol{\Sigma}$ is not known and has to be estimated from the data. In order to obtain the approximate BLUE $\hat{\boldsymbol{\beta}}$ of $\boldsymbol{\beta}$ and the BLUP $\hat{\boldsymbol{\gamma}}$ of $\boldsymbol{\gamma}$, we need to use the estimate $\hat{\boldsymbol{\Sigma}}$ of $\boldsymbol{\Sigma}$ in the formulas for $\hat{\boldsymbol{\beta}}$ and $\hat{\boldsymbol{\gamma}}$. The problem of estimating $\boldsymbol{\Sigma}$ which involves estimation of the variance components will be discussed later.

The following result gives necessary and sufficient conditions for $\hat{\boldsymbol{\gamma}}$ to be BLUP of $\boldsymbol{\gamma}$. Its proof is left as an exercise.

Lemma 11.10.1. *For the mixed model, $\hat{\boldsymbol{\gamma}}$ is BLUP for γ if and only if for any $\boldsymbol{b} \in \mathbb{R}^m$,*

(i) $E\left[\boldsymbol{b}^T\hat{\boldsymbol{\gamma}} - \boldsymbol{b}^T\boldsymbol{\gamma}\right] = 0$ *and*

(ii) $Cov\left[\boldsymbol{b}^T\hat{\boldsymbol{\gamma}} - \boldsymbol{b}^T\boldsymbol{\gamma}, \boldsymbol{l}^T\boldsymbol{Y}\right] = 0$ *for any $\boldsymbol{l} \in \boldsymbol{R}^n$ satisfying the condition $\boldsymbol{X}^T\boldsymbol{l} = \boldsymbol{0}$.*

Now let us check if $\hat{\boldsymbol{\gamma}} = \boldsymbol{DZ}^T\boldsymbol{\Sigma}^{-1}\left(\boldsymbol{Y} - \boldsymbol{X}\hat{\boldsymbol{\beta}}\right)$, where $\hat{\boldsymbol{\beta}}$ is the BLUE, is indeed the BLUP of $\boldsymbol{\gamma}$. Since $\hat{\boldsymbol{\beta}}$ is unbiased for $\boldsymbol{\beta}$, Condition (i) of Lemma 11.10.1 holds. If Condition (ii) of this lemma also holds, then $\hat{\boldsymbol{\gamma}}$ is the BLUP of $\boldsymbol{\gamma}$. Denoting $\boldsymbol{X}\hat{\boldsymbol{\beta}} = \boldsymbol{RY}$ where $\boldsymbol{R} = \boldsymbol{X}(\boldsymbol{X}^T\boldsymbol{\Sigma}^{-1}\boldsymbol{X})^{-1}\boldsymbol{X}^T\boldsymbol{\Sigma}^{-1}$, we have $\hat{\boldsymbol{\gamma}} = \boldsymbol{DZ}^T\boldsymbol{\Sigma}^{-1}(\boldsymbol{I} - \boldsymbol{R})\boldsymbol{Y}$. Therefore for any $\boldsymbol{l} \in \mathbb{R}^n$ with $\boldsymbol{X}^T\boldsymbol{l} = \boldsymbol{0}$, we have

$$\begin{aligned}\operatorname{Cov}\left[\boldsymbol{b}^T\hat{\boldsymbol{\gamma}} - \boldsymbol{b}^T\boldsymbol{\gamma}, \boldsymbol{l}^T\boldsymbol{Y}\right] &= \operatorname{Cov}\left[\boldsymbol{b}^T\hat{\boldsymbol{\gamma}}, \boldsymbol{l}^T\boldsymbol{Y}\right] - \operatorname{Cov}\left[\boldsymbol{b}^T\boldsymbol{\gamma}, \boldsymbol{l}^T\boldsymbol{Y}\right]\\ &= \operatorname{Cov}\left[\boldsymbol{b}^T\hat{\boldsymbol{\gamma}}, \boldsymbol{l}^T\boldsymbol{Y}\right] - \boldsymbol{b}^T\boldsymbol{D}\boldsymbol{Z}^T\boldsymbol{l}\\ &= \operatorname{Cov}\left[\boldsymbol{b}^T\boldsymbol{D}\boldsymbol{Z}^T\boldsymbol{\Sigma}^{-1}(\boldsymbol{I}-\boldsymbol{X}\boldsymbol{R})\boldsymbol{Y}, \boldsymbol{l}^T\boldsymbol{Y}\right] - \boldsymbol{b}^T\boldsymbol{D}\boldsymbol{Z}^T\boldsymbol{l}\\ &= \boldsymbol{b}^T\boldsymbol{D}\boldsymbol{Z}^T\boldsymbol{\Sigma}^{-1}(\boldsymbol{I}-\boldsymbol{X}\boldsymbol{R})\boldsymbol{\Sigma}\boldsymbol{l} - \boldsymbol{b}^T\boldsymbol{D}\boldsymbol{Z}^T\boldsymbol{l} = -\boldsymbol{b}^T\boldsymbol{D}\boldsymbol{Z}^T\boldsymbol{\Sigma}^{-1}\boldsymbol{X}\boldsymbol{R}\boldsymbol{\Sigma}\boldsymbol{l}.\end{aligned}$$

Since $\boldsymbol{X}^T\boldsymbol{l} = \boldsymbol{0}$, we have

$$\boldsymbol{R}\boldsymbol{\Sigma}\boldsymbol{l} = \boldsymbol{X}\left(\boldsymbol{X}^T\boldsymbol{\Sigma}^{-1}\boldsymbol{X}\right)^{-1}\boldsymbol{X}^T\boldsymbol{\Sigma}^{-1}\boldsymbol{\Sigma}\boldsymbol{l} = \left(\boldsymbol{X}^T\boldsymbol{\Sigma}^{-1}\boldsymbol{X}\right)^{-1}\boldsymbol{X}^T\boldsymbol{l} = \boldsymbol{0}.$$

This shows that Condition (ii) of Lemma 11.10.1 holds and hence $\hat{\boldsymbol{\gamma}}$ is the BLUP of $\boldsymbol{\gamma}$.

11.10.2 Mixed Model Equations

We have already seen that the expressions for the BLUE $\hat{\boldsymbol{\beta}}$ of $\boldsymbol{\beta}$ and the BLUP $\hat{\boldsymbol{\gamma}}$ of $\boldsymbol{\gamma}$ involve the inverse of $\operatorname{Cov}[\boldsymbol{Y}] = \boldsymbol{\Sigma}$, which is an $n \times n$ matrix, where n is the total number of observations. In many cases, $\boldsymbol{\Sigma}$ does not have a simple expression, and, if n is large, calculation of $\boldsymbol{\Sigma}^{-1}$ may be quite time consuming even on a modern computer. Are there some simpler formulas for obtaining $\hat{\boldsymbol{\beta}}$ and $\hat{\boldsymbol{\gamma}}$? The answer is yes and it turns out that $\hat{\boldsymbol{\beta}}$ and $\hat{\boldsymbol{\gamma}}$ can be obtained by solving the so-called *Mixed Model Equations,* which require inversions of matrices whose dimensions are much smaller than that of $\boldsymbol{\Sigma}$. Let us first see what these equations are, provide intuitive arguments which lead to them, and then show that their solutions are the BLUE and the BLUP.

Mixed model equations take the form (assuming that $\boldsymbol{D} = \operatorname{Cov}[\boldsymbol{\gamma}]$ is positive definite)

$$\begin{pmatrix}\boldsymbol{X}^T\boldsymbol{X} & \boldsymbol{X}^T\boldsymbol{Z}\\ \boldsymbol{Z}^T\boldsymbol{X} & \sigma_0^2\boldsymbol{D}^{-1} + \boldsymbol{Z}^T\boldsymbol{Z}\end{pmatrix}\begin{pmatrix}\boldsymbol{\beta}\\ \boldsymbol{\gamma}\end{pmatrix} = \begin{pmatrix}\boldsymbol{X}^T\boldsymbol{Y}\\ \boldsymbol{Z}^T\boldsymbol{Y}\end{pmatrix}. \tag{9}$$

Theorem 11.10.2.

(a) *The mixed model equations have a unique solution.*
(b) *Let $\hat{\boldsymbol{\beta}}$ and $\hat{\boldsymbol{\gamma}}$ be the solution to the mixed model equations. Then $\hat{\boldsymbol{\beta}}$ is the BLUE for $\boldsymbol{\beta}$ and $\hat{\boldsymbol{\gamma}}$ is the BLUP for $\boldsymbol{\gamma}$.*

The proof of this theorem is postponed to Section 11.10.3. We now give a simple example to show how the mixed model equations can be used in the one-factor random-effects model.

Example 11.10.6. Consider a one-factor random-effects model described in Example 11.10.1. Here $\boldsymbol{X}$ is an $n \times 1$ vector of 1's, where $n = n_1 + \cdots + n_k$ is the total number of observations. If $\pi_i = n_i/(\sigma^2/\sigma_1^2 + n_i)$, then we show below that the BLUE and BLUP for μ and γ_i are given by (denoting the random effects by $\{\gamma_i\}$ instead of $\{\alpha_i\}$)

$$\hat{\mu} = \sum(1-\pi_i)n_i\bar{Y}_{i\cdot}\Big/\sum(1-\pi_i)n_i, \quad \hat{\gamma}_i = \pi_i(\bar{Y}_{i\cdot} - \hat{\mu}).$$

For a balanced model, we have $\hat{\mu} = \bar{Y}_{..}$ and $\hat{\gamma}_i = \pi(\bar{Y}_{i.} - \bar{Y}_{..})$ with $\pi = n_0/(\sigma^2/\sigma_1^2 + n_0)$, where $n_0 = n_1 = \cdots = n_k$. In this case, the BLUE of μ does not depend on the variance components σ_1^2 and σ^2. Also note that for a balanced fixed-effects model with the constraint $\sum \gamma_i = 0$, the least squares estimate of γ_i is $\bar{Y}_{i.} - \bar{Y}_{...}$. The BLUP of γ_i in the random-effects model is obtained by shrinking $\bar{Y}_{i.} - \bar{Y}_{..}$ toward zero (since $0 < \pi < 1$). If $\sigma_1 \to \infty$, then $\pi \to 1$ and the BLUP is then the same as the BLUE of γ_i for the fixed effects case. In a sense then, the mixed model is the same as the fixed-effects model when $\sigma_1 = \infty$.

Let us now see how we can obtain the estimates given above using the mixed model equations. Denote $Y_{..} = n\bar{Y}_{..}$, and $Y_{i.} = n_i\bar{Y}_{i.}$, $i = 1, \ldots, k$. Note that in this case, $\boldsymbol{\Sigma} = \sigma_1^2\boldsymbol{Z}\boldsymbol{Z}^T + \sigma^2\boldsymbol{I}$ with $\boldsymbol{D} = \sigma_1^2\boldsymbol{I}$. So we have $\boldsymbol{X}^T\boldsymbol{X} = n\boldsymbol{I}, \boldsymbol{X}^T\boldsymbol{Z} = (n_1, \ldots, n_k)$, and $\boldsymbol{Z}^T\boldsymbol{Z}$ is a $k \times k$ diagonal matrix with diagonal elements $n_1, \ldots, n_k$. So the mixed model equations are

$$\begin{pmatrix} n & n_1 & \cdots & n_k \\ n_1 & \sigma^2/\sigma_1^2 + n_1 & 0 & 0 \\ \cdot & 0 & & 0 \\ \cdot & \cdot & \cdot & \cdot \\ \cdot & \cdot & \cdot & \cdot \\ n_k & 0 & 0 & \sigma^2/\sigma_1^2 + n_k \end{pmatrix} \begin{pmatrix} \mu \\ \gamma_1 \\ \cdot \\ \cdot \\ \cdot \\ \gamma_k \end{pmatrix} = \begin{pmatrix} Y_{..} \\ Y_{1.} \\ \cdot \\ \cdot \\ \cdot \\ Y_{k.} \end{pmatrix},$$

or

$$n\mu + n_1\gamma_1 + \cdots + n_k\gamma_k = Y_{..}, \text{ and}$$
$$n_i\mu + \left(\sigma^2/\sigma_1^2 + n_i\right)\gamma_i = Y_{i.}, \quad i = 1, \ldots, k.$$

The last k equations can be written as

$$\pi_i\mu + \gamma_i = \pi_i\bar{Y}_{i.}, \text{ ie, } \mu = \pi_i(\bar{Y}_{i.} - \mu), \quad i = 1, \ldots, k.$$

Substitute γ_i by $\pi_i(\bar{Y}_{i.} - \mu)$ in the first equation and solve for μ to obtain

$$\hat{\mu} = \sum(1 - \pi_i)n_i\bar{Y}_{i.} \Big/ \sum(1 - \pi_i)n_i \text{ and } \hat{\gamma}_i = \pi_i(\bar{Y}_{i.} - \hat{\mu}).$$

11.10.3 Motivation for Mixed Model Equations

We now provide a motivation for the mixed model equations under the assumption of normality. Let us assume that $\boldsymbol{\gamma} \sim N_q(\boldsymbol{0}, \boldsymbol{D})$, $\boldsymbol{\varepsilon} \sim N_n(\boldsymbol{0}, \sigma^2\boldsymbol{I})$, and that $\boldsymbol{\varepsilon}$ and $\boldsymbol{\gamma}$ are independent. The main idea is to treat $\boldsymbol{\gamma}$ as a parameter even though it is random. Note that conditional on $\boldsymbol{\gamma}$, $\boldsymbol{Y} \sim N_n(\boldsymbol{X\beta} + \boldsymbol{Z\gamma}, \sigma^2\boldsymbol{I})$. The joint pdf of $\boldsymbol{Y}$ and $\boldsymbol{\gamma}$ can be written as

$$\begin{aligned} f_{\boldsymbol{Y},\boldsymbol{\gamma}}(\boldsymbol{y}, \boldsymbol{\gamma}) &= f_{\boldsymbol{Y}|\boldsymbol{\gamma}}(\boldsymbol{y}|\boldsymbol{\gamma})f_{\boldsymbol{\gamma}}(\boldsymbol{\gamma}) \\ &= c\exp\left[-\|\boldsymbol{y} - \boldsymbol{X\beta} - \boldsymbol{Z\gamma}\|^2/\left(2\sigma^2\right)\right]\exp\left(-\boldsymbol{\gamma}^T\boldsymbol{D}^{-1}\boldsymbol{\gamma}/2\right), \end{aligned}$$

where the constant c depends on σ and $\boldsymbol{D}$. If we treat $\boldsymbol{\gamma}$ as nonrandom, and try to estimate $\boldsymbol{\beta}$ and $\boldsymbol{\gamma}$, then the likelihood function is

$$L(\boldsymbol{\beta}, \boldsymbol{\gamma}) = c\exp\left[-\|\boldsymbol{Y} - \boldsymbol{X\beta} - \boldsymbol{Z\gamma}\|^2/\left(2\sigma^2\right)\right]\exp\left(-\boldsymbol{\gamma}^T\boldsymbol{D}^{-1}\boldsymbol{\gamma}/2\right).$$

Maximizing L with respect to $\boldsymbol{\beta}$ and $\boldsymbol{\gamma}$ is equivalent to minimizing $-2\log L$ with respect to $\boldsymbol{\beta}$ and $\boldsymbol{\gamma}$. Differentiating $-2\log L$ with respect to $\boldsymbol{\beta}$ and $\boldsymbol{\gamma}$ and equating the derivatives to zero lead to the following equations

$$\partial(-2\log L)/\partial\boldsymbol{\beta} = 0, \text{ ie, } \boldsymbol{X}^T\boldsymbol{X}\boldsymbol{\beta} + \boldsymbol{X}^T\boldsymbol{Z}\boldsymbol{\gamma} = \boldsymbol{X}^T\boldsymbol{Y}, \text{ and}$$
$$\partial(-2\log L)/\partial\boldsymbol{\gamma} = 0, \text{ ie, } \boldsymbol{Z}^T\boldsymbol{X} + \left(\boldsymbol{Z}^T\boldsymbol{Z} + \sigma^2\boldsymbol{D}^{-1}\right)\boldsymbol{\gamma} = \boldsymbol{Z}^T\boldsymbol{Y},$$

and these are the mixed model equations given in Eq. (9).

The proof of Theorem 11.10.2 requires the following result which can be derived from the Sherman-Morrison formula (Section B.1). The proof of this lemma is left as an exercise.

Lemma 11.10.2.

(a) $\boldsymbol{\Sigma}^{-1} = \sigma^{-2}\left[\boldsymbol{I} - \boldsymbol{Z}(\boldsymbol{Z}^T\boldsymbol{Z} + \sigma^2\boldsymbol{D}^{-1})^{-1}\boldsymbol{Z}^T\right]$

(b) $\boldsymbol{\Sigma}^{-1} = \sigma^{-2}\boldsymbol{Z}(\boldsymbol{Z}^T\boldsymbol{Z} + \sigma^2\boldsymbol{D}^{-1})^{-1}\boldsymbol{D}^{-1}$,

(c) $(\boldsymbol{Z}^T\boldsymbol{Z} + \sigma^2\boldsymbol{D}^{-1})^{-1}\boldsymbol{Z}^T = \boldsymbol{D}\boldsymbol{Z}^T\boldsymbol{\Sigma}^{-1}$.

Proof of Theorem 11.10.2. Part (a) is not difficult to check and it is left as an exercise.

In order prove part (b), it is enough to show that the solutions of $\hat{\boldsymbol{\beta}}$ and $\hat{\boldsymbol{\gamma}}$ of the mixed model equations are the same as those in Theorem 11.10.1.

If $\hat{\boldsymbol{\beta}}$ and $\hat{\boldsymbol{\gamma}}$ are the solutions of the mixed model equations, then

$$\boldsymbol{Z}^T\boldsymbol{X}\hat{\boldsymbol{\beta}} + \left(\boldsymbol{Z}^T\boldsymbol{Z} + \sigma^2\boldsymbol{D}^{-1}\right)\hat{\boldsymbol{\gamma}} = \boldsymbol{Z}^T\boldsymbol{Y}, \text{ and hence}$$
$$\hat{\boldsymbol{\gamma}} = \left(\boldsymbol{Z}^T\boldsymbol{Z} + \sigma^2\boldsymbol{D}^{-1}\right)^{-1}\boldsymbol{Z}^T\left(\boldsymbol{Y} - \boldsymbol{X}\hat{\boldsymbol{\beta}}\right) = \boldsymbol{D}\boldsymbol{Z}^T\boldsymbol{\Sigma}^{-1}\left(\boldsymbol{Y} - \boldsymbol{X}\hat{\boldsymbol{\beta}}\right),$$

using part (c) of Lemma 11.10.2. Thus, $\hat{\boldsymbol{\gamma}}$ has the same form as in Theorem 11.10.1 and hence this would be the BLUP if we can show that the solution $\hat{\boldsymbol{\beta}}$ of the mixed model equations is indeed the BLUE of $\boldsymbol{\beta}$.

The first set of equations in the mixed model equations are

$$\boldsymbol{X}^T\boldsymbol{X}\hat{\boldsymbol{\beta}} + \boldsymbol{X}^T\boldsymbol{Z}\hat{\boldsymbol{\gamma}} = \boldsymbol{X}^T\boldsymbol{Y}.$$

Substituting the expression of $\hat{\boldsymbol{\gamma}}$ in this equation we have

$$\boldsymbol{X}^T\boldsymbol{X}\hat{\boldsymbol{\beta}} + \boldsymbol{X}^T\boldsymbol{Z}\boldsymbol{D}\boldsymbol{Z}^T\boldsymbol{\Sigma}^{-1}\left(\boldsymbol{Y} - \boldsymbol{X}\hat{\boldsymbol{\beta}}\right) = \boldsymbol{X}^T\boldsymbol{Y}, \text{ or}$$
$$\boldsymbol{X}^T\left[\boldsymbol{I} - \boldsymbol{Z}\boldsymbol{D}\boldsymbol{Z}^T\boldsymbol{\Sigma}^{-1}\right]\boldsymbol{X}\hat{\boldsymbol{\beta}} = \boldsymbol{X}^T\left[\boldsymbol{I} - \boldsymbol{Z}\boldsymbol{D}\boldsymbol{Z}^T\boldsymbol{\Sigma}^{-1}\right]\boldsymbol{Y}.$$

Since

$$\boldsymbol{I} - \boldsymbol{Z}\boldsymbol{D}\boldsymbol{Z}^T\boldsymbol{\Sigma}^{-1} = \boldsymbol{I} - \left[\boldsymbol{\Sigma} - \sigma^2\boldsymbol{I}\right]\boldsymbol{\Sigma}^{-1} = \sigma^2\boldsymbol{\Sigma}^{-1},$$

we have

$$\boldsymbol{X}^T\sigma^2\boldsymbol{\Sigma}^{-1}\boldsymbol{X}\hat{\boldsymbol{\beta}} = \boldsymbol{X}^T\sigma^2\boldsymbol{\Sigma}^{-1}\boldsymbol{Y}, \text{ or } \hat{\boldsymbol{\beta}} = \left(\boldsymbol{X}^T\boldsymbol{\Sigma}^{-1}\boldsymbol{X}\right)^{-1}\boldsymbol{X}^T\boldsymbol{\Sigma}^{-1}\boldsymbol{Y}.$$

The expression is the same as in Theorem 11.10.1. Thus the solution $\hat{\boldsymbol{\beta}}$ of the mixed model equations is the BLUE of β. □

11.10.4 Estimation of Variance Components

For the model given in Eq. (7)

$$\begin{aligned}\mathrm{E}[\boldsymbol{Y}] &= \boldsymbol{X\beta}, \text{ and}\\ \boldsymbol{\Sigma} = \mathrm{Cov}[\boldsymbol{Y}] &= \sigma_1^2 \boldsymbol{Z}_1\boldsymbol{Z}_1^T + \cdots + \sigma_r^2 \boldsymbol{Z}_r\boldsymbol{Z}_r^T + \sigma_0^2 I.\end{aligned}$$

Note that we have changed the notations a bit and now σ^2 is denoted by σ_0^2. We now focus on the problem of estimating the variance components $\sigma_0^2, \sigma_1^2, \ldots, \sigma_r^2$.

There are a number of well-known methods for estimating the variance components including

(i) Henderson's method III,
(ii) Maximum likelihood,
(iii) Restricted maximum likelihood (REML), and
(iv) MINQUE (minimum norm quadratic unbiased estimation).

Here we discuss only the first three methods. Computer packages such as R can be used to estimate the variance components using these procedures. Henderson's method and MINQUE do not require any distributional assumptions, whereas the maximum likelihood and the REML methods require the assumptions of normality. It should be pointed out that these methods may not produce the same estimates of the variance components. In unbalanced cases, the estimates obtained by employing any of these procedures usually do not have explicit expressions. Detailed discussion on all these methods can be found in the book "Linear Models" by Searle [60] (Chapters 9–11).

Henderson's Method III

Suppose that we want to estimate one of the variance components, say σ_r^2. Let $\boldsymbol{Q}$ be the projection on the column space of the augmented matrix $[\boldsymbol{X}, \boldsymbol{Z}_1, \ldots, \boldsymbol{Z}_r]$ and let $\boldsymbol{Q}_r$ be the projection on the column space of the matrix $[\boldsymbol{X}, \boldsymbol{Z}_1, \ldots, \boldsymbol{Z}_{r-1}]$. Then $\boldsymbol{QY}$ is the vector of fitted values when we fit the model $\boldsymbol{Y} = \boldsymbol{X\beta} + \boldsymbol{Z}_1\boldsymbol{\gamma}_1 + \cdots + \boldsymbol{Z}_r\boldsymbol{\gamma}_r + \boldsymbol{\varepsilon}$ pretending that $\boldsymbol{\gamma}_1, \ldots, \boldsymbol{\gamma}_r$ are nonrandom unknown parameters. Similarly $\boldsymbol{Q}_r\boldsymbol{Y}$ is the vector of fitted values when we fit the model $\boldsymbol{Y} = \boldsymbol{X\beta} + \boldsymbol{Z}_1\boldsymbol{\gamma}_1 + \cdots + \boldsymbol{Z}_{r-1}\boldsymbol{\gamma}_{r-1} + \boldsymbol{\varepsilon}$ pretending that $\boldsymbol{\gamma}_1, \ldots, \boldsymbol{\gamma}_{r-1}$ are nonrandom parameters. Note that an unbiased estimate of σ_0^2 is given by $\hat{\sigma}_0^2 = \|\boldsymbol{Y} - \boldsymbol{QY}\|^2/(n - rank(\boldsymbol{Q}))$. Henderson's estimate of σ_r^2 is given by

$$\hat{\sigma}_r^2 = \left[\|\boldsymbol{QY} - \boldsymbol{Q}_r\boldsymbol{Y}\|^2 - trace(\boldsymbol{Q} - \boldsymbol{Q}_r)\hat{\sigma}_0^2\right] \Big/ trace\left(\boldsymbol{Z}_r^T(\boldsymbol{I} - \boldsymbol{Q}_r)\boldsymbol{Z}_r\right). \tag{10}$$

Since $\sigma_r^2 > 0$, its estimate is usually taken to be $\max(\hat{\sigma}_r^2, 0)$.

Let us now see how this estimate is derived. First of all note that $(\boldsymbol{Q} - \boldsymbol{Q}_r)\boldsymbol{X\beta} = \boldsymbol{0}$ and $(\boldsymbol{Q} - \boldsymbol{Q}_r)\boldsymbol{Z}_i = \boldsymbol{0}$, $i = 1, \ldots, r-1$. Hence

$$\begin{aligned}(\boldsymbol{Q} - \boldsymbol{Q}_r)\boldsymbol{Y} &= (\boldsymbol{Q} - \boldsymbol{Q}_r)(\boldsymbol{X\beta} + \boldsymbol{Z}_1\boldsymbol{\gamma}_1 + \cdots + \boldsymbol{Z}_r\boldsymbol{\gamma}_r + \boldsymbol{\varepsilon})\\ &= (\boldsymbol{Q} - \boldsymbol{Q}_r)\boldsymbol{Z}_r\boldsymbol{\gamma}_r + (\boldsymbol{Q} - \boldsymbol{Q}_r)\boldsymbol{\varepsilon}.\end{aligned}$$

Then

$$\begin{aligned}
\mathrm{E}\left[\|\boldsymbol{QY}-\boldsymbol{Q}_r\boldsymbol{Y}\|^2\right] &= \mathrm{E}\left[\|(\boldsymbol{Q}-\boldsymbol{Q}_r)\boldsymbol{Z}_r\boldsymbol{\gamma}_r+(\boldsymbol{Q}-\boldsymbol{Q}_r)\boldsymbol{\varepsilon}\|^2\right]\\
&= \mathrm{E}\left[\|(\boldsymbol{Q}-\boldsymbol{Q}_r)\boldsymbol{Z}_r\boldsymbol{\gamma}_r\|^2\right]+\mathrm{E}\left[\|(\boldsymbol{Q}-\boldsymbol{Q}_r)\boldsymbol{\varepsilon}\|^2\right]\\
&= \sigma_r^2\, trace\left((\boldsymbol{Q}-\boldsymbol{Q}_r)\boldsymbol{Z}_r\boldsymbol{Z}_r^T(\boldsymbol{Q}-\boldsymbol{Q}_r)^T\right)+\sigma_0^2\, trace\left((\boldsymbol{Q}-\boldsymbol{Q}_r)(\boldsymbol{Q}-\boldsymbol{Q}_r)^T\right)\\
&= \sigma_r^2\, trace\left(\boldsymbol{Z}_r^T(\boldsymbol{Q}-\boldsymbol{Q}_r)^T(\boldsymbol{Q}-\boldsymbol{Q}_r)\boldsymbol{Z}_r\right)+\sigma_0^2\, trace\left((\boldsymbol{Q}-\boldsymbol{Q}_r)(\boldsymbol{Q}-\boldsymbol{Q}_r)^T\right).
\end{aligned}$$

Now use the facts that $\boldsymbol{Q}$ and $\boldsymbol{Q}_r$ are symmetric and $\boldsymbol{Q}-\boldsymbol{Q}_r$ is also a projection, and consequently, $(\boldsymbol{Q}-\boldsymbol{Q}_r)^T(\boldsymbol{Q}-\boldsymbol{Q}_r)=\boldsymbol{Q}-\boldsymbol{Q}_r$ and $(\boldsymbol{Q}-\boldsymbol{Q}_r)(\boldsymbol{Q}-\boldsymbol{Q}_r)^T=\boldsymbol{Q}-\boldsymbol{Q}_r$. So we have

$$\begin{aligned}
\mathrm{E}\left[\|\boldsymbol{QY}-\boldsymbol{Q}_r\boldsymbol{Y}\|^2\right] &= \sigma_r^2\, trace\left(\boldsymbol{Z}_r^T(\boldsymbol{Q}-\boldsymbol{Q}_r)\boldsymbol{Z}_r\right)+\sigma_0^2\, trace(\boldsymbol{Q}-\boldsymbol{Q}_r)\\
&= \sigma_r^2\, trace\left(\boldsymbol{Z}_r^T(I-\boldsymbol{Q}_r)\boldsymbol{Z}_r\right)+\sigma_0^2\, trace(\boldsymbol{Q}-\boldsymbol{Q}_r).
\end{aligned}$$

The last step is justified since $\boldsymbol{Z}_r^T\boldsymbol{QZ}_r=\boldsymbol{Z}_r^T\boldsymbol{Q}^T\boldsymbol{QZ}_r=\boldsymbol{Z}_r^T\boldsymbol{Z}_r$. From the last expression we now see that an unbiased estimate of σ_r^2 is of the form given above.

Except in the case of balanced models, Henderson's estimates for the variance components do not have nice forms. Here we give an example for the one-factor case.

Example 11.10.7. Consider a one-factor (random) ANOVA model as in Example 11.10.6. Here

$$\begin{aligned}
\|\boldsymbol{Y}-\boldsymbol{QY}\|^2 &= \sum_i\sum_j\left(Y_{ij}-\bar{Y}_{i.}\right)^2 = SSE, \text{ and}\\
\|\boldsymbol{QY}-\boldsymbol{Q}_1\boldsymbol{Y}\|^2 &= \sum_i\sum_j(\bar{Y}_{i.}-\bar{Y}_{..})^2=\sum n_i(\bar{Y}_{i.}-\bar{Y}_{..})^2 = SSTR.
\end{aligned}$$

Assuming that $n=n_1+\cdots+n_k$ is the total number of observations, we have $n-rank(\boldsymbol{Q})=n-k$ and $trace(\boldsymbol{Q}-\boldsymbol{Q}_1)=rank(\boldsymbol{Q})-rank(\boldsymbol{Q}_1)=k-1$. So an unbiased estimate of σ_0^2 is given by

$$\hat{\sigma}_0^2=\|\boldsymbol{Y}-\boldsymbol{QY}\|^2/(n-rank(\boldsymbol{Q}))=SSE/(n-k)=MSE.$$

Now let $\bar{\gamma}=\sum n_i\gamma_i/n$. Then

$$\begin{aligned}
\mathrm{E}\left[\|\boldsymbol{QY}-\boldsymbol{Q}_1\boldsymbol{Y}\|^2\right] &= \mathrm{E}[SSTR]\\
&= \mathrm{E}\left[\sum n_i(\gamma_i-\bar{\gamma}+\bar{\varepsilon}_{i.}-\varepsilon_{..})^2\right]\\
&= \left(n-\sum n_i^2/n\right)\sigma_1^2+(k-1)\sigma_0^2.
\end{aligned}$$

So an unbiased estimate of σ_1^2 is given by

$$\hat{\sigma}_1^2=\left[SSTR-(k-1)\hat{\sigma}_0^2\right]\Big/\left[n-\sum n_i^2/n\right]$$

$$= \left[SSTR - (k-1)MSE\right] \Big/ \left[n - \sum n_i^2/n\right]$$
$$= (k-1)\left[MSTR - MSE\right] \Big/ \left[n - \sum n_i^2/n\right].$$

In the balanced case, $n = n_0 a$ where $n_i \equiv n_0$ for all i. It is easy to check that the Henderson's estimate of σ_1^2 is then given by $\hat{\sigma}_1^2 = [MSTR - MSE]/n_0$.

Maximum Likelihood

The maximum likelihood method (under the assumption of joint normality of $\boldsymbol{\gamma}_1, \ldots, \boldsymbol{\gamma}_r$ and $\boldsymbol{\varepsilon}$) jointly estimates $\boldsymbol{\beta}$ and the variance components. We rewrite the mixed linear model given in Eq. (7) as

$$\boldsymbol{Y} = \boldsymbol{X\beta} + \sum_{i=0}^{r} \boldsymbol{Z}_i \boldsymbol{\gamma}_i, \tag{11a}$$

$$\mathrm{E}[\boldsymbol{Y}] = \boldsymbol{X\beta}, \text{ and } \boldsymbol{\Sigma}(\boldsymbol{\theta}) = \mathrm{Cov}[\boldsymbol{Y}] = \sum_{i=0}^{r} \sigma_i^2 \boldsymbol{Z}_i \boldsymbol{Z}_i^T, \tag{11b}$$

where $\boldsymbol{Z}_0 = \boldsymbol{I}$, $\boldsymbol{\gamma}_0 = \boldsymbol{\varepsilon}$, and $\boldsymbol{\theta} = \left(\sigma_0^2, \sigma_1^2, \ldots, \sigma_r^2\right)^T$. So we have $\boldsymbol{Y} \sim N_n\left(\boldsymbol{X\beta}, \boldsymbol{\Sigma}(\boldsymbol{\theta})\right)$. The likelihood therefore is

$$L(\boldsymbol{\beta}, \boldsymbol{\theta}) = c\left[1/|\boldsymbol{\Sigma}(\boldsymbol{\theta})|^{1/2}\right] \exp\left[-(\boldsymbol{Y} - \boldsymbol{X\beta})^T \boldsymbol{\Sigma}(\boldsymbol{\theta})^{-1} (\boldsymbol{Y} - \boldsymbol{X\beta})/2\right],$$

where $|\boldsymbol{\Sigma}(\boldsymbol{\theta})| = \det\left[\boldsymbol{\Sigma}(\boldsymbol{\theta})\right]$ and $c > 0$ is a constant that does not depend on $\boldsymbol{\beta}$ and $\boldsymbol{\theta}$. Maximizing the likelihood with respect to $\boldsymbol{\beta}$ and $\boldsymbol{\theta}$ is equivalent to minimizing $-2\log L$ with respect to $\boldsymbol{\beta}$ and $\boldsymbol{\theta}$. Note that

$$-2\log L(\boldsymbol{\beta}, \boldsymbol{\theta}) = -2\log(c) + (\boldsymbol{Y} - \boldsymbol{X\beta})^T \boldsymbol{\Sigma}(\boldsymbol{\theta})^{-1} (\boldsymbol{Y} - \boldsymbol{X\beta}) + \log(|\boldsymbol{\Sigma}(\boldsymbol{\theta})|).$$

We need to differentiate $-2\log L$ with respect to $\boldsymbol{\beta}$ and θ_i, $i = 0, \ldots, r$, and equate the derivatives to zero. Calculations will show that

$$\partial(-2\log L)/\partial\boldsymbol{\beta} = 0, \text{ ie, } \boldsymbol{X}^T \boldsymbol{\Sigma}(\boldsymbol{\theta})^{-1} \boldsymbol{X\beta} = \boldsymbol{X}^T \boldsymbol{\Sigma}(\boldsymbol{\theta})^{-1} \boldsymbol{Y}, \text{ and}$$
$$\partial(-2\log L)/\partial\theta_i = 0, \text{ ie,}$$
$$(\boldsymbol{Y} - \boldsymbol{X\beta})^T \left[\partial\boldsymbol{\Sigma}(\boldsymbol{\theta})^{-1}/\partial\theta_i\right] (\boldsymbol{Y} - \boldsymbol{X\beta}) + \partial\left(\log|\boldsymbol{\Sigma}(\boldsymbol{\theta})|\right)/\partial\theta_i = 0, \quad i = 0, \ldots, r.$$

Note that the first set of equations are the same as the normal equations (for estimating $\boldsymbol{\beta}$) described in Section 11.10.1. The second set of equations need simplifications. Noting that $\partial\boldsymbol{\Sigma}(\boldsymbol{\theta})/\partial\theta_i = \boldsymbol{Z}_i \boldsymbol{Z}_i^T$ and using results from Section B.5, we have

$$\partial\boldsymbol{\Sigma}(\boldsymbol{\theta})^{-1}/\partial\theta_i = -\boldsymbol{\Sigma}(\boldsymbol{\theta})^{-1}\left[\partial\boldsymbol{\Sigma}(\boldsymbol{\theta})/\partial\theta_i\right]\boldsymbol{\Sigma}(\boldsymbol{\theta})^{-1}$$
$$= -\boldsymbol{\Sigma}(\boldsymbol{\theta})^{-1} \boldsymbol{Z}_i \boldsymbol{Z}_i^T \boldsymbol{\Sigma}(\boldsymbol{\theta})^{-1}, \text{ and}$$
$$\partial\left(\log|\boldsymbol{\Sigma}(\boldsymbol{\theta})|\right)/\partial\theta_i = trace\left(\boldsymbol{\Sigma}(\boldsymbol{\theta})^{-1} \partial\boldsymbol{\Sigma}(\boldsymbol{\theta})/\partial\theta_i\right)$$
$$= trace\left(\boldsymbol{Z}_i^T \boldsymbol{\Sigma}(\boldsymbol{\theta})^{-1} \boldsymbol{Z}_i\right).$$

Thus the likelihood equation involving derivative with respect to θ_i turns out to be

$$\begin{aligned}
0 &= (\boldsymbol{Y}-\boldsymbol{X\beta})^T\left[\partial\boldsymbol{\Sigma}(\boldsymbol{\theta})^{-1}/\partial\theta_i\right](\boldsymbol{Y}-\boldsymbol{X\beta}) + \partial(\log|\boldsymbol{\Sigma}(\boldsymbol{\theta})|)/\partial\theta_i \\
&= -(\boldsymbol{Y}-\boldsymbol{X\beta})^T\boldsymbol{\Sigma}(\boldsymbol{\theta})^{-1}\boldsymbol{Z}_i\boldsymbol{Z}_i^T\boldsymbol{\Sigma}(\boldsymbol{\theta})^{-1}(\boldsymbol{Y}-\boldsymbol{X\beta}) + trace\left(\boldsymbol{Z}_i^T\boldsymbol{\Sigma}(\boldsymbol{\theta})^{-1}\boldsymbol{Z}_i\right) \\
&= -\|\boldsymbol{Z}_i^T\boldsymbol{\Sigma}(\boldsymbol{\theta})^{-1}(\boldsymbol{Y}-\boldsymbol{X\beta})\|^2 + trace\left(\boldsymbol{Z}_i^T\boldsymbol{\Sigma}(\boldsymbol{\theta})^{-1}\boldsymbol{Z}_i\right).
\end{aligned}$$

So the likelihood equations are

$$\boldsymbol{X}^T\boldsymbol{\Sigma}(\boldsymbol{\theta})^{-1}\boldsymbol{X\beta} = \boldsymbol{X}^T\boldsymbol{\Sigma}(\boldsymbol{\theta})^{-1}\boldsymbol{Y}, \tag{12a}$$

$$trace\left(\boldsymbol{Z}_i^T\boldsymbol{\Sigma}(\boldsymbol{\theta})^{-1}\boldsymbol{Z}_i\right) = \|\boldsymbol{Z}_i^T\boldsymbol{\Sigma}(\boldsymbol{\theta})^{-1}(\boldsymbol{Y}-\boldsymbol{X\beta})\|^2, \quad i=0,\ldots,r. \tag{12b}$$

These equations have no explicit solutions except in some balanced cases and iterative methods are used to solve them numerically.

Restricted Maximum Likelihood

REML is a variant of the maximum likelihood method whereby the issue of estimation of $\boldsymbol{\beta}$ is entirely bypassed and the focus is entirely on estimating the variance components. In this method, the estimate of $\boldsymbol{\theta} = (\sigma_0^2,\ldots,\sigma_r^2)^T$ is obtained by solving the equations

$$trace\left(\boldsymbol{Z}_i^T\boldsymbol{\Sigma}(\boldsymbol{\theta})^{-1}(\boldsymbol{I}-\boldsymbol{M}(\boldsymbol{\theta}))\boldsymbol{Z}_i\right) = \|\boldsymbol{Z}_i^T\boldsymbol{\Sigma}(\boldsymbol{\theta})^{-1}(\boldsymbol{I}-\boldsymbol{M}(\boldsymbol{\theta}))\boldsymbol{Y}\|^2, \quad i=0,\ldots,r, \tag{13}$$

where $\boldsymbol{M}(\boldsymbol{\theta}) = \boldsymbol{X}\left(\boldsymbol{X}^T\boldsymbol{\Sigma}(\boldsymbol{\theta})^{-1}\boldsymbol{X}\right)^{-1}\boldsymbol{X}^T\boldsymbol{\Sigma}(\boldsymbol{\theta})^{-1}$. These equations usually have no explicit solutions and iterative methods are employed in numerical computations. Once an estimate $\hat{\boldsymbol{\theta}}$ of $\boldsymbol{\theta}$ is available, $\boldsymbol{\beta}$ can be estimated by solving the approximate normal equations

$$\boldsymbol{X}^T\boldsymbol{\Sigma}\left(\hat{\boldsymbol{\theta}}\right)^{-1}\boldsymbol{X\beta} = \boldsymbol{X}^T\boldsymbol{\Sigma}\left(\hat{\boldsymbol{\theta}}\right)^{-1}\boldsymbol{Y}.$$

We will provide a justification for the REML equations. Since $rank(\boldsymbol{X}) = p$, we can find a matrix $\boldsymbol{B}$ of order $(n-p)\times n$ which has rank $n-p$ and it satisfies the equation $\boldsymbol{BX} = \boldsymbol{0}$. Let $\tilde{\boldsymbol{Y}} = \boldsymbol{BY}$, $\tilde{\boldsymbol{Z}}_i = \boldsymbol{BZ}_i$, $i=0,\ldots,r$, premultiplying both sides of Eq. (11a) leads to a modified model

$$\tilde{\boldsymbol{Y}} = \sum_{i=0}^{r}\tilde{\boldsymbol{Z}}_i\boldsymbol{\gamma}_i, \text{ and}$$

$$\text{Cov}\left[\tilde{\boldsymbol{Y}}\right] = \sum_{i=0}^{r}\sigma_i^2\tilde{\boldsymbol{Z}}_i\tilde{\boldsymbol{Z}}_i^T = \tilde{\boldsymbol{\Sigma}}(\boldsymbol{\theta}), \text{ say.}$$

Since $\tilde{\boldsymbol{Y}} \sim N_{n-p}\left(\boldsymbol{0}, \tilde{\boldsymbol{\Sigma}}(\boldsymbol{\theta})\right)$, the likelihood (based on $\tilde{\boldsymbol{Y}}$) is

$$L(\boldsymbol{\theta}) = c\left[1/|\tilde{\boldsymbol{\Sigma}}(\boldsymbol{\theta})|^{1/2}\right]\exp\left[-\tilde{\boldsymbol{Y}}^T\tilde{\boldsymbol{\Sigma}}(\boldsymbol{\theta})^{-1}\tilde{\boldsymbol{Y}}/2\right],$$

where the constant $c > 0$ does not depend on $\boldsymbol{\theta}$. In order to obtain the MLE for $\boldsymbol{\theta}$ in this case, we differentiate $-2\log L$ with respect to θ_i, $i=0,\ldots,r$, and equate the derivatives to zero. Then, as in Eq. (12b), we have

$$trace\left(\tilde{Z}_i^T \tilde{\Sigma}(\theta)^{-1} \tilde{Z}_i\right) = \|\tilde{Z}_i^T \tilde{\Sigma}(\theta)^{-1} \tilde{Y}\|^2, \quad i = 0, \ldots, r.$$

Since $\tilde{\boldsymbol{Z}}_i = \boldsymbol{B}\boldsymbol{Z}_i$ and $\tilde{\boldsymbol{\Sigma}}(\boldsymbol{\theta}) = \boldsymbol{B}\boldsymbol{\Sigma}(\boldsymbol{\theta})\boldsymbol{B}^T$, the *i*th equation is

$$trace\left(\boldsymbol{Z}_i^T \boldsymbol{B}^T \left(\boldsymbol{B}\boldsymbol{\Sigma}(\boldsymbol{\theta})\boldsymbol{B}^T\right)^{-1} \boldsymbol{B}\boldsymbol{Z}_i\right) = \|\boldsymbol{Z}_i^T \boldsymbol{B}^T \left(\boldsymbol{B}\boldsymbol{\Sigma}(\boldsymbol{\theta})\boldsymbol{B}^T\right)^{-1} \boldsymbol{B}\boldsymbol{Y}\|^2.$$

The important fact is that the matrix $\boldsymbol{B}^T \left(\boldsymbol{B}\boldsymbol{\Sigma}(\boldsymbol{\theta})\boldsymbol{B}^T\right)^{-1} \boldsymbol{B}$ does not depend on the choice $\boldsymbol{B}$ as long as $\boldsymbol{B}\boldsymbol{X} = \boldsymbol{0}$ and $rank(\boldsymbol{B}) = n - rank(\boldsymbol{X})$. The following result turns out to be true.

Lemma 11.10.3. *Assume that* $\boldsymbol{B}\boldsymbol{X} = \boldsymbol{0}$ *and* $rank(\boldsymbol{B}) = n - rank(\boldsymbol{X})$. *Let* $\boldsymbol{M}(\boldsymbol{\theta}) = \boldsymbol{X}\left(\boldsymbol{X}^T \boldsymbol{\Sigma}(\boldsymbol{\theta})^{-1} \boldsymbol{X}\right)^{-1} \boldsymbol{X}^T \boldsymbol{\Sigma}(\boldsymbol{\theta})^{-1}$. *Then*

$$\boldsymbol{B}^T \left(\boldsymbol{B}\boldsymbol{\Sigma}(\boldsymbol{\theta})\boldsymbol{B}^T\right)^{-1} \boldsymbol{B} = \boldsymbol{\Sigma}(\boldsymbol{\theta})^{-1}(\boldsymbol{I} - \boldsymbol{M}(\boldsymbol{\theta})).$$

Using Lemma 11.10.3, we can show that the REML equations are

$$trace\left(\boldsymbol{Z}_i^T \boldsymbol{\Sigma}(\boldsymbol{\theta})^{-1}(\boldsymbol{I} - \boldsymbol{M}(\boldsymbol{\theta}))\boldsymbol{Z}_i\right) = \|\boldsymbol{Z}_i^T \boldsymbol{\Sigma}(\boldsymbol{\theta})^{-1}(\boldsymbol{I} - \boldsymbol{M}(\boldsymbol{\theta}))\boldsymbol{Y}\|^2, \quad i = 1, \ldots, r.$$

Proof of Lemma 11.10.3. Simplifying the notations, writing $\boldsymbol{\Sigma}$ instead of $\boldsymbol{\Sigma}(\boldsymbol{\theta})$ and denoting $\boldsymbol{R} = \boldsymbol{\Sigma}^{1/2}\boldsymbol{B}^T$, where $\boldsymbol{\Sigma}^{1/2}$ is a symmetric square root of $\boldsymbol{\Sigma}$, we have

$$\boldsymbol{B}^T \left(\boldsymbol{B}\boldsymbol{\Sigma}\boldsymbol{B}^T\right)^{-1} \boldsymbol{B} = \boldsymbol{\Sigma}^{-1/2}\boldsymbol{R}\left(\boldsymbol{R}^T\boldsymbol{R}\right)^{-1}\boldsymbol{R}^T\boldsymbol{\Sigma}^{-1/2}.$$

Now note that $\boldsymbol{R}(\boldsymbol{R}^T\boldsymbol{R})^{-1}\boldsymbol{R}^T$ is a projection matrix. Where does this matrix project onto? Since $\boldsymbol{B}\boldsymbol{X} = \boldsymbol{0}$ and $rank(\boldsymbol{B}) = n - rank(\boldsymbol{X})$, $\mathcal{M}(\boldsymbol{R})$ is the same as the orthogonal complement of the column space of the matrix $\boldsymbol{\Sigma}^{-1/2}\boldsymbol{X}$. Now the expression of the matrix that projects on the column space of $\boldsymbol{\Sigma}^{-1/2}\boldsymbol{X}$ is given by $\boldsymbol{\Sigma}^{-1/2}\boldsymbol{X}(\boldsymbol{X}^T\boldsymbol{\Sigma}^{-1}\boldsymbol{X})^{-1}\boldsymbol{X}^T\boldsymbol{\Sigma}^{-1/2}$. So $\boldsymbol{I} - \boldsymbol{\Sigma}^{-1/2}\boldsymbol{X}(\boldsymbol{X}^T\boldsymbol{\Sigma}^{-1}\boldsymbol{X})^{-1}\boldsymbol{X}^T\boldsymbol{\Sigma}^{-1/2}$ projects on the orthogonal complement of $\boldsymbol{\Sigma}^{-1/2}\boldsymbol{X}$. Hence we must have

$$\boldsymbol{R}\left(\boldsymbol{R}^T\boldsymbol{R}\right)^{-1}\boldsymbol{R}^T = \boldsymbol{I} - \boldsymbol{\Sigma}^{-1/2}\boldsymbol{X}\left(\boldsymbol{X}^T\boldsymbol{\Sigma}^{-1}\boldsymbol{X}\right)^{-1}\boldsymbol{X}^T\boldsymbol{\Sigma}^{-1/2}.$$

Consequently,

$$\begin{aligned}
\boldsymbol{B}^T \left(\boldsymbol{B}\boldsymbol{\Sigma}\boldsymbol{B}^T\right)^{-1} \boldsymbol{B} &= \boldsymbol{\Sigma}^{-1/2}\boldsymbol{R}\left(\boldsymbol{R}^T\boldsymbol{R}\right)^{-1}\boldsymbol{R}^T\boldsymbol{\Sigma}^{-1/2} \\
&= \boldsymbol{\Sigma}^{-1/2}\left(\boldsymbol{I} - \boldsymbol{\Sigma}^{-1/2}\boldsymbol{X}\left(\boldsymbol{X}^T\boldsymbol{\Sigma}^{-1}\boldsymbol{X}\right)^{-1}\boldsymbol{X}^T\boldsymbol{\Sigma}^{-1/2}\right)\boldsymbol{\Sigma}^{-1/2} \\
&= \boldsymbol{\Sigma}^{-1}(\boldsymbol{I} - \boldsymbol{M}),
\end{aligned}$$

where $\boldsymbol{M} = \boldsymbol{X}(\boldsymbol{X}^T\boldsymbol{\Sigma}^{-1}\boldsymbol{X})^{-1}\boldsymbol{X}^T\boldsymbol{\Sigma}^{-1}$. This completes the proof. □

11.11 Inference: Examples From Mixed Models

Here we present estimation methods in a few balanced models since the expressions for the estimates in the unbalanced cases involve complicated and cumbersome notations. Except in Example 11.11.3, where the maximum likelihood method is considered, Henderson's method is used throughout to obtain the estimates of the variance components.

Example 11.11.1 (One-Factor Random Effects). We consider one-factor balanced ANOVA random-effects models as in Examples 11.10.1, 11.10.6, and 11.10.7, except now we assume that $\{\gamma_i\}$ are iid $N(0, \sigma_1^2)$ and $\{\varepsilon_{ij}\}$ are iid $N(0, \sigma_0^2)$. We have already see in Examples 11.10.6 and 11.10.7 that the BLUE for μ is $\hat{\mu} = \bar{Y}_{..}$,

$$\mathrm{E}[SSE] = (n-k)\sigma_0^2, \text{ where } n = n_0 k,$$
$$\mathrm{E}[SSTR] = (k-1)n_0\sigma_1^2 + (k-1)\sigma_0^2,$$
$$\mathrm{E}[MSE] = \sigma_0^2, \ \mathrm{E}[MSTR] = n_0\sigma_1^2 + \sigma_0^2,$$

and unbiased estimate of σ_0^2 and σ_1^2 are

$$\hat{\sigma}_0^2 = MSE \text{ and } \hat{\sigma}_1^2 = (MSTR - MSE)/n_0.$$

Direct calculation will show that $\hat{\mu} = \bar{Y}_{..} \sim N(\mu, (n_0\sigma_1^2 + \sigma_0^2)/n)$. So an estimate of $\mathrm{Var}[\hat{\mu}]$ is then given by $s^2(\hat{\mu}) = MSTR/n$. Note that $SSTR/(n_0\sigma_1^2 + \sigma_0^2) \sim \chi_{k-1}^2$ and $(\hat{\mu} - \mu)/\sqrt{(n_0\sigma_1^2 + \sigma_0^2)/n} \sim N(0, 1)$. Since $\hat{\mu} = \bar{Y}_{..}$ is independent of $SSTR$ and hence of $MSTR$,

$$(\hat{\mu} - \mu)/s(\hat{\mu}) = \frac{(\hat{\mu} - \mu)\Big/\sqrt{(n_0\sigma_1^2 + \sigma_0^2)/n}}{\sqrt{SSTR\Big/\left[(n_0\sigma_1^2 + \sigma_0^2)(k-1)\right]}} \sim t_{k-1}$$

and this fact now can be used to construct a confidence interval for μ. Unlike in the fixed effect case, MSTR is being used to estimate $\mathrm{Var}[\hat{\mu}]$ and, consequently, the df for the t-distribution is now $k-1$ instead of $n-k$.

In some cases one may want to test if H_0: $\sigma_1^2 = 0$ vs H_1: $\sigma_1^2 \neq 0$. The F-statistic for this is $F = MSTR/MSE$ and $F \sim F_{k-1,n-k}$ under H_0.

Example 11.11.2. Consider the one-factor random-effects model as in the last example and we want to construct a confidence interval for the intraclass correlation coefficient $\rho = \sigma_1^2/(\sigma_1^2 + \sigma_0^2)$, which is also the proportion of variability in the response explained by the random factor. Clearly an estimate of ρ is given by $\hat{\rho} = \hat{\sigma}_1^2/(\hat{\sigma}_1^2 + \hat{\sigma}_0^2)$, where the expressions for $\hat{\sigma}_1^2$ and $\hat{\sigma}_0^2$ are as given in the last example. Let us denote $MSTR/MSE$ by F^*. A confidence interval for ρ with confidence coefficient $1 - \alpha$ is given by $[L^*, U^*]$, where

$L^* = L/(1+L)$, $U^* = U/(1+U)$, and

$$L = (1/n_0)\left[\frac{F^*}{F_{k-1,n-k,\alpha/2}} - 1\right] \text{ and } U = (1/n_0)\left[\frac{F^*}{F_{k-1,n-k,1-\alpha/2}} - 1\right].$$

Let us see why this is indeed a confidence interval of ρ with confidence coefficient $1-\alpha$. Note that $SSTR/\left(n_0\sigma_1^2+\sigma_0^2\right) \sim \chi^2_{k-1}$, $SSE/\sigma_0^2 \sim \chi^2_{n-k}$, and that $SSTR$ and SSE are independent. Hence the random variable

$$F = \frac{\left[SSTR/\left(n_0\sigma_1^2+\sigma_0^2\right)\right]/(k-1)}{\left[SSE/\sigma_0^2\right]/(n-k)} = \left[\sigma_0^2/\left(n_0\sigma_1^2+\sigma_0^2\right)\right]F^*$$

has an F-distribution with $df = (k-1, n-k)$. Denote $c_1 = F_{k-1,n-k,1-\alpha/2}$ and $c_2 = F_{k-1,n-k,\alpha/2}$. Then

$$\begin{aligned}
1-\alpha &= P\left[c_1 \le F \le c_2\right] \\
&= P\left[c_1/F^* \le \sigma_0^2/\left(n_0\sigma_1^2+\sigma_0^2\right) \le c_2/F^*\right] \\
&= P\left[F^*/c_2 \le n_0\sigma_1^2/\sigma_0^2 + 1 \le F^*/c_1\right] \\
&= P\left[(F^*/c_2 - 1)/n_0 \le \sigma_1^2/\sigma_0^2 \le (F^*/c_1 - 1)/n_0\right] \\
&= P\left[L \le \sigma_1^2/\sigma_0^2 \le U\right] \\
&= P\left[L/(1+L) \le \sigma_1^2/\left(\sigma_1^2+\sigma_0^2\right) \le U/(1+U)\right] \\
&= P\left[L^* \le \rho \le U^*\right].
\end{aligned}$$

Example 11.11.3 (One-Factor Random-Effects Model). The setup here is the same as in the last two examples, but the estimates of the variance components are obtained using the maximum likelihood method. Here $\boldsymbol{X}$ is n-dim vector of 1's, $\beta = \mu$ is a scalar, $\boldsymbol{Z}_0 = \boldsymbol{I}$, and the matrix $\boldsymbol{Z}_1 = \begin{pmatrix} \mathbf{1} & \mathbf{0} & \cdots & \mathbf{0} \\ \mathbf{0} & \mathbf{1} & \cdots & \mathbf{0} \\ \vdots & \vdots & & \vdots \\ \mathbf{0} & \mathbf{0} & \cdots & \mathbf{1} \end{pmatrix}$ is $n \times k$, where $\mathbf{1}$ is the n_0-dim vector of 1's. Since $\boldsymbol{\Sigma}(\boldsymbol{\theta}) = \sigma_1^2\boldsymbol{Z}_1\boldsymbol{Z}_1^T + \sigma_0^2\boldsymbol{I}$, it can be checked using the Sherman-Morrison formula (Section B.1) that

$$\begin{aligned}
&\boldsymbol{\Sigma}(\boldsymbol{\theta})^{-1} = \sigma_0^{-2}\left[\boldsymbol{I} - \pi n_0^{-1}\boldsymbol{Z}_1\boldsymbol{Z}_1^T\right], \text{ where } \pi = n_0\sigma_1^2/\left(n_0\sigma_1^2+\sigma_0^2\right), \\
&\boldsymbol{X}^T\boldsymbol{\Sigma}(\boldsymbol{\theta})^{-1}\boldsymbol{X} = \sigma_0^{-2}n(1-\pi), \text{ and} \\
&\boldsymbol{X}^T\boldsymbol{\Sigma}(\boldsymbol{\theta})^{-1}(\boldsymbol{Y}-\boldsymbol{X}\beta) = \sigma_0^{-2}(1-\pi)(Y_{..} - n\mu).
\end{aligned}$$

Thus the first likelihood Eq. (12a)

$$\begin{aligned}
\boldsymbol{X}^T\boldsymbol{\Sigma}(\boldsymbol{\theta})^{-1}\boldsymbol{X} &= \boldsymbol{X}^T\boldsymbol{\Sigma}(\boldsymbol{\theta})^{-1}(\boldsymbol{Y}-\boldsymbol{X}\beta), \text{ ie,} \\
\sigma_0^{-2}(1-\pi)(Y_{..} - n\mu) &= \sigma_0^{-2}n(1-\pi)
\end{aligned}$$

leads to the usual estimate of μ (ie, $\hat{\mu} = \bar{Y}_{..}$). In order to obtain the maximum likelihood estimates of σ_0^2 and σ_1^2, more calculations are needed. The following can be checked (left as an exercise)

$$\begin{aligned}
trace\left(Z_0^T \Sigma(\theta)^{-1} Z_0\right) &= \sigma_0^{-2} n\left(1 - \pi n_0^{-1}\right), \\
trace\left(Z_1^T \Sigma(\theta)^{-1} Z_1\right) &= \sigma_0^{-2} n(1-\pi), \\
\| Z_0^T \Sigma(\theta)^{-1}\left(Y - X\hat{\beta}\right)\|^2 &= \sigma_0^{-4} \sum_i \sum_j \left[Y_{ij} - \bar{Y}_{..} - \pi(\bar{Y}_{i.} - \bar{Y}_{..})\right]^2 \\
&= \sigma_0^{-4}\left[SSE + (1-\pi)^2 SSTR\right], \text{ and} \\
\| Z_1^T \Sigma(\theta)^{-1}\left(Y - X\hat{\beta}\right)\|^2 &= \sigma_0^{-4} n_0 (1-\pi)^2 SSTR.
\end{aligned}$$

Using the above-simplified expressions, the likelihood Eq. (12b) in this case are

$$\begin{aligned}
\sigma_0^{-2} n\left(1 - \pi n_0^{-1}\right) &= \sigma_0^{-4}\left[SSE + (1-\pi)^2 SSTR\right], \text{ and} \\
\sigma_0^{-2} n(1-\pi) &= \sigma_0^{-4} n_0 (1-\pi)^2 SSTR.
\end{aligned}$$

Solutions to these equations lead to the maximum likelihood estimates of the variance components

$$\hat{\sigma}_0^2 = MSE \quad \text{and} \quad \hat{\sigma}_1^2 = \frac{SSTR/k - MSE}{n_0} = \frac{[(k-1)/k]MSTR - MSE}{n_0}.$$

This estimate of σ_1^2 is slightly different from Henderson's estimate given in Example 11.10.7.

Example 11.11.4 (Two-Factor ANOVA, Both Factors Random). The setup here is the same as in Example 11.10.2, but we now assume that it is a balanced ANOVA (ie, $n_{ij} = n_0$ for all i and j). It can be checked that the *BLUE* of μ here is $\hat{\mu} = \bar{Y}_{...}$. Recall that the sums of squares are (Example 11.6.4)

$$\begin{aligned}
SSA &= n_0 b \sum_i (\bar{Y}_{i..} - \bar{Y}_{...})^2, \quad SSB = n_0 a \sum_j \left(\bar{Y}_{.j.} - \bar{Y}_{...}\right)^2, \\
SSAB &= n_0 \sum_i \sum_j \left(\bar{Y}_{ij.} - \bar{Y}_{i..} - \bar{Y}_{.j.} + \bar{Y}_{...}\right)^2, \text{ and} \\
SSE &= \sum_i \sum_j \sum_k \left(Y_{ijk} - \bar{Y}_{ij.}\right)^2.
\end{aligned}$$

It can be shown that

$$\begin{aligned}
\mathrm{E}[MSA] &= n_0 b \sigma_1^2 + n_0 \sigma_3^2 + \sigma_0^2, \\
\mathrm{E}[MSB] &= n_0 a \sigma_2^2 + n_0 \sigma_3^2 + \sigma_0^2, \\
\mathrm{E}[MSAB] &= n_0 \sigma_3^2 + \sigma_0^2, \quad \text{and} \quad \mathrm{E}[MSE] = \sigma_0^2.
\end{aligned}$$

So the unbiased estimates of the variance components are

$$\hat{\sigma}_0^2 = MSE, \quad \hat{\sigma}_3^2 = (MSAB - MSE)/n_0,$$
$$\hat{\sigma}_1^2 = (MSA - MSAB)/(n_0 b), \quad \text{and} \quad \hat{\sigma}_2^2 = (MSB - MSAB)/(n_0 a).$$

Note that $\hat{\mu} = \overline{Y}_{...}$ is unbiased for μ and

$$\text{Var}[\hat{\mu}] = \left(n_0 b\sigma_1^2 + n_0 a\sigma_2^2 + n_0\sigma_3^2 + \sigma_0^2\right)/n,$$

where $n = n_0 ab$. So an unbiased estimate of $\text{Var}[\hat{\mu}]$ is given by $s^2(\hat{\mu}) = (MSA + MSB - MSAB)/n$.

Example 11.11.5 (Two-Factor ANOVA: Factor A Fixed, Factor B Random). We now consider a two-factor balanced mixed-effects model as in Example 11.10.3.

For this model,

$$\text{E}\left[Y_{ijk}\right] = \mu + \alpha_i \quad \text{and} \quad \text{Var}\left[Y_{ijk}\right] = \sigma_1^2 + \sigma_2^2 + \sigma_0^2.$$

It can be shown that the expected values of the mean squares are

$$\text{E}[MSA] = n_0 b \sum(\alpha_i - \bar{\alpha})^2/(a-1) + n_0\sigma_2^2 + \sigma_0^2,$$
$$\text{E}[MSB] = n_0 a\sigma_1^2 + n_0\sigma_2^2 + \sigma_0^2,$$
$$\text{E}[MSAB] = n_0\sigma_2^2 + \sigma_0^2, \quad \text{and} \quad \text{E}[MSE] = \sigma_0^2.$$

So the unbiased estimates of the variance components are

$$\hat{\sigma}_0^2 = MSE, \quad \hat{\sigma}_2^2 = (MSAB - MSE)/n_0, \text{ and}$$
$$\hat{\sigma}_1^2 = (MSB - MSAB)/(n_0 a).$$

If we want to test the null hypothesis of no effect of factor A (ie, H_0: $\alpha_1 = \cdots = \alpha_a$ vs H_1: not all α_i's are equal), then the test statistic is $F^* = MSA/MSAB$ with $df = (a-1, (a-1)(b-1))$.

For this model, the BLUE of $\mu_i = \mu + \alpha_i$ is given by $\bar{Y}_{i..}$. It can be shown that $\text{E}[\bar{Y}_{i..}] = \mu + \alpha_i$ and

$$\text{Var}[\bar{Y}_{i..}] = \left(n_0\sigma_1^2 + n_0\sigma_2^2 + \sigma_0^2\right)/(n_0 b).$$

So an unbiased estimate of $\text{Var}[\bar{Y}_{i..}]$ is given by

$$s^2(\bar{Y}_{i..}) = [(MSB - MSAB)/a + MSAB]/(n_0 b).$$

Note that

$$(MSB - MSAB)/a + MSAB = (1/a)MSB + (1 - 1/a)MSAB.$$

The last quantity is nonnegative since it is a weighted average of *MSB* and *MSAB*. Also note that

$$\begin{aligned}(MSB - MSAB)/a + MSAB &= (1/a)MSB + (1 - 1/a)MSAB \\ &= [(b-1)MSB + (a-1)(b-1)MSAB]/[a(b-1)] \\ &= [SSB + SSAB]/[a(b-1)] \\ &= n_0 \sum\sum\left(\bar{Y}_{ij\cdot} - \bar{Y}_{i\cdot\cdot}\right)^2/[a(b-1)].\end{aligned}$$

We denote the sum of squares $n_0 \sum_i \sum_j (\bar{Y}_{ij\cdot} - \bar{Y}_{i\cdot\cdot})^2$ by $SSB(A)$ (nested ANOVA case in Example 11.6.6). This sum of squares has $df = a(b-1)$ and we denote $SSB(A)/[a(b-1)]$ by $MSB(A)$. Hence we have

$$s^2(\bar{Y}_{i\cdot\cdot}) = MSB(A)/(n_0 b).$$

Since $\bar{Y}_{i\cdot\cdot}$ is independent of $SSB(A)$ and hence of $MSB(A)$, $(\bar{Y}_{i\cdot\cdot} - \mu_i)/s(\bar{Y}_{i\cdot\cdot}) \sim t_{a(b-1)}$ and this fact can be used to construct a confidence interval for μ_i.

If we want to estimate a contrast $\theta = \sum c_i\alpha_i$ of $\{\alpha_i\}$, then the BLUE is $\hat{\theta} = \sum c_i\bar{Y}_{i\cdot\cdot}$. It can be shown that

$$\mathrm{E}\left[\hat{\theta}\right] = \theta \text{ and } \mathrm{Var}\left[\hat{\theta}\right] = \left(\sum c_i^2\right)\left(n_0\sigma_2^2 + \sigma_0^2\right)/(n_0 b).$$

An unbiased estimate of $\mathrm{Var}\left[\hat{\theta}\right]$ is

$$s^2\left(\hat{\theta}\right) = \left(\sum c_i^2\right)MSAB/(n_0 b),$$

and since $\{\bar{Y}_{i\cdot\cdot}, i = 1, \ldots, a\}$ are independent of *SSAB* and hence of *MSAB*, $\hat{\theta}$ is also independent of *MSAB*. Consequently, $\left(\hat{\theta} - \theta\right)/s\left(\hat{\theta}\right) \sim t_{(a-1)(b-1)}$, and this fact can be used to construct a confidence interval for θ.

Remark 11.11.1. An alternative modeling scheme for the interactions in the last example assumes that

(i) for any j, $\sum_{i=1}^{a}(\alpha\beta)_{ij} = 0$,
(ii) for any i, $(\alpha\beta)_{i1}, \ldots, (\alpha\beta)_{ib}$ are iid $N(0, \sigma_2^2)$,
(iii) $\mathrm{Cov}[(\alpha\beta)_{ij}, (\alpha\beta)_{i'j}] = -\sigma_2^2/(a-1)$, for any $i \neq i', j = 1, \ldots, b$.

In this framework $\left\{(\alpha\beta)_{ij}\right\}$ are no longer iid and the estimates of the variance components may be different from what are given above.

Exercises

11.1. Consider a one-factor study with k levels as in Example 11.2.3.

(a) Express the ANOVA model in a regression setting by creating $k-1$ indicator variables for the factor levels. Relate the parameters of this regression model to those of the ANOVA model.

(b) Obtain the parameter estimates of the regression model in part (a) and their standard errors.

11.2. Verify the expression of $\mathrm{E}[MSTR]$ given in Example 11.4.2.

11.3. Consider the two-factor ANOVA model given in Example 11.4.4.

(a) Verify the expressions for $\mathrm{E}[SSA]$, $\mathrm{E}[SSB]$, and $\mathrm{E}[SSAB]$ as given in Example 11.4.4.

(b) Check that $\mathrm{E}[MSAB]=\sigma^2$ if and only if $(\alpha\beta)_{ij}=0$ for all i and j. Similarly, check that $\mathrm{E}[MSA]=\sigma^2$ if and only if $\alpha_i=0$ for all i, and $\mathrm{E}[MSB]=\sigma^2$ if and only if $\beta_j=0$ for all j.

11.4. Consider a real-valued response variable Y and two independent variables X_1 and X_2. Let $L(Y|X_1)$, $L(Y|X_2)$, and $L(Y|X_1,X_2)$ be the best linear predictors of Y given X_1, Y given X_2, and Y given X_1, X_2, respectively. Partial correlation between Y and X_2 given X_1 is defined as $\rho_{YX_2|X_1}=\mathrm{Corr}[Y-L(Y|X_1), X_2-L(X_2|X_1)]$.
Show that $\rho^2_{YX_2|X_1}=\left[\mathrm{E}\{Y-L(Y|X_1)\}^2-\mathrm{E}\{Y-L(Y|X_1,X_2)\}^2\right]/\mathrm{E}\{Y-L(Y|X_1)\}^2$.

11.5. Suppose a Gauss-Markov model is of the form $\boldsymbol{Y}=\boldsymbol{X}_1\boldsymbol{\beta}_1+\boldsymbol{X}_2\boldsymbol{\beta}_2+\boldsymbol{\varepsilon}$, $\boldsymbol{\varepsilon}\sim N_n(0,\sigma^2 I)$, where $\boldsymbol{X}_1$ and $\boldsymbol{X}_2$ are of full rank, and the augmented matrix $[\boldsymbol{X}_1\ \boldsymbol{X}_2]$ is also of full rank. Let $\boldsymbol{Q}_1=\boldsymbol{X}_1^T(\boldsymbol{X}_1^T\boldsymbol{X}_1)^{-1}\boldsymbol{X}_1^T$ and $\tilde{\boldsymbol{X}}_2=(I-\boldsymbol{Q}_1)\boldsymbol{X}_2$.

(a) Show that the least squares estimate of $\boldsymbol{\beta}_2$ is given by $\hat{\boldsymbol{\beta}}_2=\left(\tilde{\boldsymbol{X}}_2^T\tilde{\boldsymbol{X}}_2\right)^{-1}\tilde{\boldsymbol{X}}_2\boldsymbol{Y}$.

(b) Find the distribution of $\hat{\boldsymbol{\beta}}_2$ and use this to test H_0: $\boldsymbol{\beta}_2=0$ vs H_1: $\boldsymbol{\beta}_2\neq 0$.

11.6. Consider a Gauss-Markov model $\boldsymbol{Y}=\boldsymbol{X}_1\boldsymbol{\beta}_1+\boldsymbol{X}_2\boldsymbol{\beta}_2+\boldsymbol{\varepsilon}$, $\boldsymbol{\varepsilon}\sim N_n(0,\sigma^2 I)$. We are interested testing H_0: $\boldsymbol{\beta}_2=0$ vs H_1: $\boldsymbol{\beta}_2\neq 0$. Let SSE_F and SSE_R be the residual sums of squares for the full and the reduced model (ie, the model under H_0). The coefficient of partial determination is defined to be $R^2_{YX_2|X_1}=(SSE_R-SSE_F)/SSE_R$.

(a) Express $R^2_{YX_2|X_1}$ as a function of the F-statistic for testing H_0: $\boldsymbol{\beta}_2=0$ vs H_1: $\boldsymbol{\beta}_2\neq 0$.

(b) Use the result in part (a) to describe the distribution of $R^2_{YX_2|X_1}$ under H_0. [*Hint*: If $U\sim\chi^2_p$, $V\sim\chi^2_q$ and U and V are independent, then $U/(U+V)\sim Beta(p/2,q/2)$.]

11.7. Assume that for a two-factor study with one observation for all treatment combinations, the appropriate model is $Y_{ij}=\mu+\alpha_i+\beta_j+\varepsilon_{ij}$ with the usual constraints on $\{\alpha_i\}$ and $\{\beta_j\}$. Here $\{Y_{ij}\}$ are the observations and $\{\varepsilon_{ij}\}$ are iid $N(0,\sigma^2)$.

(a) Use the Scheffé method to obtain simultaneous confidence intervals for all contrasts in $\{\alpha_i\}$.

(b) Use Tukey's method to obtain simultaneous confidence intervals for all pairwise difference of $\{\alpha_i\}$. How would you obtain simultaneous confidence intervals for all pairwise differences of $\{\alpha_i\}$ and all pairwise differences of $\{\beta_j\}$, using Tukey's method, so that the family confidence is at least $1-\alpha$?

11.8. Prove Lemma 11.7.2.

11.9. For the ANCOVA model with one factor and one covariate as given in Example 11.7.1, use the Scheffé method to obtain simultaneous confidence intervals for all contrasts of $\{\alpha_i\}$, where $\{\alpha_i\}$ are the factor effects.

11.10. Let $\boldsymbol{Y}$ be $n \times 1$ observation vector and assume that $\boldsymbol{Y} = \boldsymbol{\mu} + \boldsymbol{\varepsilon}$, where the vector $\boldsymbol{\varepsilon}$ consists of iid observations with mean 0 and variance σ^2. Consider a model of the form $\boldsymbol{Y} = \boldsymbol{X}_k\boldsymbol{\beta}_k + \boldsymbol{\varepsilon}$, where $\boldsymbol{X}_k$ is a $n \times p_k$ matrix of rank p_k, which is being fitted to the data $\boldsymbol{Y}$ and let $\hat{\boldsymbol{\mu}}_k = \boldsymbol{X}_k\hat{\boldsymbol{\beta}}_k$ where $\hat{\boldsymbol{\beta}}_k$ is the least squares estimator $\boldsymbol{\beta}_k$. The expected value of the squared distance between $\boldsymbol{\mu}$ and $\hat{\boldsymbol{\mu}}_k$ is $D_k = \mathrm{E}\left[\|\boldsymbol{\mu} - \hat{\boldsymbol{\mu}}_k\|^2\right]$.

(a) Let $\hat{D}_k = FPE(k) - n\sigma^2$, where the expression of Akaike's FPE is given in Section 8.1 of this chapter. Is $\hat{D}_k$ an unbiased estimate of D_k? If not find its bias and find the condition under which $\left\{\mathrm{E}\left[\hat{D}_k\right] - D_k\right\}/D_k \to 0$ as $n \to \infty$.

(b) Let $\hat{D}_k = MAL(k) - n\sigma^2$, where the expression of Mallows' criterion is given in Section 8.1. Suppose we have a class of models $\boldsymbol{Y} = \boldsymbol{X}_k\boldsymbol{\beta}_k + \boldsymbol{\varepsilon}$, $k = 1, \ldots, K$, where all the $K-1$ models are nested in the Kth model (ie, $\mathcal{M}(\boldsymbol{X}_k) \subset \mathcal{M}(\boldsymbol{X}_K)$). In Mallows' criterion let $\hat{\sigma}^2$ be the MSE of the Kth model. Is $\hat{D}_k$ an unbiased estimate of D_k? If not find its bias and find the condition under which $\left\{\mathrm{E}\left[\hat{D}_k\right] - D_k\right\}/D_k \to 0$ as $n \to \infty$.

(c) In this part assume that $\boldsymbol{\varepsilon} \sim N_n(0, \sigma^2 I)$. Then the MLE of σ^2 under the model $\boldsymbol{Y} = \boldsymbol{X}_k\boldsymbol{\beta}_k + \boldsymbol{\varepsilon}$ is $\hat{\sigma}_k^2 = \|\boldsymbol{Y} - \boldsymbol{X}_k\hat{\boldsymbol{\beta}}_k\|^2/n$. If $\tilde{\boldsymbol{Y}}$ is an independent copy of $\boldsymbol{Y}$ but is independent of it, then the AIC is an estimate of $-2\mathrm{E}\left[\log f\left(\tilde{\boldsymbol{Y}}, \hat{\boldsymbol{\theta}}_k\right)\right]$, where the expectation is over $\boldsymbol{Y}$ and $\tilde{\boldsymbol{Y}}$, and $\hat{\boldsymbol{\theta}}_k$ is the vector of $\hat{\boldsymbol{\beta}}_k$ and $\hat{\sigma}_k^2$ stacked vertically. Recall that $AIC(k) = -2\log f(\boldsymbol{Y}, \hat{\boldsymbol{\theta}}_k) + 2p_k$, and denote $-2\log f(\tilde{\boldsymbol{Y}}, \hat{\boldsymbol{\theta}}_k)$ by L_k. If $\boldsymbol{\mu} = \boldsymbol{X}_k\boldsymbol{\beta}_k$ for some $\boldsymbol{\beta}_k$ (ie, the model being fitted is the correct one), then using asymptotic expansion as $n \to \infty$ and $p_k/n \to 0$, obtain an approximation of $\mathrm{E}[AIC(k)] - \mathrm{E}[L_k]$, which is the bias of $AIC(k)$ in estimating $\mathrm{E}[L_k]$.

11.11. Consider a model $\boldsymbol{Y} = \sum_{j=1}^{K} \beta_j \boldsymbol{Z}_j + \boldsymbol{\varepsilon}$, where $\boldsymbol{Y}$ is n -dim observation vector, the vectors $\{\boldsymbol{Z}_j\}$ are mutually orthogonal with $\|\boldsymbol{Z}_j\|^2 = n$, and $\boldsymbol{\varepsilon} \sim N_n(0, \sigma^2 I)$. Assume that $\beta_j \neq 0, j = 1, \ldots, k^* < n$, and $\beta_j = 0, j = k^*+1, \ldots, K$. Consider the submodels $\boldsymbol{Y} = \boldsymbol{X}_k\boldsymbol{\beta}_k + \boldsymbol{\varepsilon}$, where $\boldsymbol{X}_k = [\boldsymbol{Z}_1, \ldots, \boldsymbol{Z}_k]$ and $\boldsymbol{\beta}_k = (\beta_1, \ldots, \beta_k)^T$, $k = 1, \ldots, K$. Consider a model selection criterion of the form $F_k = \|\boldsymbol{Y} - \boldsymbol{X}_k\hat{\boldsymbol{\beta}}_k\|^2 + ck\hat{\sigma}^2, c > 0, k = 1, \ldots, K$, where $\hat{\boldsymbol{\beta}}_k$ is the least squares estimate of $\boldsymbol{\beta}_k$ for the kth model and $\hat{\sigma}^2$ is the MSE of the largest model under consideration (ie, the Kth model). The values of $F_1, F_2, \ldots$ are calculated sequentially and let $\hat{k}$ be the index so that $F_k - F_{k+1}$ becomes nonnegative for the first time (ie, F_k is strictly decreasing in k, $k = 1, \ldots, \hat{k}$ and $F_{\hat{k}+1} \geq F_{\hat{k}}$).

(a) If we test H_{k0}: $\beta_k = 0$ vs H_{k1}: $\beta_k \neq 0$, $k = 1, \ldots, K$, then consider the t-statistic $t_k = \hat{\beta}_k / s\left(\hat{\beta}_k\right)$, where $\hat{\beta}_k = \mathbf{Z}_k^T \mathbf{Y}/n$ and $s^2\left(\hat{\beta}_k\right) = \hat{\sigma}^2/n$. Then given a critical value, H_{k0} is rejected or accepted depending on whether $|t_k|$ is larger than the critical value or not. Show that $|t_k| > c^{1/2}$ for $1 \leq k \leq \hat{k}$ and $\left|t_{\hat{k}+1}\right| \leq c^{1/2}$, if model selection is done by using the criterion function $\{F_k\}$.

(b) If $c = \log n$, then prove that $P\left[\hat{k} = k^*\right] \to 1$ as $n \to \infty$. [This proves that a BIC-type criterion is capable of consistent model selection.]

11.12. Prove Lemma 11.10.1.

11.13. Prove Lemma 11.10.2.

11.14. In a one-factor random-effects model, compare the BLUP $\tilde{\alpha}_1$ of α_1 to the naive predictor $\hat{\alpha}_1 = \bar{Y}_{i\cdot} - \bar{Y}_{\cdot\cdot}$.

(a) Obtain the distributions of $\tilde{\alpha}_1 - \alpha_1$ and $\hat{\alpha}_1 - \alpha_1$.

(b) Compare the errors $\mathrm{E}\left[\left(\tilde{\alpha}_1 - \alpha_1\right)^2\right]$ and $\mathrm{E}\left[\left(\hat{\alpha}_1 - \alpha_1\right)^2\right]$.

(c) Obtain the proportional reduction in the errors, that is, $\theta = \left\{\mathrm{E}\left[\left(\hat{\alpha}_1 - \alpha_1\right)^2\right] - \mathrm{E}\left[\left(\tilde{\alpha}_1 - \alpha_1\right)^2\right]\right\} / \mathrm{E}\left[\left(\hat{\alpha}_1 - \alpha_1\right)^2\right]$ and examine it as $\sigma_1^2 \to 0$ or ∞ (assuming σ^2 to be fixed).

11.15. In a repeated measures design each of the randomly selected m subjects (factor A) is assigned to k levels of a treatment (factor B). A reasonable model is thus

$$Y_{ij} = \mu + \alpha_i + \beta_j + \varepsilon_{ij}, \quad j = 1, \ldots, k, \quad i = 1, \ldots, m,$$

where the subject effects $\{\alpha_i\}$ are iid $N(0, \sigma_\alpha^2)$, treatment effects $\{\beta_j\}$ are nonrandom with $\sum \beta_j = 0$, $\{\varepsilon_{ij}\}$ are iid $N(0, \sigma^2)$, and $\{\alpha_i\}$ are independent of $\{\varepsilon_{ij}\}$ Define $SSA = m\sum(\bar{Y}_{i\cdot} - \bar{Y}_{\cdot\cdot})^2$, $SSB = k\sum(\bar{Y}_{\cdot j} - \bar{Y}_{\cdot\cdot})^2$, and $SSE = \sum\sum(Y_{ij} - \bar{Y}_{i\cdot} - \bar{Y}_{\cdot j} + \bar{Y}_{\cdot\cdot})^2$.

(a) Find the mean, variance, and covariances of $\{Y_{ij}\}$.

(b) Obtain $\mathrm{E}[SSA]$, $\mathrm{E}[SSB]$, and $\mathrm{E}[SSE]$.

(c) Use the results in part (b) to obtain unbiased estimates of σ^2 and σ_α^2.

(d) If someone ignores the subject effect and uses a model of the form $Y_{ij} = \mu + \beta_j + \varepsilon_{ij}$ and obtains an MSE based on this model. Is this MSE an unbiased estimate of σ^2? If not find its bias.

11.16. Consider the repeated measures design as in the previous exercise. It is of interest to obtain a BLUP of α_1. The structure of the Mixed Model Equations suggest that the BLUP is a linear function of $\{\bar{Y}_{i\cdot}\}$ and $\{\bar{Y}_{\cdot j}\}$.

(a) Assuming that σ_α^2 and σ^2 are known, find the BLUP $\tilde{\alpha}_1$ of α_1.

(b) Find the distribution of $\tilde{\alpha}_1 - \alpha_1$.

11.17. Verify the expressions for $\mathrm{E}[SSTR]$ and $\mathrm{E}[MSTR]$ given in Example 11.11.1.

11.18. Verify the expressions of $\mathrm{E}[MSA]$, $\mathrm{E}[MSB]$, and $\mathrm{E}[MSAB]$ in Example 11.11.4.

11.19. Verify the expressions of $E[MSA]$, $E[MSB]$, and $E[MSAB]$ in Example 11.11.5.

11.20. Consider a balanced two-factor ANOVA model in which factor A is fixed and factor B is random as in Example 11.11.5. Let μ_i, θ, $\hat{\mu}_i$, and $\hat{\theta}$ be as in that example.

(a) Prove that $(\hat{\mu}_i - \mu_i)/s(\hat{\mu}_i) \sim t_{a(b-1)}$, where $s^2(\hat{\mu}_i) = MSB(A)/(n_0 b)$.

(b) Prove that $\left(\hat{\theta} - \theta\right)/s\left(\hat{\theta}\right) \sim t_{(a-1)(b-1)}$, where $s^2\left(\hat{\theta}\right) = (\sum c_i^2)MSAB/(n_0 b)$.

12

Multivariate Analysis

12.1 Introduction

Multivariate analysis is an area of statistics which deals with observations that are vector valued. Almost all univariate statistical methods have their multivariate counterparts. For instance, when comparing two species of the same animal, various measures such as height, length, tail length, etc., may be measured. One can then compare these two species using a multivariate version of two-sample t-test. Fisher's famous Iris data set contains four measurements for each of the three species: petal length, petal width, sepal length, and sepal width. In order to compare the three species, a multivariate analog of analysis of variance has been developed. If there is a new observation vector (of unknown species) with four measurements, then allocation of this observation vector to one of the species is known as the problem of classification.

Another class of procedures has been developed for multivariate data which deal with dimensionality reduction. If many measurements are taken on children where each measurement is a measure of intelligence, then it is often the case that these various measures are correlated with each other. If there are 20 measurements for each child, it may be reasonable to look for a few summaries which contain most of the information. These summaries are often expressed as appropriate linear combinations of the measurements. This class of methods is known as principal components and factor analyses.

We describe these methods in a systematic manner starting with a few technical results on the Wishart distribution, which is a multivariate generalization of the chi-squared distribution.

12.2 Wishart Distribution

If $\boldsymbol{Y}_1, \ldots, \boldsymbol{Y}_k$ are iid $N_p(\boldsymbol{0}, \boldsymbol{\Sigma})$ and $\boldsymbol{M} = \sum_{i=1}^{k} \boldsymbol{Y}_i \boldsymbol{Y}_i^T$, then we say that $\boldsymbol{M}$ has a (central) Wishart distribution with $df = k$ and the scale matrix $\boldsymbol{\Sigma}$, and we write $\boldsymbol{M} \sim W_p(k, \boldsymbol{\Sigma})$. If the means $\{\boldsymbol{\mu}_i\}$ of $\{\boldsymbol{Y}_i\}$ are not necessarily equal to $\boldsymbol{0}$, then $\boldsymbol{M}$ is said to have a noncentral Wishart distribution $W_p(k, \boldsymbol{\Sigma}, \boldsymbol{\Delta})$, where $\boldsymbol{\Delta} = (1/2)\boldsymbol{\Sigma}^{-1/2} \sum_{i=1}^{k} \boldsymbol{\mu}_i \boldsymbol{\mu}_i^T \boldsymbol{\Sigma}^{-1/2}$ is the noncentrality matrix. Here we discuss only the central Wishart distribution and some of its properties. Some of the results stated below will be proved later in this chapter and further details on the theory can be found in the book by Mardia et al. [61]. In multivariate

Theory and Methods of Statistics. http://dx.doi.org/10.1016/B978-0-12-802440-9.00012-6

analysis, we often deal with positive definite matrices and, for brevity of notation, we sometimes abbreviate "positive definite" by "pd."

(1) If $\boldsymbol{M} \sim W_p(k, \boldsymbol{\Sigma})$, then $\boldsymbol{BMB}^T \sim W_m(k, \boldsymbol{B\Sigma B}^T)$ for any $m \times p$ matrix $\boldsymbol{B}$.

(2) If $\boldsymbol{M}_1 \sim W_p(k_1, \boldsymbol{\Sigma})$, $\boldsymbol{M}_2 \sim W_p(k_2, \boldsymbol{\Sigma})$, and $\boldsymbol{M}_1$ and $\boldsymbol{M}_2$ are independent, then $\boldsymbol{M}_1 + \boldsymbol{M}_2 \sim W_p(k_1 + k_2, \boldsymbol{\Sigma})$.

(3) If $\boldsymbol{M} \sim W_p(k, \boldsymbol{\Sigma})$ and $\boldsymbol{\Sigma}$ is pd, then $P[\boldsymbol{M} \text{ is pd}] = 0$ if $k < p$, and $P[\boldsymbol{M} \text{ is pd}] = 1$ if $k \geq p$.

(4) If $\boldsymbol{M} \sim W_p(k, \boldsymbol{\Sigma})$, then $\boldsymbol{a}^T\boldsymbol{M}\boldsymbol{a}/\boldsymbol{a}^T\boldsymbol{\Sigma}\boldsymbol{a} \sim \chi^2_k$ if $\boldsymbol{a}^T\boldsymbol{\Sigma}\boldsymbol{a} \neq 0$, where $\boldsymbol{a}$ is in $\mathbb{R}^p$.

(5) If $\boldsymbol{M} \sim W_p(k, \boldsymbol{\Sigma})$, $k \geq p$, and $\boldsymbol{\Sigma}$ is pd, then $\boldsymbol{a}^T\boldsymbol{\Sigma}^{-1}\boldsymbol{a}/\boldsymbol{a}^T\boldsymbol{M}^{-1}\boldsymbol{a} \sim \chi^2_{k-p+1}$, where $\boldsymbol{a}$ is in $\mathbb{R}^p$.

(6) Let $\boldsymbol{Y}_1, \ldots, \boldsymbol{Y}_n$ be iid $N_p(\boldsymbol{\mu}, \boldsymbol{\Sigma})$. Define

$$\overline{\boldsymbol{Y}} = \frac{1}{n}\sum \boldsymbol{Y}_i, \quad \boldsymbol{S} = \frac{1}{n-1}\sum(\boldsymbol{Y}_i - \overline{\boldsymbol{Y}})(\boldsymbol{Y}_i - \overline{\boldsymbol{Y}})^T.$$

Then $\overline{\boldsymbol{Y}}$ and $\boldsymbol{S}$ are independent, and

$$\overline{\boldsymbol{Y}} \sim N_p(\boldsymbol{\mu}, n^{-1}\boldsymbol{\Sigma}), \quad (n-1)\boldsymbol{S} \sim W_p(n-1, \boldsymbol{\Sigma}).$$

(7) Let $\boldsymbol{Y}_1, \ldots, \boldsymbol{Y}_{n_1}$ be iid $N_p(\boldsymbol{\mu}_1, \boldsymbol{\Sigma})$, $\boldsymbol{Z}_1, \ldots, \boldsymbol{Z}_{n_2}$ be iid $N_p(\boldsymbol{\mu}_2, \boldsymbol{\Sigma})$, and assume that the samples $\{\boldsymbol{Y}_i\}$ and $\{\boldsymbol{Z}_j\}$ are independent. Define

$$\begin{aligned}
\boldsymbol{S}_1 &= \frac{1}{n_1 - 1}\sum(\boldsymbol{Y}_i - \overline{\boldsymbol{Y}})(\boldsymbol{Y}_i - \overline{\boldsymbol{Y}})^T, \\
\boldsymbol{S}_2 &= \frac{1}{n_2 - 1}\sum(\boldsymbol{Z}_j - \overline{\boldsymbol{Z}})(\boldsymbol{Z}_j - \overline{\boldsymbol{Z}})^T, \text{ and} \\
\boldsymbol{S}_{\text{pooled}} &= \frac{1}{n_1 + n_2 - 2}[(n_1 - 1)\boldsymbol{S}_1 + (n_2 - 1)\boldsymbol{S}_2].
\end{aligned}$$

Then $\overline{\boldsymbol{Y}} - \overline{\boldsymbol{Z}}$ and $\boldsymbol{S}_{\text{pooled}}$ are independent, and

$$\begin{aligned}
&\overline{\boldsymbol{Y}} - \overline{\boldsymbol{Z}} \sim N_p(\boldsymbol{\mu}_1 - \boldsymbol{\mu}_2, (1/n_1 + 1/n_2)\boldsymbol{\Sigma}), \\
&(n_1 + n_2 - 2)\boldsymbol{S}_{\text{pooled}} \sim W_p(n_1 + n_2 - 2, \boldsymbol{\Sigma}).
\end{aligned}$$

(8) Let $\boldsymbol{Y}_1, \ldots, \boldsymbol{Y}_n$, $\overline{\boldsymbol{Y}}$, and $\boldsymbol{S}$ be the same as in Property (6) above. The rv $T^2 = n(\overline{\boldsymbol{Y}} - \boldsymbol{\mu})^T\boldsymbol{S}^{-1}(\overline{\boldsymbol{Y}} - \boldsymbol{\mu})$ is called Hotelling's T^2-statistic and it is distributed as $\frac{(n-1)p}{n-p}F_{p,n-p}$, where $F_{p,n-p}$ has an F-distribution (central) with $df = (p, n-p)$.

(9) Let $\overline{\boldsymbol{Y}} - \overline{\boldsymbol{Z}}$ and $\boldsymbol{S}_{\text{pooled}}$ be the same as in Property (7), and consider the following two-sample Hotelling's T^2-statistic

$$T^2 - (1/n_1 + 1/n_2)^{-1}(\overline{\boldsymbol{Y}} - \overline{\boldsymbol{Z}} - (\mu_1 - \mu_2))^T\boldsymbol{S}^{-1}_{\text{pooled}}(\overline{\boldsymbol{Y}} - \overline{\boldsymbol{Z}} - (\mu_1 - \mu_2)).$$

This two-sample Hotelling's T^2 is distributed as $\frac{(n_1+n_2-2)p}{n_1+n_2-p-1}F_{p,n_1+n_2-p-1}$.

(10) If $k \geq p$ and $\boldsymbol{\Sigma}$ is pd, the Wishart distribution has the pdf

$$f(\boldsymbol{M}) = \frac{|\boldsymbol{M}|^{(k-p-1)/2}}{2^{kp/2}\pi^{p(p-1)/4}|\boldsymbol{\Sigma}|^{k/2}} \frac{\exp\left[-trace(\boldsymbol{\Sigma}^{-1}\boldsymbol{M})/2\right]}{\prod_{i=1}^{p} \Gamma\left[(k+1-i)/2\right]}$$

where $\boldsymbol{M}$ varies over pd matrices.

12.3 The Role of Multivariate Normal Distribution

We write $\boldsymbol{Y} \sim (\boldsymbol{\mu}, \boldsymbol{\Sigma})$ to mean that the p-dim random vector $\boldsymbol{Y}$ has a mean $\boldsymbol{\mu}$ and covariance matrix $\boldsymbol{\Sigma}$. Note that $\boldsymbol{Y}$ is not necessarily normally distributed in this notation.

12.3.1 Mahalanobis Distance

If $\boldsymbol{Y} \sim (\boldsymbol{\mu}, \boldsymbol{\Sigma})$, then the Mahalanobis distance between $\boldsymbol{Y}$ and $\boldsymbol{\mu}$ is defined to be $\Delta^2(\boldsymbol{Y}, \boldsymbol{\mu}) = (\boldsymbol{Y} - \boldsymbol{\mu})^T \boldsymbol{\Sigma}^{-1}(\boldsymbol{Y} - \boldsymbol{\mu})$. Similarly, if $\boldsymbol{Y}_1 \sim (\boldsymbol{\mu}_1, \boldsymbol{\Sigma})$ and $\boldsymbol{Y}_2 \sim (\boldsymbol{\mu}_2, \boldsymbol{\Sigma})$, then $\Delta^2(\boldsymbol{Y}_1, \boldsymbol{Y}_2) = (\boldsymbol{Y}_1 - \boldsymbol{Y}_2)^T \boldsymbol{\Sigma}^{-1}(\boldsymbol{Y}_1 - \boldsymbol{Y}_2)$. Note that Δ^2 is well defined only if $\boldsymbol{\Sigma}$ is pd. It may be worthwhile to point out that the positive square root of Δ^2 is a distance on $\mathbb{R}^p$ (and not Δ^2).

An important property of Δ^2 is that it is invariant under nonsingular linear transformations. Let $\boldsymbol{X}_1 = \boldsymbol{a} + \boldsymbol{B}\boldsymbol{Y}_1$, $\boldsymbol{X}_2 = \boldsymbol{a} + \boldsymbol{B}\boldsymbol{Y}_2$, where $\boldsymbol{a}$ is $p \times 1$, $\boldsymbol{B}$ is $p \times p$ and is nonsingular. Then $\Delta^2(\boldsymbol{X}_1, \boldsymbol{X}_2) = \Delta^2(\boldsymbol{Y}_1, \boldsymbol{Y}_2)$. Mahalanobis distance comes up naturally in multivariate analysis. For instance, if $\boldsymbol{Y}_1, \ldots, \boldsymbol{Y}_n$ are iid $(\boldsymbol{\mu}, \boldsymbol{\Sigma})$, then $\Delta^2(\overline{\boldsymbol{Y}}, \boldsymbol{\mu}) = n(\overline{\boldsymbol{Y}} - \boldsymbol{\mu})^T \boldsymbol{\Sigma}^{-1}(\overline{\boldsymbol{Y}} - \boldsymbol{\mu})$. If we want to test H_0: $\boldsymbol{\mu} = \boldsymbol{\mu}_0$, then we may use the Mahalanobis distance between $\boldsymbol{Y}$ and $\boldsymbol{\mu}_0$, that is, $\Delta^2(\boldsymbol{Y}, \boldsymbol{\mu}_0) = n(\overline{\boldsymbol{Y}} - \boldsymbol{\mu}_0)^T \boldsymbol{\Sigma}^{-1}(\overline{\boldsymbol{Y}} - \boldsymbol{\mu}_0)$, as a test statistic (assuming that $\boldsymbol{\Sigma}$ is known). If $\boldsymbol{\Sigma}$ is unknown and an estimate $\hat{\boldsymbol{\Sigma}}$ of $\boldsymbol{\Sigma}$ is available, then $\Delta^2(\boldsymbol{Y}, \boldsymbol{\mu}_0)$ can be approximated by $n(\overline{\boldsymbol{Y}} - \boldsymbol{\mu}_0)^T \hat{\boldsymbol{\Sigma}}^{-1}(\overline{\boldsymbol{Y}} - \boldsymbol{\mu}_0)$.

In the univariate case we often assume normality. However, except for prediction intervals, almost all the inference are approximately valid without normality of the population as long as the sample size n is large. The same is also true in the multivariate case as long as n is large relative to p.

12.3.2 Multivariate Central Limit Theorem

If $\boldsymbol{Y}_1, \ldots, \boldsymbol{Y}_n$ are iid with mean $\boldsymbol{\mu}$ and covariance matrix $\boldsymbol{\Sigma}$, then by the multivariate central limit theorem (Section A.4),

$$\sqrt{n}(\overline{\boldsymbol{Y}} - \boldsymbol{\mu}) \xrightarrow{\mathcal{D}} N_p(\boldsymbol{0}, \boldsymbol{\Sigma}) \text{ as } n \to \infty.$$

A consequence of this result is that

$$n(\overline{\boldsymbol{Y}} - \boldsymbol{\mu})^T \boldsymbol{\Sigma}^{-1}(\overline{\boldsymbol{Y}} - \boldsymbol{\mu}) \xrightarrow{\mathcal{D}} \chi_p^2 \text{ as } n \to \infty.$$

The sample covariance matrix $\boldsymbol{S} = \frac{1}{n-1}\sum(\boldsymbol{Y}_i - \overline{\boldsymbol{Y}})(\boldsymbol{Y}_i - \overline{\boldsymbol{Y}})^T$ is an unbiased estimate of $\boldsymbol{\Sigma}$. Since $\boldsymbol{S}$ is a consistent estimator of $\boldsymbol{\Sigma}$, we have

$$T^2 = n(\overline{\boldsymbol{Y}} - \boldsymbol{\mu})^T \boldsymbol{S}^{-1}(\overline{\boldsymbol{Y}} - \boldsymbol{\mu}) \xrightarrow{\mathcal{D}} \chi_p^2.$$

Property (8) in Section 12.2 states that when the population is normal,

$$T^2 \sim \frac{(n-1)p}{n-p} F_{p,n-p}.$$

Since the rv $F_{p,n-p}$ can be written as

$$F_{p,n-p} = \frac{W_1/p}{W_2/(n-p)},$$

where $W_1 \sim \chi_p^2$, $W_2 \sim \chi_{n-p}^2$, and W_1 and W_2 are independent, we have

$$\frac{(n-1)p}{n-p} F_{n,n-p} = \frac{n-1}{n-p} \frac{W_1}{W_2/(n-p)}.$$

When $n \to \infty$, $(n-1)/(n-p) \to 1$ and $W_2/(n-p) \xrightarrow{P} 1$, and therefore

$$T^2 \sim \frac{(n-1)p}{n-p} F_{p,n-p} \xrightarrow{\mathcal{D}} \chi_p^2.$$

12.3.3 Checking Normality

A simple indication of multivariate normality is normality of each of the p component variables. Even though this may be enough in most cases, it is important to note that the normality of the marginal distributions does not imply multivariate normality. We now discuss a strategy for checking multivariate normality when we have iid p-dim observations $\boldsymbol{Y}_1, \ldots, \boldsymbol{Y}_n$ from a population.

If the population is indeed normal, then the quantities $d_j^2 = (\boldsymbol{Y}_j - \overline{\boldsymbol{Y}})^T \boldsymbol{S}^{-1}(\boldsymbol{Y}_j - \overline{\boldsymbol{Y}})$ should be approximately iid χ_p^2. So for a given data set, we can calculate these deviances $\{d_j^2\}$ and plot them against the corresponding quantiles of the χ_p^2 distribution. If the population is multivariate normal, we expect the plot to be approximately linear. Here are the steps.

(a) Order d_j^2 from the smallest to the largest: $d_{(1)}^2 \leq \cdots \leq d_{(n)}^2$.

(b) Obtain the chi-squared plot, that is, plot $\{d_{(j)}^2\}$ against $\left\{\chi_p^2((j-0.5)/n)\right\}$, where $\chi_p^2((j-0.5)/n)$ is the $(j-0.5)/n$-quantile of the χ_p^2 distribution.

We recommend the following steps for checking multivariate normality on the basis of a data set.

Step I. Check if each of the individual p variable is univariate normal.
Step II. Check if the chi-squared plot of the deviances $\{d_j^2\}$ is linear.

Steps I and II do not guarantee multivariate normality. However, for practical purposes, these two steps are often enough for checking normality.

12.3.4 Sampling From a Normal Population

Let $\boldsymbol{Y}_1, \ldots, \boldsymbol{Y}_n$ be iid $N_p(\boldsymbol{\mu}, \boldsymbol{\Sigma})$. As before, let $\boldsymbol{S}$ (Property (6) in Section 12.2) be the sample covariance matrix. Then $\overline{\boldsymbol{Y}}$ and $\boldsymbol{S}$ are unbiased estimates of $\boldsymbol{\mu}$ and $\boldsymbol{\Sigma}$, respectively. It turns out the maximum likelihood estimate (MLE) of $\boldsymbol{\mu}$ is $\overline{\boldsymbol{Y}}$. However, the MLE of $\boldsymbol{\Sigma}$ (proved below) is

$$\tilde{\boldsymbol{S}} = \frac{1}{n} \sum (\boldsymbol{Y}_i - \overline{\boldsymbol{Y}})(\boldsymbol{Y}_i - \overline{\boldsymbol{Y}})^T = \frac{n-1}{n} \boldsymbol{S},$$

which is not an unbiased estimator for $\boldsymbol{\Sigma}$. We summarize the above and a bit more in the following result, the proof of which is given in Section 12.3.5.

Theorem 12.3.1. *Let $\overline{\boldsymbol{Y}}$ and $\boldsymbol{S}$ be the sample mean and sample covariance matrix based on n independent observations from $N_p(\boldsymbol{\mu}, \boldsymbol{\Sigma})$. Then:*

(a) *The MLE of $\boldsymbol{\mu}$ and $\boldsymbol{\Sigma}$ are $\overline{\boldsymbol{Y}}$ and $\tilde{\boldsymbol{S}}$, respectively.*
(b) *Sufficient statistics for $(\boldsymbol{\mu}, \boldsymbol{\Sigma})$ are $(\overline{\boldsymbol{Y}}, \boldsymbol{S})$.*

12.3.5 Sampling Distributions

Results given in the following theorem are important for inference when sampling from a multivariate normal population.

Theorem 12.3.2. *Let $\boldsymbol{Y}_1, \ldots, \boldsymbol{Y}_n$ be iid from $N_p(\boldsymbol{\mu}, \boldsymbol{\Sigma})$. Then the following are true:*

(a) $\sqrt{n}(\overline{\boldsymbol{Y}} - \boldsymbol{\mu}) \sim N_p(\boldsymbol{0}, \boldsymbol{\Sigma})$.
(b) $(n-1)\boldsymbol{S} \sim W_p(n-1, \boldsymbol{\Sigma})$.
(c) *$\overline{\boldsymbol{Y}}$ and $\boldsymbol{S}$ are independent.*
(d) $n(\overline{\boldsymbol{Y}} - \boldsymbol{\mu})^T \boldsymbol{\Sigma}^{-1} (\overline{\boldsymbol{Y}} - \boldsymbol{\mu}) \sim \chi^2_p$.
(e) $T^2 = n(\overline{\boldsymbol{Y}} - \boldsymbol{\mu})^T \boldsymbol{S}^{-1} (\overline{\boldsymbol{Y}} - \boldsymbol{\mu}) \sim \frac{(n-1)p}{n-p} F_{p,n-p}$.

Proof of Theorem 12.3.1.

(a) The likelihood function $L = L(\boldsymbol{\mu}, \boldsymbol{\Sigma})$ is given by

$$L = \left(1/\sqrt{2\pi}\right)^{np} (1/|\boldsymbol{\Sigma}|)^{n/2} \exp\left[-(1/2) \sum (\boldsymbol{Y}_i - \boldsymbol{\mu})^T \boldsymbol{\Sigma}^{-1} (\boldsymbol{Y}_i - \boldsymbol{\mu})\right].$$

Maximizing L with respect to $\boldsymbol{\mu}$ and $\boldsymbol{\Sigma}$ is equivalent to minimizing $-2 \log L$ with respect to $\boldsymbol{\mu}$ and $\boldsymbol{\Sigma}$. Since

$$\begin{aligned}\sum (\boldsymbol{Y}_i - \boldsymbol{\mu})^T \boldsymbol{\Sigma}^{-1} (\boldsymbol{Y}_i - \boldsymbol{\mu}) &= \sum (\boldsymbol{Y}_i - \overline{\boldsymbol{Y}})^T \boldsymbol{\Sigma}^{-1} (\boldsymbol{Y}_i - \overline{\boldsymbol{Y}}) \\ &\quad + n(\overline{\boldsymbol{Y}} - \boldsymbol{\mu})^T \boldsymbol{\Sigma}^{-1} (\overline{\boldsymbol{Y}} - \boldsymbol{\mu}) \\ &= n\, trace(\boldsymbol{\Sigma}^{-1} \tilde{\boldsymbol{S}}) + n(\overline{\boldsymbol{Y}} - \boldsymbol{\mu})^T \boldsymbol{\Sigma}^{-1} (\overline{\boldsymbol{Y}} - \boldsymbol{\mu}),\end{aligned}$$

we have

$$-2\log L = n\,trace(\boldsymbol{\Sigma}^{-1}\tilde{\mathbf{S}}) + n(\overline{\mathbf{Y}} - \boldsymbol{\mu})^T\boldsymbol{\Sigma}^{-1}(\overline{\mathbf{Y}} - \boldsymbol{\mu}) + n\log(|\boldsymbol{\Sigma}|) + np\log(2\pi).$$

Clearly, if we minimize $-2\log L$ with respect to $\boldsymbol{\mu}$, the minimum occurs at $\boldsymbol{\mu} = \overline{\mathbf{Y}}$. So

$$-2\log L(\overline{\mathbf{Y}}, \boldsymbol{\Sigma}) = n\,trace(\boldsymbol{\Sigma}^{-1}\tilde{\mathbf{S}}) + n\log(|\boldsymbol{\Sigma}|) + np\log(2\pi).$$

In order to show that $\tilde{\mathbf{S}}$ is indeed the MLE of $\boldsymbol{\Sigma}$, it is enough to show that the quantity $trace(\boldsymbol{\Sigma}^{-1}\tilde{\mathbf{S}}) + \log(|\boldsymbol{\Sigma}|)$ is minimized at $\boldsymbol{\Sigma} = \tilde{\mathbf{S}}$. Now

$$\begin{aligned}
trace(\boldsymbol{\Sigma}^{-1}\tilde{\mathbf{S}}) + \log(|\boldsymbol{\Sigma}|) &= trace(\boldsymbol{\Sigma}^{-1}\tilde{\mathbf{S}}) - \log(|\boldsymbol{\Sigma}^{-1}|)\\
&= trace\left(\tilde{\mathbf{S}}^{1/2}\boldsymbol{\Sigma}^{-1}\tilde{\mathbf{S}}^{1/2}\right) - \log\left(\left|\tilde{\mathbf{S}}^{1/2}\boldsymbol{\Sigma}^{-1}\tilde{\mathbf{S}}^{1/2}\right|\right) + \log\left(|\tilde{\mathbf{S}}|\right)\\
&= trace(\mathbf{R}) - \log(|\mathbf{R}|) + \log(|\tilde{\mathbf{S}}|),
\end{aligned}$$

where $\tilde{\mathbf{S}}^{1/2}$ is a symmetric square root of $\tilde{\mathbf{S}}$ and $\mathbf{R} = \tilde{\mathbf{S}}^{1/2}\boldsymbol{\Sigma}^{-1}\tilde{\mathbf{S}}^{1/2}$. Let $\lambda_1, \ldots, \lambda_p$ be the eigenvalues of $\mathbf{R}$. Then

$$trace(\mathbf{R}) - \log(|\mathbf{R}|) + \log(|\tilde{\mathbf{S}}|) = \sum \lambda_i - \sum \log\lambda_i + \log(|\tilde{\mathbf{S}}|).$$

The last quantity is minimized when $\lambda_1 = \cdots = \lambda_p = 1$, that is, $\mathbf{R} = \mathbf{I}$. Consequently, $\hat{\boldsymbol{\Sigma}} = \tilde{\mathbf{S}}$.

(b) Follows from the factorization theorem.

□

Proof of Theorem 12.3.2. The proof of part (a) is obvious. The proofs of parts (b) and (c) mirror their univariate counterparts. Part (d) follows from part (a). We now present the proof of the result in part (e).

Note that $R_1 = n(\overline{\mathbf{Y}} - \boldsymbol{\mu})^T\boldsymbol{\Sigma}^{-1}(\overline{\mathbf{Y}} - \boldsymbol{\mu}) \sim \chi^2_p$. Let $\mathbf{d} = \sqrt{n}(\overline{\mathbf{Y}} - \boldsymbol{\mu})$. Now we will use Property (5) of Wishart distribution from Section 12.2. Since $\overline{\mathbf{Y}}$ and $\mathbf{S}$ are independent, conditionally on $\overline{\mathbf{Y}}$,

$$R_2 = \mathbf{d}^T\boldsymbol{\Sigma}^{-1}\mathbf{d}/[\mathbf{d}^T((n-1)\mathbf{S})^{-1}\mathbf{d}] \sim \chi^2_{n-1-p+1}.$$

Since this conditional distribution does not depend on $\overline{\mathbf{Y}}$, we conclude that $R_2 \sim \chi^2_{n-p}$ unconditionally, and, R_2 is independent of $\overline{\mathbf{Y}}$ and hence of R_1. Therefore,

$$T^2 = (n-1)\frac{R_1}{R_2} = \frac{(n-1)p}{n-p}\,\frac{R_1/p}{R_2/(n-p)} \sim \frac{(n-1)p}{n-p}F_{p,n-p}.$$

□

12.4 One-Sample Inference

Let $\mathbf{Y}_1, \ldots, \mathbf{Y}_n$ be iid $N_p(\boldsymbol{\mu}, \boldsymbol{\Sigma})$. It is of interest to obtain the estimates of $\boldsymbol{\mu}$, $\boldsymbol{\Sigma}$, and confidence regions for $\boldsymbol{\mu}$. In some cases we may be interested in testing

I. H_0: $\boldsymbol{\mu} = \boldsymbol{\mu}_0$ against H_1: $\boldsymbol{\mu} \neq \boldsymbol{\mu}_0$ ($\boldsymbol{\mu}_0$ known).
II. H_0: $\boldsymbol{\psi} = \boldsymbol{\psi}_0$ against H_1: $\boldsymbol{\psi} \neq \boldsymbol{\psi}_0$, where $\boldsymbol{\psi} = \boldsymbol{D}\boldsymbol{\mu}$, $\boldsymbol{D}$ is an $m \times p$ matrix of rank $m \leq p$, and $\boldsymbol{D}$ and $\boldsymbol{\psi}_0$ are known.

There are many examples of the second testing problem. A particular characteristic of the precision instruments produced by a company is considered to be important and the company takes a random sample of n instruments. The characteristic is measured by four engineers and thus there is a vector of four measurements for each instrument. If $\boldsymbol{\mu} = (\mu_1, \ldots, \mu_4)^T$ is the mean vector, we may be interested in testing if these four measurements are same on the average, that is, H_0: $\mu_1 = \mu_2 = \mu_3 = \mu_4$. This can be restated as a testing problem given in (II) with

$$\boldsymbol{\psi} = \begin{pmatrix} 1 & 0 & 0 & -1 \\ 0 & 1 & 0 & -1 \\ 0 & 0 & 1 & -1 \end{pmatrix} \begin{pmatrix} \mu_1 \\ \mu_2 \\ \mu_3 \\ \mu_4 \end{pmatrix} = \begin{pmatrix} \mu_1 - \mu_4 \\ \mu_2 - \mu_4 \\ \mu_3 - \mu_4 \end{pmatrix} \text{ and } \boldsymbol{\psi}_0 = \begin{pmatrix} 0 \\ 0 \\ 0 \end{pmatrix}.$$

12.4.1 Confidence Ellipsoid for $\boldsymbol{\mu}$

Consider the random ellipsoid

$$A = \{\boldsymbol{u}\colon n(\overline{\boldsymbol{Y}} - \boldsymbol{u})^T \boldsymbol{S}^{-1}(\overline{\boldsymbol{Y}} - \boldsymbol{u}) \leq c_\alpha\},$$

where $c_\alpha = \frac{(n-1)p}{n-p} F_{p,n-p,\alpha}$ and $F_{p,n-p,\alpha}$ is the $(1-\alpha)$-quantile of the F-distribution with $df = (p, n-p)$. This method has been already discussed in the context of Linear models (Section 11.6.1 in Chapter 11). Since $P_{\boldsymbol{\mu},\boldsymbol{\Sigma}}[\boldsymbol{\mu}$ is in $A] = 1-\alpha$, the set A is called the confidence ellipsoid for $\boldsymbol{\mu}$ with confidence coefficient $1-\alpha$. When $p = 2$ or 3, it is not difficult to get a plot of this ellipsoid. However, it is not possible to visualize this ellipsoid when $p \geq 4$. In order to obtain confidence intervals for the individual components of $\boldsymbol{\mu}$, one can consider the confidence shadows. However, this may sometimes lead to an inefficient method for simultaneous inference. We discuss two methods for constructing confidence intervals for linear combinations of $\boldsymbol{\mu}$.

12.4.2 Simultaneous Confidence Intervals

Following the ideas given in Section 11.6.1 of Chapter 11, we present two methods for constructing simultaneous confidence intervals for linear combinations of the mean vector $\boldsymbol{\mu}$. Proofs are not given since they are the same as in Chapter 11.

(i) *Scheffé method*: Simultaneous confidence intervals for all linear combinations $\boldsymbol{l}^T\boldsymbol{\mu}$, $\boldsymbol{l}$ in $\mathbb{R}^p$, with a family confidence coefficient of $1-\alpha$ are

$$\boldsymbol{l}^T\boldsymbol{\mu}\colon \boldsymbol{l}^T\overline{\boldsymbol{Y}} \pm \sqrt{c_\alpha}\, s(\boldsymbol{l}^T\overline{\boldsymbol{Y}}),$$

where $s^2(\boldsymbol{l}^T\overline{\boldsymbol{Y}}) = \boldsymbol{l}^T\boldsymbol{S}\boldsymbol{l}/n$ and $c_\alpha = \frac{(n-1)p}{n-p} F_{p,n-p,\alpha}$.

(ii) *Bonferroni method*: Simultaneous confidence intervals for m linear combinations $\boldsymbol{l}_1^T\boldsymbol{\mu},\ldots,\boldsymbol{l}_m^T\boldsymbol{\mu}$ with a family confidence coefficient at least $1-\alpha$ are

$$\boldsymbol{l}_i^T\boldsymbol{\mu}\colon \boldsymbol{l}_i^T\overline{\boldsymbol{Y}} \pm Bs(\boldsymbol{l}_i^T\overline{\boldsymbol{Y}}), \quad i=1,\ldots,m,$$

where $B = t_{n-1,\alpha/(2m)}$.

As discussed in Chapter 11, the Bonferroni method may be inefficient if m is not small since the multiplier B increases as m increases.

12.4.3 Hypothesis Testing

We now consider the two hypothesis testing problems involving the mean vector $\boldsymbol{\mu}$ mentioned at the beginning of this section.

I. Suppose we want to test $H_0\colon \boldsymbol{\mu}=\boldsymbol{\mu}_0$ against $H_1\colon \boldsymbol{\mu}\neq\boldsymbol{\mu}_0$ ($\boldsymbol{\mu}_0$ known). Then the test statistic is Hotelling's T^2 statistic introduced earlier: $T^2 = n(\overline{\boldsymbol{Y}}-\boldsymbol{\mu}_0)^T\boldsymbol{S}^{-1}(\overline{\boldsymbol{Y}}-\boldsymbol{\mu}_0)$. Under H_0, $T^2 \sim \frac{(n-1)p}{n-p}F_{p,n-p}$. We may reject H_0 at level α if $T^2 > c_\alpha$, where $c_\alpha = \frac{(n-1)p}{n-p}F_{p,n-p,\alpha}$.

II. Let $\boldsymbol{\psi} = \boldsymbol{D}\boldsymbol{\mu}$, where $\boldsymbol{D}$ is a known $m\times p$ matrix of rank $m\le p$. Suppose we want to test $H_0\colon \boldsymbol{\psi}=\boldsymbol{\psi}_0$ against $H_1\colon \boldsymbol{\psi}\neq\boldsymbol{\psi}_0$, where $\boldsymbol{\psi}_0$ is known. Then the appropriate test statistic is

$$T^2 = n(\hat{\boldsymbol{\psi}}-\boldsymbol{\psi}_0)^T(\boldsymbol{DSD}^T)^{-1}(\hat{\boldsymbol{\psi}}-\boldsymbol{\psi}_0).$$

Under H_0, $T^2 \sim \frac{(n-1)m}{n-m}F_{m,n-m}$. So we reject H_0 at level α if $T^2 > \frac{(n-1)m}{n-m}F_{m,n-m,\alpha}$.

The second testing problem is the same as the first if we take the observation vectors to be $\boldsymbol{W}_1 = \boldsymbol{D}\boldsymbol{Y}_1,\ldots,\boldsymbol{W}_n=\boldsymbol{D}\boldsymbol{Y}_n$. Note that $\hat{\boldsymbol{\psi}} = \overline{\boldsymbol{W}} = \boldsymbol{D}\overline{\boldsymbol{Y}} \sim N_m(\boldsymbol{D}\boldsymbol{\mu},\boldsymbol{D}\boldsymbol{\Sigma}\boldsymbol{D}^T)$ and an unbiased estimate of $\boldsymbol{D}\boldsymbol{\Sigma}\boldsymbol{D}^T$ is $\boldsymbol{DSD}^T$, where $\boldsymbol{S}$ is the sample covariance matrix based on the sample $\boldsymbol{Y}_1,\ldots,\boldsymbol{Y}_n$.

12.4.4 Likelihood Ratio Test

Let us denote $(\boldsymbol{\mu},\boldsymbol{\Sigma})$ by $\boldsymbol{\theta}$. The likelihood function is

$$L(\boldsymbol{\theta}) = \left(1/\sqrt{2\pi}\right)^{np}(1/|\boldsymbol{\Sigma}|)^{n/2}\exp\left[-(1/2)\sum(\boldsymbol{Y}_i-\boldsymbol{\mu})^T\boldsymbol{\Sigma}^{-1}(\boldsymbol{Y}_i-\boldsymbol{\mu})\right].$$

The likelihood ratio statistic for testing H_0 is

$$\lambda = \frac{\max_{\boldsymbol{\theta}\text{ in }H_0} L(\boldsymbol{\theta})}{\max_{\boldsymbol{\theta}} L(\boldsymbol{\theta})} = \frac{\max\{L(\boldsymbol{\theta})\colon \boldsymbol{\theta}\text{ in the reduced model}\}}{\max\{L(\boldsymbol{\theta})\colon \boldsymbol{\theta}\text{ in the full model}\}},$$

where the "reduced model" is the same as the "model under H_0." Asymptotic theory for likelihood ratio tests tells us, under H_0,

$$-2\log\lambda \xrightarrow{\mathcal{D}} \chi_t^2$$

as $n \to \infty$, where

$$\begin{aligned} t &= (\text{\# of parameters estimated under the full model}) \\ &\quad - (\text{\# of parameters estimated under } H_0) \\ &= p. \end{aligned}$$

For large n, we reject H_0 at level α if $-2\log\lambda > \chi^2_{p,\alpha}$. It should be noted that the chi-square approximation for $-2\log\lambda$ is valid asymptotically, and it is possible to obtain the exact distribution of $-2\log\lambda$ using Theorem 12.4.1 stated below.

What is the relation between the likelihood ratio test statistic and Hotelling's T^2? The following result provides the answer when the true mean $\boldsymbol{\mu}$ is in a small neighborhood of $\boldsymbol{\mu}_0$.

Theorem 12.4.1. *Consider the problem of testing* H_0: $\boldsymbol{\mu} = \boldsymbol{\mu}_0$ *vs* H_1: $\boldsymbol{\mu} \neq \boldsymbol{\mu}_0$ *on the basis of an iid sample* $\boldsymbol{Y}_1, \ldots, \boldsymbol{Y}_n$ *from* $N_p(\boldsymbol{\mu}, \boldsymbol{\Sigma})$.

(a) *The likelihood ratio test statistic for this testing problem is* $\lambda = |\tilde{\mathbf{S}}|^{n/2}/|\mathbf{S}_0|^{n/2}$, *where* $\mathbf{S}_0 = n^{-1}\sum(\boldsymbol{Y}_i - \boldsymbol{\mu}_0)(\boldsymbol{Y}_i - \boldsymbol{\mu}_0)^T$ *and* $\tilde{\mathbf{S}}$ *is as given in Theorem 12.3.1.*
(b) *Let* $T^2 = n(\overline{\boldsymbol{Y}} - \boldsymbol{\mu}_0)^T \boldsymbol{S}^{-1}(\overline{\boldsymbol{Y}} - \boldsymbol{\mu}_0)$. *Under* $P_{\boldsymbol{\mu},\boldsymbol{\Sigma}}$, *where* $\boldsymbol{\mu}$ *is in the set* $A_n = \{\boldsymbol{u}: \|\boldsymbol{u} - \boldsymbol{\mu}_0\| \le c_n\}$, $c_n \to 0$ *as* $n \to \infty$,

$$-2\log\lambda = T^2 + o_P(1).$$

Remark 12.4.1. An examination of the proof of Theorem 12.4.1 shows that if c_n is of order $n^{-1/2}$, then the $o_P(1)$ in the last theorem can be replaced by $O_P(1/n)$. Hence, under H_0: $\boldsymbol{\mu} = \boldsymbol{\mu}_0$, or under a contiguous alternative of the form $\boldsymbol{\mu} = \boldsymbol{\mu}_0 + n^{-1/2}\boldsymbol{\delta}$, we have $-2\log\lambda = T^2 + O_P(1/n)$.

The following important identity will be used in the proof of Theorem 12.4.1

$$\mathbf{S}_0 = \tilde{\mathbf{S}} + (\overline{\boldsymbol{Y}} - \boldsymbol{\mu}_0)(\overline{\boldsymbol{Y}} - \boldsymbol{\mu}_0)^T,$$

where $\boldsymbol{S}_0$ is as given in the theorem.

Proof of Theorem 12.4.1.

(a) We use the expression for the likelihood function $L(\boldsymbol{\mu}, \boldsymbol{\Sigma})$ given in the proof of Theorem 12.3.1. Note that the MLEs of $\boldsymbol{\mu}$ and $\boldsymbol{\Sigma}$ are obtained by maximizing $L(\boldsymbol{\mu}, \boldsymbol{\Sigma})$ over $\boldsymbol{\mu}$ and $\boldsymbol{\Sigma}$. Since the MLEs of $\boldsymbol{\mu}$ and $\boldsymbol{\Sigma}$ are $\overline{\boldsymbol{Y}}$ and $\tilde{\mathbf{S}}$, respectively, we have

$$\begin{aligned} \max_{\boldsymbol{\mu},\boldsymbol{\Sigma}} L(\boldsymbol{\mu}, \boldsymbol{\Sigma}) &= L(\overline{\boldsymbol{Y}}, \tilde{\mathbf{S}}) \\ &= \left(1/\sqrt{2\pi}\right)^{np} (1/|\tilde{\mathbf{S}}|)^{n/2} \exp\left[-(1/2)\sum(\boldsymbol{Y}_i - \overline{\boldsymbol{Y}})^T \tilde{\mathbf{S}}^{-1}(\boldsymbol{Y}_i - \overline{\boldsymbol{Y}})\right] \\ &= \left(1/\sqrt{2\pi}\right)^{np} \left(1/|\tilde{\mathbf{S}}|\right)^{n/2} \exp(-pn/2). \end{aligned}$$

When $\boldsymbol{\mu} = \boldsymbol{\mu}_0$, we may write the likelihood as

$$L(\boldsymbol{\mu}_0, \boldsymbol{\Sigma}) = \left(1/\sqrt{2\pi}\right)^{np} (1/|\boldsymbol{\Sigma}|)^{n/2} \exp\left[-(1/2)\sum(\boldsymbol{Y}_i - \boldsymbol{\mu}_0)^T \boldsymbol{\Sigma}^{-1}(\boldsymbol{Y}_i - \boldsymbol{\mu}_0)\right]$$
$$= \left(1/\sqrt{2\pi}\right)^{npnp} (1/|\boldsymbol{\Sigma}|)^{n/2} \exp\left[-(n/2)trace(\boldsymbol{\Sigma}^{-1}\boldsymbol{S}_0)\right].$$

The arguments (Proof of Theorem 12.3.1) employed in obtaining the MLE of $\boldsymbol{\Sigma}$, when $\boldsymbol{\mu}$ is unknown, can be used when $\boldsymbol{\mu} = \boldsymbol{\mu}_0$. Basically the same arguments show that $L(\boldsymbol{\mu}_0, \boldsymbol{\Sigma})$ is maximized at $\boldsymbol{\Sigma} = \boldsymbol{S}_0$ where

$$\boldsymbol{S}_0 = \frac{1}{n}\sum(\boldsymbol{Y}_i - \boldsymbol{\mu}_0)(\boldsymbol{Y}_i - \boldsymbol{\mu}_0)^T.$$

Consequently,

$$\max_{\boldsymbol{\mu}=\boldsymbol{\mu}_0,\boldsymbol{\Sigma}} L(\boldsymbol{\mu}, \boldsymbol{\Sigma}) = \max_{\boldsymbol{\Sigma}} L(\boldsymbol{\mu}_0, \boldsymbol{\Sigma})$$
$$= \left(1/\sqrt{2\pi}\right)^{np} (1/|\boldsymbol{S}_0|)^{n/2} \exp\left(-(n/2)trace(\boldsymbol{S}_0^{-1}\boldsymbol{S}_0)\right)$$
$$= \left(1/\sqrt{2\pi}\right)^{np} (1/|\boldsymbol{S}_0|)^{n/2} \exp(-np/2).$$

Hence the likelihood ratio statistic for the testing problem is

$$\lambda = \frac{\max_{\boldsymbol{\mu}=\boldsymbol{\mu}_0,\boldsymbol{\Sigma}} L(\boldsymbol{\mu}, \boldsymbol{\Sigma})}{\max_{\boldsymbol{\mu},\boldsymbol{\Sigma}} L(\boldsymbol{\mu}, \boldsymbol{\Sigma})} = \frac{|\tilde{\boldsymbol{S}}|^{n/2}}{|\boldsymbol{S}_0|^{n/2}}.$$

(b) Using the identity stated before the beginning of the proof of this theorem, we can express the likelihood ratio statistic λ as

$$\lambda = \frac{1}{\left|\tilde{\boldsymbol{S}}^{-1/2}\boldsymbol{S}_0\tilde{\boldsymbol{S}}^{-1/2}\right|} = \frac{1}{|\boldsymbol{I} + \boldsymbol{b}\boldsymbol{b}^T|},$$

where $\boldsymbol{b} = \tilde{\boldsymbol{S}}^{-1/2}(\overline{\boldsymbol{Y}} - \boldsymbol{\mu}_0)$.

It is not difficult to check that the matrix $\boldsymbol{I} + \boldsymbol{b}\boldsymbol{b}^T$ has an eigenvalue equal to 1 with multiplicity $p - 1$, and the remaining eigenvalue is $1 + \|\boldsymbol{b}\|^2$. Since the determinant of a matrix is the product of its eigenvalues, we get

$$|\boldsymbol{I} + \boldsymbol{b}\boldsymbol{b}^T| = 1 + \|\boldsymbol{b}\|^2 = 1 + (\overline{\boldsymbol{Y}} - \boldsymbol{\mu}_0)^T\tilde{\boldsymbol{S}}^{-1}(\overline{\boldsymbol{Y}} - \boldsymbol{\mu}_0) = 1 + \frac{1}{n-1}T^2.$$

Since $\boldsymbol{S} = \boldsymbol{\Sigma} + o_P(1)$ and $\overline{\boldsymbol{Y}} = \boldsymbol{\mu} + O_P(n^{-1/2})$ (as $n \to \infty$),

$$T^2 \le n\|\overline{\boldsymbol{Y}} - \boldsymbol{\mu}_0\|^2\|\boldsymbol{S}\| \le n\left[\|(\overline{\boldsymbol{Y}} - \boldsymbol{\mu})\| + \|(\boldsymbol{\mu} - \boldsymbol{\mu}_0)\|\right]^2\|\boldsymbol{S}\| = o_P(n)$$

on the set $A_n = \{\boldsymbol{\mu}: \|\boldsymbol{\mu} - \boldsymbol{\mu}_0\| \le c_n\}$, and this implies $T^2/(n-1) = o_P(1)$ on the set A_n. Hence

$$-2\log\lambda = n\log(1 + T^2/(n-1)) = T^2 + o_P(1).$$

□

12.5 Two-Sample Problem

Suppose that we have two independent samples from two multivariate normal populations with different mean vectors, but the same covariance matrix. Let $\boldsymbol{Y}_{1j}, j = 1, \ldots, n_1$, be iid $N_p(\boldsymbol{\mu}_1, \boldsymbol{\Sigma})$ and $\boldsymbol{Y}_{2j}, j = 1, \ldots, n_2$ be iid $N_p(\boldsymbol{\mu}_2, \boldsymbol{\Sigma})$. We assume that the samples $\{\boldsymbol{Y}_{1j}\}$ and $\{\boldsymbol{Y}_{2j}\}$ are independent. We address the following two issues:

(a) test for H_0: $\boldsymbol{\mu}_1 = \boldsymbol{\mu}_2$ against H_1: $\boldsymbol{\mu}_1 \neq \boldsymbol{\mu}_2$.
(b) confidence statements for $\boldsymbol{\mu}_1 - \boldsymbol{\mu}_2$.

12.5.1 Estimation

The MLEs for $\boldsymbol{\mu}_1$ and $\boldsymbol{\mu}_2$ are

$$\hat{\boldsymbol{\mu}}_1 = \overline{\boldsymbol{Y}}_{1\cdot} = (1/n_1)\sum \boldsymbol{Y}_{1j}, \text{ and } \hat{\boldsymbol{\mu}}_2 = \overline{\boldsymbol{Y}}_{2\cdot} = (1/n_2)\sum \boldsymbol{Y}_{2j}.$$

An unbiased estimate of $\boldsymbol{\Sigma}$ is

$$\boldsymbol{S} = \boldsymbol{S}_{\text{pooled}} = (n_1 + n_2 - 2)^{-1}\left[(n_1 - 1)\boldsymbol{S}_1 + (n_2 - 1)\boldsymbol{S}_2\right],$$

where $\boldsymbol{S}_1$ and $\boldsymbol{S}_2$ are the sample covariance matrices on the basis of $\{\boldsymbol{Y}_{1j}: j =, 1, \ldots, n_1\}$ and $\{\boldsymbol{Y}_{2j}: j = 1, \ldots, n_2\}$, respectively, that is,

$$\boldsymbol{S}_i = (n_i - 1)^{-1}\sum_j (\boldsymbol{Y}_{ij} - \overline{\boldsymbol{Y}}_{i\cdot})(\boldsymbol{Y}_{ij} - \overline{\boldsymbol{Y}}_{i\cdot})^T, \quad i = 1, 2.$$

As in the univariate two-sample problem, the $\boldsymbol{S} = \boldsymbol{S}_{\text{pooled}}$ is a better estimator of $\boldsymbol{\Sigma}$ than $\boldsymbol{S}_1$ or $\boldsymbol{S}_2$.

The following result is useful for inference in two-sample problems.

Theorem 12.5.1. *Let us denote $\boldsymbol{\mu}_1 - \boldsymbol{\mu}_2$ and its MLE $\hat{\boldsymbol{\mu}}_1 - \hat{\boldsymbol{\mu}}_2$ by $\boldsymbol{\delta}$ and $\hat{\boldsymbol{\delta}}$, respectively. The following hold:*

(a) $(\hat{\boldsymbol{\mu}}_1, \hat{\boldsymbol{\mu}}_2, \boldsymbol{S})$ *is sufficient for* $(\boldsymbol{\mu}_1, \boldsymbol{\mu}_2, \boldsymbol{\Sigma})$.
(b) $\hat{\boldsymbol{\mu}}_1 \sim N_p(\boldsymbol{\mu}_1, (1/n_1)\boldsymbol{\Sigma})$, $\hat{\boldsymbol{\mu}}_2 \sim N_p(\boldsymbol{\mu}_2, (1/n_2)\boldsymbol{\Sigma})$, *and* $\hat{\boldsymbol{\delta}} \sim N_p(\boldsymbol{\delta}, (1/n_1 + 1/n_2)\boldsymbol{\Sigma})$.
(c) $(n_1 + n_2 - 2)\boldsymbol{S} \sim W_p(n_1 + n_2 - 2, \boldsymbol{\Sigma})$.
(d) $\hat{\boldsymbol{\mu}}_1, \hat{\boldsymbol{\mu}}_2$, *and* $\boldsymbol{S}$ *are independent.*
(e) *The two-sample Hotelling T^2-statistic*

$$T^2 = (1/n_1 + 1/n_2)^{-1}(\hat{\boldsymbol{\delta}} - \boldsymbol{\delta})^T \boldsymbol{S}^{-1}(\hat{\boldsymbol{\delta}} - \boldsymbol{\delta})$$

has the same distribution as $\frac{(n_1+n_2-2)p}{n_1+n_2-p-1}F_{p,n_1+n_2-p-1}$.

When $n_1 + n_2 \to \infty$, the two-sample Hotelling's $T^2 \xrightarrow{\mathcal{D}} \chi^2_p$.

Proof of Theorem 12.5.1. We only provide a proof of parts (c) and (e) since the rest are not difficult to establish.

(c) Note that $(n_1 - 1)\boldsymbol{S}_1 \sim W_p(n_1 - 1, \boldsymbol{\Sigma})$, $(n_2 - 1)\boldsymbol{S}_2 \sim W_p(n_2 - 1, \boldsymbol{\Sigma})$ and that $\boldsymbol{S}_1$ and $\boldsymbol{S}_2$ are independent. Hence

$$(n_1 + n_2 - 2)\boldsymbol{S} = (n_1 - 1)\boldsymbol{S}_1 + (n_2 - 1)\boldsymbol{S}_2 \sim W_p(n_1 + n_2 - 2, \boldsymbol{\Sigma}).$$

(e) We will argue as in the one-sample case. Let

$$\boldsymbol{d} = (1/n_1 + 1/n_2)^{-1/2}(\hat{\boldsymbol{\delta}} - \boldsymbol{\delta}),$$

$$R_1 = (1/n_1 + 1/n_2)^{-1}(\hat{\boldsymbol{\delta}} - \boldsymbol{\delta})^T \boldsymbol{\Sigma}^{-1}(\hat{\boldsymbol{\delta}} - \boldsymbol{\delta}), \text{ and}$$

$$R_2 = \boldsymbol{d}^T \boldsymbol{\Sigma}^{-1}\boldsymbol{d}/\{\boldsymbol{d}^T((n_1 + n_2 - 2)\boldsymbol{S})^{-1}\boldsymbol{d}\}.$$

From part (a), $R_1 \sim \chi^2_p$. From part (d), $\hat{\boldsymbol{\mu}}_1$ and $\hat{\boldsymbol{\mu}}_2$ are independent of $\boldsymbol{S}$ and hence conditionally on $\hat{\boldsymbol{\mu}}_1$ and $\hat{\boldsymbol{\mu}}_2$, $R_2 \sim \chi^2_{n_1+n_2-p-1}$ (by Property (5) in Section 12.2). Since this conditional distribution does not depend on $\hat{\boldsymbol{\mu}}_1$ and $\hat{\boldsymbol{\mu}}_2$, we conclude that $R_2 \sim \chi^2_{n_1+n_2-p-1}$ unconditionally, and R_1 and R_2 are independent.
Therefore,

$$\begin{aligned} T^2 &= (n_1 + n_2 - 2)\frac{R_1}{R_2} \\ &= \frac{(n_1 + n_2 - 2)p}{n_1 + n_2 - p - 1}\frac{R_1/p}{R_2/(n_1 + n_2 - p - 1)} \\ &\sim \frac{(n_1 + n_2 - 2)p}{n_1 + n_2 - p - 1}F_{p,n_1+n_2-p-1}. \end{aligned}$$

□

12.5.2 Hypothesis Testing

We want to test H_0: $\boldsymbol{\mu}_1 = \boldsymbol{\mu}_2$ against H_1: $\boldsymbol{\mu}_1 \neq \boldsymbol{\mu}_2$ at a level of significance α. Consider the following Hotelling's T^2 -statistic

$$T^2 = (1/n_1 + 1/n_2)^{-1}(\hat{\boldsymbol{\mu}}_1 - \hat{\boldsymbol{\mu}}_2)^T \boldsymbol{S}^{-1}(\hat{\boldsymbol{\mu}}_1 - \hat{\boldsymbol{\mu}}_2).$$

Under H_0, T^2 has the same distribution as $\frac{(n_1+n_2-2)p}{n_1+n_2-p-1}F_{p,n_1+n_2-p-1}$. So we reject H_0 at level α if $T^2 > c_\alpha$ where

$$c_\alpha = \frac{(n_1 + n_2 - 2)p}{n_1 + n_2 - p - 1}F_{p,n_1+n_2-p-1,\alpha}.$$

12.5.3 Confidence Ellipsoid for $\boldsymbol{\mu}_1 - \boldsymbol{\mu}_2$

Consider the ellipsoid

$$A = \{\boldsymbol{\delta}: (1/n_1 + 1/n_2)^{-1}(\hat{\boldsymbol{\mu}}_1 - \hat{\boldsymbol{\mu}}_2 - \boldsymbol{\delta})^T \boldsymbol{S}^{-1}(\hat{\boldsymbol{\mu}}_1 - \hat{\boldsymbol{\mu}}_2 - \boldsymbol{\delta}) \leq c_\alpha\}.$$

Since $P_{\boldsymbol{\mu}_1,\boldsymbol{\mu}_2,\boldsymbol{\Sigma}}[\boldsymbol{\mu}_1 - \boldsymbol{\mu}_2 \text{ is in } A] = 1 - \alpha$, A is a confidence ellipsoid for $\boldsymbol{\mu}_1 - \boldsymbol{\mu}_2$ with confidence coefficient $1 - \alpha$.

12.5.4 Simultaneous Confidence Intervals

(a) *Scheffé method*: Simultaneous confidence intervals for all linear combinations $\boldsymbol{l}^T\boldsymbol{\delta}$, $\boldsymbol{l}$ in $\mathbb{R}^p$ and $\boldsymbol{\delta} = \boldsymbol{\mu}_1 - \boldsymbol{\mu}_2$, with a family confidence coefficient $1-\alpha$ are given by $\boldsymbol{l}^T\hat{\boldsymbol{\delta}} \pm \sqrt{c_\alpha} s(\boldsymbol{l}^T\hat{\boldsymbol{\delta}})$, where $\hat{\boldsymbol{\delta}} = \hat{\boldsymbol{\mu}}_1 - \hat{\boldsymbol{\mu}}_2$ and $s^2(\boldsymbol{l}^T\hat{\boldsymbol{\delta}}) = (1/n_1 + 1/n_2)\boldsymbol{l}^T\boldsymbol{S}\boldsymbol{l}$.

(b) *Bonferroni method*: Simultaneous confidence intervals for $\boldsymbol{l}_1^T\boldsymbol{\delta}, \ldots, \boldsymbol{l}_m^T\boldsymbol{\delta}$ with a family confidence coefficient of at least $1-\alpha$ are

$$\boldsymbol{l}_i^T\boldsymbol{\delta}\colon \boldsymbol{l}_i^T\hat{\boldsymbol{\delta}} \pm t_{n_1+n_2-2,\alpha/(2m)} s(\boldsymbol{l}_i^T\hat{\boldsymbol{\delta}}), \quad i = 1, \ldots, m.$$

12.6 One-Factor MANOVA

Suppose that we have k multivariate normal populations with possibly different mean vectors, but the same covariance matrix. Let $\{\boldsymbol{Y}_{ij}\colon j = 1, \ldots, n_i\}$ be iid $N_p(\boldsymbol{\mu}_i, \boldsymbol{\Sigma})$, $i = 1, \ldots, k$. We may write the one-factor MANOVA model as

$$\boldsymbol{Y}_{ij} = \boldsymbol{\mu}_i + \boldsymbol{\varepsilon}_{ij}, \quad j = 1, \ldots, n_i, \quad i = 1, \ldots, k,$$

where $\{\boldsymbol{\varepsilon}_{ij}\}$ are iid $N_p(\boldsymbol{0}, \boldsymbol{\Sigma})$. We can also rewrite the above as a factor-effect model

$$\boldsymbol{Y}_{ij} = \boldsymbol{\mu} + \boldsymbol{\alpha}_i + \boldsymbol{\varepsilon}_{ij},$$

where $\boldsymbol{\mu} = \sum(n_i/n)\boldsymbol{\mu}_i$, $\boldsymbol{\alpha}_i = \boldsymbol{\mu}_i - \boldsymbol{\mu}$ and $n = n_1 + \cdots + n_k$ is the total number of observation vectors.

The following issues are of interest:

(a) test H_0: $\boldsymbol{\mu}_1 = \cdots = \boldsymbol{\mu}_k$ against H_1: not all $\boldsymbol{\mu}_i$'s are the same.

(b) confidence statements about $\boldsymbol{\mu}_i$'s and $\boldsymbol{\alpha}_i$'s.

12.6.1 Estimation

MLEs for $\boldsymbol{\mu}_1, \ldots, \boldsymbol{\mu}_k$ are $\hat{\boldsymbol{\mu}}_i = \overline{\boldsymbol{Y}}_{i.} = (1/n_i)\sum_j \boldsymbol{Y}_{ij}$, $i = 1, \ldots, k$. An unbiased estimate of the covariance matrix $\boldsymbol{\Sigma}$ is

$$\boldsymbol{S} = \frac{1}{n-k}[(n_1-1)\boldsymbol{S}_1 + \cdots + (n_k-1)\boldsymbol{S}_k], \text{ where}$$
$$\boldsymbol{S}_i = \frac{1}{n_i-1}\sum_j(\boldsymbol{Y}_{ij} - \overline{\boldsymbol{Y}}_{i.})(\boldsymbol{Y}_{ij} - \overline{\boldsymbol{Y}}_{i.})^T$$

is the sample covariance matrix on the basis of the ith sample, $i = 1, \ldots, k$. The MLEs for $\boldsymbol{\mu}$ and $\boldsymbol{\alpha}_i$ are

$$\hat{\boldsymbol{\mu}} = \overline{\boldsymbol{Y}}_{..}, \quad \hat{\boldsymbol{\alpha}}_i = \hat{\boldsymbol{\mu}}_i - \hat{\boldsymbol{\mu}} = \overline{\boldsymbol{Y}}_{i.} - \overline{\boldsymbol{Y}}_{...}$$

As in one- and two-sample cases, the MLE of $\boldsymbol{\Sigma}$

$$\tilde{\boldsymbol{S}} = \frac{1}{n}[(n_1-1)\boldsymbol{S}_1 + \cdots + (n_k-1)\boldsymbol{S}_k]$$

is a biased estimator.

The following result is useful for inference in one-factor MANOVA.

Theorem 12.6.1. *Let $\hat{\mu}_1, \ldots, \hat{\mu}_k$ be the sample means based on independent random samples from $N_p(\mu_1, \Sigma), \ldots, N_p(\mu_k, \Sigma)$, and let $\mathbf{S}$ be the pooled covariance matrix. Then the following hold:*

(a) $(\hat{\mu}_1, \ldots, \hat{\mu}_k, \mathbf{S})$ *is sufficient for* $(\mu_1, \ldots, \mu_k, \Sigma)$.
(b) $\overline{Y}_{i.} \sim N_p(\mu_i, (1/n_i)\Sigma)$.
(c) $\hat{\mu} \sim N_p(\mu, (1/n)\Sigma)$.
(d) $\hat{\alpha}_i \sim N_p(\alpha_i, (1/n_i - 1/n)\Sigma)$.
(e) $\hat{\mu}_1, \ldots, \hat{\mu}_k$ *and* $\mathbf{S}$ *are all independent.*
(f) $(n-k)\mathbf{S} \sim W_p(n-k, \Sigma)$.
(g) *When* $\mu_1 = \cdots = \mu_k$, $\sum n_i(\overline{Y}_{i.} - \overline{Y}_{..})(\overline{Y}_{i.} - \overline{Y}_{..})^T \sim W_p(k-1, \Sigma)$.

The proof of this theorem is not given. The results in it can be obtained using arguments similar to one- and two-sample cases given above, and by borrowing appropriate results from univariate analysis of variance.

12.6.2 Hypothesis Testing in One-Factor MANOVA

Suppose we wish to test H_0: $\mu_1 = \cdots = \mu_k$ against H_1: not all μ_i's are the same. This is equivalent to testing H_0: $\alpha_1 = \cdots = \alpha_k = \mathbf{0}$ against H_1: not all α_i's are equal to $\mathbf{0}$. We define a few matrices analogous to the various sums of squares in the univariate case. Total sum of squares and products (SSP), between group SSP and within group SSP, as well as their corresponding degrees of freedom are given below

$$\boldsymbol{T} = \sum_i \sum_j (\boldsymbol{Y}_{ij} - \overline{\boldsymbol{Y}}_{..})(\boldsymbol{Y}_{ij} - \overline{\boldsymbol{Y}}_{..})^T, \quad df = n-1,$$

$$\boldsymbol{B} = \sum n_i(\overline{\boldsymbol{Y}}_{i.} - \overline{\boldsymbol{Y}}_{..})(\overline{\boldsymbol{Y}}_{i.} - \overline{\boldsymbol{Y}}_{..})^T, \quad df = k-1, \text{ and}$$

$$\boldsymbol{W} = \sum_i \sum_j (\boldsymbol{Y}_{ij} - \overline{\boldsymbol{Y}}_{i.})(\boldsymbol{Y}_{ij} - \overline{\boldsymbol{Y}}_{i.})^T, \quad df = n-k.$$

It is fairly easy to check that $\boldsymbol{T} = \boldsymbol{B} + \boldsymbol{W}$.

Theorem 12.6.2. *Consider the problem of testing H_0: $\alpha_1 = \cdots = \alpha_k = \mathbf{0}$ against H_1: not all α_i's are equal to $\mathbf{0}$, in one-factor MANOVA. The likelihood ratio test statistic is*

$$\lambda = \left\{\frac{|\boldsymbol{W}|}{|\boldsymbol{T}|}\right\}^{n/2} = \Lambda^{n/2},$$

where $\Lambda = \frac{|\boldsymbol{W}|}{|\boldsymbol{T}|}$ is called Wilks' lambda.

The proof of this result will be given below in Section 12.6.4, but we describe the inference procedures first. The exact distribution of the likelihood ratio statistic λ is complicated even under H_0. In some cases we know the exact distributional results as given below. However, for the general case, we may use the asymptotic theory. From the theory of likelihood ratio tests, under H_0,

$$-2\log\lambda = -n\log\Lambda \xrightarrow{\mathcal{D}} \chi^2_{p(k-1)}$$

as $n \to \infty$. In order to obtain a better asymptotic approximation, Bartlett made a suitable modification to the test statistic. According to his modification, under H_0,

$$-[n-1-(p+k)/2]\log \Lambda \xrightarrow{\mathcal{D}} \chi^2_{p(k-1)},$$

as $n \to \infty$. If α is the given level of significance, we may reject H_0 at level α if

$$-[n-1-(p+k)/2]\log \Lambda > \chi^2_{p(k-1),\alpha}.$$

12.6.3 Simultaneous Confidence Intervals

If we are interested in constructing simultaneous confidence intervals for α_{il}, $l = 1, \ldots, p$, $i = 1, \ldots, k$, then the Bonferroni method yields the intervals

$$\alpha_{il}\text{: } \hat{\alpha}_{il} \pm t_{n-k,\alpha/(2pk)} s(\hat{\alpha}_{il}),$$

where $\hat{\alpha}_{il}$ is the lth component of $\hat{\boldsymbol{\alpha}}_i$ and $s^2(\hat{\alpha}_{il}) = (1/n_i - 1/n)s_{ll}$ and s_{ll} is the lth diagonal element of $\boldsymbol{S}$. Similarly if we want to construct simultaneous confidence intervals for all pairwise differences $\alpha_{il} - \alpha_{i'l}$, $1 \le i \ne i' \le k$, $l = 1, \ldots, p$, with a family confidence coefficient of at least $1-\alpha$, then the intervals are

$$\alpha_{il} - \alpha_{i'l}\text{: } \hat{\alpha}_{il} - \hat{\alpha}_{i'l} \pm t_{n-k,\alpha/[pk(k-1)]} s(\hat{\alpha}_{il} - \hat{\alpha}_{i'l}),$$

where $\hat{\alpha}_{il} - \hat{\alpha}_{i'l}$ is the lth component of $\hat{\boldsymbol{\alpha}}_i - \hat{\boldsymbol{\alpha}}_{i'} = \overline{\boldsymbol{Y}}_{i\cdot} - \overline{\boldsymbol{Y}}_{i'\cdot}$, and

$$s^2(\hat{\alpha}_{il} - \hat{\alpha}_{i'l}) = (1/n_i + 1/n_{i'})s_{ll}.$$

Clearly, the Bonferroni method may lead to wide intervals if p (and/or k) is not small. If k is not large, we may use one-factor ANOVA models to construct Scheffé- or Tukey-type (in the balanced case) confidence intervals for $\alpha_{il} - \alpha_{i'l}$, $1 \le i \ne i' \le k$, for each $l = 1, \ldots, p$, so that the family confidence coefficient is at least $1-\alpha$.

12.6.4 Exact Distributions of Wilks' Lambda

In general it is not easy to obtain the exact distribution of Wilks' Λ statistic and one usually uses the asymptotic distribution with Bartlett corrections. However, there are some special cases where exact distributions have been obtained and some of these are presented in following table:

$p = 1$	$k \ge 2$	$\frac{n-k}{k-1}\frac{1-\Lambda}{\Lambda} \sim F_{k-1,n-k}$
$p = 2$	$k \ge 2$	$\frac{n-k-1}{k-1}\frac{1-\sqrt{\Lambda}}{\sqrt{\Lambda}} \sim F_{2(k-1),2(n-k-1)}$
$p \ge 1$	$k = 2$	$\frac{n-p-1}{p}\frac{1-\Lambda}{\Lambda} \sim F_{p,n-p-1}$
$p \ge 1$	$k = 3$	$\frac{n-p-2}{p}\frac{1-\sqrt{\Lambda}}{\sqrt{\Lambda}} \sim F_{2p,2(n-p-2)}$

Proof of Theorem 12.6.2. Note that the likelihood is

$$L(\boldsymbol{\mu}_1,\ldots,\boldsymbol{\mu}_k,\boldsymbol{\Sigma}) = \left(1/\sqrt{2\pi}\right)^{pn}(1/|\boldsymbol{\Sigma}|)^{n/2}\exp\left[-\sum_i\sum_j(\boldsymbol{Y}_{ij}-\boldsymbol{\mu}_i)^T\boldsymbol{\Sigma}^{-1}(\boldsymbol{Y}_{ij}-\boldsymbol{\mu}_i)/2\right].$$

Under H_0: $\boldsymbol{\mu}_1 = \cdots = \boldsymbol{\mu}_k$, the MLE for the common mean $\boldsymbol{\mu}$ is $\hat{\boldsymbol{\mu}} = \overline{\boldsymbol{Y}}_{..}$ and the MLE for $\boldsymbol{\Sigma}$ is

$$\mathbf{S}_0 = (1/n)\sum_i\sum_j(\boldsymbol{Y}_{ij}-\overline{\boldsymbol{Y}}_{..})(\boldsymbol{Y}_{ij}-\overline{\boldsymbol{Y}}_{..})^T.$$

For the general case, the MLE for $\boldsymbol{\mu}_i$ is $\hat{\boldsymbol{\mu}}_i = \overline{\boldsymbol{Y}}_{i.}$ and the MLE of $\boldsymbol{\Sigma}$ is

$$\tilde{\mathbf{S}} = (1/n)\sum_i\sum_j(\boldsymbol{Y}_{ij}-\overline{\boldsymbol{Y}}_{i.})(\boldsymbol{Y}_{ij}-\overline{\boldsymbol{Y}}_{i.})^T.$$

Following the argument similar to the one- and two-sample cases, we can show that

$$\begin{aligned}\max_{\boldsymbol{\mu}_1=\cdots=\boldsymbol{\mu}_k,\boldsymbol{\Sigma}} L(\boldsymbol{\mu}_1,\ldots,\boldsymbol{\mu}_k,\boldsymbol{\Sigma}) &= L\left(\overline{\boldsymbol{Y}}_{..},\mathbf{S}_0\right)\\ &= \left(1/\sqrt{2\pi}\right)^{pn}(1/|\mathbf{S}_0|)^{n/2}\exp(-pn/2),\ \text{and}\\ \max_{\boldsymbol{\mu}_1,\ldots,\boldsymbol{\mu}_k,\boldsymbol{\Sigma}} L(\boldsymbol{\mu}_1,\ldots,\boldsymbol{\mu}_k,\boldsymbol{\Sigma}) &= L\left(\overline{\boldsymbol{Y}}_{1.},\ldots,\overline{\boldsymbol{Y}}_{k.},\tilde{\mathbf{S}}\right)\\ &= \left(1/\sqrt{2\pi}\right)^{pn}(1/|\tilde{\mathbf{S}}|)^{n/2}\exp(-pn/2).\end{aligned}$$

So the likelihood ratio test statistic for testing equality of the means is

$$\lambda = \frac{\max_{\boldsymbol{\mu}_1=\cdots=\boldsymbol{\mu}_k,\boldsymbol{\Sigma}} L(\boldsymbol{\mu}_1,\ldots,\boldsymbol{\mu}_k,\boldsymbol{\Sigma})}{\max_{\boldsymbol{\mu}_1,\ldots,\boldsymbol{\mu}_k,\boldsymbol{\Sigma}} L(\boldsymbol{\mu}_1,\ldots,\boldsymbol{\mu}_k,\boldsymbol{\Sigma})} = \left(\frac{|\tilde{\mathbf{S}}|}{|\mathbf{S}_0|}\right)^{n/2} = \left(\frac{|\boldsymbol{W}|}{|\boldsymbol{T}|}\right)^{n/2}.$$

The last equality holds since $\mathbf{S}_0 = (1/n)\boldsymbol{T}$ and $\tilde{\mathbf{S}} = (1/n)\boldsymbol{W}$.

□

12.6.5 More Tests for One-Factor MANOVA

In Section 12.6.2, we have discussed the likelihood ratio test for H_0: $\boldsymbol{\alpha}_1 = \cdots = \boldsymbol{\alpha}_k = \mathbf{0}$ against H_1: not all $\boldsymbol{\alpha}_i$'s are $\mathbf{0}$. There are three other well-known tests, and computer packages routinely report them. These are

(a) Lawley-Hotelling trace: $trace(\boldsymbol{BW}^{-1})$,
(b) Pillai's trace: $trace(\boldsymbol{BT}^{-1})$, and
(c) Roy's largest root: the largest eigenvalue of $\boldsymbol{BT}^{-1}$.

For each of these tests, we reject H_0 if the corresponding statistic is larger than a threshold value. It turns out that all these four test statistics can be written as functions of the eigenvalues of $\boldsymbol{BT}^{-1}$. Let $\hat{\lambda}_1 \geq \cdots \geq \hat{\lambda}_p$ be the eigenvalues of $\boldsymbol{B}$ with respect to $\boldsymbol{T}$ (ie, the eigenvalues of $\boldsymbol{T}^{-1/2}\boldsymbol{BT}^{-1/2}$, where $\boldsymbol{T}^{-1/2}$ is symmetric). Then we have

(a) likelihood ratio: $\left\{\prod_{i=1}^{p}(1-\hat{\lambda}_i)\right\}^{n/2}$,
(b) Lawley-Hotelling trace: $\sum_{i=1}^{p}\hat{\lambda}_i/(1-\hat{\lambda}_i)$,
(c) Pillai's trace: $\sum_{i=1}^{p}\hat{\lambda}_i$, and
(d) Roy's largest root: $\hat{\lambda}_1$.

Remark 12.6.1. Since the rank of the matrix $\boldsymbol{B}$ is $s = \min(p, k-1)$, the number of nonzero generalized eigenvalues of $\boldsymbol{B}$ with respect to $\boldsymbol{T}$ is equal to s. Thus $\hat{\lambda}_j > 0, j = 1, \ldots, s$, and the remaining $\hat{\lambda}_j$'s are 0.

Interpretation of Tests in MANOVA

When testing H_0: $\boldsymbol{\alpha}_1 = \cdots = \boldsymbol{\alpha}_k = \mathbf{0}$, the reduced and the full MANOVA models are

$$\boldsymbol{Y}_{ij} = \boldsymbol{\mu} + \boldsymbol{\varepsilon}_{ij} \text{ (reduced)},$$
$$\boldsymbol{Y}_{ij} = \boldsymbol{\mu} + \boldsymbol{\alpha}_i + \boldsymbol{\varepsilon}_{ij} \text{ (full)}.$$

Now if we look at $\{\boldsymbol{e}^T\boldsymbol{Y}_{ij}\}$, $\boldsymbol{e} \in \mathbb{R}^p$, then we get the reduced and full models for the one-factor ANOVA case

$$\boldsymbol{e}^T\boldsymbol{Y}_{ij} = \boldsymbol{e}^T\boldsymbol{\mu} + \boldsymbol{e}^T\boldsymbol{\varepsilon}_{ij} \text{ (reduced)},$$
$$\boldsymbol{e}^T\boldsymbol{Y}_{ij} = \boldsymbol{e}^T\boldsymbol{\mu} + \boldsymbol{e}^T\boldsymbol{\alpha}_i + \boldsymbol{e}^T\boldsymbol{\varepsilon}_{ij} \text{ (full)}.$$

Since the coefficient of determination $R^2(\boldsymbol{e})$ is the proportional reduction in the residual sum of squares from the reduced to the full model (following the terminology used in Chapter 11), we have

$$R^2(\boldsymbol{e}) = \frac{SSE_R - SSE_F}{SSE_R} = \frac{\sum n_i(\boldsymbol{e}^T\overline{\boldsymbol{Y}}_{i.} - \boldsymbol{e}^T\overline{\boldsymbol{Y}}_{..})^2}{\sum_i\sum_j(\boldsymbol{e}^T\boldsymbol{Y}_{ij} - \boldsymbol{e}^T\overline{\boldsymbol{Y}}_{..})^2} = \frac{\boldsymbol{e}^T\boldsymbol{B}\boldsymbol{e}}{\boldsymbol{e}^T\boldsymbol{T}\boldsymbol{e}}. \tag{1}$$

Let $\hat{\lambda}_1 \geq \cdots \geq \hat{\lambda}_p$ be the generalized eigenvalues of $\boldsymbol{B}$ with respect to $\boldsymbol{T}$, that is, $\hat{\lambda}_1, \ldots, \hat{\lambda}_p$ are the eigenvalues of $\boldsymbol{T}^{-1/2}\boldsymbol{B}\boldsymbol{T}^{-1/2}$ with the corresponding eigenvectors $\hat{\boldsymbol{u}}_1, \ldots, \hat{\boldsymbol{u}}_p$ (Section B.4). If we denote $\hat{\boldsymbol{e}}_i = \boldsymbol{T}^{-1/2}\hat{\boldsymbol{u}}_i$, then $R^2(\hat{\boldsymbol{e}}_i) = \hat{\lambda}_i$, $i = 1, \ldots, p$.

Clearly, $R^2(\boldsymbol{e})$ is maximized at $\boldsymbol{e} = \hat{\boldsymbol{e}}_1$ and $R^2(\hat{\boldsymbol{e}}_1) = \hat{\lambda}_1$. The next largest value of $R^2(\boldsymbol{e})$ is obtained by maximizing it over $\boldsymbol{e}$ subject to the constraint $\boldsymbol{e}^T\boldsymbol{T}\hat{\boldsymbol{e}}_1 = 0$. This maximum is attained at $\boldsymbol{e} = \hat{\boldsymbol{e}}_2$ and $R^2(\hat{\boldsymbol{e}}_2) = \lambda_2$. This argument can be carried out further and it shows that $\{R^2(\hat{\boldsymbol{e}}_i)\}$ are the same as the generalized eigenvalues $\{\hat{\lambda}_i\}$ of $\boldsymbol{B}$ with respect to $\boldsymbol{T}$.
We can now express the four test statistics given above for the one-factor MANOVA in terms of $R^2(\hat{\boldsymbol{e}}_1), \ldots, R^2(\hat{\boldsymbol{e}}_p)$,

(a) likelihood ratio: $\left\{\prod_{i=1}^{p}(1 - R^2(\hat{\boldsymbol{e}}_i))\right\}^{n/2}$,
(b) Lawley-Hotelling trace: $\sum_{i=1}^{p} R^2(\hat{\boldsymbol{e}}_i)/\left[1 - R^2(\hat{\boldsymbol{e}}_i)\right]$,
(c) Pillai trace: $\sum_{i=1}^{p} R^2(\hat{\boldsymbol{e}}_i)$, and
(d) Roy's largest root: $R^2(\hat{\boldsymbol{e}}_1)$.

12.7 Two-Factor MANOVA

Let us consider a two-factor balanced MANOVA model

$$\boldsymbol{Y}_{ijk} = \boldsymbol{\mu} + \boldsymbol{\alpha}_i + \boldsymbol{\beta}_j + (\boldsymbol{\alpha\beta})_{ij} + \boldsymbol{\varepsilon}_{ijk},$$

$k = 1, \ldots, n_0, j = 1, \ldots, b, i = 1, \ldots, a$, where $\{\boldsymbol{\varepsilon}_{ijk}\}$ are iid $N_p(\mathbf{0}, \boldsymbol{\Sigma})$,

$$\sum \boldsymbol{\alpha}_i = \mathbf{0}, \quad \sum \boldsymbol{\beta}_j = \mathbf{0},$$
$$\sum_i (\boldsymbol{\alpha\beta})_{ij} = \mathbf{0} \text{ for all } j, \text{ and } \sum_j (\boldsymbol{\alpha\beta})_{ij} = \mathbf{0} \text{ for all } i.$$

For notational simplicity we write $\boldsymbol{\gamma}_{ij}$ for $(\boldsymbol{\alpha\beta})_{ij}$. The vectors $\{\boldsymbol{\alpha}_i\}$, $\{\boldsymbol{\beta}_j\}$, and $\{\boldsymbol{\gamma}_{ij}\}$ are the main effects of factor A, main effects of factor B, and interaction effects, respectively. The total number of observation vectors is $n = n_0ab$.

12.7.1 Estimation

The MLEs of $\boldsymbol{\mu}$, $\boldsymbol{\alpha}_i$, $\boldsymbol{\beta}_j$, and $\boldsymbol{\gamma}_{ij}$ are

$$\hat{\boldsymbol{\mu}} = \overline{\boldsymbol{Y}}_{\ldots}, \quad \hat{\boldsymbol{\alpha}}_i = \overline{\boldsymbol{Y}}_{i..} - \overline{\boldsymbol{Y}}_{\ldots}, \quad \hat{\boldsymbol{\beta}}_j = \overline{\boldsymbol{Y}}_{.j.} - \overline{\boldsymbol{Y}}_{\ldots}, \text{ and}$$
$$\hat{\boldsymbol{\gamma}}_{ij} = \overline{\boldsymbol{Y}}_{ij.} - \overline{\boldsymbol{Y}}_{i..} - \overline{\boldsymbol{Y}}_{.j.} + \overline{\boldsymbol{Y}}_{\ldots}.$$

We now write down the sums of squares and products matrices along with their degrees of freedom,

$$\boldsymbol{SSP}_{\text{tot}} = \sum_i \sum_j \sum_k (\boldsymbol{Y}_{ijk} - \overline{\boldsymbol{Y}}_{\ldots})(\boldsymbol{Y}_{ijk} - \overline{\boldsymbol{Y}}_{\ldots})^T, \quad df = n - 1$$
$$\boldsymbol{SSP}_A = n_0 b \sum \hat{\boldsymbol{\alpha}}_i \hat{\boldsymbol{\alpha}}_i^T, \quad df = a - 1,$$
$$\boldsymbol{SSP}_B = n_0 a \sum \hat{\boldsymbol{\beta}}_j \hat{\boldsymbol{\beta}}_j^T, \quad df = b - 1,$$
$$\boldsymbol{SSP}_{AB} = n_0 \sum_i \sum_j \hat{\boldsymbol{\gamma}}_{ij} \hat{\boldsymbol{\gamma}}_{ij}^T, \quad df = (a-1)(b-1),$$
$$\boldsymbol{SSP}_{\text{res}} = \sum_i \sum_j \sum_k (\boldsymbol{Y}_{ijk} - \overline{\boldsymbol{Y}}_{ij.})(\boldsymbol{Y}_{ijk} - \overline{\boldsymbol{Y}}_{ij.})^T, \quad df = n - ab.$$

The following lemma lists some useful facts.

Lemma 12.7.1. *For a two-factor MANOVA, the following hold:*

(a) $\boldsymbol{SSP}_{\text{tot}} = \boldsymbol{SSP}_A + \boldsymbol{SSP}_B + \boldsymbol{SSP}_{AB} + \boldsymbol{SSP}_{\text{res}}$.
(b) $df(\boldsymbol{SSP}_{\text{tot}}) = df(\boldsymbol{SSP}_A) + df(\boldsymbol{SSP}_B) + df(\boldsymbol{SSP}_{AB}) + df(\boldsymbol{SSP}_{\text{res}})$.
(c) $\boldsymbol{SSP}_A$, $\boldsymbol{SSP}_B$, $\boldsymbol{SSP}_{AB}$, *and* $\boldsymbol{SSP}_{\text{res}}$ *are independent.*
(d) $\boldsymbol{SSP}_{\text{res}} \sim W_p(n - ab, \boldsymbol{\Sigma})$.
(e) *An unbiased estimate of* $\boldsymbol{\Sigma}$ *is* $\frac{1}{n-ab}\boldsymbol{SSP}_{\text{res}}$.

12.7.2 Hypothesis Testing in Two-Factor MANOVA

We will write down here only the likelihood ratio tests. However, as in the one-factor MANOVA case, there are other tests such as those by Pillai, Lawley-Hotelling, and Roy, and computer packages routinely report them.

Test for Interactions

Suppose we wish to test H_0: $\boldsymbol{\gamma}_{ij} = \mathbf{0}$ for all i and j, against H_1: not all $\boldsymbol{\gamma}_{ij}$ are zero.

Wilks' lambda for this test is $\Lambda = \frac{|\boldsymbol{SSP}_{\text{res}}|}{|\boldsymbol{SSP}_{AB}+\boldsymbol{SSP}_{\text{res}}|}$ and the likelihood ratio test statistic is $\lambda = \Lambda^{n/2}$. Under H_0, using Bartlett's modification we have

$$-\left\{ab(n_0-1) - \frac{p+1-(a-1)(b-1)}{2}\right\}\log\Lambda \xrightarrow{\mathcal{D}} \chi^2_{(a-1)(b-1)p},$$

as $n \to \infty$. We can use this result to obtain the critical value or the p-value in order to carry out the test.

Test for the Main Effects of Factor A

If we wish to test H_0: $\boldsymbol{\alpha}_i = \mathbf{0}$ for all i, against H_1: not all $\boldsymbol{\alpha}_i$ are zero, Wilks' lambda criterion is $\Lambda = \frac{|\boldsymbol{SSP}_{\text{res}}|}{|\boldsymbol{SSP}_{A}+\boldsymbol{SSP}_{\text{res}}|}$ and the likelihood ratio test statistic is $\lambda = \Lambda^{n/2}$.

Once again, we can use the following asymptotic result to carry out this test. Under H_0, using Bartlett's approximation we have

$$-\left\{ab(n_0-1) - \frac{p+1-(a-1)}{2}\right\}\log\Lambda \xrightarrow{\mathcal{D}} \chi^2_{(a-1)p},$$

as $n \to \infty$.

Test for the Main Effects of Factor B

If we wish to test H_0: $\boldsymbol{\beta}_j = \mathbf{0}$ for all j, against H_1: not all $\boldsymbol{\beta}_j$ are zero, Wilks' lambda criterion is $\Lambda = \frac{|\boldsymbol{SSP}_{\text{res}}|}{|\boldsymbol{SSP}_{B}+\boldsymbol{SSP}_{\text{res}}|}$ and the likelihood ratio is $\lambda = \Lambda^{n/2}$. It turns out that under H_0 (using Bartlett's modification),

$$-\left\{ab(n_0-1) - \frac{p+1-(b-1)}{2}\right\}\log\Lambda \xrightarrow{\mathcal{D}} \chi^2_{(b-1)p},$$

as $n \to \infty$.

12.7.3 Simultaneous Confidence Intervals

Simultaneous confidence intervals with a family confidence coefficient of at least $1-\alpha$ for all the pairwise differences in the main effect of factor A are given by

$$\alpha_{il} - \alpha_{i'l}\colon \hat{\alpha}_{il} - \hat{\alpha}_{i'l} + Bs(\hat{\alpha}_{il} - \hat{\alpha}_{i'l}), \quad 1 \le i \ne i' \le a, \ \ l = 1, \ldots, p,$$

where, $B = t_{n-ab,\alpha/[pa(a-1)]}$, $s(\hat{\alpha}_{il} - \hat{\alpha}_{i'l}) = \sqrt{2s_{ll}/(n_0 b)}$, and s_{ll} is the lth diagonal element of $\boldsymbol{S}$.

Simultaneous confidence intervals with a family confidence coefficient of at least $1-\alpha$ for all the pairwise differences in the main effect of factor B are given by

$$\beta_{jl}-\beta_{j'l}\colon \hat{\beta}_{jl}-\hat{\beta}_{j'l}+Bs(\hat{\beta}_{jl}-\hat{\beta}_{j'l}),\quad 1\le j\ne j'\le b,\ \ l=1,\ldots,p,$$

where $B=t_{n-ab,\alpha/[pb(b-1)]}$ and $s(\hat{\beta}_{jl}-\hat{\beta}_{j'l})=\sqrt{2s_{ll}/(n_0 a)}$.

Simultaneous confidence intervals with a family confidence coefficient of at least $1-\alpha$ for all the pairwise differences in the mean response are given by

$$\mu_{ijl}-\mu_{i'j'l}\colon \hat{\mu}_{ijl}-\hat{\mu}_{i'j'l}+Bs(\hat{\mu}_{ijl}-\hat{\mu}_{i'j'l}),$$
$$1\le i\ne i'\le a,\ \ 1\le j\ne j'\le b,\ \ (i,j)\ne(i',j'),\ \ l=1,\ldots,p,$$

where $B=t_{n-ab,\alpha/[pab(ab-1)]}$ and $s(\hat{\mu}_{ijl}-\hat{\mu}_{i'j'l})=\sqrt{2s_{ll}/n_0}$. [Note that $\hat{\boldsymbol{\mu}}_{ij}=\overline{\boldsymbol{Y}}_{ij\cdot\cdot}$.]

The simultaneous confidence intervals given above are expected to be wide. Therefore, we may use Tukey's method for pairwise comparisons for each of the p univariate ANOVA models with a confidence level of $1-\alpha/p$ so that the overall confidence level is at least $1-\alpha$.

12.8 Multivariate Linear Model

In the usual linear model framework, the response for each of the n observations is real valued. Thus the observed vector of responses $\boldsymbol{Y}$ is n-dim, and if the design matrix $\boldsymbol{X}$ is $n\times k$, then the Gauss-Markov model is written as $\boldsymbol{Y}=\boldsymbol{X\beta}+\boldsymbol{\varepsilon}$, where $\boldsymbol{\beta}$ is an unknown vector of parameters and $\boldsymbol{\varepsilon}$ is an $n\times 1$ vector of mutually uncorrelated random errors with mean zero and common variance σ^2 (Section 11.1 in Chapter 11). If each of the n observations is p-dim, then we have an $n\times p$ matrix $\boldsymbol{Y}$ of observed responses which can modeled by a generalization of the framework described in Section 11.1 of Chapter 11 resulting in a multivariate linear model. We have also seen in Chapter 11 that the Gauss-Markov setup and its extensions include regression, analysis of variance, analysis of covariance, random- and mixed-effect models as special cases, and the same is true for their multivariate counterparts.

In the multivariate case, we have p columns of $n\times 1$ observation vectors $\boldsymbol{Y}_1,\ldots,\boldsymbol{Y}_p$, and for each observed vector of responses, the model is

$$\boldsymbol{Y}_j=\boldsymbol{X\beta}_j+\boldsymbol{\varepsilon}_j,\quad j=1,\ldots,p,$$

where $\boldsymbol{\beta}_j$ is a k-dim unknown vector of parameters and $\boldsymbol{\varepsilon}_j$ is an n-dim vector of mutually uncorrelated mean zero random errors. Thus we have

$$[\boldsymbol{Y}_1,\ldots,\boldsymbol{Y}_p]=\boldsymbol{X}\big[\boldsymbol{\beta}_1,\ldots,\boldsymbol{\beta}_p\big]+[\boldsymbol{\varepsilon}_1,\ldots,\boldsymbol{\varepsilon}_p],\ \text{or}$$
$$\boldsymbol{Y}=\boldsymbol{X\beta}+\boldsymbol{\varepsilon},$$

where $\boldsymbol{Y}$ is $n\times p$ with columns $\boldsymbol{Y}_1,\ldots,\boldsymbol{Y}_p$, $\boldsymbol{X}$ is $n\times k$, $\boldsymbol{\beta}$ is $k\times p$ with columns $\boldsymbol{\beta}_1,\ldots,\boldsymbol{\beta}_p$, and $\boldsymbol{\varepsilon}$ is $n\times p$ with columns $\boldsymbol{\varepsilon}_1,\ldots,\boldsymbol{\varepsilon}_p$. This is the multivariate linear model and it has the same formal structure as in Eq. (1) in Chapter 11. It is possible to analyze each of the p linear models separately, but this strategy has a drawback since the p observations in each row of the response matrix may be correlated, that is, the p elements in each row of $\boldsymbol{\varepsilon}$ may be correlated, and a procedure which analyzes the p linear models separately, fails to take into account this dependence. A joint analysis of these p models is therefore preferable. In the subsequent discussion we assume that $\boldsymbol{X}^T\boldsymbol{X}$ is nonsingular.

In the multivariate linear model, we often assume that the rows of $\boldsymbol{\varepsilon}$ are iid $N_p(\boldsymbol{0}, \boldsymbol{\Sigma})$. The goal is to estimate the matrices of unknown parameters $\boldsymbol{\beta}$ and $\boldsymbol{\Sigma}$, carry out inferential procedures such as tests of hypotheses and construction of confidence intervals, and make predictions whenever necessary. We do not discuss random- and mixed-effect cases here. We write down a few basic results on the estimation of $\boldsymbol{\beta}$ and $\boldsymbol{\Sigma}$, and the distributions of the estimates.

12.8.1 Estimation of β and Σ

The normal equations and least squares estimate of $\boldsymbol{\beta}$ are similar to those in the univariate case, and they are

$$\boldsymbol{X}^T\boldsymbol{X}\hat{\boldsymbol{\beta}} = \boldsymbol{X}^T\boldsymbol{Y}, \quad \hat{\boldsymbol{\beta}} = (\boldsymbol{X}^T\boldsymbol{X})^{-1}\boldsymbol{X}^T\boldsymbol{Y}.$$

In order to see why the same equations come up, we may consider rewriting the model. Postmultiplying both sides of the multivariate linear model by $\boldsymbol{\Sigma}^{-1/2}$ leads to

$$\begin{aligned}\left[\boldsymbol{R}_1, \ldots, \boldsymbol{R}_p\right] &= \boldsymbol{X}\left[\boldsymbol{\gamma}_1, \ldots, \boldsymbol{\gamma}_p\right] + \left[\boldsymbol{\delta}_1, \ldots, \boldsymbol{\delta}_p\right], \text{ or}\\ \boldsymbol{R} &= \boldsymbol{X}\boldsymbol{\gamma} + \boldsymbol{\delta},\end{aligned}$$

where $\boldsymbol{R}_j = \boldsymbol{Y}_j\boldsymbol{\Sigma}^{-1/2}, \boldsymbol{\gamma}_j = \boldsymbol{\beta}_j\boldsymbol{\Sigma}^{-1/2}, \boldsymbol{\delta}_j = \boldsymbol{\varepsilon}_j\boldsymbol{\Sigma}^{-1/2}, j = 1, \ldots, p$. Under normality (ie, the rows of $\boldsymbol{\varepsilon}$ are iid $N_p(\boldsymbol{0}, \boldsymbol{\Sigma})$), the error vectors $\boldsymbol{\delta}_1, \ldots, \boldsymbol{\delta}_p$ are mutually independent, and each $\boldsymbol{\delta}_j$ has n entries which are iid with mean 0 and variance 1. Even if the assumption of normality is not valid, $\boldsymbol{\delta}_1, \ldots, \boldsymbol{\delta}_p$ are mutually uncorrelated, and we can minimize

$$\sum \|\boldsymbol{R}_j - \boldsymbol{X}\boldsymbol{\gamma}_j\|^2$$

with respect to $\boldsymbol{\gamma}_1, \ldots, \boldsymbol{\gamma}_p$ in order to get the least squares estimates, which lead to the following p normal equations

$$\begin{aligned}\boldsymbol{X}^T\boldsymbol{X}\boldsymbol{\gamma}_j &= \boldsymbol{X}^T\boldsymbol{R}_j, \quad \text{ie, } \boldsymbol{X}^T\boldsymbol{X}\boldsymbol{\beta}_j\boldsymbol{\Sigma}^{-1/2} = \boldsymbol{X}^T\boldsymbol{Y}_j\boldsymbol{\Sigma}^{-1/2}, \quad \text{ie,}\\ \boldsymbol{X}^T\boldsymbol{X}\boldsymbol{\beta}_j &= \boldsymbol{X}^T\boldsymbol{Y}_j, \quad j = 1, \ldots, p, \text{ or}\\ \boldsymbol{X}^T\boldsymbol{X}\boldsymbol{\beta} &= \boldsymbol{X}^T\boldsymbol{Y}, \quad \text{ie, } \hat{\boldsymbol{\beta}} = (\boldsymbol{X}^T\boldsymbol{X})^{-1}\boldsymbol{X}^T\boldsymbol{Y}.\end{aligned}$$

Since

$$\begin{aligned}\sum \|\boldsymbol{R}_j - \boldsymbol{X}\boldsymbol{\gamma}_j\|^2 &= trace((\boldsymbol{R} - \boldsymbol{X}\boldsymbol{\gamma})^T(\boldsymbol{R} - \boldsymbol{X}\boldsymbol{\gamma}))\\ &= trace((\boldsymbol{Y} - \boldsymbol{X}\boldsymbol{\beta})^T(\boldsymbol{Y} - \boldsymbol{X}\boldsymbol{\beta})\boldsymbol{\Sigma}^{-1}),\end{aligned}$$

it follows that the normal equations are obtained by minimizing $trace((\boldsymbol{Y} - \boldsymbol{X}\boldsymbol{\beta})^T(\boldsymbol{Y} - \boldsymbol{X}\boldsymbol{\beta})\boldsymbol{\Sigma}^{-1})$ with respect to $\boldsymbol{\beta}$.

An unbiased estimate of $\boldsymbol{\Sigma}$ is given by

$$\boldsymbol{S} = \frac{1}{n-k}(\boldsymbol{Y} - \boldsymbol{X}\hat{\boldsymbol{\beta}})^T(\boldsymbol{Y} - \boldsymbol{X}\hat{\boldsymbol{\beta}}).$$

As in the univariate case, the estimated response $\hat{\boldsymbol{Y}} = \boldsymbol{X}\hat{\boldsymbol{\beta}}$ and the residuals $\hat{\boldsymbol{\varepsilon}} = \boldsymbol{Y} - \boldsymbol{X}\hat{\boldsymbol{\beta}}$ are uncorrelated, and are independent when the rows of $\boldsymbol{\varepsilon}$ are iid $N_p(\boldsymbol{0}, \boldsymbol{\Sigma})$, in which case, $\boldsymbol{S}$ and $\hat{\boldsymbol{\beta}}$ are independent as well.

12.8.2 Properties of the Estimates of β and Σ

For the rest of this section, we assume that the rows of $\boldsymbol{\varepsilon}$ are iid $N_p(\boldsymbol{0}, \boldsymbol{\Sigma})$. We have the following result which can be used for tests of hypotheses and construction of confidence intervals.

Theorem 12.8.1.

(a) *The MLEs of $\boldsymbol{\beta}$ and $\boldsymbol{\Sigma}$ are*

$$\hat{\boldsymbol{\beta}} = (\boldsymbol{X}^T\boldsymbol{X})^{-1}\boldsymbol{X}^T\boldsymbol{Y}, \quad \tilde{\boldsymbol{S}} = \frac{1}{n}(\boldsymbol{Y} - \boldsymbol{X}\hat{\boldsymbol{\beta}})^T(\boldsymbol{Y} - \boldsymbol{X}\hat{\boldsymbol{\beta}}).$$

(b) *$(\hat{\boldsymbol{\beta}}, \boldsymbol{S})$ are sufficient for $(\boldsymbol{\beta}, \boldsymbol{\Sigma})$.*
(c) *$\hat{\boldsymbol{Y}} = \boldsymbol{X}\hat{\boldsymbol{\beta}}$ and $\hat{\boldsymbol{\varepsilon}} = \boldsymbol{Y} - \boldsymbol{X}\hat{\boldsymbol{\beta}}$ are independent.*
(d) *$\hat{\boldsymbol{\beta}}$ and $\boldsymbol{S}$ are independent.*
(e) *For any vector $\boldsymbol{b}$ in $\mathbb{R}^p$, $\hat{\boldsymbol{\beta}}\boldsymbol{b} \sim N_k(\boldsymbol{\beta}\boldsymbol{b}, \boldsymbol{b}^T\boldsymbol{\Sigma}\boldsymbol{b}(\boldsymbol{X}^T\boldsymbol{X})^{-1})$.*
(f) *$(n-k)\boldsymbol{S} \sim W_p(n-k, \boldsymbol{\Sigma})$.*

12.9 Principal Components Analysis

Principal components analysis is a widely used method in multivariate analysis, and its goal is to reduce the dimensionality of the data with as little loss of information as possible. We first describe the basic ideas behind principal components, and discuss estimation issues later. Let $\boldsymbol{Y}$ be a p-dim random vector with mean $\boldsymbol{\mu}$ and covariance matrix $\boldsymbol{\Sigma}$. In some cases we assume that the diagonal elements of $\boldsymbol{\Sigma}$ are 1 which means that $\boldsymbol{\Sigma}$ is a correlation matrix.

If $\{\boldsymbol{e}_1, \ldots, \boldsymbol{e}_p\}$ is an orthonormal basis of $\mathbb{R}^p$, then

$$\boldsymbol{Y} - \boldsymbol{\mu} = \sum_{j=1}^{p}\left[\boldsymbol{e}_j^T(\boldsymbol{Y} - \boldsymbol{\mu})\right]\boldsymbol{e}_j, \text{ and}$$

$$\|\boldsymbol{Y} - \boldsymbol{\mu}\|^2 = \sum_{j=1}^{p}\left[\boldsymbol{e}_j^T(\boldsymbol{Y} - \boldsymbol{\mu})\right]^2.$$

Note that $\boldsymbol{e}_j^T(\boldsymbol{Y} - \boldsymbol{\mu})$ is a random variable with mean 0 and variance $\boldsymbol{e}_j^T\boldsymbol{\Sigma}\boldsymbol{e}_j$. The total variability of $\boldsymbol{Y}$ is

$$\sum_{j=1}^{p} \mathrm{Var}[Y_j] = \sum_{j=1}^{p} \mathrm{E}[Y_j - \mu_j]^2 = \mathrm{E}\left\{\sum_{j=1}^{p} (Y_j - \mu_j)^2\right\}$$
$$= \mathrm{E}[\|\boldsymbol{Y} - \boldsymbol{\mu}\|^2] = trace(\boldsymbol{\Sigma}).$$

On the other hand

$$\sum_{j=1}^{p} \mathrm{Var}[Y_j] = \mathrm{E}[\|\boldsymbol{Y} - \boldsymbol{\mu}\|^2] = \sum_{j=1}^{p} \mathrm{E}\left[\boldsymbol{e}_j^T(\boldsymbol{Y} - \boldsymbol{\mu})\right]^2 = \sum_{j=1}^{p} \mathrm{Var}[\boldsymbol{e}_j^T \boldsymbol{Y}].$$

Now if it happens that for some $k < p$ (and hopefully k is small),

$$\mathrm{E}[\|\boldsymbol{Y} - \boldsymbol{\mu}\|^2] \approx \sum_{j=1}^{k} \mathrm{E}\left[\boldsymbol{e}_j^T(\boldsymbol{Y} - \boldsymbol{\mu})\right]^2, \text{ and}$$

$$\sum_{j=k+1}^{p} \mathrm{E}\left[\boldsymbol{e}_j^T(\boldsymbol{Y} - \boldsymbol{\mu})\right]^2 \text{ is small,}$$

then the total variability of $\boldsymbol{Y}$ is explainable (to a large extent) by the variability of k random variables $\boldsymbol{e}_1^T\boldsymbol{Y}, \dots, \boldsymbol{e}_k^T\boldsymbol{Y}$. Ideally, the reduction in dimensionality is substantial if k is small in comparison to p. Let $\lambda_1 \geq \cdots \geq \lambda_p$ be the eigenvalues of $\boldsymbol{\Sigma}$ with $\boldsymbol{u}_1, \dots, \boldsymbol{u}_p$, the corresponding orthonormal eigenvectors. From the properties of eigenvalues and eigenvectors (Section B.2) we have that

$$\max\left\{\boldsymbol{e}_1^T\boldsymbol{\Sigma}\boldsymbol{e}_1 + \cdots + \boldsymbol{e}_k^T\boldsymbol{\Sigma}\boldsymbol{e}_k\text{: } \boldsymbol{e}_1, \dots, \boldsymbol{e}_k \text{ orthonormal}\right\} = \lambda_1 + \cdots + \lambda_k,$$

and the maximum is attained at $\boldsymbol{e}_1 = \boldsymbol{u}_1, \dots, \boldsymbol{e}_k = \boldsymbol{u}_k$. So

$$\max\left\{\mathrm{Var}[\boldsymbol{e}_1^T\boldsymbol{Y}] + \cdots + \mathrm{Var}[\boldsymbol{e}_k^T\boldsymbol{Y}]\text{: } \boldsymbol{e}_1, \dots, \boldsymbol{e}_k \text{ orthonormal}\right\}$$
$$= \mathrm{Var}[\boldsymbol{u}_1^T\boldsymbol{Y}] + \cdots + \mathrm{Var}[\boldsymbol{u}_k^T\boldsymbol{Y}] = \lambda_1 + \cdots + \lambda_k\,.$$

Remark 12.9.1.

(a) The random variable $\boldsymbol{u}_1^T(\boldsymbol{Y} - \boldsymbol{\mu})$ is called the first principal component of Y, $\boldsymbol{u}_2^T(\boldsymbol{Y} - \boldsymbol{\mu})$ is the second principal component, and so on. The vectors $\boldsymbol{u}_1, \boldsymbol{u}_2, \dots$ are sometimes called the loading vectors.

(b) Principal components and the eigenvalues are not unit free. For this reason, in many cases one may prefer to carry out principal components analysis on the standardized variables which means that $\boldsymbol{\Sigma}$ is the correlation matrix.

(c) Note that $\pi_k = (\lambda_1 + \cdots + \lambda_k)/(\lambda_1 + \cdots + \lambda_p)$ is the proportion of the total variability of $\boldsymbol{Y}$ explained by the first k principal components. We can expect k to be small when the variables are strongly correlated. When the variables are mutually uncorrelated, principal components analysis is not useful.

Here is a summary of the properties of principal components.

Lemma 12.9.1. *Let* $Z_s = \boldsymbol{u}_s^T(\boldsymbol{Y} - \boldsymbol{\mu})$, $s = 1, 2, \dots$ *Then:*

(a) $E[Z_s] = 0$.
(b) $Var[Z_s] = \lambda_s$.
(c) $Cov[Z_s, Z_{s'}] = 0$ *if* $s \neq s'$.
(d) $Corr[Y_l, Z_s] = \sqrt{\lambda_s} u_{ls}/\sqrt{\sigma_{ll}}$, *where* Y_l *is the lth component of* $\boldsymbol{Y}$, u_{ls} *is the lth element of the vector* $\boldsymbol{u}_s$ *and* σ_{ll} *is the lth diagonal element of* $\boldsymbol{\Sigma}$.

12.9.1 Regression Interpretation of Principal Components

Let $Z_1 = \boldsymbol{e}_1^T(\boldsymbol{Y} - \boldsymbol{\mu}), \ldots, Z_k = \boldsymbol{e}_k^T(\boldsymbol{Y} - \boldsymbol{\mu})$ be k linear functions of the random vector $\boldsymbol{Y}$ such that the random variables $Z_1, \ldots, Z_k$ are mutually uncorrelated. Now consider the problem of predicting Y_j, the jth component of $\boldsymbol{Y}$, from $Z_1, \ldots, Z_k$ using a linear regression and let τ_j^2 be the prediction error. Then $\sum_{j=1}^p \tau_j^2$ is the total prediction error for predicting $Y_1, \ldots, Y_p$ (each separately) using $Z_1, \ldots, Z_k$. If $\sum_{j=1}^p \tau_j^2$ is quite small, then we may conclude that the information in the vector $\boldsymbol{Y}$ can be well summarized by $Z_1, \ldots, Z_k$. It turns out that $\sum_{j=1}^p \tau_j^2$ is minimized when $Z_1, \ldots, Z_k$ are the first k principal components of $\boldsymbol{Y}$.

Let us now justify this regression interpretation. Note that $Y_j - \mu_j, Z_1, \ldots, Z_k$ have zero means, and hence the intercept term for the regression of $Y_j - \mu_j$ on $Z_1, \ldots, Z_k$ is zero. Clearly,

$$\tau_j^2 = \min_{\beta_{j1},\ldots,\beta_{jk}} \mathrm{E}\left\{Y_j - \mu_j - \sum_{s=1}^{k} \beta_{js} Z_s\right\}^2.$$

Since $Z_1, \ldots, Z_k$ are assumed to be mutually uncorrelated, the solution to this minimization problem is given by $\beta_{js}^* = \mathrm{Cov}[Y_j, Z_s]/\mathrm{Var}[Z_s]$, $1 \leq s \leq k$, and consequently

$$\tau_j^2 = \mathrm{Var}[Y_j] - \sum_{s=1}^{k} \mathrm{Cov}[Y_j, Z_s]^2/\mathrm{Var}[Z_s].$$

Let $\boldsymbol{\sigma}_1, \ldots, \boldsymbol{\sigma}_p$ be the column vectors of the covariance matrix $\boldsymbol{\Sigma}$, that is, $\boldsymbol{\Sigma} = [\boldsymbol{\sigma}_1, \ldots, \boldsymbol{\sigma}_p]$. Then $\mathrm{Cov}[Y_j, Z_s] = \boldsymbol{e}_s^T \boldsymbol{\sigma}_j$, $\mathrm{Var}[Z_s] = \boldsymbol{e}_s^T \boldsymbol{\Sigma} \boldsymbol{e}_s$, and we have

$$\tau_j^2 = \mathrm{Var}[Y_j] - \sum_{s=1}^{k} \frac{(\boldsymbol{e}_s^T v_j)^2}{\boldsymbol{e}_s^T \boldsymbol{\Sigma} \boldsymbol{e}_s}.$$

Hence

$$\sum_{j=1}^{p} \tau_j^2 = \sum_{j=1}^{p} \mathrm{Var}[Y_j] - \sum_{j=1}^{p} \sum_{s=1}^{k} \frac{(\boldsymbol{e}_s^T \boldsymbol{\sigma}_j)^2}{\boldsymbol{e}_s^T \boldsymbol{\Sigma} \boldsymbol{e}_s}$$

$$= \sum_{j=1}^{p} \mathrm{Var}[Y_j] - \sum_{s=1}^{k} \sum_{j=1}^{p} \frac{(\boldsymbol{e}_s^T \boldsymbol{\sigma}_j)^2}{\boldsymbol{e}_s^T \boldsymbol{\Sigma} \boldsymbol{e}_s}.$$

Since $\sum_{j=1}^{p} \text{Var}[Y_j] = trace(\boldsymbol{\Sigma})$ and

$$\sum_{j=1}^{p} (\boldsymbol{e}_s^T \boldsymbol{\sigma}_j)^2 = \boldsymbol{e}_s^T \sum_{j=1}^{p} \boldsymbol{\sigma}_j \boldsymbol{\sigma}_j^T \boldsymbol{e}_s = \boldsymbol{e}_s^T \boldsymbol{\Sigma}^2 \boldsymbol{e}_s,$$

we have

$$\sum_{j=1}^{p} \tau_j^2 = trace(\boldsymbol{\Sigma}) - \sum_{s=1}^{k} \frac{\boldsymbol{e}_s^T \boldsymbol{\Sigma}^2 \boldsymbol{e}_s}{\boldsymbol{e}_s^T \boldsymbol{\Sigma} \boldsymbol{e}_s}.$$

Minimizing $\sum_{j=1}^{p} \tau_j^2$ with respect to $\boldsymbol{e}_1, \ldots, \boldsymbol{e}_k$ subject to the constrains $\boldsymbol{e}_s^T \boldsymbol{\Sigma} \boldsymbol{e}_{s'} = \text{Cov}[Z_s, Z_{s'}] = 0, 1 \leq s \neq s' \leq k$, is equivalent to maximizing

$$\sum_{s=1}^{k} \frac{\boldsymbol{e}_s^T \boldsymbol{\Sigma}^2 \boldsymbol{e}_s}{\boldsymbol{e}_s^T \boldsymbol{\Sigma} \boldsymbol{e}_s} \tag{2}$$

with respect to $\boldsymbol{e}_1, \ldots, \boldsymbol{e}_k$ subject to the constraints $\boldsymbol{e}_s^T \boldsymbol{\Sigma} \boldsymbol{e}_{s'} = 0, 1 \leq s \neq s' \leq k$. This a generalized eigenvalue problem described in Section B.4. If $\lambda_1 \geq \lambda_2 \geq \cdots$ are the generalized eigenvalues of $\boldsymbol{A} = \boldsymbol{\Sigma}^2$ with respect to $\boldsymbol{B} = \boldsymbol{\Sigma}$, the maximum value of Eq. (2) is given by $\lambda_1 + \cdots + \lambda_k$. Since $\lambda_1, \lambda_2, \ldots$ are the eigenvalues of $\boldsymbol{B}^{-1/2} \boldsymbol{A} \boldsymbol{B}^{-1/2} = \boldsymbol{\Sigma}^{-1/2} \boldsymbol{\Sigma}^2 \boldsymbol{\Sigma}^{-1/2} = \boldsymbol{\Sigma}$ with the corresponding orthonormal eigenvectors $\boldsymbol{u}_1, \boldsymbol{u}_2, \ldots$, the maximum of Eq. (2) occurs at $\boldsymbol{e}_s = \boldsymbol{u}_s, s = 1, \ldots, k$.

This proves that among all the k mutually uncorrelated linear functions of $\boldsymbol{Y}$, the principal components $Z_1 = \boldsymbol{u}_1^T(\boldsymbol{Y} - \boldsymbol{\mu}), \ldots, Z_k = \boldsymbol{u}_k^T(\boldsymbol{Y} - \boldsymbol{\mu})$ are the best linear predictors of $Y_1, \ldots, Y_p$.

12.9.2 Estimation of Principal Components

Till now we have discussed the concept of principal components for a random vector from a theoretical standpoint. We now take up the issue of estimating them from the data. Let $\boldsymbol{Y}_1, \ldots, \boldsymbol{Y}_n$ be iid with mean vector $\boldsymbol{\mu}$ and covariance matrix $\boldsymbol{\Sigma}$. Note that we have made no assumption on the distribution of $\boldsymbol{Y}_i$ (such as normality).

Even though the covariance matrix of the standardized variables is the correlation matrix, for notational simplicity, we use $\boldsymbol{\Sigma}$ to denote both the covariance and correlation matrices. Similarly we use $\boldsymbol{S}$ to denote both the sample covariance and correlation matrices. Note that if we want to find the principal components of $\boldsymbol{Y} - \boldsymbol{\mu}$, then we deal with the covariance matrix of $\boldsymbol{Y}$. Whereas if we want to find the principal components of $\left[diag(\boldsymbol{\Sigma})\right]^{-1/2}(\boldsymbol{Y} - \boldsymbol{\mu})$ (ie, the standardized variables), we deal with the correlation matrix.

As before, let $\lambda_1 \geq \lambda_2 \geq \cdots$ be the eigenvalues of $\boldsymbol{\Sigma}$ with $\boldsymbol{u}_1, \boldsymbol{u}_2, \ldots$ as the corresponding orthonormal eigenvectors. Similarly, let $\hat{\lambda}_1 \geq \hat{\lambda}_2 \geq \cdots$ be the eigenvalues of $\boldsymbol{S}$ with the corresponding orthonormal eigenvectors $\hat{\boldsymbol{u}}_1, \hat{\boldsymbol{u}}_2, \ldots$. It is easy to guess that $\hat{\lambda}_j$ estimates λ_j

and $\hat{\boldsymbol{u}}_j$ estimates $\pm \boldsymbol{u}_j$, $j = 1, \ldots, p$. The reason for "$\pm$" is that if $\boldsymbol{u}_j$ is an eigenvector of $\boldsymbol{\Sigma}$ with the eigenvalue λ_j, then $-\boldsymbol{u}_j$ is also an eigenvector of $\boldsymbol{\Sigma}$ with the same eigenvalue λ_j. Estimated principal components are $\hat{Z}_s = \hat{\boldsymbol{u}}_s^T(\boldsymbol{Y} - \overline{\boldsymbol{Y}}), s = 1, \ldots, p$. Recall that the population (or theoretical) principal components have zero means and are uncorrelated. The estimated (sample) principal components have similar properties. Define the scores of the sth principal component to be $\hat{Z}_{is} = \hat{\boldsymbol{u}}_s^T(\boldsymbol{Y}_i - \overline{\boldsymbol{Y}})$, $i = 1, \ldots, n$. A result analogous to Lemma 12.9.1 holds for the sample principal components when E[·], Var[·], etc., are replaced by the sample mean, sample variance, etc., and this is left as an exercise. In particular,

$$\sum_{i=1}^{n} \hat{Z}_{is}/n = 0, \quad \sum_{i=1}^{n} \hat{Z}_{is}^2/(n-1) = \hat{\lambda}_s, \text{ and}$$
$$\sum_{i=1}^{n} \hat{Z}_{is}\hat{Z}_{is'}/(n-1) = 0, \quad s \neq s'.$$

Scree Plot

The ratio $\hat{\lambda}_k / \sum \hat{\lambda}_j = \hat{\lambda}_k / trace(\boldsymbol{S})$ estimates the proportion of total variability of $\boldsymbol{Y}$ explained by the kth principal component. For this reason, it is useful to plot $\hat{\lambda}_k / trace(\boldsymbol{S})$ against k. This plot graphically displays how the eigenvalues decay. It is also useful to plot the cumulative ratio $\hat{\pi}_k = (\hat{\lambda}_1 + \cdots + \hat{\lambda}_k)/(\hat{\lambda}_1 + \cdots + \hat{\lambda}_p)$ against k. This gives us an idea of how much dimensionality reduction is possible since $\hat{\pi}_k$ is an estimate of $\pi_k = (\lambda_1 + \cdots + \lambda_k)/(\lambda_1 + \cdots + \lambda_p)$, the proportion of the total variability of $\boldsymbol{Y}$ explained by the first k principal components.

12.9.3 Asymptotic Results in Principal Components Analysis

Asymptotic distributions of sample eigenvalues and sample eigenvectors are somewhat complicated. It is important to point out that the asymptotic distributions depend on the distribution of the multivariate population from which the observations are taken. This is quite unlike the limit theorems for the estimated mean vectors and associated statistics. Here we only deal with the case when the eigenvalues of the population covariance matrix $\boldsymbol{\Sigma}$ are distinct. When some of the eigenvalues have multiplicities larger than 1, the asymptotic distributions of the sample eigenvalues and sample eigenvectors are not even normal (they are mixtures of normals). We do not discuss such cases here.

As before, let the spectral decompositions of $\boldsymbol{\Sigma}$ and sample covariance matrix $\boldsymbol{S}$ be

$$\boldsymbol{\Sigma} = \sum_{j=1}^{p} \lambda_j \boldsymbol{u}_j \boldsymbol{u}_j^T \text{ and } \boldsymbol{S} = \sum_{j=1}^{p} \hat{\lambda}_j \hat{\boldsymbol{u}}_j \hat{\boldsymbol{u}}_j^T,$$

where $\lambda_1 \geq \lambda_2 \geq \cdots$ and $\hat{\lambda}_1 \geq \hat{\lambda}_2 \geq \cdots$. We assume all the eigenvalues are distinct (ie, $\lambda_1 > \lambda_2 > \cdots > \lambda_p$). For continuous distributions, it can be shown that the sample eigenvalues $\hat{\lambda}_1, \ldots, \hat{\lambda}_p$ are distinct with probability 1.

The jth principal component of $\boldsymbol{Y}$ is $Z_j = \boldsymbol{u}_j^T(\boldsymbol{Y} - \boldsymbol{\mu})$. We know that $\mathrm{E}[Z_j] = 0$ and $\mathrm{Var}[Z_j] = \lambda_j$. We also know that $Z_1, \ldots, Z_p$ are mutually uncorrelated. If the population of $\boldsymbol{Y}$ is normal then $Z_1, \ldots, Z_p$ are independent and $Z_j \sim N(0, \lambda_j), j = 1, \ldots, p$.

We will write down the asymptotic distributions of the sample eigenvalues and sample eigenvectors. These results can be proved using the perturbation theory of matrices which is outside the scope of this book.

Theorem 12.9.1. *Let $\boldsymbol{Y}_1, \ldots, \boldsymbol{Y}_n$ be iid with mean $\boldsymbol{\mu}$ and covariance matrix $\boldsymbol{\Sigma}$, and assume that the eigenvalues of $\boldsymbol{\Sigma}$ are distinct. All the results below are true assuming that $n \to \infty$.*

(a) *$\sqrt{n}(\hat{\boldsymbol{\lambda}} - \boldsymbol{\lambda}) \xrightarrow{\mathcal{D}} N_p(\boldsymbol{0}, \boldsymbol{W})$, where element (j, k) of $\boldsymbol{W}$ is given by $Cov[Z_j^2, Z_k^2]$, where $\boldsymbol{\lambda}$ and $\hat{\boldsymbol{\lambda}}$ are p-dim vectors of the eigenvalues $\lambda_1, \ldots \lambda_p$ and their estimates.*

(b) *If the population is normal (ie, $\{\boldsymbol{Y}_i\}$ are iid $N_p(\boldsymbol{\mu}, \boldsymbol{\Sigma})$), then the matrix $\boldsymbol{W}$ in part (a) is a diagonal matrix whose jth diagonal element is given by $2\lambda_j^2$.*

(c) *For any $1 \le i \le k$, $\sqrt{n}(\hat{\boldsymbol{u}}_i \pm \boldsymbol{u}_i) \xrightarrow{\mathcal{D}} N_p(\boldsymbol{0}, \boldsymbol{R}_i)$, where*

$$\boldsymbol{R}_i = \sum_{j \neq i} \sum_{k \neq i} \frac{Cov[Z_j Z_i, Z_k Z_i]}{(\lambda_j - \lambda_i)(\lambda_k - \lambda_i)} \boldsymbol{u}_j \boldsymbol{u}_k^T.$$

(d) *If the population is normal, then the matrix $\boldsymbol{R}_i$ in part (c) has the simplified form*

$$\boldsymbol{R}_i = \lambda_i \sum_{j \neq i} \frac{\lambda_j}{(\lambda_j - \lambda_i)^2} \boldsymbol{u}_j \boldsymbol{u}_j^T.$$

The results given above allow us to construct confidence intervals for the eigenvalues.

Remark 12.9.2. Even though $\{Z_j\}$ are mutually uncorrelated, the same is not necessarily true for $\{Z_j^2\}$. If the population is normal, $\{Z_j^2\}$ are mutually independent, and hence mutually uncorrelated. Thus, under normality, $\mathrm{Cov}[Z_j^2, Z_k^2] = 0, j \neq k$, and $\mathrm{Cov}[Z_j^2, Z_k^2] = \mathrm{Var}[Z_j^2] = 2\lambda_j^2$ when $j = k$. Therefore, part (b) of the last result follows from part (a). When $j \neq i$ and $k \neq i$,

$$\mathrm{Cov}[Z_j Z_i, Z_k Z_i] = \mathrm{E}[Z_j Z_k Z_i^2] - \mathrm{E}\,[Z_j Z_i]\boldsymbol{E}[Z_k Z_i] = \mathrm{E}[Z_j Z_k Z_i^2].$$

Under normality, when j, k, and i are all distinct, $\mathrm{E}[Z_j Z_k Z_i^2] = 0$. When $j = k \neq i$, then $\mathrm{E}[Z_j Z_k Z_i^2] = \mathrm{E}[Z_j^2]\mathrm{E}[Z_i^2] = \lambda_j \lambda_i$. This shows that part (d) of the last theorem follows from part (c).

When the population is normal, estimation of $\boldsymbol{W}$ or $\boldsymbol{R}_i$ is rather easy since $\hat{\boldsymbol{W}}$ is a diagonal matrix with diagonal elements $2\hat{\lambda}_1^2, \ldots, 2\hat{\lambda}_p^2$, and

$$\hat{\boldsymbol{R}}_i = \hat{\lambda}_i \sum_{j \neq i} \frac{\hat{\lambda}_j}{(\hat{\lambda}_j - \hat{\lambda}_i)^2} \hat{\boldsymbol{u}}_j \hat{\boldsymbol{u}}_j^T.$$

However, if the population is not normal, then we need to estimate these matrices ($\boldsymbol{W}$ and $\boldsymbol{R}_i$) using the principal component scores. For the ith principal component, the scores are $\{\hat{Z}_{ti} = \boldsymbol{u}_i^T(\boldsymbol{Y}_t - \overline{\boldsymbol{Y}}) \colon t = 1, \ldots, n\}$. Estimate of element (j, k) of $\boldsymbol{W}$ is

$$\begin{aligned}\hat{W}_{jk} &= \text{sample covariance of } \{(\hat{Z}_{tj}^2, \hat{Z}_{tk}^2): t = 1, \ldots, n\} \\ &= (n-1)^{-1}\left\{\sum_{t=1}^{n} \hat{Z}_{tj}^2 \hat{Z}_{tk}^2 - n\hat{\lambda}_j\hat{\lambda}_k\right\}.\end{aligned}$$

Similarly

$$\hat{\boldsymbol{R}}_i = \sum_{j\neq i}\sum_{k\neq i} \frac{v(j,k,i)}{(\hat{\lambda}_j - \hat{\lambda}_i)(\hat{\lambda}_k - \hat{\lambda}_i)} \hat{\boldsymbol{u}}_j \hat{\boldsymbol{u}}_k^T,$$

where $v(j,k,i)$ is the sample estimate of $\mathrm{Cov}[Z_jZ_i, Z_kZ_i]$ and is given by

$$v(j,k.i) = (n-1)^{-1} \sum_{t=1}^{n} \hat{Z}_{tj}\hat{Z}_{tk}\hat{Z}_{ti}^2.$$

Confidence Interval for λ_j

An approximate confidence interval for λ_j with a confidence coefficient $1-\alpha$ is given by $\hat{\lambda}_j \pm z_{\alpha/2}\sqrt{\hat{W}_{jj}/n}$, where $z_{\alpha/2}$ is the $(1-\alpha/2)$-quantile of the standard normal distribution. Note that $\hat{W}_{jj} = 2\hat{\lambda}_j^2$ for the normal case, whereas $\hat{W}_{jj}$ is the sample variance of $\{\hat{Z}_{tj}^2: t = 1, \ldots, n\}$ in the general case. Sometimes it may be preferable to construct a confidence interval for $\log\lambda_j$. By the delta method,

$$\sqrt{n}(\log\hat{\lambda}_j - \log\lambda_j) \xrightarrow{\mathcal{D}} N(0, W_{jj}/\lambda_j^2).$$

If the population is normal, then the natural logarithm is a variance stabilizing transformation since $W_{jj}/\lambda_j^2 = 2$, and in such a case, an approximate confidence interval for $\log\lambda_j$ with confidence coefficient $1-\alpha$ is given by $\log\hat{\lambda}_j \pm z_{\alpha/2}\sqrt{2/n}$. In the general case, an approximate confidence interval for $\log\lambda_j$ is $\log\hat{\lambda}_j \pm z_{\alpha/2}\sqrt{\hat{W}_{jj}/(n\hat{\lambda}_j^2)}$.

Estimation of $\pi_k = (\lambda_1 + \cdots + \lambda_k)/(\lambda_1 + \cdots + \lambda_p)$

Recall that π_k is the proportion of variability of $\boldsymbol{Y}$ explained by the first k principal components. An estimate of π_k is

$$\begin{aligned}\hat{\pi}_k &= (\hat{\lambda}_1 + \cdots + \hat{\lambda}_k)/(\hat{\lambda}_1 + \cdots + \hat{\lambda}_p) \\ &= (\hat{\lambda}_1 + \cdots + \hat{\lambda}_k)/trace(\boldsymbol{S}).\end{aligned}$$

In order to obtain a confidence interval for π_k, we need to find the asymptotic distribution of $\hat{\pi}_k$. Let

$$g(\boldsymbol{\lambda}) = (\lambda_1 + \cdots + \lambda_k)/(\lambda_1 + \cdots + \lambda_p)$$

and let $\boldsymbol{g}_1(\boldsymbol{\lambda})$ be the vector of partial derivatives of g. By the delta method,

$$\sqrt{n}(\hat{\pi}_k - \pi_k) \xrightarrow{\mathcal{D}} N(0, \boldsymbol{g}_1(\boldsymbol{\lambda})^T \boldsymbol{W} \boldsymbol{g}_1(\boldsymbol{\lambda})) \text{ as } n \to \infty.$$

12.10 Factor Analysis

Let $\boldsymbol{Y}$ be a p-dim vector with mean $\boldsymbol{\mu}$ and covariance matrix $\boldsymbol{\Sigma}$. A factor model with k common factors ($k \le p$) is

$$\boldsymbol{Y} - \boldsymbol{\mu} = \boldsymbol{L}\boldsymbol{f} + \boldsymbol{\varepsilon} \text{ or } \boldsymbol{Y} = \boldsymbol{\mu} + \boldsymbol{L}\boldsymbol{f} + \boldsymbol{\varepsilon}, \tag{3}$$

where $\boldsymbol{L}$ is a $p \times k$ matrix of factor loadings (nonrandom), $\boldsymbol{f}$ is a $k \times 1$ vector of common factors (random), and $\boldsymbol{\varepsilon}$ is a $p \times 1$ vector of specific factors (random). We assume that

$$\mathrm{E}[\boldsymbol{f}] = \boldsymbol{0},\ \mathrm{Cov}[\boldsymbol{f}] = \boldsymbol{I},\ \mathrm{E}[\boldsymbol{\varepsilon}] = \boldsymbol{0},\ \mathrm{Cov}[\boldsymbol{\varepsilon}] = \boldsymbol{\Psi},\ \text{and } \mathrm{Cov}[\boldsymbol{f}, \boldsymbol{\varepsilon}] = \boldsymbol{0}.$$

Here $\boldsymbol{\Psi}$ is assumed to be a diagonal matrix with positive diagonal elements $\psi_1, \ldots, \psi_p$.

Note that if the factor model is correct, then $\mathrm{E}[\boldsymbol{Y}] = \boldsymbol{\mu}$ and $\mathrm{Cov}[\boldsymbol{Y}] = \boldsymbol{L}\boldsymbol{L}^T + \boldsymbol{\Psi}$. The goal of factor analysis is to approximate the covariance matrix $\boldsymbol{\Sigma}$ by a matrix of the form $\boldsymbol{L}\boldsymbol{L}^T + \boldsymbol{\Psi}$, where $\boldsymbol{L}$ is $p \times k$ and $\boldsymbol{\Psi}$ is a diagonal matrix with positive diagonal elements. In other words, factor analysis can be viewed as a dimensionality reduction method, and, in order for this method to be useful, k should be as small as possible (at least in comparison to p).

Let Y_i be the ith component of $\boldsymbol{Y}$. If the factor model is correct, then $\mathrm{E}[Y_i] = \mu_i$ and

$$\mathrm{Var}[Y_i] = l_{i1}^2 + \cdots + l_{ik}^2 + \psi_i = h_i^2 + \psi_i, \quad i = 1, \ldots, p,$$

where $h_i^2 = l_{i1}^2 + \cdots + l_{ik}^2$ is called the "communality" and ψ_i is called the specific variance.

It is important to note that the factor model in Eq. (3) is not identifiable. If $\boldsymbol{G}$ is a $k \times k$ orthogonal matrix and $\tilde{\boldsymbol{L}} = \boldsymbol{L}\boldsymbol{G}$, then we can write

$$\boldsymbol{Y} - \boldsymbol{\mu} = \boldsymbol{L}\boldsymbol{f} + \boldsymbol{\varepsilon} = \tilde{\boldsymbol{L}}\tilde{\boldsymbol{f}} + \boldsymbol{\varepsilon},$$

where $\tilde{\boldsymbol{f}} = \boldsymbol{G}^T\boldsymbol{f}$. Note that $\mathrm{E}[\tilde{\boldsymbol{f}}] = \boldsymbol{0}$ and $\mathrm{Cov}[\tilde{\boldsymbol{f}}] = \boldsymbol{I}$, and

$$\mathrm{Cov}[\boldsymbol{Y}] = \boldsymbol{L}\boldsymbol{L}^T + \Psi = \tilde{\boldsymbol{L}}\tilde{\boldsymbol{L}}^T + \boldsymbol{\Psi}.$$

12.10.1 Estimation of L and Ψ

Let $\boldsymbol{Y}_1, \ldots, \boldsymbol{Y}_n$ be a random sample from a population with mean $\boldsymbol{\mu}$ and covariance matrix $\boldsymbol{\Sigma}$. As usual the mean vector is estimated by the sample mean. The goal is to estimate $\boldsymbol{L}$ and $\boldsymbol{\Psi}$ from the data assuming that the factor model given in Eq. (3) is appropriate (ie, $\mathrm{Cov}[\boldsymbol{Y}] = \boldsymbol{L}\boldsymbol{L}^T + \boldsymbol{\Psi}$). The following two methods are widely used for estimating $\boldsymbol{L}$ and $\boldsymbol{\Psi}$:

(a) principal components,
(b) maximum likelihood.

Principal Factor Analysis

Let $\hat{\lambda}_1 \ge \cdots \ge \hat{\lambda}_p$ be the eigenvalues of the sample covariance matrix $\boldsymbol{S}$ with the corresponding orthonormal eigenvectors $\hat{\boldsymbol{u}}_1, \ldots, \hat{\boldsymbol{u}}_p$. Then the $p \times k$ matrix

$\hat{\boldsymbol{L}} = [\hat{\lambda}_1^{1/2}\hat{\boldsymbol{u}}_1, \ldots, \hat{\lambda}_k^{1/2}\hat{\boldsymbol{u}}_k]$ is taken to be an estimate of $\boldsymbol{L}$, that is, the columns of $\hat{\boldsymbol{L}}$ are $\hat{\boldsymbol{l}}_j = \hat{\lambda}_j^{1/2}\hat{\boldsymbol{u}}_j, j = 1, \ldots, k$, and hence $\hat{\boldsymbol{L}}\hat{\boldsymbol{L}}^T = \sum_{j=1}^k \hat{\lambda}_j\hat{\boldsymbol{u}}_j\hat{\boldsymbol{u}}_j^T$. Here

$$\hat{h}_i^2 = \hat{l}_{i1}^2 + \cdots + \hat{l}_{ik}^2, \quad \hat{\psi}_i = s_{ii} - \hat{h}_i^2,$$

where s_{ii} is the ith diagonal element of the sample covariance matrix $\boldsymbol{S}$. An estimate of the proportion of total variability of $\boldsymbol{Y}$ explained by the first factor is $\|\hat{\boldsymbol{l}}_1\|^2/trace(\boldsymbol{S}) = \hat{\lambda}_1/trace(\boldsymbol{S})$. In general, an estimate of the proportion of total variability of $\boldsymbol{Y}$ explained by the jth factor is $\hat{\lambda}_j/trace(\boldsymbol{S}), j = 1, \ldots, k$.

If the sample correlation matrix is used in the analysis instead of the sample covariance matrix, then $\hat{\boldsymbol{L}} = [\hat{\lambda}_1^{1/2}\hat{\boldsymbol{u}}_1, \ldots, \hat{\lambda}_k^{1/2}\hat{\boldsymbol{u}}_k]$, where $\hat{\lambda}_1 \geq \hat{\lambda}_2 \geq \cdots$ are the eigenvalues of the sample correlation matrix with the corresponding normalized eigenvectors $\hat{\boldsymbol{u}}_1, \hat{\boldsymbol{u}}_2, \ldots$. In this case, $\hat{h}_i^2 = \hat{l}_{i1}^2 + \cdots + \hat{l}_{ik}^2$, $\hat{\psi}_i = 1 - \hat{h}_i^2$, and the proportion of the total variability explained by the jth factor is $\hat{\lambda}_j/p, j = 1, 2, \ldots$.

Maximum Likelihood

This method typically assumes that $\boldsymbol{Y}_1, \ldots, \boldsymbol{Y}_n$ are iid $N_p(\boldsymbol{\mu}, \boldsymbol{\Sigma})$. The maximum likelihood estimates of $\boldsymbol{L}$ and $\boldsymbol{\Psi}$ are obtained by maximizing the likelihood under the constraint that $\boldsymbol{L}\boldsymbol{\Psi}^{-1}\boldsymbol{L}$ is a diagonal matrix. This constraint eliminates the problem of nonuniqueness of the factor loading matrix $\boldsymbol{L}$. Explicit expression for the estimates are not available. However, computer packages such R or MATLAB can be used to obtain the estimates of $\boldsymbol{L}$ and $\boldsymbol{\Psi}$. Estimate of the proportion of the total variability explained by the jth factor is $\|\hat{\boldsymbol{l}}_j\|^2/trace(\boldsymbol{S}), j = 1, \ldots, k$, where $\hat{\boldsymbol{l}}_1, \ldots, \hat{\boldsymbol{l}}_k$ are the columns of $\hat{\boldsymbol{L}}$.

If the correlation matrix is used in the analysis, the estimate of the proportion of the total variability explained by the jth factor is $\|\hat{\boldsymbol{l}}_j\|^2/p, j = 1, \ldots, k$.

How Many Factors

Since we are trying to estimate $\boldsymbol{\Sigma}$ by a matrix of the form $\boldsymbol{L}\boldsymbol{L}^T + \boldsymbol{\Psi}$, a reasonable way to determine k is to examine the residual matrix $\boldsymbol{S} - (\hat{\boldsymbol{L}}\hat{\boldsymbol{L}}^T + \hat{\boldsymbol{\Psi}})$. If all the elements of the residual matrix are small, then we may assume that $\boldsymbol{L}\boldsymbol{L}^T + \boldsymbol{\Psi}$ is a good approximation to $\boldsymbol{\Sigma}$. A more formal way to decide the number of factors is to minimize an AIC-type criterion over k. However, we do not address that issue here.

Factor Rotation

If $\hat{\boldsymbol{L}}$ is an estimate of the factor loading matrix $\boldsymbol{L}$ and $\hat{\boldsymbol{L}}_* = \hat{\boldsymbol{L}}\boldsymbol{G}$, where $\boldsymbol{G}$ is a $k \times k$ orthogonal matrix, then $\hat{\boldsymbol{L}}\hat{\boldsymbol{L}}^T + \hat{\boldsymbol{\Psi}} = \hat{\boldsymbol{L}}_*\hat{\boldsymbol{L}}_*^T + \hat{\boldsymbol{\Psi}}$. So we may take $\hat{\boldsymbol{L}}_*$ to be also a valid estimate of $\boldsymbol{L}$. It has been suggested that one should choose $\hat{\boldsymbol{L}}_*$ so that for each column of $\hat{\boldsymbol{L}}_*$, some of the elements are relatively large and the rest are small. Kaizer's Varimax Rotation tries to achieve this. The procedure is described below.

Let $\tilde{l}_{*ij} = \hat{l}_{*ij}/\hat{h}_i$, $j = 1, \ldots, k$, $i = 1, \ldots, p$, where $\hat{l}_{*ij}$ is the jth element in the ith row of $\hat{\boldsymbol{L}}_*$. Now consider the $p \times k$ matrix $\tilde{\boldsymbol{L}}_*$ whose elements are given by $\tilde{l}_{*ij}$'s. Note that each row of $\tilde{\boldsymbol{L}}_*$ has length 1 since for any $1 \leq i \leq p$, $\sum_{j=1}^{k} \tilde{l}^2_{*ij} = 1$. We consider a criterion which maximizes the variability of the squares of the elements in each column of $\tilde{\boldsymbol{L}}_*$. For $1 \leq j \leq k$, consider a measure of the variability of $\left\{\tilde{l}^2_{*1j}, \ldots, \tilde{l}^2_{*pj}\right\}$

$$(1/p) \sum_{i=1}^{p} \left\{ \tilde{l}^4_{*ij} - \left(\sum_i \tilde{l}^2_{*ij}/p \right)^2 \right\}.$$

Now add these measures over $j = 1, \ldots, k$, to obtain the following criterion

$$\begin{aligned} Q &= \sum_{j=1}^{k} (1/p) \sum_{i=1}^{p} \left\{ \tilde{l}^4_{*ij} - \left(\sum_i \tilde{l}^2_{*ij}/p \right)^2 \right\} \\ &= \sum_{j=1}^{k} \left[\text{variance of } \left\{ \tilde{l}^2_{*ij};\ i = 1, \ldots, p \right\} \right]. \end{aligned}$$

We can maximize Q in order to obtain $\hat{\boldsymbol{L}}_*$ which is a rotated version of $\hat{\boldsymbol{L}}$. There is no explicit expression for the rotated matrix, but one can obtain this estimate by using a computer package.

12.10.2 Prediction of Common Factors

Prediction of common factors in a factor analysis setting is similar to prediction in a random- or mixed-effect model discussed in Section 11.10 of Chapter 11. Assume that the factor model in Eq. (3) holds, $\boldsymbol{\mu}$, $\boldsymbol{L}$, and $\boldsymbol{\Psi}$ are known (or estimated using past observations), then the goal is to predict $\boldsymbol{f}$ when $\boldsymbol{Y}$ is observed. We restrict ourselves to predictors which are linear functions of $\boldsymbol{Y}$. A predictor $\hat{\boldsymbol{f}}$ of $\boldsymbol{f}$ is called unbiased if $\mathrm{E}[\hat{\boldsymbol{f}} - \boldsymbol{f}] = \boldsymbol{0}$. In the following discussion we assume that $\boldsymbol{L}$ has rank k, and the diagonal elements of $\boldsymbol{\Psi}$ are positive. Since we can center $\boldsymbol{Y}$ by subtracting the mean $\boldsymbol{\mu}$, we assume that $\boldsymbol{\mu} = \boldsymbol{0}$ in the subsequent discussion.

If we ignore that $\boldsymbol{f}$ is random and minimize the ordinary least squares criterion $\|\boldsymbol{Y} - \boldsymbol{L}\boldsymbol{f}\|$ with respect to $\boldsymbol{f}$, we get a predictor of the form $\hat{\boldsymbol{f}}^{(1)} = (\boldsymbol{L}^T\boldsymbol{L})^{-1}\boldsymbol{L}^T\boldsymbol{Y}$. Similarly, we may try to obtain a weighted least squares predictor of $\boldsymbol{f}$ as follows. If we premultiply both sides of Eq. (3) by $\boldsymbol{\Psi}^{-1/2}$, we get model of the form $\tilde{\boldsymbol{Y}} = \boldsymbol{L}\boldsymbol{f} + \tilde{\boldsymbol{\varepsilon}}$, where $\tilde{\boldsymbol{Y}} = \boldsymbol{\Psi}^{-1/2}\boldsymbol{Y}$, $\tilde{\boldsymbol{L}} = \boldsymbol{\Psi}^{-1/2}\boldsymbol{L}$, and $\tilde{\boldsymbol{\varepsilon}} = \boldsymbol{\Psi}^{-1/2}\boldsymbol{\varepsilon}$. Noting that $\mathrm{E}[\tilde{\boldsymbol{\varepsilon}}] = \boldsymbol{0}$ and $\mathrm{Cov}[\tilde{\boldsymbol{\varepsilon}}] = \boldsymbol{I}$, we may minimize $\|\tilde{\boldsymbol{Y}} - \tilde{\boldsymbol{L}}\boldsymbol{f}\|^2 = (\boldsymbol{Y} - \boldsymbol{L}\boldsymbol{f})^T\boldsymbol{\Psi}^{-1}(\boldsymbol{Y} - \boldsymbol{L}\boldsymbol{f})$ with respect to $\boldsymbol{f}$, and this leads to another linear predictor $\hat{\boldsymbol{f}}^{(2)} = (\boldsymbol{L}^T\boldsymbol{\Psi}^{-1}\boldsymbol{L})^{-1}\boldsymbol{L}^T\boldsymbol{\Psi}^{-1}\boldsymbol{Y}$. Both $\hat{\boldsymbol{f}}^{(1)}$ and $\hat{\boldsymbol{f}}^{(2)}$ are unbiased linear predictors of $\boldsymbol{f}$, but they are not the best in terms of prediction error.

As in Section 11.10 of Chapter 11, we may define the concept of best linear unbiased predictor as follows.

Definition 12.10.1. A linear function $\hat{\boldsymbol{f}}$ of $\boldsymbol{Y}$ is called a best linear unbiased predictor of $\boldsymbol{f}$ if

(i) $\hat{\boldsymbol{f}}$ is an unbiased predictor of $\boldsymbol{f}$, that is, $\mathrm{E}\left[\hat{\boldsymbol{f}} - \boldsymbol{f}\right] = \boldsymbol{0}$.

(ii) For any $\boldsymbol{a} \in \mathbb{R}^k$, $\mathrm{E}\left[\boldsymbol{a}^T\hat{\boldsymbol{f}} - \boldsymbol{a}^T\boldsymbol{f}\right]^2 \le \mathrm{E}\left[\boldsymbol{l}^T\boldsymbol{Y} - \boldsymbol{a}^T\boldsymbol{f}\right]^2$ for all linear unbiased predictors $\boldsymbol{l}^T\boldsymbol{Y}$ of $\boldsymbol{a}^T\boldsymbol{f}$, $\boldsymbol{l} \in \mathbb{R}^p$.

Arguments used in Sections 11.10.1 and 11.10.2 lead to the following best linear predictor

$$\hat{\boldsymbol{f}}^{(3)} = \left(\boldsymbol{L}^T\boldsymbol{\Psi}^{-1}\boldsymbol{L} + \boldsymbol{I}\right)^{-1}\boldsymbol{L}^T\boldsymbol{\Psi}^{-1}\boldsymbol{Y}.$$

We leave it to the reader to prove that the best linear unbiased predictor of $\boldsymbol{f}$ is unique and is equal to $\hat{\boldsymbol{f}}^{(3)}$ as given above.

12.11 Classification and Discrimination

Suppose that we have k populations (each p-dim) with means $\boldsymbol{\mu}_1, \ldots, \boldsymbol{\mu}_k$ and the same covariance matrix $\boldsymbol{\Sigma}$. If we have an observation vector $\boldsymbol{y}$ from one of these populations, then the goal of the classification problem is to guess which population $\boldsymbol{y}$ comes from. If $\boldsymbol{y}$ is closer to $\boldsymbol{\mu}_i$ than all other means, then a reasonable guess is that $\boldsymbol{y}$ comes from population i. It turns out that this intuition is also mathematically valid. Recall that Mahalanobis distance between $\boldsymbol{y}$ and $\boldsymbol{\mu}_i$ is $\Delta^2(\boldsymbol{y}, \boldsymbol{\mu}_i) = (\boldsymbol{y} - \boldsymbol{\mu}_i)^T\boldsymbol{\Sigma}^{-1}(\boldsymbol{y} - \boldsymbol{\mu}_i)$. So a reasonable rule is: allocate $\boldsymbol{y}$ to population i if

$$\begin{aligned} &\Delta^2(\boldsymbol{y}, \boldsymbol{\mu}_i) < \Delta^2(\boldsymbol{y}, \boldsymbol{\mu}_j), \quad \text{for all } j \neq i, \text{ ie,} \qquad (4)\\ &-2\boldsymbol{\mu}_i^T\boldsymbol{\Sigma}^{-1}\boldsymbol{y} + \boldsymbol{\mu}_i^T\boldsymbol{\Sigma}^{-1}\boldsymbol{\mu}_i < -2\boldsymbol{\mu}_j^T\boldsymbol{\Sigma}^{-1}\boldsymbol{y} + \boldsymbol{\mu}_j^T\boldsymbol{\Sigma}^{-1}\boldsymbol{\mu}_j, \quad \text{for all } j \neq i. \end{aligned}$$

This is called a linear discriminant rule since the criterion for discrimination between the populations depends linearly on $\boldsymbol{y}$.

In some cases, the prior probabilities $\{\pi_1, \ldots, \pi_k\}$ of the populations need to be taken into account. For such a case, the rule given above is modified: now we allocate $\boldsymbol{y}$ to population i if

$$\Delta^2(\boldsymbol{y}, \boldsymbol{\mu}_i) - 2\log\pi_i < \Delta^2(\boldsymbol{y}, \boldsymbol{\mu}_j) - 2\log\pi_j, \quad \text{for all } j \neq i. \qquad (5)$$

Note that the rule defined by the inequalities given in (5) is also a linear discriminant rule, and the rule in (4) is a special case of the rule in (5) when the prior is noninformative, that is, when $\pi_1 = \pi_2 = \cdots = \pi_k = 1/k$.

Note. If the k populations have different covariance matrices, then the rules defined by the inequalities in (4) and (5) need to be modified, and, in such a case, we are led to what is known as a quadratic discriminant rule. We discuss this issue below.

12.11.1 Bayes' Rule for Classification

Let us assume that the chance that the random vector $\boldsymbol{Y}$ is from population i (with pdf f_i) is π_i, $i = 1, \ldots, k$, where $\pi_1 + \cdots + \pi_k = 1$. Let J be a discrete variable taking values $1, \ldots, k$, with $P[J = i] = \pi_i$. In this framework, the conditional pdf of $\boldsymbol{Y}$ given $J = i$ is f_i. The marginal pdf of $\boldsymbol{Y}$ is $f(\boldsymbol{y}) = \pi_1 f_1(\boldsymbol{y}) + \cdots + \pi_k f_k(\boldsymbol{y})$, and the conditional probability of $J = i$ given $\boldsymbol{Y} = \boldsymbol{y}$ is $P[J = i | \boldsymbol{Y} = \boldsymbol{y}] = \pi_i f_i(\boldsymbol{y})/f(\boldsymbol{y})$. Given an observation vector $\boldsymbol{Y} = \boldsymbol{y}$, the decision rule is to allocate $\boldsymbol{y}$ to population i if

$$
\begin{aligned}
P[J = i|\boldsymbol{Y} = \boldsymbol{y}] > P[J = j|\boldsymbol{Y} = \boldsymbol{y}], \quad &\text{for all } j \neq i, \text{ ie,} \\
\pi_i f_i(\boldsymbol{y}) > \pi_j f_j(\boldsymbol{y}), \quad &\text{for all } j \neq i.
\end{aligned} \tag{6}
$$

This is known as the Bayes' rule for classification, and it is the "best" rule as will be discussed below.

12.11.2 The Normal Case

We now discuss the case where the populations are $N_p(\boldsymbol{\mu}_i, \boldsymbol{\Sigma}_i) i = 1, \ldots, k$. If the populations have the same covariance matrix, that is, $\boldsymbol{\Sigma}_1 = \cdots = \boldsymbol{\Sigma}_k = \boldsymbol{\Sigma}$, then we are led to the linear discriminant rule. Otherwise, we have a quadratic discriminant rule.

Case I: $\boldsymbol{\Sigma}_1 = \cdots = \boldsymbol{\Sigma}_k = \boldsymbol{\Sigma}$.

The pdf of the jth population is

$$f_j(\boldsymbol{y}) = \left(1/\sqrt{2\pi}\right)^p (1/|\boldsymbol{\Sigma}|^{1/2}) \exp\left[-(1/2)(\boldsymbol{y} - \boldsymbol{\mu}_j)^T \boldsymbol{\Sigma}^{-1} (\boldsymbol{y} - \boldsymbol{\mu}_j)\right].$$

According to the Bayes' rule given in Eq. (6), allocate $\boldsymbol{y}$ to population i if, for all $j \neq i$,

$$
\begin{gathered}
\pi_i f_i(\boldsymbol{y}) > \pi_j f_j(\boldsymbol{y}), \text{ ie,} \\
-2 \log f_i(\boldsymbol{y}) - 2 \log \pi_i < -2 \log f_j(\boldsymbol{y}) - 2 \log \pi_j, \text{ ie,} \\
(\boldsymbol{y} - \boldsymbol{\mu}_i)^T \boldsymbol{\Sigma}^{-1} (\boldsymbol{y} - \boldsymbol{\mu}_i) - 2 \log \pi_i < (\boldsymbol{y} - \boldsymbol{\mu}_j)^T \boldsymbol{\Sigma}^{-1} (\boldsymbol{y} - \boldsymbol{\mu}_j) - 2 \log \pi_j.
\end{gathered}
$$

This is precisely the rule defined by the inequalities given in (5). If $\pi_1, \ldots, \pi_k$ are unknown, one often assumes that the prior is noninformative, that is, $\pi_1 = \cdots = \pi_k = 1/k$. In such a case, we are led to the rule given in (4).

Case II: $\boldsymbol{\Sigma}_1, \ldots, \boldsymbol{\Sigma}_k$ not the same.

The pdf of the jth population is

$$f_j(\boldsymbol{y}) = \left(1/\sqrt{2\pi}\right)^p \left(1/|\boldsymbol{\Sigma}_j|^{1/2}\right) \exp\left[-(1/2)(\boldsymbol{y} - \boldsymbol{\mu}_j)^T \boldsymbol{\Sigma}_j^{-1} (\boldsymbol{y} - \boldsymbol{\mu}_j)\right].$$

According to the Bayes' rule given in Eq. (6) allocate $\boldsymbol{y}$ to population i if, for all $j \neq i$,

$$
\begin{aligned}
&\pi_i f_i(\boldsymbol{y}) > \pi_j f_j(\boldsymbol{y}), \text{ ie,} \\
&-2 \log f_i(\boldsymbol{y}) - 2 \log \pi_i < -2 \log f_j(\boldsymbol{y}) - 2 \log \pi_j, \text{ ie,} \\
&(\boldsymbol{y} - \boldsymbol{\mu}_i)^T \boldsymbol{\Sigma}_i^{-1} (\boldsymbol{y} - \boldsymbol{\mu}_i) + \log |\boldsymbol{\Sigma}_i| - 2 \log \pi_i \\
&\quad < (\boldsymbol{y} - \boldsymbol{\mu}_j)^T \boldsymbol{\Sigma}_j^{-1} (\boldsymbol{y} - \boldsymbol{\mu}_j) + \log |\boldsymbol{\Sigma}_j| - 2 \log \pi_j.
\end{aligned}
$$

This is called the quadratic discriminant rule. Note that unlike the case $\boldsymbol{\Sigma}_1 = \cdots = \boldsymbol{\Sigma}_k$, the quadratic terms involving $\boldsymbol{y}$ do not cancel out.

12.11.3 Sample Estimates

Suppose that we have n_i observations from population i, $i = 1, \ldots, k$. Then we can estimate the population means $\boldsymbol{\mu}_1, \ldots, \boldsymbol{\mu}_k$. Let $\boldsymbol{S}_i$ be the sample covariance matrix on the basis of n_i observations from population i.

Linear Discriminant Rule

In order to apply the linear discriminant rule, we need an estimate of $\boldsymbol{\Sigma}$ in addition to the estimates of $\boldsymbol{\mu}_1, \ldots, \boldsymbol{\mu}_k$. Recall that an unbiased estimate of $\boldsymbol{\Sigma}$ is given by

$$\boldsymbol{S} = \boldsymbol{S}_{\text{pooled}} = \frac{1}{n-k}[(n_1 - 1)\boldsymbol{S}_1 + \cdots + (n_k - 1)\boldsymbol{S}_k],$$

where $n = n_1 + \cdots + n_k$ is the total number of observation vectors.

If $\pi_1, \ldots, \pi_k$ are known (or if they are estimated), then the discriminant rule is: allocate $\boldsymbol{y}$ to population i if, for all $j \neq i$

$$(\boldsymbol{y} - \hat{\boldsymbol{\mu}}_i)^T \boldsymbol{S}^{-1}(\boldsymbol{y} - \hat{\boldsymbol{\mu}}_i) - 2\log\pi_i < (\boldsymbol{y} - \hat{\boldsymbol{\mu}}_j)^T \boldsymbol{S}^{-1}(\boldsymbol{y} - \hat{\boldsymbol{\mu}}_j) - 2\log\pi_j, \text{ ie,}$$

$$-2\hat{\boldsymbol{\mu}}_i^T \boldsymbol{S}^{-1}\boldsymbol{y} + \hat{\boldsymbol{\mu}}_i^T \boldsymbol{S}^{-1}\hat{\boldsymbol{\mu}}_i - 2\log\pi_i < -2\hat{\boldsymbol{\mu}}_j^T \boldsymbol{S}^{-1}\boldsymbol{y} + \hat{\boldsymbol{\mu}}_j^T \boldsymbol{S}^{-1}\hat{\boldsymbol{\mu}}_j - 2\log\pi_j.$$

This rule is simplified when the prior is noninformative since the terms involving $\{\log\pi_i\}$ cancel out.

Remark 12.11.1. In some cases it may be possible to estimate $\pi_1, \ldots, \pi_k$. Suppose the observations in the sample are $(J_t, \boldsymbol{Y}_t)$, $t = 1, \ldots, n$, where J_t are iid. In such a case, $n_i = \{\#\text{ of } J_t = i\}$ is random, and $(n_1, \ldots, n_k)$ is *Multinomial*$(n; \pi_1, \ldots, \pi_k)$. We can then use the estimate $\hat{\pi}_j = n_j/n$ (or $\hat{\pi}_j = (n_j + 1/2)/(n + 1/2)$) of π_j in the classification rule.

Quadratic Discriminant Rule

The quadratic discriminant rule is: allocate $\boldsymbol{y}$ to population i if, for all $j \neq i$,

$$(\boldsymbol{y} - \hat{\boldsymbol{\mu}}_i)^T \boldsymbol{S}_i^{-1}(\boldsymbol{y} - \hat{\boldsymbol{\mu}}_i) + \log|\boldsymbol{S}_i| - 2\log\pi_i$$
$$< (\boldsymbol{y} - \hat{\boldsymbol{\mu}}_j)^T \boldsymbol{S}_j^{-1}(\boldsymbol{y} - \hat{\boldsymbol{\mu}}_j) + \log|\boldsymbol{S}_j| - 2\log\pi_j.$$

One big drawback of the quadratic discriminant rule is that we need to estimate $\boldsymbol{\Sigma}_1, \ldots, \boldsymbol{\Sigma}_k$ which is equivalent to estimating $kp(p + 1)/2$ parameters of the covariance matrices. This can lead to inefficiencies especially if $p(p + 1)/2$ is not small in comparison to $\min\{n_1, \ldots, n_k\}$. A plausible remedy is to estimate $\boldsymbol{\Sigma}_i$ by shrinking $\boldsymbol{S}_i$ toward $\boldsymbol{S}$, the pooled estimate constructed in the linear discriminant rule (ie, estimate $\boldsymbol{\Sigma}_i$ by $\hat{\boldsymbol{\Sigma}}_i = (1-\alpha_i)\boldsymbol{S}_i + \alpha_i\boldsymbol{S}$, $0 \leq \alpha_i \leq 1$). One can then carry out a quadratic discriminant rule using $\{\hat{\boldsymbol{\Sigma}}_i\}$ instead of

$\{\boldsymbol{S}_i\}$. The constants $\{\alpha_i\}$ need to be estimated from the data, and methods for doing this include:

(i) minimizing an AIC-type criterion, and
(ii) minimizing a cross-validation type criterion.

12.11.4 Probability of Misclassification

Let $\boldsymbol{Y}$ be a p-dim random vector which comes from one of the k populations with pdf's $f_1, \ldots, f_k$. As before assume that $\pi_1, \ldots, \pi_k$ are the prior probabilities which may or may not be noninformative. If $d(\boldsymbol{Y})$ is a function from $\mathbb{R}^p$ to the set $\{1, \ldots, k\}$, then it is called a classifier or a classification function. If $d(\boldsymbol{Y}) = i$, then the classifier d allocates $\boldsymbol{Y}$ to population i. Now if $\boldsymbol{Y}$ is actually from population i, but the classifier allocates it to population j (ie, $d(\boldsymbol{Y}) = j, j \neq i$), then there is a misclassification. Given that $\boldsymbol{Y}$ is from population i, the probability of misclassification is

$$P[d(\boldsymbol{Y}) \neq i | J = i] = \sum_{j \neq i} P[d(\boldsymbol{Y}) = j | J = i].$$

So the total probability of misclassification for the classifier is

$$\begin{aligned} P[d(\boldsymbol{Y}) \neq J] &= \sum_{i=1}^{k} P[d(\boldsymbol{Y}) \neq i | J = i] P[J = i] \\ &= \sum_{i=1}^{k} P[d(\boldsymbol{Y}) \neq i | J = i] \pi_i. \end{aligned}$$

The following result states that the classification rule defined by the inequalities given in (6) has the smallest total probability of misclassification among all classifiers.

Theorem 12.11.1. *Let d^* be the Bayes' rule for classification defined by the inequalities given in (6) that is,*

$$d^*(\boldsymbol{y}) = i \text{ if } \pi_i f_i(\boldsymbol{y}) > \pi_j f_j(\boldsymbol{y}) \text{ for all } j \neq i, i = 1, \ldots, k.$$

If $\mathcal{D}$ is the set of all classifiers from $\mathbb{R}^p$ to $\{1, \ldots, k\}$, then $\min_{d \in \mathcal{D}} P[d(\boldsymbol{Y}) \neq J] = P[d^*(\boldsymbol{Y}) \neq J]$.

Remark 12.11.2. Note that when $\pi_i f_i(\boldsymbol{y}) = \pi_j f_j(\boldsymbol{y})$ for some $i \neq j$, there may be an ambiguity in how to classify $\boldsymbol{y}$. If $\boldsymbol{Y}$ has a continuous distribution, then $P\big[\pi_i f_i(\boldsymbol{Y}) = \pi_j f_j(\boldsymbol{Y})\big] = 0$ for any $j \neq i$. For this reason, the definition of d^* as given above is adequate for continuous distributions.

Proof of Theorem 12.11.1. Since $P[d(\boldsymbol{Y}) \neq J] = 1 - P[d(\boldsymbol{Y}) = J]$, it is enough to show that for any classification rule d,

$$P[d(\boldsymbol{Y}) = J] \leq P[d^*(\boldsymbol{Y}) = J].$$

Note that $P[d(\boldsymbol{Y}) = J] = \mathrm{E}[P\{d(\boldsymbol{Y}) = J | \boldsymbol{Y}\}]$. We will show that for classification rule d

$$P[d(\boldsymbol{Y}) = J | \boldsymbol{Y}] \leq P[d^*(\boldsymbol{Y}) = J | \boldsymbol{Y}].$$

In the calculation of the conditional probability $P[d(\boldsymbol{Y}) = J|\boldsymbol{Y}]$, we may assume that $d(\boldsymbol{Y})$ is fixed since $\boldsymbol{Y}$ is fixed. Hence

$$P[d(\boldsymbol{Y}) = J|\boldsymbol{Y}] = \sum_{i=1}^{k} I(d(\boldsymbol{Y}) = i)P[J = i|\boldsymbol{Y}],$$

where I is the indicator function such that $I(u = v) = 1$ if $u = v$ and $= 0$ otherwise. Assume that the maximum of $P[J = i|\boldsymbol{Y}]$, over $i = 1, \ldots, k$, is attained at i^*. Hence by definition, $d^*(\boldsymbol{y}) = i^*$. So $I(d^*(\boldsymbol{Y}) = i)$ is equal to 0 or 1 depending on whether $i \neq i^*$ or $i = i^*$. Hence

$$\begin{aligned} P[d(\boldsymbol{Y}) = J|\boldsymbol{Y}] &= \sum_{i=1}^{k} I(d(\boldsymbol{Y}) = i)P[J = i|\boldsymbol{Y}] \\ &\leq \sum_{i=1}^{k} I(d(\boldsymbol{Y}) = i)P[J = i^*|\boldsymbol{Y}] = P[J = i^*|\boldsymbol{Y}] \\ &= \sum_{i=1}^{k} I(d^*(\boldsymbol{Y}) = i)P[J = i|\boldsymbol{Y}] = P[d^*(\boldsymbol{Y}) = J]. \end{aligned}$$

This concludes the proof of this result.

□

12.11.5 Classification: Fisher's Method

This method does not require the normality assumption and is flexible enough to provide nonlinear classification rules. Suppose that we have n_i observations vectors $\{\boldsymbol{Y}_{ij}\colon j = 1, \ldots, n_i\}$ from population i, $i = 1, \ldots, k$. If $\boldsymbol{e}$ is in $\mathbb{R}^p$, then we have a one-factor ANOVA model

$$\boldsymbol{e}^T\boldsymbol{Y}_{ij} = \boldsymbol{e}^T\boldsymbol{\mu}_i + \boldsymbol{e}^T\boldsymbol{\varepsilon}_{ij}, \quad j = 1, \ldots, n_i, \quad i = 1, \ldots, k,$$

and $R^2(\boldsymbol{e})$, the coefficient of determination (given in Eq. (1)), is

$$R^2(\boldsymbol{e}) = \frac{\boldsymbol{e}^T\boldsymbol{B}\boldsymbol{e}}{\boldsymbol{e}^T\boldsymbol{T}\boldsymbol{e}} = \frac{\boldsymbol{e}^T\boldsymbol{B}\boldsymbol{e}}{\boldsymbol{e}^T\boldsymbol{B}\boldsymbol{e} + \boldsymbol{e}^T\boldsymbol{W}\boldsymbol{e}} = \frac{(\boldsymbol{e}^T\boldsymbol{B}\boldsymbol{e})/(\boldsymbol{e}^T\boldsymbol{W}\boldsymbol{e})}{(\boldsymbol{e}^T\boldsymbol{B}\boldsymbol{e})/(\boldsymbol{e}^T\boldsymbol{W}\boldsymbol{e}) + 1},$$

where $\boldsymbol{B}$, $\boldsymbol{W}$, and $\boldsymbol{T}$ are the between group, within group, and total SSP matrices, respectively. Maximizing $R^2(\boldsymbol{e})$ with respect to $\boldsymbol{e}$ leads to a generalized eigenvalue problem (Section B.4). Now maximizing $R^2(\boldsymbol{e})$ is equivalent to maximizing the ratio $\boldsymbol{e}^T\boldsymbol{B}\boldsymbol{e}/\boldsymbol{e}^T\boldsymbol{W}\boldsymbol{e}$ which in turn is equivalent to maximizing $\boldsymbol{e}^T\boldsymbol{B}\boldsymbol{e}/\boldsymbol{e}^T\boldsymbol{S}\boldsymbol{e}$, where $\boldsymbol{S} = (n-k)^{-1}\boldsymbol{W}$ is the pooled covariance matrix ($n = n_1 + \cdots + n_k$).

Let $\hat{\lambda}_1 \geq \hat{\lambda}_2 \geq \cdots$ be the eigenvalues of $\boldsymbol{B}$ with respect to $\boldsymbol{S}$ (ie, these are the eigenvalues of $\boldsymbol{S}^{-1/2}\boldsymbol{B}\boldsymbol{S}^{-1/2}$). Let $\hat{\boldsymbol{u}}_1, \hat{\boldsymbol{u}}_2, \ldots$ be the corresponding orthonormal eigenvectors of $\boldsymbol{S}^{-1/2}\boldsymbol{B}\boldsymbol{S}^{-1/2}$. If $\hat{\boldsymbol{e}}_j = \boldsymbol{S}^{-1/2}\hat{\boldsymbol{u}}_j$, $j = 1, 2 \ldots$, then $\hat{\boldsymbol{e}}_j^T\boldsymbol{S}\hat{\boldsymbol{e}}_j = 1$ for all j, and $\hat{\boldsymbol{e}}_i^T\boldsymbol{S}\hat{\boldsymbol{e}}_j = 0$ whenever $i \neq j$. Since $s = rank(\boldsymbol{B}) = \min(k-1, p)$, $\hat{\lambda}_1 \geq \cdots \geq \hat{\lambda}_s > 0$ and $\hat{\lambda}_{s+1} = \cdots = \hat{\lambda}_p = 0$. We will call $Z_1 = \hat{\boldsymbol{e}}_1^T\boldsymbol{Y}$ the first discriminant, $Z_2 = \hat{\boldsymbol{e}}_2^T\boldsymbol{Y}$ the second discriminant, and so on. The vector of discriminants $\boldsymbol{Z}$ is given by $(\hat{\boldsymbol{e}}_1^T\boldsymbol{Y},\ldots,\hat{\boldsymbol{e}}_s^T\boldsymbol{Y})^T$. If we write $\boldsymbol{Z}_{ij} = (\hat{\boldsymbol{e}}_1^T\boldsymbol{Y}_{ij},\ldots,\hat{\boldsymbol{e}}_s^T\boldsymbol{Y}_{ij})^T$, then we can write an approximate MANOVA model

$$\boldsymbol{Z}_{ij} = \tilde{\boldsymbol{\theta}}_i + \boldsymbol{\delta}_{ij}, \quad j = 1,\ldots,n_i, \quad i = 1,\ldots,k,$$

where $\tilde{\boldsymbol{\theta}}_i = (\hat{\boldsymbol{e}}_1^T\boldsymbol{\mu}_i,\ldots,\hat{\boldsymbol{e}}_s^T\boldsymbol{\mu}_i)^T$, and for each i, sample means and sample covariance matrix of $\{\boldsymbol{Z}_{ij}\}$ are $\{\tilde{\boldsymbol{\theta}}_i\}$ and $\boldsymbol{I}$, respectively.

Method of Classification

Suppose $\boldsymbol{y}$ is from one of the k populations and we need to classify it. Let $\boldsymbol{z} = (\hat{\boldsymbol{e}}_1^T\boldsymbol{y},\ldots,\hat{\boldsymbol{e}}_s^T\boldsymbol{y})^T$ be the vector of discriminant scores of $\boldsymbol{y}$. Let $\hat{\boldsymbol{\theta}}_i = (\hat{\boldsymbol{e}}_1^T\hat{\boldsymbol{\mu}}_i,\ldots,\hat{\boldsymbol{e}}_s^T\hat{\boldsymbol{\mu}}_i)^T$. Allocate $\boldsymbol{y}$ to population i if

$$\|\boldsymbol{z} - \hat{\boldsymbol{\theta}}_i\|^2 < \|\boldsymbol{z} - \hat{\boldsymbol{\theta}}_j\|^2 \quad \text{for all } j \neq i.$$

If it is decided to use q discriminants ($q < s$), then we create a vector $\boldsymbol{z} = (\hat{\boldsymbol{e}}_1^T\boldsymbol{y},\ldots,\hat{\boldsymbol{e}}_q^T\boldsymbol{y})^T$ using the first q discriminants and let $\hat{\boldsymbol{\theta}}_i = (\hat{\boldsymbol{e}}_1^T\hat{\boldsymbol{\mu}}_i,\ldots,\hat{\boldsymbol{e}}_q^T\hat{\boldsymbol{\mu}}_i)^T$ be the corresponding sample means. Then the rule is: allocate $\boldsymbol{y}$ to population i if

$$\|\boldsymbol{z} - \hat{\boldsymbol{\theta}}_i\|^2 < \|\boldsymbol{z} - \hat{\boldsymbol{\theta}}_j\|^2 \quad \text{for all } j \neq i.$$

In practice, the decision to use the first q discriminants is usually based on how close the ratio $\hat{\pi}_q = (\hat{\lambda}_1+\cdots+\hat{\lambda}_q)/(\hat{\lambda}_1+\cdots+\hat{\lambda}_s)$ is to 1, and this approach makes intuitive sense since $\hat{\pi}_q$ is an estimate of the proportion of variability in $\boldsymbol{Y}$ explained by the first q discriminants.

Connection to Linear Discriminant Rule

The linear discriminant rule is not necessarily the same as Fisher's. These two rules are the same if all the discriminants are used (ie, number of discriminants is equal to $s = rank(\boldsymbol{B})$) in Fisher's method and the prior is noninformative.

Lemma 12.11.1. *Fisher's classification rule with $s = rank(\boldsymbol{B})$ discriminants is equivalent to the linear discriminant rule with a noninformative prior (ie, $\pi_1 = \cdots = \pi_k = 1/k$).*

Proof of Lemma 12.11.1. Recall that $\hat{\lambda}_1 \geq \hat{\lambda}_2 \geq \cdots$ are the eigenvalues of $\boldsymbol{S}^{-1/2}\boldsymbol{B}\boldsymbol{S}^{-1/2}$ with the corresponding orthonormal eigenvectors $\hat{\boldsymbol{u}}_1, \hat{\boldsymbol{u}}_2, \ldots$. Set $\hat{\boldsymbol{e}}_l = \boldsymbol{S}^{-1/2}\hat{u}_l$, $l = 1,\ldots,p$. Since $\hat{\lambda}_l = 0$ for $l \geq s+1$, we have for any $l \geq s+1$

$$0 = \hat{\boldsymbol{u}}_l^T\boldsymbol{S}^{-1/2}\boldsymbol{B}\boldsymbol{S}^{-1/2}\hat{\boldsymbol{u}}_l = \hat{\boldsymbol{e}}_l^T\boldsymbol{B}\hat{\boldsymbol{e}}_l = \sum_{1\leq i\leq k} n_i(\hat{\boldsymbol{e}}_l^T\overline{\boldsymbol{Y}}_{i.} - \hat{\boldsymbol{e}}_l^T\overline{\boldsymbol{Y}}_{..})^2.$$

Hence, $\hat{\boldsymbol{e}}_l^T\overline{\boldsymbol{Y}}_{i.} = \hat{\boldsymbol{e}}_l^T\overline{\boldsymbol{Y}}_{..}$ whenever $l \geq s+1$. For $1 \leq i \leq k$, and any $\boldsymbol{y}$ in $\mathbb{R}^p$,

$$\begin{aligned}
&(\boldsymbol{y} - \overline{\boldsymbol{Y}}_{i.})^T\boldsymbol{S}^{-1}(\boldsymbol{y} - \overline{\boldsymbol{Y}}_{i.}) \\
&= \|\boldsymbol{S}^{-1/2}(\boldsymbol{y} - \overline{\boldsymbol{Y}}_{i.})\|^2 \\
&= \sum_{l=1}^{p}\{\hat{\boldsymbol{u}}_l^T\boldsymbol{S}^{-1/2}(\boldsymbol{y} - \overline{\boldsymbol{Y}}_{i.})\}^2 \quad (\{\hat{\boldsymbol{u}}_1,\ldots,\hat{\boldsymbol{u}}_p\} \text{ is an orthonormal basis of } \mathbb{R}^p)
\end{aligned}$$

$$= \sum_{l=1}^{p} \{\hat{\boldsymbol{e}}_l^T(\boldsymbol{y} - \overline{\boldsymbol{Y}}_{i\cdot})\}^2 = \sum_{1 \le l \le s} \{\hat{\boldsymbol{e}}_l^T(\boldsymbol{y} - \overline{\boldsymbol{Y}}_{i\cdot})\}^2 + \sum_{s+1 \le l \le p} \{\hat{\boldsymbol{e}}_l^T(\boldsymbol{y} - \overline{\boldsymbol{Y}}_{i\cdot})\}^2$$

$$= \sum_{l=1}^{s} \{\hat{\boldsymbol{e}}_l^T(\boldsymbol{y} - \overline{\boldsymbol{Y}}_{i\cdot})\}^2 + \sum_{l=s+1}^{p} \{\hat{\boldsymbol{e}}_l^T(\boldsymbol{y} - \overline{\boldsymbol{Y}}_{\cdot\cdot})\}^2$$

$$= \|\boldsymbol{z} - \hat{\boldsymbol{\theta}}_i\|^2 + \sum_{l=s+1}^{p} \{\hat{\boldsymbol{e}}_l^T(\boldsymbol{y} - \overline{\boldsymbol{Y}}_{\cdot\cdot})\}^2,$$

where $\boldsymbol{z} = (\hat{\boldsymbol{e}}_1^T\boldsymbol{y}, \ldots, \hat{\boldsymbol{e}}_s^T\boldsymbol{y})^T$ and $\hat{\boldsymbol{\theta}}_i = (\hat{\boldsymbol{e}}_1^T\hat{\boldsymbol{\mu}}_i, \ldots, \hat{\boldsymbol{e}}_s^T\hat{\boldsymbol{\mu}}_i)^T$. Note that the second sum in the last expression does not depend on i. So comparing $(\boldsymbol{y} - \overline{\boldsymbol{Y}}_{i\cdot})^T\boldsymbol{S}^{-1}(\boldsymbol{y} - \overline{\boldsymbol{Y}}_{i\cdot})$ to $(\boldsymbol{y} - \overline{\boldsymbol{Y}}_{i'\cdot})^T\boldsymbol{S}^{-1}(\boldsymbol{y} - \overline{\boldsymbol{Y}}_{i'\cdot})$, $i \neq i'$, is equivalent to comparing $\|\boldsymbol{z} - \hat{\boldsymbol{\theta}}_i\|^2$ to $\|\boldsymbol{z} - \hat{\boldsymbol{\theta}}_{i'}\|^2$.

This argument shows that, for the noninformative prior, the linear discriminant rule is equivalent to Fisher's rule if the number of discriminants used by Fisher's method is equal to $rank(\boldsymbol{B})$.

□

12.12 Canonical Correlation Analysis

Canonical correlation analysis is a descriptive method that seeks to obtain measures of association between two sets of multivariate observations. Let $\boldsymbol{X}$ be $q \times 1$ and $\boldsymbol{Y}$ be $p \times 1$ random vectors with means $\boldsymbol{\mu}_X$ and $\boldsymbol{\mu}_Y$, respectively. Assume that the covariance matrix of $\begin{pmatrix}\boldsymbol{X}\\ \boldsymbol{Y}\end{pmatrix}$ is $\begin{pmatrix}\boldsymbol{\Sigma}_{11} & \boldsymbol{\Sigma}_{12}\\ \boldsymbol{\Sigma}_{21} & \boldsymbol{\Sigma}_{22}\end{pmatrix}$,

$$\boldsymbol{\Sigma}_{11} = \mathrm{Cov}[\boldsymbol{X}], \boldsymbol{\Sigma}_{22} = \mathrm{Cov}[\boldsymbol{Y}], \text{ and } \boldsymbol{\Sigma}_{12} = \mathrm{Cov}[\boldsymbol{X}, \ \boldsymbol{Y}] = \boldsymbol{\Sigma}_{21}^T.$$

Assume that $\boldsymbol{\Sigma}_{11}$ and $\boldsymbol{\Sigma}_{22}$ are nonsingular.

The goal is to find linear functions $\boldsymbol{a}^T\boldsymbol{X}$ and $\boldsymbol{b}^T\boldsymbol{Y}$, $\boldsymbol{a} \in \mathbb{R}^q$ and $\boldsymbol{b} \in \mathbb{R}^p$, which maximize the correlation between $\boldsymbol{a}^T\boldsymbol{X}$ and $\boldsymbol{b}^T\boldsymbol{Y}$. Clearly, we can rescale $\boldsymbol{a}$ and $\boldsymbol{b}$ so that $\mathrm{Var}[\boldsymbol{a}^T\boldsymbol{X}] = \mathrm{Var}[\boldsymbol{b}^T\boldsymbol{Y}] = 1$ (ie, $\boldsymbol{a}^T\boldsymbol{\Sigma}_{11}\boldsymbol{a} = \boldsymbol{b}^T\boldsymbol{\Sigma}_{22}\boldsymbol{b} = 1$). In such a case, $\mathrm{Corr}[\boldsymbol{a}^T\boldsymbol{X}, \boldsymbol{b}^T\boldsymbol{Y}] = \boldsymbol{a}^T\boldsymbol{\Sigma}_{12}\boldsymbol{b}$.

If $\mathrm{Corr}[\boldsymbol{a}^T\boldsymbol{X}, \boldsymbol{b}^T\boldsymbol{Y}] \le 0$, then the correlation between $-\boldsymbol{a}^T\boldsymbol{X}$ and $\boldsymbol{b}^T\boldsymbol{Y}$ is nonnegative. This shows that maximizing the absolute value of $\mathrm{Corr}[\boldsymbol{a}^T\boldsymbol{X}, \boldsymbol{b}^T\boldsymbol{Y}]$ is equivalent to maximizing $\mathrm{Corr}[\boldsymbol{a}^T\boldsymbol{X}, \boldsymbol{b}^T\boldsymbol{Y}]$.

Let $\lambda_1 \ge \lambda_2 \ge \cdots$ be the eigenvalues of the matrix $\boldsymbol{\Sigma}_{11}^{-1/2}\boldsymbol{\Sigma}_{12}\boldsymbol{\Sigma}_{22}^{-1}\boldsymbol{\Sigma}_{21}\boldsymbol{\Sigma}_{11}^{-1}$ with $\boldsymbol{u}_1, \boldsymbol{u}_2, \ldots$ as the corresponding orthonormal eigenvectors. Let

$$\rho_i = \lambda_i^{1/2}, \quad \boldsymbol{a}_i = \boldsymbol{\Sigma}_{11}^{-1/2}\boldsymbol{u}_i, \text{ and } \quad \boldsymbol{b}_i = \rho_i^{-1}\boldsymbol{\Sigma}_{22}^{-1}\boldsymbol{\Sigma}_{21}\boldsymbol{a}_i, \tag{7}$$

$i = 1, \ldots, k = \min(p, q)$. Then $\rho_1 \ge \rho_2 \ge \cdots$ are called the canonical correlations, $\boldsymbol{a}_1^T\boldsymbol{X}$, $\boldsymbol{a}_2^T\boldsymbol{X}, \ldots$ are called the canonical variates of $\boldsymbol{X}$, and $\boldsymbol{b}_1^T\boldsymbol{Y}$, $\boldsymbol{b}_2^T\boldsymbol{Y}, \ldots$ are called the canonical variates of $\boldsymbol{Y}$. Detailed arguments are given in Section 12.12.4.

The following result summarizes the key ideas.

Theorem 12.12.1. *Let* $\{\rho_i\}$, $\{\boldsymbol{a}_i\}$, *and* $\{\boldsymbol{b}_i\}$ *be as given in Eq. (7). Then the following hold:*

(a) $Var[\boldsymbol{a}_i^T\boldsymbol{X}] = Var[\boldsymbol{b}_i^T\boldsymbol{Y}] = 1, i = 1, \ldots, k.$
(b) *When* $i \neq j$, $Cov[\boldsymbol{a}_i^T\boldsymbol{X}, \boldsymbol{a}_j^T\boldsymbol{X}] = 0$, $Cov[\boldsymbol{b}_i^T\boldsymbol{Y}, \boldsymbol{b}_j^T\boldsymbol{Y}] = 0$, *and* $Cov[\boldsymbol{a}_i^T\boldsymbol{X}, \boldsymbol{b}_j^T\boldsymbol{Y}] = 0$.
(c) $Corr[\boldsymbol{a}_i^T\boldsymbol{X}, \boldsymbol{b}_i^T\boldsymbol{Y}] = \rho_i, i = 1, \ldots, k.$

Proof of this result is given in Section 12.12.4. We now write down two more results on canonical correlations.

Theorem 12.12.2. *Let* $\{\rho_i\}$, $\{\boldsymbol{a}_i\}$, *and* $\{\boldsymbol{b}_i\}$ *be as in Theorem 12.12.1. Then:*

(a) $\max_{\boldsymbol{a}^T\boldsymbol{\Sigma}_{11}\boldsymbol{a}=1,\, \boldsymbol{b}^T\boldsymbol{\Sigma}_{22}\boldsymbol{b}=1} Corr[\boldsymbol{a}^T\boldsymbol{X}, \boldsymbol{b}^T\boldsymbol{Y}] = Corr[\boldsymbol{a}_1^T\boldsymbol{X}, \boldsymbol{b}_1^T\boldsymbol{Y}] = \rho_1.$
(b) *Under the constraints* $\boldsymbol{a}^T\boldsymbol{\Sigma}_{11}\boldsymbol{a}_i$, $i = 1, \ldots, r-1$,
$\max_{\boldsymbol{a}^T\boldsymbol{\Sigma}_{11}\boldsymbol{a}=1,\, \boldsymbol{b}^T\boldsymbol{\Sigma}_{22}\boldsymbol{b}=1} Corr[\boldsymbol{a}^T\boldsymbol{X}, \boldsymbol{b}^T\boldsymbol{Y}] = Corr[\boldsymbol{a}_r^T\boldsymbol{X}, \boldsymbol{b}_r^T\boldsymbol{Y}] = \rho_r, r = 2, \ldots, k.$

Proof of Theorem 12.12.2 is not given since it is basically a restatement of Theorem B.4.1 given in Appendix B. The following result on invariance of canonical correlations under nonsingular linear transformations is left as an exercise.

Lemma 12.12.1 (Invariance). *If* $\tilde{\boldsymbol{X}} = \boldsymbol{c} + \boldsymbol{U}\boldsymbol{X}$ *and* $\tilde{\boldsymbol{Y}} = \boldsymbol{d} + \boldsymbol{V}\boldsymbol{Y}$, *where* $\boldsymbol{U}$ *is* $q \times q$ *and nonsingular,* $\boldsymbol{V}$ *is* $p \times p$ *and nonsingular, and* $\boldsymbol{c}$, $\boldsymbol{d}$, $\boldsymbol{U}$, *and* $\boldsymbol{V}$ *are nonrandom, then the canonical correlations between* $\tilde{\boldsymbol{X}}$ *and* $\tilde{\boldsymbol{Y}}$ *are the same as those between* $\boldsymbol{X}$ *and* $\boldsymbol{Y}$. *Moreover, if* $\{\boldsymbol{a}_i\}$ *and* $\{\boldsymbol{b}_i\}$ *are the canonical vectors of* $\boldsymbol{X}$ *and* $\boldsymbol{Y}$, *then* $\tilde{\boldsymbol{a}}_i = \{\boldsymbol{U}^{-1}\}^T\boldsymbol{a}_i$ *and* $\tilde{\boldsymbol{b}}_i = \{\boldsymbol{V}^{-1}\}^T\boldsymbol{b}_i$ *are the canonical vectors of* $\tilde{\boldsymbol{X}}$ *and* $\tilde{\boldsymbol{Y}}$, *respectively,* $i = 1, \ldots, k$.

12.12.1 Sample Estimates

Let $(\boldsymbol{X}_t, \boldsymbol{Y}_t)$, $t = 1, \ldots, n$, be the n pairs of vector observations, and let $\boldsymbol{S}_{11}$ and $\boldsymbol{S}_{22}$ be the sample covariance matrices of $\{\boldsymbol{X}_t\}$ and $\{\boldsymbol{Y}_t\}$, respectively. Set $\boldsymbol{S}_{12} = \frac{1}{n-1}\sum(\boldsymbol{X}_t - \bar{\boldsymbol{X}})(\boldsymbol{Y}_t - \bar{\boldsymbol{Y}})^T$ and $\boldsymbol{S}_{21} = \boldsymbol{S}_{12}^T$.

Let $\hat{\lambda}_1 \geq \hat{\lambda}_2 \geq \cdots$ be the eigenvalues of the matrix $\boldsymbol{S}_{11}^{-1/2}\boldsymbol{S}_{12}\boldsymbol{S}_{22}^{-1}\boldsymbol{S}_{21}\boldsymbol{S}_{11}^{-1}$ with $\hat{\boldsymbol{u}}_1, \hat{\boldsymbol{u}}_2, \ldots$ as the corresponding orthonormal eigenvectors. Let

$$\hat{\rho}_i = \hat{\lambda}_i^{1/2}, \quad \hat{\boldsymbol{a}}_i = \boldsymbol{S}_{11}^{-1/2}\hat{\boldsymbol{u}}_i, \text{ and } \quad \hat{\boldsymbol{b}}_i = \hat{\rho}_i^{-1}\boldsymbol{S}_{22}^{-1}\boldsymbol{S}_{21}\hat{\boldsymbol{a}}_i,$$

$i = 1, \ldots, k = \min(p, q)$. Then

$\hat{\rho}_1 \geq \hat{\rho}_2 \geq \cdots$ are the estimated canonical correlations,
$\hat{\boldsymbol{a}}_1^T\boldsymbol{X}, \hat{\boldsymbol{a}}_2^T\boldsymbol{X}, \ldots$ are the estimated canonical variates of $\boldsymbol{X}$, and
$\hat{\boldsymbol{b}}_1^T\boldsymbol{Y}, \hat{\boldsymbol{b}}_2^T\boldsymbol{Y}, \ldots$ are the estimated canonical variates of $\boldsymbol{Y}$.

12.12.2 Test for $\Sigma_{12} = \mathbf{0}$

Are $\boldsymbol{X}$ and $\boldsymbol{Y}$ uncorrelated? The likelihood ratio statistic for testing H_0: $\boldsymbol{\Sigma}_{12} = \mathbf{0}$ against H_1: $\boldsymbol{\Sigma}_{12} \neq \mathbf{0}$ is

$$\lambda = |I - S_{22}^{-1} S_{21} S_{11}^{-1} S_{12}|^{n/2} = \{(1 - \hat{\rho}_1^2) \cdots (1 - \hat{\rho}_k^2)\}^{n/2} = \Lambda^{n/2},$$

where $\Lambda = (1 - \hat{\rho}_1^2) \cdots (1 - \hat{\rho}_k^2)$ is Wilks' lambda. Under H_0,

$$-[n - (p + q + 3)/2)] \log \Lambda \xrightarrow{\mathcal{D}} \chi^2_{pq},$$

as $n \to \infty$.

If we want to test that only $m < k$ of the population canonical correlations are non-negative, then the test is based on the statistic $\Lambda = (1 - \hat{\rho}_{m+1}^2) \cdots (1 - \hat{\rho}_k^2)$. Then under H_0: $\rho_{m+1} = \cdots = 0$,

$$-[n - (p + q + 3)/2] \log \Lambda \xrightarrow{\mathcal{D}} \chi^2_{(p-m)(q-m)},$$

as $n \to \infty$.

Example 12.12.1. Scores of $n = 88$ students in five subjects are given in Mardia et al. [61]. The subjects are Mechanics (C), Vectors (C), Algebra (O), Analysis (O), and Statistics (O).

Here "C" and "O" stand for closed- and open-book examinations. The goal is to find canonical correlations between open- and closed-book scores. Let $\boldsymbol{X}$ be the vector closed-book marks and $\boldsymbol{Y}$ be the vector of open-book marks. The sample covariance matrix is

$$S = \begin{pmatrix} 302.3 & 125.8 & 100.4 & 105.1 & 116.1 \\ & 170.9 & 84.2 & 93.6 & 97.9 \\ & & 111.6 & 110.8 & 120.5 \\ & & & 217.9 & 153.8 \\ & & & & 294.4 \end{pmatrix}.$$

Here $q = 2$ and $p = 3$, and so $k = \min(q, p) = 2$.

The canonical correlations are $\hat{\rho}_1 = 0.6630$ and $\hat{\rho}_2 = 0.0412$. The first canonical scores are

$$\hat{\boldsymbol{a}}_1^T \boldsymbol{X} = 0.0260X_1 + 0.0518X_2, \text{ and } \hat{\boldsymbol{b}}_1^T \boldsymbol{Y} = 0.0824Y_1 + 0.0081Y_2 + 0.0035Y_3.$$

12.12.3 Cross-Classified Data and Canonical Correlation

Consider the 5×5 contingency table given below on the occupational status of $n = 3497$ father-son pairs. Data sets of this type have been analyzed by many authors in order to investigate issues of intergenerational mobility. Note that occupational status is a qualitative variable and its numbering of 1 through 5 is purely descriptive. Can we assign numerical values for father's and son's status so that they can be treated as "quantitative" variables? This issue will be addressed here using canonical correlation analysis.

Suppose that we have an $r \times c$ contingency table with cell counts $\{n_{ij}\}$ and the total is $n = \sum\sum n_{ij}$. For $i = 1, \ldots, r$, and $j = 1, \ldots, c$, create the following indicator variables, $t = 1, \ldots, n$,

$$X_{ti} = \begin{cases} 1 & \text{if individual } t \text{ belongs to the } i\text{th row category} \\ 0 & \text{otherwise} \end{cases}$$

$$Y_{tj} = \begin{cases} 1 & \text{if individual } t \text{ belongs to the } j\text{th column category} \\ 0 & \text{otherwise.} \end{cases}$$

Thus the observations are $(\boldsymbol{X}_t, \boldsymbol{Y}_t)$, $t = 1, \ldots, n$, where $\boldsymbol{X}_t = (X_{t1}, \ldots, X_{tr})^T$ and $\boldsymbol{Y}_t = (Y_{t1}, \ldots, Y_{tc})^T$. Suppose the numerical values for father's status and son's status are $a_1, \ldots, a_r$ and $b_1, \ldots, b_c$, respectively (here $r = c = 5$). We wish to find $\{a_j\}$ and $\{b_j\}$ so that the correlation between father's and son's status is maximized and this is done by employing canonical correlation analysis. Since $\boldsymbol{X}$ has only one nonzero component, $\boldsymbol{a}^T\boldsymbol{X}$ takes one of the r values $a_1, \ldots, a_r$. Similarly, $\boldsymbol{b}^T\boldsymbol{Y}$ takes one of the c values $b_1, \ldots, b_c$. Thus a father-son pair has a bivariate score (a_i, b_j), if the father is in the ith row category and the son is in the jth column category. We can now find $\boldsymbol{a} = (a_1, \ldots, a_r)^T$ and $\boldsymbol{b} = (b_1, \ldots, b_c)^T$ to maximize $\text{Corr}[\boldsymbol{a}^T\boldsymbol{X}, \quad \boldsymbol{b}^T\boldsymbol{Y}]$.

Social mobility data: $n = 3497$

Father's Status	Subject's Status					Total	Percent
	1	2	3	4	5		
1	50	45	8	18	8	129	3.7
2	28	174	84	154	55	495	14.2
3	11	78	110	223	96	516	14.8
4	14	150	185	714	447	1510	43.2
5	0	42	72	320	411	845	24.2
Total	103	489	459	1429	1017	3497	
Percent	2.9	13.9	13.1	40.9	29.1		

Classes: 1 = professional, 2 = intermediate, 3 = skilled, 4 = semiskilled, 5 = unskilled.

The first canonical correlation is $\hat{\rho}_1 = 0.504$. The canonical scores are given below

Father's Status	1	2	3	4	5
	0	3.15	4.12	4.55	4.96

Son's Status	1	2	3	4	5
	0	3.34	4.49	4.87	5.26

Scores seem to be increasing for both the father and the son. Father's scores seem to be correlated to son's scores. Social Classes 1 and 2 seem to be more distinct from one another than other adjacent social classes, both for the son's and the father's.

12.12.4 Technical Notes

Derivation of Canonical Correlations

The following lemma is useful and it is similar to a result in Section B.1.

Lemma 12.12.2. *Let $\boldsymbol{w}$ be in $\mathbb{R}^m$. Then $\max_{\|\boldsymbol{a}\|=1} \boldsymbol{w}^T\boldsymbol{a} = \|\boldsymbol{w}\|$, and the maximum is attained at $\boldsymbol{a} = \boldsymbol{w}/\|\boldsymbol{w}\|$.*

Now let us find the canonical correlations and the corresponding canonical vectors. Since $\boldsymbol{a}^T\boldsymbol{\Sigma}_{11}\boldsymbol{a} = \boldsymbol{b}^T\boldsymbol{\Sigma}_{22}\boldsymbol{b} = 1$, we have

$$\text{Corr}[\boldsymbol{a}^T\boldsymbol{X}, \boldsymbol{b}^T\boldsymbol{Y}] = \text{Cov}[\boldsymbol{a}^T\boldsymbol{X}, \boldsymbol{b}^T\boldsymbol{Y}] = \boldsymbol{a}^T\boldsymbol{\Sigma}_{12}\boldsymbol{b}.$$

Making a change of vector $\boldsymbol{v} = \boldsymbol{\Sigma}_{22}^{1/2}\boldsymbol{b}$, we have

$$\begin{aligned}\max_{\boldsymbol{b}^T\boldsymbol{\Sigma}_{22}\boldsymbol{b}=1} \text{Corr}[\boldsymbol{a}^T\boldsymbol{X}, \boldsymbol{b}^T\boldsymbol{Y}) &= \max_{\|\boldsymbol{v}\|=1} \boldsymbol{a}^T\boldsymbol{\Sigma}_{12}\boldsymbol{\Sigma}_{22}^{-1/2}\boldsymbol{v}\\ &= \|\boldsymbol{\Sigma}_{22}^{-1/2}\boldsymbol{\Sigma}_{21}\boldsymbol{a}\| = (\boldsymbol{a}^T\boldsymbol{\Sigma}_{12}\boldsymbol{\Sigma}_{22}^{-1}\boldsymbol{\Sigma}_{21}\boldsymbol{a})^{1/2},\end{aligned}$$

and this maximum occurs at $\boldsymbol{v} = \boldsymbol{\Sigma}_{22}^{-1/2}\boldsymbol{\Sigma}_{21}\boldsymbol{a}/\|\boldsymbol{\Sigma}_{22}^{-1/2}\boldsymbol{\Sigma}_{21}\boldsymbol{a}\|$, that is, at $\boldsymbol{b} = \boldsymbol{\Sigma}_{22}^{-1}\boldsymbol{\Sigma}_{21}\boldsymbol{a}/\|\boldsymbol{\Sigma}_{22}^{-1/2}\boldsymbol{\Sigma}_{21}\boldsymbol{a}\|$.

Now making a change of vector $\boldsymbol{u} = \boldsymbol{\Sigma}_{11}^{1/2}\boldsymbol{a}$, we get

$$\max_{\boldsymbol{a}^T\boldsymbol{\Sigma}_{11}\boldsymbol{a}=1} \boldsymbol{a}^T\boldsymbol{\Sigma}_{12}\boldsymbol{\Sigma}_{22}^{-1}\boldsymbol{\Sigma}_{21}\boldsymbol{a} = \max_{\|\boldsymbol{u}\|=1} \boldsymbol{u}^T\boldsymbol{\Sigma}_{11}^{-1/2}\boldsymbol{\Sigma}_{12}\boldsymbol{\Sigma}_{22}^{-1}\boldsymbol{\Sigma}_{21}\boldsymbol{\Sigma}_{11}^{-1/2}\boldsymbol{u}.$$

Let $\lambda_1 \geq \lambda_2 \geq \cdots$ be the eigenvalues of $\boldsymbol{\Sigma}_{11}^{-1/2}\boldsymbol{\Sigma}_{12}\boldsymbol{\Sigma}_{22}^{-1}\boldsymbol{\Sigma}_{21}\boldsymbol{\Sigma}_{11}^{-1/2}$ with $\boldsymbol{u}_1, \boldsymbol{u}_2, \ldots$ as the corresponding orthonormal eigenvectors. So

$$\rho_i = \lambda_i^{1/2}, \quad \boldsymbol{a}_i = \boldsymbol{\Sigma}_{11}^{-1/2}\boldsymbol{u}_i, \text{ and } \quad \boldsymbol{b}_i = \boldsymbol{\Sigma}_{22}^{-1}\boldsymbol{\Sigma}_{21}\boldsymbol{a}_i/\|\boldsymbol{\Sigma}_{22}^{-1/2}\boldsymbol{\Sigma}_{21}\boldsymbol{a}_i\|.$$

Note that

$$\|\boldsymbol{\Sigma}_{22}^{-1/2}\boldsymbol{\Sigma}_{21}\boldsymbol{a}_i\|^2 = \boldsymbol{a}_i^T\boldsymbol{\Sigma}_{12}\boldsymbol{\Sigma}_{22}^{-1}\boldsymbol{\Sigma}_{21}\boldsymbol{a}_i = \boldsymbol{u}_i^T\boldsymbol{\Sigma}_{11}^{-1/2}\boldsymbol{\Sigma}_{12}\boldsymbol{\Sigma}_{22}^{-1}\boldsymbol{\Sigma}_{21}\boldsymbol{\Sigma}_{11}^{-1/2}\boldsymbol{u}_i = \rho_i^2.$$

Thus we have $\boldsymbol{b}_i = \rho_i^{-1}\boldsymbol{\Sigma}_{22}^{-1}\boldsymbol{\Sigma}_{21}\boldsymbol{a}_i$.

Proof of Theorem 12.12.1.

(a) It is easy to see that $\text{Var}[\boldsymbol{a}_i^T\boldsymbol{X}] = \boldsymbol{a}_i^T\boldsymbol{\Sigma}_{11}\boldsymbol{a}_i = \boldsymbol{u}_i^T\boldsymbol{u}_i = 1$.
Note that

$$\begin{aligned}\text{Var}[\boldsymbol{b}_i^T\boldsymbol{Y}] &= \boldsymbol{b}_i^T\boldsymbol{\Sigma}_{22}\boldsymbol{b}_i = \rho_i^{-2}\boldsymbol{a}_i^T\boldsymbol{\Sigma}_{12}\boldsymbol{\Sigma}_{22}^{-1}\boldsymbol{\Sigma}_{12}\boldsymbol{a}_i\\ &= \rho_i^{-2}\boldsymbol{u}_i^T\boldsymbol{\Sigma}_{11}^{-1/2}\boldsymbol{\Sigma}_{12}\boldsymbol{\Sigma}_{22}^{-1}\boldsymbol{\Sigma}_{21}\boldsymbol{\Sigma}_{11}^{-1/2}\boldsymbol{u}_i = \rho_i^{-2}\rho_i^2 = 1.\end{aligned}$$

(b) It is fairly easy to see that $\boldsymbol{a}_i^T\boldsymbol{X}$ and $\boldsymbol{a}_j^T\boldsymbol{X}$ are uncorrelated for $i \neq j$ since

$$\text{Cov}[\boldsymbol{a}_i^T\boldsymbol{X}, \boldsymbol{a}_j^T\boldsymbol{X}] = \boldsymbol{a}_i^T\boldsymbol{\Sigma}_{11}\boldsymbol{a}_j = \boldsymbol{u}_i^T\boldsymbol{u}_j = 0.$$

Note that when $i \neq j$, we have

$$\begin{aligned}\text{Cov}[\boldsymbol{b}_i^T\boldsymbol{Y}, \boldsymbol{b}_j^T\boldsymbol{Y}] &= \boldsymbol{b}_i^T\boldsymbol{\Sigma}_{22}\boldsymbol{b}_j = \rho_i^{-1}\rho_j^{-1}\boldsymbol{a}_i^T\boldsymbol{\Sigma}_{12}\boldsymbol{\Sigma}_{22}^{-1}\boldsymbol{\Sigma}_{21}\boldsymbol{a}_j\\ &= \rho_i^{-1}\rho_j^{-1}\boldsymbol{u}_i^T\boldsymbol{\Sigma}_{11}^{-1/2}\boldsymbol{\Sigma}_{12}\boldsymbol{\Sigma}_{22}^{-1}\boldsymbol{\Sigma}_{21}\boldsymbol{\Sigma}_{11}^{-1/2}\boldsymbol{u}_j = 0.\end{aligned}$$

The last step follows from the fact that the eigenvectors $\boldsymbol{u}_1, \boldsymbol{u}_2, \ldots$ are orthonormal. Similarly

$$\text{Cov}[\boldsymbol{a}_i^T \boldsymbol{X}, \boldsymbol{b}_j^T \boldsymbol{Y}] = \boldsymbol{a}_i^T \boldsymbol{\Sigma}_{12} \boldsymbol{b}_j = \rho_j^{-1} \boldsymbol{a}_i^T \boldsymbol{\Sigma}_{12} \boldsymbol{\Sigma}_{22}^{-1} \boldsymbol{\Sigma}_{21} \boldsymbol{a}_j = 0.$$

The last step follows from the intermediate step in the proof for $\text{Cov}[\boldsymbol{b}_i^T \boldsymbol{Y}, \boldsymbol{b}_j^T \boldsymbol{Y}] = 0$.

(c) Note that

$$\begin{aligned}
\text{Corr}[\boldsymbol{a}_i^T \boldsymbol{X}, \boldsymbol{b}_i^T \boldsymbol{Y}] &= \text{Cov}[\boldsymbol{a}_i^T \boldsymbol{X}, \boldsymbol{b}_i^T \boldsymbol{Y}] \\
&= \boldsymbol{a}_i^T \boldsymbol{\Sigma}_{12} \boldsymbol{b}_i = \rho_i^{-1} \boldsymbol{a}_i^T \boldsymbol{\Sigma}_{12} \boldsymbol{\Sigma}_{22}^{-1} \boldsymbol{\Sigma}_{21} \boldsymbol{a}_i \\
&= \rho_i^{-1} \boldsymbol{u}_i^T \boldsymbol{\Sigma}_{11}^{-1/2} \boldsymbol{\Sigma}_{12} \boldsymbol{\Sigma}_{22}^{-1} \boldsymbol{\Sigma}_{21} \boldsymbol{\Sigma}_{11}^{-1/2} \boldsymbol{u}_i = \rho_i^{-1} \rho_i^2 = \rho_i.
\end{aligned}$$

□

Exercises

12.1. For this question, you may use the following facts: if $X \sim \chi^2_m$, then $\text{E}[X] = m$ and $\text{E}[1/X] = 1/(m-2)$, $m > 2$

(a) Show that $\text{E}[F_{u,v}] = \frac{v}{v-2}$ where $F_{u,v}$ has an F-distribution with $df = (u, v)$, $v > 2$. [*Hint*: $F_{u,v} = \frac{R_1/u}{R_2/v}$, where $R_1 \sim \chi^2_u$, $R_2 \sim \chi^2_v$, and R_1 and R_2 are independent.]

(b) Let $\boldsymbol{M} \sim W_p(k, \boldsymbol{\Sigma})$, where $\boldsymbol{\Sigma}$ is positive definite and $k > p + 1$. Show that $\text{E}[\boldsymbol{M}] = k\boldsymbol{\Sigma}$.

(c) Let $\boldsymbol{M}$ be as in part (b). Show that $\text{E}[\boldsymbol{M}^{-1}] = c\boldsymbol{\Sigma}^{-1}$ for some constant $c > 0$, and find an explicit expression for this constant. [*Hint*: Use Property (5) of Wishart distribution given in Section 12.2].

12.2. Consider a repeated measures study in which, for each of the n randomly selected subjects, an attribute (such as growth) is recorded at times $t_1, \ldots, t_p$. Thus for the ith individual, the vector of measurements is $\boldsymbol{Y}_i = (Y_{i1}, \ldots, Y_{ip})^T$. For each of the three popular models given below, β_0 is a constant, $\{\alpha_i\}$ are iid $N(0, \sigma^2_\alpha)$, $\{\delta_{ij}\}$ are iid $N(0, \sigma^2_\delta)$:

(i) $Y_{ij} = \beta_0 + \alpha_i + \tau_j + \delta_{ij}$, τ_j's are time effects (fixed) with $\sum \tau_j = 0$.

(ii) $Y_{ij} = \beta_0 + \alpha_i + \beta_1 t_j + \delta_{ij}$, β_1 is the slope (constant).

(iii) $Y_{ij} = \beta_0 + \alpha_i + \beta_{i1} t_j + \delta_{ij}$, $\{\beta_{i1}\}$ are iid random slopes with $\beta_{i1} \sim N(\beta_1, \sigma^2_\beta)$.

It is understood that $\{\alpha_i\}$, $\{\beta_{i1}\}$, $\{\delta_{ij}\}$ are all mutually independent. For each of the three models above, $\boldsymbol{Y}_1, \ldots, \boldsymbol{Y}_n$ are iid $N_p(\boldsymbol{\mu}, \boldsymbol{\Sigma})$. Explicitly obtain the elements of $\boldsymbol{\mu}$ and $\boldsymbol{\Sigma}$ in each of the three cases.

12.3. Let $\boldsymbol{Y} \sim N_p(\boldsymbol{\mu}, \boldsymbol{\Sigma})$ and $\boldsymbol{u}$ is a vector in $\mathbb{R}^p$. Let $Z = \boldsymbol{u}^T(\boldsymbol{Y} - \boldsymbol{\mu})/\sqrt{\boldsymbol{u}^T \boldsymbol{\Sigma} \boldsymbol{u}}$.

(a) If $\boldsymbol{u}$ is nonrandom, then show that $Z \sim N(0, 1)$.

(b) Now assume that $\boldsymbol{u}$ is a random vector which is independent of $\boldsymbol{Y}$ and $P[\boldsymbol{u}^T \boldsymbol{\Sigma} \boldsymbol{u} = 0] = 0$. Show that $Z \sim N(0, 1)$ and Z is independent of $\boldsymbol{u}$.

12.4. Suppose that the growth of n randomly selected children are observed at p different time points and let $\boldsymbol{Y}_i$ be the p-dim vector of observed growths for the ith child. Assume that $\boldsymbol{\varepsilon}_i = \boldsymbol{Y}_i - \text{E}[\boldsymbol{Y}_i]$ are iid $N_p(\boldsymbol{0}, \boldsymbol{\Sigma})$, $\boldsymbol{\Sigma}$ unknown. It is also assumed that the mean growth at time t can be modeled by $\beta_0 + \beta_1 t + \cdots + \beta_d t^d$, where $\{\beta_j\}$ are unknown and need to be estimated.

(a) If growth of each of the n children are observed at times $t_1, \ldots, t_p$, then $\mathrm{E}[\boldsymbol{Y}_i] = \boldsymbol{D}\beta$ and obtain an explicit expression for the matrix $\boldsymbol{D}$. Explicitly write down $-2 \times$ log-likelihood.

(b) If the observation time points for these children are not necessarily the same, that is, the growth of the ith child is measured at times t_{ij}, $j = 1, \ldots, p$, then $\mathrm{E}[\boldsymbol{Y}_i] = \boldsymbol{D}_i\beta$ and obtain an explicit expression for $\boldsymbol{D}_i$. Explicitly write down $-2 \times$ log-likelihood.

(c) Show that the likelihood equations (ie, the minimizers of $-2 \times$ log-likelihood) for part (a) are

$$\boldsymbol{\beta} = (\boldsymbol{D}^T\boldsymbol{\Sigma}^{-1}\boldsymbol{D})^{-1}\boldsymbol{D}^T\boldsymbol{\Sigma}^{-1}\overline{\boldsymbol{Y}}, \quad \boldsymbol{\Sigma} = n^{-1}\sum(\boldsymbol{Y}_i - \boldsymbol{D}\boldsymbol{\beta})(\boldsymbol{Y}_i - \boldsymbol{D}\boldsymbol{\beta})^T.$$

[Estimates of $\boldsymbol{\beta}$ and $\boldsymbol{\Sigma}$ are obtained by iterations starting with some reasonable estimates of $\boldsymbol{\beta}$ and $\boldsymbol{\Sigma}$. For instance one may start with an initial estimate of $\boldsymbol{\beta}$ as $\hat{\boldsymbol{\beta}}^0 = (\boldsymbol{D}^T\boldsymbol{D})^{-1}\boldsymbol{D}^T\overline{\boldsymbol{Y}}$ and use this in the expression of $\boldsymbol{\Sigma}$ (in the likelihood equations) to get an intial estimate $\hat{\boldsymbol{\Sigma}}^0$ of $\boldsymbol{\Sigma}$.]

(d) Show that the likelihood equations (ie, the minimizers of $-2 \times$ log-likelihood) for part (b) are

$$\boldsymbol{\beta} = \left(\sum \boldsymbol{D}_i^T\boldsymbol{\Sigma}^{-1}\boldsymbol{D}_i\right)^{-1}\left(\sum \boldsymbol{D}_i^T\boldsymbol{\Sigma}^{-1}\overline{\boldsymbol{Y}}\right),$$
$$\boldsymbol{\Sigma} = n^{-1}\sum(\boldsymbol{Y}_i - \boldsymbol{D}_i\boldsymbol{\beta})(\boldsymbol{Y}_i - \boldsymbol{D}_i\boldsymbol{\beta})^T.$$

12.5. Let $\boldsymbol{Y}_1, \ldots, \boldsymbol{Y}_n$ be iid $N_p(\boldsymbol{\mu}, \boldsymbol{\Sigma})$, but assume that $\boldsymbol{\Sigma}$ is known. The statistic $T_0^2 = n(\bar{\boldsymbol{Y}} - \boldsymbol{\mu}_0)^T\boldsymbol{\Sigma}^{-1}(\bar{\boldsymbol{Y}} - \boldsymbol{\mu}_0)$ may be used for testing $H_0\colon \boldsymbol{\mu} = \boldsymbol{\mu}_0$ against $H_1\colon \boldsymbol{\mu} \neq \boldsymbol{\mu}_0$.

(a) Find the distribution of T_0^2 under H_0. If H_0 is not true, what is the distribution of T_0^2?

(b) Show that $\mathrm{E}\left[T_0^2\right] = p + n(\boldsymbol{\mu} - \boldsymbol{\mu}_0)^T\boldsymbol{\Sigma}^{-1}(\boldsymbol{\mu} - \boldsymbol{\mu}_0)$, when the true mean is $\boldsymbol{\mu}$.

12.6. Let $\boldsymbol{Y}_1, \ldots, \boldsymbol{Y}_n$ be iid $N_p(\boldsymbol{\mu}, \boldsymbol{\Sigma})$. Consider the problem of testing $H_0\colon \boldsymbol{\mu} = \boldsymbol{\mu}_0$ against $H_1\colon \boldsymbol{\mu} \neq \boldsymbol{\mu}_0$. Then the Hotelling's T^2 statistic is $T^2 = n(\overline{\boldsymbol{Y}} - \boldsymbol{\mu}_0)^T$ $\boldsymbol{S}^{-1}(\overline{\boldsymbol{Y}} - \boldsymbol{\mu}_0)$, where $\boldsymbol{S}$ is the sample covariance matrix. For $\boldsymbol{a}$ in $\mathbb{R}^p$, we denote $\boldsymbol{Y}(\boldsymbol{a}) = \boldsymbol{a}^T\boldsymbol{Y}$, $\mu(\boldsymbol{a}) = \boldsymbol{a}^T\boldsymbol{\mu}$, $\varepsilon(\boldsymbol{a}) = \boldsymbol{a}^T\boldsymbol{\varepsilon}$, $\mu_0(\boldsymbol{a}) = \boldsymbol{a}^T\boldsymbol{\mu}_0$, and $s(\boldsymbol{a})^2 = \boldsymbol{a}^T\boldsymbol{S}\boldsymbol{a}$.

(a) Consider the univariate model $Y_i(\boldsymbol{a}) = \mu(\boldsymbol{a}) + \varepsilon_i(\boldsymbol{a})$. Then for the problem of testing $H_0\colon \mu(\boldsymbol{a}) = \mu_0(\boldsymbol{a})$ against $H_1\colon \mu(\boldsymbol{a}) \neq \mu_0(\boldsymbol{a})$, we may use the t-statistic $t(\boldsymbol{a}) = \sqrt{n}(\overline{Y}(\boldsymbol{a}) - \mu_0(\boldsymbol{a}))/s(\bar{Y}(\boldsymbol{a}))$, where $s^2(\bar{Y}(\boldsymbol{a})) = \boldsymbol{a}^T\boldsymbol{S}\boldsymbol{a}/n$. Show that $\max_{\boldsymbol{a}\neq\boldsymbol{0}} t(\boldsymbol{a})^2 = T^2$.

(b) Show that $\mathrm{E}[T^2] = \frac{p(n-1)}{n-p-2}$ when H_0 is true.

(c) Show that $\mathrm{E}[T^2] = \frac{n(n-1)}{n-p-2}(\boldsymbol{\mu} - \boldsymbol{\mu}_0)^T\boldsymbol{\Sigma}^{-1}(\boldsymbol{\mu} - \boldsymbol{\mu}_0) + \frac{p(n-1)}{n-p-2}$. Show that $\mathrm{E}[T^2] = \frac{p(n-1)}{n-p-2}$ when and only when H_0 is true.

12.7. Let $\boldsymbol{Y}_1, \ldots, \boldsymbol{Y}_n$ be iid $N_p(\boldsymbol{\mu}, \boldsymbol{\Sigma})$. Let $\theta_i = \mu_{i+2} - 2\mu_{i+1} + \mu_i$, $i = 1, \ldots, p-2$. Suppose that we are interested in testing H_0: $\boldsymbol{\theta} = \boldsymbol{0}$ against H_1: $\boldsymbol{\theta} \neq \boldsymbol{0}$, where $\boldsymbol{\theta}$ is the $(p-2)$-dim vector consisting of elements $\theta_1, \ldots, \theta_{p-2}$. [This kind of testing may be important in growth analysis where we may want to test if μ_i's are linear in i.]

(a) Express $\boldsymbol{\theta}$ as a linear function $\boldsymbol{D\mu}$ of $\boldsymbol{\mu}$ by finding a matrix $\boldsymbol{D}$ explicitly.

(b) Find the MLE $\hat{\boldsymbol{\theta}}$ of $\boldsymbol{\theta}$. Obtain the distribution of this estimate. Obtain an unbiased estimate of the covariance matrix of $\hat{\boldsymbol{\theta}}$.

(c) Obtain an appropriate statistic for testing H_0 against H_1, and find the distribution of this statistic under H_0.

12.8. Let $\boldsymbol{Y}_1, \ldots, \boldsymbol{Y}_n$ be iid $N_{2p}(\boldsymbol{\mu}, \boldsymbol{\Sigma})$. Let $\boldsymbol{\mu} = \binom{\boldsymbol{\mu}_1}{\boldsymbol{\mu}_2}$ where $\boldsymbol{\mu}_1$ and $\boldsymbol{\mu}_2$ are p-dim. We are interested in inference on $\boldsymbol{\theta} = \boldsymbol{\mu}_1 - \boldsymbol{\mu}_2$. This type of issue comes up in the case of paired observations. Assume that $\boldsymbol{\Sigma}$ is of the form $\begin{pmatrix} \boldsymbol{\Sigma}_{11} & \boldsymbol{\Sigma}_{12} \\ \boldsymbol{\Sigma}_{21} & \boldsymbol{\Sigma}_{22} \end{pmatrix}$, where $\boldsymbol{\Sigma}_{11}$, $\boldsymbol{\Sigma}_{12}$, $\boldsymbol{\Sigma}_{21}$, and $\boldsymbol{\Sigma}_{22}$ are $p \times p$ matrices. Similarly the sample covariance matrix $\boldsymbol{S}$ can be written as $\begin{pmatrix} \boldsymbol{S}_{11} & \boldsymbol{S}_{12} \\ \boldsymbol{S}_{21} & \boldsymbol{S}_{22} \end{pmatrix}$, where $\boldsymbol{S}_{11}$, $\boldsymbol{S}_{12}$, $\boldsymbol{S}_{21}$, and $\boldsymbol{S}_{22}$ are $p \times p$ matrices.

(a) Express $\boldsymbol{\theta}$ in the form $\boldsymbol{D\mu}$ and find the matrix $\boldsymbol{D}$ explicitly. Find the MLE of $\boldsymbol{\theta}$ and then find the distribution of this estimate. Obtain the parameters of this distribution explicitly in terms of $\boldsymbol{\theta}$, $\boldsymbol{\Sigma}_{11}$, $\boldsymbol{\Sigma}_{12}$, $\boldsymbol{\Sigma}_{21}$, and $\boldsymbol{\Sigma}_{22}$.

(b) Suppose you want to test H_0: $\boldsymbol{\theta} = \boldsymbol{0}$ against H_1: $\boldsymbol{\theta} \neq \boldsymbol{0}$. Find the appropriate Hotelling's T^2-statistic and obtain its distribution under H_0. Modify this test statistic appropriately if $\boldsymbol{\Sigma}$ were known. Find the distribution of this modified statistic under H_0.

(c) Obtain simultaneous confidence intervals for $\theta_1, \ldots, \theta_m$ with family confidence of at least $1 - \alpha$ using the Bonferroni approach. Obtain an explicit expression for the standard error of the estimate of each θ_i as a combination of the elements of the matrices $\boldsymbol{S}_{11}$, $\boldsymbol{S}_{12}$, $\boldsymbol{S}_{21}$, and $\boldsymbol{S}_{22}$.

(d) In some cases it is of interest to test H_0: $\theta_1 + \cdots + \theta_m \leq 0$ against H_1: $\theta_1 + \cdots + \theta_m > 0$ at a level of significance α, where $m \leq p$. Obtain an appropriate statistic for this testing problem and state the decision rule.

12.9. Consider a one-factor MANOVA model $\boldsymbol{Y}_{ij} = \boldsymbol{\mu}_i + \boldsymbol{\varepsilon}_{ij}$, $j = 1, \ldots, n_i$, $i = 1, \ldots, k$, where $\{\boldsymbol{\varepsilon}_{ij}\}$ are iid $N_p(\boldsymbol{0}, \boldsymbol{\Sigma})$. Let $\boldsymbol{B}$, $\boldsymbol{W}$, and $\boldsymbol{T}$ denote the between group, within group, and total SSP matrices, respectively.

(a) Show that the matrices $\boldsymbol{B}$, $\boldsymbol{W}$, and $\boldsymbol{T}$ are nonnegative definite.

(b) Show that $|\boldsymbol{W}| \leq |\boldsymbol{T}|$.

(c) Assume that $\boldsymbol{T}$ is positive definite with probability 1. Show that $|\boldsymbol{W}| = |\boldsymbol{T}|$ when and only when $\overline{\boldsymbol{Y}}_{i\cdot} = \overline{\boldsymbol{Y}}_{\cdot\cdot}$ for $i = 1, \ldots, k$. [*Hint for parts (b) and (c)*: Look at $\boldsymbol{T}^{-1/2}\boldsymbol{W}\boldsymbol{T}^{-1/2}$.]

(d) Let $\boldsymbol{\mu} = \sum (n_i/n)\boldsymbol{\mu}_i$ and $\boldsymbol{\alpha}_i = \boldsymbol{\mu}_i - \boldsymbol{\mu}$, where $n = n_1 + \cdots + n_k$. Let $\hat{\boldsymbol{\alpha}}_i = \overline{\boldsymbol{Y}}_{i\cdot} - \overline{\boldsymbol{Y}}_{\cdot\cdot}$. Show that $\overline{\boldsymbol{Y}}_{\cdot\cdot} \sim N_p(\boldsymbol{\mu}, n^{-1}\boldsymbol{\Sigma})$ and $\hat{\boldsymbol{\alpha}}_i \sim N_p(\boldsymbol{\alpha}_i, (n_i^{-1} - n^{-1})\boldsymbol{\Sigma})$.

(e) Let $\boldsymbol{\theta} = \sum c_i \boldsymbol{\alpha}_i$ be a contrast, that is, $\{c_i\}$ are real numbers and they satisfy the constraint $\sum c_i = 0$. Let $\hat{\boldsymbol{\theta}} = \sum c_i \hat{\boldsymbol{\alpha}}_i$. Show that $\hat{\boldsymbol{\theta}} \sim N_p(\boldsymbol{\theta}, \sum c_i^2/n_i \boldsymbol{\Sigma})$.

(f) First prove that $\mathrm{E}[\boldsymbol{B}] = (k-1)\boldsymbol{\Sigma} + \sum m_i \boldsymbol{\alpha}_i \boldsymbol{\alpha}_i^T$.
Then show that $\mathrm{E}[\boldsymbol{B}] = (k-1)\boldsymbol{\Sigma}$ when and only when $\boldsymbol{\mu}_1 = \cdots = \boldsymbol{\mu}_k$.

12.10. Consider a one-factor MANOVA model as in Exercise 12.9. Consider the transformed data $\boldsymbol{Z}_{ij} = \boldsymbol{a} + \boldsymbol{A}\boldsymbol{Y}_{ij}$, where $\boldsymbol{a}$ is a vector in $\mathbb{R}^p$ and $\boldsymbol{A}$ is a nonsingular matrix of order pxp.

(a) Write down a one-factor MANOVA model for $\{\boldsymbol{Z}_{ij}\}$, that is, $\boldsymbol{Z}_{ij} = \boldsymbol{\theta}_i + \boldsymbol{\delta}_{ij}$, where $\boldsymbol{\delta}_{ij}$'s have zero means. Find the distribution of $\{\boldsymbol{\delta}_{ij}\}$. Express $\boldsymbol{\theta}_i$ and the parameters of the distribution of $\boldsymbol{\delta}_{ij}$ explicitly in terms of $\boldsymbol{\mu}_i$, $\boldsymbol{a}$, $\boldsymbol{A}$, and $\boldsymbol{\Sigma}$.

(b) Show that the between group, within group, and total SSP matrices for the transformed data $\{\boldsymbol{Z}_{ij}\}$ are $\boldsymbol{ABA}^T$, $\boldsymbol{AWA}^T$, and $\boldsymbol{ATA}^T$, respectively.

(c) Consider the problem of testing H_0: $\boldsymbol{\theta}_1 = \cdots = \boldsymbol{\theta}_k$ against H_1: not all $\boldsymbol{\theta}_i$'s are the same. Show that the Wilks' lambda, Pillai trace, and Roy's largest root statistics obtained on the basis of the data $\{\boldsymbol{Z}_{ij}\}$ are the same as those for $\{\boldsymbol{Y}_{ij}\}$.

(d) When $k = 2$, show that the Lawley-Hotelling statistic $trace(\boldsymbol{BW}^{-1})$ is equal to cT^2, for some constant $c > 0$, where T^2 is the two-sample T^2-statistic for testing H_0: $\boldsymbol{\mu}_1 = \boldsymbol{\mu}_2$ vs H_1: $\boldsymbol{\mu}_1 \neq \boldsymbol{\mu}_2$. Find the constant c.

12.11. Let $\boldsymbol{Y}_1, \ldots, \boldsymbol{Y}_n$ be iid p-dim random vectors with mean vector $\boldsymbol{\mu}$ and covariance matrix $\boldsymbol{\Sigma}$ whose eigenvalues are $\lambda_1 \geq \lambda_2 \geq \cdots$ with the corresponding orthonormal eigenvectors $\boldsymbol{u}_1, \boldsymbol{u}_2, \ldots$. Let $\hat{\lambda}_1 \geq \hat{\lambda}_2 \geq \cdots$ be the eigenvalues of the sample covariance matrix $\boldsymbol{S}$ with the corresponding orthonormal eigenvectors $\hat{\boldsymbol{u}}_1, \hat{\boldsymbol{u}}_2 \ldots$. The sample values of the jth principal component are $\hat{Z}_{tj} = \hat{\boldsymbol{u}}_j^T(\boldsymbol{Y}_t - \overline{\boldsymbol{Y}})$, $t = 1, \ldots, n$.

(a) Show that the sample mean and sample variance of $\{\hat{Z}_{tj}$: $t = 1, \ldots, n\}$ are zero and $\hat{\lambda}_j$, respectively.

(b) Show that the sample correlation of $\{(\hat{Z}_{ti}, \hat{Z}_{tj})$: $t = 1, \ldots, n\}$, $i \neq j$, is equal to zero.

For parts (c) and (d) assume that the population is normal and that the eigenvalues $\lambda_1, \ldots, \lambda_p$ are distinct.

(c) Let $\pi_k = (\lambda_1 + \cdots + \lambda_k)/trace(\boldsymbol{\Sigma})$ and $\hat{\pi}_k = (\hat{\lambda}_1 + \cdots + \hat{\lambda}_k)/trace(\boldsymbol{S})$, $k < p$. Let $\boldsymbol{\Sigma}_k = \sum_{j=1}^k \lambda_j \boldsymbol{u}_j \boldsymbol{u}_j^T$. Show that $\sqrt{n}(\hat{\pi}_k - \pi_k) \xrightarrow{D} N(0, 2\tau_k^2)$, where $\tau_k^2 = \left[(1-\pi_k)^2\, trace(\boldsymbol{\Sigma}_k^2) + \pi_k^2\, trace\{(\boldsymbol{\Sigma} - \boldsymbol{\Sigma}_k)^2\}\right]/trace(\boldsymbol{\Sigma})^2$.

(d) Let θ and $\hat{\theta}$ be the geometric means of $\{\lambda_1, \ldots, \lambda_p\}$ and $\{\hat{\lambda}_1, \ldots, \hat{\lambda}_p\}$, respectively. Show that $\sqrt{n}(\hat{\theta} - \theta) \xrightarrow{D} N(0, 2\theta^2/p)$.

12.12. Consider the factor model given in Eq. (3) and assume that $\boldsymbol{L}$ and $\boldsymbol{\Psi}$ are known and $\boldsymbol{\mu} = \boldsymbol{0}$. Suppose it is desired to predict the vector of common factors $\boldsymbol{f}$ when $\boldsymbol{Y}$ is given and this problem examines the three predictors $\hat{\boldsymbol{f}}^{(1)}$, $\hat{\boldsymbol{f}}^{(2)}$, and $\hat{\boldsymbol{f}}^{(3)}$ given in Section 12.10.2.

(a) Show that $\hat{\boldsymbol{f}}^{(3)}$ is the best linear unbiased predictor of $\boldsymbol{f}$.

(b) For any $\boldsymbol{a} \in \mathbb{R}^k$, explicitly calculate $\mathrm{E}\left[\boldsymbol{a}^T\hat{\boldsymbol{f}}^{(1)} - \boldsymbol{a}^T\boldsymbol{f}\right]^2$, $\mathrm{E}\left[\boldsymbol{a}^T\hat{\boldsymbol{f}}^{(2)} - \boldsymbol{a}^T\boldsymbol{f}\right]^2$, and $\mathrm{E}\left[\boldsymbol{a}^T\hat{\boldsymbol{f}}^{(3)} - \boldsymbol{a}^T\boldsymbol{f}\right]^2$, and rank these three predictors $\hat{\boldsymbol{f}}^{(1)}$, $\hat{\boldsymbol{f}}^{(2)}$, and $\hat{\boldsymbol{f}}^{(3)}$ from the worst to the best.

12.13. Let $\boldsymbol{Y} \sim N_p(\boldsymbol{\mu}, \boldsymbol{\Sigma})$ and assume that the factor model given in Eq. (3) holds, that is, $\boldsymbol{Y} = \boldsymbol{\mu} + \boldsymbol{L}\boldsymbol{f} + \boldsymbol{\varepsilon}$, where $\boldsymbol{L}$ is $p \times k$, $k < p$, $\boldsymbol{f} \sim N_k(\boldsymbol{0}, I)$, $\boldsymbol{\varepsilon} \sim N_p(\boldsymbol{0}, \boldsymbol{\Psi})$, where $\boldsymbol{\Psi}$ is diagonal, and $\boldsymbol{f}$ and $\boldsymbol{\varepsilon}$ are assumed to be independent. Suppose it is desired to predict $\boldsymbol{f}$ when $\boldsymbol{Y}$ is given and the matrices $\boldsymbol{L}$ and $\boldsymbol{\Psi}$ are known and $\boldsymbol{\mu} = \boldsymbol{0}$.

(a) Find the conditional distribution of $\boldsymbol{f}$ given $\boldsymbol{Y}$ and show that the mean of this conditional distribution is $\hat{\boldsymbol{f}}^{(3)}$ as given in Section 12.10.2.

(b) Using the joint distribution of $\boldsymbol{Y}$ and $\boldsymbol{f}$, obtain an analog of the mixed model equations given in Section 11.10.2 of Chapter 11 (assuming that $\boldsymbol{L}$ and $\boldsymbol{\Psi}$ are known and $\boldsymbol{\mu} = \boldsymbol{0}$) and show that $\hat{\boldsymbol{f}}^{(3)}$ is the solution of this equation.

12.14. Let $\boldsymbol{Y}$ be from one of the populations $N_p(\boldsymbol{\mu}_1, \boldsymbol{\Sigma})$ (population 1) or $N_p(\boldsymbol{\mu}_2, \boldsymbol{\Sigma})$ (population 2) with probabilities π_1 and π_2, respectively, with $\pi_1 + \pi_2 = 1$. Consider a linear discriminant rule with prior probabilities π_1 and π_2. Assume that $\boldsymbol{\mu}_1$, $\boldsymbol{\mu}_2$, $\boldsymbol{\Sigma}$, and π_1 are known. Denote $(\boldsymbol{\mu}_1 - \boldsymbol{\mu}_2)^T\boldsymbol{\Sigma}^{-1}(\boldsymbol{\mu}_1 - \boldsymbol{\mu}_2)$ and $\log(\pi_1/\pi_2)$ by δ^2 and c, respectively.

(a) Let $R = (\boldsymbol{\mu}_1 - \boldsymbol{\mu}_2)^T\boldsymbol{\Sigma}^{-1}(\boldsymbol{Y} - \overline{\boldsymbol{\mu}})$, where $\overline{\boldsymbol{\mu}} = (\boldsymbol{\mu}_1 + \boldsymbol{\mu}_2)/2$. Show that the linear discriminant rule allocates $\boldsymbol{Y}$ to population 1 if $R > -c$, and allocates $\boldsymbol{Y}$ to population 2 if the inequality is reversed.

(b) Show that R is $N(\delta^2/2, \delta^2)$ if $\boldsymbol{Y}$ is from population 1, and R is $N(-\delta^2/2, \delta^2)$ if $\boldsymbol{Y}$ is from population 2.

(c) Show that the probability of misclassification is equal to $\Phi(-\delta/2 - c/\delta)$ if $\boldsymbol{Y}$ is from population 1, and is equal to $\Phi(-\delta/2 + c/\delta)$ when $\boldsymbol{Y}$ is from population 2. [Here Φ is the cdf of the standard normal distribution.]

(d) If $\boldsymbol{Y}$ is from one of the two populations (with probability π_1 from population 1 and with probability π_2 from population 2) and it is classified using a linear discriminant rule, find the probability of misclassifying $\boldsymbol{Y}$, in terms of δ, π_1, and π_2.

12.15. Prove Theorem 12.12.2.

12.16. Prove Lemma 12.12.1.

13 Time Series

13.1 Introduction

In the previous chapters, except for random-effects models in Chapter 11, all the different types of statistical modeling and procedures are concerned with data sets consisting of independent observations. In practice, however, there are many cases when the assumption of independence is not tenable and this is particularly true when the observations are recorded over time and/or over space (geographical locations). A simple example for such a data set is daily records of average levels of ozone concentration in the air at various locations in a particular geographical region. This chapter is only concerned with data sets consisting of observations recorded over time and such observations are usually dependent. For instance, when unemployment rates are recorded over months, observation in a particular month depends on the employment levels in the previous months. Methodologies developed for the investigation of such data sets are called time series methods. Here are a few examples of time series data:

(a) Annual precipitation at Lake Michigan in the last 75 years.
(b) Annual temperature anomalies (ie, average yearly temperature minus a base value) in the last 150 years.
(c) Monthly unemployment in the United States in the last 50 years.
(d) Monthly electricity sales to the residential sector in the United States in the last 50 years.
(e) EEG data (used in diagnosing patients) at a particular skull location.

We may denote the observations as $\{Y_t\colon\ t = 1, \ldots, n\}$ where the time unit may be a year or a month or a week or even fraction of a second depending on the particular problem at hand. An actual examination of the data in the first example indicates that the annual precipitation $\{Y_t\colon t = 1, \ldots, n = 75\}$ tends to fluctuate around a constant value. In the second example, there is an overall increase in the annual temperature anomalies $\{Y_t\colon\ t = 1, \ldots, n = 150\}$ over the last 150 years and thus it is reasonable to model Y_t as a smooth part (trend) plus a rough part (random errors which are identically distributed, but not necessarily independent). An examination of the monthly electricity sales series $\{Y_t\colon\ t = 1, \ldots, n = 12 \times 50 = 600\}$ would reveal an overall increase over time. Moreover,

Theory and Methods of Statistics. http://dx.doi.org/10.1016/B978-0-12-802440-9.00013-8

there is a seasonal factor (ie, January sales in consecutive years tend to be similar, February sales in consecutive years tend to be similar, and so on). There may also be a cyclical factor (about 9–12 years) as the sales may dip a little after economic recessions. Thus Y_t may be modeled as a sum of the trend, the seasonal effect, the cyclical effect, and a mean zero random error. Thus a general model for Y_t for three of the five examples given above may be

$$\text{Example (a): } Y_t = \mu + X_t,$$
$$\text{Example (b): } Y_t = \mu_t + X_t,$$
$$\text{Example (d): } \log Y_t = \mu_t + S_t + C_t + X_t,$$

where μ_t is the trend, $\{S_t\}$ are the seasonal effects, $\{C_t\}$ are the cycles, and $\{X_t\}$ are the random errors which are identically distributed but may not be independent. Note that in Example (d), the series is transformed by the natural logarithm in order achieve constant variability over time and this will be discussed below. The trend can be modeled as a smooth function by nonparametric methods. Seasonals $\{S_t\}$ may be modeled by various methods including linear combination of sines and cosines. If the period of cycles are known, they may also be modeled by sines and cosines. Apart from the above-mentioned methods for modeling the trend, the seasonals, and the cycles, there are also probabilistic modeling schemes. Forecasting is one of the important goals in time series analysis. Forecasting at time $t + h$, $h \geq 1$, requires estimates of μ_{t+h}, S_{t+h}, C_{t+h}, and X_{t+h} which can be added to get a forecast value of Y_{t+h} as $\hat{Y}_{t+h} = \hat{\mu}_{t+h} + \hat{S}_{t+h} + \hat{C}_{t+h} + \hat{X}_{t+h}$.

In the fifth example above, the observations are combination of waves (alpha, beta, theta, delta, etc.), and the weights of the combinations vary depending on whether the subject is normal or has a disease such as epilepsy. For instance, theta waves are in the frequency range of 3–8 Hz and are present in diseased patients, whereas alpha waves vary in the frequency range of 8–13 Hz and are present in normal individuals without any external stimulus. The goal of the analysis here in not prediction, but to determine the combination of different waves in the observed series for a particular patient and this is done by using the spectral analysis which is described in the last section of this chapter.

This chapter is concerned with understanding and analysis of the rough part $\{X_t\}$. The main assumptions are

(i) $\{X_t\}$ are identically distributed,
(ii) for any t, the correlation between X_t and X_{t+h}, $h \geq 0$, depends only on h, and
(iii) the correlation between X_t and X_{t+h} is negligible when h is large.

A series with properties (i) and (ii) is called *stationary*, and property (iii) is a statistical necessity, which allows forecasting and consistent estimation for parametric models based on observed data sets.

Types of Nonstationary Series

There may be many sources of nonstationarity, but we briefly point out three of them:

(i) unequal means over time,
(ii) unequal variance over time, and
(iii) inappropriate time scale.

Often, a reasonable way to view unequal means is to treat it as a smooth function of time (trend). For some series, such as daily records of sulfur dioxide levels in the air, sharp changes may correspond to unusual events such as volcano eruptions, and may need to be incorporated into the trend.

In the analysis of financial time series data such as rate of return on stocks, the main focus of investigation is on modeling the variance which changes over time. In other cases, unequal variance can sometimes be remedied by transforming the data using a Box-Cox transformation

$$Y_t(\lambda) = \begin{cases} \frac{Y_t^\lambda - 1}{\lambda} & \lambda \neq 0 \\ \log(Y_t) & \lambda = 0. \end{cases}$$

If a transformation turns out to be useful, then analysis and modeling are done on the transformed series. For instance, a logarithmic transformation for monthly electricity sales (Example (d) above) is appropriate for achieving equal variance.

For some series, such as a signal from bird chirping, attempts are made to plot, understand, and analyze the data on a time scale $H(t)$ (a nonlinear function of time t) on which the series may be stationary. Typically H is unknown and needs to be estimated.

A Simple Method for Extracting the Stationary Part

Consider a series of the form $Y_t = m_t + S_t + X_t$, where $\{m_t\}$ is the smooth trend, $\{S_t\}$ are the seasonal effects with period s, and $\{X_t\}$ is the stationary part. As mentioned above, in some cases approximate equal variance may be achieved by employing Box-Cox transformations. When the seasonal effects are absent (ie, the model is $Y_t = m_t + X_t$), a popular method for the analysis of such series employs the integrated autoregressive-moving average (ARIMA) models. When both the trend and the seasonals are present, then such a series can be analyzed by using what is known as the integrated seasonal autoregressive-moving average (seasonal ARIMA) model. A good detailed description of these models can be found in the book by Box et al. [62].

Here we briefly describe a simple regression method for estimating m_t and S_t, and this procedure works well for some series. Since $X_t = Y_t - (m_t + S_t)$, we can obtain an approximation of X_t if we can obtain estimates of m_t and S_t. For the sake of identifiability, let us assume that $S_{t-r+1} + \cdots + S_t = 0$ for any t, where r is the seasonal order. In order to simplify the discussion, let us assume that we are dealing with quarterly data (ie, $r = 4$). If we model the trend m_t by a polynomial of degree d, then

$$m_t = \beta_0 + \beta_1 t + \cdots + \beta_d t^d.$$

In order to account for the seasonal effects, create variables, I_{t1}, I_{t2}, and I_{t3} as

$$I_{t1} = \begin{cases} 1 & \text{if time } t \text{ is Quarter 1} \\ -1 & \text{if time } t \text{ is Quarter 4}, \\ 0 & \text{otherwise} \end{cases} \quad I_{t2} = \begin{cases} 1 & \text{if time } t \text{ is Quarter 2} \\ -1 & \text{if time } t \text{ is Quarter 4}, \\ 0 & \text{otherwise} \end{cases}$$

$$I_{t3} = \begin{cases} 1 & \text{if time } t \text{ is Quarter 3} \\ -1 & \text{if time } t \text{ is Quarter 4} \\ 0 & \text{otherwise.} \end{cases}$$

Thus $m_t + S_t$ can be modeled as

$$\beta_0 + \beta_1 t + \cdots + \beta_d t^d + \theta_1 I_{t1} + \theta_2 I_{t2} + \theta_3 I_{t3},$$

and the unknown parameters $\beta_0, \ldots, \beta_d$ and θ_1, θ_2, and θ_3 can be estimated by minimizing the least squares criterion

$$\sum_{t=1}^{n} \left[Y_t - \beta_0 - \beta_1 t - \cdots - \beta_d t^d - \theta_1 I_{t1} - \theta_2 I_{t2} - \theta_3 I_{t3} \right]^2.$$

Once the estimates of these parameters are obtained, one can get estimates $\hat{m}_t$ and $\hat{S}_t$ of m_t and S_t. Then an estimate of X_t is given by $\hat{X}_t = Y_t - (\hat{m}_t + \hat{S}_t)$.

It is important to note that the seasonal fluctuations $\{S_t\}$ are assumed to be nonrandom in this discussion, which may not always be appropriate.

13.2 Concept of Stationarity

In order to develop an appropriate mathematical framework, it is usually assumed that the data $\{X_1, \ldots, X_n\}$ is a finite section of an infinite series $\{X_t: \ -\infty < t < \infty\}$. Throughout this chapter, we use the terms "series" and "process" to mean a sequence of random variables. A series $\{X_t\}$ is called *strictly stationary* if $\{X_t, \ldots, X_{t+k}\}$ has the same distribution as $\{X_{t+h}, \ldots, X_{t+h+k}\}$ for any t and $h \geq 0$. This notion of stationarity is usually too strict to be useful in applications since it is difficult to verify it in practice. A weaker version, also known as covariance stationarity, assumes that $\text{Cov}[X_t, X_{t+h}]$ depend only on h. It is easy to see that strict stationarity implies weak stationarity. From now on, we assume that $\{X_t\}$ has mean μ and $\gamma(h) = \text{Cov}[X_t, X_{t+h}]$ for any t and h. Since $\gamma(-h) = \text{Cov}[X_{t+h}, X_t]$ and $\text{Cov}[X_t, X_{t+h}] = \text{Cov}[X_{t+h}, X_t]$, it thus follows that $\gamma(-h) = \gamma(h)$ for any integer h. The correlation between X_t and X_{t+h} can be easily seen to be $\rho(h) = \gamma(h)/\gamma(0)$ and this quantity also does not depend on t. The sequences $\{\gamma(h)\}$ and $\{\rho(h)\}$ are called the autocovariance function and the autocorrelation function, respectively.

A time series is *stationary Gaussian* if the joint distribution of $\{X_{t+1}, \ldots, X_{t+p}\}$, for any t and $p \geq 1$, has a p-dim multivariate normal distribution with the mean vector $(\mu, \ldots, \mu)^T$

and covariance matrix $\boldsymbol{\Gamma}_p$ with elements $\gamma(k-j)$, $1 \leq j, k \leq p$. It can be shown that a covariance-stationary Gaussian time series is also "strictly" stationary.

Definition 13.2.1. A series $\{X_t\}$ is called (weakly) stationary if, for any t, $\mathrm{E}[X_t] = \mu$ and $\mathrm{Cov}[X_t, X_{t+h}]$ depends only on h. Its autocovariance $\{\gamma(h)\}$ and autocorrelation functions $\{\rho(h)\}$ are $\gamma(h) = \mathrm{Cov}[X_t, X_{t+h}]$ and $\rho(h) = \mathrm{Corr}[X_t, X_{t+h}] = \gamma(h)/\gamma(0)$.

We now present a few examples of stationary series and one example of a nonstationary series.

Example 13.2.1 (White Noise). Let $\{X_t\}$ be iid with mean μ and variance σ^2, then it is stationary with

$$\gamma(h) = \begin{cases} \sigma^2 & \text{if } h = 0 \\ 0 & \text{if } h \neq 0 \end{cases} \text{ and } \quad \rho(h) = \begin{cases} 1 & \text{if } h = 0 \\ 0 & \text{if } h \neq 0 \end{cases}.$$

If $\{X_t\}$ are iid with $\mu = 0$, then such a series is called white noise. Some authors dispense with the iid assumption and call a series white noise if $\{X_t\}$ have zero mean, identical variances, and are mutually uncorrelated.

Example 13.2.2 (Moving Average Process). If a series $\{X_t\}$ can be written as

$$X_t - \mu = \varepsilon_t + \theta\varepsilon_{t-1},$$

where $\{\varepsilon_t\}$ are iid with mean 0 and variance σ^2, then it is called a moving average process of order 1. For such a series, $\mathrm{E}[X_t] = \mu$, and the autocovariances are

$$\begin{aligned}
\gamma(0) &= \mathrm{Var}[X_t] = \mathrm{Var}[\varepsilon_t] + \mathrm{Var}[\theta\varepsilon_{t-1}] = (1+\theta^2)\sigma^2, \\
\gamma(1) &= \mathrm{Cov}[X_t, X_{t-1}] = \mathrm{Cov}[\varepsilon_t + \theta\varepsilon_{t-1}, \varepsilon_{t-1} + \theta\varepsilon_{t-2}] \\
&= \mathrm{Cov}[\theta\varepsilon_{t-1}, \varepsilon_{t-1}] = \theta\sigma^2, \text{ and} \\
\gamma(h) &= 0, \quad h \geq 2.
\end{aligned}$$

Clearly, the autocorrelations are

$$\rho(1) = \theta/(1+\theta^2), \quad \rho(h) = 0, \ h \geq 2.$$

A process more general than the above is

$$X_t - \mu = \varepsilon_t + \theta_1\varepsilon_{t-1} + \cdots + \theta_q\varepsilon_{t-q},$$

where $\{\varepsilon_t\}$ are mean 0 iid variables with common variance σ^2. This is known as a moving average process of order q and is denoted by *MA*(q). For this series, $\mathrm{E}[X_t] = \mu$, and the autocovariances and autocorrelations are

$$\begin{aligned}
\gamma(h) &= \sigma^2 \sum_{j=0}^{q-h} \theta_j\theta_{j+h}, \quad 0 \leq h \leq q, \text{ with } \theta_0 = 1, \\
\gamma(h) &= 0, \quad h \geq q+1, \\
\rho(h) &= \sum_{j=0}^{q-h} \theta_j\theta_{j+h} \Bigg/ \sum_{j=0}^{q} \theta_j^2, \quad 0 \leq h \leq q, \text{ and} \\
\rho(h) &= 0, \quad h \geq q+1.
\end{aligned}$$

Example 13.2.3 (Autoregressive Process). If a series $\{X_t\}$ is representable as

$$X_t - \mu = \phi(X_{t-1} - \mu) + \varepsilon_t,$$

where $\{\varepsilon_t\}$ are iid with mean 0 and variance σ^2, then it is called an autoregressive process of order 1. For this series, $\mathrm{E}[X_t] = \mu$. In order to obtain the variance of X_t, note that $\mathrm{Cov}[X_{t-1}, \varepsilon_t] = 0$, and hence

$$\gamma(0) = \mathrm{Var}[X_t] = \mathrm{Var}[\phi X_{t-1}] + \mathrm{Var}[\varepsilon_t] = \phi^2\gamma(0) + \sigma^2, \text{ and}$$
$$\gamma(0) = (1 - \phi^2)^{-1}\sigma^2.$$

The condition $\gamma(0) = \mathrm{Var}[X_t] > 0$ requires that $\phi^2 < 1$ (ie, $|\phi| < 1$) and this will be shown later to be the condition for the series to be stationary. Autocovariances of this series can be obtained by using a recursive procedure. Since X_{t-1} is uncorrelated with ε_t, we have

$$\begin{aligned}\gamma(1) &= \mathrm{Cov}[X_t, X_{t-1}] = \mathrm{Cov}[\phi X_{t-1} + \varepsilon_t, X_{t-1}] \\ &= \phi\mathrm{Cov}[X_{t-1}, X_{t-1}] = \phi\gamma(0).\end{aligned}$$

Similarly, noting that ε_t is uncorrelated with X_{t-h} for $h \geq 1$, we have

$$\begin{aligned}\gamma(h) &= \mathrm{Cov}[X_t, X_{t-h}] = \mathrm{Cov}[\phi X_{t-1} + \varepsilon_t, X_{t-h}] \\ &= \phi\mathrm{Cov}[X_{t-1}, X_{t-h}] = \phi\gamma(h-1).\end{aligned}$$

Thus for $h = 2, \ldots,$ we have

$$\begin{aligned}\gamma(2) &= \phi\gamma(1) = \phi^2\gamma(0), \\ \gamma(3) &= \phi\gamma(2) = \phi^3\gamma(0), \ldots.\end{aligned}$$

This argument shows that for any $h \geq 0$

$$\gamma(h) = \phi^h\gamma(0) \text{ with } \gamma(0) = (1 - \phi^2)^{-1}\sigma^2, \text{ and } \rho(h) = \phi^h.$$

Since $|\phi| < 1$, $\rho(h)$ converges to zero exponentially as $h \to \infty$.

A general version of the simple autoregressive process is

$$X_t - \mu = \phi_1(X_{t-1} - \mu) + \cdots + \phi_p(X_{t-p} - \mu) + \varepsilon_t,$$

where $\{\varepsilon_t\}$ are iid with mean 0 and variance σ^2. This is an autoregressive process of order p with mean $\mathrm{E}[X_t] = \mu$ and it is denoted by *AR*(p). Conditions on $\phi_1, \ldots, \phi_p$ needed for the process $\{X_t\}$ to be stationary will be discussed later. In the *AR*(p) case, it is not possible to obtain explicit expressions for the autocovariance or autocorrelation functions though computer packages such as R can be used to obtain them for given values of $\phi_1, \ldots, \phi_p$ and σ^2. It should be pointed out that, under the condition of stationarity, the autocorrelation function $\rho(h)$ converges rapidly (exponentially) to zero as $h \to \infty$.

Example 13.2.4 (Autoregressive-Moving Average Process). A series $\{X_t\}$ which has both *AR*(p) and *MA*(q) parts is called an *ARMA*(p, q) process, and is representable in the form

$$X_t - \mu = \phi_1(X_{t-1} - \mu) + \cdots + \phi_p(X_{t-p} - \mu)$$
$$+ \varepsilon_t + \theta_1\varepsilon_{t-1} + \cdots + \theta_q\varepsilon_{t-q},$$

where $\{\varepsilon_t\}$ are iid with mean 0 and variance σ^2. Conditions on $\phi_1, \ldots, \phi_p$ are needed to guarantee stationarity as in the *AR*(*p*) case. In general there are no explicit forms for the autocovariances and autocorrelations even though computer packages can be used to obtain them. As in the *AR*(*p*) case, under appropriate conditions, the autocorrelation $\rho(h)$ converges to zero exponentially as $h \to \infty$. The error terms $\{\varepsilon_t\}$ in AR, MA, or ARMA series are sometimes called *innovations*.

Example 13.2.5 (Random Walk). A mean zero *AR*(1) series with $\phi = 1$ is called a random walk and it has the form $X_t = X_{t-1} + \varepsilon_t$, $t = 1, 2, \ldots$, where $\{\varepsilon_t\}$ are iid with mean 0 and variance σ^2. It is easy to see that $X_t = X_0 + \varepsilon_1 + \cdots + \varepsilon_t$ and assuming that $\mathrm{E}[X_0] = \mu$, we have

$$\mathrm{E}[X_t] = \mu \quad \text{and} \quad \mathrm{Cov}[X_t, X_{t+h}] = \mathrm{Var}[X_0] + t\sigma^2, \quad h \geq 0.$$

Clearly this series is not stationary since $\mathrm{Cov}[X_t, X_{t+h}]$ depends on t. Random walks are sometimes used for modeling the trend of a nonstationary time series.

13.2.1 Representation of the Autocovariance Function

We begin with an important result on the representation of the covariance function $\{\gamma(h): -\infty < h < \infty\}$ of a stationary series. Under the assumption that $\sum_{h=-\infty}^{\infty}|\gamma(h)| < \infty$, the function

$$f(w) = \sum_{h=-\infty}^{\infty} \exp[-2\pi ihw]\gamma(h), \quad h \geq 0, \tag{1}$$

is well defined on $[-1/2, 1/2]$, where $i = \sqrt{-1}$. This function f is called the spectral density function of the series $\{X_t\}$ with the autocovariance function $\{\gamma(h)\}$. It is periodic with period 1 (ie, f can be defined for any real w and $f(w) = f(w+1)$). More detailed discussion on spectral density function will be given in a later section.

Since $\gamma(h) = \gamma(-h)$ and $\exp[-2\pi ihw] + \exp[2\pi ihw] = 2\cos(2\pi hw)$, we have

$$\begin{aligned} f(w) &= \sum_{h=-\infty}^{-1} \exp[-2\pi ihw]\gamma(h) + \gamma(0) + \sum_{h=1}^{\infty} \exp[-2\pi ihw]\gamma(h) \\ &= \gamma(0) + \sum_{h=1}^{\infty}\big\{\exp[2\pi ihw] + \exp[-2\pi ihw]\big\}\gamma(h) \\ &= \gamma(0) + 2\sum_{h=1}^{\infty} \cos(2\pi hw)\gamma(h). \end{aligned}$$

Thus the spectral density function is a trigonometric series with autocovariances $\{\gamma(h)\}$ as the coefficients. It is symmetric about 0 (ie, $f(-w) = f(w)$), and is nonnegative (to be shown below). What is the connection between the autocovariance function

$\{\gamma(h)\}$ and the spectral density function? If we know the spectral density function f, then we can get the autocovariances as

$$\gamma(h) = \int_{-1/2}^{1/2} \exp[2\pi i h w] f(w) dw, \; h \geq 0.$$

In order to verify that f is nonnegative, consider the arguments given below. The discrete cosine and sine transforms of $\{X_t - \mu, t = 1, \ldots, n\}$ are

$$\tilde{X}_{c,n}(w) = n^{-1/2} \sum_{t=1}^{n} (X_t - \mu) \cos(2\pi w t), \text{ and}$$
$$\tilde{X}_{s,n}(w) = n^{-1/2} \sum_{t=1}^{n} (X_t - \mu) \sin(2\pi w t),$$

where $|w| \leq 1/2$. Then

$$\begin{aligned}
\mathrm{E}\left[\tilde{X}_{c,n}(w)\right]^2 + \mathrm{E}\left[\tilde{X}_{s,n}(w)\right]^2 &= n^{-1} \sum_{1 \leq s,t \leq n} \cos(2\pi w s) \cos(2\pi w t) \mathrm{Cov}[X_s, X_t] \\
&\quad + n^{-1} \sum_{1 \leq s,t \leq n} \sin(2\pi w s) \sin(2\pi w t) \mathrm{Cov}[X_s, X_t] \\
&= n^{-1} \sum_{1 \leq s,t \leq n} \cos(2\pi w(t-s)w) \mathrm{Cov}[X_s, X_t] \\
&= n^{-1} \sum_{1 \leq s,t \leq n} \cos(2\pi w(t-s)w) \gamma(t-s).
\end{aligned}$$

It is not difficult to check that, for any $-(n-1) \leq h \leq n-1$, the number of (s, t) pairs when $t - s = h$, $1 \leq s, t \leq n$, equals $n - |h|$, and therefore

$$\begin{aligned}
\mathrm{E}\left[\tilde{X}_{c,n}(w)\right]^2 + \mathrm{E}\left[\tilde{X}_{s,n}(w)\right]^2 &= n^{-1} \sum_{h=-n+1}^{n-1} (n - |h|) \cos(2\pi h w) \gamma(h) \\
&= \sum_{h=-n+1}^{n-1} (1 - |h|/n) \cos(2\pi h w) \gamma(h).
\end{aligned}$$

Clearly, the last sum is nonnegative, and since $\sum |\gamma(h)| < \infty$, this sum converges to $\sum_{h=-\infty}^{\infty} \cos(2\pi h w) \gamma(h) = f(w)$ by the Dominated Convergence Theorem. Thus $f(w) \geq 0$ for any $|w| \leq 1/2$.

Example 13.2.6. Let $\{X_t\}$ be iid with mean 0 and variance σ^2. Then $\gamma(h) = \sigma^2$ if $h = 0$ and $\gamma(h) = 0$ if $h \neq 0$. For this case, the spectral density function is $f(w) = \sigma^2$ for any $w \in [-1/2, 1/2]$.

Example 13.2.7. Let $\{X_t\}$ be a mean zero *MA*(1) series, that is, $X_t = \varepsilon_t + \theta\varepsilon_{t-1}$ where $\{\varepsilon_t\}$ are iid with mean 0 and variance σ^2. We have already seen that

$$\gamma(h) = \begin{cases} (1+\theta^2)\sigma^2 & h = 0 \\ \theta\sigma^2 & h = 1. \\ 0 & h \geq 2 \end{cases}$$

Since $\gamma(h) = \gamma(-h)$ for any integer h, the spectral density function is

$$\begin{aligned} f(w) &= \gamma(0) + 2\sum_{h=1}^{\infty} \cos(2\pi hw)\gamma(h) \\ &= (1+\theta^2)\sigma^2 + 2\theta\sigma^2\cos(2\pi w) \\ &= \sigma^2\left[1 + \theta^2 + 2\theta\cos(2\pi w)\right]. \end{aligned}$$

Example 13.2.8. Let $\{X_t\}$ be a stationary *AR*(1) series as described in Example 13.2.3. We have already seen that for this sequence $\gamma(h) = \phi^h\gamma(0)$, $h \geq 0$, and $\gamma(0) = (1-\phi^2)^{-1}\sigma^2$. So the spectral density function of the *AR*(1) series is

$$\begin{aligned} f(w) &= \gamma(0) + 2\sum_{h=1}^{\infty} \cos(2\pi hw)\gamma(h) \\ &= \gamma(0) + 2\sum_{h=1}^{\infty} \phi^h\gamma(0)\cos(2\pi hw) \\ &= \gamma(0)\left[1 + 2\sum_{h=1}^{\infty} \phi^h\cos(2\pi hw)\right]. \end{aligned}$$

Denoting $z = \exp(-2\pi iw)$ and noting that $2\cos(2\pi hw) = z^h + \bar{z}^h$, where $\bar{z} = \exp(2\pi iw)$ is the complex conjugate of z, we have

$$\begin{aligned} 1 + 2\sum_{h=1}^{\infty} \phi^h\cos(2\pi hw) &= 1 + \sum_{h=1}^{\infty} \phi^h\left(z^h + \bar{z}^h\right) \\ &= \sum_{h=0}^{\infty} (\phi z)^h + \sum_{h=0}^{\infty} (\phi\bar{z})^h - 1 = (1-\phi z)^{-1} + (1-\phi\bar{z})^{-1} - 1 \\ &= (1-\phi^2)\frac{1}{1+\phi^2 - 2\phi\cos(2\pi w)}, \end{aligned}$$

where the last step is obtained after some simple algebra. Hence we have

$$\begin{aligned} f(w) &= \gamma(0)(1-\phi^2)\frac{1}{1+\phi^2 - 2\phi\cos(2\pi w)} \\ &= \sigma^2\frac{1}{1+\phi^2 - 2\phi\cos(2\pi w)}, \end{aligned}$$

where the justification of the last step follows from the fact that $\gamma(0) = (1-\phi^2)^{-1}\sigma^2$.

Remark 13.2.1. There is a more general result on the representation of the autocovariance function which does not require a restrictive condition like $\sum|\gamma(h)| < \infty$. It can be shown (Bochner's Theorem) that there exists a nonincreasing function F (called the spectral distribution function) on $[-1/2, 1/2]$ such that $\gamma(h) = \int_{-1/2}^{1/2} \exp(2\pi ihw)dF(w)$ for any integer h. When F is differentiable, its derivative f is the spectral density function. Further details can be found in the books by Brockwell and Davis [63] and Gikhman and Skorokhod [64].

13.2.2 Linear Time Series

In this book we only consider linear time series, which, as the same suggests, are linear combinations of mean zero variables $\{\varepsilon_t\}$ of the form

$$X_t = \mu + \sum_{j=-\infty}^{\infty} a_j \varepsilon_{t-j},$$

where $\sum a_j^2 < \infty$, and $\{\varepsilon_t\}$ are mutually uncorrelated with mean 0 and variance σ^2. For this series, the mean and the autocovariance function are

$$\mathrm{E}[X_t] = \mu, \quad \gamma(h) = \sigma^2 \sum_{j=-\infty}^{\infty} a_j a_{j+h}.$$

Such a representation exists if the spectral density f is integrable. If $\{X_t\}$ is stationary Gaussian, then $\{\varepsilon_t\}$ are iid $N(0, \sigma^2)$ variables.

A stationary series $\{X_t\}$ is said to have a causal representation if it can be written as

$$X_t = \mu + \sum_{j=0}^{\infty} a_j \varepsilon_{t-j},$$

where $\{\varepsilon_t\}$ are mutually uncorrelated mean zero rv's with common variance σ^2. If $\int_{-1/2}^{1/2} \log(f(w))dw > -\infty$, then the series admits a causal representation. As before, if $\{X_t\}$ is stationary Gaussian, then $\{\varepsilon_t\}$ are iid $N(0, \sigma^2)$ variables. A detailed discussion on these representations can be found in Chapter 5 of the book by Gikhman and Skorokhod [64].

In practice, it is usually assumed that $\{\varepsilon_t\}$ are iid with mean 0 and variance σ^2 even if they are not normally distributed. Particular examples of linear time series are ARMA processes. There are other time series that are nonlinear such as the bilinear series

$$X_t = \mu + \sum_{j=-\infty}^{\infty} a_j \varepsilon_{t-j} + \sum_{-\infty<j,k<\infty} b_{jk} \varepsilon_{t-j} \varepsilon_{t-k}$$

with $\sum|a_j| < \infty$ and $\sum|b_{jk}| < \infty$, but such series are not discussed in this chapter.

From now on, it will be assumed that $\{X_t\}$ *is a linear time series with* $\{\varepsilon_t\}$ *iid mean zero rv's with finite variance* σ^2, that is,

$$X_t = \mu + \sum_{j=-\infty}^{\infty} a_j \varepsilon_{t-j}, \quad \text{with} \sum_{j=-\infty}^{\infty} \left| a_j \right| < \infty, \tag{2}$$

$$\{\varepsilon_t\} \text{ iid}, \quad \mathrm{E}[\varepsilon_t] = 0, \text{ and } \quad \mathrm{Var}\,[\varepsilon_t] = \sigma^2.$$

This series $\{X_t\}$ is stationary as discussed above.

Notations

The following notations will be used throughout this chapter. For any stationary series with autocovariance function γ and for any positive integer h, $\boldsymbol{\Gamma}_h$ will denote the $h \times h$ matrix whose element (j, k) is given by $\gamma(j-k)$ and $\boldsymbol{\gamma}_h$ will denote the $h \times 1$ vector with elements $\gamma(1), \dots, \gamma(h)$. Similarly, $\hat{\boldsymbol{\Gamma}}_h$ and $\hat{\boldsymbol{\gamma}}_h$ will denote the estimates of $\boldsymbol{\Gamma}_h$ and $\boldsymbol{\gamma}_h$ when the autocovariances are estimated based on the available data. We note that $\boldsymbol{\Gamma}_h$ is the covariance matrix of $(X_{t+1}, \dots, X_{t+h})$ and it is nonnegative definite.

13.2.3 Time Reversibility for Linear Prediction

If $\{X_t\}$ is stationary with autocovariance function γ, then we show below that the coefficients associated with the best linear predictor (forecast) of X_t based on $X_{t-1}, \dots, X_{t-h}$ are the same as those of the best linear predictor (backcast) of X_{t-h-1} based on $X_{t-h}, \dots, X_{t-1}$. This result is true for any $h \geq 1$. In order to simplify the notations, we assume that $\{X_t\}$ has zero mean, since we can always subtract the mean μ from X_t to achieve this.

A linear predictor $\phi_1 X_{t-1} + \cdots + \phi_h X_{t-h}$ of X_t has the prediction error $PE^{(f)} = \mathrm{E}[X_t - \phi_1 X_{t-1} - \cdots - \phi_h X_{t-h}]^2$. Since $X_t, X_{t-1}, \dots, X_{t-h}$ have zero means, there is no need to include an intercept term in the formula for the predictor. The best linear predictor can be obtained by minimizing $PE^{(f)}$ with respect to $\phi_1, \dots, \phi_h$. As in the case of linear models, we can differentiate $PE^{(f)}$ with respect to $\phi_j, j = 1, \dots, h$, and equate the derivatives to zero, which leads to the normal equations. When the derivative of $PE^{(f)}$ with respect to ϕ_j is set to 0, we have

$$\begin{aligned} -2\mathrm{E}[X_t - \phi_1 X_{t-1} - \cdots - \phi_h X_{t-h}]X_{t-j} &= 0, \text{ ie,} \\ -\gamma(j) + \phi_1 \gamma(j-1) + \cdots + \phi_h \gamma(j-h) &= 0, \text{ ie,} \\ \phi_1 \gamma(j-1) + \cdots + \phi_h \gamma(j-h) &= \gamma(j), \end{aligned}$$

$j = 1, \dots, h$. These normal equations can be written in the matrix form as

$$\boldsymbol{\Gamma}_h \boldsymbol{\phi} = \boldsymbol{\gamma}_h, \tag{3}$$

where $\boldsymbol{\phi} = (\phi_1, \dots, \phi_h)^T$, and they are also known as the *Yule-Walker equations*. Moreover, the prediction error of the best linear predictor with coefficients $\phi_1, \dots, \phi_h$ satisfying the Yule-Walker equations is given by

$$
\begin{aligned}
\mathrm{E}[X_t-\phi_1X_{t-1}-\cdots-\phi_hX_{t-h}]^2 &= \mathrm{E}[X_t^2]-2\mathrm{E}[X_t(\phi_1X_{t-1}+\cdots+\phi_hX_{t-h})] \\
&\quad+\mathrm{E}\big[\phi_1X_{t-1}+\cdots+\phi_hX_{t-h}\big]^2 \\
&= \gamma(0)-2\boldsymbol{\phi}^T\boldsymbol{\gamma}_h+\boldsymbol{\phi}^T\boldsymbol{\Gamma}_h\boldsymbol{\phi}=\gamma(0)-\boldsymbol{\phi}^T\boldsymbol{\gamma}_h \\
&= \gamma(0)-\phi_1\gamma(1)-\cdots-\phi_h\gamma(h).
\end{aligned}
$$

A linear predictor $\phi_1X_{t-h}+\phi_2X_{t-h+1}+\cdots+\phi_hX_{t-1}$ of X_{t-h-1} has the prediction error $PE^{(b)}=\mathrm{E}[X_{t-h-1}-\phi_1X_{t-h}-\cdots-\phi_hX_{t-1}]^2$. In order to obtain the best linear predictor we need to differentiate $PE^{(b)}$ with respect to $\phi_1,\ldots,\phi_h$ and equate the derivatives to 0. Setting the derivative of $PE^{(b)}$ with respect to ϕ_j to 0, we have

$$
\begin{aligned}
-2\mathrm{E}[(X_{t-h-1}-\phi_1X_{t-h}-\cdots-\phi_hX_{t-1})X_{t-h-1+j}] &= 0, \text{ ie,} \\
-\gamma(j)+\phi_1\gamma(j-1)+\cdots+\phi_h\gamma(j-h) &= 0, \text{ ie,} \\
\phi_1\gamma(j-1)+\cdots+\phi_h\gamma(j-h) &= \gamma(j),
\end{aligned}
$$

$j=1,\ldots,h$. These normal equations in the matrix form are exactly the same as in Eq. (3). Since the normal equations for forecasting (ie, linear prediction of X_t based on $X_{t-1},\ldots,X_{t-h}$) and backcasting (ie, linear prediction of X_{t-h-1} based on $X_{t-h},\ldots,X_{t-1}$) are the same, the vectors of coefficients $\boldsymbol{\Gamma}_h^{-1}\boldsymbol{\gamma}_h$ are also the same. Using the arguments given above, it is fairly easy to verify that the prediction error for the best linear predictor for backcasting X_{t-h-1} is exactly the same as that for forecasting X_t. Thus we arrive at an important result.

Lemma 13.2.1. *Let $\{X_t\}$ be a mean zero stationary series.*

(a) *For any positive integer h, if $\hat{X}_t^{(f)}=\phi_1X_{t-1}+\cdots+\phi_hX_{t-h}$ is the best linear predictor of X_t based on $X_{t-1},\ldots,X_{t-h}$, then $\hat{X}_{t-h-1}^{(b)}=\phi_1X_{t-h}+\phi_2X_{t-h+1}+\cdots+\phi_hX_{t-1}$ is the best linear predictor of X_{t-h-1} based on $X_{t-h},\ldots,X_{t-1}$.*

(b) *The vector of coefficients $\boldsymbol{\phi}$ of the best predictor described in part (a) is given by the solution of the normal equations (ie, the Yule-Walker equations) $\boldsymbol{\Gamma}_h\boldsymbol{\phi}=\boldsymbol{\gamma}_h$.*

(c) *The prediction errors $E\big[X_t-\hat{X}_t^{(f)}\big]^2$ and $E\big[X_{t-h-1}-\hat{X}_{t-h-1}^{(b)}\big]^2$ are the same and they are given by $\gamma(0)-\phi_1\gamma(1)-\cdots-\phi_h\gamma(h)$.*

13.3 Estimation of the Mean and the Autocorrelation Function

Suppose that $X_1,\ldots,X_n$ are observations from a stationary series with mean μ and covariance function γ. Some useful descriptive statistics for an initial analysis involve estimation of the mean, the autocovariance function γ, the autocorrelation function ρ, etc. These are the topics of discussion in this section.

13.3.1 Estimation of the Mean

An estimate of μ is given by $\hat{\mu}=\bar{X}_n=n^{-1}\sum_{t=1}^{n}X_t$. It is fairly easy to see that $\hat{\mu}$ is an unbiased estimator of μ. In order to obtain the standard error of this estimate, one needs to calculate the variance of $\hat{\mu}$. Note that

$$\text{Var}[\hat{\mu}] = n^{-2} \sum_{s,t=1}^{n} \text{Cov}[X_s, X_t] = n^{-2} \sum_{s,t=1}^{n} \gamma(s-t).$$

The argument used in Section 13.2.1 to demonstrate the nonnegativeness of the spectral density function can be employed to show that

$$\tau_n^2 = n\text{Var}[\hat{\mu}] = n^{-1} \sum_{h=-(n-1)}^{n-1} (n - |h|)\gamma(h) = \sum_{h=-n+1}^{n-1} (1 - |h|/n)\gamma(h)$$
$$\to \sum_{h=-\infty}^{\infty} \gamma(h) := \tau^2,$$

assuming that $\sum |\gamma(h)| < \infty$. Since $\gamma(-h) = \gamma(h)$, we may also rewrite τ_n^2 and τ^2 as

$$\tau_n^2 = n\text{Var}[\hat{\mu}] = \sum_{h=-n+1}^{n-1} (1 - |h|/n)\gamma(h) = \gamma(0) + 2\sum_{h=1}^{n-1}(1 - h/n)\gamma(h),$$
$$\tau^2 = \sum_{h=-\infty}^{\infty} \gamma(h) = \gamma(0) + 2\sum_{h=1}^{\infty} \gamma(h).$$

If it is assumed that $\gamma(h)$ converges rapidly to zero as $h \to \infty$, a condition satisfied by many series such as *ARMA*(p,q), then one may ignore $\gamma(h)$, $h > L$, for a suitably chosen integer L which may depend on n (eg, $L \approx \sqrt{n}$). In such a case, a reasonable estimate of τ_n^2 is

$$\hat{\tau}_n^2 = n^{-1}\hat{\gamma}(0) + 2\sum_{h=1}^{L}(1 - h/n)\hat{\gamma}(h) \approx n^{-1}\left[\hat{\gamma}(0) + 2\sum_{h=1}^{L}\hat{\gamma}(h)\right],$$

where $\hat{\gamma}(h)$ is an estimate of $\gamma(h)$. It turns out that the central limit theorem (CLT) holds for $\hat{\mu} = \bar{X}_n$ under reasonable conditions.

Theorem 13.3.1. *Assume that* $X_1, \ldots, X_n$ *are observations from a linear stationary series* $\{X_t\}$ *as given in Eq. (2). Then as* $n \to \infty$, $\sqrt{n}(\hat{\mu} - \mu) \xrightarrow{\mathcal{D}} N(0, \tau^2)$, *where* $\tau^2 = \lim_{n\to\infty} n\text{Var}[\hat{\mu}]$.

This result now can be used to construct an approximate confidence interval for μ. Let $\hat{\tau}_n$ be as given above. Assuming that $\hat{\tau}_n/\tau_n - 1 \xrightarrow{P} 0$, an approximate confidence interval for μ with confidence coefficient $1 - \alpha$ is $\hat{\mu} \pm z_{\alpha/2}\hat{\tau}_n/\sqrt{n}$.

A proof of the above result on the asymptotic normality of $\hat{\mu} = \bar{X}_n$ uses the CLT for the sample mean of m-dependent rv's (Theorem 3.3.3) and it involves careful details as it requires verifying technical conditions, and the details are given in Brockwell and Davis [63]. However, the basic idea behind the proof is simple and can be summarized in the following steps. Let $W_n = \sqrt{n}(\bar{X}_n - \mu)$.

I. The truncated rv's $\{X_{m,t}\}$

$$X_{m,t} = \mu + \sum_{j=-m}^{m} a_j \varepsilon_{t-j},$$

are $2m$-dependent with mean μ where m is a positive integer.

II. Let $\bar{X}_{m,n} = n^{-1} \sum_{t=1}^{n} X_{m,t}$ and $W_{m,n} = \sqrt{n}\left(\bar{X}_{m,n} - \mu\right)$. Since $\{X_{m,t}\}$ is $2m$-dependent, the CLT holds for $W_{m,n}$ (Theorem 3.3.3). Thus $W_{m,n} \xrightarrow{\mathcal{D}} Z_m \sim N(0, \tau(m)^2)$, where $\tau^2(m) = \lim_{n\to\infty} \mathrm{Var}\left[W_{m,n}\right]$.

III. Show that for any $\delta > 0$,

$$\limsup_{n\to\infty} P\left[\left|W_{m,n} - W_n\right| > \delta\right] \to 0$$

as $m \to \infty$. A sufficient condition for this, via Chebychev's inequality, is

$$\limsup_{n\to\infty} \mathrm{E}\left[\left|W_{m,n} - W_n\right|^2\right] \to 0$$

as $m \to \infty$.

IV. If $\tau^2(m) = \lim_{n\to\infty} \mathrm{Var}\left[W_{m,n}\right] \to \tau^2 > 0$, then $W_n \xrightarrow{D} Z \sim N(0, \tau^2)$.

The four steps given above can be written down in a general framework which does not require $\{X_t\}$ to be a linear stationary series. The general framework is useful since this result can also be applied for obtaining asymptotic normality of estimates of the autocovariances and autocorrelations.

Theorem 13.3.2. *Let $X_1, \ldots, X_n$ be observations from an infinite sequence $\{X_t\}$ of rv's with common mean μ and let $W_n = \sqrt{n}(\bar{X}_n - \mu)$, where $\bar{X}_n = n^{-1} \sum_{t=1}^{n} X_t$. Assume that for any positive integer m, there exists an m-dependent series $\{X_{m,t}\}$ with common mean μ_m and let $W_{m,n} = \sqrt{n}(\bar{X}_{m,n} - \mu_m)$, where $\bar{X}_{m,n} = n^{-1} \sum_{t=1}^{n} X_{m,t}$. Then $W_n \xrightarrow{D} Z \sim N(0, \tau^2)$ if the following conditions hold*

(i) *for every positive integer m, $W_{m,n} \xrightarrow{\mathcal{D}} Z_m \sim N(0, \tau(m)^2)$, where $\tau^2(m) = \lim_{n\to\infty} \mathrm{Var}\left[W_{m,n}\right]$,*

(ii) *for any $\delta > 0$, $\limsup_{n\to\infty} P\left[\left|W_{m,n} - W_n\right| > \delta\right] \to 0$ as $m \to \infty$,*

(iii) *$\tau^2(m)$ converges to a constant $\tau^2 > 0$ as $m \to \infty$.*

Let us briefly examine why this theorem is true. For any real number z and $\delta > 0$,

$$\begin{aligned} P[W_n \le z] &= P\left[W_n \le z, \left|W_n - W_{m,n}\right| \le \delta\right] \\ &\quad + P\left[W_n \le z, \left|W_n - W_{m,n}\right| > \delta\right] \\ &\le P\left[W_{m,n} \le z + \delta\right] + P\left[\left|W_n - W_{m,n}\right| > \delta\right]. \end{aligned}$$

Denoting $\Delta(m, \delta) = \limsup_{n\to\infty} P\left[\left|W_n - W_{m,n}\right| > \delta\right]$, we have

$$\begin{aligned} \limsup_{n\to\infty} P[W_n \le z] &\le \limsup_{n\to\infty} P\left[W_{m,n} \le z + \delta\right] + \Delta(m, \delta) \\ &= P[Z_m \le z + \delta] + \Delta(m, \delta). \end{aligned}$$

Since $\tau(m)^2 \to \tau^2$, letting $m \to \infty$ we have

$$\limsup_{n\to\infty} P[W_n \le z] \le P[Z \le z + \delta], \text{ and thus}$$
$$\limsup_{n\to\infty} P[W_n \le z] \le P[Z \le z],$$

since $\delta > 0$ is arbitrary.

A similar argument will show that

$$\limsup_{n\to\infty} P[W_n > z] \le P[Z > z], \text{ and hence}$$
$$\liminf_{n\to\infty} P[W_n \le z] \ge P[Z \le z].$$

Since

$$P[Z \le z] \le \liminf_{n\to\infty} P[W_n \le z] \le \limsup_{n\to\infty} P[W_n \le z] \le P[Z \le z],$$

we conclude that $\lim_{n\to\infty} P[W_n \le z] = P[Z \le z]$.

13.3.2 Estimation of Autocovariance and Autocorrelation Functions

Estimates of $\gamma(h) = \text{Cov}[X_t, X_{t+h}]$ and $\rho(h) = \text{Corr}\left[X_t, X_{t+h}\right]$, $h = 0, 1, \ldots$ are

$$\hat{\gamma}(h) = n^{-1} \sum_{t=1}^{n-h} (X_t - \bar{X})(X_{t+h} - \bar{X}) \quad \text{and} \quad \hat{\rho}(h) = \hat{\gamma}(h)/\hat{\gamma}(0). \tag{4}$$

This estimate of $\gamma(h)$ is not unbiased as will be clear soon. For notational convenience, we assume that $\text{E}[X_t] = 0$ since $X_t - \bar{X}$ is the same as $Y_t - \bar{Y}$, where $Y_t = X_t - \mu$ is the centered X_t. Simple algebra shows that

$$\begin{aligned}
\hat{\gamma}(h) &= n^{-1} \sum_{t=1}^{n-h} (X_t - \bar{X})(X_{t+h} - \bar{X}) \\
&= n^{-1} \sum_{t=1}^{n-h} X_t X_{t+h} - \bar{X} n^{-1} \sum_{t=1}^{n-h} X_{t+h} - \bar{X} n^{-1} \sum_{t=1}^{n-h} X_t + n^{-1}(n-h)\bar{X}^2 \\
&= n^{-1} \sum_{t=1}^{n-h} X_t X_{t+h} - (1 + h/n)\bar{X}^2 + \bar{X}\left[n^{-1} \sum_{t=1}^{h} X_t + n^{-1} \sum_{t=n-h+1}^{n} X_t\right].
\end{aligned}$$

We have already seen from the last section that $\text{Var}[\bar{X}] = \tau_n^2/n = O(n^{-1})$ assuming that $\sum|\gamma(h)| < \infty$. Since $\text{E}[X_t] = 0$, $\text{E}[\bar{X}^2] = \text{Var}[\bar{X}]$. It is not difficult to verify that the expectation of the last term in the last displayed equation is $O(n^{-3/2})$. Since $\text{E}\left[X_t X_{t+h}\right] = \gamma(h)$, we have

$$\begin{aligned}E[\hat{\gamma}(h)] &= n^{-1}(n-h)\gamma(h) + \tau_n^2/n + O(n^{-3/2})\\ &= \gamma(h) - (h/n)\gamma(h) + \tau_n^2/n + O(n^{-3/2}).\end{aligned}$$

Expression for τ_n^2/n given in the last section shows that $(h/n)\gamma(h)$ is not cancelled by τ_n^2/n and thus the bias of $\hat{\gamma}(h)$ is $O(h/n)$.

From the preceding arguments, we get the following simple result.

Lemma 13.3.1. *Let $\{X_t\}$ be stationary with the autocovariance function $\{\gamma(h)\}$ which satisfies the condition $\sum|\gamma(h)| < \infty$. Then*

$$\begin{aligned}\hat{\gamma}(h) &= n^{-1}\sum_{t=1}^{n-h}(X_t-\mu)(X_{t+h}-\mu) - (\bar{X}-\mu)^2 + R_1\\ &= n^{-1}\sum_{t=1}^{n}(X_t-\mu)(X_{t+h}-\mu) + R_2,\end{aligned}$$

where $E|R_1| = O(n^{-3/2})$ and $E|R_2| = O(n^{-1})$.

Why is the biased estimate of $\gamma(h)$ used in practice? The reason is that the $n \times n$ matrix $\hat{\boldsymbol{\Gamma}}_n = ((\hat{\gamma}(j-k)))_{n\times n}$, which is an estimate of the covariance matrix $\boldsymbol{\Gamma}_n$, is nonnegative definite. Had we used an unbiased estimate of $\gamma(h)$, then this nonnegative definiteness property is not guaranteed to be preserved. An unbiased estimate of $\gamma(h)$ is

$$\tilde{\gamma}(h) = (n-h-1)^{-1}\sum_{t=1}^{n-h}(X_t-\bar{X}_1)(X_{t+h}-\bar{X}_2), \text{ with}$$

$$\bar{X}_1 = (n-h)^{-1}\sum_{t=1}^{n-h}X_t \text{ and } \bar{X}_2 = (n-h)^{-1}\sum_{t=h+1}^{n}X_t.$$

Asymptotic distributions of the estimated autocovariances and autocorrelations are known. We first write down the joint distribution of $\hat{\rho}(1), \ldots, \hat{\rho}(h)$.

Theorem 13.3.3. *Let $\{X_t\}$ be a linear stationary series as given in Eq. (2) with the extra assumption $E[\varepsilon_t^4] < \infty$. Let $\boldsymbol{\rho}_h$ be the $h \times 1$ vector with elements $\rho(1), \ldots, \rho(h)$, and similarly let $\hat{\boldsymbol{\rho}}_h$ be the $h \times 1$ vector with elements $\hat{\rho}(1), \ldots, \hat{\rho}(h)$, where $\hat{\rho}(j)$'s are as in Eq. (4). Then $\sqrt{n}(\hat{\boldsymbol{\rho}} - \boldsymbol{\rho}) \xrightarrow{\mathcal{D}} N_h(\mathbf{0}, \mathbf{W})$ as $n \to \infty$, where element (j,k) of the matrix $\mathbf{W}$ is*

$$w_{jk} = \sum_{l=-\infty}^{\infty} e(j,l)e(k,l), \text{ with}$$

$$e(j,l) = [\rho(l+j) + \rho(l-j) - 2\rho(j)\rho(l)]\big/\sqrt{2}.$$

If $\{X_t\}$ are iid, then $\rho(j) = 0$ for any $j \geq 1$, $\mathbf{W}$ is the $h \times h$ identity matrix, and, $\sqrt{n}\hat{\rho}(j)$, $j = 1, \ldots, h$, are asymptotically iid $N(0,1)$ variables.

Note that if the series $\{X_t\}$ are iid, then $e(j,l)$ is equal to 1 if $l = j$ and is 0 otherwise. In such a case, the matrix $\mathbf{W}$ in the above theorem the identity. So if $\{X_t\}$ are iid, then $\sqrt{n}\hat{\rho}(j)$, $j = 1, 2, \ldots$ are asymptotically iid $N(0,1)$ rv's. The estimated autocorrelation function is widely used to check if a series is white noise, that is, to check if $\rho(h) = 0$ when $h \neq 0$. In practice, one usually plots $\hat{\rho}(h)$ against h to check if $\hat{\rho}(h)$ is outside the range $0 \pm 2/\sqrt{n}$ to investigate if $\rho(h) \neq 0$. Such a plot is called the ACF plot.

The proof of the above theorem on the asymptotic distribution of the estimated autocorrelations follows from the asymptotic normality of the estimated autocovariances and this is stated below.

Theorem 13.3.4. *Let $\{X_t\}$ be a linear stationary series as given in Eq. (2) with the extra assumption $E[\varepsilon_t^4] < \infty$. Let $\boldsymbol{\delta}_h$ be the $(h+1)$-dim vector with elements $\gamma(0), \ldots, \gamma(h)$ and, similarly, let $\hat{\boldsymbol{\delta}}_h$ be the vector of $\hat{\gamma}(0), \ldots, \hat{\gamma}(h)$. Then $\sqrt{n}(\hat{\boldsymbol{\delta}}_h - \boldsymbol{\delta}_h) \xrightarrow{\mathcal{D}} N_{h+1}(\mathbf{0}, \boldsymbol{V})$ where element (j,k) of $\boldsymbol{V}$ is*

$$\left[E(\varepsilon_t^4) - 3\sigma^4\right]\gamma(j)\gamma(k) + \sum_{l=-\infty}^{\infty} f(j,l)f(k,l), \text{ with}$$

$$\sigma^2 = E[\varepsilon_t^2],\ f(j,l) = \left[\gamma(l+j) + \gamma(l-j)\right]\Big/\sqrt{2}.$$

The proof of the above result uses the following steps and Theorem 13.3.2 given above. The details are quite long, and are given in the book by Brockwell and Davis [63].

As in the analysis of bias of $\hat{\gamma}(h)$, we may use the centered variable $X_t - \mu$, but we continue to denote it by X_t with the understanding that $E[X_t] = 0$.

I. Define $\tilde{\gamma}(j) = n^{-1}\sum_{t=1}^{n} X_t X_{t+j}$. Then, by Lemma 13.3.1, $\hat{\gamma}(j) = \tilde{\gamma}(j) + O_P(n^{-1})$ since $\bar{X} = O_P(n^{-1/2})$.

II. It is enough to prove the CLT for $\sum_{j=0}^{h} c_j \hat{\gamma}(j)$ where $c_0, \ldots, c_h$ are constants (Cramér-Wold device). Since $\hat{\gamma}(j) = \tilde{\gamma}(j) + O_P(n^{-1}), j = 0, \ldots, h$, it is enough to prove the CLT for $\sum_{j=0}^{h} c_j \tilde{\gamma}(j)$.

III. Denoting $S_t = X_t(c_0 X_t + \cdots + c_h X_{t+h})$, we have

$$\sum_{j=0}^{h} c_j \tilde{\gamma}(j) = n^{-1}\sum_{t=1}^{n} X_t(c_0 X_t + \cdots + c_h X_{t+h}) = n^{-1}\sum_{t=1}^{n} S_t.$$

IV. Let $X_{m,t} = \sum_{j=-m}^{m} a_j \varepsilon_{t-j}$, where m is a positive integer and

$$S_{m,t} = X_{m,t}(c_0 X_{m,t} + \cdots + c_h X_{m,t+h}).$$

Note that $\{X_{m,t}\}$ and $\{S_{m,t}\}$ are $2m$- and $(h+2m)$-dependent sequences of rv's, respectively. Moreover, $E[S_t]$ is the same for all t and the same is true for $E[S_{m,t}]$. Let

$$\bar{S}_n = n^{-1}\sum_{t=1}^{n} S_t, \quad \bar{S}_{m,n} = n^{-1}\sum_{t=1}^{n} S_{m,t},$$

$$\theta = E[S_t], \quad \theta_m = E[S_{m,t}],$$

$$W_n = \sqrt{n}\left(\bar{S}_n - \theta\right), \quad \text{and} \quad W_{m,n} = \sqrt{n}\left(\bar{S}_{m,n} - \theta_m\right).$$

V. For any positive integer m, show that $W_{m,n} \xrightarrow{\mathcal{D}} Z_m \sim N(0, \nu_m^2)$ as $n \to \infty$ (Theorem 3.4.3).

VI. Show that $\lim_{m\to\infty} \Delta(m,\delta) = 0$, where $\Delta(m,\delta) = \limsup_{n\to\infty} P\left[|W_n - W_{m,n}| > \delta\right]$.

VII. Show that $\nu_m \to \nu^2$ as $m \to \infty$, where $\nu^2 = \sum_{0\le j,k\le h} c_j c_k v_{jk}$ (v_{jk} is element (j,k) of the matrix $\boldsymbol{V}$).

13.3.3 Diagnostics

Diagnostic tools are often employed for an observed series $X_1,\ldots,X_n$ in order to investigate if there is trend, if the assumption of equal variance is reasonable, if the sequence is iid, if the assumption of normality is reasonable, etc. Before we discuss the diagnostic methods, let us introduce another descriptive measure called the partial autocorrelation function (PACF) which is commonly used in the analysis of stationary time series data. PACF of order h is the partial correlation between X_t and X_{t-h} given $X_{t-1},\ldots,X_{t-h+1}$. More formally, if $X_t^{(f)}$ is the best linear predictor of X_t based on $X_{t-1},\ldots,X_{t-h+1}$ and $X_{t-h}^{(b)}$ is the best linear predictor of X_{t-h} based on $X_{t-h+1},\ldots,X_{t-1}$, then the partial autocorrelation between X_t and X_{t-h} is $\pi(h) = \text{Corr}\left[X_t - X_t^{(f)}, X_{t-h} - X_{t-h}^{(b)}\right]$, $h \ge 2$. By definition, $\pi(1) = \rho(1)$. In a later section, partial autocorrelations will be discussed in detail along with appropriate formulas that can be used for computations. If a series is $AR(p)$, then $\pi(h) = 0$ when $h \ge p+1$, and if $\hat{\pi}(h)$ is the estimator of $\pi(h)$, then $\sqrt{n}\hat{\pi}(h)$ is approximately distributed as $N(0,1)$ for large n.

We now discuss some graphical and formal diagnostic procedures.

(a) Plot of the series against time reveals if there is a trend or if the assumption of equal variance (across time) is reasonable. In some cases, the problem of unequal variance can be remedied by an appropriate Box-Cox transformation of the observed series.

(b) In order to check if $\{X_t\}$ are iid, one may plot the estimated autocorrelations (the ACF plot) along with $0 \pm 2/\sqrt{n}$ bars in order to assess if the autocorrelations are close to zero. ACF plot of the estimated residuals $\{\hat{\varepsilon}_t\}$ after fitting an ARMA model can also be used to assess appropriateness of the model. For instance, if an $AR(p)$ model is fitted, the estimated residuals are

$$\hat{\varepsilon}_t = X_t - \bar{X} - \sum_{j=1}^{p} \hat{\phi}_j (X_{t-j} - \bar{X}), \quad t = p+1,\ldots,n,$$

where $\bar{X}$ is the sample mean, and $\hat{\phi}_1,\ldots,\hat{\phi}_p$ are the estimated autoregressive parameters. If $AR(p)$ provides an appropriate description of the data, then $\{\hat{\varepsilon}_t\}$ are approximately iid and its ACF plot would indicate this. Similar logic can be used when fitting an $MA(q)$ or an $ARMA(p,q)$ model to check its adequacy.

(c) There are a number of formal tests for checking if the observations are iid. We may want to test, for a given positive integer h, H_0: $\rho(1) = \cdots = \rho(h) = 0$ against H_1: at least one of $\rho(1), \ldots, \rho(h)$ is nonzero. We mention two tests: Portmanteau and Ljung-Box. The test statistics are

$$Q = n\sum_{j=1}^{h} \hat{\rho}(j)^2 \quad \text{and} \quad Q_{LB} = n(n+2)\sum_{j=1}^{h} \hat{\rho}(j)^2/(n-j),$$

where $\{\hat{\rho}(j)\}$ are the estimated autocorrelations as given in Eq. (4). Under H_0, each of the two statistics (Q and Q_{LB}) has an approximate chi-squared distribution with h degrees of freedom. So we can reject the null hypothesis at level α if $Q > \chi^2_{h,\alpha}$ or $Q_{LB} > \chi^2_{h,\alpha}$, where $\chi^2_{h,\alpha}$ is the $(1-\alpha)$-quantile of the chi-squared distribution with h degrees of freedom. These tests are based on the asymptotic result given in Theorem 13.3.3. According to this theorem, if the series consists of iid observations, then $\sqrt{n}\hat{\rho}(j), j = 1, \ldots, h$, are approximately iid $N(0, 1)$ for large n. Ljung-Box test provides a small sample correction to the Portmanteau test by obtaining a better estimate of the asymptotic variance of $\hat{\rho}(j)$. In Portmanteau test, $\text{Var}[\hat{\rho}(j)]$ is approximated by $1/n$, whereas Ljung-Box uses the approximation $(n-j)/[n(n+2)]$.

(d) One can examine the histogram of the data to check if the assumption of normality is justifiable.

(e) A plot of $\hat{\pi}(h)$ against h along with $0 \pm 2/\sqrt{n}$ bars is known as the PACF plot and it can be used to make an assessment if $\pi(h)$ is substantially different from zero. This plot is useful in guessing the order of an autoregressive model since $\pi(h) = 0, h \geq p+1$, for an $AR(p)$ model. It is a common practice to use ACF and PACF plots in the initial analysis of the data.

It is useful to keep in mind that for an $MA(q)$ model, autocorrelations of lag $q+1$ or higher are all zero. Similarly, for an $AR(p)$ model, partial autocorrelations of order $p+1$ or higher are zero. So the ACF plot is useful for an initial guess of the order of a moving average model and the PACF plot is useful in guessing the order of an autoregressive model. For instance, if the autocorrelations of lag 3 or higher are all negligible, then an $MA(2)$ model may provide a reasonable description of the data. If the partial autocorrelations of lag 3 or higher are negligible, then $AR(2)$ may be a reasonable model for the data. It should be pointed, however, out that if the true model is $MA(q)$, the asymptotic mean of $\hat{\rho}(h)$ is 0 for $h \geq q+1$, but the asymptotic variance of $\hat{\rho}(h)$ is $\sum_{j=-q}^{q} \rho(j)^2/n$ and not $1/n$ and thus the $\pm 2/\sqrt{n}$ bounds are not necessarily equal to $\pm 2SE[\hat{\rho}(h)]$. Nevertheless, the ACF plot can be a useful graphical method for assessing if some moving average model is reasonable.

The ACF of a stationary autoregressive series, and under the condition of invertibility (discussed later), the PACF of a moving average series decrease rapidly with lag h. In general, for an ARMA series both the ACF and PACF decrease rapidly. The following table provides a summary.

Model	AR(p)	MA(q)	ARMA(p, q)
ACF	Tails off	Cuts off after lag q	Tails off
PACF	Cuts off after lag p	Tails off	Tails off

In order to get a good predictive model, one should use a criterion such as AIC to select an appropriate model. Nevertheless, the use of ACF and PACF plots may sometimes lead to a reasonable predictive model.

13.3.4 Notation of Backshift Operator

The notation of backshift operator makes the description of ARMA processes convenient. The backshift operator B is defined as $BX_t = X_{t-1}$. Note that $X_{t-k} = B^k X_t$ for any k with the understanding that $B^0 X_t = X_t$. A mean zero $AR(1)$ process can be written as

$$X_t = \phi_1 BX_t + \varepsilon_t, \text{ or } X_t - \phi_1 BX_t = \varepsilon_t, \text{ or } (1 - \phi_1 B)X_t = \varepsilon_t.$$

Similarly, a mean zero $AR(p)$ process can be written as

$$(1 - \phi_1 B - \cdots - \phi_p B^p)X_t = \varepsilon_t, \text{ or } \phi(B)X_t = \varepsilon_t,$$
$$\text{with } \phi(B) = 1 - \phi_1 B - \cdots - \phi_p B^p.$$

If a series $\{X_t\}$ with mean μ is $AR(p)$, then

$$\phi(B)(X_t - \mu) = \varepsilon_t, \text{ where } \phi(B) = 1 - \phi_1 B - \cdots - \phi_p B^p.$$

Similarly, an $MA(q)$ series $\{X_t\}$ with mean μ can be expressed as

$$X_t - \mu = \varepsilon_t + \theta_1 \varepsilon_{t-1} + \cdots + \theta_q \varepsilon_{t-q} = \theta(B)\varepsilon_t, \text{ where}$$
$$\theta(B) = 1 + \theta_1 B + \cdots + \theta_q B^q.$$

An $ARMA(p, q)$ series $\{X_t\}$ with mean μ can be written as

$$\phi(B)(X_t - \mu) = \theta(B)\varepsilon_t.$$

As mentioned earlier, the error terms $\{\varepsilon_t\}$ are often called *innovations* in the time series literature.

13.4 Partial Autocorrelation Function (PACF)

As discussed in the last section, the partial autocorrelation (PACF) plot is widely used in the analysis of stationary data as it aids in the preliminary identification of autoregressive models. Let us recall that, for any $h \geq 2$, partial autocorrelation $\pi(h)$ of order h of a stationary series $\{X_t\}$ is defined to be the partial correlation between X_t and X_{t-h} given $X_{t-1}, \ldots, X_{t-h+1}$. When $h = 1$, $\pi(1)$ is defined to be $\mathrm{Corr}[X_t, X_{t-1}]$. If $\hat{X}^{(f)}_{h-1,t}$ is the best linear predictor (forecast) of X_t based on $X_{t-1}, \ldots, X_{t-h+1}$ and $\hat{X}^{(b)}_{h-1,t-h}$ is the best linear predictor (backcast) of X_{t-h} based on $X_{t-h+1}, \ldots, X_{t-1}$, then for any $h \geq 2$, the partial autocorrelation of order h is

$$\pi(h) = \text{Corr}\left[X_t - \hat{X}^{(f)}_{h-1,t}, X_{t-h} - \hat{X}^{(b)}_{h-1,t-h}\right].$$

In this section, we discuss partial autocorrelations in some detail. We show that $\pi(h) = \phi_{h,h}$, where $\phi_{h,h}$ is the coefficient associated with X_{t-h} in the expression of the best linear predictor of X_t based on $X_{t-1}, \ldots, X_{t-h}$. As a by-product of the discussion, we also obtain an important recursion formula (known as *Durbin-Levinson* recursions) which relates the autoregressive coefficients of the *AR*(h) fit to those of *AR*($h-1$) fit.

We first write down a basic result.

Lemma 13.4.1. *If* $\{X_t\}$ *is a stationary AR(p) series, then* $\pi(h) = 0$ *for* $h \geq p+1$.

It is fairly easy to see why this lemma is true. Assume that $E[X_t] = 0$. Since the series is assumed to be *AR*(p) with autoregressive coefficients $\phi_1, \ldots, \phi_p$, for any $h \geq p+1$, the best linear predictor of X_t based on $X_{t-1}, \ldots, X_{t-h+1}$ is $\hat{X}^{(f)}_{h-1,t} = \phi_1 X_{t-1} + \cdots + \phi_p X_{t-p}$ and $X_t - \hat{X}^{(f)}_{h-1,t} = \varepsilon_t$. If $\hat{X}^{(b)}_{h-1,t-h}$ is the best linear predictor of X_{t-h} based on $X_{t-h+1}, \ldots, X_{t-1}$, then the remainder $X_{t-h} - \hat{X}^{(b)}_{h-1,t-h}$ is a linear function of $X_{t-1}, \ldots, X_{t-h}$. Since ε_t is uncorrelated with $X_{t-1}, \ldots, X_{t-h}$, we can conclude that $X_t - \hat{X}_{h-1,t} = \varepsilon_t$ is uncorrelated with $X_{t-h} - \hat{X}^{(b)}_{h-1,t-h}$ and thus $\pi(h) = 0$.

13.4.1 Expression for $\pi(h)$ and Durbin-Levinson Iterations

Let $\{X_t\}$ be a mean zero stationary series which is not necessarily an autoregressive series of any finite order. Let $\phi_{h,1}, \ldots, \phi_{h,h}$ be the coefficients of the best linear predictor of X_t based on $X_{t-1}, \ldots, X_{t-h}$ (ie, $\hat{X}^{(f)}_{h,t} = \phi_{h,1} X_{t-1} + \cdots + \phi_{h,h} X_{t-h}$). The arguments here are given in terms of the theoretical autocovariances which can be replaced by their sample estimates for numerical computations based on the available data.

By Lemma 13.2.1

$$\hat{X}^{(f)}_{h-1,t} = \phi_{h-1,1} X_{t-1} + \cdots + \phi_{h-1,h-1} X_{t-h+1},$$
$$\hat{X}^{(b)}_{h-1,t-h} = \phi_{h-1,1} X_{t-h+1} + \cdots + \phi_{h-1,h-1} X_{t-1},$$

and the mean square error in forecasting X_t by $\hat{X}^{(f)}_{h-1,t}$ is the same as the error in backcasting X_{t-h} by $\hat{X}^{(b)}_{h-1,t-h}$. This common prediction error is

$$\begin{aligned} PE(h-1) &= E\left[X_t - \hat{X}^{(f)}_{h,t}\right]^2 = E\left[X_{t-h} - \hat{X}^{(b)}_{h-1,t-h}\right]^2 \\ &= \gamma(0) - \left[\phi_{h-1,1}\gamma(1) + \cdots + \phi_{h-1,h-1}\gamma(h)\right]. \end{aligned}$$

The identity $\pi(h) = \phi_{h,h}$ and Durbin-Levinson iterative formula follow by equating the coefficients of two different but equivalent expressions of the best linear predictor of X_t based on $X_{t-1}, \ldots, X_{t-h}$. This basic idea is detailed the following observations.

(i) Any linear predictor of X_t based on $X_{t-1}, \ldots, X_{t-h}$ can be expressed as

$$L = a_1 X_{t-1} + \cdots + a_{h-1} X_{t-h+1} + a_h \delta_{h-1,t-h},$$

where $\delta_{h-1,t-h} = X_{t-h} - \hat{X}^{(b)}_{h-1,t-h}$.

(ii) Since $\delta_{h-1,t-h}$ is uncorrelated with $X_{t-1}, \ldots, X_{t-h+1}$, the prediction error $\mathrm{E}[X_t - L]^2$, where L is as in (i), is minimized at

$$\begin{aligned} a_1^* &= \phi_{h-1,1}, \ldots, a_{h-1}^* = \phi_{h-1,h-1}, \text{ and} \\ a_h^* &= \mathrm{Cov}[X_t, \delta_{h-1,t-h}]/\mathrm{Var}[\delta_{h-1,t-h}] \\ &= \mathrm{Cov}\left[X_t - \hat{X}^{(f)}_{h-1,t}, X_{t-h} - \hat{X}^{(b)}_{h-1,t-h}\right] \Big/ \mathrm{Var}\left[X_{t-h} - \hat{X}^{(b)}_{h-1,t-h}\right] \\ &= \mathrm{Corr}\left[X_t - \hat{X}^{(f)}_{h-1,t}, X_{t-h} - \hat{X}^{(b)}_{h-1,t-h}\right] = \pi(h), \end{aligned}$$

where the last two steps follow from the fact that
$\mathrm{E}\left[X_t - \hat{X}^{(f)}_{h-1,t}\right]^2 = \mathrm{E}\left[X_{t-h} - \hat{X}^{(b)}_{h-1,t-h}\right]^2$.

(iii) The coefficients of the best linear predictor $\hat{X}^{(f)}_{h,t}$ are given in (ii) above, and using Lemma 13.2.1 we have

$$\begin{aligned} \hat{X}^{(f)}_{h,t} &= a_1^* X_{t-1} + \cdots + a_{h-1}^* X_{t-h+1} + a_h^* \delta_{h-1,t-h} \\ &= \phi_{h-1,1} X_{t-1} + \cdots + \phi_{h-1,h-1} X_{t-h+1} \\ &\quad + \pi(h)\left[X_{t-h} - \phi_{h-1,1} X_{t-h+1} + \cdots + \phi_{h-1,h-1} X_{t-1}\right] \\ &= [\phi_{h-1,1} - \pi(h)\phi_{h-1,h-1}] X_{t-1} + \cdots \\ &\quad + [\phi_{h-1,h-1} - \pi(h)\phi_{h-1,1}] X_{t-h+1} + \pi(h) X_{t-h}. \end{aligned}$$

(iv) The expression for the best linear predictor $\hat{X}^{(f)}_{h,t}$ of X_t based on $X_{t-1}, \ldots, X_{t-h}$ is

$$\phi_{h,1} X_{t-1} + \cdots + \phi_{h,h-1} X_{t-h+1} + \phi_{h,h} X_{t-h},$$

and this should be the same as $\hat{X}^{(f)}_{h,t}$ in (iii). Equating the coefficients of $X_{t-1}, \ldots, X_{t-h}$ in these two expressions, we have

$$\begin{aligned} \phi_{h,h} &= \pi(h) = a_h^* \\ &= \mathrm{Cov}[X_t, \delta_{h-1,t-h}]/\mathrm{Var}[\delta_{h-1,t-h}] \\ &= \frac{\gamma(h) - \phi_{h-1,1}\gamma(h-1) - \cdots - \phi_{h-1,h-1}\gamma(1)}{\gamma(0) - \phi_{h-1,1}\gamma(1) - \cdots - \phi_{h-1,h-1}\gamma(h-1)}, \text{ and} \\ \phi_{h,j} &= \phi_{h-1,j} - \phi_{h,h}\phi_{h-1,h-j}, \quad j = 1, \ldots, h-1. \end{aligned}$$

From Step (iv), we have $\phi_{h,h} = \pi(h)$. The expressions in (iv), which relate the $AR(h)$ coefficients to $AR(h-1)$, provide a recursion formula for calculating the autoregressive coefficients, which is known as the *Durbin-Levinson* algorithm. In order to obtain the estimate $\hat{\pi}(h)$ of $\pi(h)$, it is enough to fit an $AR(h)$ model and solve the Yule-Walker equations with $\{\hat{\gamma}(j)\}$ in place of $\{\gamma(j)\}$. Also note that the Durbin-Levinson algorithm provides an iterative scheme for solving the Yule-Walker equations starting from AR(1). We thus arrive at the following important results.

Theorem 13.4.1. *If $\{X_t\}$ is a mean zero stationary series, and if $\phi_{h,1}X_{t-1} + \cdots + \phi_{h,h}X_{t-h}$ is the best linear predictor of X_t based on $X_{t-1}, \ldots, X_{t-h}$, then the partial autocorrelation of order $h \geq 2$ is given by $\pi(h) = \phi_{h,h}$. If $\{X_t\}$ is stationary AR(p), then $\sqrt{n}\hat{\pi}(h) \xrightarrow{\mathcal{D}} N(0,1)$ as $n \to \infty$ for $h \geq p+1$.*

Theorem 13.4.2 (Durbin-Levinson Iterations). *For a mean zero stationary series $\{X_t\}$, if $\phi_{h,1}X_{t-1} + \cdots + \phi_{h,h}X_{t-h}$ is the best linear predictor of X_t based on $X_{t-1}, \ldots, X_{t-h}$, then the coefficients $\phi_{h,1}, \ldots, \phi_{h,h}$ are related to those of the best linear predictor of X_t based on $X_{t-1}, \ldots, X_{t-h-1}$ as given in observation (iv) above.*

We now provide a justification of the result on the asymptotic normality of $\hat{\pi}(h)$. We will see in Section 13.8.1 that when $\{X_t\}$ is *AR(p)*, then $\sqrt{n}(\hat{\boldsymbol{\phi}}_h - \boldsymbol{\phi}_h) \xrightarrow{\mathcal{D}} N_h(\mathbf{0}, \sigma^2\boldsymbol{\Gamma}_h^{-1})$, $h \geq p+1$, where $\boldsymbol{\phi}_h = \boldsymbol{\Gamma}_h^{-1}\boldsymbol{\gamma}_h$ and $\hat{\boldsymbol{\phi}}_h$ is the solution of the Yule-Walker equations $\hat{\boldsymbol{\Gamma}}_h\boldsymbol{\phi} = \hat{\boldsymbol{\gamma}}_h$. Thus, $\sqrt{n}(\hat{\pi}(h) - \pi(h)) \xrightarrow{\mathcal{D}} N\left(0, \sigma^2\boldsymbol{\Gamma}_h^{-1}(h,h)\right)$, where $\boldsymbol{\Gamma}_h^{-1}(h,h)$ is the last element of the matrix $\boldsymbol{\Gamma}_h^{-1}$. From matrix algebra it is known that

(i) $\boldsymbol{\Gamma}_h^{-1}(h,h) = |\boldsymbol{\Gamma}_{h-1}|/|\boldsymbol{\Gamma}_h|$, where $|\cdot|$ denotes the determinant,

(ii) $\boldsymbol{\Gamma}_h = \begin{pmatrix} \boldsymbol{\Gamma}_{h-1} & \boldsymbol{\gamma}_{h-1} \\ \boldsymbol{\gamma}_{h-1}^T & \gamma(0) \end{pmatrix}$ and $|\boldsymbol{\Gamma}_h| = |\boldsymbol{\Gamma}_{h-1}|\left[\gamma(0) - \boldsymbol{\gamma}_{h-1}^T\boldsymbol{\Gamma}_{h-1}^{-1}\boldsymbol{\gamma}_{h-1}\right]$,

(iii) $\boldsymbol{\Gamma}_h^{-1}(h,h) = 1\Big/\left[\gamma(0) - \boldsymbol{\gamma}_{h-1}^T\boldsymbol{\Gamma}_{h-1}^{-1}\boldsymbol{\gamma}_{h-1}\right]$, and

(iv) from Lemma 13.2.1, $\gamma(0) - \boldsymbol{\gamma}_{h-1}^T\boldsymbol{\Gamma}_{h-1}^{-1}\boldsymbol{\gamma}_{h-1} = \mathrm{E}\left[X_t - X_{h-1,t}^{(f)}\right]^2$.

If $\{X_t\}$ is stationary *AR(p)*, then $X_{h-1,t}^{(f)} = X_{p,t}^{(f)}$ when $h \geq p+1$ and

$$\mathrm{E}\left[X_t - X_{h-1,t}^{(f)}\right]^2 = \mathrm{E}\left[X_t - X_{p,t}^{(f)}\right]^2 = \gamma(0) - \boldsymbol{\gamma}_p^T\boldsymbol{\Gamma}_p^{-1}\boldsymbol{\gamma}_p,$$

Stationarity of $\{X_t\}$ implies

$$\begin{aligned} \gamma(0) &= \mathrm{Var}[X_t] = \mathrm{Var}\left[\phi_1X_{t-1} + \cdots + \phi_pX_{t-p} + \varepsilon_t\right] \\ &= \mathrm{Var}\left[\phi_1X_{t-1} + \cdots + \phi_pX_{t-p}\right] + \mathrm{Var}[\varepsilon_t] \\ &= \boldsymbol{\phi}^T\boldsymbol{\Gamma}_p\boldsymbol{\phi} + \sigma^2 = \boldsymbol{\gamma}_p^T\boldsymbol{\Gamma}_p^{-1}\boldsymbol{\gamma}_p + \sigma^2, \text{ and hence} \\ \sigma^2 &= \gamma(0) - \boldsymbol{\gamma}_p^T\boldsymbol{\Gamma}_p^{-1}\boldsymbol{\gamma}_p, \text{ and} \end{aligned}$$

$$\sigma^2\boldsymbol{\Gamma}_h^{-1}(h,h) = \sigma^2\Big/\left[\gamma(0) - \boldsymbol{\gamma}_{h-1}^T\boldsymbol{\Gamma}_{h-1}^{-1}\boldsymbol{\gamma}_{h-1}\right] = 1$$

for any $h \geq p+1$. This shows that $\sqrt{n}\hat{\pi}(h) \xrightarrow{\mathcal{D}} N(0,1)$ as $n \to \infty$ for $h \geq p+1$.

13.5 Causality and Invertibility

This section addresses a number of important mathematical issues and theoretical results, but their proofs are not given. For theoretical details including proofs, the readers may consult the book by Brockwell and Davis [63].

A series $\{X_t\}$ is called causal if it has an $MA(\infty)$ representation, that is

$$X_t - \mu = \sum_{j=0}^{\infty} \psi_j \varepsilon_{t-j}, \text{ with } \psi_0 = 1, \tag{5}$$

where $\{\varepsilon_t\}$ are mean zero iid with common variance σ^2 and $\sum |\psi_j| < \infty$. Any causal series is stationary since for any $h \geq 0$,

$$\text{Cov}[X_t, X_{t+h}] = \sigma^2 \sum_{j=0}^{\infty} \psi_j \psi_{j+h} = \gamma(h),$$

depends only on h. In general, explicit expressions for these ψ weights are difficult to obtain (unless it is an $MA(q)$ model) even though iterative formulas are available. Fortunately, packages such as R will calculate these ψ coefficients for an $ARMA(p,q)$ model.

A series $\{X_t\}$ is invertible if it can be written as an $AR(\infty)$ series, that is, it has the representation

$$X_t - \mu = \sum_{j=1}^{\infty} \pi_j (X_{t-j} - \mu) + \varepsilon_t, \tag{6}$$

where $\{\varepsilon_t\}$ are iid with mean 0 and common variance σ^2. Clearly, any $AR(p)$ model is invertible. However, an invertible model need not be stationary. For instance if $X_t = X_{t-1} + \varepsilon_t$, where $\{\varepsilon_t\}$ are iid mean 0 with variance σ^2, then $\{X_t\}$ is invertible but is not stationary (Example 13.2.5). Except for autoregressive models, there are no simple expressions for $\{\pi_j\}$ for $MA(q)$ or $ARMA(p,q)$ models. However, a computing package such as R can be used to obtain them.

Mathematical conditions for invertibility and causality will be discussed later. An invertible expression is useful for obtaining the forecast values, whereas a causal representation makes it easy to obtain the variance of forecasts.

Example 13.5.1. A mean zero $AR(1)$ series $\{X_t\}$ may be rewritten as

$$\begin{aligned}
X_t &= \phi X_{t-1} + \varepsilon_t = \phi(\phi X_{t-2} + \varepsilon_{t-1}) + \varepsilon_t \\
&= \phi^2 X_{t-2} + \phi \varepsilon_{t-1} + \varepsilon_t \\
&= \phi^2 (\phi X_{t-3} + \varepsilon_{t-2}) + \phi \varepsilon_{t-1} + \varepsilon_t \\
&= \phi^3 X_{t-3} + \phi^2 \varepsilon_{t-2} + \phi \varepsilon_{t-1} + \varepsilon_t.
\end{aligned}$$

We may repeat this argument to get

$$X_t = \phi^r X_{t-r} + \phi^{r-1}\varepsilon_{t-r+1} + \cdots + \phi\varepsilon_{t-1} + \varepsilon_t,$$

for any positive integer r. If $|\phi| < 1$, then $\phi^r \to 0$ and hence $\phi^r X_{t-r} \xrightarrow{P} 0$ as $r \to \infty$. Thus we may reexpress X_t as

$$X_t = \sum_{j=0}^{\infty} \phi^j \varepsilon_{t-j},$$

which is a causal representation of the series $\{X_t\}$ with $\psi_j = \phi^j$.

Example 13.5.2. A mean zero $MA(1)$ series may be rewritten as

$$\varepsilon_t = X_t - \theta\varepsilon_{t-1} = X_t - \theta(X_{t-1} - \theta\varepsilon_{t-2}) = X_t - \theta X_{t-1} + \theta^2\varepsilon_{t-2}.$$

Repeating this argument we have, for any positive integer r,

$$\varepsilon_t = X_t - \theta X_{t-1} + \cdots + (-\theta)^r X_{t-r} + (-\theta)^{r+1}\varepsilon_{t-r-1}.$$

If $|\theta| < 1$, then $(-\theta)^{r+1}\varepsilon_{t-r-1} \xrightarrow{P} 0$ as $r \to \infty$ and thus we have an invertible representation for the $MA(1)$ series

$$\varepsilon_t = X_t - \theta X_{t-1} + \theta^2 X_{t-2} + \cdots, \text{ and hence}$$

$$X_t = \theta X_{t-1} - \theta^2 X_{t-2} + \cdots + \varepsilon_t = \sum_{j=1}^{\infty} \pi_j X_{t-j} + \varepsilon_t,$$

with $\pi_j = (-1)^{j-1}\theta^j$.

Example 13.5.3. Consider an $ARMA(2,2)$ model with $\phi_1 = 0.8$, $\phi_2 = -0.15$, $\theta_1 = 0.6$, $\theta_2 = 0.08$. Using R, we have obtained the first 12 values of $\{\psi_j\}$ starting with $\psi_1 = 1.400$ are

$$[1.400, 1.050, 0.630, 0.347, 0.184, 0.094, 0.050, 0.024, 0.012, 0.006, 0.003, 0.002].$$

Note that the values of ψ_j become small for large j and this is typical for any ARMA series satisfying the condition of stationarity as will be seen later. Using R, we can get an invertible expression, and the first 12 values of $\{\pi_j\}$ starting with $\pi_1 = 1.400$ are

$$[1.400, -0.910, 0.434, -0.188, 0.078, -0.032, 0.013, -0.005, 0.002, -0.001, 0.0003, -0.0001].$$

The values of π_j rapidly approach 0 as j increases.

13.5.1 Usefulness of Invertible Representation

If we know the weights $\{\pi_j\}$, then it is easy to do the forecasting from the data $\{X_1, \ldots, X_n\}$ assuming that $\pi_j \to 0$ as $j \to \infty$, which is generally true for ARMA processes under appropriate conditions. For notational simplicity let us assume that $\mu = 0$. Then we can write this series as

$$X_t = \varepsilon_t + \pi_1 X_{t-1} + \pi_2 X_{t-2} + \cdots.$$

If we have the entire past $\{X_t: -\infty < t \leq n\}$, then the forecasted value of X_{n+1} is

$$\hat{X}_{n+1} = \pi_1 X_n + \pi_2 X_{n-1} + \pi_3 X_{n-2} + \cdots,$$

If X_{n+1} were known, then forecast of X_{n+2} would be

$$\pi_1 X_{n+1} + \pi_2 X_n + \pi_3 X_{n-1} + \cdots.$$

Since X_{n+1} is unknown, then we can substitute it by $\hat{X}_{n+1}$ (justification given in Lemma 13.6.1), leading to the forecasted value of X_{n+2} as

$$\hat{X}_{n+2} = \pi_1 \hat{X}_{n+1} + \pi_2 X_n + \pi_3 X_{n-1} + \cdots.$$

This method can now be replicated to forecast X_{n+3}, X_{n+4}, etc.

Typically μ is not equal to zero, but obtaining the forecasts is not difficult with the known π values. For instance, the forecasting formula for X_{n+1} is given by

$$\hat{X}_{n+1} - \mu = \pi_1(X_n - \mu) + \pi_2(X_{n-1} - \mu) + \pi_3(X_{n-2} - \mu) + \cdots.$$

These expressions for forecasts assume that the entire past $\{X_t: -\infty < t \leq n\}$ is known. However, if $\pi_j \to 0$ rapidly as $j \to \infty$, which is the case for ARMA processes under the condition of invertibility, then the terms involving X_t, $t \leq 0$, may be ignored since the associated π-coefficients are negligible and thus the approximate forecasts are linear functions of the available data $X_1, \ldots, X_n$.

13.5.2 Usefulness of Causal Representation

Any practical approach to forecasting is incomplete without addressing the issue of prediction limits (or prediction intervals) of these forecasts. If the observed series is $\{X_1, \ldots, X_n\}$, then the h step ahead forecast is denoted by $\hat{X}_{n+h}$. The forecast error is $X_{n+h} - \hat{X}_{n+h}$, which is not known since X_{n+h} is unknown. For all the cases we consider in this chapter, the mean of the forecast error is equal to zero (or close to zero when the parameters of the model are estimated). The mean square error of the forecast error is denoted by

$$\begin{aligned}\sigma^2(h) &= \text{Var}[(X_{n+h} - \hat{X}_{n+h})|X_1, \ldots, X_n] \\ &= \text{E}[(X_{n+h} - \hat{X}_{n+h})^2|X_1, \ldots, X_n].\end{aligned}$$

Thus a prediction interval for X_{n+h} with confidence coefficient $1 - \alpha$ is $\hat{X}_{n+1} \pm z_{\alpha/2}\sigma(h)$ if the series is assumed to be stationary Gaussian.

13.5.3 Important Technical issues

Here is a summary of the technical issues for $AR(p)$, $MA(q)$, and $ARMA(p, q)$ models which will be discussed in this section.

1. For $AR(p)$ models, we need conditions on the autoregressive coefficients $\phi_1, \ldots, \phi_p$ in order to guarantee stationarity.

2. Moving average models are not in general identifiable. Thus for $MA(q)$ models, we need conditions on the moving average coefficients $\theta_1, \ldots, \theta_q$ in order to guarantee identifiability.
3. For $ARMA(p, q)$ models,
   (i) the coefficients in the AR part must satisfy constrains to guarantee stationarity (as in (1) above),
   (ii) the MA coefficients must satisfy constraints in order to guarantee identifiability of the model (as in (2) above),
   (iii) conditions on the AR and the MA coefficients are needed in order to guarantee "nonredundancy." This issue will be discussed in soon.

Nonuniqueness of Moving Average Models

Moving average models are not unique. For instance, consider an $MA(1)$ series $X_t - \mu = \varepsilon_t + \theta\varepsilon_{t-1}$, where $\{\varepsilon_t\}$ is white noise with variance σ^2. Under the assumption of normality (ie, $\{\varepsilon_t\}$ are normally distributed), any stationary series is completely characterized by the mean and autocovariances. Thus if two sequences have the same mean and autocovariance functions, they are equally good descriptions of the data, that is, they provide the same fit and they have the same predictive performances. Consider the following two models

$$X_t - \mu = \varepsilon_t + \theta\varepsilon_{t-1}, \quad X_t - \mu = \varepsilon'_t + (1/\theta)\varepsilon'_{t-1},$$

where $\theta \neq 0$, $\{\varepsilon_t\}$ is white noise with variance σ^2, and $\{\varepsilon'_t\}$ is white noise with variance $\theta^2\sigma^2$. Note that we only observe the data $\{X_t\}$, not $\{\varepsilon_t\}$ or $\{\varepsilon'_t\}$. Both models have the same mean μ. All the autocovariances of lag 2 or higher are zero for both models. For the first model

$$\gamma(0) = (1+\theta^2)\sigma^2, \; \gamma(1) = \theta\sigma^2, \text{ and } 0 = \gamma(2) = \gamma(3) = \cdots.$$

For the second model,

$$\gamma(0) = [1 + (1/\theta)^2](\theta^2\sigma^2) = (1+\theta^2)\sigma^2,$$
$$\gamma(1) = (1/\theta)(\theta^2\sigma^2) = \theta\sigma^2, \text{ and } 0 = \gamma(2) = \gamma(3) = \cdots.$$

So both the models have identical mean and autocovariance structures. Hence they will provide identical fits and predictions. This nonuniqueness, or lack of identifiability, is problematic since there are multiple "correct" models.

How is this issue resolved? If the value of θ is larger than 1 in magnitude and $\mathrm{Var}[\varepsilon_t] = \sigma^2$, then we may as well consider the second model for which coefficient associated with ε_{t-1} is $1/\theta$ whose magnitude is less than 1. Similarly if the moving average coefficient $(1/\theta)$ of the second model is larger than 1 in magnitude, then we may decide to use the first model for which the coefficient would be smaller than 1 in magnitude. Thus we can always choose a model whose moving average coefficient is no larger than 1 in magnitude and this is what is done in practice.

The same issue of nonidentifiability comes up for general $MA(q)$ models. One can restrict attention to those models with appropriate conditions on the MA parameters $\theta_1, \ldots, \theta_q$ needed for identifiably and this is what is usually done.

Redundancy Issue for *ARMA*(*p*, *q*) Models

If a series $\{X_t\}$ is white noise (ie, $X_t = \varepsilon_t$, where $\{\varepsilon_t\}$ is white noise), then subtracting $0.5X_{t-1} = 0.5\varepsilon_{t-1}$ from this series we have

$$X_t - 0.5X_{t-1} = \varepsilon_t - 0.5\varepsilon_{t-1}, \quad \text{ie, } X_t = 0.5X_{t-1} + \varepsilon_t - 0.5\varepsilon_{t-1}.$$

Now it seems that the series $\{X_t\}$ is $ARMA(1, 1)$, whereas in reality it is a white noise. As a matter of fact we can rewrite X_t as

$$X_t = \phi X_{t-1} + \varepsilon_t - \phi\varepsilon_{t-1},$$

for any $-1 < \phi < 1$. Once again it looks as if $\{X_t\}$ is $ARMA(1, 1)$; however, there is a redundant parameter ϕ. Also note that the number of such redundant models is infinite. In general if $\theta = -\phi$ in an $ARMA(1, 1)$ model $X_t = \phi X_{t-1} + \varepsilon_t + \theta\varepsilon_{t-1}$, then there is redundancy. When $\theta \neq \phi$ then this redundancy is no longer present. The same issue needs to be addressed for general ARMA models and constraints on the parameters are needed in order to avoid any redundancy.

Condition for Stationarity for *AR*(*p*) Models

If $\{X_t\}$ follows an $AR(p)$ model, then it is stationary (and causal) if it can be written in the form Eq. (5). Conditions on the autoregressive coefficients $\phi_1, \ldots, \phi_p$ are needed to guarantee that the series can be written in this form. Let $\phi(z) = 1 - \phi_1 z - \cdots - \phi_p z^p$ be a polynomial in z, where z is complex. This polynomial has p roots which can be real or complex valued.

Lemma 13.5.1. *If the absolute values of all the roots of the polynomial $\phi(z)$ of an $AR(p)$ series are larger than 1, then the series is stationary.*

When $p = 1$, the root of the polynomial is $1/\phi_1$. The condition that the absolute value of $1/\phi_1$ is larger than 1 is equivalent to the condition $-1 < \phi_1 < 1$. In this case, the autocorrelation function is $\rho(h) = \phi_1^h$, $h = 0, 1, \ldots$, and $\rho(h)$ converges to zero exponentially fast as h increases (Example 13.2.3).

For $p = 2$, the condition on ϕ_1 and ϕ_2 for stationarity is a bit more complicated. The condition is: $-1 < \phi_2 < 1$ and $-1 < \phi_1/(1 - \phi_2) < 1$, that is, (ϕ_1, ϕ_2) is inside the triangle $\Delta = \{(u_1, u_2): \; -1 < u_2 < 1, -1 < u_1/(1 - u_2) < 1\}$. The roots of $\phi(z)$ are reciprocals of the roots of $g(z) = z^2 - \phi_1 z - \phi_2$. The roots of $g(z)$ are $(1/2)\left[\phi_1 \pm \sqrt{\phi_1^2 + 4\phi_2}\right]$, and they must be smaller than 1 in magnitude for the $AR(2)$ series to be stationary. In the $AR(2)$ case, it can be shown that $\rho(h) = \phi_1\rho(h-1) + \phi_2\rho(h-2)$, $h = 2, 3, \ldots$. When the roots r_1 and r_2 of $g(z)$ are real and distinct, the theory of difference equations tells us that the autocorrelations $\rho(h)$ behave like $c_1 r_1^h + c_2 r_2^h$, where c_1 and c_2 are real. When $r_1 = r_2$, that is, $\phi_1^2 + 4\phi_2 = 0$, then $\rho(h) = c_1 r_1^h + c_2 h r_1^h$, c_1, c_2 real. Thus the autocorrelation function $\{\rho(h)\}$ decays exponentially in h when the roots of $g(z)$ are real.

When r_1 and r_2 are complex, the $AR(2)$ can model series with pseudo-cyclical behavior. In this case, $\phi_1^2+4\phi_2 < 0$, r_2 is the complex conjugate of r_1 and $\rho(h)$ is of the form $c_1 r_1^h + c_2 r_2^h$, where c_1 and c_2 are complex. Calculations show that $\rho(h) = |r_1|^h [sign(\phi_1)]^h \sin(2\pi Ch + D)/\sin(D)$, where $sign(\phi_1) = 1$ if $\phi_1 > 0$ and $= -1$ if $\phi_1 < 0$, and C and D are constants. Since $|r_1| = |r_2| < 1$, $\rho(h)$ decays exponentially in a sinusoidal fashion.

In general, the autocorrelations of an $AR(p)$ process decay as a mixture of exponentials or as damped (sinusoidal) exponentials.

Identifiability of *MA*(q) Models

Consider the polynomial $\theta(z) = 1 + \theta_1 z + \cdots + \theta_q z^q$, where $\theta_1, \ldots, \theta_q$ are the parameters of the $MA(q)$ model. This polynomial has q roots which can be real or complex valued.

Lemma 13.5.2. *If the absolute values of all the roots of the polynomial $\theta(z)$ of an $MA(q)$ series are larger than or equal to 1, then the series is identifiable.*

For the $MA(1)$ model, this translates into the condition that $-1 \leq \theta_1 \leq 1$. For $MA(2)$ this condition is equivalent to the condition that $(-\theta_1, -\theta_2)$ is in the triangle Δ as in the $AR(2)$ case, except that $(-\theta_1, -\theta_2)$ is allowed to be on the boundaries of the triangle.

Lemma 13.5.3. *If for an $MA(q)$ or $ARMA(p, q)$ series, the absolute values of all the roots of the polynomial $\theta(z)$ are larger than 1, then the series is invertible.*

For the $MA(1)$ model, the conditions for identifiability and invertibility are $-1 \leq \theta_1 \leq 1$ and $-1 < \theta_1 < 1$, respectively. For $MA(2)$ the model is invertible if $(-\theta_1, -\theta_2)$ is in the triangle Δ as in the $AR(2)$ case. Thus the condition for invertibility guarantees identifiability.

Stationarity, Invertibility, and Nonredundancy of *ARMA*(*p*, *q*) Models

We have already seen that for an $ARMA(1, 1)$ series, there is no redundancy when $\theta \neq -\phi$. For this case the roots of $\phi(z)$ and $\theta(z)$ are $1/\phi$ and $-1/\theta$, respectively. Thus nonredundancy is achieved when $\phi(z)$ and $\theta(z)$ have no common root. This same condition is true for the general ARMA model.

Lemma 13.5.4. *For an $ARMA(p, q)$ series, if $\phi(z)$ and $\theta(z)$ have no common root, then the model is nonredundant.*

For the general ARMA model we need the following:

(a) roots of $\phi(z)$ are larger than 1 in magnitude (condition for stationarity),
(b) roots of $\theta(z)$ are larger than or equal to 1 (condition for identifiability), and
(c) the roots of $\phi(z)$ are distinct from the roots of $\theta(z)$ (condition for nonredundancy).

These are summarized in the following result.

Lemma 13.5.5. *An $ARMA(p, q)$ series is stationary, identifiable, and nonredundant if the following conditions hold:*

(a) *the roots of* $\phi(z)$ *are larger than 1 in magnitude,*
(b) *the roots of* $\theta(z)$ *are larger than or equal to 1 in magnitude,*
(c) $\phi(z)$ *and* $\theta(z)$ *have no common roots.*
For invertibility, in addition to stationarity and nonredundancy, condition (b) needs to be replaced by the stronger condition
(b′) *the roots of* $\theta(z)$ *are larger than 1 in magnitude.*

13.6 Forecasting

This section deals with the issues of forecasting, and obtaining prediction intervals for an ARMA series $\{X_t\}$. Here the parameters associated with the ARMA model are assumed to be known, but they need to be estimated in practice, and the issue of estimation will be discussed later. Suppose that observations are $X_1, \ldots, X_n$ and we wish to forecast $X_{n+1}, X_{n+2}, \ldots$, then a linear predictor of X_{n+h} is of the form $a_0 + a_1X_1 + \cdots + a_nX_n$. If all the variables are centered (ie, they are subtracted by their means), then we may take $a_0 = 0$. For notational convenience, we consider the issue of forecasting a centered series and we continue to denote the centered series as $\{X_t\}$.

If we wish to predict X_{n+1} using a linear predictor based on $\boldsymbol{X}_n = (X_1, \ldots, X_n)^T$ for the $AR(1)$ case which is modeled as $X_{n+1} = \phi X_n + \varepsilon_{n+1}$, the best linear predictor of X_{n+1} is $\hat{X}_{n+1} = \phi X_n$. If X_{n+1} were available, then the best linear predictor of X_{n+2} based on $\boldsymbol{X}_{n+1} = (X_1, \ldots, X_n, X_{n+1})$ would be ϕX_{n+1}. Now X_{n+1} is unobserved, but a predicted value $\hat{X}_{n+1}$ of X_{n+1} is available. Thus substituting $\hat{X}_{n+1}$ in place of X_{n+1}, we can obtain a linear predictor of X_{n+2} based on $\boldsymbol{X}_n$ as $\hat{X}_{n+2} = \phi\hat{X}_{n+1} = \phi^2 X_n$. We can similarly obtain a linear predictor of X_{n+h} based on $X_1, \ldots, X_n$ as $\hat{X}_{n+h} = \phi^h X_n$. Is $\hat{X}_{n+h} = \phi^h X_n$ the best linear predictor of X_{n+h} based on $X_1, \ldots, X_n$? The answer is yes.

Let $\hat{X}_{n+h}$ denote the best linear predictor of X_{n+h} based on $X_1, \ldots, X_n$. In general, it is easy to obtain an expression for the best predictor of X_{n+h} based on $X_1, \ldots, X_{n+h-1}$ and in this expression we can substitute $\hat{X}_{n+1}, \ldots, \hat{X}_{n+h-1}$ for $X_{n+1}, \ldots, X_{n+h-1}$ in order to obtain the best linear predictor of X_{n+h} based on $X_1, \ldots, X_n$. This is based on a rather simple argument as outlined below.

If Y is a rv and $\boldsymbol{W}_1, \boldsymbol{W}_2$ are two random vectors, then the best predictors of Y given $\boldsymbol{W}_1$, and given $\boldsymbol{W}_2$ and $\boldsymbol{W}_1$ are $\mathrm{E}[Y \mid \boldsymbol{W}_1\}$ and $\mathrm{E}[Y \mid \boldsymbol{W}_2, \boldsymbol{W}_1]$, respectively. The law of iterative expectations tells us $\mathrm{E}[Y \mid \boldsymbol{W}_1] = \mathrm{E}[\mathrm{E}\{Y \mid \boldsymbol{W}_2, \boldsymbol{W}_1\} \mid \boldsymbol{W}_1]$. Is this result true for best linear predictors? The answer is yes if Y, $\boldsymbol{W}_1$, $\boldsymbol{W}_2$ are jointly normally distributed, since the best predictor is the same as the best linear predictor. However, the result is true more generally in the sense described in the result below. Let the best linear predictors of Y given $\boldsymbol{W}_1$, and Y given $\boldsymbol{W}_1$, $\boldsymbol{W}_2$ be denoted by $L(Y \mid \boldsymbol{W}_1)$ and $L(Y \mid \boldsymbol{W}_2, \boldsymbol{W}_1)$, respectively. We can then write the following result.

Lemma 13.6.1. *Let* $L(\cdot \mid \cdot)$ *be the best linear predictor as described above.*

(a) *L is linear in the response in the sense that for any rv's* Y_1 *and* Y_2,

$$L(Y_1 + Y_2 \mid \boldsymbol{W}_1) = L(Y_1 \mid \boldsymbol{W}_1) + L(Y_2 \mid \boldsymbol{W}_1).$$

(b) *L satisfies the iterative formula* $L(Y\,|\,\mathbf{W}_1) = L[L(Y\,|\,\mathbf{W}_2, \mathbf{W}_1)\,|\,\mathbf{W}_1]$.
(c) *If* $\mathbf{W}_1$ *and* $\mathbf{W}_2$ *are uncorrelated, that is,* $Cov[\mathbf{W}_1, \mathbf{W}_2] = \mathbf{0}$, *then*

$$L[L(Y\,|\,\mathbf{W}_2)\,|\,\mathbf{W}_1] = 0 \quad \textit{and} \quad L(Y\,|\,\mathbf{W}_2, \mathbf{W}_1) = L(Y\,|\,\mathbf{W}_2) + L(Y\,|\,\mathbf{W}_1).$$

This proof of this lemma is left as an exercise. It can be used to justify replacing $X_{n+1}, \ldots, X_{n+h-1}$ by $\hat{X}_{n+1}, \ldots, \hat{X}_{n+h-1}$ in the expression for the best linear predictor of X_{n+h} given $X_1, \ldots, X_{n+h-1}$ in order to obtain the best linear predictor of X_{n+h} given $X_1, \ldots, X_n$.

13.6.1 Forecasting an *AR*(*p*) Series

Forecasting with an $AR(p)$ model with autoregressive coefficients $\phi_1, \ldots, \phi_p$ is quite simple as it has a regression form. Suppose that observations are $X_1, \ldots, X_n$ and we wish to forecast $X_{n+1}, X_{n+2}, \ldots$. The best linear predictor of X_{n+h} based on $X_1, \ldots, X_n$ will be denoted by $\hat{X}_{n+h}$. Since

$$X_t = \phi_1 X_{t-1} + \cdots + \phi_p X_{t-p} + \varepsilon_t,$$

where $\{\varepsilon_t\}$ are iid with mean 0 and variance σ^2, the best linear predictor of X_{n+1} based on $X_1, \ldots, X_n$ is

$$\hat{X}_{n+1} = \phi_1 X_n + \cdots + \phi_p X_{n+1-p}.$$

If X_{n+1} were known the best linear predictor based on $X_1, \ldots, X_{n+1}$ is $\phi_1 X_{n+1} + \phi_2 X_n + \cdots + \phi_p X_{n+2-p}$. In Lemma 13.6.1, use $Y = X_{n+2}$, $\mathbf{W}_1 = (X_1, \ldots, X_n)^T$ and $W_2 = X_{n+1}$ to obtain

$$\begin{aligned}
\hat{X}_{n+2} &= L(Y\,|\,\mathbf{W}_1) = L[L\{Y\,|\,W_2, \mathbf{W}_1\}\,|\,\mathbf{W}_1] \\
&= L\big[\phi_1 X_{n+1} + \phi_2 X_n + \cdots + \phi_p X_{n+2-p}\,|\,\mathbf{W}_1\big] \\
&= \phi_1 L(X_{n+1}\,|\,\mathbf{W}_1) + \phi_2 X_n + \cdots + \phi_p X_{n+2-p} \\
&= \phi_1 \hat{X}_{n+1} + \phi_2 X_n + \cdots + \phi_p X_{n+2-p}.
\end{aligned}$$

We have used linearity of L (part (a) of Lemma 13.6.1) and the fact that $L(X_t\,|\,\mathbf{W}_1) = X_t$ for any $t = 1, \ldots, n$.

If we denote $\hat{X}_t = X_t$, $t = 1, \ldots, n$, then the argument used above can be employed to show that for any $h \geq 1$,

$$\hat{X}_{n+h} = \phi_1 \hat{X}_{n+h-1} + \cdots + \phi_p \hat{X}_{n+h-p}.$$

13.6.2 Forecasting an *MA*(*q*) Series

For a mean zero $MA(q)$ series with coefficients $\theta_1, \ldots, \theta_q$, X_{n+h} is uncorrelated with X_n when $h \geq q+1$. Thus $\hat{X}_{n+h} = 0$ for $h = q+1, q+2, \ldots$. We discuss below how to find the formula for forecasting $\hat{X}_{n+1}, \ldots, \hat{X}_{n+q}$. As will be clear in the subsequent discussion, unlike in the $AR(p)$ case where the forecasted value $\hat{X}_{n+h}$ depends only on $X_n, \ldots, X_{n+1-p}$, here the forecasted value of X_{n+h}, $1 \leq h \leq q$, depends on the entire available past $X_n, \ldots, X_1$.

For an invertible series given in Eq. (6), the forecasting formula is simple if the entire past $\{X_t: -\infty < t \leq n\}$ were available. Absolute summability of $\{\pi_j\}$ guarantees that $\pi_j \to 0$ as $j \to \infty$, and for an invertible *MA*(q) model, π_j decays exponentially fast as j increases. So when predicting X_{n+h} based on $X_1, \ldots, X_n$, we may simply truncate the π series at $j = n$ thus approximating the process by an *AR*(p) sequence with $p = n$, and then use the methods associated with forecasting an autoregressive process as given in Section 13.6.1.

We now discuss how to carry out forecasting without having to obtain the values of $\{\pi_j\}$ assuming that the series is invertible. Note that

$$X_{n+1} = \varepsilon_{n+1} + \theta_1\varepsilon_n + \cdots + \theta_q\varepsilon_{n+1-q}.$$

For the moment assume that in addition to the observations $\boldsymbol{X}_n = (X_1, \ldots, X_n)^T$ we also have $\boldsymbol{\varepsilon}_0 = (\varepsilon_{-q+1}, \ldots, \varepsilon_0)^T$, and we predict X_{n+1} based on $\boldsymbol{\varepsilon}_0$ and $\boldsymbol{X}_n$. Any linear combination of $\boldsymbol{\varepsilon}_0$ and $\boldsymbol{X}_n$ can be rewritten as a linear combination of $\varepsilon_{-q+1}, \ldots, \varepsilon_n$, and vice versa. Thus the best linear predictor of X_{n+1} based on $\boldsymbol{\varepsilon}_0$ and $\boldsymbol{X}_n$ is given by

$$\hat{X}_{n+1} = \theta_1\varepsilon_n + \cdots + \theta_q\varepsilon_{n+1-q}.$$

In order to forecast X_{n+2}, note that

$$X_{n+2} = \varepsilon_{n+2} + \theta_1\varepsilon_{n+1} + \theta_2\varepsilon_n + \cdots + \theta_q\theta_{n+2-q},$$

and $(\varepsilon_{n+2}, \varepsilon_{n+1})$ is uncorrelated with $\boldsymbol{W}_1 = (\boldsymbol{\varepsilon}_0^T, X_1, \ldots, X_n)^T$. Thus the best linear predictor of X_{n+2} given $\boldsymbol{W}_1$ is

$$\hat{X}_{n+2} = \theta_2\varepsilon_n + \cdots + \theta_q\theta_{n+2-q}.$$

A similar argument will show that

$$\begin{aligned}
\hat{X}_{n+3} &= \theta_3\varepsilon_n + \cdots + \theta_q\varepsilon_{n+3-q}, \\
\hat{X}_{n+h} &= \theta_h\varepsilon_n + \cdots + \theta_q\varepsilon_{n+h-q}, \quad 1 \leq h \leq q, \text{ and} \\
\hat{X}_{n+h} &= 0, \quad h > q.
\end{aligned}$$

The forecasts $\hat{X}_{n+h}$, $h = 1, \ldots, q$, depend on knowing $\varepsilon_n, \varepsilon_{n-1}, \ldots, \varepsilon_{n+1-q}$. We now point out how to obtain these from $\boldsymbol{\varepsilon}_0$ and $\boldsymbol{X}_n$. We may obtain ε_1 and ε_2 as

$$\begin{aligned}
\varepsilon_1 &= X_1 - (\theta_1\varepsilon_0 + \cdots + \theta_q\varepsilon_{1-q}), \text{ and} \\
\varepsilon_2 &= X_2 - (\theta_1\varepsilon_1 + \cdots + \theta_q\varepsilon_{2-q}).
\end{aligned}$$

Continuing this way, once we have $\varepsilon_1, \ldots, \varepsilon_t$, we may obtain

$$\varepsilon_{t+1} = X_{t+1} - (\theta_1\varepsilon_t + \cdots + \theta_q\varepsilon_{t+1-q}), \quad t = 1, \ldots, n-1.$$

It is important to note that only the data $\boldsymbol{X}_n = (X_1, \ldots, X_n)$ is available and not $\boldsymbol{\varepsilon}_0$. Even though $\varepsilon_n, \ldots, \varepsilon_{n+1-q}$ are linear combinations of $\boldsymbol{\varepsilon}_0$ and $\boldsymbol{X}_n$, often one takes $\boldsymbol{\varepsilon}_0 = \boldsymbol{0}$ since the coefficients associated with $\boldsymbol{\varepsilon}_0$ in these linear combinations are negligible if the assumption of invertibility is valid. Without the assumption of invertibility, it is still possible to obtain the forecasts, but that issue will not be discussed here.

Example 13.6.1. Consider a mean zero $MA(1)$ series with the moving average coefficient θ. Assuming ε_0 to be known, we can get

$$\varepsilon_1 = X_1 - \theta\varepsilon_0, \quad \varepsilon_2 = X_2 - \theta\varepsilon_1, \ldots, \quad \varepsilon_n = X_n - \theta\varepsilon_{n-1}.$$

Using the arguments given in Example 13.5.2, we have

$$\varepsilon_n = X_n - \theta X_{n-1} + \theta^2 X_{n-2} + \cdots + (-1)^{n-1}\theta^{n-1}X_1 + (-1)^n\theta^n\varepsilon_0.$$

The forecasted value $\hat{X}_{n+1}$ of X_{n+1} based on $\varepsilon_0, X_1, \ldots, X_n$ is

$$\begin{aligned}\hat{X}_{n+1} &= \theta\varepsilon_n \\ &= \theta\Big[X_n - \theta X_{n-1} + \theta^2 X_{n-2} + \cdots + (-1)^{n-1}\theta^{n-1}X_1\Big] + (-1)^n\theta^{n+1}\varepsilon_0.\end{aligned}$$

The term $\theta^{n+1}\varepsilon_0$ is negligible if $|\theta| < 1$ (ie, the series is invertible) and thus $\hat{X}_{n+1}$ is approximately a linear combination of $X_1, \ldots, X_n$.

13.6.3 Forecasting an *ARMA* Series

If $\{X_t\}$ is mean zero $ARMA(p,q)$, then the method for forecasting X_{n+h} based on $\boldsymbol{X}_n = (X_1, \ldots, X_n)^T$ combines the methods given for AR and MA models. If the series in invertible, then we can obtain, in principle, the (approximate) best linear predictors of X_{n+h}, $h = 1, 2, \ldots$, based on $X_1, \ldots, X_n$, by approximating $\{X_t\}$ by an $AR(p)$ process with $p = n$.

For the moment assume that $\boldsymbol{\varepsilon}_p = (\varepsilon_{p+1-q}, \ldots, \varepsilon_p)^T$ is known. Then we can obtain $\varepsilon_{p+1}, \ldots, \varepsilon_n$ as linear combinations of $\boldsymbol{\varepsilon}_p$ and $\boldsymbol{X}_n$ as will be shown below.

Since

$$\begin{aligned}X_{n+1} &= \phi_1 X_n + \cdots + \phi_p X_{n+1-p} + \varepsilon_{n+1} + \theta_1\varepsilon_n + \cdots + \theta_q\varepsilon_{n+1-q} \\ &= \phi_1 X_n + \cdots + \phi_p X_{n+1-p} + \theta_1\varepsilon_n + \cdots + \theta_q\varepsilon_{n+1-q} + \varepsilon_{n+1},\end{aligned}$$

the best linear predictor $\hat{X}_{n+1}$ of X_{n+1} is given by

$$\hat{X}_{n+1} = \phi_1 X_n + \cdots + \phi_p X_{n+1-p} + \theta_1\varepsilon_n + \cdots + \theta_q\varepsilon_{n+1-q}.$$

Since

$$\begin{aligned}X_{n+2} = \phi_1 X_{n+1} &+ \phi_2 X_n + \cdots + \phi_p X_{n+2-p} \\ &+ \theta_1\varepsilon_{n+1} + \theta_2\varepsilon_n + \cdots + \theta_q\varepsilon_{n+2-q} + \varepsilon_{n+2},\end{aligned}$$

the best linear predictor of X_{n+2} would be

$$\phi_1 X_{n+1} + \phi_2 X_n + \cdots + \phi_p X_{n+2-p} + \theta_1\varepsilon_{n+1} + \theta_2\varepsilon_n + \cdots + \theta_q\varepsilon_{n+2-q},$$

if X_{n+1}, ε_{n+1} were known. Following Lemma 13.6.1 we can now replace X_{n+1} by $\hat{X}_{n+1}$ and ε_{n+1} by $\hat{\varepsilon}_{n+1} = 0$ (since ε_{n+1} is uncorrelated with $\boldsymbol{\varepsilon}_p$ and $X_1, \ldots, X_n$). The best linear predictor of X_{n+2} given $\boldsymbol{\varepsilon}_p$ and $X_1, \ldots, X_n$ is

$$\hat{X}_{n+2} = \phi_1\hat{X}_{n+1} + \phi_2 X_n + \cdots + \phi_p X_{n+2-p} + \theta_2\varepsilon_n + \cdots + \theta_q\varepsilon_{n+2-q}.$$

A similar argument will show that

$$\hat{X}_{n+h} = \phi_1 \hat{X}_{n+h-1} + \cdots + \phi_p \hat{X}_{n+h-p} + \theta_h \varepsilon_n + \cdots + \theta_q \varepsilon_{n+h-q}, \quad h = 1, \ldots, q,$$

$$\hat{X}_{n+h} = \phi_1 \hat{X}_{n+h-1} + \cdots + \phi_p \hat{X}_{n+h-p}, \quad h > q,$$

with the understanding that $\hat{X}_t = X_t$ for $t = 1, \ldots, n$.

The forecasts $\hat{X}_{n+h}$, $h \geq 1$, depend on $\varepsilon_{n+1-q}, \ldots, \varepsilon_n$. We need to obtain their values when $\boldsymbol{\varepsilon}_p$ and $X_1, \ldots, X_n$ are available, and this can be done iteratively starting with ε_{p+1}

$$\varepsilon_{p+1} = X_{p+1} - \phi_1 X_p - \cdots - \phi_p X_1 - (\theta_1 \varepsilon_p + \cdots + \theta_q \varepsilon_{p+1-q}),$$

$$\varepsilon_{p+2} = X_{p+2} - \phi_1 X_{p+1} - \cdots - \phi_p X_2 - (\theta_1 \varepsilon_{p+1} + \cdots + \theta_q \varepsilon_{p+2-q}),$$

and, when $\varepsilon_{p+1}, \ldots, \varepsilon_t$, $t \geq p+1$, are available, we can find

$$\varepsilon_{t+1} = X_{t+1} - \phi_1 X_t - \cdots - \phi_p X_{t+1-p} - (\theta_1 \varepsilon_t + \cdots + \theta_q \varepsilon_{t+1-q}).$$

Thus $\varepsilon_n, \ldots, \varepsilon_{n+1-q}$ and hence $\hat{X}_{n+h}$, $h \geq 1$, are linear combinations of $\boldsymbol{\varepsilon}_p$ and $X_1, \ldots, X_n$. Under the conditions of invertibility, the coefficients associated with $\varepsilon_p, \ldots, \varepsilon_{p+1-q}$ are negligible. In practice, often $\varepsilon_p, \ldots, \varepsilon_{p+1-q}$ are taken to be zeros.

Example 13.6.2. Let us assume that we have an *ARMA*(1, 1) process $X_t = \phi X_{t-1} + \varepsilon_t + \theta \varepsilon_{t-1}$, $\phi \neq -\theta$, and we want to predict $X_{n+1}, X_{n+2}, \ldots$ using the observations $X_1, \ldots, X_n$. If ε_1 were available, then generate $\varepsilon_2, \varepsilon_n, \ldots, \varepsilon_n$ as

$$\varepsilon_2 = X_2 - (\phi X_1 + \theta \varepsilon_1),\ \varepsilon_3 = X_3 - (\phi X_2 + \theta \varepsilon_2), \ldots, \varepsilon_n = X_n - (\phi X_{n-1} + \theta \varepsilon_{n-1})$$

Thus predicted values of X_{n+h}, $h \geq 1$, are

$$\hat{X}_{n+1} = \phi X_n + \theta \varepsilon_n,\ \hat{X}_{n+2} = \phi \hat{X}_{n+1} + \theta \hat{\varepsilon}_{n+1} = \phi \hat{X}_{n+1}, \text{ and}$$

$$\hat{X}_{n+h+1} = \phi \hat{X}_{n+h},\ h \geq 2.$$

Remark 13.6.1. For the *MA*(*q*) and *ARMA*(*p*, *q*) cases, it is generally assumed that the series is invertible. However, the condition of invertibility is more of a convenience than necessity. Even without the condition of invertibility, it is still possible to forecast X_{n+h}, $h \geq 1$, using the best linear predictor $a_1 X_n + \cdots + a_{n-1} X_1$, but unlike in the invertible case, the coefficients $\{a_j\}$ may not converge to 0 rapidly. For instance, for the mean zero *MA*(1) case if we have $\theta = -1$, then the best linear predictor of X_{n+1} is of the form

$$\hat{X}_{n+1} = \sum_{j=0}^{n-1} a_j X_{n-j}, \text{ where } a_j = -1 + (j+1)/(n+1).$$

Note that the coefficients $\{a_j\}$ increase linearly from $a_0 = -n/(n+1)$ to $a_{n-1} = -1/(n+1)$.

13.6.4 Standard Error for Prediction

Calculations of standard errors of predictions are needed to construct confidence bounds. Let $\sigma^2(h) = \mathrm{E}[(X_{n+h} - \hat{X}_{n+h})^2 \,|\boldsymbol{X}_n]$ be the mean square error for predicting X_{n+h} based on the data $\boldsymbol{X}_n = (X_1, \ldots, X_n)^T$. Once again we assume that the parameters of the ARMA model are known and formulas for $\sigma^2(h)$ depend on the unknown parameters. In practice these parameters are replaced by their sample estimates. Under the assumption of normality, a prediction interval for X_{n+h} with confidence coefficient $1-\alpha$ is given by $\hat{X}_{n+h} \pm z_{\alpha/2}\sigma(h)$.

The formulas for $\sigma^2(h)$ can be obtained easily for a series with causal representation

$$X_t = \sum_{j=0}^{\infty} \psi_j \varepsilon_{t-j}, \text{ with } \psi_0 = 1,$$

where $\{\varepsilon_t\}$ are iid with mean 0 and variance σ^2, and $\sum|\psi_j| < \infty$. If we assume that the entire past $X_n = \{X_t\colon -\infty < t \le n\}$ is known, then it is equivalent to knowing the entire past of $\{\varepsilon_t\colon -\infty < t \le n\}$ of the errors. If a predictor of X_{n+h} is a linear combination of $\{X_t\colon -\infty < t \le n\}$, then it is a linear combination of $\{\varepsilon_t\colon -\infty < t \le n\}$. Under appropriate conditions of stationarity and invertibility, the dependence on $\{\varepsilon_t\colon -\infty < t \le 0\}$ is negligible. Since

$$X_{n+1} = \sum_{j=1}^{\infty} \psi_j \varepsilon_{n+1-j} + \varepsilon_{n+1},$$

the best linear predictor $\hat{X}_{n+1}$ of X_{n+1} is $\sum_{j=1}^{\infty} \psi_j \varepsilon_{n+1-j}$, and the mean square error of prediction is

$$\sigma^2(1) = \mathrm{E}\left[\left(X_{n+1} - \hat{X}_{n+1}\right)^2 \middle| \boldsymbol{X}_n\right]^2 = \mathrm{E}\left[\varepsilon_{n+1}^2\right] = \sigma^2.$$

The best linear predictor of X_{n+2} given X_n is

$$\begin{aligned} \hat{X}_{n+2} &= \psi_2\varepsilon_n + \psi_3\varepsilon_{n-1} + \cdots, \text{ and} \\ X_{n+2} - \hat{X}_{n+2} &= [\varepsilon_{n+2} + \psi_1\varepsilon_{n+1} + \psi_2\varepsilon_n + \cdots] \\ &\quad - [\psi_2\varepsilon_n + \psi_3\varepsilon_{n-1} + \cdots] \\ &= \varepsilon_{n+2} + \psi_1\varepsilon_{n+1}. \end{aligned}$$

Hence the mean square error for predicting X_{n+2} by $\hat{X}_{n+2}$ is

$$\begin{aligned} \sigma^2(2) &= \mathrm{E}\left[\left(X_{n+2} - \hat{X}_{n+2}\right)^2 \middle| \boldsymbol{X}_n\right] = \mathrm{E}\left[(\varepsilon_{n+2} + \psi_1\varepsilon_{n+1})^2 \middle| \boldsymbol{X}_n\right] \\ &= \sigma^2 + \psi_1^2\sigma^2 = \left(1 + \psi_1^2\right)\sigma^2. \end{aligned}$$

A very similar argument will show that the best linear predictor of X_{n+h} based on the entire past and the mean square error of prediction are

$$\begin{aligned} \hat{X}_{n+h} &= \psi_h\varepsilon_n + \psi_{h+1}\varepsilon_{n-1} + \cdots, \\ X_{n+h} - \hat{X}_{n+h} &= \varepsilon_{n+h} + \psi_1\varepsilon_{n+h-1} + \cdots + \psi_{h-1}\varepsilon_{n+1}, \text{ and} \end{aligned}$$

$$\sigma^2(h) = \mathrm{E}\left[\left(X_{n+h} - \hat{X}_{n+h}\right)^2 \middle| \boldsymbol{X}_n\right]$$
$$= \left(1 + \psi_1^2 + \cdots + \psi_{h-1}^2\right)\sigma^2.$$

Note that $\sigma^2(h) \to \sigma^2 \sum_{j=0}^{\infty} \psi_j^2$ as $h \to \infty$, where $\psi_0 = 1$. The limit of $\sigma^2(h)$ is a constant as we assume that $\{\psi_j\}$ is absolutely summable.

13.7 ARIMA Models and Forecasting

If a series $\{Y_t\}$ is nonstationary in the mean in the sense that there is a trend, then the first difference $\{X_t = Y_t - Y_{t-1}\}$ or the second difference $\{X_t = Y_t - 2Y_{t-1} + Y_{t-2}\}$ may behave like a stationary series, and an ARMA model may be used for the differenced series $\{X_t\}$. Thus if the dth-order difference of the sequence $\{Y_t\}$ follows an *ARMA*(p, q) model, then we say that the series $\{Y_t\}$ follows an *ARIMA*(p, d, q) model [ARIMA stands for "integrated autoregressive-moving average"].

Let us briefly discuss forecasting when $d = 1$, which is often used in practice. Let $X_t = Y_t - Y_{t-1}$. If the observations are $Y_1, \ldots, Y_n$, then we have the differenced values $X_2 = Y_2 - Y_1, \ldots, X_n = Y_n - Y_{n-1}$. If $\{X_t\}$ is modeled by *ARMA*(p, q), we can obtain forecasts $\hat{X}_{n+h}$, $h = 1, 2, \ldots$. Since

$$Y_{n+h} = X_{n+h} + \cdots + X_{n+1} + Y_n,$$

the forecast of Y_{n+h} is

$$\hat{Y}_{n+h} = \hat{X}_{n+h} + \cdots + \hat{X}_{n+1} + Y_n.$$

The formula for the mean square error for prediction is a bit more complicated in comparison to the stationary case. Following the arguments in Section 13.6.4,

$$\hat{X}_{n+l} - X_{n+l} = \sum_{j=1}^{l} \psi_{l-j}\varepsilon_{n+j}, \text{ where } \psi_0 = 1, \text{ and}$$

$$\hat{Y}_{n+h} - Y_{n+h} - \sum_{l=1}^{h} (\hat{X}_{n+l} - X_{n+l})$$
$$= \sum_{l=1}^{h}\sum_{j=1}^{l} \psi_{l-j}\varepsilon_{n+j} = \sum_{j=1}^{h}\left(\sum_{l=0}^{h-j} \psi_l\right)\varepsilon_{n+j},$$

where $\{\varepsilon_t\}$ are the error terms in ARMA model for $\{X_t\}$. Thus the mean square error for predicting Y_{n+h} is

$$\sigma^2(h) = \mathrm{E}\left[(\hat{Y}_{n+h} - Y_{n+h})^2 | Y_1, \ldots, Y_n\right] = \sigma^2 \sum_{j=1}^{h}\left(\sum_{l=0}^{h-j} \psi_l\right)^2.$$

Unlike in the stationary case (Section 13.6.4), $\sigma^2(h)$ here increases linearly with h as the expression above shows. Thus the prediction interval for Y_{n+h} can be wide unless h is small.

13.8 Parameter Estimation

We now discuss how to estimate the parameters of ARMA models based on the data $X_1, \ldots, X_n$ and write down the asymptotic distributions of the parameter estimates. For the Gaussian series, the maximum likelihood method can be used to estimate the parameters. However, the actual implementation may always not be easy due to the dependence of the observations and, often appropriate approximations to the likelihood are used in order ease the computation. Details can be found in the book by Box et al. [62]. Here we will basically focus on the least squares type methods. We begin with estimation of the parameters of AR models since it is simpler than the MA or ARMA models. For the discussion below, we assume that the series $\{X_t\}$ is stationary, invertible, and identifiable.

13.8.1 Parameter Estimation: *AR*(*p*) Models

If a series $\{X_t\}$ is $AR(p)$, we may estimate the parameters by the method of least squares since it can rewritten as

$$X_t = \phi_0 + \phi_1 X_{t-1} + \cdots + \phi_p X_{t-p} + \varepsilon_t, \text{ where}$$
$$\phi_0 = \mu - (\phi_1 + \cdots + \phi_p)\mu.$$

Thus one may minimize $\sum_{t=p+1}^{n} [X_t - \phi_0 - \phi_1 X_{t-1} - \cdots - \phi_p X_{t-p}]^2$ with respect to $\phi_0, \ldots, \phi_p$ in order to obtain their least squares estimates. However, numerically more stable estimates are obtained via Yule-Walker equations. Recall that the theoretical Yule-Walker equations $\boldsymbol{\Gamma}_p \boldsymbol{\phi} = \boldsymbol{\gamma}_p$, $\boldsymbol{\phi} = (\phi_1, \ldots, \phi_p)^T$, as given in Eq. (3) are obtained by minimizing $Q = \mathrm{E}[(X_t - \mu) - \phi_1(X_{t-1} - \mu) - \cdots - \phi_p(X_{t-p} - \mu)]^2$ with respect to $\phi_1, \ldots, \phi_p$. If $\{\gamma(j)\}$ in Yule-Walker equations are replaced by their empirical estimates $\{\hat{\gamma}(j)\}$, then we get the empirical version of the equations $\hat{\boldsymbol{\Gamma}}_p \boldsymbol{\phi} = \hat{\boldsymbol{\gamma}}_p$, that is, $\hat{\boldsymbol{\phi}} = \hat{\boldsymbol{\Gamma}}_p^{-1} \hat{\boldsymbol{\gamma}}_p$. The Yule-Walker estimate $\hat{\boldsymbol{\phi}}$ can also be obtained by padding the data $X_1, \ldots, X_n$ as follows. Create $\tilde{X}_t$, $t = -p+1, \ldots, n+p$, where $\tilde{X}_t = X_t - \bar{X}$, $t = 1, \ldots, n$ and $\tilde{X}_t = 0$ otherwise. Then one can minimize the least squares criterion

$$\sum_{t=1}^{n+p} (\tilde{X}_t - \phi_1 \tilde{X}_{t-1} - \cdots \phi_{t-p} \tilde{X}_{t-p})^2$$

with respect to $\phi_1, \ldots, \phi_p$ and the normal equations are the same as the Yule-Walker equations $\hat{\boldsymbol{\Gamma}}_p \boldsymbol{\phi} = \hat{\boldsymbol{\gamma}}_p$.

Theorem 13.8.1. *Let $\hat{\boldsymbol{\phi}}$ be the solution of the Yule-Walker equations $\hat{\boldsymbol{\Gamma}}_p\boldsymbol{\phi} = \hat{\boldsymbol{\gamma}}_p$, where $\hat{\boldsymbol{\Gamma}}$ and $\hat{\boldsymbol{\gamma}}_p$ are estimates of $\boldsymbol{\Gamma}_p$ and $\boldsymbol{\gamma}_p$ based on the observations $X_1, \ldots, X_n$ from an AR(p) process. Assuming that $E[\varepsilon_t^4] < \infty$, $\sqrt{n}(\hat{\boldsymbol{\phi}} - \boldsymbol{\phi}) \xrightarrow{\mathcal{D}} N_p(0, \sigma^2\boldsymbol{\Gamma}_p^{-1})$.*

Remark 13.8.1. The above result is also true if the true model is *AR*(*p*) but we fit an autoregressive model of order higher than p. For $h > p$, let $\boldsymbol{\phi}_h = \begin{pmatrix}\boldsymbol{\phi}\\ \mathbf{0}\end{pmatrix}$, where $\boldsymbol{\phi}$ is the vector of autoregressive parameters of the *AR*(*p*) model and $\mathbf{0}$ is a $(h-p)$-dim vector of zeros. Let $\hat{\boldsymbol{\phi}}_h$ be the solution of the Yule-Walker equations for estimating the parameters of an *AR*(*h*) model, that is, $\hat{\boldsymbol{\phi}}_h = \hat{\boldsymbol{\Gamma}}_h^{-1}\hat{\boldsymbol{\gamma}}_h$. Then $\sqrt{n}(\hat{\boldsymbol{\phi}}_h - \boldsymbol{\phi}_h) \xrightarrow{\mathcal{D}} N_h(\mathbf{0}, \sigma^2\boldsymbol{\Gamma}_h^{-1})$ assuming that $E[\varepsilon_t^4] < \infty$.

This result can be used to carry out inferences on $\phi_1, \ldots, \phi_p$ including construction of confidence intervals and deciding if an autoregressive term can be dropped from the model.

An outline of the proof will be given below. One may heuristically guess the asymptotic result since the *AR*(*p*) model can be reexpressed as a Gauss-Markov model and one may use the distributional results of the least squares estimates. However, applying the distributional results of the Gauss-Markov model for the autoregressive case requires justifications due to dependence of the observations. We provide an outline of the main arguments used in the derivation of the asymptotic normality of $\hat{\boldsymbol{\phi}}$.

Since $\{X_t\}$ can be centered by subtracting the mean μ, we assume that $E[X_t] = 0$. The main idea behind the proof is to decompose $\hat{\boldsymbol{\gamma}}_p$ so that

$$\hat{\boldsymbol{\gamma}}_p = \hat{\boldsymbol{\Gamma}}_p\boldsymbol{\phi} + n^{-1/2}\boldsymbol{\delta} + O_P(n^{-1}),$$

where $\boldsymbol{\delta}$ is a p-dim vector whose jth element is $\delta_j = n^{-1/2}\sum_{t=1}^{n} X_t\varepsilon_{t+j}$. For the moment assume that $\boldsymbol{\delta} \xrightarrow{\mathcal{D}} N_p(\mathbf{0}, \sigma^2\boldsymbol{\Gamma}_p)$. Then

$$\begin{aligned}\hat{\boldsymbol{\Gamma}}_p(\hat{\boldsymbol{\phi}} - \boldsymbol{\phi}) &= \hat{\boldsymbol{\gamma}}_p - \hat{\boldsymbol{\Gamma}}_p\boldsymbol{\phi} = \left[\hat{\boldsymbol{\Gamma}}_p\boldsymbol{\phi} + n^{-1/2}\boldsymbol{\delta} + O_P(n^{-1})\right] - \hat{\boldsymbol{\Gamma}}_p\boldsymbol{\phi}\\ &= n^{-1/2}\boldsymbol{\delta} + O_P(n^{-1}).\end{aligned}$$

It follows from Theorem 13.3.4 that $\hat{\gamma}(h) - \gamma(h) \xrightarrow{P} 0$, $h = 0, \ldots, p-1$, and hence $\hat{\boldsymbol{\Gamma}}_p - \boldsymbol{\Gamma}_p \xrightarrow{P} 0$ and $\hat{\boldsymbol{\Gamma}}_p^{-1} - \boldsymbol{\Gamma}_p^{-1} \xrightarrow{P} 0$. Therefore,

$$n^{1/2}(\hat{\boldsymbol{\phi}} - \boldsymbol{\phi}) = \hat{\boldsymbol{\Gamma}}_p^{-1}\boldsymbol{\delta} + O_P(n^{-1/2}) \xrightarrow{\mathcal{D}} N_p(\mathbf{0}, \sigma^2\boldsymbol{\Gamma}_p^{-1}).$$

Let us briefly examine why the asymptotic normality of $\boldsymbol{\delta}$ is valid. It is fairly easy to check that

$$\begin{aligned}&E\left[\delta_j\right] = 0,\\ &E\left[\delta_j\delta_k\right] = n^{-1}\sum_{t=1}^{n} E\left[X_tX_{t+k-j}\right] = \gamma(k-j),\ j \le k, \text{ and}\\ &\text{Cov}[\boldsymbol{\delta}] = \boldsymbol{\Gamma}_p.\end{aligned}$$

In order to establish the asymptotic normality of $\boldsymbol{\delta}$, use the Cramér-Wold device, that is, establish the asymptotic normality of a linear function $c_1\delta_1 + \cdots + c_p\delta_p = \boldsymbol{c}^T\boldsymbol{\delta}$ of $\boldsymbol{\delta}$, where $\boldsymbol{c} = (c_1, \ldots, c_p)^T$ is a vector of constants. Now, denoting $S_t = X_t(c_1\varepsilon_{t+1} + \cdots + c_p\varepsilon_{t+p})$, we have

$$\boldsymbol{c}^T\boldsymbol{\delta} = n^{-1/2}\sum_{t=1}^{n} X_t(c_1\varepsilon_{t+1} + \cdots + c_p\varepsilon_{t+p}) = n^{-1/2}\sum_{t=1}^{n} S_t$$

Since $\{X_t\}$ is stationary and has a causal representation, $X_t = \sum_{j=0}^{\infty}\psi_j\varepsilon_{t-j}$ with $\psi_0 = 0$ and $\sum|\psi_j| < \infty$. For any positive integer m, let $X_{m,t} = \sum_{j=0}^{m}\psi_j\varepsilon_{t-j}$ and $S_{m,t} = X_{m,t}(c_1\varepsilon_{t+1} + \cdots + c_p\varepsilon_{t+p})$. Since $\{\varepsilon_t\}$ are iid, the process $\{S_{m,t}\}$ is $(m+p)$-dependent and we can use Theorem 13.3.2 (details omitted) in order to establish asymptotic normality of $\boldsymbol{\delta}$.

Let us now see why the decomposition of $\hat{\boldsymbol{\gamma}}_p$ given above is valid. Since we are assuming that $\mathrm{E}[X_t] = 0$, $X_t = \boldsymbol{\phi}^T\boldsymbol{X}_{t-1} + \varepsilon_t$ where $\boldsymbol{X}_{t-1} = (X_{t-1}, \ldots, X_{t-p})^T$. Using Lemma 13.3.1, the jth element of $\hat{\boldsymbol{\gamma}}_p$ is

$$\begin{aligned}
\hat{\gamma}(j) &= n^{-1}\sum_{t=1}^{n} X_tX_{t+j} + O_P(n^{-1})\\
&= n^{-1}\sum_{t=1}^{n} X_t\left(\boldsymbol{\phi}^T\boldsymbol{X}_{t+j-1} + \varepsilon_{t+j}\right) + O_P(n^{-1})\\
&= \sum_{l=1}^{p}\phi_l n^{-1}\sum_{t=1}^{n} X_tX_{t+j-l} + n^{-1}\sum_{t=1}^{n} X_t\varepsilon_{t+j} + O_P(n^{-1})\\
&= \sum_{l=1}^{p}\phi_l\left[\hat{\gamma}(j-l) + O_P(n^{-1})\right] + n^{-1/2}\delta_j + O_P(n^{-1})\\
&= \sum_{l=1}^{p}\phi_l\hat{\gamma}(j-l) + n^{-1/2}\delta_j + O_P(n^{-1}), \qquad h = 1, \ldots, p.
\end{aligned}$$

In the matrix notations, these equations can be written as

$$\hat{\boldsymbol{\gamma}}_p = \hat{\boldsymbol{\Gamma}}_p\boldsymbol{\phi} + n^{-1/2}\boldsymbol{\delta} + O_P(n^{-1}).$$

13.8.2 Parameter Estimation: *MA*(*q*) Models

Parameter estimation for moving average models is more complicated in comparison to autoregressive models and no closed form solution is available. If the series $\{X_t\}$ has mean μ and we have observations $X_1, \ldots, X_n$, we can estimate μ by the sample mean $\bar{X}$ and then continue to do the analysis based on $\{X_t - \bar{X}\}$. For this reason, in order to make the notations simple, we assume that the series has zero mean. We now describe the least squares and the maximum likelihood methods for estimating $\theta_1, \ldots, \theta_q$.

Let us follow the ideas used in forecasting *MA*(*q*) models. If $\boldsymbol{\varepsilon}_0 = (\varepsilon_{-q+1}, \ldots, \varepsilon_0)^T$ were known, given $\boldsymbol{\theta} = (\theta_1, \ldots, \theta_q)$, we may calculate ε_t's starting with

$$\varepsilon_1(\boldsymbol{\theta}, \boldsymbol{\varepsilon}_0) = X_1 - (\theta_1\varepsilon_0 + \cdots + \theta_q\varepsilon_{-q+1}),$$

and then iteratively compute, when $\varepsilon_{-q+1}, \ldots, \varepsilon_{t-1}$ are available, ε_t as

$$\varepsilon_t(\boldsymbol{\theta}, \boldsymbol{\varepsilon}_0) = X_t - (\theta_1 \varepsilon_{t-1} + \cdots + \theta_q \varepsilon_{t-q}), \quad t = 2, \ldots, n.$$

It is important to note that each ε_t is a linear combination of $X_t, \ldots, X_1$ and $\boldsymbol{\varepsilon}_0$. For the $MA(1)$ case, denoting θ_1 by θ, we have

$$\varepsilon_t(\theta, \varepsilon_0) = X_t - \theta X_{t-1} + \cdots + (-\theta)^{t-1} X_1 + (-\theta)^t \varepsilon_0.$$

In the invertible case, that is, $|\theta| < 1$, dependence of $\varepsilon_t(\theta, \varepsilon_0)$ on ε_0 is negligible when t is not small since $(-\theta)^t \to 0$ rapidly. For the general $MA(q)$ case, the same is true when it is invertible.

In order to obtain the estimate of $\boldsymbol{\theta}$, we may minimize $\sum_{t=1}^n \varepsilon_t(\boldsymbol{\theta}, \boldsymbol{\varepsilon}_0)^2$, with respect to $\boldsymbol{\theta}$. This estimate $\hat{\boldsymbol{\theta}}$ depends on $\boldsymbol{\varepsilon}_0$, even though this dependence is negligible in the invertible case as pointed out before. How can one implement this estimation method in practice? The following are among many options:

(i) take $\boldsymbol{\varepsilon}_0 = \mathbf{0}$,
(ii) obtain the MLE, and
(iii) obtain a modified MLE.

The first option is often used and is reasonable in the invertible case. In order to see the second and the third options let us note that the joint pdf of $\boldsymbol{\varepsilon}_0$ and $X_1, \ldots, X_n$, under normality (ie, $\{\varepsilon_t\}$ are iid $N(0, \sigma^2)$), is

$$f(\boldsymbol{x}, \boldsymbol{\varepsilon}_0, \boldsymbol{\theta}, \sigma) = \left(\sqrt{2\pi}\sigma\right)^{-n-q} \exp\left[-\frac{1}{2\sigma^2}\sum_{t=1}^n \varepsilon_t(\boldsymbol{\theta}, \boldsymbol{\varepsilon}_0)^2 - \frac{1}{2\sigma^2}\|\boldsymbol{\varepsilon}_0\|^2\right],$$

where $\boldsymbol{x} = (x_1, \ldots, x_n)^T$. In order to obtain the MLE (option (ii)), one needs to maximize the likelihood after $\boldsymbol{\varepsilon}_0$ has been integrated out, that is, maximize

$$f(\boldsymbol{X}, \boldsymbol{\theta}, \sigma) = \int f(\boldsymbol{X}, \boldsymbol{\theta}, \boldsymbol{\varepsilon}_0, \sigma) g(\boldsymbol{\varepsilon}_0, \sigma) d\boldsymbol{\varepsilon}_0,$$

where g is the marginal pdf of $\boldsymbol{\varepsilon}_0$ and $\boldsymbol{X} = (X_1, \ldots, X_n)^T$. Clearly this integration is rather difficult since $\sum_{t=1}^n \varepsilon_t(\boldsymbol{\theta}, \boldsymbol{\varepsilon}_0)^2$ does not have a simple expression involving $\boldsymbol{\varepsilon}_0$. Another option is to approximate this integral $f(\boldsymbol{X}, \boldsymbol{\theta}, \sigma)$ by an average of M iid copies of $\boldsymbol{\varepsilon}_0$, that is, approximate $f(\boldsymbol{X}, \boldsymbol{\theta}, \sigma)$ by

$$f_M(\boldsymbol{X}, \boldsymbol{\theta}, \sigma) = M^{-1} \sum_{j=1}^M f(\boldsymbol{X}, \boldsymbol{\varepsilon}_{0j}, \boldsymbol{\theta}, \sigma),$$

where $\boldsymbol{\varepsilon}_{0j}$ are iid as $\boldsymbol{\varepsilon}_0$. Mathematically $f_M(\boldsymbol{x}, \boldsymbol{\theta}, \sigma)$ converges to $f(\boldsymbol{x}, \boldsymbol{\theta}, \sigma)$ in probability as $M \to \infty$. In practice, $M = n$ should be adequate.

In the third option, one may try to maximize $f(\boldsymbol{X}, \boldsymbol{\theta}, \boldsymbol{\varepsilon}_0, \sigma)$ with respect to $\boldsymbol{\theta}, \sigma$, and $\boldsymbol{\varepsilon}_0$, an idea used in the derivation of mixed model equations in Chapter 11. Thus one may minimize

$$-2\log f(\boldsymbol{X}, \boldsymbol{\theta}, \boldsymbol{\varepsilon}_0, \sigma) = \frac{1}{\sigma^2}\sum_{t=1}^{n} \varepsilon_t(\boldsymbol{\theta}, \boldsymbol{\varepsilon}_0)^2 + \frac{1}{\sigma^2}\|\boldsymbol{\varepsilon}_0\|^2 + (n+q)\log(2\pi\sigma^2),$$

with respect to $\boldsymbol{\theta}$, σ, and $\boldsymbol{\varepsilon}_0$. Note that the first two terms in the last expression have the same denominator and hence in order to obtain estimates of $\boldsymbol{\theta}$ and $\boldsymbol{\varepsilon}_0$ one needs to minimize the penalized criterion

$$\sum_{t=1}^{n} \varepsilon_t(\boldsymbol{\theta}, \boldsymbol{\varepsilon}_0)^2 + \|\boldsymbol{\varepsilon}_0\|^2$$

with respect to $\boldsymbol{\theta}$ and $\boldsymbol{\varepsilon}_0$.

For the least squares method, once we have obtained $\hat{\boldsymbol{\theta}}$, the estimate of σ^2 can be obtained as $\hat{\sigma}^2 = (n)^{-1}\sum \varepsilon_t(\hat{\boldsymbol{\theta}}, \boldsymbol{\varepsilon}_0)^2$, where either $\boldsymbol{\varepsilon}_0 = \mathbf{0}$ or it is estimated by optimizing a penalized criterion as described above.

The following asymptotic result holds for the estimate $\hat{\boldsymbol{\theta}}$ of $\boldsymbol{\theta}$. In order to describe the covariance matrix of the asymptotic distribution of $\hat{\boldsymbol{\theta}}$, we will adopt some simple notations as in Brockwell and Davis [63]. Let $\{D_t\}$ be a mean zero *AR*(q) process of the form $\theta(B)D_t = \delta_t$, where $\theta(B) = 1 + \theta_1 B + \cdots + \theta_q B^q$ and $\{\delta_t\}$ are iid with mean 0 and variance σ^2. Let $\boldsymbol{V} = \text{Cov}[\boldsymbol{D}_t]$, where $\boldsymbol{D}_t = [D_1, \ldots, D_q]^T$.

Theorem 13.8.2. *Let $X_1, \ldots, X_n$ be observations from a Gaussian MA(q) series which is invertible. Then as $n \to \infty$, $\sqrt{n}(\hat{\boldsymbol{\theta}} - \boldsymbol{\theta}) \xrightarrow{\mathcal{D}} N_q(\mathbf{0}, \sigma^2\mathbf{V}^{-1})$, where* **V** *is as given above and $\hat{\boldsymbol{\theta}}$ is estimated using any of the methods described above.*

This result makes it possible to obtain the standard errors of the estimates $\hat{\theta}_1, \ldots, \hat{\theta}_q$ and construct confidence intervals for $\theta_1, \ldots, \theta_p$ or carry out tests of hypotheses.

13.8.3 Parameter Estimation: *ARMA*(*p*, *q*) Models

As in the *MA*(q) case, parameter estimates for ARMA models do not have closed-form expressions. There are many methods for the estimation of parameters and many textbooks on time series provide details of these methods. As in the *MA*(q) case, we can center the observations by the sample mean so that we can assume the mean of the series to be zero. Here we outline a simple least squares type method extending some of the ideas outlined for the *AR*(p) and *MA*(q) cases. Details can be found in the well-known book on time series by Box et al. [62].

We assume that $\{X_t\}$ is mean zero stationary Gaussian *ARMA*(p, q) series which is causal and invertible. If $\boldsymbol{\varepsilon}_p = (\varepsilon_{p-q+1}, \ldots, \varepsilon_p)^T$ were available, then for given values of $\boldsymbol{\phi}$ and $\boldsymbol{\theta}$, we can obtain

$$\varepsilon_{p+1}(\boldsymbol{\phi}, \boldsymbol{\theta}, \boldsymbol{\varepsilon}_p) = (X_{p+1} - \phi_1 X_p - \cdots - \phi_p X_1) - (\theta_1 \varepsilon_p + \cdots + \theta_q \varepsilon_{p-q+1}).$$

Once $\varepsilon_{p+1}(\boldsymbol{\phi}, \boldsymbol{\theta}, \boldsymbol{\varepsilon}_p), \ldots, \varepsilon_{t-1}(\boldsymbol{\phi}, \boldsymbol{\theta}, \boldsymbol{\varepsilon}_p)$ are obtained, then it is possible to calculate

$$\varepsilon_t(\boldsymbol{\phi}, \boldsymbol{\theta}, \boldsymbol{\varepsilon}_p) = (X_t - \phi_1 X_{t-1} - \cdots - \phi_p X_{t-p}) - (\theta_1 \varepsilon_{t-1} + \cdots + \theta_q \varepsilon_{t-q}),$$

$t = p+2, \ldots, n$. Following the ideas in the $MA(q)$ case, in order to obtain estimates $\hat{\boldsymbol{\phi}}$ and $\hat{\boldsymbol{\theta}}$, we may minimize $\sum_{t=p+1}^{n} \varepsilon_t(\boldsymbol{\phi}, \boldsymbol{\theta}, \boldsymbol{\varepsilon}_p)^2$, with $\boldsymbol{\varepsilon}_p = \mathbf{0}$, with respect to $\boldsymbol{\phi}$ and $\boldsymbol{\theta}$, or we may minimize

$$\sum_{t=p+1}^{n} \varepsilon_t(\boldsymbol{\phi}, \boldsymbol{\theta}, \boldsymbol{\varepsilon}_p)^2 + \|\varepsilon_p\|^2,$$

with respect to $\boldsymbol{\phi}$, $\boldsymbol{\theta}$, and $\boldsymbol{\varepsilon}_p$.

As described in the $AR(p)$ estimation case, we may also used the idea of padding the data at the beginning and at the end to obtain $\{\tilde{X}_t : t = -p+1, \ldots, n+p\}$ and then taking $\boldsymbol{\varepsilon}_0 = (\varepsilon_{-q+1}, \ldots, \varepsilon_0)^T$, we can obtain $\varepsilon_t(\boldsymbol{\phi}, \boldsymbol{\theta}, \boldsymbol{\varepsilon}_0)$, $t = 1, \ldots, n+p$, and then obtain the estimates $\hat{\boldsymbol{\phi}}$ and $\hat{\boldsymbol{\theta}}$ by minimizing $\sum_{t=p+1}^{n} \varepsilon_t(\boldsymbol{\phi}, \boldsymbol{\theta}, \boldsymbol{\varepsilon}_0)^2$, with $\boldsymbol{\varepsilon}_0 = \mathbf{0}$, or by minimizing the penalized criterion

$$\sum_{t=1}^{n+p} \varepsilon_t(\boldsymbol{\phi}, \boldsymbol{\theta}, \boldsymbol{\varepsilon}_0)^2 + \|\boldsymbol{\varepsilon}_0\|^2$$

with respect to $\boldsymbol{\phi}$, $\boldsymbol{\theta}$, and $\boldsymbol{\varepsilon}_0$.

An estimate of σ^2 is

$$\hat{\sigma}^2 = (n)^{-1} \sum_{t=p+1}^{n} \varepsilon_t(\hat{\boldsymbol{\phi}}, \hat{\boldsymbol{\theta}}, \hat{\boldsymbol{\varepsilon}}_0)^2,$$

where $\hat{\boldsymbol{\varepsilon}}_0$ is $\mathbf{0}$ or is obtained by optimizing a penalized criterion as described above.

We now write down the asymptotic distribution of $(\hat{\boldsymbol{\phi}}, \hat{\boldsymbol{\theta}})$. Let $\{C_t\}$ and $\{D_t\}$ be mean zero $AR(p)$ and $AR(q)$ series

$$\phi(B)C_t = \delta_t, \quad \theta(B)D_t = \delta_t,$$

where $\phi(B) = 1 - \phi_1 B - \cdots - \phi_p B^p$, $\theta(B) = 1 + \theta_1 B + \cdots + \theta_q B^q$ and $\{\delta_t\}$ are iid with mean 0 and variance σ^2. Let $\boldsymbol{R} = [C_1, \ldots, C_p, D_1, \ldots, D_q]$ and $\boldsymbol{V} = \mathrm{Cov}[\boldsymbol{R}]$.

Theorem 13.8.3. *Let $X_1, \ldots, X_n$ be observations from a stationary Gaussian $ARMA(p, q)$ series which is invertible and nonredundant. Let $\boldsymbol{\beta}$ the $(p+q) \times 1$ vector of $\boldsymbol{\phi}$ and $\boldsymbol{\theta}$ stacked vertically and, similarly, let $\hat{\boldsymbol{\beta}}$ be the stacked vector of $\hat{\boldsymbol{\phi}}$ and $\hat{\boldsymbol{\theta}}$ which are obtained using any of the least squares methods outlined above. Then $\sqrt{n}(\hat{\boldsymbol{\beta}} - \boldsymbol{\beta}) \xrightarrow{\mathcal{D}} N_{p+q}(\mathbf{0}, \sigma^2 \boldsymbol{V}^{-1})$, where $\boldsymbol{V}$ is described above.*

Approximate standard errors of the estimates of $\hat{\boldsymbol{\phi}}$ and $\hat{\boldsymbol{\theta}}$ can be obtained using this theorem.

13.9 Selection of an Appropriate ARMA model

As in any statistical method, model selection is an important part of time series analysis. As in the case of linear models, it is possible to select an appropriate ARMA model by using a criterion such as AIC or BIC. We also introduce another widely used criterion known as AICC which provides a small sample correction to AIC. For an $ARMA(p, q)$ model, the total number of estimated parameters (excluding σ^2) is $k = p + q + 1$. As in the case of linear models (Section 11.8.3 in Chapter 11), the *AIC*, *AICC*, and *BIC* values with $k = p + q + 1$, can be written as

$$AIC(k) = -2\log L(\hat{\mu}_k, \hat{\phi}_k, \hat{\theta}_k, \hat{\sigma}_k) + 2k, \quad BIC(k) = -2\log L(\hat{\mu}_k, \hat{\boldsymbol{\phi}}_k, \hat{\boldsymbol{\theta}}_k, \hat{\sigma}_k) + [\log(n)]k,$$
$$AICC(k) = AIC(k) + 2k(k+1)/(n-k-1),$$

where L is the likelihood, and $(\hat{\mu}_k, \hat{\boldsymbol{\phi}}_k, \hat{\boldsymbol{\theta}}_k, \hat{\sigma}_k)$ is the MLE of $(\mu, \boldsymbol{\phi}, \boldsymbol{\theta}, \sigma)$ under the *ARMA*(p, q) model. Statistical computing packages such as R or MATLAB can calculate the values of these criteria for an ARMA model given a data set. As usual, an appropriate model is selected by minimizing the criterion of choice.

13.10 Spectral Analysis

Analysis of time series has two important aspects

(a) model fitting and forecasting, and
(b) understanding of the hidden periodicities.

We have discussed the first aspect in the previous sections, and we now discuss the second which involves a deeper analysis of the spectral density function defined in Eq. (1).

The basis of spectral analysis is an important mathematical result which states that any stationary time series (causal or noncausal) can be approximated by linear combinations of sines and cosines with random coefficients. Toward this end, let us define an important quantity: frequency. A frequency is a real number between 0 and 1/2. For the monthly energy (electricity) data mentioned at the beginning of this chapter, once the trend is estimated and subtracted from the data, the detrended series has a similar pattern of behavior every 12 time points (months). Energy consumptions in January are similar, energy consumptions in March are similar, and so on. In such a case, we can say that energy consumption has an "important" frequency at $w = 1/12$. A plot of the annual sunspots recorded over the last $n = 313$ years reveals that the peaks are occurring between 8 and 12 years. However, unlike in the energy data (which is seasonal), the times of the peaks are not fixed in the sunspot series. There are 28.5 cycles in the series. So the series tends to behave similarly every $313/28.5 = 10.98$ years (on the average). Thus there is a peak at frequency at $w = 1/10.98 = 0.091$.

Consider a series $X_t = A\cos(2\pi wt) + B\sin(2\pi wt)$, $0 \le w \le 1/2$, where A and B are mutually uncorrelated rv's with mean 0 and variance σ^2. Then $\mathrm{E}[X_t] = 0$, $\mathrm{Var}[X_t] = \sigma^2$, and the series $\{X_t\}$ is stationary since $\mathrm{Cov}[X_t, X_{t+h}] = \sigma^2\cos(2\pi wh)$ depends only on h. Note that $\mathrm{Corr}[X_t, X_{t+h}] = \cos(2\pi hw) = 1$ whenever hw is a positive integer. Thus if $1/w$ is an integer, then the series repeats itself at every $1/w$ time points, that is, $X_t = X_{t+h}$ when h is an integer multiple of $1/w$. However, even if $1/w$ is not a rational number, the correlation $\cos(2\pi wh)$ is high whenever hw is close to an integer. We call this series an elementary periodic series with frequency w and variance σ^2.

Now consider a series X_t which is a sum of M such elementary periodic series with distinct frequencies $w_1, \ldots, w_M$ and variances $\sigma_1^2, \ldots, \sigma_M^2$, that is,

$$X_t = \sum_{j=1}^{M} \{A_j \cos(2\pi w_j t) + B_j \sin(2\pi w_j t)\}, \tag{7}$$

where $\{A_j\}$, $\{B_j\}$ are mean zero mutually uncorrelated rv's with $\mathrm{Var}[A_j] = \mathrm{Var}[B_j] = \sigma_j^2$. It is also assumed that $\{A_j\}$ and $\{B_j\}$ are also uncorrelated to each other. Then it can be seen that $\{X_t\}$ is stationary with

$$\mathrm{E}[X_t] = 0, \ \mathrm{Var}[X_t] = \sigma_1^2 + \cdots + \sigma_M^2, \text{ and}$$

$$\mathrm{Cov}[X_t, X_{t+h}] = \sum_{j=1}^{M} \sigma_j^2 \cos(2\pi w_j h) = \gamma(h). \tag{8}$$

Remarks

(a) If X_t has mean μ, then the representation above is valid with $X_t - \mu$ on the left-hand side of Eq. (7). From now on we assume the mean to be equal to zero since we can always carry out spectral analysis after subtracting the mean from the series.

(b) For the monthly electricity consumption data $\{Y_t = \log(\text{sales})\}$, we have briefly discussed the model $Y_t = m_t + S_t + X_t$, where $\{X_t\}$ is stationary (Section 13.1). There are many approaches to the analysis of such a data set. We may estimate the trend m_t and the seasonal effect S_t, and then subtract them from Y_t in order to get an estimate of the stationary part $\{X_t\}$. However, there is another way of modeling this. We can subtract the trend only and the remainder, that is, $S_t + X_t$ can be often considered stationary, especially if $\{S_t\}$ is deemed to be stochastic. Thus if a sequence has no trend, but has seasonality whose variance does not depend on time t, then the sequence itself can be considered stationary.

(c) In some trivial cases, M may be small. But in general M is large.

(d) The goal of spectral analysis is to find σ_j's. Since $\mathrm{Var}[X_t] = \sigma_1^2 + \cdots + \sigma_M^2$, the contribution of the jth elementary periodic series to this variance (at frequency w_j) is σ_j^2. It is of interest to find out which frequencies contribute more to this variability than the others.

(e) If M is large (mathematically $M \to \infty$) and almost all σ_j's are not equal to zero, then these σ_j's need to be small so that $\mathrm{Var}[X_t] = \sigma_1^2 + \cdots + \sigma_M^2$ remains finite as $M \to \infty$. This can be done if $\sigma_j^2 = O(1/M)$. In Section 13.2.1, it is written that if $\sum_{h=-\infty}^{\infty} |\gamma(h)| < \infty$, then the spectral density is given by $f(w) = \sum_{h=-\infty}^{\infty} \gamma(h) \exp(-2\pi i h w)$. However, for the series in Eq. (7), $\sum_{h=-\infty}^{\infty} |\gamma(h)|$ is not finite and hence we may consider a truncated version $f_M(w) = \sum_{h=-M}^{M} \gamma(h) \exp(-2\pi i h w)$. It is left as an exercise for the reader to verify that

$$f_M(w) = [(2M+1)/2]\sigma_j^2 \quad \text{if } w = j/(2M+1), \ j = 1, \ldots, M.$$

In order for f_M to be approximately equal to a bounded spectral density f for large M, we need to have $\sigma_j^2 \approx [2/(2M+1)]f(w_j)$ when $w_j = j/(2M+1)$ and one of the goals of the spectral analysis is to obtain an estimate of the spectral density f based on the available data $X_1, \ldots, X_n$.

13.10.1 Representation of a Stationary Series

If $\{X_t\}$ is a mean zero stationary series, can one approximate it as in Eq. (7)? The answer is yes if the underlying spectral density function f as defined in Eq. (1) is square integrable. The arguments given here are heuristic and detailed proofs can be found in Gikhman and Skorokhod [64]. Consider the following random functions

$$Z_{1,M}(w) = \sum_{s=-M}^{M} \frac{\sin(2\pi sw)}{2\pi s} X_s \quad \text{and} \quad Z_{2,M}(w) = \sum_{s=-M}^{M} \frac{1-\cos(2\pi sw)}{2\pi s} X_s,$$

with the understanding that when $s = 0$, $\sin(2\pi sw)/(2\pi s) = w$ and $[1-\cos(2\pi sw)]/(2\pi s) = 0$. Then it is easy to check that for $-M \le t \le M$,

$$X_t = 2\int_0^{1/2} \cos(2\pi tw)dZ_{1,M}(w) + 2\int_0^{1/2} \sin(2\pi tw)dZ_{2,M}(w).$$

Now consider the limiting random functions Z_1 of $Z_{1,M}$ and Z_2 of $Z_{2,M}$ as $M \to \infty$

$$Z_1(w) = \sum_{s=-\infty}^{\infty} \frac{\sin(2\pi sw)}{2\pi s} X_s \quad \text{and} \quad Z_2(w) = \sum_{s=-\infty}^{\infty} \frac{1-\cos(2\pi sw)}{2\pi s} X_s,$$

which exist for $0 \le w < 1/2$ in the sense that $E[Z_1(w)^2] < \infty$ and $E[Z_2(w)^2] < \infty$. Let us denote $R_{t,M} = \int_0^{1/2} \cos(2\pi tw)dZ_{1,M}(w)$. For any given t, the sequence $\{R_{t,N}\}$ is Cauchy in the mean square sense, that is, $E[(R_{t,M} - R_{t,N})^2] \to 0$ and $M, N \to \infty$. Thus a limit of $R_{t,M}$ exists in the mean square sense as $M \to \infty$ and the limit is denoted by

$$\int_0^{1/2} \cos(2\pi tw)dZ_1(w).$$

A similar argument can be used to show that the limit of $\int_0^{1/2} \sin(2\pi tw)dZ_{2,M}(w)$ exists in the mean square sense as $M \to \infty$ and the limit is denoted by

$$\int_0^{1/2} \sin(2\pi tw)dZ_2(w).$$

Thus we can represent the time series $\{X_t\}$ as

$$X_t = 2\int_0^{1/2} \cos(2\pi tw)dZ_1(w) + 2\int_0^{1/2} \sin(2\pi tw)dZ_2(w). \tag{9}$$

The random function Z_1 has the *orthogonal increment* property, that is, if $w_1 \neq w_2$, then $Z_1(w_1 + \delta) - Z_1(w_1)$ and $Z_1(w_2 + \delta) - Z_1(w_2)$, $\delta > 0$, are uncorrelated if the intervals $(w_1, w_1 + \delta]$ and $(w_2, w_2 + \delta]$ are disjoint. Moreover, $E[\{Z_1(w + \delta) - Z_1(w)\}^2] \approx (\delta/2)f(w)$ when $\delta > 0$ is small enough. The random function Z_2 also has the same property as Z_1. Additionally, $Z_1(w_1)$ is uncorrelated with $Z_2(w_2)$ for any w_1, w_2. The representation of X_t as given above in Eq. (9) is known as the *Cramér Representation* of a stationary series.

Now consider approximating the integral $2\int_0^{1/2}\cos(2\pi tw)dZ_1(w)$ by a finite sum as follows. For a positive integer M, denote $w_j = j/(2M+1)$, $I_j = (w_{j-1}, w_j]$, $Z_1(I_j) = Z_1(w_j) - Z_1(w_{j-1})$, $j = 1, \ldots, M$. Ignoring the integral over $(w_M, 1/2)$, which is reasonable when M is large, we have

$$2\int_0^{1/2}\cos(2\pi tw)dZ_1(w) \approx 2\sum_{j=1}^{M}\int_{I_j}\cos(2\pi tw)dZ_1(w)$$
$$\approx 2\sum_{j=1}^{M}\cos(2\pi tw_j)Z_1(I_j) = \sum_{j=1}^{M}A_j\cos(2\pi tw_j),$$

with $A_j = 2Z_1(I_j)$. Due to the orthogonal increment property of the random function Z_1, the rv's $\{A_j\}$ are mutually uncorrelated and

$$\text{Var}[A_j] = 4\text{Var}[Z_1(I_j)] \approx 4[1/\{2(2M+1)\}]f(w_j) = [2/(2M+1)]f(w_j).$$

Similarly we can approximate

$$2\int_0^{1/2}\sin(2\pi tw)dZ_2(w) \approx \sum_{j=1}^{M}B_j\cos(2\pi tw_j),$$

where $B_j = 2Z_2(I_j)$. Here $\{B_j\}$ are also mutually uncorrelated with $\text{Var}[B_j] \approx [2/(2M+1)]f(w_j)$. Moreover, $\{A_j\}$ and $\{B_j\}$ are also uncorrelated with each other. Thus combining all the arguments above we have

$$X_t \approx \sum_{j=1}^{M}\{A_j\cos(2\pi w_j t) + B_j\sin(2\pi w_j t)\},$$

when M is large.

Remark 13.10.1. The Cramér representation of X_t as given in Eq. (9) is true more generally under weaker conditions than given here (Chapter 5 in Gikhman and Skorokhod [64]), but we will not concern ourselves with such mathematical details.

13.10.2 Periodogram

Consider the series in Eq. (7) with $w_j = j/n, j = 1, \ldots, M$, where M is the largest integer for which $M/n < 1/2$. We can estimate A_j and B_j from the data $\{X_1, \ldots, X_n\}$ using the method of least squares

$$\hat{A}_j = \frac{\sum_{t=1}^{n}X_t\cos(2\pi w_j t)}{\sum_{t=1}^{n}\cos^2(2\pi w_j t)},\quad \hat{B}_j = \frac{\sum_{t=1}^{n}X_t\sin(2\pi w_j t)}{\sum_{t=1}^{n}\sin^2(2\pi w_j t)},\quad \text{or}$$
$$\hat{A}_j = (2/n)\sum_{t=1}^{n}X_t\cos(2\pi w_j t),\quad \hat{B}_j = (2/n)\sum_{t=1}^{n}X_t\sin(2\pi w_j t).$$

The last equalities hold since it can be shown that

$$\sum_{t=1}^{n} \cos^2(2\pi w_j t) = n/2 \quad \text{and} \quad \sum_{t=1}^{n} \sin^2(2\pi w_j t) = n/2.$$

The quantity $P(w_j) = \hat{A}_j^2 + \hat{B}_j^2$ is called the scaled periodogram and its rescaled version

$$I(w_j) = (n/4)P(w_j)$$

is called the periodogram. The main use of the periodogram $I(w_j)$ is as an estimator of $f(w_j)$, the spectral density function at frequency $w_j = j/n$. There is a related quantity called the discrete Fourier transform of the data

$$\begin{aligned} d(w_j) &= n^{-1/2} \sum_{t=1}^{n} X_t \exp(-2\pi i w_j t) \\ &= n^{-1/2} \sum_{t=1}^{n} X_t \cos(2\pi w_j t) - i n^{-1/2} \sum_{t=1}^{n} X_t \sin(2\pi w_j t) \\ &:= \tilde{X}_{c,n}(w_j) - i\tilde{X}_{s,n}(w_j) \end{aligned}$$

where $i = \sqrt{-1}$ is the imaginary number, and $\tilde{X}_{c,n}(w_j)$ and $\tilde{X}_{s,n}(w_j)$ are the discrete cosine and sine transforms introduced in Section 13.2.1. The connection between the discrete Fourier transform and the periodogram is

$$I(w_j) = \left|d(w_j)\right|^2 = \tilde{X}_{c,n}^2 + \tilde{X}_{s,n}^2.$$

In the general case when the series $\{X_t\}$ may not have zero mean, the definitions of $\tilde{X}_{c,n}$, $\tilde{X}_{s,n}$, d, and I are based on centered observations $\{X_t - \bar{X}\}$, that is,

$$\begin{aligned} \tilde{X}_{c,n}(w_j) &= n^{-1/2} \sum_{t=1}^{n} (X_t - \bar{X}) \cos(2\pi w_j t), \\ \tilde{X}_{s,n}(w_j) &= n^{-1/2} \sum_{t=1}^{n} (X_t - \bar{X}) \sin(2\pi w_j t), \\ d(w_j) &= \tilde{X}_{c,n}(w_j) - i\tilde{X}_{s,n}(w_j),\ I(w_j) = \left|d(w_j)\right|^2 = \tilde{X}_{c,n}^2 + \tilde{X}_{s,n}^2, \end{aligned}$$

and the periodogram values are obtained at frequencies $w_j = j/n, j = 1, \ldots, M$. We should note that the periodogram values at frequencies $\{w_j\}$ as defined in the last displayed identities can be, and are usually, calculated when observations $X_1, \ldots, X_n$ from a stationary series are available.

If $\{X_t\}$ is a mean zero Gaussian series, then clearly $\tilde{X}_{c,n}(w_j)$ and $\tilde{X}_{s,n}(w_j)$ are normally distributed since each of them is a linear function of $X_1, \ldots, X_n$. However, asymptotic normality of $\tilde{X}_{c,n}(w_j)$ and $\tilde{X}_{s,n}(w_j)$ hold if the observations are from a linear process as given in Eq. (2) using arguments outlined in the proof of Theorem 13.3.1.

Theorem 13.10.1. *Let $\{X_t\}$ be a mean zero linear process as given in Eq. (2) and assume that $0 < w_j = j/n < 1/2, j = 1, \ldots, M$. As $n \to \infty$,*

(a) $\tilde{X}_n(w_j) = (\tilde{X}_{c,n}(w_j), \tilde{X}_{s,n}(w_j))^T \xrightarrow{\mathcal{D}} N_2(\mathbf{0}, (1/2)f(w_j)\boldsymbol{I}), j = 1, \ldots, M$, *where* $\boldsymbol{I}$ *is the* 2×2 *identity matrix, and* $\tilde{X}_n(w_1), \ldots, \tilde{X}_n(w_M)$ *are asymptotically independent,*

(b) $I(w_j)$ *is approximately distributed as* $(\xi_j/2)f(w_j)$, *where* $\xi_j \sim \chi_2^2, j = 1, \ldots, M$, *and* $I(w_1), \ldots, I(w_M)$ *are approximately independent.*

13.10.3 Estimation of the Spectral Density

Since $I(w_j)$ is approximately distributed as $(\xi_j/2)f(w_j)$, where $\xi_j \sim \chi_2^2$, and $\mathrm{E}[I(w_j)]$ is approximately equal to $f(w_j)$ (noting that $\mathrm{E}[\xi_j] = 2$). Thus $I(w_j)$ is almost an unbiased estimator of $f(w_j)$. However, it is not a consistent estimator since $\mathrm{Var}[I(w_j)]$ is approximately equal to $\mathrm{Var}[\xi_j/2]f(w_j)^2 = f(w_j)^2$ and this variance does not converge to zero as the sample size $n \to \infty$. In order to construct a consistent estimator of the spectral density at any point $w \in (0, 1/2)$, we may use a weighted average of $I(w_j)$ for w_j in a small neighborhood of w. Assuming that the spectral density function f is smooth, such a weighted average should lead to a better estimate of $f(w)$ since $I(w_j)$'s are asymptotically independent. Toward this purpose, we may use the kernel method for regression discussed in Chapter 9. Let K be a pdf on $[-1, 1]$ and, as in Chapter 9, assume that

$$\text{(i) } K \text{ is symmetric about } 0, \quad \text{(ii) } \int zK(z)dz = 0.$$

In the discussion below it is assumed that periodogram values at frequencies 0 and 1/2 are excluded, so the total number of frequencies M may be smaller than $[n/2]$.

We can now obtain a nonparametric estimate of $f(w)$, $w \in (0, 1/2)$, with kernel K and bandwidth $h_n \to 0$ as

$$\hat{f}(w) = \frac{\sum_{j=1}^{M} K(h_n^{-1}(w - w_j))I(w_j)}{\sum_{j=1}^{M} K(h_n^{-1}(w - w_j))}.$$

The bias and variance properties of the kernel estimate $\hat{f}(w)$ are similar to those in Chapter 9. Writing $K_j = K(h_n^{-1}(w - w_j))$ and $K_+ = \sum K_j$, we have

$$\begin{aligned}\mathrm{E}[\hat{f}(w)] - f(w) &\approx \sum K_j \mathrm{E}[\xi_j/2]f(w_j)/K_+ - f(w)\\ &= \sum K_j f(w_j)/K_+ - f(w) = \sum K_j[f(w_j) - f(w)]/K_+.\end{aligned}$$

If f is twice differentiable and f'' is continuous, then a two term Taylor expansion yields

$$f(w_j) - f(w) = (w_j - w)f'(w) + (1/2)(w_j - w)^2 f''(w_j'),$$

where w_j' lies between w and w_j. Employing arguments similar to the ones used in Chapter 9, we get the bias of $\hat{f}(w)$ as

$$\begin{aligned}E[\hat{f}(w)] - f(w) &\approx \sum K_j[f(w_j) - f(w)]/K_+ \\ &= (1/2)h_n^{-2} f''(w) \int z^2 K(z) dz[1 + o(1)] \\ &= (1/2)h_n^2 f''(w)\mu_2(K)[1 + o(1)],\end{aligned}$$

where $\mu_2(K) = \int z^2 K(z) dz$.

Approximate independence of $I(w_j)$'s may be used to calculate the variance of $\hat{f}(w)$ and thus we have

$$\begin{aligned}\text{Var}[\hat{f}(w)] &\approx \sum K_j^2 \text{Var}[\xi_j/2] f(w_j)^2/K_+^2 = \sum K_j^2 f(w_j)^2/K_+^2 \\ &\approx \left[\sum K_j^2/K_+^2\right] f(w)^2, \qquad (10)\end{aligned}$$

where the last step is justified since $f(w_j)$ is approximately equal to $f(w)$ by continuity when $|w_j - w| \leq h_n$. Once again, employing arguments similar to the ones used in Chapter 9, we have

$$\text{Var}[\hat{f}(w)] \approx (nh_n)^{-1} f(w)^2 \int K^2(z) dz = (Mh_n)^{-1} f(w)^2 \|K\|^2,$$

where $\|K\|^2 = \int K^2(z) dz$.

Since the mean square error of any estimator is the sum of its variance and square of its bias, the mean square error of $\hat{f}(w)$ is approximately given by

$$(nh_n)^{-1} f(w)^2 \|K\|^2 + (1/4)h_n^4 f''(w)^2 \mu_2(K)^2.$$

The last expression is convex in h_n and it is minimized at $h_n^* = c_1 n^{-1/5}$, where $c_1 = \left[f(w)\|K\|/\{f''(w)\mu_2(K)\}\right]^{2/5}$. The minimum mean square error of $\hat{f}(w)$ (at $h_n = h_n^*$) is approximately equal to $c_2 n^{-4/5}$, where c_2 is a constant that depends $f(w)$, $f''(w)$, $\|K\|$, and $\mu_2(K)$.

Remark 13.10.2.

(a) It is possible to obtain asymptotic normality of $\hat{f}(w)$ as given in Chapter 9 and the results are similar.

(b) As discussed in Section 9.6 of Chapter 9, a drawback of the kernel density or kernel regression estimates is that near the boundary points of the independent variable, the bias may be of order h_n and not h_n^2. However, in the case of spectral density estimation, this does not pose a problem since the spectral density f is periodic and symmetric about 0, so one can obtain a periodogram estimate at point $-w_j$, where $0 < w_j < 1/2$, by taking $I(-w_j) = I(w_j)$. Even though $I(0) = 0$ (follows from the formula), we can obtain an estimate $\hat{f}(0)$ of $f(0)$ using the kernel method since the periodogram values at negative frequencies can be obtained as mentioned above. For a frequency near zero, say at $w = h_n/2$, all the values of $I(w_j)$, $|w_j - w| \leq h_n$, are now available (substituting $\hat{f}(0)$ for the periodogram at frequency 0) and we can obtain an estimate $\hat{f}(w)$ of $f(w)$ using the kernel method. Thus the bias of $\hat{f}(w)$ is of order h_n^2 when w is close to zero. However, it should be pointed out that when w is close to 0,

the variance of $\hat{f}(w)$ is different from the formula given in Eq. (10) and it needs to be recalculated since $\hat{f}(w)$ is no longer a weighted average of approximately independent rv's.
Similar strategies can be used to estimate $f(1/2)$ and $f(w)$ when w is close to 1/2. Since f is periodic on $[-1/2, 1/2]$ and symmetric about 0, for any $1/2 < w < 1$, $f(w) = f(w-1) = f(1-w)$ and the same is also true for the periodogram values. For any $n/2 < j < n$, we can get $I(j/n) = I(1 - j/n)$. Thus a kernel estimate of $f(w)$ for w near 1/2 does not have any inadequacy in terms of inflated bias, but its variance needs to be recalculated as it is not the same as the formula given in Eq. (10).

(c) As in any nonparametric method, one needs to obtain an estimate of the bandwidth in a data dependent manner. One may apply the method of cross-validation as outlined in Chapters 9 and 11 for this purpose. It should be pointed out that in the context of kernel regression discussed in Chapter 9, the theoretical justification of the method of cross-validation relies on the assumption that the data consist of iid observations $(Y_j, X_j), j = 1, \ldots, n$. A kernel estimate of the spectral density $f(w)$ is based on the data $\{(I(w_j), w_j), j = 1, \ldots, M\}$, which are not iid. Even though w_j's are nonrandom, the use of cross-validation does not pose a problem since w_j's are equally spaced.

13.10.4 Linear Filtering

For a series $\{X_t\}$, it is sometimes of interest to study the behavior of the first difference or a running weighted average of the series such as

$$(a)\ Z_t = X_t - X_{t-1},$$
$$(b)\ Z_t = (1/2)X_t + (1/2)X_{t-1}, \text{ and}$$
$$(c)\ Z_t = (1/3)X_t + (1/3)X_{t-1} + (1/3)X_{t-2}.$$

For each of the three cases above, $\{Z_t\}$ is a linear combination of $\{X_t\}$. A linear combination of $\{X_t\}$ is called a filtered series of $\{X_t\}$. It turns out that there is a nice formula connecting the spectral density of the original series to that of the filtered series when $\{X_t\}$ is stationary. Denoting the spectral density functions of $\{X_t\}$ and $\{Z_t\}$ by f_X and f_Z, respectively, the spectral density functions for $\{Z_t\}$ for (a) and (b) are (justifications given below)

(a) $f_Z(w) = [2 - 2\cos(2\pi w)] f_X(w)$, and
(b) $f_Z(w) = (1/2)[1 + \cos(2\pi w)] f_X(w)$.

In each case, the spectral density of Z_t is equal to the spectral density of X_t times a weight function. Note that in (a), the weight function is zero at $w = 0$ and it monotonically increases to the value of 4 at $w = 1/2$. In other words, the higher the frequency, the higher is the weight, indicating that the first difference of $\{X_t\}$ is a rougher series than $\{X_t\}$. For the second case, the weight function $(1/2)[1 + \cos(2\pi w)]$ equals 1 at $w = 0$ and then it decreases to zero at $w = 1/2$. This indicates that the running average $\{Z_t\}$ of $\{X_t\}$ is smoother than $\{X_t\}$ since, for this case, higher frequencies have lower weights.

Consider the filtered series of a mean zero stationary series $\{X_t\}$,

$$Z_t = \sum_{j=-\infty}^{\infty} \psi_j X_{t-j},$$

where $\{\psi_j\}$ are constants satisfying the condition $\sum_{j=-\infty}^{\infty} |\psi_j| < \infty$. If γ_X is the autocovariance function of the series $\{X_t\}$, then

$$\begin{aligned} \mathrm{E}[Z_t^2] &= \sum_{-\infty<j,k<\infty} \psi_j \psi_k \operatorname{Cov}[X_{t-j}, X_{t-k}] \\ &= \sum_{-\infty<j,k<\infty} \psi_j \psi_k \gamma_X(j-k) < \infty, \end{aligned}$$

since γ_X is bounded and $\sum_{j=-\infty}^{\infty} |\psi_j| < \infty$. The fact that $\{Z_t\}$ is stationary follows from the fact that

$$\begin{aligned} \operatorname{Cov}[Z_t, Z_{t+h}] &= \sum_{-\infty<j,k<\infty} \psi_j \psi_k \operatorname{Cov}[X_{t-j}, X_{t+h-k}] \\ &= \sum_{-\infty<j,k<\infty} \psi_j \psi_k \gamma_X(h+j-k) \end{aligned}$$

depends only on h. Thus we conclude that $\{Z_t\}$ is stationary with the autocovariance function

$$\gamma_Z(h) = \sum_{-\infty<j,k<\infty} \psi_j \psi_k \gamma_X(h+j-k). \tag{11}$$

Now let us look at the spectral density function of the series $\{Z_t\}$ which is

$$\begin{aligned} f_Z(w) &= \sum_{h=-\infty}^{\infty} \exp(-2\pi ihw) \gamma_Z(h) \\ &= \sum_{h=-\infty}^{\infty} \exp(-2\pi ihw) \sum_{-\infty<j,k<\infty} \psi_j \psi_k \gamma_X(h+j-k) \\ &= \sum_{h=-\infty}^{\infty} \sum_{-\infty<j,k<\infty} [\psi_j \exp(2\pi ijw)][\psi_k \exp(-2\pi ikw)] \\ &\quad \times [\exp(-2\pi i(h+j-k)w) \gamma_X(h+j-k)] \\ &= \sum_{-\infty<j,k<\infty} [\psi_j \exp(2\pi ijw)][\psi_k \exp(-2\pi ikw)] \\ &\quad \times \sum_{h=-\infty}^{\infty} [\exp(-2\pi i(h+j-k)w) \gamma_X(h+j-k)]. \end{aligned}$$

Writing $l = h + j - k$, we see that

$$\sum_{h=-\infty}^{\infty} [\exp(-2\pi i(h+j-k)w)\gamma_X(h+j-k)]$$
$$= \sum_{l=-\infty}^{\infty} \exp(-2\pi ilw)\gamma_X(l) = f_X(w).$$

Hence

$$\begin{aligned} f_Z(w) &= \sum_{-\infty<j,k<\infty} [\psi_j \exp(2\pi ijw)][\psi_k \exp(-2\pi ikw)]f_X(w) \\ &= \left| \sum_{j=-\infty}^{\infty} \psi_j \exp(-2\pi ijw) \right|^2 f_X(w) \\ &= |\Psi(w)|^2 f_X(w), \text{ where} \end{aligned}$$

$$\Psi(w) = \sum_{j=-\infty}^{\infty} \psi_j \exp(-2\pi iwj) \tag{12}$$

is called the frequency response function. Thus we arrive at the following important result.

Lemma 13.10.1. *Let $\{X_t\}$ be a mean zero stationary series with autocovariance function γ_X and spectral density function f_X. Consider the filtered series $Z_t = \sum_{j=-\infty}^{\infty} \psi_j X_{t-j}$ where $\{\psi_j\}$ are constants satisfying the condition $\sum_{j=-\infty}^{\infty} |\psi_j| < \infty$.*

(a) *The series $\{Z_t\}$ is mean zero stationary with the autocovariance function $\{\gamma_Z(h)\}$ given in Eq. (11).*

(b) *The spectral density function of $\{Z_t\}$ is $f_Z(w) = |\Psi(w)|^2 f_X(w)$, where $\Psi(w)$ is the frequency response function given in Eq. (12).*

Example 13.10.1. Let $Z_t = X_t - X_{t-1}$, where $\{X_t\}$ is mean zero stationary. In this case $\psi_0 = 1$, $\psi_1 = -1$, and $\psi_j = 0$ when $j \neq 0$ and $j \neq 1$. Then the frequency response function is

$$\begin{aligned} \Psi(w) &= \sum_{j=-\infty}^{\infty} \psi_j \exp(-2\pi iwj) \\ &= (1)\exp(-2\pi iw0) + (-1)\exp(-2\pi iw) \\ &= 1 - \exp(-2\pi iw) = 1 - \cos(2\pi w) + i\sin(2\pi w), \text{ and} \\ |\Psi(w)|^2 &= [1 - \cos(2\pi w)]^2 + \sin^2(2\pi w) = 2 - 2\cos(2\pi w). \end{aligned}$$

Hence the spectral density of $\{Z_t\}$ is

$$f_Z(w) = [2 - 2\cos(2\pi w)]f_X(w).$$

Example 13.10.2. Let $Z_t = (1/2)X_t + (1/2)X_{t-1}$, where $\{X_t\}$ is mean zero stationary. Then $\psi_0 = \psi_1 = 1/2$ and $\psi_j = 0$ otherwise. The frequency response function is

$$\begin{aligned}\Psi(w) &= \sum_{j=-\infty}^{\infty} \psi_j \exp(-2\pi iwj) \\ &= (1/2)\exp(-2\pi iw0) + (1/2)\exp(-2\pi iw) \\ &= (1/2)\big[1 + \exp(-2\pi iw)\big] \\ &= (1/2)\big[1 + \cos(2\pi w) - i\sin(2\pi w)\big], \text{ and} \\ |\Psi(w)|^2 &= (1/4)[1 + \cos(2\pi w)]^2 + (1/4)\sin^2(2\pi w) \\ &= (1/4)[2 + 2\cos(2\pi w)] = (1/2)[1 + \cos(2\pi w)].\end{aligned}$$

Thus the spectral density function of $\{Z_t\}$ is

$$f_Z(w) = (1/2)[1 + \cos(2\pi w)]f_X(w).$$

Example 13.10.3. Let $Z_t = (X_t + \cdots + X_{t-L+1})/L$, where $\{X_t\}$ is mean zero and stationary. Then $\psi_j = 1/L$ when $j = 0, \ldots, L-1$, and $= 0$ otherwise. The frequency response function is

$$\begin{aligned}\Psi(w) &= \sum_{j=-\infty}^{\infty} \psi_j \exp(-2\pi iwj) = (1/L)\sum_{j=0}^{L-1} \exp(-2\pi iwj) \\ &= (1/L)\frac{1 - \exp(-2\pi iwL)}{1 - \exp(-2\pi iw)}.\end{aligned}$$

Since

$$\begin{aligned}|1 - \exp(-2\pi iwL)|^2 &= |1 - \cos(2\pi wL) + i\sin(2\pi wL)|^2 \\ &= |1 - \cos(2\pi wL)|^2 + |\sin(2\pi wL)|^2 \\ &= 2 - 2\cos(2\pi wL), \text{ and similarly} \\ |1 - \exp(-2\pi iw)|^2 &= 2 - 2\cos(2\pi w),\end{aligned}$$

we have

$$\begin{aligned}|\Psi(w)|^2 &= (1/L^2)\frac{|1 - \exp(-2\pi iwL)|^2}{|1 - \exp(-2\pi iw)|^2} \\ &= (1/L^2)\frac{2 - 2\cos(2\pi wL)}{2 - 2\cos(2\pi w)} = (1/L^2)\frac{1 - \cos(2\pi wL)}{1 - \cos(2\pi w)}.\end{aligned}$$

The spectral density function of $\{Z_t\}$ is

$$f_Z(w) = (1/L^2)\frac{1 - \cos(2\pi wL)}{1 - \cos(2\pi w)} f_X(w).$$

13.10.5 Spectral Density for ARMA

In this section we use Lemma 13.10.1 to obtain an explicit expression of the spectral density function of an *ARMA*(p,q) series. As in the previous sections, let $\phi_1, \ldots, \phi_p$ be the AR parameters and let $\theta_1, \ldots, \theta_q$ be the MA parameters of the *ARMA*(p, q) series. The polynomials used in the discussion on stationarity, invertibility, etc., in Section 13.5 are

$$\theta(z) = 1 + \theta_1 z + \cdots + \theta_q z^q, \quad \phi(z) = 1 - \phi_1 z - \cdots - \phi_p z^p.$$

Let us first consider a mean zero *MA*(q) series which is of the form $X_t = \varepsilon_t + \sum_{j=1}^{q} \theta_j \varepsilon_{t-j}$, where $\{\varepsilon_t\}$ are iid with mean 0 and variance σ^2. So $\{X_t\}$ is a filtered series of $\{\varepsilon_t\}$ with $\psi_0 = 1$, $\psi_j = \theta_j, j = 1, \ldots, q$, and $\psi_j = 0$ otherwise. Hence the frequency response function of $\{X_t\}$ is

$$\begin{aligned}\Psi(w) &= \sum_{j=-\infty}^{\infty} \psi_j \exp(-2\pi i w j) = 1 + \sum_{j=1}^{q} \theta_j \exp(-2\pi i w j) \\ &= \theta(z), \text{ with } z = \exp(-2\pi i w).\end{aligned}$$

Since the spectral density function of $\{\varepsilon_t\}$ is $f_\varepsilon(w) = \sigma^2$ for all w, the spectral density of $\{X_t\}$ is

$$f_X(w) = |\Psi(w)|^2 f_\varepsilon(w) = \sigma^2 |\theta(z)|^2, \text{ with } z = \exp(-2\pi i w).$$

Let us now find the spectral density function of an *AR*(p) series. For this series

$$\begin{aligned}X_t &= \phi_1 X_{t-1} + \cdots + \phi_p X_{t-p} + \varepsilon_t, \text{ ie,} \\ \varepsilon_t &= X_t - \phi_1 X_{t-1} - \cdots - \phi_p X_{t-p}.\end{aligned}$$

Thus $\{\varepsilon_t\}$ is a filtered series of $\{X_t\}$ with $\psi_0 = 1$, $\psi_j = -\phi_j, j = 1, \ldots, p$, and $\psi_j = 0$ otherwise. The frequency response function is

$$\begin{aligned}\Psi(w) &= \sum_{j=-\infty}^{\infty} \psi_j \exp(-2\pi i w j) = 1 - \sum_{j=1}^{p} \phi_j \exp(-2\pi i w j) \\ &= \phi(z), \text{ with } z = \exp(-2\pi i w).\end{aligned}$$

It then follows that

$$\begin{aligned}f_\varepsilon(w) &= |\phi(w)|^2 f_X(w), \text{ and hence} \\ f_X(w) &= \sigma^2 \frac{1}{|\phi(z)|^2}, \text{ with } z = \exp(\;2\pi i w).\end{aligned}$$

Now let us look at the *ARMA*(p, q) series

$$\begin{aligned}X_t &= \phi_1 X_{t-1} + \cdots + \phi_p X_{t-p} + \varepsilon_t + \theta_1 \varepsilon_{t-1} + \cdots + \theta_q \varepsilon_{t-q}, \text{ ie,} \\ X_t - \phi_1 X_{t-1} - \cdots - \phi_p X_{t-p} &= \varepsilon_t + \theta_1 \varepsilon_{t-1} + \cdots + \theta_q \varepsilon_{t-q}.\end{aligned}$$

The left-hand side of the last expression is a filtered series of $\{X_t\}$ with the frequency response function $\Psi_1(w)$ and the right-hand side is a filtered series $\{\varepsilon_t\}$ with frequency response function $\Psi_2(w)$, where $\Psi_1(w) = \phi(z)$ and $\Psi_2(w) = \theta(z)$ with $z = \exp(-2\pi i w)$. Since the spectral density function of the filtered series on the left equals the spectral density function on the right-hand side, we have

$$|\Psi_1(w)|^2 f_X(w) = |\Psi_2(w)|^2 f_\varepsilon(w) = \sigma^2 |\Psi_2(w)|^2, \text{ ie,}$$
$$|\theta(z)|^2 f_X(w) = \sigma^2 |\theta(z)|^2 f_\varepsilon(w).$$

Therefore, the spectral density function of an *ARMA*(p, q) series is

$$f_X(w) = \sigma^2 \frac{|\theta(z)|^2}{|\theta(z)|^2}, \text{ with } z = \exp(-2\pi i w).$$

Since

$$\begin{aligned}
\theta(z) &= 1 + \sum_{j=1}^{q} \theta_j \exp(-2\pi i j w) \\
&= 1 + \sum_{j=1}^{q} \theta_j \cos(2\pi j w) - i \sum_{j=1}^{q} \theta_j \sin(2\pi j w), \\
\phi(z) &= 1 - \sum_{j=1}^{p} \phi_j \exp(-2\pi i j w) \\
&= 1 - \sum_{j=1}^{p} \phi_j \cos(2\pi j w) + i \sum_{j=1}^{p} \phi_j \sin(2\pi j w),
\end{aligned}$$

the spectral density function f_X of an *ARMA*(p, q) series can be written as

$$f_X(w) = \sigma^2 \frac{\left[1 + \sum_{j=1}^{q} \theta_j \cos(2\pi j w)\right]^2 + \left[\sum_{j=1}^{q} \theta_j \sin(2\pi j w)\right]^2}{\left[1 - \sum_{j=1}^{p} \phi_j \cos(2\pi j w)\right] + \left[\sum_{j=1}^{p} \phi_j \sin(2\pi j w)\right]^2}.$$

Some Special Cases

I. Spectral density of *AR*(1).
When $z = \exp(-2\pi i w)$, we have

$$|\phi(z)|^2 = 1 + \phi^2 - 2\phi \cos(2\pi w).$$

Hence we get

$$f_X(w) = \sigma^2 \frac{1}{1 + \phi^2 - 2\phi \cos(2\pi w)}.$$

II. Spectral density of *AR*(2).
Using some algebra, we have

$$\begin{aligned}
|\phi(z)|^2 &= \left[1 - \phi_1 \cos(2\pi w) - \phi_2 \cos(4\pi w)\right]^2 + \left[\phi_1 \sin(2\pi j w) + \phi_2 \sin(4\pi w)\right]^2 \\
&= 1 + \phi_1^2 + \phi_2^2 + 2\phi_1(1 - \phi_2)\cos(2\pi w) - 2\phi_2 \cos(4\pi w).
\end{aligned}$$

Since $\cos(4\pi w) = 2\cos^2(2\pi w) - 1$, substituting this in the last expression, we have

$$|\phi(z)|^2 = \phi_1^2 + (1+\phi_2)^2 + 2\phi_1(1-\phi_2)\cos(2\pi w) - 4\phi_2\cos^2(2\pi w),$$

which is a quadratic polynomial in $\cos(2\pi w)$. Thus the spectral density function is

$$f_X(w) = \sigma^2 \frac{1}{\phi_1^2 + (1+\phi_2)^2 + 2\phi_1(1-\phi_2)\cos(2\pi w) - 4\phi_2\cos^2(2\pi w)}.$$

III. Spectral density of $MA(1)$.
Here

$$|\theta(z)|^2 = 1 + \theta^2 + 2\theta\cos(2\pi w), \text{ with } z = \exp(-2\pi i w),$$

and hence the spectral density is

$$f_X(w) = \sigma^2[1 + \theta^2 + 2\theta\cos(2\pi w)].$$

IV. Spectral density of $ARMA(1,1)$.
If the parameters of an $ARMA(1,1)$ series are ϕ and θ, then

$$|\theta(z)|^2 = 1 + \theta^2 + 2\theta\cos(2\pi w), \text{ and}$$
$$|\phi(z)|^2 = 1 + \phi^2 - 2\phi\cos(2\pi w), \text{ with } z = \exp(-2\pi i w).$$

Thus the spectral density of an $ARMA(1,1)$ series is

$$f_X(w) = \sigma^2 \frac{1 + \theta^2 + 2\theta\cos(2\pi w)}{1 + \phi^2 - 2\phi\cos(2\pi w)}.$$

Exercises

In all the problems below, the autocorrelations and autocovariances of stationary series are denoted by $\{\gamma(h)\}$ and $\{\rho(h)\}$, respectively. For any ARMA series, it is understood that the mean and variance of the innovations $\{\varepsilon_t\}$ are 0 and σ^2, respectively.

13.1. Let $\{X_{1t}\}, \ldots, \{X_{kt}\}$ be k independent stationary series with autocovariance functions $\gamma_1, \ldots, \gamma_k$. Then show that $W_t - c_1X_{t1} + \cdots + c_kX_{tk}$, where $c_1, \ldots, c_k$ are constants, is also stationary. Find the autocovariances and autocorrelations of the series $\{W_t\}$ in terms of the autocovariance and autocorrelation functions of the series $\{X_{1t}\}, \ldots, \{X_{kt}\}$.

13.2. For each of the following models, determine if it is stationary and invertible. It is understood that $\{\varepsilon_t\}$ are iid with mean 0 and variance σ^2.

(i) $X_t = 6 + \varepsilon_t + 1.2\varepsilon_{t-1}$.
(ii) $X_t = -5 + \varepsilon_t + 0.6\varepsilon_{t-1} + 0.7\varepsilon_{t-2}$.
(iii) $X_t = 3 + 0.5X_{t-1} + \varepsilon_t = 0.4\varepsilon_{t-1}$.
(iv) $X_t = 9 + 0.7X_{t-1} + 0.6X_{t-1} + \varepsilon_t$.
(v) $X_t = 2 - 0.5X_{t-1} - 0.4X_{t-2} + \varepsilon_t + 0.3\varepsilon_{t-1} + 0.6\varepsilon_{t-2}$.

13.3. Assume that $\{X_t\}$ follows a stationary *ARMA*(1, 1) model with the autoregressive and moving average parameters ϕ and θ, respectively, with $\phi \neq -\theta$.

(a) Show that for any positive integer r, one may write X_t as

$$X_t = \phi^r X_{t-r} + \sum_{j=0}^{r-1} \psi_j \varepsilon_{t-j} + \phi^{r-1}\theta\varepsilon_{t-r},$$

where $\psi_0 = 1$ and $\psi_j = (\phi + \theta)\phi^{j-1}, j = 1, \ldots$.

(b) Use the result in part (a) to argue that $X_t = \sum_{j=0}^{\infty} \psi_j \varepsilon_{t-j}$, where $\{\psi_j\}$ are as given in part (a).

(c) Show that $\text{Var}[X_t] = \left[1 + (\phi + \theta)^2(1 - \phi^2)^{-1}\right]\sigma^2$.

(d) Show that the autocovariance function of $\{X_t\}$ is $\gamma(h) = (\phi + \theta)(1 + \phi\theta)(1 - \phi^2)^{-1}\sigma^2, h = 1, 2, \ldots$.

13.4. Let $\{X_t\}$ be a stationary *ARMA*(1, 1) series with mean μ. Assuming that ϕ, θ, and σ^2 are known, use the results in Exercise 13.3 to obtain an estimate of $\tau_n^2 = \text{Var}[\hat{\mu}]$ in terms of ϕ, θ, σ^2, and n, where $\hat{\mu} = n^{-1}\sum_{t=1}^{n} X_t$ is an estimate of μ based on the available data $X_1, \ldots, X_n$.

13.5. Let $\{X_t\}$ be stationary (not necessarily *AR*(p)) with mean μ. Denote $\rho(1)$ by ϕ.

(a) Show that the best linear predictor (forecast) of X_t from X_{t-1} is $X_t^{(f)} = \mu + \phi(X_{t-1} - \mu)$.

(b) Show that the best linear predictor (backcast) of X_{t-2} from X_{t-1} is $X_{t-2}^{(b)} = \mu + \phi(X_{t-1} - \mu)$.

(c) Show that $\text{Var}\left[X_t - X_t^{(f)}\right] = (1 + \phi^2)\gamma(0) - 2\phi\gamma(1)$.

(d) Show that $\text{Var}\left[X_{t-2} - X_{t-2}^{(f)}\right] = (1 + \phi^2)\gamma(0) - 2\phi\gamma(1)$.

(e) Show that $\text{Cov}\left[X_t - X_t^{(f)}, X_{t-2} - X_{t-2}^{(b)}\right] = \gamma(2) - 2\phi\gamma(1) + \phi^2\gamma(0)$.

(f) Show that the partial correlation between X_t and X_{t-2} given X_{t-1} is given by $[\gamma(2) - 2\phi\gamma(1) + \phi^2\gamma(0)]/[(1 + \phi^2)\gamma(0) - 2\phi\gamma(1)]$.

13.6. Suppose that when fitting an *AR*(p) model to the data $X_1, \ldots, X_n$ from a stationary series $\{X_t\}$ with mean μ, the data are expanded by padding with $2p$ extra values the sample mean $\bar{X}$ at the beginning and at the end. In particular, let $\tilde{X}_t, t = -p + 1, \ldots, n + p$ be such that $\tilde{X}_t = X_t$ if $1 \leq t \leq n$ and $\tilde{X}_t = \bar{X}$ otherwise. Show that the normal equations obtained by minimizing $\sum_{t=1}^{n+p}\left(Y_t - \phi_1 Y_{t-1} - \cdots - \phi_p Y_{t-p}\right)^2$, where $Y_t = \tilde{X}_t - \bar{X}$, with respect to $\phi_1, \ldots, \phi_p$ are the same as the Yule-Walker equations.

13.7. For an invertible *MA*(1) model obtain the partial correlation function $\{\pi(h)\}$ and express it as a function of the moving average parameter θ.

13.8. Let $\{X_t\}$ be a stationary, invertible, and nonredundant *ARMA*(1, 1) series with zero mean, and AR and MA parameters ϕ and θ.

(a) Obtain an invertible representation of $\{X_t\}$ as Eq. (6) by finding $\{\psi_j\}$ explicitly in terms of ϕ and θ.

(b) If the $ARMA(1,1)$ series $\{X_t\}$ has mean zero and its invertible representation is $X_{t+1} = \sum_{j=0} \pi_j X_{t-j} + \varepsilon_t$, then consider predicting X_{n+1} using $\hat{X}_{n+1} = \pi_1 X_n + \cdots + \pi_n X_1$. Find the mean square error of prediction $\mathrm{E}[\hat{X}_{n+1} - X_{n+1}]^2$.

13.9. In Section 13.5 (after Lemma 13.5.1), it has been pointed out that the autocorrelation function of a stationary $AR(2)$ series is representable in terms of the roots z_1, z_2 of the equation $g(z) = 0$, where $g(z) = z^2 - \phi_1 z - \phi_2$. Condition of stationarity requires that (ϕ_1, ϕ_2) is inside the triangular region $\Delta = \{(u_1, u_2)\colon |u_2| < 1, |u_1/(1-u_2)| < 1\}$.
Show that (ϕ_1, ϕ_2) is inside Δ if and only if $|z_1| < 1$ and $|z_2| < 1$.

13.10. Let $\{X_t\}$ be an $AR(2)$ series with autoregressive coefficients ϕ_1 and ϕ_2.

(a) Show that $|\phi_2| < 1$.

(b) Show that $\rho(1) = \phi_1/(1-\phi_2)$.

(c) $\sigma^2 = (1 - \phi_1^2 - \phi_2^2)\gamma(0) - 2\phi_1\phi_2\gamma(1)$.

(d) Show that $\rho(h) = \phi_1\rho(h-1) + \phi_2\rho(h-2)$, $h \geq 2$.

13.11. For an $AR(1)$ series $\{X_t\}$ with autoregressive coefficient $-1 < \phi < 1$, show that the prediction error for predicting X_{n+h} by its best linear predictor based on $X_1, \ldots, X_n$ is $\sigma^2(h) = \sigma^2 \frac{1-\phi^{2h}}{1-\phi^2}$.

13.12. **(a)** Let $\{X_t\}$ be a mean zero time series following an $MA(1)$ model. Let $\hat{X}_t$ be the forecasted value of X_t based on the past $X_{t-1}, X_{t-2}, \ldots$, and $\hat{X}_{t+1}$ be the forecasted value of X_{t+1} based on the past $X_t, X_{t-1}, \ldots$ Show that $\hat{X}_{t+1} = \theta(X_t - \hat{X}_t)$. [Here θ is the moving average parameter.]

(b) The series $\{Y_t\}$ follows an $ARIMA(0,1,1)$ model and assume that the series $\{X_t\}$, the first difference of $\{Y_t\}$, has zero mean. Let $\hat{Y}_t$ be the forecasted value of Y_t based on the past $Y_{t-1}, Y_{t-2}, \ldots$, and $\hat{Y}_{t+1}$ be the forecasted value of Y_{t+1} based on the past $Y_t, Y_{t-1}, \ldots$. Show that $\hat{Y}_{t+1} = Y_t + \theta(Y_t - \hat{Y}_t)$, where $-1 < \theta < 1$ is the moving average coefficient.

13.13. If $\{X_t\}$ follows an $MA(q)$ model, then show that the asymptotic variance of $\hat{\rho}(h)$ is $\sum_{j=-q}^{q} \rho(j)^2/n$, for any $h \geq q+1$.

13.14. **(a)** If $\{X_t\}$ is stationary $AR(1)$ and the parameter ϕ is estimated using the Yule-Walker method based on a sample $X_1, \ldots, X_n$. Obtain the asymptotic distribution of $\sqrt{n}(\hat{\phi} - \phi)$ and explicitly obtain the parameters of this distribution in terms of ϕ and σ^2.

(b) Repeat part (a) for the parameter estimate $\hat{\theta}$ of θ for an $MA(1)$ series.

(c) If $\{X_t\}$ is stationary $ARMA(1,1)$ and it is invertible and nonredundant. Let $\hat{\phi}$ and $\hat{\theta}$ be the estimates of ϕ and θ using the methods in Section 13.8.3. Find the joint asymptotic distribution of $\sqrt{n}(\hat{\phi} - \phi)$ and $\sqrt{n}(\hat{\theta} - \theta)$, and explicitly obtain the parameters of this distribution in terms of ϕ, θ, and σ^2.

13.15. Prove Lemma 13.6.1.

13.16. Let $\{X_t\}$ be as in Eq. (7). Show that $\{X_t\}$ is stationary with the autocovariance function $\{\gamma(h)\}$, where $\gamma(h) = \sum_{j=1}^{M} \sigma_j^2 \cos(2\pi w_j h)$.

13.17. Let $\{X_t\}$ be stationary and define $W_t = (X_{t-2} + 2X_{t-1} + 3X_t + 2X_{t+1} + X_{t+2})$.

(a) Find the frequency response function $\Psi(w)$.

(b) Obtain the spectral density function of $\{W_t\}$ in terms of the spectral density function of $\{X_t\}$.

(c) Plot the square of the absolute value of the frequency response function.

13.18. For a stationary series $\{X_t\}$, its second difference is $W_t = X_t - 2X_{t-1} + X_{t-2}$.

(a) Find the frequency response function $\Psi(w)$.

(b) Obtain the spectral density function of $\{W_t\}$ in terms of the spectral density function of $\{X_t\}$.

(c) Plot the square of the absolute value of the frequency response function.

Appendix A
Results From Analysis and Probability

A.1 Some Important Results in Integration Theory

Theorem A.1.1 (Lebesgue Dominated Convergence). *Let $\{f_n\}$ be a sequence of integrable functions on $\mathcal{X}$. If*

(i) $\lim_{n\to\infty} f_n(x) = f(x)$ *a.e. in $\mathcal{X}$, that is, for all $x \notin \mathcal{S}$ where $\int_{\mathcal{S}} dx = 0$, and*
(ii) *there is an integrable function g on $\mathcal{X}$ such that $|f_n(x)| \le g(x)$ for all n and for all $x \in \mathcal{X}$,*

then

$$\lim_{n\to\infty} \int_{\mathcal{X}} f_n(x)\, dx = \int_{\mathcal{X}} \lim_{n\to\infty} f_n(x)\, dx = \int_{\mathcal{X}} f(x)\, dx.$$

Proof. See Royden [65, p. 88]. □

The nest two theorems follow from the Dominated Convergence Theorem in a straightforward manner.

Theorem A.1.2 (Monotone Convergence). *Let $\{f_n\}$ be a sequence of nonnegative functions on $\mathcal{X}$ such that $0 \le f_1(x) \le f_2(x) \le \cdots$ and let $\lim_{n\to\infty} f_n(x) = f(x)$ where f is integrable on $\mathcal{X}$. Then*

$$\lim_{n\to\infty} \int_{\mathcal{X}} f_n(x) = \int_{\mathcal{X}} f(x)\, dx.$$

Theorem A.1.3 (Differentiation Under Integration). *Let $f(x,t)$ for $(x,t) \in [a,b] \times [c,d]$ be such that*

(i) *$f(x,t)$ is an integrable function of x on $[a,b]$ for each $t \in [c,d]$,*
(ii) *the partial derivative $\partial f/\partial t$ exists and is bounded on $[a,b] \times [c,d]$.*

Then

$$\frac{d}{dt} \int_a^b f(x,t)\, dx = \int_a^b \frac{\partial f(x,t)}{\partial t}\, dx.$$

Theorem A.1.4 (Fubini). *If $f(x,y)$ is integrable on $\mathcal{X} \times \mathcal{Y}$, then*

$$\iint_{\mathcal{X}\times\mathcal{Y}} f(x,y)\,dx\,dy = \int_{\mathcal{X}}\left[\int_{\mathcal{Y}} f(x,y)\,dy\right]dx = \int_{\mathcal{Y}}\left[\int_{\mathcal{X}} f(x,y)\,dx\right]dy,$$

that is, the double integral can be evaluated interactively either way.

A.2 Convex Functions

Definition A.2.1. A real-valued function f on an interval $(a,b) \subset \mathbb{R}$, or more generally on $(a_1,b_1) \times \cdots \times (a_k,b_k) \subset \mathbb{R}^k$, is said to be convex if for any x_1, x_2 in its domain and for any $0 < \lambda < 1$,

$$f(\lambda x_1 + (1-\lambda)x_2) \leq \lambda f(x_1) + (1-\lambda)f(x_2).$$

The function is strictly convex if the above inequality is a strict inequality for all such x_1, x_2 and λ.

Geometrically, a function is convex if the straight line joining any two points on its graph lies entirely above the graph.

A twice differentiable function f on $(a,b) \subset R$ is convex iff $f''(x) \geq 0$ for all $a < x < b$. More generally, a function f on $(a_1,b_1) \times \cdots \times (a_k,b_k) \subset R^k$ for which all second partial derivatives exist and are finite, is convex iff the Hessian (ie, the matrix of second partial derivatives) is nonnegative definite.

From the above definition it follows by induction, that if X is an rv taking values $x_1,\ldots,x_r$ in (a,b) with $P[X = x_i] = \lambda_i, i = 1,\ldots,r$ with $\sum_{i=1}^r \lambda_i = 1$, then

$$f(\mathrm{E}[X]) = f\left(\sum_{i=1}^r \lambda_i x_i\right) \leq \sum_{i=1}^r \lambda_i f(x_i) = \mathrm{E}[f(X)].$$

Obviously, this inequality also holds for a random vector X taking values in $(a_1,b_1) \times \cdots \times (a_k,b_k) \subset R^k$.

The following theorem asserts that this inequality holds for arbitrary random vector X with finite expectation.

Theorem A.2.1 (Jensen's Inequality). *If f is a convex function on $I = (a_1,b_1) \times \cdots \times (a_k,b_k) \subset R^k$ and $\mathbf{X}$ is a k-dim random vector with $P[\mathbf{X} \in I] = 1$, and with finite expectation, then*

$$f(E[\mathbf{X}]) \leq E[f(\mathbf{X})].$$

Moreover the above inequality is strict iff f is strictly convex, unless $\mathbf{X}$ is a constant with probability 1.

Proof. The proof depends on the Supporting Hyperplane Theorem for convex sets. See Ferguson [1, p. 76]. □

An Application

Consider the function $f(x, y) = -x^p y^{1-p}$, $x > 0$, $y > 0$, and $0 < p < 1$, which can be shown to be convex on the first quadrant on R^2 by verifying that its Hessian is nonnegative definite. The following theorem is now proved by an application of Jensen's Inequality.

Theorem A.2.2 (Hölder's Inequality). *If X and Y are positive rv's with finite means and* $0 \le p \le 1$, *then*

$$E\left[X^p Y^{1-p}\right] \le (E[X])^p (E[Y])^{1-p}.$$

Taking $X = U^2$ *and* $Y = V^2$ *and* $p = 1/2$, *the Cauchy-Schwarz inequality follows as a special case.*

A.3 Stieltjes Integral

Let f and g be real-valued functions on an interval $[a, b]$. The Stieltjes integral of f with respect to g on $[a, b]$, denoted by $\int_a^b f(x)\ dg(x)$ or simply $\int_a^b f\ dg$ is a generalization of the Riemann integral $\int_a^b f(x)\ dx$. As in Riemann integration, we need the concept of partitions.

Definition A.3.1.

(i) A partition of $[a, b]$ is a finite set of real numbers $P = \{x_0, x_1, \ldots, x_n\}$ where $a = x_0 < x_1 < \cdots < x_n = b$, of which $[x_{i-1}, x_i]$ are segments with length $\Delta x_i = x_i - x_{i-1}$, and $\Delta P = \max\{\Delta x_i, i = 1, \ldots, n\}$ is the norm of P.

(ii) A partition $Q = \{y_0, y_1, \ldots, y_m\}$ is a refinement of $P = \{x_0, x_1, \ldots, x_n\}$ if $P \subset Q$, in which case, $\Delta Q \le \Delta P$.

(iii) A partition $Q = \{\xi_1, \ldots, \xi_n\}$ is an intermediate partition of $P = \{x_0, x_1, \ldots, x_n\}$ if $x_{i-1} \le \xi_i \le x_i$ for all i.

(iv) For real-valued functions f and g on $[a, b]$, a partition $P = \{x_0, x_1, \ldots, x_n\}$ of $[a, b]$ and an intermediate partition $Q = \{\xi_1, \ldots, \xi_n\}$ of P, the Stieltjes sum of f with respect to g on $[a, b]$ corresponding to P and Q is defined as

$$S(f, g, P, Q) = \sum_{i=1}^{n} f(\xi_i)\Delta g_i, \quad \text{where } \Delta g_i = g(x_i) - g(x_{i-1}).$$

[This generalizes the Riemann Sum $\sum_{i=1}^{n} f(\xi_i)\Delta x_i$, where $\Delta x_i = x_i - x_{i-1}$.]

(v) The Stieltjes integral of f with respect to g on $[a, b]$ is defined as a number $\int_a^b f(x)\ dg(x)$ having the property that for every $\varepsilon > 0$ there exists a $\delta > 0$ such that

$$\left| S(f, g, P, Q) - \int_a^b f\ dg \right| < \varepsilon$$

for all P with $\Delta P < \delta$ and for all intermediate partitions Q of P.

[Definition (v) is a formal way of saying, $\int_a^b f\,dg = \lim_{\Delta P \to 0} S(f, g, P, Q)$.]

As in Riemann integration,

$$\int_b^a f dg = -\int_a^b f\,dg, \quad \text{for all } a < b, \text{ and}$$
$$\int_a^a f dg = 0, \quad \text{by convention.}$$

Existence

The Stieltjes integral $\int_a^b f\,dg$ exists if f is continuous and g is nondecreasing on $[a, b]$. More generally, $\int_a^b$ exists if f has at most a finite number of discontinuities, g is of bounded variation (as defined below) and f and g have no common discontinuity on $[a, b]$. For practical purposes, this generality will suffice.

Definition A.3.2 (Functions of Bounded Variation). The variation of f on $[a, b]$ corresponding to a partition $P = \{x_0, x_1, \ldots, x_n\}$

$$V^P(f) = \sum_{i=1}^{n} |f(x_i) - f(x_{i-1})|.$$

Obviously, if Q is a refinement of P, then $V^P(f) \le V^Q(f)$. If the set $\{V^P(f)$: P is a partition of $[a, b]\}$ is bounded, then f is called a function of bounded variation on $[a, b]$ and $V^{[a,b]}(f) = \sup_P V^P(f)$ is the total variation of f on $[a, b]$.

If f is differentiable on $[a, b]$ with $|f'(x)| \le M$ for all $x \in [a, b]$, then f is of bounded variation on $[a, b]$ and $V^{[a,b]}(f) \le M(b-a)$. More generally, f is of bounded variation iff it is the difference of two nondecreasing functions. Indeed, for a function of bounded variation if we let

$$v(x; f) = V^{[a,x]}(f) \quad \text{for } a < x \le b \text{ and } v(a; f) = 0,$$

then the function $v(x; f)$ called the total variation function of f and the function $r(x; f) = v(x; f) - f(x)$ called the residual function of f, on $[a, b]$, are both nondecreasing and $f = v - r$.

Properties of Stieltjes Integrals

1. If f is Riemann integrable and g has continuous derivative g' on $[a, b]$, then the Riemann integral $\int_a^b f(x)g'(x)\,dx$ and the Stieltjes integral $\int_a^b f(x)\,dg(x)$ both exit and are equal.

2. (a) If $\int_a^b f_i\, dg$, $i = 1, 2$, exist, then so does $\int_a^b (k_1 f_1 + k_2 f_2)\, dg$ for k_1 and k_2 constants and

$$\int_a^b (k_1 f_1 + k_2 f_2)\, dg = k_1 \int_a^b f_1\, dg + k_2 \int_a^b f_2\, dg,$$

and (a′) if $\int_a^b f dg_i$, $i = 1, 2$, exist, then so does $\int_a^b f d(k_1 g_1 + k_2 g_2)$ for k_1, k_2 constants and

$$\int_a^b f\, d(k_1 g_1 + k_2 g_2) = k_1 \int_a^b f\, dg_1 + k_2 \int_a^b f\, dg_2.$$

(b) If $\int_a^b f\, dg$ exists, then for $a < c < b$, $\int_a^c f\, dg$ and $\int_c^b f\, dg$ exist and

$$\int_a^b f\, dg = \int_a^c f\, dg + \int_c^b f\, dg.$$

(c) If $\int_a^b f\, dg$ exists, then $\int_c^d f\, dg$ exists for any $[c, d] \subset [a, b]$.

Theorem A.3.1 (Integration by Parts). *If $\int_a^b f\, dg$ exists, then $\int_a^b g\, df$ also exists and*

$$\int_a^b f\, dg = f(b)g(b) - f(a)g(a) - \int_a^b g\, df.$$

4. *Change of variable.* Suppose that $\int_a^b f\, dg$ exists, h is a strictly increasing and continuous function on $[p, q]$ with $h(p) = a$ and $h(q) = b$. Then for $F = f \circ h$ and $G = g \circ h$ on $[p, q]$, $\int_p^q F\, dG$ exists and is equal to $\int_a^b f\, dg$.

Riemann-Stieltjes (R-S) integral. If in the Stieltjes Sum $S(f, g, P, Q)$ we replace $f(\xi_i)$ by $m_i = \inf_{x \in [x_{i-1}, x_i]} f(x)$ or $M_i = \sup_{x \in [x_{i-1}, x_i]} f(x)$, then the resulting sums, denoted by $\underline{RS}(f, g, P)$ and $\overline{RS}(f, g, P)$, respectively, are called the Lower and the Upper R-S sums which are generalizations of the Lower and Upper Riemann Sums. Since $\underline{RS}(f, g, P) \le \overline{RS}(f, g, Q)$ for all partitions P, Q of $[a, b]$, we have

$$\underline{\int_a^b} f\, dg := \sup_P \underline{RS}(f, g, P) \le \inf_P \overline{RS}(f, g, P) := \overline{\int_a^b} f\, dg.$$

If $\underline{\int_a^b} f\, dg$ and $\overline{\int_a^b} f\, dg$, called the lower and upper R-S integrals are equal, then the common value is called the R-S integral of f with respect to g on $[a, b]$.

The Stieltjes integral and the Riemann-Stieltjes integral both exist and are equal if f has at most a finite number of discontinuities, g is nondecreasing, and f and g do not have any common discontinuity on $[a, b]$.

Expected value of a random variable. The expected value of an rv with cdf F is defined as

$$\mathrm{E}_F[X] = \int x\, dF(x).$$

More generally, $E_F[g(X)] = \int g(x)\, dF(x)$ where g has at most a finite number of discontinuities.

If F is differentiable, $F' = f$ is the pdf of X, and $\int x\, dF(x) = \int xf(x)\, dx$. If F increases only by jumps at $x_1, x_2, \ldots$ with jump size $f(x_i)$ at x_i, then X is discrete with pmf $f(x_i), i = 1, 2, \ldots$, and $\int x\, dF(x) = \sum_i x_i f(x_i)$.

If $P[X \geq 0] = 1$, then we have the following alternative expression for $E_F[X] < \infty$, using integration by parts:

$$\begin{aligned} E_F[X] &= \int_0^\infty x\, dF(x) = -\int_0^\infty x\, d[1 - F(x)] \\ &= \int_0^\infty [1 - F(x)]\, dx = \int_0^\infty P[X > x]\, dx, \end{aligned}$$

because $x[1 - F(x)] \leq \int_x^\infty yF(y)$ which converges to 0 as $x \to \infty$ by virtue of $E_F[X] < \infty$.

Empirical cdf. For a random sample $(X_1, \ldots, X_n)$ from F, the function

$$F_n(x) = n^{-1} \sum_{i=1}^{n} I_{(-\infty, x]}(X_i),$$

which increases by jumps of size $1/n$ at $X_1, \ldots, X_n$ is called the empirical cdf. It follows that

$$\int g(x)\, dF_n(x) = n^{-1} \sum_{i=1}^{n} g(X_i) = \text{Sample mean of } g(X_1), \ldots, g(X_n).$$

We often want to deal with $n^{-1} \sum_{i=1}^{n} g(X_i) - E_F[g(X)]$, the difference between the sample mean of $g(X_1), \ldots, g(X_n)$ and its expected value. This can be represented as

$$\begin{aligned} n^{-1} \sum_{i=1}^{n} g(X_i) - E_F[g(X)] &= \int g(x)\, dF_n(x) - \int g(x)\, dF(x) \\ &= \int g(x) d[F_n(x) - F(x)]. \end{aligned}$$

A.4 Characteristic Function, Weak Law of Large Number, and Central Limit Theorem

Theorem A.4.1 (Helly-Bray Theorem). $X_n \xrightarrow{\mathcal{L}} X$ *implies* $E[g(X_n)] \to E[g(X)]$ *for all bounded and a.e. continuous functions* g.

Proof. See Breiman [42, p. 160]. □

Note. Let D_g denote the set of discontinuity points of g. If $P[X \in D_g] = 0$, then g is a.e. continuous.

Characteristics Function

Definition A.4.1. The characteristic function (cf) of an rv X with cdf F, or the cf of F is defined to be

$$\varphi(t) = \varphi_X(t) = \mathrm{E}[\mathrm{e}^{itX}] = \int_{-\infty}^{\infty} \mathrm{e}^{itx}\, dF(x) = \int_{-\infty}^{\infty} \cos(tx)\, dF(x) + i \int \sin(tx)\, dF(x),$$

where t is real and $i = \sqrt{-1}$. In general, $\varphi(t)$ is complex-valued, with

$$\varphi(0) = 1, |\varphi(t)| \leq \int_{-\infty}^{\infty} dF(x) = 1 \text{ for all } t, \text{ and } \varphi(-t) = \overline{\varphi(t)} \quad \text{for all } t,$$

where $\bar{z}$ denotes the complex conjugate of a complex number z.

If f is symmetrically distributed (about 0) (ie, $X \overset{\mathcal{D}}{=} -X$), then

$$\overline{\varphi(t)} = \varphi(-t) = \mathrm{E}[\mathrm{e}^{i(-t)X}] = \mathrm{E}[^{it(-X)}] = \mathrm{E}[\mathrm{e}^{itX}] = \varphi(t).$$

Thus the cf of a symmetric rv is real-valued function of t.

By dominated convergence, $\varphi(t)$ is continuous and if $m_k = \mathrm{E}\,[X^k]$ exists, then $\varphi(t)$ is k-times differentiable; moreover, we can differentiate under the integral sign, that is,

$$\varphi^{(r)}(t) = i^r \int_{-\infty}^{\infty} x^r \mathrm{e}^{itx}\, dF(x) \text{ for } 0 \leq r \leq k, \text{ and}$$

$$\varphi^{(r)}(0) = i^r \int_{-\infty}^{\infty} x^r\, dF(x) = i^r \mathrm{E}[X^r] = i^r m_r, \quad 0 \leq r \leq k.$$

In the neighborhood of $t = 0$, we have the McLaurin Series

$$\varphi(t) = 1 + \sum_{r=1}^{k} (m_r/r!)(it)^r + o(t^k), \quad \text{as } t \to 0.$$

Special Cases

1. Let Φ denote the cdf of $N(0, 1)$. Then the cf of Φ is

$$\varphi(t) = \int_{-\infty}^{\infty} \mathrm{e}^{itx}\, d\Phi(x) = \mathrm{e}^{-t^2/2}.$$

2. Let F denote the cdf of an rv X with $P[X = c] = 1$ (ie, $F(x) = 0$ for $x < c$ and $F(x) = 1$ for $x \geq c$). Then the cf of F is $\varphi(t) = \mathrm{e}^{itc}$.

Properties of cf

1. If φ_X is the cf of X, then $\varphi_{aX+b}(t) = \mathrm{e}^{itb}\varphi_X(at)$ for constant a, b.
2. If $X_1, \ldots, X_n$ are independent, then $\varphi_{\sum_{i=1}^n X_i}(t) = \prod_{i=1}^n \varphi_{X_i}(t)$.

In particular, if $X_1, \ldots, X_n$ are iid as X, then

$$\varphi_{\sum_{i=1}^n X_i/\sqrt{n}}(t) = \{\varphi_X(t/\sqrt{n})\}^n.$$

We now state the following two fundamental theorems in the theory of characteristic functions, for the proof of which we refer to Cramér [18, p. 93–8].

Theorem A.4.2 (Uniqueness of Characteristic Functions on $\mathbb{R}$). *If $(a-h, a+h)$ is a continuity interval of a cdf F (ie, $a \pm h$ are continuity points of F), and if φ is the cf of F, then*

$$F(a+h) - F(a-h) = \lim_{T\to\infty} \frac{1}{\pi} \int_{-T}^{T} \frac{\sin(ht)}{t} e^{-ita} \varphi(t)\, dt.$$

Consequently, the cf φ determines the cdf F.

Theorem A.4.3 (Continuity of Characteristic Functions on R). *If $\{F_n\}$ is a sequence of cdf's and $\{\varphi_n\}$ is the corresponding sequence of cf's, then $F_n \to F$ (at all continuity points of F) iff there exists a φ which is continuous at $t = 0$, such that $\varphi_n(t) \to \varphi(t)$ for all t. Moreover, if there is such a φ, then it is the cf of F.*

Note. The Uniqueness and Continuity Theorems for Characteristic Functions also extend to R^k. See Cramér [18, p. 100–3].

Remark A.4.1. Combining the Helly-Bray Theorem with the Continuity Theorem for Characteristic Functions, we now conclude that the following are equivalent:

1. $X_n \xrightarrow{\mathcal{L}} X$.
2. $F_n(x) \to F(x)$ at all continuity points of F.
3. $\mathrm{E}[g(X_n)] \to \mathrm{E}[g(X)]$ for all bounded and a.e. continuous functions g.
4. $\varphi_n(t) = \mathrm{E}\left[e^{itX_n}\right] \to \varphi(t) = \mathrm{E}\left[e^{itX}\right]$ for all t.

Applications of the Continuity Theorem for Characteristic Functions

Theorem A.4.4. *If $X_n \xrightarrow{\mathcal{L}} X$ and g is a a.e. continuous function, then $g(X_n) \xrightarrow{\mathcal{L}} g(X)$.*

Proof. Since $\cos(tg(x))$ and $\sin(tg(x))$ are a.e. continuous functions of x for every t, it follows from the Helly-Bray Theorem that

$$\begin{aligned}\varphi_{g(X_n)}(t) &= \mathrm{E}\left[e^{itg(X_n)}\right] = \mathrm{E}\left[\cos(tg(X_n))\right] + i\mathrm{E}[\sin(tg(X_n))] \\ &\to \mathrm{E}\left[\cos(tg(X))\right] + i\mathrm{E}[\sin(tg(X))] = \mathrm{E}\left[e^{itg(X)}\right] = \varphi_{g(X)}(t).\end{aligned}$$

Hence $g(X_n) \xrightarrow{\mathcal{L}} g(X)$ by the Continuity Theorem. □

We now prove Theorems 3.2.1–3.2.3 of Chapter 3.

Theorem A.4.5 (Weak Law of Large Numbers (Khinchine)). *If $X_1, X_2, \ldots$ are iid as X with $E[X] = \mu$, then $\bar{X}_n = n^{-1}\sum_{i=1}^n X_i \xrightarrow{\mathcal{P}} \mu$.*

Proof. The cf of $\bar{X}_n$ is

$$\varphi_{\bar{X}_n}(t) = \varphi_{n^{-1}\sum_{i=1}^n X_i}(t) = \{\varphi_X(t/n)\}^n = \left[1 + i(t/n)\mu + o(1/n)\right]^n$$

by McLaurin's series expansion of $\varphi_X(t/n)$ about 0. Hence

$$\lim_{n\to\infty} \varphi_{\bar{X}_n}(t) = \lim_{n\to\infty}\left[1 + i(t/n)\mu + o(1/n)\right]^n = \mathrm{e}^{it\mu},$$

which is the cf of an rv X_0 which takes the value μ with probability 1. Hence $\bar{X}_n \xrightarrow{\mathcal{L}} X_0$ by the Continuity Theorem, so that for all $\varepsilon > 0$,

$$P[\bar{X}_n \le \mu - \varepsilon] \to P[X_0 \le \mu - \varepsilon] = 0 \text{ and}$$
$$P[\bar{X}_n \le \mu + \varepsilon] \to P[X_0 \le \mu + \varepsilon] = 1.$$

Thus, for any $\varepsilon > 0$,

$$\lim_{n\to\infty} P[|\bar{X}_n - \mu| \le \varepsilon] = \lim_{n\to\infty} P[\bar{X}_n \le \mu + \varepsilon] - \lim_{n\to\infty} P\big[\bar{X}_n \le \mu - \varepsilon\big] = 1 = 0 = 1.$$

□

Theorem A.4.6 (Central Limit Theorem (Lindeberg-Lévy)). *If $X_1, X_2, \ldots$ are iid as X with $E[X] = \mu$ and $Var[X] = \sigma^2$, then*

$$Z_n = \frac{1}{\sigma\sqrt{n}}\sum_{j=1}^n (X_j - \mu) \xrightarrow{\mathcal{L}} Z \sim N(0,1), \ ie, \quad \frac{1}{\sqrt{n}}\sum_{j=1}^n (X_j - \mu) \xrightarrow{\mathcal{L}} \sigma Z \sim N(0, \sigma^2).$$

Proof. Let $(X_j - \mu)/\sigma = Y_j$. Then $Y_1, Y_2, \ldots$ are iid as $Y = (X - \mu)/\sigma$ with $\mathrm{E}[Y] = 0$ and $\mathrm{E}[Y^2] = 1$. Then

$$\begin{aligned}
\varphi_{Z_n}(t) &= \varphi_{n^{-1/2}\sum_{j=1}^n Y_j}(t) = \left\{\varphi_Y(t/\sqrt{n})\right\}^n \\
&= \left[1 + i(t/\sqrt{n})0 + (1/2)i^2(t/\sqrt{n})^2 1 + o(1/n)\right]^n \\
&= \left[1 - t^2/(2n) + o(1/n)\right]^n \to \mathrm{e}^{-t^2/2} = \varphi_Z(t).
\end{aligned}$$

Hence $Z_n \xrightarrow{\mathcal{L}} Z$ by the Continuity Theorem. □

Theorem A.4.7 (The Cramér-Wold Device). *Let $\{\boldsymbol{X}_n\}$ be k-dim random vectors such that $\boldsymbol{a}^T\boldsymbol{X}_n \xrightarrow{\mathcal{L}} \boldsymbol{a}^T\boldsymbol{X}$ for all $\boldsymbol{a} \in \mathbb{R}^k$. Then $\boldsymbol{X}_n \xrightarrow{\mathcal{L}} \boldsymbol{X}$.*

Proof. Since $\boldsymbol{a}^T\boldsymbol{X}_n \xrightarrow{\mathcal{L}} \boldsymbol{a}^T\boldsymbol{X}$ for all $\boldsymbol{a} \in \mathbb{R}^k$,

$$\varphi_{\boldsymbol{a}^T\boldsymbol{X}_n}(t) = \mathrm{E}[\mathrm{e}^{it(\boldsymbol{a}^T\boldsymbol{X}_n)}] \to \varphi_{\boldsymbol{a}^T\boldsymbol{X}}(t) = \mathrm{E}[\mathrm{e}^{it(\boldsymbol{a}^T\boldsymbol{X})}], \quad \text{for all } t \in R \text{ and } \boldsymbol{a} \in \mathbb{R}^k.$$

But $\{t\boldsymbol{a}\colon t \in \mathbb{R}, \boldsymbol{a} \in \mathbb{R}^k\} = \{\boldsymbol{s}\colon \boldsymbol{s} \in \mathbb{R}^k\}$, and therefore,

$$\varphi_{\boldsymbol{X}_n}(\boldsymbol{s}) = \mathrm{E}\Big[\mathrm{e}^{i\boldsymbol{s}^T\boldsymbol{X}_n}\Big] \to \mathrm{E}\Big[\mathrm{e}^{i\boldsymbol{s}^T\boldsymbol{X}}\Big] = \varphi_{\boldsymbol{X}}(\boldsymbol{s}) \quad \text{for all } \boldsymbol{s} \in \mathbb{R}^k.$$

Hence $\boldsymbol{X}_n \xrightarrow{\mathcal{L}} \boldsymbol{X}$ by the Continuity Theorem for Characteristic Functions on $\mathbb{R}^k$. □

Theorem A.4.8 (Multivariate Central Limit Theorem). *If $\boldsymbol{X}_1, \boldsymbol{X}_2, \ldots$ are iid as $\boldsymbol{X}$ in $\mathbb{R}^k$ with the mean vector $E[\boldsymbol{X}] = \boldsymbol{\mu}$ and the covariance matrix $E[(\boldsymbol{X} - \boldsymbol{\mu})(\boldsymbol{X} - \boldsymbol{\mu})^T] = \boldsymbol{\Sigma}$ which is positive definite, then*

$$n^{-1/2} \sum_{j=1}^{n} (\boldsymbol{X}_j - \boldsymbol{\mu}) \xrightarrow{\mathcal{L}} \boldsymbol{W} \sim N_k(\boldsymbol{0}, \boldsymbol{\Sigma}).$$

Proof. For $\boldsymbol{a} \in \mathbb{R}^k$, let $Y_j = \boldsymbol{a}^T \boldsymbol{X}_j$ and $Y = \boldsymbol{a}^T \boldsymbol{X}$. Then $Y_1, Y_2, \ldots$ are iid as Y where $\mathrm{E}[Y] = \boldsymbol{a}^T \boldsymbol{\mu}$ and $\mathrm{Var}[Y] = \boldsymbol{a}^T \boldsymbol{\Sigma} \boldsymbol{a}$, so that

$$n^{-1/2} \sum_{j=1}^{n} (\boldsymbol{a}^T \boldsymbol{X}_j - \boldsymbol{a}^T \boldsymbol{\mu}) \xrightarrow{\mathcal{L}} Z_{\boldsymbol{a}} \sim N(0, \boldsymbol{a}^T \boldsymbol{\Sigma} \boldsymbol{a})$$

by the univariate Central Limit Theorem. On the other hand, if $\boldsymbol{W} \sim N_k(\boldsymbol{0}, \boldsymbol{\Sigma})$, then $\boldsymbol{a}^T \boldsymbol{W} \sim N(0, \boldsymbol{a}^T \boldsymbol{\Sigma} \boldsymbol{a})$. Thus

$$\boldsymbol{a}^T \left\{ n^{-1/2} \sum_{j=1}^{n} (\boldsymbol{X}_j - \boldsymbol{\mu}) \right\} \xrightarrow{\mathcal{L}} \boldsymbol{a}^T \boldsymbol{W} \quad \text{for all } \boldsymbol{a} \in \mathbb{R}^k,$$

and so the theorem follows by the Cramér-Wold device. □

A.5 Weak Convergence of Probabilities on C[0,1]

A metric on a nonempty set S is a function d: $S \times S \to R$ such that (i) $d(x, y) \geq 0$ and $d(x, y) = 0$ iff $x = y$, (ii) $d(x, y) = d(y, x)$, (iii) $d(x, y) \leq d(x, z) + d(y, z)$ for all $x, y, z \in S$. We call (S, d) a metric space.

For each $x \in S$ and $\varepsilon > 0$, the set $S_\varepsilon(x) = \{y\colon d(x, y) < \varepsilon\}$ is the open sphere of radius ε centered at x. A set $G \subset S$ is open $\iff$ for each $x \in G$, $S_\varepsilon(x) \subset G$ for some $\varepsilon > 0$ $\iff$ G is a union of open spheres.

In S, a sequence $\{x_n\}$ converges to x if $\lim_{n\to\infty} d(x_n, x) = 0$. The boundary of $A \subset S$ is $\partial A = \{x \in S\colon x$ is a limit point of sequences in both A and $A^c\}$.

A collection of open sets $\{G_\lambda\colon \lambda \in \Lambda\}$ is an open covering of $A \subset S$ if $A \subset \cup_{\lambda\in\Lambda} G_\lambda$. A set $K \subset S$ is compact if for every open covering of K, there is a finite subcovering.

Let $\{P_n\}$ and $\{P\}$ be probabilities on $(S, \mathcal{S})$ where $\mathcal{S}$ is the Borel σ-field, that is, the smallest σ-field of subsets of S which includes all open sets and let $C(S)$ be the set of all bounded continuous functions f: $S \to \mathrm{R}$.

The sequence $\{P_n\}$ converges weakly to P iff $\int f \, dP_n \to \int f \, dP$ for all $f \in C(S)$. This is denoted by $P_n \xrightarrow{w} P$. Equivalently, $P_n \xrightarrow{w} P$ iff $P_n(A) \to P(A)$ for all $A \in \mathcal{S}$ for which $P(\partial A) = 0$. The weak limit is unique (ie, if $P_n \xrightarrow{w} P$ and $P_n \xrightarrow{w} Q$), then $P = Q$.

A transformation g: $(S, \mathcal{S}) \to (S', \mathcal{S}')$ is measurable if $g^{-1}(B) \in \mathcal{S}$ for all $B \in \mathcal{S}'$, in which case, for $\{P_n\}$ and P on $(S, \mathcal{S})$, $\{P_n g^{-1}\}$ and $P g^{-1}$ are induced probabilities on $(S', \mathcal{S}')$, where

$$P_n g^{-1}(B) = P_n(g^{-1}(B)) \text{ and } P g^{-1}(B) = P(g^{-1}(B)) \quad \text{for all } B \in \mathcal{S}'.$$

A function g: $(S, d) \to (S', d')$ is continuous iff $x_n \to x$ in $S \Longrightarrow g(x_n) \to g(x)$ in S'.

Theorem A.5.1 (Continuous Mapping Theorem). *If* $P_n \xrightarrow{w} P$ *on* $(S, \mathcal{S})$ *and if* g: $(S, d) \to (S', d')$ *is continuous, then* $P_n g^{-1} \xrightarrow{w} Pg^{-1}$ *on* $(S', \mathcal{S}')$. *More generally, the theorem holds if* $P(D_g) = 0$ *where* D_g *is the set of discontinuity points of* g.

The continuous mapping theorem stated above is a generalization of Theorem 3.2.5(III) dealing with probability distributions of k-dim rv's.

Let $C = C[0, 1]$ be the set of all continuous functions on $[0, 1]$ with the metric $d(x, y) = \sup_{0 \le t \le 1} |x(t) - y(t)|$ and let $\mathcal{C}$ denote the Borel σ-field in C. On $(C, \mathcal{C})$, we now consider the weak convergence of $\{P_n\}$ to P, which holds under two conditions, namely, convergence of finite-dimensional distributions (fdd) and "tightness" of the sequence $\{P_n\}$. For notational simplicity, we describe these conditions in terms of random elements $X_n(\cdot)$ of $(C, \mathcal{C}, P_n)$ and $X(\cdot)$ of $(C, \mathcal{C}, P)$.

Definition A.5.1. For each positive integer k and $t_1 < \cdots < t_k$ in $[0, 1]$, the distributions of $(X_n(t_1), \ldots, X_n(t_k))$ and $(X(t_1), \ldots, X(t_k))$ are called fdd of $X_n(\cdot)$ and $X(\cdot)$, respectively.

Definition A.5.2. A family of probabilities Π on $(S, \mathcal{S})$ is tight if for every $\varepsilon > 0$, there exists a compact set K such that $P(K) > 1 - \varepsilon$ for all $P \in \Pi$. In particular, a sequence of probabilities $\{P_n\}$ on $(C, \mathcal{C})$ is tight if for every $\varepsilon > 0$, there exists a compact set K in C such that $P[X_n(\cdot) \in K] > 1 - \varepsilon$ for all n.

Notation

We are writing $P[X_n(\cdot) \in K]$ for $P_n(K)$.

Theorem A.5.2. *Let* $\{X_n(\cdot), n = 1, 2, \ldots\}$ *and* $X(\cdot)$ *denote random elements of* $(C, \mathcal{C}, P_n)$ *and* $(C, \mathcal{C}, P)$, *respectively. Then* $P_n \xrightarrow{w} P$ *or equivalently,* $X_n \xrightarrow{w} X$ *if*

- ***(i)*** *the fdd's of* $\{X_n(\cdot)\}$ *converge to those of* $X(\cdot)$,
- ***(iia)*** $\{X_n(0)\}$ *is tight (ie,* $X_n(0) = O_P(1)$*), and*
- ***(iib)*** *there exist constants* $\gamma \ge 0$ *and* $\alpha > 1$ *and a nondecreasing, continuous function* F *on* $[0, 1]$ *such that*

$$P\big[|X_n(t_2) - X_n(t_1)| \ge \lambda\big] \le \frac{1}{\lambda^\gamma} |F(t_2) - F(t_1)|^\alpha$$

holds for all t_1, t_2 *and* n, *and all* $\lambda > 0$ *(see [43, p. 95–6]).*

Appendix B
Basic Results From Matrix Algebra

This appendix lists some basic definitions, formulas, and results for vectors and matrices which are used in this book. We begin with some simple definitions and elementary results.

B.1 Some Elementary Facts

It is known from the theory of matrices that the number of linearly independent rows of a matrix $\boldsymbol{A}$ equals the number of linearly independent columns, and the rank of $\boldsymbol{A}$ (denoted by $rank(\boldsymbol{A})$) is defined to be the number of linearly independent rows of $\boldsymbol{A}$ (or the number of linearly independent columns). For any vector $\boldsymbol{x}$, $\boldsymbol{x}^T\boldsymbol{x}$ will be denoted by $\|\boldsymbol{x}\|^2$, which equals the square of the length of $\boldsymbol{x}$. A matrix $\boldsymbol{A}$ of order $n \times m$ is said to have a full rank if $rank(\boldsymbol{A}) = \min(n, m)$.

For any $n \times n$ matrix $\boldsymbol{A}$, its quadratic form is defined to be $q(\boldsymbol{x}) = \boldsymbol{x}^T\boldsymbol{A}\boldsymbol{x}$, where $\boldsymbol{x} \in \mathbb{R}^n$. If $\boldsymbol{A}$ is not symmetric, then $q(\boldsymbol{x})$ may also be written as $\boldsymbol{x}^T\tilde{\boldsymbol{A}}\boldsymbol{x}$, where $\tilde{\boldsymbol{A}} = (1/2)(\boldsymbol{A} + \boldsymbol{A}^T)$ is the symmetrized version of $\boldsymbol{A}$.

Definition B.1.1. All the matrices in this definition are assumed to be square of order n (ie, the matrices have n rows and n columns).

- **(a)** Trace of a matrix is defined to be the sum of its diagonal elements (ie, $trace(\boldsymbol{A}) = \sum_{i=1}^{n} a_{ii}$).
- **(b)** The determinant $\boldsymbol{A}$ (denoted by $|\boldsymbol{A}|$) is defined to be $\sum_{\pi}(-1)^{\pi} a_{i,\pi(i)}$, where the sum is over all permutations π of $\{1, \ldots, n\}$, and $(-1)^{\pi}$ equals 1 or -1 depending on whether π is a positive or a negative permutation.
- **(c)** A symmetric matrix $\boldsymbol{A}$ is called nonnegative definite if its quadratic form $\boldsymbol{x}^T\boldsymbol{A}\boldsymbol{x} \geq 0$ for any $\boldsymbol{x} \in \mathbb{R}^n$. If $\boldsymbol{x}^T\boldsymbol{A}\boldsymbol{x} > 0$ for all $0 \neq \boldsymbol{x} \in \mathbb{R}^n$, then $\boldsymbol{A}$ is called a positive definite matrix.
- **(d)** A matrix $\boldsymbol{A}$ is said to be orthogonal if its rows are orthonormal (ie, the row vectors are orthogonal to each other and each has unit length). Consequently, $\boldsymbol{A}\boldsymbol{A}^T = \boldsymbol{I}$. It is easy to check that $\boldsymbol{A}$ is nonsingular, $\boldsymbol{A}^T = \boldsymbol{A}^{-1}$ and $\boldsymbol{A}^T\boldsymbol{A} = \boldsymbol{I}$. Since $\boldsymbol{A}^T\boldsymbol{A} = \boldsymbol{I}$, columns of $\boldsymbol{A}$ are orthonormal.

Here are some important results on rank and trace of matrices.

Lemma B.1.1.

(a) *For any matrix* $\mathbf{A}$ *of order* $n \times m$, $rank(\mathbf{A}\mathbf{A}^T) = rank(\mathbf{A})$.
(b) *If* $\mathbf{A}$ *and* $\mathbf{B}$ *are of order* $n \times m$, *then* $rank(\mathbf{A}+\mathbf{B}) \le rank(\mathbf{A}) + rank(\mathbf{B})$.
(c) *If* $\mathbf{A}$ *and* $\mathbf{B}$ *are of order* $n \times m$ *and* $m \times k$, *respectively, then* $rank(\mathbf{A}\mathbf{B}) \le \min[rank(\mathbf{A}), rank(\mathbf{B})]$.
(d) *If* $\mathbf{A}$ *and* $\mathbf{B}$ *are of order* $n \times n$, *then* $trace(\mathbf{A}+\mathbf{B}) = trace(\mathbf{A}) + trace(\mathbf{B})$.
(e) *If* $\mathbf{A}$ *and* $\mathbf{B}$ *are of order* $n \times m$ *and* $m \times n$, *then* $trace(\mathbf{A}\mathbf{B}) = trace(\mathbf{B}\mathbf{A})$.
(f) *If* $\mathbf{A}$ *and* $\mathbf{B}$ *are of order* $n \times n$, *then* $|\mathbf{A}\mathbf{B}| = |\mathbf{A}|\,|\mathbf{B}|$.

A class of formulas known as the *Sherman-Morrison Formulas* are quite useful in inverting matrices.

Theorem B.1.1.

(a) *If* $\boldsymbol{a} \in \mathbb{R}^n$, *then* $(\boldsymbol{I} + \boldsymbol{a}\boldsymbol{a}^T)^{-1} = \boldsymbol{I} - (1 + \|\boldsymbol{a}\|^2)^{-1}\boldsymbol{a}\boldsymbol{a}^T$.
(b) *If* $\boldsymbol{a} \in \mathbb{R}^n$, $\|\boldsymbol{a}\| \neq 1$, *then* $(\boldsymbol{I} - \boldsymbol{a}\boldsymbol{a}^T)^{-1} = \boldsymbol{I} + (1 - \|\boldsymbol{a}\|^2)^{-1}\boldsymbol{a}\boldsymbol{a}^T$.
(c) *If* $\mathbf{A}$ *is of order* $n \times m$, *then* $(\boldsymbol{I} + \boldsymbol{A}\boldsymbol{A}^T)^{-1} = \boldsymbol{I} - \boldsymbol{A}(\boldsymbol{I} + \boldsymbol{A}^T\boldsymbol{A})^{-1}\boldsymbol{A}^T$.
(d) *If* $\mathbf{A}$ *is* $n \times m$ *and* $\mathbf{B}$ *is a positive definite matrix of order* $n \times n$, *then* $(\boldsymbol{B} + \boldsymbol{A}\boldsymbol{A}^T)^{-1} = \boldsymbol{B}^{-1} - \boldsymbol{B}^{-1}\boldsymbol{A}(\boldsymbol{I} + \boldsymbol{A}^T\boldsymbol{B}^{-1}\boldsymbol{A})^{-1}\boldsymbol{A}^T\boldsymbol{B}^{-1}$.

The following is the Cauchy-Schwarz inequality for the matrices.

Theorem B.1.2. *Let* $\boldsymbol{a}$ *be an n-dim vector and* $\mathbf{A}$ *be a positive definite matrix of order* $n \times n$.

(a) *For any* $\boldsymbol{x} \in \mathbb{R}^n$, $|\boldsymbol{a}^T\boldsymbol{x}|^2 \le \|\boldsymbol{a}\|^2\|\boldsymbol{x}\|^2$. *Moreover,* $\sup\{|\boldsymbol{a}^T\boldsymbol{x}|^2/\|\boldsymbol{x}\|^2 : \boldsymbol{x} \in \mathbb{R}^n\} = \|\boldsymbol{a}\|^2$ *and this supremum is attained at* $\boldsymbol{x} = \boldsymbol{a}$.
(b) *For any* $\boldsymbol{x} \in \mathbb{R}^n$, $|\boldsymbol{a}^T\boldsymbol{x}|^2 \le [\boldsymbol{a}^T\boldsymbol{A}^{-1}\boldsymbol{a}][\boldsymbol{x}^T\boldsymbol{A}\boldsymbol{x}]$. *Moreover,* $\sup\{|\boldsymbol{a}^T\boldsymbol{x}|^2/[\boldsymbol{x}^T\boldsymbol{A}\boldsymbol{x}] : \boldsymbol{x} \in \mathbb{R}^n\} = \boldsymbol{a}^T\boldsymbol{A}^{-1}\boldsymbol{a}$ *and this supremum is attained at* $\boldsymbol{x} = \boldsymbol{A}^{-1}\boldsymbol{a}$.

B.2 Eigenvalues and Eigenvectors

For a square matrix $\boldsymbol{A}$ of order n, if there exists a scalar (may be complex) and a vector $\boldsymbol{x}$ (may be complex) such that $\boldsymbol{A}\boldsymbol{x} = \lambda\boldsymbol{x}$, then λ is called an eigenvalue of $\boldsymbol{A}$ with the corresponding eigenvector $\boldsymbol{x}$. The following result lists some basic properties of eigenvalues.

Lemma B.2.1.

(a) *If* $\mathbf{A}$ *is symmetric, then all its eigenvalues and eigenvectors are real.*
(b) *If* $\mathbf{A}$ *is nonsingular, then all its eigenvalues are nonzero.*
(c) *If* λ *is an eigenvalue of symmetric matrix* $\mathbf{A}$ *with the corresponding eigenvector* $\boldsymbol{x}$, *then for any positive integer r,* λ^r *is an eigenvalue of* $\mathbf{A}^r$ *with eigenvector* $\boldsymbol{x}$
(d) *The nonzero eigenvalues of* $\mathbf{A}^T\mathbf{A}$, *where* $\mathbf{A}$ *is a matrix of order* $n \times m$, *are the same as those of* $\mathbf{A}\mathbf{A}^T$.
(e) *The eigenvalues of a nonnegative definite (positive definite) matrix* $\mathbf{A}$ *are nonnegative (positive).*

The following is an important result that is widely used in Linear Models and Multivariate Analysis.

Theorem B.2.1 (Spectral Decomposition Theorem). *If $\boldsymbol{A}$ is a symmetric matrix of order $n \times n$, then there exist an $n \times n$ orthogonal matrix $\boldsymbol{U}$ with columns $\boldsymbol{u}_1, \ldots, \boldsymbol{u}_n$ (ie, $\boldsymbol{U} = [\boldsymbol{u}_1, \ldots, \boldsymbol{u}_n]$) and a diagonal matrix $\boldsymbol{\Lambda}$ of order $n \times n$ with diagonal elements $\lambda_1, \ldots, \lambda_n$ such that*

$$\boldsymbol{A} = \boldsymbol{U}\boldsymbol{\Lambda}\boldsymbol{U}^T = \sum_{j=1}^{n} \lambda_j \boldsymbol{u}_j \boldsymbol{u}_j^T.$$

Here $\{\lambda_j\}$ are the eigenvalues of $\boldsymbol{A}$ with the corresponding orthonormal eigenvectors $\{\boldsymbol{u}_j\}$.

Is there an analog of the Spectral Decomposition Theorem for an arbitrary matrix $\boldsymbol{A}$ is of order $n \times m$? The answer is yes. Positive square roots of the eigenvalues of $\boldsymbol{A}^T\boldsymbol{A}$ are called the *singular values* of $\boldsymbol{A}$. Since $\boldsymbol{A}^T\boldsymbol{A}$ is nonnegative definite, its eigenvalues are nonnegative and thus the square roots of the eigenvalues of $\boldsymbol{A}^T\boldsymbol{A}$ are real. Clearly, if $\boldsymbol{A}$ is symmetric, then $\boldsymbol{A}^T\boldsymbol{A} = \boldsymbol{A}^2$ with eigenvalues $\{\lambda_j^2\}$, where $\{\lambda_j\}$ are the eigenvalues of $\boldsymbol{A}$, and the singular values of $\boldsymbol{A}$ are $\{|\lambda_j|\}$.

Theorem B.2.2 (Singular Value Decomposition). *Let $\boldsymbol{A}$ be a matrix of order $n \times m$, $m \leq n$. There exist an $n \times m$ suborthogonal matrix $\boldsymbol{U}$ (ie, the columns of $\boldsymbol{U}$ are orthonormal), an $m \times m$ orthogonal matrix $\boldsymbol{V}$, and an $m \times m$ diagonal matrix $\boldsymbol{\Lambda}$ with nonnegative diagonal entries $\lambda_1, \ldots, \lambda_m$ such that*

$$\boldsymbol{A} = \boldsymbol{U}\boldsymbol{\Lambda}\boldsymbol{V}^T = \sum_{j=1}^{m} \lambda_j \boldsymbol{u}_j \boldsymbol{v}_j^T,$$

where $\{\boldsymbol{u}_j\}$ and $\{\boldsymbol{v}_j\}$ are the columns of $\boldsymbol{U}$ and $\boldsymbol{V}$, respectively.

It is important to note that $\{\lambda_j\}$ in the Singular Value Decomposition of $\boldsymbol{A}$ are the singular values of $\boldsymbol{A}$, and $\{\lambda_j^2\}$ are the eigenvalues of the matrix $\boldsymbol{A}^T\boldsymbol{A}$.

We finally write down two important formulas for the trace and determinant of matrices.

Lemma B.2.2. *Let $\boldsymbol{A}$ be a symmetric matrix of order $n \times n$ with eigenvalues $\{\lambda_j, j = 1, \ldots, n\}$. Then,*

$$trace(\boldsymbol{A}) = \sum_{j=1}^{n} \lambda_j \quad and \quad |\boldsymbol{A}| = \prod_{j=1}^{n} \lambda_j.$$

The following result on the optimization of quadratic forms involves the eigenvalues and eigenvectors.

Theorem B.2.3. *Let $\lambda_1 \geq \cdots \geq \lambda_n$ be the eigenvalues of a symmetric matrix $\boldsymbol{A}$ of order $n \times n$ with the corresponding orthonormal eigenvectors $\boldsymbol{u}_1, \ldots, \boldsymbol{u}_n$. It is understood that $\boldsymbol{x}$, $\boldsymbol{x}_1, \ldots$ written below are in $\mathbb{R}^n$. Then the following hold:*

(a) $\sup\{\boldsymbol{x}^T\boldsymbol{A}\boldsymbol{x}: \|\boldsymbol{x}\| = 1\} = \sup_{\boldsymbol{x}\neq 0} \frac{\boldsymbol{x}^T\boldsymbol{A}\boldsymbol{x}}{\boldsymbol{x}^T\boldsymbol{x}} = \lambda_1$, *and this supremum is attained at* $\boldsymbol{x} = \boldsymbol{u}_1$.

(b) $\inf\{\boldsymbol{x}^T\boldsymbol{A}\boldsymbol{x}: \|\boldsymbol{x}\| = 1\} = \inf_{\boldsymbol{x}\neq 0} \frac{\boldsymbol{x}^T\boldsymbol{A}\boldsymbol{x}}{\boldsymbol{x}^T\boldsymbol{x}} = \lambda_n$, *and this infimum is attained at* $\boldsymbol{x} = \boldsymbol{u}_n$.

(c) *For* $1 < m \leq n$, *we have*

$$\begin{aligned}
&\sup\left\{\boldsymbol{x}_1^T\boldsymbol{A}\boldsymbol{x}_1 + \cdots + \boldsymbol{x}_m^T\boldsymbol{A}\boldsymbol{x}_m: \boldsymbol{x}_1, \ldots, \boldsymbol{x}_m \text{ are orthonormal}\right\} \\
&= \sup\left\{\frac{\boldsymbol{x}_1^T\boldsymbol{A}\boldsymbol{x}_1}{\boldsymbol{x}_1^T\boldsymbol{x}_1} + \cdots + \frac{\boldsymbol{x}_m^T\boldsymbol{A}\boldsymbol{x}_m}{\boldsymbol{x}_m^T\boldsymbol{x}_m}: \boldsymbol{x}_1 \neq \boldsymbol{0}, \ldots, \boldsymbol{x}_m \neq \boldsymbol{0} \text{ are orthogonal}\right\} \\
&= \lambda_1 + \cdots + \lambda_m,
\end{aligned}$$

and this supremum is attained at $\boldsymbol{x}_1 = \boldsymbol{u}_1, \ldots, \boldsymbol{x}_m = \boldsymbol{u}_m$.

B.3 Functions of Symmetric Matrices

If $\boldsymbol{A}$ is a symmetric matrix of order $n \times n$, then for any real-valued function f whose domain contains the eigenvalues of $\boldsymbol{A}$, it is possible to define the corresponding function of $\boldsymbol{A}$ using the Spectral Decomposition Theorem. More formally, let f be a real-valued function with domain $D \subset \mathbb{R}$, and let $\{\lambda_j: j = 1, \ldots, n\}$ be the eigenvalues of $\boldsymbol{A}$ with the corresponding orthonormal eigenvector $\{\boldsymbol{u}_j\}$. If the eigenvalues of $\boldsymbol{A}$ are inside the set D, then the matrix function $f(\boldsymbol{A})$ is defined to be $f(\boldsymbol{A}) = \sum_{j=1}^n f(\lambda_j)\boldsymbol{u}_j\boldsymbol{u}_j^T$. Here are some examples that are useful in Linear Models and Multivariate Analysis.

I. (Square root of a matrix) Let $\boldsymbol{A}$ be nonnegative definite and let $f(u) = u^{1/2}$, $u \geq 0$ (ie, $D = [0, \infty)$). Then

$$\boldsymbol{A}^{1/2} = f(\boldsymbol{A}) = \sum_{j=1}^n f(\lambda_j)\boldsymbol{u}_j\boldsymbol{u}_j^T = \sum_{j=1}^n \lambda_j^{1/2}\boldsymbol{u}_j\boldsymbol{u}_j^T.$$

Clearly, $\boldsymbol{A}^{1/2}$ is symmetric and it is fairly easy to check that $\boldsymbol{A}^{1/2}\boldsymbol{A}^{1/2} = \boldsymbol{A}$.

II. (Inverse of a matrix) If $\boldsymbol{A}$ is nonsingular, then all its eigenvalues are nonzero and let $f(u) = u^{-1}$, $u \neq 0$ (ie, $D = \mathbb{R} - \{0\}$). Then

$$\boldsymbol{A}^{-1} = f(\boldsymbol{A}) = \sum_{j=1}^n \lambda_j^{-1}\boldsymbol{u}_j\boldsymbol{u}_j^T.$$

It is fairly easy to check that $\boldsymbol{A}^{-1}$ is indeed the inverse of the matrix $\boldsymbol{A}$.

III. (Inverse of square root of a matrix) Let $\boldsymbol{A}$ be positive definite and let $f(u) = u^{-1/2}$, $u > 0$ (ie, $D = (0, \infty)$). Then

$$\boldsymbol{A}^{-1/2} = f(\boldsymbol{A}) = \sum_{j=1}^n \lambda_j^{-1/2}\boldsymbol{u}_j\boldsymbol{u}_j^T.$$

It is clear that $\boldsymbol{A}^{-1/2}$ is symmetric, and it is easy to verify that $\boldsymbol{A}^{-1/2}\boldsymbol{A}^{-1/2} = \boldsymbol{A}^{-1}$.

B.4 Generalized Eigenvalues

Let $\boldsymbol{A}$ and $\boldsymbol{B}$ be symmetric matrices of order $n \times n$ where $\boldsymbol{B}$ is positive definite. We say that λ is a (generalized) eigenvalue of $\boldsymbol{A}$ with respect to $\boldsymbol{B}$ if there is a vector $\boldsymbol{l}$ in $\mathbb{R}^n$ such that $\boldsymbol{Al} = \lambda \boldsymbol{Bl}$. Premultiplying both sides by $\boldsymbol{B}^{-1}$ we get $\boldsymbol{B}^{-1}\boldsymbol{Al} = \lambda \boldsymbol{l}$. In other words, if λ is an eigenvalue of $\boldsymbol{A}$ with respect to $\boldsymbol{B}$, then λ is also an eigenvalue of $\boldsymbol{B}^{-1}\boldsymbol{A}$, and the converse is also true. Similarly we can show that if λ is an eigenvalue of $\boldsymbol{A}$ with respect to $\boldsymbol{B}$, then λ is also an eigenvalue of $\boldsymbol{AB}^{-1}$ and of $\boldsymbol{B}^{-1/2}\boldsymbol{AB}^{-1/2}$, where $\boldsymbol{B}^{-1/2}$ is symmetric. The following result summarizes these observations.

Lemma B.4.1. *Eigenvalues of* $\boldsymbol{AB}^{-1}$, $\boldsymbol{B}^{-1}\boldsymbol{A}$, *and* $\boldsymbol{B}^{-1/2}\boldsymbol{AB}^{-1/2}$ *are the same.*

Let $\lambda_1 \geq \cdots \geq \lambda_n$ be the generalized eigenvalues of $\boldsymbol{A}$ with respect to $\boldsymbol{B}$. By Lemma B.4.1, $\lambda_1, \ldots, \lambda_p$ are also the eigenvalues of $\boldsymbol{B}^{-1/2}\boldsymbol{AB}^{-1/2}$. By the Spectral Decomposition Theorem, we have

$$\boldsymbol{B}^{-1/2}\boldsymbol{AB}^{-1/2} = \sum_{j=1}^{n} \lambda_j \boldsymbol{u}_j \boldsymbol{u}_j^T, \text{ and } \sum_{j=1}^{n} \boldsymbol{u}_j \boldsymbol{u}_j^T = \boldsymbol{I},$$

where $\{\boldsymbol{u}_j\}$ are the orthonormal eigenvector of the matrix $\boldsymbol{B}^{-1/2}\boldsymbol{AB}^{-1/2}$ corresponding to the eigenvalues $\{\lambda_j\}$. We now write down an analog of Theorem B.2.3 for the generalized eigenvalues.

Theorem B.4.1. *Let* $\boldsymbol{A}$ *and* $\boldsymbol{B}$ *be symmetric* $n \times n$ *matrices, and assume that* $\boldsymbol{B}$ *is positive definite. Let* $\lambda_1 \geq \cdots \geq \lambda_n$ *be the eigenvalues of* $\boldsymbol{B}^{-1/2}\boldsymbol{AB}^{-1/2}$ *with the corresponding orthonormal eigenvectors* $\boldsymbol{u}_1, \ldots, \boldsymbol{u}_n$. *It is understood that* $\boldsymbol{x}, \boldsymbol{x}_1, \ldots$ *written below are in* $\mathbb{R}^n$. *Then the following hold:*

(a) $\sup_{\boldsymbol{x} \neq \boldsymbol{0}} \frac{\boldsymbol{x}^T\boldsymbol{Ax}}{\boldsymbol{x}^T\boldsymbol{Bx}} = \lambda_1$, *and this supremum is attained at* $\boldsymbol{x} = \boldsymbol{B}^{-1/2}\boldsymbol{u}_1$.

(b) $\inf_{\boldsymbol{x} \neq \boldsymbol{0}} \frac{\boldsymbol{x}^T\boldsymbol{Ax}}{\boldsymbol{x}^T\boldsymbol{Bx}} = \lambda_n$, *and this infimum is attained at* $\boldsymbol{x} = \boldsymbol{B}^{-1/2}\boldsymbol{u}_n$.

(c) *For* $1 < m \leq n$, *we have*

$$\sup\left\{\frac{\boldsymbol{x}_1^T\boldsymbol{Ax}_1}{\boldsymbol{x}_1^T\boldsymbol{Bx}_1} + \cdots + \frac{\boldsymbol{x}_m^T\boldsymbol{Ax}_m}{\boldsymbol{x}_m^T\boldsymbol{Bx}_m} : \boldsymbol{x}_i^T\boldsymbol{Bx}_j = 0, 1 \leq i \neq j \leq m, \boldsymbol{x}_i \neq 0, i = 1, \ldots, m\right\}$$
$$= \lambda_1 + \cdots + \lambda_m,$$

and this supremum is attained at $\boldsymbol{x}_1 = \boldsymbol{B}^{-1/2}\boldsymbol{u}_1, \ldots, \boldsymbol{x}_m = \boldsymbol{B}^{-1/2}\boldsymbol{u}_m$.

B.5 Matrix Derivatives

In many cases one needs to differentiate the quadratic form or the trace or the determinant of a matrix. There are a number of useful formulas for such purposes. For any function $f : \mathbb{R}^n \to \mathbb{R}$, the column vector of its first order partial derivatives and the $n \times n$ matrix of second order partial derivatives (also known as the Hessian) will be denoted by $\frac{\partial}{\partial \boldsymbol{x}} f(\boldsymbol{x})$ and

$\frac{\partial^2}{\partial \boldsymbol{x} \partial \boldsymbol{x}} f(\boldsymbol{x})$, respectively. Let $\boldsymbol{A}$ be a symmetric matrix of order $n \times n$, and let $l(\boldsymbol{x}) = \boldsymbol{A}\boldsymbol{x}$ and $q(\boldsymbol{x}) = \boldsymbol{x}^T \boldsymbol{A}\boldsymbol{x}, \boldsymbol{x} \in \mathbb{R}^n$.

Lemma B.5.1. *Let $l(\boldsymbol{x})$ and $q(\boldsymbol{x})$ be defined as above. Then,*

$$\frac{\partial}{\partial \boldsymbol{x}} l(\boldsymbol{x}) = \boldsymbol{A}, \ \frac{\partial}{\partial \boldsymbol{x}} q(\boldsymbol{x}) = 2\boldsymbol{A}\boldsymbol{x}, \ \frac{\partial^2}{\partial \boldsymbol{x} \partial \boldsymbol{x}} q(\boldsymbol{x}) = 2\boldsymbol{A}.$$

For the result given below, it is assumed that the square matrix $\boldsymbol{A}(x)$ of order n is a function of a real variable x and element (i,j) of $\boldsymbol{A}(x)$ is $a_{ij}(x)$. Let $\dot{\boldsymbol{A}}(x)$ denote the matrix obtained by differentiating all elements $\boldsymbol{A}(x)$ with respect to x (ie, $\dot{\boldsymbol{A}}(x) = ((\dot{a}_{ij}(x)))$, where $\dot{a}_{ij}(x) = da_{ij}(x)/dx$). The following result presents expressions for the derivatives of $\boldsymbol{A}^{-1}(x)$ and $|\boldsymbol{A}(x)|$ with respect to x.

Lemma B.5.2. *Let $\boldsymbol{A}(x)$ be an $n \times n$ symmetric nonsingular matrix whose elements depend on a real variable x, and let $\dot{\boldsymbol{A}}(x)$ be the matrix obtained by differentiating the elements of $\boldsymbol{A}(x)$ with respect to x. Then,*

$$\frac{d}{dx}\boldsymbol{A}^{-1}(x) = -\boldsymbol{A}^{-1}(x)\dot{\boldsymbol{A}}(x)\boldsymbol{A}^{-1}(x),$$

$$\frac{d}{dx}|\boldsymbol{A}(x)| = |\boldsymbol{A}(x)| \, trace(\dot{\boldsymbol{A}}(x)\boldsymbol{A}^{-1}(x)).$$

It is fairly easy to check the first equality. Since $\boldsymbol{I} = \boldsymbol{A}^{-1}(x)\boldsymbol{A}(x)$, differentiating both sides with respect to x, we have

$$\begin{aligned} \boldsymbol{0} = \frac{d}{dx}\boldsymbol{I} &= \frac{d}{dx}\boldsymbol{A}^{-1}(x)\boldsymbol{A}(x) \\ &= \left[\frac{d}{dx}\boldsymbol{A}^{-1}(x)\right]\boldsymbol{A}(x) + \boldsymbol{A}^{-1}(x)\left[\frac{d}{dx}\boldsymbol{A}(x)\right] \\ &= \left[\frac{d}{dx}\boldsymbol{A}^{-1}(x)\right]\boldsymbol{A}(x) + \boldsymbol{A}(x)^{-1}\dot{\boldsymbol{A}}(x). \end{aligned}$$

Postmultiply by $\boldsymbol{A}^{-1}(x)$ on both sides to get the result.

The proof of the second equality is a bit more involved than that of the first. We outline a proof via spectral decomposition of $\boldsymbol{A}(x) = \sum_{j=1}^{n} \lambda_j(x)\boldsymbol{u}_j(x)\boldsymbol{u}_j^T(x)$, where for each x, $\{\boldsymbol{u}_j(x)\colon j = 1, \ldots, n\}$ are orthonormal. Denote the derivatives of $\lambda_j(x)$ and $\boldsymbol{u}_j(x)$ by $\dot{\lambda}_j(x)$ and $\dot{\boldsymbol{e}}_j(x)$, respectively. Since $|\boldsymbol{A}(x)| = \lambda_1(x) \cdots \lambda_n(x)$, we have

$$\begin{aligned} \frac{d}{dx}|\boldsymbol{A}(x)| &= \frac{d}{dx}[\lambda_1(x)\cdots\lambda_n(x)] = \sum_{j=1}^{n}[\lambda_1(x)\cdots\lambda_n(x)]\left[\dot{\lambda}_j(x)/\lambda_j(x)\right] \\ &= |\boldsymbol{A}(x)|\sum_{j=1}^{n}\dot{\lambda}_j(x)/\lambda_j(x). \end{aligned}$$

The second equality would hold if we can establish that

$$trace(\dot{\boldsymbol{A}}(x)\boldsymbol{A}^{-1}(x)) = \sum_{j=1}^{n}\dot{\lambda}_j(x)/\lambda_j(x).$$

Since $\boldsymbol{A}(x) = \sum_{j=1}^{n} \lambda_j(x)\boldsymbol{u}_j(x)\boldsymbol{u}_j^T(x)$, differentiating both sides with respect to x we have

$$\begin{aligned}\dot{\boldsymbol{A}}(x) &= \sum_{j=1}^{n} \dot{\lambda}_j(x)\boldsymbol{u}_j(x)\boldsymbol{u}_j^T(x) + \sum_{j=1}^{n} \lambda_j(x)\dot{\boldsymbol{u}}_j(x)\boldsymbol{u}_j^T(x) + \sum_{j=1}^{n} \lambda_j(x)\boldsymbol{u}_j(x)\dot{\boldsymbol{u}}_j^T(x)\\ &:= \boldsymbol{B}_1(x) + \boldsymbol{B}_2(x) + \boldsymbol{B}_3(x).\end{aligned}$$

The result now follows nothing that

$$\begin{aligned}\boldsymbol{A}^{-1}(x) &= \sum_{j=1}^{n} \lambda_j(x)^{-1}\boldsymbol{u}_j(x)\boldsymbol{u}_j^T(x),\\ trace(\boldsymbol{B}_1(x)\boldsymbol{A}(x)^{-1}) &= \sum_{j=1}^{n} \dot{\lambda}_j(x)/\lambda_j(x), \text{ and}\\ trace(\boldsymbol{B}_2(x)\boldsymbol{A}(x)^{-1}) = trace(\boldsymbol{B}_3(x)\boldsymbol{A}(x)^{-1}) &= \sum_{j=1}^{n} \boldsymbol{u}_j^T(x)\dot{\boldsymbol{u}}_j(x) = 0,\end{aligned}$$

where the last step is justified as $\boldsymbol{u}_j^T(x)\dot{\boldsymbol{u}}_j(x) = 0$ for all j, which can be verified by differentiating both sides of the identity $\boldsymbol{u}_j^T(x)\boldsymbol{u}_j(x) = 1$ with respect to x.

B.6 Orthogonal Projection

For a matrix $\boldsymbol{A}$ of order $n \times m$, we denote its column space $\{\boldsymbol{Ax}: \boldsymbol{x} \in \mathbb{R}^m\}$ by $\mathcal{M}(\boldsymbol{A})$. The orthogonal complement of $\mathcal{M}(\boldsymbol{A})$, denoted by $\mathcal{M}(\boldsymbol{A})^{\perp}$, is the set $\{\boldsymbol{y} \in \mathbb{R}^n: \boldsymbol{y}^T\boldsymbol{u} = 0$ for any $\boldsymbol{u} \in \mathcal{M}(\boldsymbol{A})\}$. A square matrix $\boldsymbol{A}$ of order n is called idempotent if $\boldsymbol{A}^2 = \boldsymbol{A}$. A symmetric matrix $\boldsymbol{A}$ is called a (orthogonal) projection matrix if it is symmetric and idempotent. It is fairly easy to see that if $\boldsymbol{A}$ is a projection matrix, then so is $\boldsymbol{I} - \boldsymbol{A}$. Since $(\boldsymbol{I} - \boldsymbol{A})\boldsymbol{A} = \boldsymbol{A} - \boldsymbol{A}^2 = \boldsymbol{A} - \boldsymbol{A} = \boldsymbol{0}$, it follows that $\mathcal{M}(\boldsymbol{A})^{\perp} = \mathcal{M}(\boldsymbol{I} - \boldsymbol{A})$. If λ is an eigenvalue of $\boldsymbol{A}$, then λ^2 is an eigenvalue of $\boldsymbol{A}^2$. Since $\boldsymbol{A} = \boldsymbol{A}^2$, we have $\lambda = \lambda^2$ and thus $\lambda = 0$ or 1. The following lists a few important properties of a projection matrix.

Theorem B.6.1. *Let* $\boldsymbol{A}$ *be a* $n \times n$ *projection matrix. Let* $\mathcal{M}(\boldsymbol{A}) = \{\boldsymbol{Ax}: \boldsymbol{x} \in \mathbb{R}^n\}$ *be the column space of* $\boldsymbol{A}$. *The following hold:*

(a) $\boldsymbol{I} - \boldsymbol{A}$ *is a projection matrix.*
(b) *All the eigenvalues of* $\boldsymbol{A}$ *are either 0 or 1.*
(c) $trace(\boldsymbol{A}) = rank(\boldsymbol{A})$.
(d) $\mathcal{M}(\boldsymbol{A})^{\perp} = \mathcal{M}(\boldsymbol{I} - \boldsymbol{A})$.
(e) *If* $\boldsymbol{B}$ *is an* $n \times n$ *projection matrix and* $\mathcal{M}(\boldsymbol{B}) \subset \mathcal{M}(\boldsymbol{A})$, *then* $\boldsymbol{A} - \boldsymbol{B}$ *is a projection on* $\mathcal{M}(\boldsymbol{A}) \cap \mathcal{M}(\boldsymbol{B})^{\perp}$ *and* $\boldsymbol{AB} = \boldsymbol{B}$ *(ie,* $(\boldsymbol{I} - \boldsymbol{A})\boldsymbol{B} = \boldsymbol{0}$*).*

Suppose that $\boldsymbol{A}$ is of order $n \times m$, with rank $m \le n$. Given a vector $\boldsymbol{y} \in \mathbb{R}^n$, how do we find a vector in $\mathcal{M}(\boldsymbol{A})$ that is closest to $\boldsymbol{y}$? Clearly, this is equivalent to minimizing $\|\boldsymbol{y} - \boldsymbol{Ax}\|^2$ with respect to $\boldsymbol{x} \in \mathbb{R}^m$, and if a minimum is attained at $\boldsymbol{x} = \boldsymbol{x}^*$, then $\hat{\boldsymbol{y}} = \boldsymbol{Ax}^*$ is the element in $\mathcal{M}(\boldsymbol{A})$ that is closest to $\boldsymbol{y}$. It is fairly easy to verify that $\boldsymbol{x}^* = (\boldsymbol{A}^T\boldsymbol{A})^{-1}\boldsymbol{A}^T\boldsymbol{y}$ and $\hat{\boldsymbol{y}} = \boldsymbol{A}(\boldsymbol{A}^T\boldsymbol{A})^{-1}\boldsymbol{A}^T\boldsymbol{y} = \boldsymbol{Q}_A\boldsymbol{y}$, where $\boldsymbol{Q}_A = \boldsymbol{A}(\boldsymbol{A}^T\boldsymbol{A})^{-1}\boldsymbol{A}^T$. It is also easy to check that $\boldsymbol{Q}_A$ is a projection matrix and $\mathcal{M}(\boldsymbol{A}) = \{\boldsymbol{Q}_A\boldsymbol{u} : \boldsymbol{u} \in \mathbb{R}^n\}$.

B.7 Distribution of Quadratic Forms

In this section, all the matrices $\boldsymbol{A}$, $\boldsymbol{A}_1$, etc., associated with quadratic forms of $\boldsymbol{Y}$, where $\boldsymbol{Y} \sim N_n(\boldsymbol{\mu}, \boldsymbol{I})$, are assumed to be symmetric of order $n \times n$. If $\boldsymbol{A}$ is a projection matrix of rank $p \leq n$, then it has exactly p eigenvalues equal to 1 (and the rest are 0), and let $\boldsymbol{u}_1, \ldots, \boldsymbol{u}_p$ be the corresponding orthonormal eigenvectors. Then

$$\|\boldsymbol{A}\boldsymbol{Y}\|^2 = \boldsymbol{Y}^T\boldsymbol{A}\boldsymbol{Y} = \sum_{j=1}^{p}\left(\boldsymbol{u}_j^T\boldsymbol{Y}\right)^2 := \sum_{j=1}^{p} W_j^2.$$

Since $\{\boldsymbol{u}_j\}$ are orthonormal, $\{W_j = \boldsymbol{u}_j^T\boldsymbol{Y}, j = 1, \ldots, p\}$ are independent with $W_j \sim N(\boldsymbol{u}_j^T\boldsymbol{\mu}, 1)$. Results from Section 2.2.9 in Chapter 2 tell us $\|\boldsymbol{A}\boldsymbol{Y}\|^2 = \sum_{j=1}^{p} W_j^2 \sim \chi_p^2(\delta^2)$, where $\delta^2 = (1/2)\sum_{j=1}^{p}\left(\boldsymbol{u}_j^T\boldsymbol{\mu}\right)^2 = (1/2)\boldsymbol{\mu}^T\boldsymbol{A}\boldsymbol{\mu} = (1/2)\|\boldsymbol{A}\boldsymbol{\mu}\|^2$.

It turns out that the converse is also true, that is, if $\boldsymbol{Y}^T\boldsymbol{A}\boldsymbol{Y} \sim \chi_p^2(\delta^2)$, then $\boldsymbol{A}$ must be a projection matrix of rank p. In order to see this, let us assume that $\boldsymbol{A}$ has rank q and its nonzero eigenvalues and the corresponding normalized eigenvectors are $\lambda_1, \ldots, \lambda_q$ and $\boldsymbol{u}_1, \ldots, \boldsymbol{u}_q$, respectively. Then

$$\boldsymbol{Y}^T\boldsymbol{A}\boldsymbol{Y} = \sum_{j=1}^{q} \lambda_j W_j^2,$$

where $\{W_j = \boldsymbol{u}_j^T\boldsymbol{Y}, j = 1, \ldots, q\}$ are independent with $W_j \sim N(\boldsymbol{u}_j^T\boldsymbol{\mu}, 1)$. Since $W_j^2 \sim \chi_1^2(\delta_j^2)$ with $\delta_j^2 = (1/2)\left(\boldsymbol{u}_j^T\boldsymbol{\mu}\right)^2$, and $\{W_j : j = 1, \ldots, q\}$ are independent, the characteristic function (cf) of $\boldsymbol{Y}^T\boldsymbol{A}\boldsymbol{Y}$ is the product of the cf's of $\lambda_j W_j^2, j = 1, \ldots, q$. And this product of cf's must be equal the cf of $\chi_p^2(\delta^2)$, since $\boldsymbol{Y}^T\boldsymbol{A}\boldsymbol{Y} \sim \chi_p^2(\delta^2)$ by assumption. An examination of this equality of the characteristic functions shows (details not given here) that p must be equal to q and $\lambda_1 = \cdots = \lambda_q = 1$. This proves that $\boldsymbol{A}$ is a projection matrix of rank p and we have the following result.

Lemma B.7.1. *If $\boldsymbol{Y} \sim N_n(\boldsymbol{\mu}, \boldsymbol{I})$, then $\boldsymbol{Y}^T\boldsymbol{A}\boldsymbol{Y} \sim \chi_p^2(\delta^2)$ if and only if $\boldsymbol{A}$ is a projection matrix of rank p.*

We now discuss a more general result. Let $\boldsymbol{A}_1$ and $\boldsymbol{A}_2$ be two projection matrices of ranks p_1 and p_2, respectively, and $\boldsymbol{A}_1\boldsymbol{A}_2 = 0$. We can therefore find orthonormal vectors $\{\boldsymbol{u}_j : j = 1, \ldots, p_1 + p_2\}$ such that

$$A_1 = \sum_{j=1}^{p_1} \boldsymbol{u}_j \boldsymbol{u}_j^T \text{ and } A_2 = \sum_{j=p_1+1}^{p_1+p_2} \boldsymbol{u}_j \boldsymbol{u}_j^T.$$

Since $\{W_j = \boldsymbol{u}_j^T \boldsymbol{Y}, j = 1, \ldots, p_1 + p_2\}$ are independent with $W_j \sim N(\boldsymbol{u}_j^T \boldsymbol{\mu}, 1)$, we can conclude that

$$\|A_1 Y\|^2 = Y^T A_1 Y = \sum_{j=1}^{p_1} W_j^2 \sim \chi^2_{p_1}(\delta_1^2),$$

$$\|A_2 Y\|^2 = Y^T A_2 Y = \sum_{j=p_1+1}^{p_1+p_2} W_j^2 \sim \chi^2_{p_2}(\delta_2^2),$$

and that $Y^T A_1 Y$ and $Y^T A_2 Y$ are independent, where $\delta_1^2 = (1/2)\boldsymbol{\mu}^T A_1 \boldsymbol{\mu}$ and $\delta_2^2 = (1/2)\boldsymbol{\mu}^T A_2 \boldsymbol{\mu}$. Moreover, $Y^T(A_1 + A_2)Y \sim \chi^2_{p_1+p_2}(\delta_1^2 + \delta_2^2)$.

It turns out that a converse of this is also true as given in the following result. The proofs of the next two results use ideas similar to the ones given above and details can be found in Rao [66].

Lemma B.7.2. *Let $A = A_1 + A_2$ and assume that*

(i) $Y^T A Y \sim \chi^2_p(\delta^2)$,
(ii) $Y^T A_1 Y \sim \chi^2_{p_1}(\delta_1^2)$, *and*
(iii) $P[Y^T A_2 Y \geq 0] = 1$.

Then A_2 is a projection matrix of rank $p - p_1$ and $A_1 A_2 = \boldsymbol{0}$. Consequently, $Y^T A_2 Y \sim \chi^2_{p-p_1}(\delta^2 - \delta_1^2)$ and $Y^T A_1 Y$ is independent of $Y^T A_2 Y$.

Lemma B.7.3. *Assume that $Y^T A_1 Y \sim \chi^2_{p_1}(\delta_1^2), \ldots, Y^T A_r Y \sim \chi^2_{p_r}(\delta_j^2)$. Then a necessary and sufficient condition that $Y^T A_1 Y, \ldots, Y^T A_r Y$ are independent is that $A_j A_k = \boldsymbol{0}$ for all $j \neq k$, in which case, $Y^T(A_1 + \cdots + A_r)Y \sim \chi^2_{p_1+\cdots+p_r}(\delta_1^2 + \cdots + \delta_r^2)$.*

Bibliography

1. Ferguson T. *Mathematical Statistics*. New York: Academic Press; 1967.
2. Lehmann E. *Testing Statistical Hypotheses*. New York: John Wiley; 1959.
3. Lehmann E. *Theory of Point Estimation*. New York: John Wiley; 1983.
4. Hájek J, Šidák Z. *Theory of Rank Tests*. New York: Academic Press; 1967.
5. Gosset W. On the probable error of a mean. Biometrika 1908;6:1–25.
6. Kolmogorov A. *Foundations of the Theory of Probability*. New York: Chelsea Publishing Company; 1933 (English Translation, 1950).
7. Hoeffding W. Probability inequalities for sums of bounded random variables. J Am Statist Assoc 1963;58:13–30.
8. Uspensky J. *Introduction to Mathematical Probability*. New York and London: McGraw Hill; 1937.
9. Doob J. *Stochastic Processes*. New York: John Wiley; 1953.
10. Hoeffding W, Robbins H. The central limit theorem for dependent random variables. Duke Math J 1948;15:773–80.
11. Stein C. Inadmissibility of the usual estimator for the mean of a multivariate normal distribution. In: Proc Third Berkeley Symp Math Statist Prob, vol. 1. Univ of Calif Press, CA; 1956. p. 197–206.
12. Hodges J, Lehmann E. Some problems in minimax point estimation. Ann Math Statist 1950;21:182–97.
13. Karlin S. Admissibility for estimation with quadratic loss. Ann Math Statist 1958;29:406–36.
14. Stein C. The admissibility of Pitman's estimator for a single location parameter. Ann Math Statist 1959;30:970–9.
15. Rao C. Information and accuracy attainable in the estimation of statistical parameters. Bull Calc Math Soc 1945;37:81–91.
16. Blackwell D. Conditional expectation and unbiased sequential estimation. Ann Math Statist 1947;18:105–10.
17. Basu D. On statistics independent of a complete sufficient statistic. Sankhya 1955;15:377–80.
18. Cramér H. *Mathematical Methods of Statistics*. Princeton, NJ: Princeton University Press; 1946.
19. Pitman E. The estimation of location and scale parameters of a continuous population of any given form. Biometrika 1939;30:391–421.
20. Fisher R. On the mathematical foundations of theoretical statistics. Phil Trans R Soc Lond Ser A 1921;222:309–68.
21. Rao C. Criteria of estimation in large samples. Sankhya 1963;25:189–206.
22. Neyman J, Pearson E. On the problem of the most efficient tests of statistical hypotheses. Phil Trans R Soc Lond Ser A 1933;231:289–337.
23. Neyman J, Pearson E. Contributions to the theory of testing statistical hypotheses. I. Unbiased critical regions of type A and type A_1. Stat Res Mem 1936;1:1–37.
24. Wald A. *Sequential Analysis*. New York: John Wiley; 1947.

25. Bahadur R. A note on the fundamental identity of sequential analysis. Ann Math Statist 1958;29:534–43.

26. Neyman J. Outline of a theory of statistical estimation based on the classical theory of probability. Phil Trans R Soc 1937;235:333–80.

27. Wald A. Note on the consistency of the maximum likelihood estimate. Ann Math Statist 1949;20:595–601.

28. Neyman J, Scott E. Consistent estimates based on partially consistent observations. Econometrika 1948;16:1–32.

29. Basu D. An inconsistency of the method of maximum likelihood. Ann Math Statist 1955;26:144–5.

30. Ferguson T. An inconsistent maximum likelihood estimate. J Am Statist Assoc 1982;77:831–4.

31. LeCam L. On some asymptotic properties of maximum likelihood estimates and related Bayes estimates. Univ Calif Publ Statist 1953;1:277–330.

32. Chernoff H, Lehmann E. The use of maximum likelihood estimates in χ^2 tests for goodness of fit. Ann Math Statist 1954;25:579–86.

33. Hoeffding W. A class of statistics with asymptotic normal distribution. Ann Math Statist 1948;19:293–325.

34. Roussas G. *Contiguity of Probability Measures*. Cambridge: Cambridge University Press; 1972.

35. Fisher R, Yates F. *Statistical Tables for Biological, Agricultural and Medical Research*. Edinburgh-London: Oliver and Boyd; 1938.

36. Hoeffding W. "Optimum" nonparametric tests. In: Proc Second Berkeley Symp Math Statist Prob, vol. 1. Univ Calif Press, CA; 1950. p. 83–92.

37. Terry M. Some rank order tests which are most powerful against specific parametric alternatives. Ann Math Statist 1952;23:346–66.

38. Chenoff H, Savage I. Asymptotic normality and efficiency of certain nonparametric test statistics. Ann Math Statist 1958;29:972–94.

39. Dvoretzky A, Kiefer J, Wolfowtz J. Asymptotic minimax character of the sample distribution function and of the classical multinomial estimator. Ann Math Statist 1956;27:642–69.

40. Doob J. Heuristic approach to the Kolmogorov-Smirnov theorems. Ann Math Statist 1949;20:393–403.

41. Freedman D. *Brownian Motion and Diffusion*. New York: Springer; 1983.

42. Breiman L. *Probability*. Reading, MA: Addison-Wesley; 1968.

43. Billingsley P. *Convergence of Probability Measures*. New York: John Wiley; 1968.

44. Wiener N. Un probleme de probabilités énombrables. Bull Soc Math de France 1924;52:569–78.

45. Rosenblatt M. Remarks on some nonparametric estimates of a density function. Ann Math Statist 1956;27:832–7.

46. Parzen E. On estimation of a probability density and mode. Ann Math Statist 1962;33:1065–76.

47. Nadaraya E. On nonparametric estimates of density functions and regression curves. Theor Prob Appl 1965;10:186–90.

48. Bhattacharya P. Estimation of a probability density function and its derivatives. Sankhya Ser A 1967;29:373–82.

49. Silverman B. *Density Estimation for Statistics and Data Analysis*. London: Chapman and Hall; 1968.

50. Fan J, Gijbels I. *Local Polynomial Modelling and its Applications*. London: Chapman-Hall; 1996.

51. Stone C. On asymptotically optimal window selection rule for kernel density estimates. Ann Statist 1984;12:1285–97.

52. Kaplan E, Meier P. Nonparametric estimation from incomplete observations. J Am Statist Assoc 1958;53:457–81.

53. Efron B. The two-sample problem with censored data. In: Proc Fifth Berkeley Symp Math Statist Prob, vol. 4. University of California Press, CA; 1967. p. 831–53.

54. von Mises R. On the asymptotic distribution of differentiable statistical functionals. Ann Math Statist 1947;18:309–48.

55. Serfling R. *Approximation Theorems in Mathematical Statistics*. New York: John Wiley; 1980.

56. Huber P. *Robust Statistics*. New York: John Wiley; 1981.

57. Fernholtz L. *von Mises Calculus for Statistical Functionals*. New York: Springer-Verlag; 1983. Lecture Notes in Statistics.

58. Bahadur R. A note on quantiles in large samples. Ann Math Statist 1966;37:577–80.

59. Efron B, Tibshirani R.J. *An Introduction to the Bootstrap*. New York: Chapman and Hall; 1993.

60. Searle SR. *Linear Models*. New York: John Wiley; 1971.

61. Mardia K, Kent J, Bibby J. *Multivariate Analysis*. New York: Academic Press; 1979.

62. Box G, Jenkins G, Reinsel G. *Time Series Analysis: Forecasting and Control*. New Jersey: Prentice-Hall; 1994.

63. Brockwell P, Davis R. *Time Series: Theory and Methods*. New York: Springer; 1991.

64. Gikhman I, Skorokhod A. *Introduction to the Theory of Random Processes*. New York: Dover; 1996.

65. Royden H. *Real Analysis*. New York: McMillan; 1968.

66. Rao C. *Linear Statistical Inference and its Applications*. New Delhi: Wiley Eastern; 1973.

67. Hájek J. A characterization of limiting distributions of regular estimates. Z Wahr Verw Geb 1970;14:323–30.

68. Inagaki N. On the limiting distribution of a sequence of estimators with uniformity property. Ann Inst Stat Math 1970;22:113–97.

69. Pitman E. *Notes on Nonparametric Statistical Inference*. New York (Mimeographed): Columbia Univ; 1948.

Index

D

P

Kolibri 360 GmbH
Ferdinand-Jühlke-Straße 7
99095 Erfurt, Deutschland
produktsicherheit@kolibri360.de